Lecture Notes in Computer Science 4975

Commenced Publication in 1973
Founding and Former Series Editors:
Gerhard Goos, Juris Hartmanis, and Jan van Leeuwen

Falai Chen   Bert Jüttler (Eds.)

Advances in Geometric Modeling and Processing

5th International Conference, GMP 2008
Hangzhou, China, April 23-25, 2008
Proceedings

 Springer

Volume Editors

Falai Chen
University of Science and Technology of China, Department of Mathematics
Hefei, Anhui 230026, China
E-mail: chenfl@ustc.edu.cn

Bert Jüttler
Johannes Kepler University, Institute of Applied Geometry
Altenberger Str. 69, 4040 Linz, Austria
E-mail: bert.juettler@jku.at

Library of Congress Control Number: 2008924624

CR Subject Classification (1998): I.3.5, I.3.7, I.4.8, G.1.2, F.2.2, I.5, G.2

LNCS Sublibrary: SL 1 – Theoretical Computer Science and General Issues

ISSN	0302-9743
ISBN-10	3-540-79245-7 Springer Berlin Heidelberg New York
ISBN-13	978-3-540-79245-1 Springer Berlin Heidelberg New York

Springer is a part of Springer Science+Business Media

springer.com

© Springer-Verlag Berlin Heidelberg 2008

Typesetting: Camera-ready by author, data conversion by Scientific Publishing Services, Chennai, India
Printed on acid-free paper   SPIN: 12255566   06/3180   5 4 3 2 1 0

Preface

Geometric Modeling and Processing (GMP) is a biennial international conference on geometric modeling, simulation and computing, which provides researchers and practitioners with a forum for exchanging new ideas, discussing new applications, and presenting new solutions. Previous GMP conferences were held in Pittsburgh (2006), Beijing (2004), Tokyo (2002), and Hong Kong (2000). This, the 5th GMP conference, was held in Hangzhou, one of the most beautiful cities in China.

GMP 2008 received 113 paper submissions, covering a wide spectrum of geometric modeling and processing, such as curves and surfaces, digital geometry processing, geometric feature modeling and recognition, geometric constraint solving, geometric optimization, multiresolution modeling, and applications in computer vision, image processing, scientific visualization, robotics and reverse engineering. Each paper was reviewed by at least three members of the program committee and external reviewers. Based on the recommendations of the reviewers, 34 regular papers were selected for oral presentation, and 17 short papers were selected for poster presentation. All selected papers are included in these proceedings.

We thank all authors, external reviewers and program committee members for their great effort and contributions, which made this conference a success. We thank the conference chairs Ron Goldman and Guozhao Wang, and the steering committee members Shimin Hu, Myung-Soo Kim, Ralph Martin, Helmut Pottmann, Kenji Shimada, Hiromasa Suzuki and Wenping Wang for their consistent advice and help. Very special thanks go to the local organizer, Ligang Liu, for his hard work in organizing the conference and in managing the online review system. Very special thanks also go to Ms. Bayer, who collected the final versions of the papers. We thank the Natural Science Foundation of China and the National Key Basic Research Program of China (Project number 2004CB318000) for their financial support. We thank Zhejiang University for hosting the conference. The Activity Group on Geometric Design and Computing of the Chinese Society of Industrial and Applied Mathematics provided valuable support to the conference. Finally, we thank all of the participants for making this conference a success.

February 2008

Falai Chen
Bert Jüttler

Conference Committee

Honorary Conference Chair

Jia-Guang Sun (Tsinghua University)

Conference Co-chairs

Ron Goldman (Rice University, USA)
Guozhao Wang (Zhejiang University, China)

Program Co-chairs

Bert Jüttler (Johannes Kepler University, Austria)
Falai Chen (University of Sciences and Technology of China, China)

GMP Steering Committee

Myung-Soo Kim (Seoul National University, Korea)
Kenji Shimada (Carnegie Mellon University, USA)
Shimin Hu (Tsinghua University, China)
Ralph Martin (Cardiff University, UK)
Helmut Pottmann (Institut für Geometrie, TU Wien, Austria)
Hiromasa Suzuki (University of Tokyo, Japan)
Wenping Wang (Hong Kong University, Hong Kong)

Program Committee

Franz Aurenhammer (Technische Universität Graz, Austria)
Chandrajit Bajaj (Univ. of Texas at Austin, USA)
Hujun Bao (Zhejiang University, China)
Alexander Belyaev (Heriot-Watt University, UK)
Wim Bronsvoort (Delft University of Technology, The Netherlands)
Stephen Cameron (Oxford University, UK)
Fuhua (Frank) Cheng (University of Kentucky, USA)
Jian-Song Deng (University of Science and Technology, China)
Gershon Elber (Technion, Israel)
Gerald Farin (Arizona State Univ., USA)
Rida T. Farouki (University of California, Davis, USA)
Anath Fischer (Technion, Israel)
Michael Floater (Sintef Applied Mathematics, Norway)

Tamas Varady (Geomagic Hungary, Hungary)
Johannes Wallner (Institut für Geometrie, TU Graz, Austria)
Charlie Wang (The Chinese University of Hong Kong)
Guojin Wang (Zhejiang University, China)
Jiaye Wang (Shandong University, China)
Michael Yu Wang (The Chinese University of Hong Kong)
Joe Warren (Rice University, USA)
Guoliang Xu (Institute of Computational Mathematics, Chinese Academy of
    Sciences (CAS))
Soji Yamakawa (Carnegie Mellon University, USA)
Xiuzi Ye (Zhejiang University, China)
Hong-Bin Zha (Peking University, China)
Caiming Zhang (Shandong University, China)
Jianmin Zheng (Nanyang Technological University)
Kun Zhou (Microsoft Research Asia, China)

Organization Committee

Ligang Liu (Chair, Zhejiang University, China)
Ruofeng Tong (Zhejiang University, China)
Xingjiang Lu (Zhejiang University, China)
Jiangyun Li (Zhejiang University, China)
Zhihao Zheng (Zhejiang University, China)
Hongxin Zhang (Zhejiang University, China)
Hongwei Lin (Zhejiang University, China)

Additional Reviewers

Abdelwahed Abbas
Andy Shiue
Antoine Bouthors
Ariel Shamir
Basile Sauvage
Bernhard Kornberger
Binhai Zhu
Chen Wenyu
Chenglei Yang
David Cohen-Steiner
Dong Yu
Emil Zagar
Esmeralda Mainar
Franca Giannini
G.P. Bonneau
Gabriel Peyre

Georgios Papaioannou
Guido Brunnett
Haijun Su
Hayong Shin
Henry Kang
Hongbo Fu
Iddo Hanniel
Ioannis Ivrissimtzis
Jiaping Wang
Jieqing Tan
Jingyi Yu
Joon Kyung Seong
Junho Kim
Junji Sone
Karthik Ramani
Katsuaki Kawachi

Kazuya Kobayashi
Kerstin Mueller
Klaus Hildebrandt
Koichi Matsuda
Konrad Polthier
Ku-Jin Kim
Kwan H. Lee
Kwanghee Ko
Li Guiqing
Liu Shengjun
Lixian Zhang
Martin Bendsoe
Martin Reuter
Miguel A. Otaduy
Min Gyu Choi
Ming Li
Oded Ziedman
Oliver Labs
Pal Benko
Peter Salvi
Pierre Alliez
Qian-Yi Zhou
Renjiang Zhang
Rhaleb Zayer
Sagi Schein
Sang-Uk Cheon
Scott Schaefer
Seung-Hyun Yoon

Shin Yoshizawa
Sigal Ar
Spyridon Vosinakis
Taek-Hee Lee
Taichi Watanabe
Takayuki Itoh
Takis Sakkalis
Tim Goodman
Tomohiro Mizoguchi
Wang Yimin
Waqar Saleem
Weiwei Xu
Xia Qi
Xiaohan Shi
Xinguo Liu
Yang Xingqiang
Yaron Lipman
Ye Duan
Yiying Tong
Yong-Jin Liu
Yongjie Zhang
Yoshiyuki Furukawa
Young Joon Ahn
Yunjin Lee
Zeyun Yu
Zhang Qin
Zhouwang Yang

Table of Contents

II   Short Papers

III   A Comment

Part I

Regular Papers

Automatic PolyCube-Maps

Juncong Lin[1], Xiaogang Jin[1], Zhengwen Fan[1], and Charlie C.L. Wang[2]

[1] State Key Lab of CAD&CG, Zhejiang University, Hangzhou, 310027, P.R. China
{linjuncong,jin,fanzhengwen}@cad.zju.edu.cn
[2] Department of Mechanical and Automation Engineering, the Chinese University of
Hong Kong, Shatin, N.T., Hong Kong
cwang@mae.cuhk.edu.hk

Abstract. We propose an automatic PolyCube-Maps construction scheme. Firstly, input mesh is decomposed into a set of feature regions, and further split into patches. Then, each region is approximated by a simple basic polycube primitive with each patch mapped to a rectangular sub-surface of the basic polycube primitive which can be parameterized independently. After that, an iterative procedure is performed to improve the parameterization quality globally. By these steps, we can obtain the polycubic parameterization result efficiently.

Keywords: parameterization, PolyCube-Maps, automatic construction, Reeb graph.

1 Introduction

Since being introduced to computer graphics as a method for mapping textures onto surfaces (ref. [1] and [2]), surface parameterization has become an active research area in the past decades.

Early work in mesh parameterization mostly focused on planar parameterization of meshes with disk-like topology. These methods generally aimed at achieving certain optimal metrics such as angular distortion, stretch and area distortion or a tradeoff between these optimal metrics. Planar parameterization is only applicable to surfaces with disk topology. Hence, when dealing with closed surfaces and surfaces with genus greater than zero, a cutting/chart generation step needs to be conducted. However, these cuts always introduce discontinuities, which will lead to artifacts in rendering or remeshing along the boundary. Finding appropriate cuts in order to minimize the artifacts is notoriously difficult.

For those applications which are sensitive to discontinuities in the parameterization, different primitives (as base domains) for parameterization are desired. Simplicial complexes, spheres and periodic planar regions with transition curves have been investigated by a number of researchers, where these methods construct a seamless parameterization of a mesh over a triangulated base complex. A popularly used non-planar base domain is simplicial complex. One of the key problems for this sort of parameterization is to obtain the complexes of the surface as the base domains, where various techniques have been presented to solve

F. Chen and B. Jüttler (Eds.): GMP 2008, LNCS 4975, pp. 3–16, 2008.
© Springer-Verlag Berlin Heidelberg 2008

the problem (e.g., [3], [4], and [5]). After determining a suitable base mesh, the original mesh can be parameterized by assigning each of its vertices to a simplex of the base domain, along with barycentric coordinates inside it. Spherical domain also attracted lots of researchers to give support on seamless and continuous parameterization of genus-0 models. Several methods (ref. [6], [7], and [8]) extended the barycentric, convex boundary planar methods to sphere to obtain a spherical parameterization. Regrettably, such an extension scheme is unstable and may collapse to a degenerate solution. An efficient and bijective alternative is suggested by multi-resolution techniques in [9] and [10], where these methods obtain an initial guess by simplifying the model until it becomes a tetrahedron, then trivially embed it on the sphere and progressively add back the vertices.

These non-planar domain (either simplex or sphere) parameterization methods did avoid the seams that existed in planar parameterization methods. However, texture mapping with these methods might fail for those triangle vertices that have parameter values on different domain triangles, because their linear interpolation is a secant triangle that falls outside the surface and the color information defined on the domain triangles of the base complex cannot be mapped to those vertices. Tarini et al. [11] proposed the PolyCube-Maps concept for seamless texture mapping, and their work opens the way to a new category of surface parameterization. The PolyCube-Maps method is inspired from the well-known cube maps that is commonly used for environment mapping and seamless texture mapping for cube like object. PolyCube-Maps use polycube rather than a single cube as the 3D texture domain. A polycube is a shape composed of axis-aligned unit cubes that are attached face to face. A polycube is also required to resemble the shape of the given mesh and to capture the large scale features so as to achieve best parameterization result. After the polycube is constructed, Tarini et al. conduct a one-to-one map between the 3D shape and the polycube by projecting points from the shape to polycube. To the best of our knowledge, it is the first one to use a quadrilateral base domain which is much more suitable for quadrilateral remeshing of the input surface and for spline fitting. Several researchers further applied PolyCube-Maps to various areas. For example, Fan et al [12] proposed a novel mesh morphing approach based on polycubic cross-parameterization. Wang et al. [13] presented the so called polycube splines with polycube map serving as its parametric domain.

1.1 Contributions

As described above, PolyCube Maps have become a new category of surface parameterization methods, and demonstrated great advantages in various applications. For a globally smooth parameterization method, the construction of the base domain and the control of the smoothness and distortion at patch boundaries are two most important ingredients. Tarni et al. provided a good solution to the second problem; however, the generation of polycube in [11] was manual which is tedious for complex models and prevents the popularization of such a new parameterization method. When compared with other non-planar parameterizations all of which have the automatic base domain construction methods, this problem becomes notable. In this paper, we present a novel solution to the first problem and provide an automatic construction method. To our knowledge, this appears to be the first approach to construct polycubes automatically.

2 Automatic PolyCube-Maps Paradigm

Fig.1 shows the flowchart of our automatic PolyCube-Maps method. Our method consists of three steps:

1. *Feature-based mesh segmentation.* Input model is segmented into parts corresponding to relevant features of the shape. We take a feature based segmentation strategy and employ embedded Reeb graphs to guide the segmentation procedure.
2. *Polycube approximation.* We create a basic polycube primitive for each subpart, and then further divide each subpart into patches which will be assigned to the faces of the basic polycube primitive later.
3. *Parameterization.* After approximation, we adopt the improved barycentric coordinate method to map all the vertices onto the corresponding faces of polycube. An iterative optimization procedure is then employed to improve the final result.

Our strategy for the mapping of the 3D shape and the polycube is more close to Wang et al. [13], which is better than the direct projection method as Tarini et al [11] did. However, we focus more on the construction of polycube to achieve a better parameterization quality while the authors of [13] made the construction of polycube to be independent of the actual geometry of 3D shape allowing different complexity and resolution for the polycube.

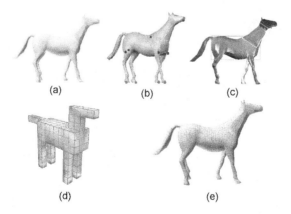

Fig. 1. Automatic PolyCube-Maps: (a) Input model, (b)Reeb graph based feature segmentation, (c) polycube approximation, (d) polycube, and (e) texture-mapping

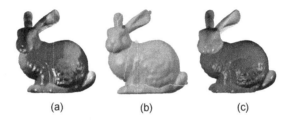

(a)   (b)   (c)

Fig. 2. Feature based mesh segmentation. (a)AGD of Bunny; (b)the embeded Reeb graph; (c)result of feature segmentation.

3   Feature Based Mesh Segmentation

Mesh segmentation plays an important role in various computer graphics applications. In the context of PolyCube-Maps generation, the key issue is to identify the genus of input mesh, so as to separate handles from the main parts and serve for the following polycube approximation step. In this paper, we adopt a feature-based method to segment input mesh based on the topological analysis of the average geodesic distance function with the help of Reeb graph. For the term *feature*, we mean large protrusions on the input mesh. Our mesh segmentation approach is akin to [14], but with revisions on critical points filtering and genus reduction.

To build an embedded Reeb graph, we need define an appropriate quotient function which will directly affect the stability properties of the topological structure. In our implementation, we choose the average geodesic distance function defined by Zhang et al. [14] as the quotient function to construct topological structures. The average geodesic distance function was firstly introduced by Hilaga et al. [15] for shape matching, and Zhang et al. proposed the variant one defined below in [14].

$$AGD_n(\mathbf{p}) := \frac{A_n(\mathbf{p})}{\min_{q \in S} A_n(\mathbf{q})}, (n = 1, 2, \ldots \infty) \tag{1}$$

where $A_n(\mathbf{p})$ is defined as

$$A_n(\mathbf{p}) = \sqrt[n]{\frac{\int_{q \in S} g^n(\mathbf{p}, \mathbf{q}) d\mathbf{q}}{Area(S)}} \tag{2}$$

$g(\mathbf{p}, \mathbf{q})$ is the geodesic distance between two points $\mathbf{p}$ and $\mathbf{q}$. We use $n = 2$ in our implementation to achieve a balance between performance and result. The term AGD is used for short in the rest of the paper for the *average geodesic distance function*. We take an approximate strategy to compute the AGD quickly. Firstly, we sample 3D points on the surface of the input mesh evenly. Then, we calculate the geodesic distances from them other than all the surface points using the method introduced by Surazhsky et al. [16]. Finally, a satisfactory approximation of AGD can be computed. After defining AGD on the surface, we follow [14] to construct the induced Reeb graph Γ by performing region-growing in the ascending order of AGD. Specifically, we add one triangle at a time in the increasing order of the AGD starting from a triangle with global minimum AGD value until the surface is covered. The AGD of a triangle $T = \{v_1, v_2, v_3\}$ is defined as $AGD(T) = \min\{AGD(v_1), AGD(v_2), AGD(v_3)\}$. The boundary of the visited region consists of a number of loops. We label the addling triangle with the following five tags:

1. Minimum: where one new boundary loop starts.
2. Maximum: where one boundary loop vanishes. This is the tip of a protrusion.
3. Splitting saddle: where one boundary loop intersects itself and splits into two.
4. Merging saddle: where two boundary loops intersect and merge into one. This signifies the formulation of a handle.
5. Regular: where the number of boundary loops does not change.

Since complex surfaces often contain many small protrusions, the embedded Reeb graph can have an excessive number of local maxima and saddle points which should be removed before practical use. In [14], Zhang et al. chose to alter the order in which triangles were added to eliminate extra local maxima and splitting saddle points. However, such a filtering scheme is inefficient and cannot thoroughly avoid adding extra splitting saddle points. In this paper, we present a new filtering scheme which is more efficient and allows users to control the size of the identifying feature. As shown in Fig.3, denoting the region ending with

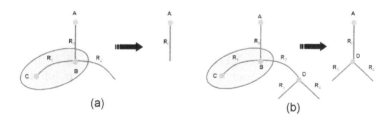

Fig. 3. Eliminate extra splitting saddle points

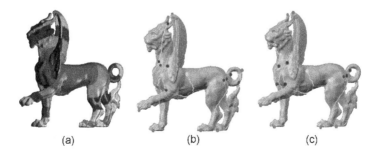

Fig. 4. Embbed Reeb graphs with different filtering constant: (a) the AGD of Feline, (b) $\gamma=0.1$, and (c) $\gamma=0.2$

B by R_0 and the other region split from B by R_2, for every region R_1 ending with a maximum C, if the starting critical point B of R_1 is a splitting saddle and satisfies

$$AGD(C) - AGD(B) < \gamma \times (AGD_{max} - AGD_{min}) \qquad (3)$$

we merge R_1 and R_2 into R_0 and delete the critical points B and C. Fig.4 shows the experimental result of our filtering scheme applied to the feline model.

Equipped with the embedded Reeb graph, we can segment input surface into meaningful parts according to the topological structure revealed. We construct a *separating cycle* [17] on the surface for every tip point **p** of a feature. Separating cycle is a closed curve on the surface that separates the feature from the remaining body. Firstly, a separating region R corresponding to **p** is constructed by performing region-growing starting from **p**. Secondly, we reduce R into its skeleton by repeatedly performing elementary collapses: treating R as a 2-complex (with boundary edges). Remove triangles with at least one boundary edge from R along with one of the boundary edges, we apply the operation until all 2-cells (triangles) are removed and the 2-complex is reduced to a 1-complex. Finally, the separating cycle γ which is neither inside nor outside the feature region is selected as the desired one and optimized.

If the input surface is an object with genus n, we need to separate the handle from the body by assigning them to different regions so as to construct a homeomorphous polycube. For a genus n surface, a loop doe not always divide the surface into two connected components. Loops with this property are associated to the elements of the first homology group which form an Abelian group with $2n$ generators, we call them *non-separating cycles* [17]. To reduce the genus size, we try to identify an appropriate non-separating cycle for every handle and cut the surface open along the cycles. Zhang et al. in [14] firstly constructed a basic loop ρ for a handle by computing two shortest paths in Γ that connect the merging saddle point $\mathbf{q}_i$ and the splitting saddle point $\mathbf{p}_i$ of the handle, then a nearby non-separating cycle γ is created for one of the passages of each basic loop ρ by performing region-growing from ρ in an increasing order of the distance function from $\mathbf{p}_i$. Since splitting saddle point is generally closer to the main part, the strategy in [14] can separate the handle from the main part well in most cases. However, it does not work for some complex cases such as the

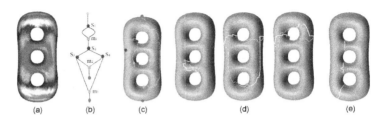

Fig. 5. Embedded Reeb graphs with different filtering constant: (a) The AGD of Feline; (b)γ=0.1; (c)γ=0.2

3-Holes model shown in Fig.5. The model has a genus 3, with 3 cycles contained in the Reeb graph. The cycles signaled by merging saddle points $\mathbf{m}_2$ and $\mathbf{m}_3$ respectively have the same splitting saddle point s_2. If we perform region-growing from s_2, we cannot reduce the size/number of genus properly. We choose to perform region-growing from s_2 at one side, and from $\mathbf{m}_3$ for another side. If the cycle in the Reeb graph is approaching a non-separating cycle (e.g. the first cycle in Fig.5(d)), Zhang's method will lead to a longer and worse non-separating cycle. We take a more complicated strategy here. Firstly, if the length of a non-separating cycle γ is smaller than ρ, we take it as the ideal non-separating cycle. Otherwise, we perform region-growing starting from the intersection point of γ and ρ to construct a new non-separating cycle ζ. Our experiment shows that we can always get a better non-separating cycle (see Fig.5(e)). Besides, after transforming the input model to an appropriate orientation and taking the $x - y$ plane as the axis plane, we hope that the ideal non-separating cycle will align with the axis plane as close as possible so that we do not stop region-growing at the first intersection point but until the nth intersection point. $n = 10$ is chosen in our implementation. Then, we choose the ideal one from the non-separating cycles constructed from n intersection points.

4   Polycube Approximation

After applying the segmentation procedure in section 3, we get the result of segmentation including feature regions, boundaries and non-separating cycles. In this section, we will build an appropriate *basic polycube primitive* (such as cube, L-shape, O-shape, U-shape) for each feature region, and determine the vertices/paths on the feature region corresponding to the vertices/edges of the basic polycube primitive. For example, we need to divide the region into 6 rectangle patches if it is approximated by a cube, so as to map to the 8 vertices/12 edges of the basic polycube primitive. To get a better initial parameterization, the boundary length of each patch should be close to the edge length of the corresponding face on the basic polycube primitive. Besides, we have to check the validity of the corresponding basic polycube primitive — it is valid if the current basic polycube primitive overlaps with existed basic polycube primitives either on the boundary or non-separating cycle, else it is invalid. For the invalid case, we need to reconstruct a basic polycube primitive for the region until we get a valid one. In the rest of this section, we will describe our algorithm for polycube approximation in detail.

4.1   Determine the Unit Length and Vertices of Region Boundary

We need to determine the size of the *unit cube* constituting the basic polycube primitive so that we can determine the number of unit cubes for region approximation. To save the texture memory, we hope to use as few cubes as possible to approximate the region with the least stretch. We choose the region boundary γ with minimum length, project it to the axis plane, and take the four ordered vertices $\mathbf{p}_0$, $\mathbf{p}_1$, $\mathbf{p}_2$, $\mathbf{p}_3$ on the projection of γ that is closest to the 4 vertices of

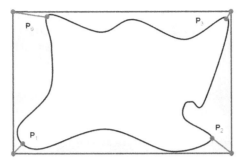

Fig. 6. Determine the vertices of region boundary

the bounding box of γ as the initial boundary vertices. Then we adjust $\mathbf{p}_2$ to divide γ into two segments ($\mathbf{p}_0\mathbf{p}_1\mathbf{p}_2$ and $\mathbf{p}_0\mathbf{p}_3\mathbf{p}_2$) with equal length. After that, we adjust $\mathbf{p}_1$ to make $\mathbf{p}_0\mathbf{p}_1 = \mathbf{p}_2\mathbf{p}_3$ and $\mathbf{p}_1\mathbf{p}_2 = \mathbf{p}_3\mathbf{p}_0$. We finally adjust $\mathbf{p}_0$, $\mathbf{p}_1$, $\mathbf{p}_2$, $\mathbf{p}_3$ simultaneously until $\mathbf{p}_0\mathbf{p}_1/\mathbf{p}_1\mathbf{p}_2 = \mathbf{p}_2\mathbf{p}_3/\mathbf{p}_3\mathbf{p}_0 = integer$ (assuming $\mathbf{p}_0\mathbf{p}_1 > \mathbf{p}_1\mathbf{p}_2$ and $\mathbf{p}_2\mathbf{p}_3 > \mathbf{p}_3\mathbf{p}_0$) and take the length of the short one as the size of unit cube. We apply the similar operations to other boundaries and make the lengths of both short and long segments to be integer of unit cube length.

4.2 Approximate the Region from Segmentation of Maximum Point

For the region corresponding to the maximum point $\mathbf{p}$, we firstly calculate the geodesic distance d from $\mathbf{p}$ to the boundary γ. If d is larger than γ's length, we use a cube to approximate the region. Otherwise, a bounding box will be adopted. As shown in Fig.7, assume the the length of $\mathbf{p}_0\mathbf{p}_1$ and $\mathbf{p}_2\mathbf{p}_3$ to be a, the length of $\mathbf{p}_1\mathbf{p}_2$ and $\mathbf{p}_3\mathbf{p}_0$ to be b, we firstly perform region-growing starting from $\mathbf{p}$ in the increasing order of the geodesic distance from $\mathbf{p}$ and stop until the geodesic distance to the maximum point $\mathbf{p}$ equals to $\sqrt{a^2 + b^2}/2$. Then we determine 4 vertices ($\mathbf{q}_0$, $\mathbf{q}_1$, $\mathbf{q}_2$, $\mathbf{q}_3$) on the boundary of the new region ζ corresponding to the 4 vertices ($\mathbf{p}_0$, $\mathbf{p}_1$, $\mathbf{p}_2$, $\mathbf{p}_3$) on γ, and the paths between the vertices pairs. Here we find the shortest path τ between γ and ζ with two end points $\mathbf{u}$, $\mathbf{v}$ on

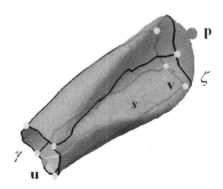

Fig. 7. Approximate the region segmented from maximum point

γ and ζ respectively, then we determine $\mathbf{q}_0$, $\mathbf{q}_1$, $\mathbf{q}_2$, $\mathbf{q}_3$ satisfying $\frac{\mathbf{vq}_0}{\mathbf{up}_0} = \frac{\mathbf{vq}_1}{\mathbf{up}_1} = \frac{\mathbf{vq}_2}{\mathbf{up}_2} = \frac{\mathbf{vq}_3}{\mathbf{up}_3}$ and find the shortest path between the vertices pairs $\mathbf{p}_0\mathbf{q}_0$, $\mathbf{p}_1\mathbf{q}_1$, $\mathbf{p}_2\mathbf{q}_2$, $\mathbf{p}_3\mathbf{q}_3$. Finally, we use a cube to approximate the region and determine the number of unit cubes contained.

4.3   Approximate the Region Segmented by Non-separating Cycle

For those regions segmented by non-separating cycle, if the two sides of non-separating cycle are both contained in the region (the torus in Fig.8), a O-shape is adopted to approximate it. Otherwise, cube, L-shape, or U-shape will be the candidates. The detailed approximation scheme is as follows.

1. Determine the shortest path s between region boundary γ and non-separating cycle ζ with two end points $\mathbf{p}$ and $\mathbf{q}$.
2. Project s onto the main plane, and locate the origin $\mathbf{o}$ of s on the plane. The distance from origin $\mathbf{o}$ to all the points on s is nearly equal. If the sector $\mathbf{poq}$ covers a quadrant with more than 60 degrees, we put a bend point in the sector locating at the intersection of s and the line that starts from origin and divides the quadrant into two equal parts (Fig.9).
3. The predetermined 4 vertices pairs on the region boundary and non-separating cycle may not be in an appropriate position. We can use the method in Section 4.2 to adjust their position.
4. If the region is with high curvature, directly connecting the vertex pairs with the shortest path between them may not be the optimum. We thus further improve the path by introducing reference rings – as shown in Fig.10(b), for every shortest path s between γ and ζ, we construct a reference ring passing every bend point by connecting the bend point of other shortest paths. Then, the bend point of the shortest path is adjusted according to the determined ratio (see Section 4.1).

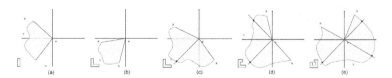

Fig. 8. Torus and polycube primitives. The blue spheres represent the corresponding non-separating cycle or region boundary, and the green ones represent the vertices on the basic polycube primitives.

Fig. 9. Determine the basic polycube primitive based on the shortest path between two boundaries

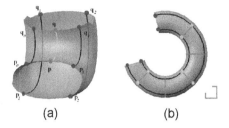

Fig. 10. Construct the boundary of mesh patch: (a) Connect the corresponding vertex pairs directly and (b) connect the corresponding vertex pair through the reference cycle. Green cycles represent the cycles at vertices while red ones are extra cycles.

4.4   Approximate Region with Bounding Box

Bounding box is employed to approximate other regions. Firstly, we construct the bounding box of the region according to the result of PCA analysis. Then, we find the 8 vertices on the region which are closest to the 8 vertices of bounding box respectively. Then the position of 8 vertices is adjusted to meet the requirement of a cube and make slight adjustment so that the geodesic distance between them meets the requirement of a cube and is an integer number of unit cube's length. Finally, we connect vertices with the shortest path and divide the region into 6 patches.

4.5   Construct Global Coordinate for the Polycube

The polycube approximation is generated by following the *neighborhood rule*. That is, by beginning with any region, the following region to be processed should be the neighbor of the processed ones. We set the starting point of the first built basic polycube primitive as the origin. Then, the position of the newly constructed basic polycube primitives can be determine according to the boundary between the existing basic polycube primitives. After all the regions are processed, we translate the whole polycube to make the starting point of every basic polycube have a positive coordinate.

5   Parameterization

After the construction of polycube, we will further map every region onto its corresponding basic polycube primitive. For every patch of a region, we parameterize it onto the corresponding facet of the basic polycube primitive using the mean value coordinates method in [18]. The patch's boundary is mapped to the rectangle in the 2D parameterization domain, and every inner point will be the mean value of the one-ring as

$$\sum_{i=1}^{k} \lambda_i v_i = v_0 \tag{4}$$

$$\lambda_i = \frac{\omega_i}{\sum_{j=1}^{k} \omega_j} (\omega_i = \frac{\tan(\alpha_{i-1}/2) + \tan(\alpha_i/2)}{\| v_i - v_0 \|}) \tag{5}$$

We construct this mapping for every region, so as to get an initial global parameterization of the whole model. However, such an initial parameterization is usually not satisfied, we therefore conduct an iterative procedure to optimize the mapping and to minimize the overall distortion of the parameterization. For each vertex v around the boundary of basic polycube's facet, the mapping of one-ring of v may cover several facets, we apply the following optimization procedure for every such vertex:

- If v and its one-ring neighbors all lie on the same plane (a plane may contain several facets from different basic polycube primitives), we recalculate its texture coordinate by equation (4) and (5) according to the texture coordinate of its one-ring neighbour. The weights ω_i can be obtained from the original input mesh. If the new assigned texture coordinate will lead to the switch of triangle, the operation will be cancelled.
- If v and its one-ring neighbors lie on two planes P_1 and P_2, we rotate all the vertices of v's one-ring lying on P_2 to P_1, and then recalculate the texture coordinate following equation (4) and (5). If this leads to the switch of triangle, the operation will also be cancelled (see Fig.11).
- No action is given for those vertices whose one-ring covers more than two planes.
- For those boundary vertices, who have not been optimized in the above steps, we adopt a relaxation scheme following [4].

We repeat the optimization procedure until no vertices need to be optimized or the maximum iteration number is reached.

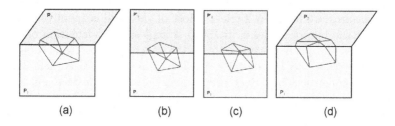

(a)   (b)   (c)   (d)

Fig. 11. Vertex slippage between two polycube planes

6   Experimental Results

We implement the algorithm on a standard PC with 2.4 GHz Pentium 4 CPU, 512MB RAM. Fig.12 gives the polycube map result of the Stanford Bunny, while Fig.13 shows the result for 3-ring torus which is a genus-3 surface. Fig.12(a) and

Fig. 12. The polycube parameterization of Bunny

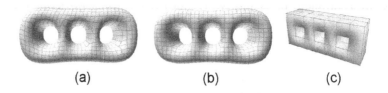

Fig. 13. The polycube parameterization of 3-Holes

Fig.13(a) show the texture map of the initial parameterization using the mean value coordinates, and Fig.12(b) and Fig.13(b) show the result after the iterative optimization procedure. It is very clear to see that the quality of the poly-cubic parameterization is greatly improved by the optimization procedure. The texel values of the texture map have been filled using a regular pattern. Fig.12(c) and Fig.13(c) are the normal maps of the Stanford Bunny and 3-ring torus respectively, where colors are used to encode unit surface normals. The time data and quality measures are listed in Table 1. Most of the time is spent on AGD calculation, for the bunny model with 15000 triangles, the calculation of geodesic distances between specified vertex to all the other vertices costs about 1 second. As we adopt about 150 samples, the total time for AGD is then about 150 seconds. The construction of Reeb graph takes 6 seconds, the segmentation and polycube approximation cost about 20 seconds, and the parameterization and optimization use about 15 seconds. The performance can be further improved by using approximate geodesic distance. For comparison purpose, we adopt the same measures as [11] to evaluate the quality of the automatic poly-cubic parameterization scheme. Areas and angle distortions are measured by integrating and normalizing the values $(\sigma_1\sigma_2 + 1/(\sigma_1\sigma_2))$ and $(\sigma_1/\sigma_2 + \sigma_2/\sigma_1)$ respectively, where σ_1 and σ_2 are the singular values of the Jacobian matrix J_Φ (see [19] for more details). The stretch metric is computed as [9]. For all measures, the optimal value is 1. Considering it is an automatic scheme, the parameterization quality is acceptable though not so excellent as the manual construction in [11]. There are still spaces for further improvement in the quality of polycube approximation, this will be taken into account in our future work.

Table 1. Statistics for Automatic Polycube Maps

Model	Triangle Number	Timing		Distortion				Stretch	
		AGD	Others	Area		Angle			
					[11]		[11]		[11]
Bunny	15000	150s	41s	1.15	1.034	1.12	1.069	0.79	0.892
3-Holes	4000	75s	25s	1.01	1.003	1.02	1.011	0.92	0.986
Horse	2000	20s	18s	1.22		1.32		0.75	

7   Conclusions

In this paper, we present an automatic method for constructing PolyCube-Maps. After going through a feature based mesh segmentation, each part of the model is approximated with an appropriate basic polycube primitive, and the whole polycube domain is created by putting these basic polycube primitives together. For every part, we further divide it into patches corresponding to the facets of its basic polycube primitive, and we parameterize these patches onto the facets using the mean value coordinate method. After going through an iterative optimization step, the final PolyCube-Maps is constructed. In the future, we will focus on improving the parameterization's quality. Besides, we will further explore the applications of this new parameterization mechanism in surface quadrangulation, and the quadrilateral base domain of polycube map makes it suitable for quadrilateral remeshing.

References

1. Bernnis, C., Vezien, J.M., Iglesias, G.: Piecewise surface flattening for nondistorted texture mapping. In: ACM SIGGRAPH, pp. 237–246 (1991)
2. Maillot, J., Yahia, H., Verroust, A.: Ineractive texture mapping. In: ACM SIGGRAPH, pp. 27–34 (1993)
3. Eck, M., DeRose, T., Duchamp, T., Hoppe, H., Lounsbery, M., Stuetzle, W.: Multiresolution analysis of arbitrary meshes. In: ACM SIGGRAPH, pp. 173–182 (1995)
4. Guskov, I., Vidimce, K., Sweldens, W., Schroder, P.: Normal meshes. In: ACM SIGGRAPH, pp. 95–102 (2000)
5. Lee, A., Sweldens, W., Schroder, P., Cowsar, L., Dobkin, D.: Maps: Multiresolution adaptive parameterization of surfaces. In: ACM SIGGRAPH, pp. 85–94 (1998)
6. Alexa, M.: Merging polyhedral shapes with scattered features. The Visual Computer 16(1), 26–37 (2000)
7. Gu, X., Yau, S.T.: Global conformal surface parameterization. In: Symposium on Geometry Processing, pp. 127–137 (2003)
8. Kobbelt, L.P., Vorsatz, J., Labisk, U., Seidel, H.-P.: A shrink-wrapping approach to remeshing polygonal surfaces. Computer Graphics Forum 18(3), 119–130 (1999)
9. Praun, E., Hoppe, H.: Spherical parameterization and remeshing. ACM Transactions on Graphics 22(3), 340–349 (2003)
10. Shapiro, A., Tal, A.: Polyhedron realization for shape transformation. The Visual Computer 14(8), 429–444 (1998)

11. Tarini, M., Hormann, K., Cignoni, P., Montani, C.: Polycube-maps. ACM Transactions on Graphics 23(3), 853–860 (2004)
12. Fan, Z., Jin, X., Feng, J., Sun, H.: Mesh morphing using polycube-based cross-parameterization. Computer Animation and Virtual Worlds 16, 499–508 (2005)
13. Wang, H., He, Y., Li, X., Gu, X., Qin, H.: Polycube splines. In: ACM Symposium on Solid and Physical Modeling, pp. 241–251 (2007)
14. Zhang, E., Mischaikow, K., Turk, G.: Feature-based surface parameterization and texture mapping. ACM Transactions on Graphics 24(1), 1–27 (2005)
15. Hilaga, M., Shinagawa, Y., Kohmura, T., Kunii, T.L.: Topology matching for fully automatic similarity estimation of 3d shapes. In: ACM SIGGRAPH, pp. 203–212 (2001)
16. Surazhsky, V., Surazhsky, T., Kirsanov, D., Gortler, S., Hoppe, H.: Fast exact and approximate geodesics on meshes. ACM Transactions on Graphics 24(3), 553–560 (2005)
17. Erickson, J., Har-Peled, S.: Optimally cutting a surface into disk. In: Symposium on Computational Geometry, pp. 244–253 (2002)
18. Floater, M.: Mean value coordinates. Computer Aided Geometric Design 20(1), 19–27 (2003)
19. Degener, P., Meseth, J., Klein, R.: An adaptable surface parameterization method. In: Proceedings 12th International Meshing Roundtable, pp. 201–213 (2003)

Bicubic G^1 Interpolation of Irregular Quad Meshes Using a 4-Split

Stefanie Hahmann[1], Georges-Pierre Bonneau[2], and Baptiste Caramiaux[1]

Laboratoire Jean Kuntzmann, CNRS UMR 5224
BP. 53, F-38041 Grenoble Cedex 9, France,
[1]Grenoble INP, [2]Université Joseph Fourier, INRIA

Abstract. We present a piecewise bi-cubic parametric G^1 spline surface interpolating the vertices of any irregular quad mesh of arbitrary topological type. While tensor product surfaces need a chess boarder parameterization they are not well suited to model surfaces of arbitrary topology without introducing singularities. Our spline surface consists of tensor product patches, but they can be assembled with G^1-continuity to model any non-tensor-product configuration. At common patch vertices an arbitrary number of patches can meet. The parametric domain is built by 4-splitting one unit square for each input quadrangular face. This key idea of our method enables to define a very low degree surface, that interpolates the input vertices and results from an explicit and local procedure : no global linear system has to be solved.

Keywords: G^1 continuity, arbitrary topology, interpolation, quad meshes, Bézier surfaces, 4-split

1 Introduction

Most existing modeling systems restrict to tensor product patches. But this type of surface is not well suited to model surfaces of arbitrary topological type. Tensor product surfaces generally need to have a chess boarder parametric domain, except torus like surfaces or singular parameterizations. However surfaces of arbitrary topology or surfaces with interesting features like suitcase corners and house corners require the existence of patch corners where three, five or more patches meet. Surfaces defined on irregular quadrilateral or triangular meshes are thus a powerful alternative to singularly parameterized tensor product surfaces.

While triangular meshes are widely used to model surfaces of arbitrary topology, quad (quadrilateral) meshes are more appropriate to model objects with symmetry axis like a simple vase or a statue for example. Furthermore modeling with quadrilateral surface patches can make profit from the powerful tools of existing modeling systems for purposes of efficient evaluation, display, intersections, or surface interrogation for example.

The present paper tackles the problem of defining a G^1-continuous surface **interpolating** the vertices of an irregular quad mesh with low degree polynomial tensor product patches. Irregular quad mesh means a mesh without tensor

F. Chen and B. Jüttler (Eds.): GMP 2008, LNCS 4975, pp. 17–32, 2008.

product configuration. Contrary to Catmull-Clark subdivision surfaces [1,9] and surface splines [15] and other works which are not designed to interpolate the given mesh, we aim to solve the *interpolation* problem in a completely general form. Several special interpolation methods exist [20,19,8,18], but they all make either restrictions on the mesh topology (vertices of valence 3,4,5 only) or on the input data, and lead to higher degree patches. Further methods deviate from standard polynomial representation [13,2]. Others solve the similar but more constraint problem of interpolation of a mesh of boundary curves [14,10] and thus need to make severe restrictions on the input data. The smoothness of interpolatory subdivision surfaces suffers from extraordinary points even with improved schemes [4,21] and don't provide an explicit polynomial representations. Others [11] only produce C^0 surfaces in order to overcome the mathematical difficult G^1 constraints. These constraints are generally replaced by simple heuristics in order to reduce the discontinuity. It is obvious that fewer constraints give degrees of freedom for shape design. Remains the problem how to use them optimally. However, these surfaces don't lie in the same continuity class as G^1 surfaces and can therefore not be compared with. It would be interesting to have convergent G^1 surfaces or even practical G^2 schemes, but this seems still to be an actual research problem [16].

In this paper a new surface scheme for G^1 interpolation of irregular quad meshes by tensor product Bézier patches is presented. The surface construction is based on a 1-to-4 split of the parameter domain for each surface patch. This technique has already been used successfully in the case interpolating triangular G^1 surfaces [6]. It is shown in this paper that thanks to the 4-split technique, the resulting surface is of very low degree(3,3) and locally defined. The interpolation scheme applies to any irregular 2-manifold surface mesh with or without boundary. The scheme is completely general in the sense that it doesn't make any restriction of the valence of the vertices, i.e. no restriction of the number of patches joining around a common vertex, and it doesn't make any restriction on the input data. Furthermore, it gives explicit formulas for all control points, no linear system has to be solved.

2 G^1 Continuity

2.1   Notations

A cubic Bézier curve is defined by $B(t) = \sum_{i=0}^{3} \boldsymbol{b}_i B_i^3(t)$, $t \in [0,1] \subset \mathbb{R}$ where the $\boldsymbol{b}_i \in \mathbb{R}^3$ are the Bézier control points and $B_i^3(t) = \binom{3}{i}(1-t)^{3-i} t^i$ are cubic Berstein polynomials. A bicubic tensor product Bézier patch is define by

$$B(u,v) = \sum_{i=0}^{3} \sum_{j=0}^{3} \boldsymbol{b}_{ij} B_i^3(u) B_j^3(v), (u,v) \in [0,1]^2 \subset \mathbb{R}^2. \tag{1}$$

Let $\mathcal{M}$ be a 2-manifold surface mesh in $\mathbb{R}^3$ with quadrilateral faces with or without boundary and of arbitrary topological genus. The mesh faces don't need

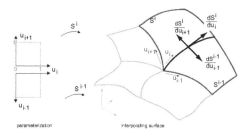

Fig. 1. Parameterization of the surface patches around a mesh vertex

to be planar, and there is no restriction on the number of faces meeting at a vertex. $\mathcal{M}$ is called an *irregular quad mesh*. The number of edges incident in a vertex is referred to as *order of a vertex*.

The problem we address is the computation of Bernstein-Bézier coefficients of a bicubic G^1 continuous surface interpolating the vertices of an irregular quad mesh. Since we want to be able to model a surface of arbitrary genus each surface patch can be seen as the image of one quadrilateral domain. Thus instead of considering a map from a regular, chess-board-like planar domain onto the surface, we consider many maps of one quadrilateral domain onto different patches of the surface.

Let us consider a mesh vertex $\boldsymbol{v} \in \mathbb{R}^3$ of order n, and its neighbouring points $\boldsymbol{v}_1, \ldots, \boldsymbol{v}_n$ ordered in a trigonometric sense. The patches around the vertex are numbered $S^1, \ldots, S^n$. The super- and subscripts $i = 1, \ldots, n$ are taken modulo n. Let $u_i \in [0, 1]$ be the parameter corresponding to the curve between $\boldsymbol{v}$ and $\boldsymbol{v}_i$, the surface patch S^i is then parameterized as illustrated in Figure 1.

2.2 General Tangent Plane Continuity

We require the interpolating surface to be polynomial and tangent plane continuous, i.e. G^1 continuous. This means that the surface is C^∞ everywhere except at the inner patch boundaries where it has continuously varying tangent planes.

Let S^i and S^{i-1} be two adjacent tensor product Bézier patches parameterized as in Figure 1. S^i and S^{i-1} join at the common boundary with tangent plane continuity, denoted G^1, if and only if there exist three scalar functions Φ_i, ν_i and μ_i such that

$$\Phi_i(u_i) \frac{\partial S^i}{\partial u_i}(u_i, 0) = \nu_i(u_i) \frac{\partial S^i}{\partial u_{i+1}}(u_i, 0) + \mu_i(u_i) \frac{\partial S^{i-1}}{\partial u_{i-1}}(0, u_i) , \qquad (2)$$

where $\nu_i(u_i) \cdot \mu_i(u_i) > 0$ (preservation of orientation, avoiding ridges) and $\frac{\partial S^i}{\partial u_i}(u_i, 0) \times \frac{\partial S^i}{\partial u_{i+1}}(u_i, 0) \neq 0$ (well defined normal vectors). Furthermore the surface patches are constructed such that the patch corners coincide with the mesh vertices (**interpolation**) and such that the patch boundaries correspond to the edges of the mesh.

2.3   Vertex Consistency Problem

At a patch corner, where n patches meet, the G^1 continuity is directly related to the twists (the second order mixed partial derivatives at a patch corner). Indeed, for polynomial patches, which lie in the continuity class C^2, both twists $\frac{\partial^2 S^i}{\partial u_i \partial u_{i+1}}(0,0)$ and $\frac{\partial^2 S^i}{\partial u_{i+1} \partial u_i}(0,0)$ are identical. Therefore the following necessary conditions of G^1-continuity must hold at the mesh vertices ($i = 1, \ldots, n$):

$$
\nu_i(0)\, \frac{\partial^2 S^i}{\partial u_i \partial u_{i+1}}(0,0) + \mu_i(0)\, \frac{\partial^2 S^{i-1}}{\partial u_{i-1} \partial u_i}(0,0)
$$

$$
= \Phi_i'(0)\, \frac{\partial S^i}{\partial u_i}(0,0) + \Phi_i(0)\, \frac{\partial^2 S^i}{\partial u_i^2}(0,0) - \nu_i'(0)\, \frac{\partial S^i}{\partial u_{i+1}}(0,0) - \mu_i'(0)\, \frac{\partial S^{i-1}}{\partial u_{i-1}}(0,0)
$$

$$
(3)
$$

Both equations (2) and (3) have to be satisfied in order to yield G^1 continuity of the surface patches around a corner vertex. These equations relate the values of the scalar functions and the first, second derivatives and the twists around a mesh vertex. Whatever the procedure is to construct the surface, either fix first the scalar function and compute then the surface control points, or choose first the surface control points and compute then the function values; in any case a serious difficulty arises: the matrices involved are generally circulant and singular when n is even. This problem is called the *twist compatibility problem*, or *vertex consistency problem*.

Our procedure consists in constructing first a network of C^2-consistent boundary curves and twists before filling-in the patches. We overcome the vertex consistency problem by computing combinations of control points which lie in the image space of these circulant matrices and thus avoid all singularity problems. This is the reason why we can deal with vertices of any valence, even or odd and we are able to give explicit formulas of the control points and don't need to solve any possibly singular system.

3   Bicubic Interpolation

In the rest of the paper we present now the new G^1 surface construction interpolating an irregular quad mesh which has the following important features:

- bicubic Bézier patches,
- local definition,
- interpolation of mesh vertices,
- 1-to-4 split of parameter domain,
- explicit formulas for all control points.

3.1   4-Split Method

The 4-split method has been introduced in [6] for triangular meshes. It can be adapted to quad meshes by 4-splitting the parameter domain at value $u_i = 1/2$. Therefore always four Bézier patches are constructed in correspondance to each mesh face. The group of four patches is referred to as *macro-patch*. The macro-patches join together with G^1 continuity. They are C^1 continuous inside.

This 4-split is the key contribution since it enables many important benefits: the resulting surface is simultaneously local and of very low degree while having a sufficient amount of degrees of freedom for shape design. The 4-split technique implies that the present surface construction method differs from all previously published methods since we are dealing with piecewise polynomial curves and surfaces instead of only one polynomial curve and surface segment.

3.2   Overview of the Algorithm

In order to satisfy the G^1-conditions between adjacent patches and at the mesh vertices (2) and (3), we first choose the scalar functions of as low degree as possible, see Section 3.3. In Section 3.4 we then compute a C^2-consistent boundary curve network and all remaining control points while always satisfying the G^1 equations. The computation of the surface control points is divided into three steps:

1.   Vertex computations: The first step constructs for all mesh vertices first and second derivatives for all incident boundary curve, and the twists for all incident patches.
2.   Edge computations: The second step constructs for all mesh edges the middle control point and all inner control points of the two adjacent patches which determine the cross-boundary tangents.
3.   Face computations: The third step constructs for all mesh faces the remaining inner control points of the corresponding macro-patch.

For all steps we will give explicit formulas of the Bézier control points allowing for a direct implementation. Due to the restricted number of pages we will not be able to give all proofs here. A complete and detailed version of this method is available as technical report [7].

3.3 G^1 Continuity Conditions and Choice of the Scalar Functions

From the general G^1 conditions, see Section 2, it can be seen that the degree of the polynomial patches can not be freely chosen. The degree is related to the degree of the scalar functions Φ_i, ν_i and μ_i since both sides of equation (2) need to be of same polynomial degree. In order to keep the degree of the patches low, the degree of the scalar functions has to be low too. Note also that the degree of Φ_i has to be of one higher than the degree of ν_i and μ_i in order to equalize the degree on both sides of equation (2). Ideally this would mean to take ν_i and μ_i constant and Φ_i linear. But then the surface construction would not be local, since $\Phi_i'(0)$ occurs in the second G^1 condition (3). In that way, the degree of the rectangular patches would not be raised. A local definition is important since the modification of one mesh vertex should not modify the whole surface but only the macro-patches around. The outside boundary of this group of macro-patches should not be affected by this modification. Therefore all computations around a vertex v should not depend on informations related to the neighbouring vertices v_i. This is the reason why in [12] Φ_i has been chosen quadratic for the construction of triangular patches. However in the present method we can make profit from the 4-split and define Φ_i piecewise linear. Thus the vertex informations can be separated between v and v_i.

Finally the choice of the scalar functions is the following for $i = 1, \ldots, n$:

$$\nu_i(u_i) \equiv \mu_i(u_i) \equiv \tfrac{1}{2} \quad \forall u_i \in [0, 1]$$

$$\Phi_i(u_i) \equiv \begin{cases} \cos(\frac{2\pi}{n})(1 - 2u_i) & \text{for } u_i \in [0, \tfrac{1}{2}] \\ -\cos(\frac{2\pi}{n_i})(2u_i - 1) & \text{for } u_i \in [\tfrac{1}{2}, 1] \end{cases} \tag{4}$$

where n_i is the order of the vertex v_i. We motivate the choice of Φ_i in the following observations: For symmetry reasons we choose $\nu_i = \mu_i = \tfrac{1}{2}$ and as simplification we introduce the notations

$$\begin{aligned} \Phi^0 &:= \Phi_i(0), \\ \Phi^1 &:= \Phi_i'(0). \end{aligned} \tag{5}$$

Therefore we have to look first at the G^1 conditions at the mesh vertices where $u_i = 0$. Condition (2) thus gives the following homogeneous system

$$\begin{bmatrix} -\Phi^0 & \tfrac{1}{2} & 0 & \cdots & 0 & \tfrac{1}{2} \\ \tfrac{1}{2} & -\Phi^0 & \tfrac{1}{2} & & & 0 \\ 0 & \ddots & \ddots & \ddots & & \vdots \\ \vdots & & & & & \tfrac{1}{2} \\ \tfrac{1}{2} & & & & \tfrac{1}{2} & -\Phi^0 \end{bmatrix} \begin{bmatrix} \frac{\partial S^1}{\partial u_1}(0,0) \\ \\ \vdots \\ \\ \frac{\partial S^n}{\partial u_n}(0,0) \end{bmatrix} = O$$

where the determinant is equal to $\Pi_{i=0}^{n} \cos(\frac{2\pi k}{n}) - \Phi^0$, see [3]. A non-trivial solution exists if and only if the determinant is equal zero, i.e. if $\Phi^0 = \cos(\frac{2\pi k}{n})$ for some integer k. We set $k = 1$ in order to ensure that the vectors $\frac{\partial S^i}{\partial u_i}(0,0)$ span a plane and are ordered properly. Thus we have $\Phi^0 = \Phi_i(0) = \cos\frac{2\pi}{n}$.

Then we have to look at the G^1 conditions at the opposite mesh vertex where $u_i = 1$. Here condition (2) implies

$$\left(\cos\frac{2\pi}{n_i}\right) \frac{\partial S^i}{\partial \bar{u}_i}(0,0) = \frac{1}{2}\frac{\partial S^i}{\partial \bar{u}_{i+1}}(0,0) + \frac{1}{2}\frac{\partial S^{i-1}}{\partial \bar{u}_{i-1}}(0,0),$$

where n_i is the order of the vertex associated with $S^i(1,0)$, and $\bar{u}_i = -u_i$, $\bar{u}_{i+1} = u_{i+1}$, $\bar{u}_{i-1} = u_{i-1}$.

Linearity and differentiation implies

$$\left(-\cos\frac{2\pi}{n_i}\right) \frac{\partial S^i}{\partial u_i}(0,0) = \frac{1}{2}\frac{\partial S^i}{\partial u_{i+1}}(0,0) + \frac{1}{2}\frac{\partial S^{i-1}}{\partial u_{i-1}}(0,0).$$

And so we deduce the value of Φ_i at $u_i = 1$ being equal $\Phi_i(1) = -\cos\frac{2\pi}{n_i}$.

Finally, if one chooses the functions Φ_i to be linear, this would imply that $\Phi^1 = \Phi_i'(0)$ in (3) depends on the order n_i of a neighbouring vertex. This would make the algorithm global instead of local, which is not acceptable. In our method, the 4-splitting of the domain rectangles enables to separate vertex information by choosing Φ_i **piecewise linear** and continuous, as defined in equation (4).

Why do we choose $\Phi_i(1/2) = 0$? First, setting a value for Φ_i at $u_i = 1/2$ ensures the locality property, because if one vertex is moved, the mid-edge value allows to compute the first derivative without opposite vertex informa-tion. Then, setting it equal 0 is justified by the observations that $n = n_i$ implies $\Phi_i(0) = -\Phi_i(1)$ and therefore the Φ_i would be a single linear function. Moreover, $n = n_i = 4$ implies $\Phi_i(u_i) = 0$ for all $u_i \in [0,1]$. This particular case corresponds to a tensor product configuration, and the C^1-continuity is guaranteed. This choice would not have been possible without 4-splitting the parameter domain.

Now the scalar functions have been chosen, the new **G^1 conditions** state as follows for $i = 1, \dots, n$:

$$\Phi_i(u_i) \frac{\partial S^i}{\partial u_i}(u_i, 0) = \frac{1}{2} \frac{\partial S^i}{\partial u_{i+1}}(u_i, 0) + \frac{1}{2} \frac{\partial S^{i-1}}{\partial u_{i-1}}(0, u_i), \quad u_i \in [0,1], \quad (6)$$

and

$$\frac{1}{2} \frac{\partial^2 S^i}{\partial u_i \partial u_{i+1}}(0,0) + \frac{1}{2} \frac{\partial^2 S^{i-1}}{\partial u_{i-1} \partial u_i}(0,0) = \Phi^1 \frac{\partial S^i}{\partial u_i}(0,0) + \Phi^0 \frac{\partial^2 S^i}{\partial u_i^2}(0,0), \quad (7)$$

with $\Phi^1 = \Phi_i'(0)$ and $\Phi^0 = \Phi_i(0) = \cos\frac{2\pi}{n}$, see (4) and (5). Condition (7) can be written in matrix form

$$T\,t = \Phi^1 d_1 + \Phi^0\, d_2 \qquad (8)$$

where

$$T = \begin{bmatrix} \frac{1}{2} & 0 & \cdots & 0 & \frac{1}{2} \\ \frac{1}{2} & \frac{1}{2} & \cdots & & 0 \\ & & \ddots & & \\ 0 & \cdots & & \frac{1}{2} & \frac{1}{2} & 0 \\ 0 & \cdots & & 0 & \frac{1}{2} & \frac{1}{2} \end{bmatrix}, \quad d_1 = \begin{bmatrix} \frac{\partial S^1}{\partial u_1}(0,0) \\ \vdots \\ \frac{\partial S^n}{\partial u_n}(0,0) \end{bmatrix}, \quad d_2 = \begin{bmatrix} \frac{\partial^2 S^1}{\partial u_1 \partial u_1}(0,0) \\ \vdots \\ \frac{\partial^2 S^n}{\partial u_n \partial u_n}(0,0) \end{bmatrix}$$

3.4   Surface Construction

First a C^2-consistent network of piecewise cubic Bézier boundary curves corre-sponding to the edges of the control mesh $\mathcal{M}$ is constructed. The end points interpolate the mesh vertices. The first and second derivatives and the twists at the end points satisfy the G^1 conditions (6) and (8). Then the G^1 continuity is established between adjacent patches along the boundary curves by computing the first inner row of control points on each side of the boundary curve. And finally the remaining Bézier control points are computed for each piecewise cu-bic macro-patch such that the resulting surface is globally G^1 continuous. The

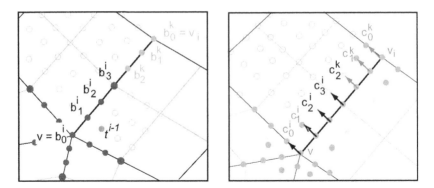

Fig. 2. Left: Boundary curve control points and twists computed around each mesh vertex. Right: Cross-boundary tangents and first inner row of control-points.

whole surface construction is governed by the requirements to get a locally defined polynomial surface of as low degree as possible and overall G^1 continuous.

The G^1-conditions and in particular the second one (8) relate the first and second derivatives of the boundary curve's end points and the twists of all patches around a mesh vertex. Since we want a locally defined surface we shall compute all these quantities around each vertex independent from all neighbouring vertices. Note that the degree of a polynomial curve interpolating first and second derivatives at the extremities should be ≥ 5. The advantage of using the 4-split is to separate these information by using only **piecewise C^1 cubic curves.**

1. Vertex neighbourhood computations. Let us consider the computations around a mesh vertex v of order n. For each boundary curve incident in v we compute only the first three control points $b_0^i, b_1^i, b_2^i \in \mathbb{R}^3$, $i = 1, \ldots, n$, of the cubic curve corresponding to the parameter interval $[0, \frac{1}{2}]$, see Figure 2 left.

Let us denote

$$
b_0 := \begin{bmatrix} b_0^1 \\ \vdots \\ b_0^n \end{bmatrix}_{n \times 3}, \quad
b_1 := \begin{bmatrix} b_1^1 \\ \vdots \\ b_1^n \end{bmatrix}_{n \times 3}, \quad
b_2 := \begin{bmatrix} b_2^1 \\ \vdots \\ b_2^n \end{bmatrix}_{n \times 3}
$$

the sets of Bézier control points around vertex v, and

$$
\bar{v} := \begin{bmatrix} v \\ \vdots \\ v \end{bmatrix}_{n \times 3}, \quad
p := \begin{bmatrix} v_1 \\ \vdots \\ v_n \end{bmatrix}_{n \times 3}
$$

where p is referred to the *vertex neighbourhood of v*. b_0, b_1, b_2 are computed such that the boundary curves interpolate the mesh vertex v and are compatible with the G^1 conditions (6) and (8). One possibility to satisfy these conditions is to compute the first and second derivatives

$$d_1 = 6(b_1 - b_0)$$
$$d_2 = 24(b_2 - 2b_1 + b_0)$$

of the boundary curve at $u_i = 0$ such that d_1 and d_2 lie in the image space of matrix (T) in equation (8) and simultaneously satisfy (6) at $u_i = 0$. It has been shown in [6,12] that control points satisfying these two conditions can be obtained as linear combinations of the vertex neighbourhood p as follows:

$$b_0 = \bar{v}$$
$$b_1 = \bar{v} + B^1 p \tag{9}$$
$$b_2 = (\gamma_0 + \gamma_1 + \tfrac{\gamma_2}{3})\bar{v} + B^2 p, \qquad \gamma_0 + \gamma_1 + \gamma_2 = 1$$

where B^1, B^2 are $n \times n$ matrices with

$$B_{ij}^1 = \tfrac{\beta}{n} \cos\left(\tfrac{2\pi(j-i)}{n}\right), \quad i, j = 1, \ldots, n$$
$$B_{ij}^2 = \tfrac{\gamma_1 \beta}{n} \cos\left(\tfrac{2\pi(j-i)}{n}\right) + \gamma_2 \begin{cases} 1/6 & \text{if } j = i-1, i+1 \\ 1/3 & \text{if } j = i \\ 0 & \text{otherwise.} \end{cases}$$

The scalar values $\beta, \gamma_0, \gamma_1, \gamma_2$ are shape parameters which can be freely chosen. β controls the magnitude of the tangent vector, and so the amplitude of the tangent plane at vertex v. A geometrical interpretation of (9) shows that the points b_1^i, form the affine image of a regular n-gon whose centroid is the point b_0. $\gamma_0, \gamma_1, \gamma_2$ control the second derivative, since b_2 is defined in [6] as $b_2 = \gamma_0 b_0 + \gamma_1 b_1 + \gamma_2(\tfrac{1}{3}b_0 + \tfrac{1}{6}v_{i-1} + \tfrac{1}{3}v_i + \tfrac{1}{6}v_{i+1})$. This particular definition of b_2 allows to prove that d_2 lies in the image space of matrix (T) in (8), and thus that the boundary curve network satisfies the G^1 conditions. See Section 4 for more discussion on the use of these parameters.

Let us now compute the twists

$$t = \begin{bmatrix} t^1 \\ \vdots \\ t^n \end{bmatrix}_{n \times 3} \qquad \text{with} \qquad t^i = \frac{\partial^2 S^i}{\partial u_i \partial u_{i+1}}(0,0).$$

In terms of Bézier control points of a bicubic tensor product surface, see (1) the twist is given by the vector $t^i = 18(b_{11} - b_{10} - b_{01} + b_{00})$ [5]. The twists can be interpreted geometrically as the vector which measures the deviation of control point b_{11} to the tangent plane in b_{00}, i.e. to the tangent plane at the mesh vertex v. Thus, once we have computed the twist, we can compute the patch control point b_{11}, the so-called twist point, see Figure 2 left.

The twists appear in G^1 condition (8). The right hand side has been constructed such that it lies in the image space of T. Thus there exist points $\tilde{b}_0, \tilde{b}_1, \tilde{b}_2$ such that

$$T\tilde{b}_0 = b_0, \quad T\tilde{b}_1 = b_1, \quad T\tilde{b}_2 = b_2.$$

Due to the simple structure of T and (9) it can now be verified easily that

$$\widetilde{b}_0^i = \tfrac{1}{2}(b_0^i + b_0^{i-1}) = b_0^i = v$$
$$\widetilde{b}_1^i = v + \sum_{j=1}^{n} \tfrac{\beta}{n}\left[\cos\left(\tfrac{2\pi(j-i)}{n}\right) + \tan\left(\tfrac{\pi}{n}\right)\sin\left(\tfrac{2\pi(j-i)}{n}\right)\right]v_j$$
$$\widetilde{b}_2^i = \gamma_0 b_0^i + \gamma_1 \widetilde{b}_1^i + \tfrac{1}{3}\gamma_2(v + v_i + v_{i+1})$$

are solutions of these three equations. Any linear combination of points lying in the image space of T does also, so is for d_1, d_2. It is now just as simple to check that b_0, b_1, b_2 satisfy the other G^1 condition (6) at $u_i = 0$ as well. Therefore it is now possible to solve (8) for the twists by computing

$$\begin{aligned}
\Phi^1 r^1 + \Phi^0 r^2 &= 6\Phi^1(b_1 - b_0) + 24\Phi^0(b_2 - 2b_1 + b_0)\\
&= (-6\Phi^1 + 24\Phi^0)b_0 + (6\Phi^1 - 48\Phi^0)b_1 + 24\Phi^0 b_2\\
&= T\Big[(-6\Phi^1 + 24(1+\gamma_0)\Phi^0)b_0 + (6\Phi^1 + (-48 + 24\gamma_1)\Phi^0)\widetilde{b}_1\\
&\quad + 8\gamma_2\Phi^0(v + v_i + v_{i+1})\Big].
\end{aligned}$$

and by inserting this expression into (8) :

$$\begin{aligned}
t_i &= 8\gamma_2\Phi^0(v_i + v_{i+1}) - 16\gamma_2\Phi^0 v\\
&\quad + \sum_{j=1}^{n} \tfrac{(6\Phi^1 + (24\gamma_1 - 48)\Phi^0)\beta}{n}\left[\cos\left(\tfrac{2\pi(j-i)}{n}\right) + \tan\left(\tfrac{\pi}{n}\right)\sin\left(\tfrac{2\pi(j-i)}{n}\right)\right]v_j.
\end{aligned}$$

(10)

To sum up, the vertex computations determine the curve control points b_0^i, b_1^i, b_2^i, and the twists around each vertex independently from its neighbours. The control points are C^2-consistent which means that they garantee the G^1-continuity and the twist compatibility of the patches at the patch corners.

2. Edge computations. The edge computations are now establishing the G^1-continuity between adjacent patches. Once again, these computations are local for each boundary curve independently of the others.

Let us consider a boundary curve between two adjacent patches S^i and S^{i-1} corresponding to the edge $\overline{vv_i}$ and let $b_0^L, b_1^L, b_2^L, b_0^R, b_1^R, b_2^R$ denote its control points computed in the previous section. The only remaining requirement on the boundary curves is to be of class C^1. Each boundary curve of the patch network is therefore a piecewise C^1 cubic curve, when setting

$$b_3^L = b_3^R = \frac{1}{2}(b_2^L + b_2^R). \tag{11}$$

G^1 continuity between adjacent patches can now be established by computing the first inner row of control points on each side of the boundary curve. These control points together with the boundary curve control points define the cross-boundary tangents. They have to satisfy the following G^1-conditions

$$\frac{\partial S^i}{\partial u_{i+1}}(u_i, 0) = \Phi_i(u_i)\frac{\partial S^i}{\partial u_i}(u_i, 0) + \Psi_i(u_i)V_i(u_i),$$

(12)

$$\frac{\partial S^{i-1}}{\partial u_{i-1}}(0, u_i) = \Phi_i(u_i)\frac{\partial S^i}{\partial u_i}(u_i, 0) - \Psi_i(u_i)V_i(u_i),$$

where Ψ_i is some scalar function and V_i some vector function.

These conditions are equivalent to (6) and (8), but they have the advantage that they are both written in terms of a single function V_i. V_i together with $\frac{\partial S^i}{\partial u_i}(u_i, 0)$ characterizes the tangent plane behaviour of the surface along the boundary curve $S^i(u_i, 0)$.

Now, Ψ_i and V_i have to be computed of as low degree as possible while satisfying (12). It is clear from these equations that the degrees of Ψ_i and V_i determine the degree of the resulting surface S^i. Thus if we are able to construct the product $\Psi_i V_i$ being cubic, left and right hand side of equations (12) would be cubic and thus the final surface would be bi-cubic.

It can now be shown that thanks to the 4-split Ψ_i can be chosen linear and V_i can be chosen piecewise C^0 quadratic. Ψ_i is given by

$$\Psi_i(u_i) = sin(\frac{2\pi}{n})(1 - u_i) + sin(\frac{2\pi}{n_i})u_i, \qquad u_i \in [0, 1] \qquad (13)$$

and the part of V_i next to vertex $\boldsymbol{v}$ parameterized over $[0, \frac{1}{2}]$ is given by

$$V_i(u_i) = \sum_{k=0}^{2} \boldsymbol{v}_k^2 B_k^2(2u_i), \qquad u_i \in [0, \frac{1}{2}]$$

where

$$\begin{aligned} \boldsymbol{v}_0 &= V^0 \boldsymbol{p}, \\ \boldsymbol{v}_1 &= V^1 \boldsymbol{p}, \end{aligned} \qquad (14)$$

with V^0, V^1 two $n \times n$ matrices:

$$V_{ij}^0 = \frac{6\beta}{n} \sin\left(\frac{2\pi(j-i)}{n}\right), \quad i, j = 1, \ldots, n,$$

$$V_{ij}^1 = \frac{1}{\psi_i^0} \left[(6\Phi^1 - 48\Phi^0 + 24\gamma_1\Phi^0)\tan(\frac{\pi}{n}) - 6\psi_i^1\right] \frac{\beta}{n} \sin\left(\frac{2\pi(j-i)}{n}\right)$$
$$+ \frac{4}{\psi_i^0}\gamma_2\Phi^0 \begin{cases} 1 & \text{if } j = i+1 \\ -1 & \text{if } j = i-1 \end{cases},$$

and $\Psi_i^1 := \Psi_i'(0) = sin(\frac{2\pi}{n}) - sin(\frac{2\pi}{n_i})$. The middle control vector $\boldsymbol{v}_2^i$ for each vector function V_i, $i = 0, \ldots, n$ can be chosen freely subject to $V_i(1/2^-) = V_i(1/2^+)$. But we choose V_i being C^1-continuous, i.e.

$$\boldsymbol{v}_2^i = \frac{1}{2}\boldsymbol{v}_1^i + \frac{1}{2}\boldsymbol{v}_1^k \qquad (15)$$

where $\boldsymbol{v}_1^k$ is known from the opposite vertex along the edge. All these values are obtained by making the cross-boundary tangents (12) compatible with the surface tangent plane and the twists at the end points which have already been determined in the previous section.

Let us know compute the cross-boundary tangents in Bézier form. $\frac{\partial S^i}{\partial u_{i+1}}(u_i, 0)$ is a piecewise C^0 cubic polynomial whose control vectors are proportional to the vectors between the first inner row of control points $\boldsymbol{c}_i$ and the boundary control

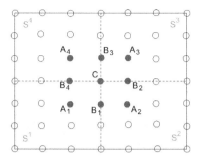

Fig. 3. Bézier control points of a piecewise bi-cubic macro-patch. The white control points ensure G^1-continuity between adjacent macro-patches. The dark control points ensure C^1-continuity inside a macro-patch.

points b_i, see Figure 2 right. The unknown control points c_j^i of the first part can be obtained by comparing the coefficients on both sides of the following identity:

$$\frac{\partial S^i}{\partial u_{i+1}}(u_i, 0) = \sum_{j=0}^{3} 6(c_j^i - b_j^i) B_j^3(2u_i)$$
$$= \Phi_i(u_i) \frac{\partial S^i}{\partial u_i}(u_i, 0) + \Psi_i(u_i) V_i(u_i), \qquad u_i \in [0, \tfrac{1}{2}]$$

$$= \cos(\tfrac{2\pi}{n})(1 - 2u_i) \sum_{j=0}^{2} 6(b_{j+1}^i - b_j^i) B_j^2(2u_i)$$
$$+ \left(\sin(\tfrac{2\pi}{n})(1 - u_i) + \sin(\tfrac{2\pi}{n_i})u_i \right) \sum_{j=0}^{2} v_j^i B_j^2(2u_i)$$

The control points of the right hand side are known from (4), (9), (13), (14), (15). The control points c_i^k from the second part of the cross-boundary tangents are known analogously from the opposite vertex.

3. Face computations. The last step of the algorithm computes the remaining control points inside each macro-patch. Remember that the 4-split produces piecewise C^1 cubic boundary curves and piecewise C^1 cubic cross-boundary tangents, which guarantees G^1 continuity of the macro-patches around all corner vertices and along the common boundary curves. Figure 3 shows all control points which are already determined as white dots. It will now be shown how to compute the remaining nine control points (dark dots) such that the four sub-patches join with C^1-continuity inside.

Following the C^1 condition for uniformly parameterized tensor product Bézier patches [5], and following the fact that the macro-patch joins its four neighbouring macro-patches with C^1-continuity at the parameter value $u_i = 1/2^1$, we get the following necessary and sufficient five conditions for the remaining nine control points:

---

[1] This point is quite crucial. Without C^1 continuity these edge mid-points would become ordinary vertices of valence 4, for which the G^1 conditions have to be applied. However by prooving that our construction automatically garantees C^1 continuity there, we don't have to worry about the C^2-consistency there. A formal proof involves (12) and the Φ function (4).

$$B_i = \frac{A_i + A_{i+1}}{2}, \qquad C = \frac{B_i + B_{i+2}}{2} = \frac{A_1 + A_2 + A_3 + A_4}{4}.$$

i=1,...,4. The subscripts are taken modulo 4, see Figure 3. It remain 4 degrees of freedom. In our implementation we choose A_1, A_2, A_3, A_4 as dof.

4   Results and Implementation Issues

The interpolation method offers several parameters as degrees of freedom, which can be used for shape design and optimization. Let us briefly discuss the possibility to fix them. There are two groups of parameters which act on different parts of the surface. First, there are three scalar values $\beta, \gamma_0, \gamma_2$ ($\gamma_1 = 1 - \gamma_0 - \gamma_2$) available for each mesh vertex. β controls the length of the first derivative vectors ($b_1 - b_0$). The γ values control the position of the second derivative control points b_2. Thanks to this intitive geometric meaning, they can be offered to the user as interactive design handles. Another possibility is to use a least-squares (LS) approximation to determine them optimally. We propose to approximate in a LS sense "ideal" boundary curves around each mesh vertex. A common heuristic to estimate these "ideal" boundary curves can be found in [17]. This leads to a small 3x3 linear system to solve. One can even reduce it to a 2x2 linear system when setting $\gamma_0 = -1$. Second, four control points inside each macro-patch are free. Once again, one can fix them in a LS sense by minimizing an energy functional or by using some heuristics. In our implementation we use a simple heuristic, the 4 points A_i are taken as linear combination of their neighbours offseted by some normal component.

The first example in Figure 4 shows how a simple cube can be interpolated. The different shapes have been obtained by varying the boundary curve's parameters β, γ_0 and γ_2. The following parameters have been used for the first cube $\beta = 0.13, \gamma_0 = -1.1, \gamma_2 = 0.1$, the second cube: $\beta = 0.33, \gamma_0 = -1.3, \gamma_2 = 0.3$, the third cube: $\beta = 0.13, \gamma_0 = -1.3, \gamma_2 = 0.3$. All vertices are of valence 3.

The second example in Figure 5 is a surface of topological genus 1, a tore. We show the resulting surface with and without the macro-patch boundary curves, the input mesh and the patch-work of all Bézier patches composing the tore.

The third example in Figure 6 is a complex surface of topological genus 2, a vase. The shape is more complex in the sense that it has varying curvature

Fig. 4. Quad mesh of a cube. Three interpolating G^1-contuous surfaces with varying paremeter sets.

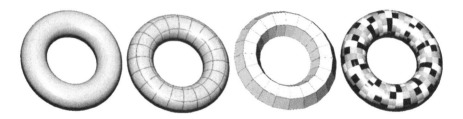

Fig. 5. Quad mesh of a tore: interpolating G^1 surface with and without boundary curves, input mesh and patch work of bicubic Bézier patches

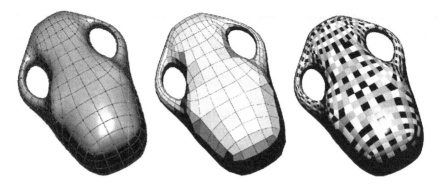

Fig. 6. Quad mesh of a vase: Resulting G^1-contuous surface with boundary curves, input mesh and patch work of bicubic Bézier patches

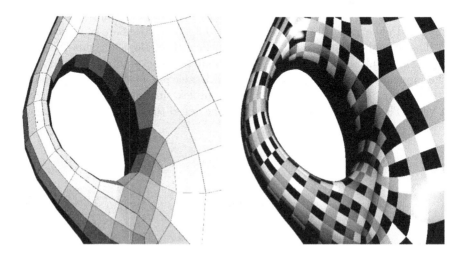

Fig. 7. Zoom on vertices with valence 3,4,5, and 6 of the vase data set

values and mesh vertices with different valences. A zoom on these vertices and the interpolating surface patches is shown in Figure 7.

5   Conclusion

We presented a method for quad mesh interpolation with G^1-continuous bi-cubic Bézier patches. The key contribution is a 1-to-4 split of the parameter domain of each patch. This enables the method to be local and of very low degree while satisfying the interpolation and the G^1-constraints. However sufficient number of degrees of freedom are available for shape design. The presentation we choose in this paper aims to facilite the direct implementation of this method. The algorithm is clearly presented in three steps. For each step we give the explicit formulas for all Bézier control points. This algorithm is completely general in the sense that it produces always a G^1 surface without any extraordinary or singular points, and it makes no restriction about the number of patches which meet around a mesh vertex.

It would further be interesting to study the possibility to get more degrees of freedom for the boundary curve's design. It is well known that their shape is mostly influencing the final shape. Future reserach will also adress the problem of fitting dense input data with that scheme.

Acknowledgments. The work was partially supported by the European Community 6-th framework program, with the Network of Excellence AIM@SHAPE (IST NoE No 506766) http://www.aimatshape.net. The model of Figures 6, 7 have been made available by C. Grimm at Washington University in Saint Louis.

References

1. Catmull, E., Clark, J.: Recursive generated b-spline surfaces on arbitrary topological meshes. Computer Aided-Design 10, 350–355 (1978)
2. Chiyokura, H., Kimura, F.: Design of solids with free-form surfaces. In: SIGGRAPH 1983: Proceedings of the 10th annual conference on Computer graphics and interactive techniques, pp. 289–298 (1983)
3. Davis, P.: Circulant Matrices. Wiley, Chichester (1979)
4. Dyn, N., Levine, D., Gregory, J.A.: A butterfly subdivision scheme for surface interpolation with tension control. ACM Trans. Graph. 9(2), 160–169 (1990)
5. Farin, G.: Curves and Surfaces for Computer Aided Geometric Design, 4th edn. Academic Press, New York (1996)
6. Hahmann, S., Bonneau, G.-P.: Triangular G1 interpolation by 4-splitting domain triangles. Computer Aided Geometric Design 17(8), 731–757 (2000)
7. Hahmann, S., Bonneau, G.-P., and Caramiaux, B.: Report on G¹ interpolation of irregular quad meshes. Technical Report IMAG RR 1087-I, Institut d'Informatique et de Mathématiques Appliquées de Grenoble, France (2007)
8. Hahn, J.: Filling polygonal holes with rectangular patches. In: Theory and practice of geometric modeling, pp. 81–91. Springer, New York (1989)

9. Halstead, M., Kass, M., DeRose, T.: Efficient, fair interpolation using catmull-clark surfaces. In: SIGGRAPH 1993: Proceedings of the 20th annual conference on Computer graphics and interactive techniques, pp. 35–44 (1993)
10. Liu, Q., Sun, T.C.: G1 interpolation of mesh curves. Computer-Aided Design 26(4), 259–267 (1994)
11. Liu, Y., Mann, S.: Approximate continuity for parametric bézier patches. In: SPM 2007: Proceedings of the 2007 ACM symposium on Solid and physical modeling, Beijing, China, pp. 315–321 (2007)
12. Loop, C.: A G^1 triangular spline surface of arbitrary topological type. Computer Aided Geometric Design 11, 303–330 (1994)
13. Loop, C., DeRose, T.D.: Generalized b-spline surfaces of arbitrary topology. In: SIGGRAPH 1990: Proceedings of the 17th annual conference on Computer graphics and interactive techniques, pp. 347–356 (1990)
14. Peters, J.: Smooth interpolation of a mesh of curves. Constructive Approx. 7, 221–246 (1991)
15. Peters, J.: C^1 surface splines. SIAM Journal of Numerical Analysis 32(2), 645–666 (1995)
16. Peters, J.: C^2 free-form surfaces of degree (3, 5). Computer Aided Geometric Design 19(2), 113–126 (2002)
17. Piper, B.: Visually smooth interpolation with triangular bézier patches. In: Farin, G. (ed.) Geometric Modeling: Algorithms and new Trends, pp. 221–233. SIAM, Philadelphia (1987)
18. Reif, U.: Biquadratic g-spline surfaces. Computer Aided Geometric Design 12(2), 193–205 (1995)
19. Sarraga, R.F.: G1 interpolation of generally unrestricted cubic bézier curves. Computer Aided Geometric Desgn 4(1-2), 23–39 (1987)
20. Van Wijk, J.J.: Bicubic patches for approximating non-rectangular control-point meshes. Computer Aided Geometric Design 3(1), 1–13 (1986)
21. Zorin, D., Schröder, P., Sweldens, W.: Interpolating subdivision for meshes with arbitrary topology. In: SIGGRAPH 1996: Proceedings of the 23rd annual conference on Computer graphics and interactive techniques, pp. 189–192 (1996)

Bounding the Distance between a Loop Subdivision Surface and Its Limit Mesh

Zhangjin Huang and Guoping Wang

School of Electronic Engineering and Computer Science,
Peking University, Beijing 100871, China
zhangjin.huang@gmail.com
http://www.graphics.pku.edu.cn

Abstract. Given a control mesh of a Loop subdivision surface, by pushing the control vertices to their limit positions, a *limit mesh* of the Loop surface is obtained. Compared with the control mesh, the limit mesh is a tighter linear approximation in general, which inscribes the limit surface. We derive an upper bound on the distance between a Loop subdivision surface patch and its limit triangle in terms of the maximum norm of the mixed second differences of the initial control vertices and a constant that depends only on the valence of the patch's extraordinary vertex. A subdivision depth estimation formula for the limit mesh approximation is also proposed.

Keywords: Loop subdivision surfaces, Limit mesh, Error bound, Subdivision depth.

1 Introduction

A subdivision surface is defined as the limit of a sequence of successively refined control meshes. In practice such as surface rendering, surface trimming and surface intersection, a linear approximation (for example, a refined control mesh) is used to substitute the smooth surface. It is natural to ask the following questions: How does one estimate the *error (distance)* between a limit surface and its linear approximation (for instance, the control mesh)? How many steps of subdivision would be necessary to satify a user-specified error tolerance?

Loop subdivision surfaces generalize the quartic three-directional box spline surfaces to triangular meshes of arbitrary topology [1]. Several works have been devoted to studying the distance between a Loop surface and its control mesh approximation [2,3,4,5,6]. These error estimation techniques can be classified into two classes.

One is the vertex based method [2,3], which measures the distance between the control vertices and their limit positions. In [2], Lanquetin et al. derived a wrong exponential bound which resulted in a improper subdivision depth formula. In [3], Wang et al. proposed a proper exponential bound with an awkward subdivision depth estimation technique. The vertex based bounds all suffer from one problem as pointed out in [3]: they may be smaller than the actual maximal distance between the limit surface and its control mesh in some cases.

F. Chen and B. Jüttler (Eds.): GMP 2008, LNCS 4975, pp. 33–46, 2008.

The other is the patch based method [4,5,6], which estimates the parametric distance between a Loop surface patch and its control mesh. Wu et al. presented an accurate error measure for adaptive subdivision and interference detection [4,5]. But their estimate is dependent on recursive subdivision, thus can not be used to pre-compute the error bound after n steps of subdivision or predict the recursion depth within a user-specified tolerance. In [6], Huang derived a bound in terms of the maximum norm of the mixed second differences of the initial control vertices, and a subdivision depth estimation approach.

The *limit mesh* of a Loop surface, obtained by pushing the control vertices of its control mesh to their limit positions, has been applied in surface reconstruction [8]. The limit mesh is usually considered a better linear approximation than the corresponding control mesh. But the approximation error for Loop surfaces has not been investigated yet. Recently, an effective approach has been proposed to bound the error of the limit mesh approximation to a Catmull-Clark subdivision surface [7]. With an analogous technique, in this paper we derive a bound on the maximum distance between a Loop surface patch and its limit triangle. The bound is expressed in terms of the maximum norm of the mixed second differences of the Loop patch's initial control mesh and a constant that depends only on the valence of the patch's extraordinary vertex.

The paper is organized as follows. Section 2 introduces some definitions and notations. In Sections 3 and 4, we derive distance bounds for regular Loop patches and extraordinary Loop patches, respectively. And a subdivision depth estimation method for the limit mesh approximation is presented in Section 5. In Section 6, we compare the limit mesh approximation with the control mesh approximation. Finally we conclude the paper with some suggestions for future work.

2   Definition and Notation

Without loss of generality, we assume the initial control mesh has been subdivided at least once, isolating the extraordinary vertices so that each face is a triangle and contains at most one extraordinary vertex.

2.1   Distance

Given a control mesh of a Loop subdivision surface $\tilde{\mathbf{S}}$, for each interior triangle $\mathbf{F}$ in the control mesh, there is a corresponding *limit triangle* $\overline{\mathbf{F}}$ in the limit mesh, and a corresponding surface patch $\mathbf{S}$ in the limit surface $\tilde{\mathbf{S}}$. Obviously, the limit triangle $\overline{\mathbf{F}}$ is the triangle formed by connecting the three corner points of $\mathbf{S}$.

The *distance* between a Loop patch $\mathbf{S}$ and the corresponding limit triangle $\overline{\mathbf{F}}$ is defined as the maximum distance between $\mathbf{S}(v, w)$ and $\overline{\mathbf{F}}(v, w)$, that is,

$$\max_{(v,w)\in\Omega} \|\mathbf{S}(v, w) - \overline{\mathbf{F}}(v, w)\| \ ,$$

where Ω is the unit triangle, $\mathbf{S}(v, w)$ is the Stam's parametrization [9] of $\mathbf{S}$ over Ω, and $\overline{\mathbf{F}}(v, w)$ is the linear parametrization of $\overline{\mathbf{F}}$ over Ω.

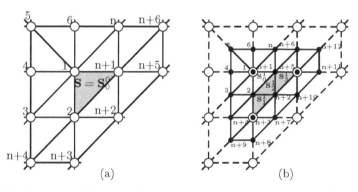

(a)   (b)

Fig. 1. (a) Ordering of the control vertices of an extraordinary patch **S** of valence n.
(b) Ordering of the new control vertices (*solid dots*) after one step of Loop subdivision.

2.2   Second Order Norm

The control mesh Π of a Loop patch **S** consists of $n + 6$ control vertices $\Pi = \{\mathbf{P}_i : i = 1, 2, \ldots, n+6\}$, where n is the valence of **F**'s only extraordinary vertex (if has, otherwise $n = 6$) and called the *valence* of the patch **S** (see Figure 1(a)).

The *second order norm* of Π, denoted $M = M^0 = M_0^0$, is defined as the maximum norm of the following $n + 9$ mixed second differences (MSDs) $\{\alpha_i : 1 \leq i \leq n+9\}$ of the $n + 6$ vertices of Π:

$$
\begin{aligned}
M = \max\{&\|\mathbf{P}_1 + \mathbf{P}_2 - \mathbf{P}_{n+1} - \mathbf{P}_3\|, \{\|\mathbf{P}_1 + \mathbf{P}_i - \mathbf{P}_{i-1} - \mathbf{P}_{i+1}\| : 3 \leq i \leq n\}, \\
&\|\mathbf{P}_1 + \mathbf{P}_{n+1} - \mathbf{P}_n - \mathbf{P}_2\|, \|\mathbf{P}_2 + \mathbf{P}_{n+1} - \mathbf{P}_1 - \mathbf{P}_{n+2}\|, \\
&\|\mathbf{P}_2 + \mathbf{P}_{n+2} - \mathbf{P}_{n+1} - \mathbf{P}_{n+3}\|, \|\mathbf{P}_2 + \mathbf{P}_{n+3} - \mathbf{P}_{n+2} - \mathbf{P}_{n+4}\|, \\
&\|\mathbf{P}_2 + \mathbf{P}_{n+4} - \mathbf{P}_{n+3} - \mathbf{P}_3\|, \|\mathbf{P}_2 + \mathbf{P}_3 - \mathbf{P}_{n+4} - \mathbf{P}_1\|, \\
&\|\mathbf{P}_{n+1} + \mathbf{P}_n - \mathbf{P}_1 - \mathbf{P}_{n+6}\|, \|\mathbf{P}_{n+1} + \mathbf{P}_{n+6} - \mathbf{P}_n - \mathbf{P}_{n+5}\|, \\
&\|\mathbf{P}_{n+1} + \mathbf{P}_{n+5} - \mathbf{P}_{n+6} - \mathbf{P}_{n+2}\|, \|\mathbf{P}_{n+1} + \mathbf{P}_{n+2} - \mathbf{P}_{n+5} - \mathbf{P}_2\|\} \\
= \max\{&\|\alpha_i\| : i = 1, \ldots, n+9\} \ .
\end{aligned}
$$

$$(1)$$

M is also called the (level-0) second order norm of the patch **S**. For a regular patch ($n = 6$), there are 15 mixed second differences.

Through subdivision we can generate $n+12$ new vertices $\mathbf{P}_i^1$, $i = 1, \ldots, n+12$ (see Figure 1(b)), which are called the level-1 control vertices of **S**. All these level-1 control vertices compose **S**'s level-1 control mesh $\Pi^1 = \{\mathbf{P}_i^1 : i = 1, 2, \ldots, n+12\}$. We use $\mathbf{P}_i^k$ and Π^k to represent the level-k control vertices and level-k control mesh of **S**, respectively, after k steps of subdivision of Π.

The level-1 control vertices form four control vertex sets $\Pi_0^1, \Pi_1^1, \Pi_2^1$ and Π_3^1, corresponding to the control meshes of the subpatches $\mathbf{S}_0^1, \mathbf{S}_1^1, \mathbf{S}_2^1$ and $\mathbf{S}_3^1$, respectively (see Figure 1b), where $\Pi_0^1 = \{P_i^1 : 1 \leq i \leq n+6\}$. The subpatch $\mathbf{S}_0^1$ is an extraordinary patch, but $\mathbf{S}_1^1, \mathbf{S}_2^1$ and $\mathbf{S}_3^1$ are regular triangular patches [9]. Following the definition in Equation (1), one can define the second order norms M_i^1

for $\mathbf{S}_i^1, i = 0, 1, 2, 3$, respectively. $M^1 = \max\{M_i^1 : i = 0, 1, 2, 3\}$ is defined as the second order norm of the level-1 control mesh Π^1. After k steps of subdivision on Π, one gets 4^k control point sets $\Pi_i^k : i = 0, 1, \ldots, 4^k - 1$ corresponding to the 4^k subpatches $\mathbf{S}_i^k : i = 0, 1, \ldots, 4^k - 1$ of $\mathbf{S}$, with $\mathbf{S}_0^k$ being the only level-k extraordinary patch (if $n \neq 6$). We denote M_i^k and M^k as the second order norms of Π_i^k and Π^k, respectively.

3   Regular Patch

In this section, both the regular Loop patch $\mathbf{S}$ and its corresponding limit triangle $\overline{\mathbf{F}}$ are first expressed in quartic triangular Bézier form. Then we bound the distance between $\mathbf{S}$ and $\overline{\mathbf{F}}$ by measuring the deviations between their corresponding Bézier points.

3.1   Quartic Triangular Bézier Form

If $\mathbf{S}$ is a regular Loop patch, then $\mathbf{S}(u, v)$ can be expressed as a quartic box spline patch defined by 12 control vertices $\mathbf{p}_i, 1 \leq i \leq 12$ (see Figure 2):

$$\mathbf{S}(v, w) = \sum_{i=1}^{12} \mathbf{p}_i N_i(v, w), \quad (v, w) \in \Omega , \qquad (2)$$

where $N_i(v, w), 1 \leq i \leq 12$ are the quartic box spline basis functions. The expressions of $N_i(v, w)$ refer to [9]. The surface is defined over the unit triangle:

$$\Omega = \{(v, w) \,|\, v \in [0, 1] \text{ and } w \in [0, 1 - v]\} .$$

We introduce the third parameter $u = 1 - v - w$ such that (u, v, w) forms a barycentric system of coordinates for the unit triangle.

$\mathbf{S}$ is a quartic triangular Bézier patch, thus $\mathbf{S}(u, v)$ can be written in terms of Bernstein polynomials [11]:

$$\mathbf{S}(u, v) = \sum_{i=1}^{15} \mathbf{b}_i B_i(v, w), \quad (v, w) \in \Omega , \qquad (3)$$

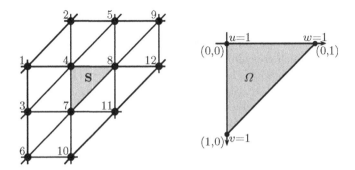

Fig. 2. Control vertices of a quartic box spline patch and their ordering

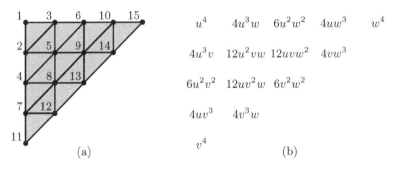

$$u^4 \qquad 4u^3w \qquad 6u^2w^2 \qquad 4uw^3 \qquad w^4$$

$$4u^3v \qquad 12u^2vw \qquad 12uvw^2 \qquad 4vw^3$$

$$6u^2v^2 \qquad 12uv^2w \qquad 6v^2w^2$$

$$4uv^3 \qquad 4v^3w$$

$$v^4$$

(a)   (b)

Fig. 3. (a) Ordering of the Bézier points of a quartic triangular Bézier patch. (b) Quartic Bernstein polynomials corresponding to the Bézier points.

where $\mathbf{b}_i, 1 \leq i \leq 15$ are the Bézier points of $\mathbf{S}$, and $B_i(v,w), 1 \leq i \leq 15$ are the quartic Bernstein polynomials (see Figure 3). The correspondence between the standard representation $B_{ijk}^4(u,v,w) = \frac{4!}{i!j!k!}u^iv^jw^k$, $i,j,k \geq 0$, $i+j+k = 4$ and $B_i(v,w), 1 \leq i \leq 15$ is $(i,j,k) \mapsto \frac{(j+k)(j+k+1)}{2} + k + 1$. With the algorithm developed in [10], we can get the 15×12 matrix T which converts from the 12 control vertices $(\mathbf{p}_i)$ to the 15 Bézier points $(\mathbf{b}_i)$ (see Appendix 1).

Note that $\overline{\mathbf{F}} = \{\mathbf{b}_1, \mathbf{b}_{11}, \mathbf{b}_{15}\}$ is the limit triangle corresponding to the center triangle $\mathbf{F} = \{\mathbf{p}_4, \mathbf{p}_7, \mathbf{p}_8\}$ (see Figures 2 and 3). The linear parametrization of $\overline{\mathbf{F}}$ is

$$\overline{\mathbf{F}}(v,w) = u\mathbf{b}_1 + v\mathbf{b}_{11} + w\mathbf{b}_{15} ,$$

where $u = 1-v-w$. By the linear precision property of the Bernstein polynomials [11], we can express $\overline{\mathbf{F}}(v,w)$ as the following quartic Bézier form:

$$\overline{\mathbf{F}}(v,w) = \sum_{i=1}^{15} \overline{\mathbf{b}}_i B_i(v,w) , \qquad (4)$$

where $\overline{\mathbf{b}}_{\frac{(j+k)(j+k+1)}{2}+k+1} = \overline{\mathbf{F}}(\frac{j}{4},\frac{k}{4})$, $j,k \geq 0$, $0 \leq j+k \leq 4$ are the Bézier points.

3.2   Bounding the Distance

According to the previous analysis, for $(v,w) \in \Omega$, it follows that

$$\|\mathbf{S}(v,w) - \overline{\mathbf{F}}(v,w)\| = \left\| \sum_{i=1}^{15}(\mathbf{b}_i - \overline{\mathbf{b}}_i)B_i(v,w) \right\| \leq \sum_{i=1}^{15} \|\mathbf{b}_i - \overline{\mathbf{b}}_i\| B_i(v,w) . \qquad (5)$$

In the following, we compute the smallest possible constants $\delta_i, i = 1, 2, \ldots, 15$ such that $\|\mathbf{b}_i - \overline{\mathbf{b}}_i\|$ is bounded by

$$\|\mathbf{b}_i - \overline{\mathbf{b}}_i\| \leq \delta_i M ,$$

where M is the second order norm of $\mathbf{S}$. It is obvious that $\delta_1 = \delta_{11} = \delta_{15} = 0$. $\mathbf{b}_i$ and $\overline{\mathbf{b}}_i, 1 \leq i \leq 15$ are the convex combinations of the control vertices $\mathbf{p}_i$,

$i = 1, 2, \ldots, 12$, and $\mathbf{b}_i - \overline{\mathbf{b}}_i$ can be expressed as a linear combination of the 15 MSDs $\alpha_l, l = 1, 2, \ldots 15$ defined in Equation (1) as follows:

$$\mathbf{b}_i - \overline{\mathbf{b}}_i = \sum_{l=1}^{15} x_l^i \alpha_l ,$$

where $x_l^i, l = 1, 2, \ldots, 15$ are undetermined real coefficients. It follows that

$$\|\mathbf{b}_i - \overline{\mathbf{b}}_i\| \leq \sum_{l=1}^{15} \|x_l^i \alpha_l\| \leq \sum_{l=1}^{15} |x_l^i| \|\alpha_l\| \leq \sum_{l=1}^{15} |x_l^i| M .$$

Therefore, to get a tight upper bound for $\|\mathbf{b}_i - \overline{\mathbf{b}}_i\|$, we solve the following constrained minimization problem:

$$\delta_i = \min \sum_{l=1}^{15} |x_l^i| \tag{6}$$
$$s.t. \quad \sum_{l=1}^{15} x_l^i \alpha_l = \mathbf{b}_i - \overline{\mathbf{b}}_i .$$

By symmetry, we only need to solve three constrained minimization problems. With the help of the symbolic computation of *Mathematica*, we have

$$\delta_1 = \delta_{11} = \delta_{15} = 0 ,$$
$$\delta_2 = \delta_3 = \delta_7 = \delta_{10} = \delta_{12} = \delta_{14} = \frac{1}{4} ,$$
$$\delta_4 = \delta_6 = \delta_{13} = \frac{1}{3} ,$$
$$\delta_5 = \delta_8 = \delta_9 = \frac{5}{12} .$$

It follows that

$$\mathcal{B}(v, w) = \sum_{i=1}^{15} \delta_i B_i(v, w) = v + w - v^2 - vw - w^2 .$$

We obtain a bound on the pointwise distance between $\mathbf{S}(v, w)$ and $\overline{\mathbf{F}}(v, w)$:

Theorem 1. *For $(v, w) \in \Omega$, we have*

$$\|\mathbf{S}(v, w) - \overline{\mathbf{F}}(v, w)\| \leq \mathcal{B}(v, w) M ,$$

where $\mathcal{B}(v, w) = v + w - v^2 - vw - w^2$ is called the distance bound function *of* $\mathbf{S}(v, w)$ *with respect to* $\overline{\mathbf{F}}(v, w)$.

By symmetry, the maximum of $\mathcal{B}(u, v), (u, v) \in \Omega$ must occur on the diagonal $\mathcal{B}(t, t) = 2t - 3t^2, 0 \leq t \leq \frac{1}{2}$. Since

$$\max_{0 \leq t \leq 1/2} 2t - 3t^2 = \frac{1}{3} = \mathcal{B}(\frac{1}{3}, \frac{1}{3}) ,$$

we have a bound on the maximal distance between $\mathbf{S}(u,v)$ and $\overline{\mathbf{F}}(u,v)$ as stated in the following theorem:

Theorem 2. *The distance between a regular Loop patch* $\mathbf{S}$ *and the corresponding limit triangle* $\overline{\mathbf{F}}$ *is bounded by*

$$\max_{(v,w)\in\Omega} \|\mathbf{S}(v,w) - \overline{\mathbf{F}}(v,w)\| \leq \frac{1}{3}M \ .$$

4   Extraordinary Patch

For an extraordinary patch, we first partition it into regular triangular sub-patches with Stam's parametrization [9], then derive a distance bound function for each regular subpatch with the technique developed in Section 3.

4.1   Stam's Parametrization

Through subdivision an extraordinary Loop patch $\mathbf{S}$ of valence n can be partitioned into an infinite sequence of regular triangular patches $\{\mathbf{S}_m^k\}, k \geq 1, m = 1, 2, 3$. If we partition the unit triangle Ω into an infinite set of tiles $\{\Omega_m^k\}, k \geq 1, m = 1, 2, 3$ (see Figure 4), accordingly, with

$$\Omega_1^k = \{(v,w) \,|\, v \in [2^{-k}, 2^{-k+1}] \text{ and } w \in [0, 2^{-k+1} - v]\} \ ,$$
$$\Omega_2^k = \{(v,w) \,|\, v \in [0, 2^{-k}] \text{ and } w \in [2^{-k} - v, 2^{-k}]\} \ ,$$
$$\Omega_3^k = \{(v,w) \,|\, v \in [0, 2^{-k}] \text{ and } w \in [2^{-k}, 2^{-k+1} - v]\} \ .$$

Each tile Ω_m^k corresponds to a box spline patch $\mathbf{S}_m^k$. And $\mathbf{S}_m^k(v,w)$ is defined over the unit triangle with the form as Equation (2). Therefore, the parametrization for $\mathbf{S}(v,w)$ is constructed as follows [9]:

$$\mathbf{S}(v,w) \,|_{\Omega_m^k} = \mathbf{S}_m^k(\tilde{v},\tilde{w}) = \mathbf{S}_m^k(\mathbf{t}_m^k(v,w)) \ ,$$

where the transformation $\mathbf{t}_m^k$ maps the tile Ω_m^k onto the unit triangle Ω.

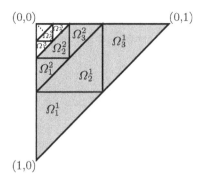

Fig. 4. Partition of the parameter domain Ω

The center triangle of $\mathbf{S}$'s control mesh is $\mathbf{F} = \{\mathbf{P}_1, \mathbf{P}_2, \mathbf{P}_{n+1}\}$ (see Figure 1), and the corresponding limit triangle is $\overline{\mathbf{F}} = \{\overline{\mathbf{P}}_1, \overline{\mathbf{P}}_2, \overline{\mathbf{P}}_{n+1}\}$, with $\overline{\mathbf{P}}_i$ being the limit point of $\mathbf{P}_i, i = 1, 2, n+1$. Let $\overline{\mathbf{F}}(u, v)$ be the linear parametrization of $\overline{\mathbf{F}}$:

$$\overline{\mathbf{F}}(v, w) = u\overline{\mathbf{P}}_1 + v\overline{\mathbf{P}}_2 + w\overline{\mathbf{P}}_{n+1}, \quad (v, w) \in \Omega . \tag{7}$$

The limit triangle $\overline{\mathbf{F}}$ can be partitioned into sub-triangles defined over Ω_m^k as follows:

$$\overline{\mathbf{F}}(v, w) \mid_{\Omega_m^k} = \widehat{\mathbf{F}}_m^k(\mathbf{t}_m^k(v, w))) .$$

Here $\widehat{\mathbf{F}}_m^k$ is the linear patch defined as

$$\widehat{\mathbf{F}}_m^k(v, w) = u\overline{\mathbf{b}}_1 + v\overline{\mathbf{b}}_{11} + w\overline{\mathbf{b}}_{15}, \quad (v, w) \in \Omega . \tag{8}$$

If the three corners of Ω_m^k are $(v_1, w_1), (v_2, w_2)$ and (v_3, w_3), which correspond to $(0, 0), (1, 0)$ and $(0, 1)$ in Ω via the transformation $\mathbf{t}_m^k$, respectively. Then

$$\overline{\mathbf{b}}_1 = \overline{\mathbf{F}}(v_1, w_1) , \quad \overline{\mathbf{b}}_{11} = \overline{\mathbf{F}}(v_2, w_2) , \quad \overline{\mathbf{b}}_{15} = \overline{\mathbf{F}}(v_3, w_3) .$$

4.2   Bounding the Distance

Similar to the analysis in Section 3, we can rewrite $\mathbf{S}_m^k(v, w)$ and $\widehat{\mathbf{F}}_m^k(v, w)$ into the quartic Bézier forms as Equations (3) and (4), respectively. Thus for $(v, w) \in \Omega_m^k$, we have

$$\|\mathbf{S}(v, w) - \overline{\mathbf{F}}(v, w)\| = \|\mathbf{S}_m^k(\tilde{v}, \tilde{w}) - \widehat{\mathbf{F}}_m^k(\tilde{v}, \tilde{w})\| \leq \sum_{i=1}^{15} \|\mathbf{b}_i - \overline{\mathbf{b}}_i\| B_i(\tilde{v}, \tilde{w}). \tag{9}$$

Notice that $\widehat{\mathbf{F}}_m^k$ is not the limit triangle of the triangular patch $\mathbf{S}_m^k$ but one portion of the extraordinary patch $\mathbf{S}$'s limit triangle $\overline{\mathbf{F}}$. So we can not use the results for $\|\mathbf{b}_i - \overline{\mathbf{b}}_i\|$ derived in Section 3 directly.

By solving 15 constrained minimization problems with the form similar to Equation (6), we have

$$\|\mathbf{b}_i - \overline{\mathbf{b}}_i\| \leq \delta_i M, \ 1 \leq i \leq 15 .$$

Consequently, it follows from Equation (9) that

$$\|\mathbf{S}_m^k(v, w) - \widehat{\mathbf{F}}_m^k(v, w)\| \leq \mathcal{B}_m^k(v, w)M, \ (v, w) \in \Omega ,$$

where the quartic Bézier function

$$\mathcal{B}_m^k(v, w) = \sum_{i=1}^{15} \delta_i B_i(v, w), \ (v, w) \in \Omega \tag{10}$$

is the *distance bound function* of $\mathbf{S}_m^k(v, w)$ with respect to $\widehat{\mathbf{F}}_m^k(v, w)$. Then the *distance bound function* of $\mathbf{S}(v, w)$ with respect to $\overline{\mathbf{F}}(v, w)$, $\mathcal{B}(v, w), (v, w) \in \Omega$, can be defined as follows:

$$\mathcal{B}(v, w) \mid_{\Omega_m^k} = \mathcal{B}_m^k(\mathbf{t}_m^k(v, w)), \ k \geq 1, m = 1, 2, 3 .$$

It is obvious that $\mathcal{B}(0,0) = \mathcal{B}(1,0) = \mathcal{B}(0,1) = 0$, and $\mathcal{B}(v,w)$ is a piecewise quartic triangular Bézier function over Ω away from $(0,0)$. Let $\beta(n) = \max_{(v,w)\in\Omega} \mathcal{B}(v,w)$, we have the following theorem on the maximal distance between $\mathbf{S}(v,w)$ and $\overline{\mathbf{F}}(v,w)$:

Theorem 3. *The distance between an extraordinary Loop patch* $\mathbf{S}$ *of valence* n *and the corresponding limit face* $\overline{\mathbf{F}}$ *is bounded by*

$$\max_{(v,w)\in\Omega} \|\mathbf{S}(v,w) - \overline{\mathbf{F}}(v,w)\| \le \beta(n)M \ , \tag{11}$$

where $\beta(n)$ is a constant that depends only on n, the valence of $\mathbf{S}$.

For $3 \le n \le 50$, we investigate the maximums of the quartic Bézier functions $\mathcal{B}_m^k(v,w), k \ge 1, m = 1,2,3$ over Ω assisted by plotting the graph of $\mathcal{B}(v,w)$. The following facts are found:

1. $\mathcal{B}(v,w)$ attains its maximum in the domain $\overline{\Omega} = \Omega \cap \{(v,w)\,|\,v+w \ge \frac{1}{4}\}$ (shaded region in Figure 4).
2. $\mathcal{B}(v,w)$ attains its maximum either on the diagonal $\mathcal{B}(t,t), t \in [0,\frac{1}{2}]$ or on the borders $\mathcal{B}(t,0)$ and $\mathcal{B}(0,t), t \in [0,1]$.

By symmetry, to compute the value of $\beta(n)$, at most four distance bound functions corresponding to $\mathbf{S}_1^1, \mathbf{S}_2^1, \mathbf{S}_1^2, \mathbf{S}_2^2$ are needed to be analyzed. Numerical results for $\beta(n), 3 \le n \le 10$ are given in Table 2. For regular Loop patches, the constant $\beta(n) = \frac{1}{3}$ is optimum. But for extraordinary Loop patches, the constants can be improved through further subdividing the subpatches $\mathbf{S}_m^k$.

5   Subdivision Depth Estimation

Before estimating subdivision depth, we investigate the recurrence relation between the second order norms of the control meshes of $\mathbf{S}$ at different levels.

5.1   Convergence Rate of Second Order Norm

If the second order norms M_0^{k+j} and M_0^j satisfy the following recurrence inequality

$$M_0^{k+j} \le r_j(n)M_0^k, \quad j \ge 0 \ , \tag{12}$$

where $r_j(n)$ is a constant which depends on n, the valence of the extraordinary vertex, and $r_0(n) \equiv 1$. We call $r_j(n)$ the *j-step convergence rate* of second order norm. The convergence rate reflects how fast the control mesh converges to the limit surface. The smaller the convergence rate is, the faster the control mesh converges. It is obvious that $r_{j+k} \le r_j r_k$.

Let $\alpha_i^k, i = 1,2,\ldots,n+9$ be the MSDs of $\Pi_0^k, k \ge 0$ defined as in Equation (1). For each $l = 1,2,\ldots,n+9$, we express α_l^{k+1} as a linear combination of α_i^k:

$$\alpha_l^{k+1} = \sum_{i=1}^{n+9} x_i^l \alpha_i^k \ ,$$

where $x_i^l, i = 1, 2, \ldots, n + 9$ are undetermined real coefficients. Then we can bound $\|\alpha_l^{k+1}\|$ by $c_l(n) M_0^k$, where $c_l(n)$ is the solution of the following constrained minimization problem

$$c_l(n) = \min \sum_{i=1}^{n+9} |x_i^l| \,,$$
$$s.t. \quad \sum_{i=1}^{n+9} x_i^l \alpha_i^k = \alpha_l^{k+1} \,. \tag{13}$$

Since $M_0^{k+1} = \max\{\|\alpha_l^{k+1}\| : 1 \leq l \leq n + 9\}$, we get an estimate for $r_1(n)$ as follows

$$r_1(n) = \max_{1 \leq l \leq n+9} c_l(n) \,.$$

By symmetry, we only need to solve at most four constrained minimization problems corresponding to $\alpha_1^{k+1} = \mathbf{P}_1^{k+1} + \mathbf{P}_2^{k+1} - \mathbf{P}_{n+1}^{k+1} - \mathbf{P}_3^{k+1}$, $\alpha_{n+1}^{k+1} = \mathbf{P}_2 + \mathbf{P}_{n+1} - \mathbf{P}_1 - \mathbf{P}_{n+2}$, $\alpha_{n+2}^{k+1} = \mathbf{P}_2 + \mathbf{P}_{n+2} - \mathbf{P}_{n+1} - \mathbf{P}_{n+3}$, and $\alpha_{n+3}^{k+1} = \mathbf{P}_2 + \mathbf{P}_{n+3} - \mathbf{P}_{n+2} - \mathbf{P}_{n+4}$, respectively. Since $\mathbf{P}_1^{k+1}$ is the extraordinary vertex, it is not surprising to find out that $c_1(n)$ is the maximum for $n > 3$. The special case is $r_1(3) = c_4(3) = 0.4375 > c_1(3) = 0.3125$. Then it follows that

$$r_1(n) = c_1(n), \quad n > 3 \,.$$

Similarly, we can estimate $r_j(n)$, $n \geq 3$, $j > 1$ by solving only one constrained minimization problem (13) with α_1^{k+1} replaced by α_1^{k+j}. Then we have

Lemma 1. *If M_0^k represents the second order norm of the level-k extraordinary subpatch $S_0^k, k \geq 0$ of valence n, then it follows that*

$$M_0^{k+j} \leq r_j(n) M_0^k, \quad j \geq 1 \,.$$

The above lemma works in a more general sense, that is, if M_0^k is replaced with M^k, the second order norm of the level-k control mesh Π^k, the estimates for $r_j(n)$ still work.

Though $r_2(n)$ can be roughly estimated as $r_1(n)^2$, in practice $r_2(n)$ may derive better results than $r_1(n)^2$ as shown in the next subsection. Table 1 shows the convergence rates $r_j(n), j = 1, 2, 3$ for $3 \leq n \leq 10$. The convergence rate $r_j(6) = (\frac{1}{4})^j$ is sharp for regular patches. $r_j(n) > r_j(6), n \neq 6$ means the control mesh near an extraordinary vertex converges to the limit surface slower than a regular mesh.

Table 1. Convergence rates $r_i(n)$, $i = 1, 2, 3$

n	3	4	5	6	7	8	9	10
$r_1(n)$	0.437500	0.382813	0.540907	0.250000	0.674025	0.705136	0.726355	0.741711
$r_2(n)$	0.082031	0.142700	0.258367	0.062500	0.372582	0.402608	0.424000	0.439960
$r_3(n)$	0.020752	0.053148	0.118899	0.015625	0.197695	0.219995	0.236377	0.248872

5.2   Subdivision Depth

Given an error tolerance $\epsilon > 0$, the *subdivision depth* of a Loop patch $\mathbf{S}$ with respect to ϵ is a positive integer d such that if the control mesh of S is recursively subdivided d times, the distance between the resulting limit mesh and $\mathbf{S}$ is smaller than ϵ.

The distance between an extraordinary Loop patch $\mathbf{S}(v, w)$ and its level-1 limit mesh can be expressed as

$$\max_{i=0,1,2,3} \max_{(v,w)\in\Omega} \|\mathbf{S}_i^1(v, w) - \overline{\mathbf{F}}_i^1(v, w)\| ,$$

where $\mathbf{S}_i^1, i = 0, 1, 2, 3$ are the level-1 subpatches of $\mathbf{S}$ as described in Section 2.2, and $\overline{\mathbf{F}}_i^1$ are the limit triangles corresponding to $\mathbf{S}_i^1, i = 0, 1, 2, 3$, respectively. It is easy to see that

$$\max_{(v,w)\in\Omega} \|\mathbf{S}_0^1(v, w) - \overline{\mathbf{F}}_0^1(v, w)\| \leq \beta(n)M_0^1 \leq \beta(n)r_1(n)M^0 ,$$

$$\max_{(v,w)\in\Omega} \|\mathbf{S}_i^1(v, w) - \overline{\mathbf{F}}_i^1(v, w)\| \leq \beta(6)M_i^1 \leq \frac{1}{3}r_1(n)M^0, \ i = 1, 2, 3 .$$

Then it follows that

$$\max_{i=0,1,2,3} \max_{(v,w)\in\Omega} \|\mathbf{S}_i^1(v, w) - \overline{\mathbf{F}}_i^1(v, w)\| \leq \max\{\beta(n), \frac{1}{3}\}r_1(n)M^0 .$$

Since $r_1(n) \geq \frac{1}{4}$ and $r_{i+1}(n) \geq \frac{1}{4}r_i(n)$, the distance between an extraordinary Loop patch $\mathbf{S}(v, w)$ and its level-k limit mesh is bounded by

$$\max_{i=0,1,\ldots,4^k-1} \max_{(v,w)\in\Omega} \|\mathbf{S}_i^k(v, w) - \overline{\mathbf{F}}_i^k(v, w)\| \leq \max\{\beta(n), \frac{1}{3}\}r_k(n)M^0 .$$

Assume $k = \lambda l_j + j, 0 \leq j \leq \lambda - 1$, then $r_k(n) \leq (r_\lambda(n))^{l_j}r_j(n)$. Let

$$\max\{\beta(n), \frac{1}{3}\}r_k(n)M^0 \leq \max\{\beta(n), \frac{1}{3}\}(r_\lambda(n))^{l_j}r_j(n)M^0 < \epsilon ,$$

then it follows that $l_j \geq \left\lceil \log_{\frac{1}{r_\lambda(n)}} \left(\frac{r_j(n)\max\{\beta(n), \frac{1}{3}\}M^0}{\epsilon} \right) \right\rceil$. Consequently, we have the following subdivision depth estimation theorem for Loop patches.

Theorem 4. *Given a Loop patch $\mathbf{S}$ of valence n and an error tolerance $\epsilon > 0$, after*

$$k = \min_{0\leq j\leq\lambda-1} \lambda l_j + j \tag{14}$$

steps of subdivision on the control mesh of $\mathbf{S}$, the distance between $\mathbf{S}$ and its level-k limit mesh is smaller than ϵ. Here,

$$l_j = \left\lceil \log_{\frac{1}{r_\lambda(n)}} \left(\frac{r_j(n)\max\{\beta(n), \frac{1}{3}\}M}{\epsilon} \right) \right\rceil, \quad 0 \leq j \leq \lambda - 1, \lambda \geq 1 .$$

In particular, for regular Loop patches, $k = \left\lceil \log_4 \left(\frac{M}{3\epsilon} \right) \right\rceil$.

6   Comparison

Both a control mesh and its corresponding limit mesh can be employed to approximate a Loop surface in practical applications. This section compares these two approximation representations within the framework of the second order difference techniques.

The distance between a Loop patch $\mathbf{S}$ of valence n and its control mesh can be bounded in terms of the second order norm M as [6]:

$$\max_{(v,w)\in\Omega} \|\mathbf{S}(v,w) - \mathbf{F}(v,w)\| \leq C_\lambda(n)M, \quad \lambda \geq 1 , \tag{15}$$

where

$$C_\lambda(n) = \beta(n)\frac{\sum_{i=0}^{\lambda-1} r_i(n)}{1 - r_\lambda(n)} .$$

Table 2 illustrates the comparison results of the constants $C_3(n)$ and $\beta(n)$ for $3 \leq n \leq 10$. It can be seen that $C_3(6) = \frac{1}{2}$ is the smallest of $C_3(n), n \geq 3$. $\beta(n) < C_3(6), n \geq 3$ means that the limit mesh approximates a Loop surface better than the corresponding control mesh in general.

Given a Loop patch $\mathbf{S}$ of valence n and an error tolerance $\epsilon > 0$, the subdivision depth estimation formula for the control mesh approximation is [6]:

$$k = \min_{0 \leq j \leq a-1} al_j + j , \tag{16}$$

where

$$l_j = \left\lceil \log_{\frac{1}{r_\lambda(n)}} \left(\frac{r_j(n)C_\lambda(n)M}{\epsilon} \right) \right\rceil, \quad 0 \leq j \leq \lambda - 1, \lambda \geq 1 .$$

For regular Loop patches, $k = \left\lceil \log_4 \left(\frac{M}{2\epsilon} \right) \right\rceil$.

Table 3 shows the comparison results of subdivision depths. The error tolerance ϵ is set to 0.0001, and the second order norm M is assumed to be 1. As can be seen from the table, the limit mesh approximation has a 20% improvement over the control mesh approximation in most of the cases.

Table 2. Comparison of $C_3(n)$ and $\beta(n)$, $3 \leq n \leq 10$

n	3	4	5	6	7	8	9	10
$C_3(n)$	0.872850	0.780397	0.858623	0.500000	0.875407	0.866176	0.856245	0.847476
$\beta(n)$	0.358813	0.350722	0.342499	0.333333	0.329252	0.332001	0.333880	0.335299

Table 3. Comparison of subdivision depths

n	3	4	5	6	7	8	9	10
control mesh	8	10	14	7	19	21	22	23
limit mesh	7	9	12	6	16	17	18	18

7   Conclusions and Future Work

In this paper we investigate the distance (error) between a Loop subdivision surface and its limit mesh. The maximal distance between a Loop patch and its limit triangle is bounded in terms of the second order norm of the initial control vertices and a constant that depends on the valence of the patch. An efficient subdivision depth estimation technique is also proposed. Test results show that a limit mesh approximates the limit surface better than the corresponding control mesh in general.

The bounds achieved are still upper bounds, not necessarily strict. In future work we hope to derive an accurate error measure with a technique similar to Wu et al's [4,5]. Besides the parametric distance, the bounds of other distances, such as the Hausdorff distance, are yet to be investigated.

Acknowledgments. We would like to thank the anonymous reviewers for their comments. This work was supported by the 973 Program of China (2004CB719403), NSF of China (60573151, 60473100), the 863 Program of China (2006AA01Z334, 2007AA01Z318), and China Postdoctoral Science Foundation (20060390359).

References

1. Loop, C.T.: Smooth Subdivision Surfaces Based on Triangles. M.S. Thesis, Department of Mathematics, University of Utah (1987)
2. Lanquetin, S., Neveu, M.: A Priori Computation of the Number of Surface Subdivision Levels. In: Proceedings of International Conference on Computer Graphics and Vision 2004 (Graphicon 2004), pp. 87–94 (2004)
3. Wang, H., Sun, H., Qin, K.: Estimating Recursion Depth for Loop Subdivision. International Journal of CAD/CAM 4(1), 11–18 (2004)
4. Wu, X., Peters, J.: Interference Detection for Subdivision Surfaces. Computer Graphics Forum 23(3), 577–585 (2004)
5. Wu, X., Peters, J.: An Accurate Error Measure for Adaptive Subdivision Surfaces. In: Proceedings of International Conference on Shape Modeling and Applications 2005 (SMI 2005), pp. 51–56 (2005)
6. Huang, Z.: Estimating Error Bounds and Subdivision Depths for Loop Subdivision Surfaces. Technical report, Peking University (2007), http://www.graphics.pku.edu.cn/hzj/pub/tr-loop-2007.pdf
7. Huang, Z., Wang, G.: Distance Between a Catmull-Clark Subdivision Surface and Its Limit Mesh. In: Proceedings of the 2007 ACM Symposium on Solid and Physical Modeling (SPM 2007), pp. 233–240 (2007)
8. Hoppe, H., DeRose, T., Duchamp, T., Halstead, M., Jin, H., McDonald, J., Schweitzer, J., Stuetzle, W.: Piecewise Smooth Surface Reconstruction. In: Proceedings of SIGGRAPH 1994, pp. 295–302. ACM, New York (2004)
9. Stam, J.: Evaluation of Loop Subdivision Surfaces. In: Subdivision for Modeling and Animation, SIGGRAPH 1999 Course Notes (1999)
10. Lai, M.J.: Fortran Subroutines For B-Nets of Box Splines on Three- and Four-Directional Meshes. Numerical Algorithms 2, 33–38 (1992)
11. Farin, G.: Curves and Surfaces for CAGD — A Practical Guide, 5th edn. Morgan Kaufmann Publishers, San Francisco (2002)

Appendix 1: Conversion Matrix T

The 15×12 matrix T which converts from the 12 control vertices of a quartic box spline surface patch to the 15 Bézier points of the corresponding quartic triangular Bézier patch is as follows:

$$T = \frac{1}{24} \begin{bmatrix}
2 & 2 & 2 & 12 & 2 & 0 & 2 & 2 & 0 & 0 & 0 & 0 \\
1 & 0 & 3 & 12 & 1 & 0 & 4 & 3 & 0 & 0 & 0 & 0 \\
0 & 1 & 1 & 12 & 3 & 0 & 3 & 4 & 0 & 0 & 0 & 0 \\
0 & 0 & 4 & 8 & 0 & 0 & 8 & 4 & 0 & 0 & 0 & 0 \\
0 & 0 & 1 & 10 & 1 & 0 & 6 & 6 & 0 & 0 & 0 & 0 \\
0 & 0 & 0 & 8 & 4 & 0 & 4 & 8 & 0 & 0 & 0 & 0 \\
0 & 0 & 3 & 4 & 0 & 1 & 12 & 3 & 0 & 0 & 1 & 0 \\
0 & 0 & 1 & 6 & 0 & 0 & 10 & 6 & 0 & 0 & 1 & 0 \\
0 & 0 & 0 & 6 & 1 & 0 & 6 & 10 & 0 & 0 & 1 & 0 \\
0 & 0 & 0 & 4 & 3 & 0 & 3 & 12 & 1 & 0 & 1 & 0 \\
0 & 0 & 2 & 2 & 0 & 2 & 12 & 2 & 0 & 2 & 2 & 0 \\
0 & 0 & 1 & 3 & 0 & 0 & 12 & 4 & 0 & 1 & 3 & 0 \\
0 & 0 & 0 & 4 & 0 & 0 & 8 & 8 & 0 & 0 & 4 & 0 \\
0 & 0 & 0 & 3 & 1 & 0 & 4 & 12 & 0 & 0 & 3 & 1 \\
0 & 0 & 0 & 2 & 2 & 0 & 2 & 12 & 2 & 0 & 2 & 2
\end{bmatrix}$$

A Carving Framework for Topology Simplification of Polygonal Meshes

Nate Hagbi and Jihad El-Sana

Computer Science Department
Ben-Gurion University
Israel

Abstract. The topology of polygonal meshes has a large impact on the performance of various geometric processing algorithms, such as rendering and collision detection algorithms. Several approaches for simplifying topology have been discussed in the literature. These methods operate locally on models, which makes their effect on topology hard to predict and analyze. Most existing methods also tend to exhibit various disturbing artifacts, such as shrinking of the input and splitting of its components. We propose a novel top-down method for topology simplification that avoids the problems common in existing methods. The method starts with a simple, genus-zero mesh that bounds the input and gradually introduces topological features by a series of carving operations. Through this process a multiresolution stream of meshes is created with increasing topologic level of detail. Following the proposed approach, we present a practical carving algorithm that is based on the Constrained Delaunay Tetrahedralization (CDT). The algorithm pretetrahedralizes the complement of the input with respect to its convex hull and then eliminates tetrahedra in a prioritized manner. We present quality results for two families of meshes that are difficult to simplify by all existing methods known to us - topologically complex and highly clustered meshes.

Keywords: model simplification, topology simplification, level-of-detail generation, shape approximation and geometric modeling.

1 Introduction

Topology simplification has been identified in the graphics literature as a critical task for various general and domain-specific applications. For example, simplifying the topology of complex models is often important for geometric simplification to achieve quality results. Topology simplification is also necessary for generating level-of-detail representations. For this purpose aggregating a large number of objects is required before simplifying each of the objects thoroughly. Topology simplification is important for efficiency reasons as well. For instance, the performance of collision-detection algorithms is known to depend heavily on the genus of objects. Topological features can also cause several undesirable artifacts, such as image-space aliasing.

F. Chen and B. Jüttler (Eds.): GMP 2008, LNCS 4975, pp. 47–61, 2008.

Several approaches have been proposed to reduce the topologic complexity of polygonal meshes. In most approaches, the simplification process has effects on topology that are hard to predict and analyze. Proximity-based methods, such as pair-collapse [10] [18] and clustering [17] [19], are based on proximity between vertices. They allow topological changes and aggregation but do not identify and remove parts of the surface that become internal. Thus, they may end up complicating topology rather than simplifying it. Such methods can also join components or do the opposite (see Figure 1). The number of components in the model can then increase and decrease unexpectedly before reaching one in a decimation process. In contrast, one would like each component of the original surface to be represented by at most one component in each generated approximation. Namely, the number of components is preferred to be monotonically decreasing or increasing through the decimation or refinement, respectively.

Another problem that is common in most existing simplification methods is shrinking of the input. In addition to reduction in the overall size of the model, shrinking causes small components to reach sub-pixel dimensions and finally disappear. A class of meshes particularly vulnerable to shrinking consists of meshes whose center is relatively highly clustered, i.e., meshes that consist of many small components with high density. In such models the vertices in the center pull the approximation toward the inside. In contrast, when viewing from the outside one would like a highly clustered model to be coarsened from the inside toward the outside. This would maintain its general shape and dimensions. This can also be motivated by a basic assumption that usually the central part of a complex model is less visible to an outside viewer. For instance, to simplify a model of a detailed motor that includes all the internal parts, it is reasonable to begin with a more aggressive simplification on the inside than the outside.

This work presents a novel method for the topology simplification of unstructured polygonal surfaces that addresses all the above issues. The method draws from the α-shape [7] and shrink-wrapping [16] concepts. The method defines a refinement scheme and operates on the input top-down. A simple genus-zero,

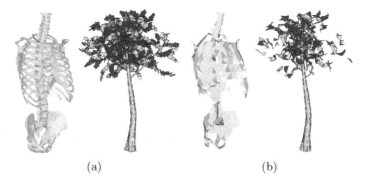

(a)   (b)

Fig. 1. The effect of the pair-collapse operator on a topologically complex model and a highly clustered model. (a) Original models (b) Pair-collapse simplified models. Note the increased number of disconnected components and shrunk appearance.

bounding object, such as the convex hull, is used as initial approximation. Topological features, such as splitting to components and tunnels, are gradually introduced into the approximation. This is done by carving parts of the complement of the input with respect to the initial approximation. This process is repeated until the input is reached. The method allows natural aggregation of multiple components and genus reduction, while maintaining the visual properties of the mesh.

The remainder of this paper is organized as follows. Section 2 reviews related topology simplification approaches, as well as several surface reconstruction and remeshing concepts. Section 3 outlines our general method and discusses design decisions regarding the implemention of a carving algorithm. Section 4 describes our carving algorithm and its implementation, which are based on the Constrained Delaunay Tetrahedralization (CDT). Section 5 presents our results. Section 6 summarizes our contribution, discusses limitations and concludes.

2   Related Work

In this section we briefly review closely related work in topology simplification, surface reconstruction and remeshing.

2.1   Topology Simplification

In various cases, topology simplifying steps have to be taken for performing aggressive geometric simplification without degenerating parts of the input and losing its basic characteristics. Meshes consisting of multiple disjoint components or scenes consisting of a large number of objects demand a different approach than simplifying each component separately. In such cases, even after each component has reached minimum-resolution, the number of minimum-resolution components can still imply a massive model size.

Rossignac and Borrel [19] imposed a global grid over the input dataset and cluster vertices into cells, which are replaced by new vertices. Later, Luebke and Erikson [17] extended this method and defined a tight octree over an input model to generate a view-dependent hierarchy. These approaches can simplify genus and reduce the number of components if the desired simplification regions fall within a grid cell.

He *et al.* [14] proposed to use a low-pass filter in the volumetric domain to simplify the genus of the input in a controlled fashion. This method is capable of aggregating disjoint components and reducing the genus of objects. However, there is no correlation between vertices of the input and vertices of the simplified mesh. In addition, meshes simplified by this method are subject to shrinking. Wood *et al.* [25] developed an algorithm for removing topological errors in isosurfaces in the form of tiny handles by sweeping the models and constructing a Reeb graph. Bischoff *et al.* [5] extract isosurfaces with controlled topology. They start with an initial estimate of the final surface and then morph the model into the final result using a topology preserving growing scheme. Zhou *et al.* [26] recently proposed a method for solid models, which repairs topological errors in the form of small surface handles. Gerstner and Pajarola [11] proposed a technique for controlled topology simplification by multiresolution isosurface extraction. Andujar *et al.* [2] have developed an automatic simplification algorithm to generate a multiresolution representation from an initial boundary representation of a polyhedral solid.

Popović and Hoppe [18], and Garland and Heckbert [10] proposed the (vertex) pair-collapse operator. Unlike edge-collapse based algorithms, pair-collapse based algorithms can join parts of a model together and therefore operate much better than edge-collapse on complex inputs consisting of numerous components [10]. However, the pair-collapse operator performs topology altering steps without analyzing the resulting topology and can demonstrate several undesirable artifacts. For instance, the method does not keep track of parts of the surface that become internal as a result of topology altering steps. Moreover, though able of aggregating disjoint components, the pair-collapse operator can also do the opposite, i.e., divide a single component (see Figure 1). Schroeder [20] introduced the vertex-split operator, which is designed to keep the mesh consistent when a topology simplifying vertex-decimation step takes place and a non-manifold neighborhood of a vertex is formed. Vertex-decimation is still, however, unable of aggregating disjoint components and does not identify and remove parts that become internal.

El-Sana and Varshney [9] roll a sphere of radius α over the surface of the input mesh. Areas not accessible to the sphere are identified and eliminated and the resulting holes are resurfaced. This approach suggests no way of aggregating disjoint components. In addition, smoothly resurfacing the eliminated parts of the surface is an intricate task. This makes the method suitable for objects with tunnels, holes and cavities, which are otherwise flat, such as mechanical parts, rather than smoothly curved, such as animal models. The operation also produces a single approximation per execution and suggests no intuitive way for automatically generating a multiresolution stream. Ju *et al.* [15] presented a guided topology editing algorithm. The resulting model is topologically consistent with a defined target shape. The algorithm allows removing and adding various topological features, while ensuring that each topological change is made by minimal modification to the source model.

2.2 Surface Reconstruction and Remeshing

Range scanning methods produce a set of points which are used to reconstruct a scanned physical object. Remeshing algorithms convert meshes into meshes with favorable properties, such as subdivision connectivity.

Edelsbrunner and Mücke [8] defined the 3D α-shape over the Delaunay tetrahedralization of a given set of points. Amenta and Bern [1] proposed the first surface reconstruction algorithm with correctness guarantee under a given sampling condition. Bernardini and Bajaj [4] proposed to use α-hulls to elegantly reconstruct a surface from an unorganized set of points that describes a mechanical model. Attene and Spagnuolo [3] proposed to construct the Delaunay tetrahedralization of the point set and then perform a sculpturing process. Gopi

et al.[13] developed a projection-based approach to determine the incident faces
on a point and used sampling criteria that guarantee a topologically correct
mesh after surface reconstruction. Recently, Giesen and John [12] extended the
Gabriel graph by associating its edges with the critical points of a dynamic system induced by the sample points. The above approaches start with a set of
points and aim at reconstructing the appropriate surface, while our approach
starts with a surface and aims to generate topology-simplified representations
for it. The result quality of the above reconstruction methods is measured with
respect to the distance of the reconstructed surface from the input point set,
while the result quality of topology simplification algorithms is measured with
respect to the input surface. Kobbelt *et al.* [16] simulate a shrink-wrap process to convert a given unstructured triangle mesh into a mesh with subdivision
connectivity. The complexity of the base meshes generated by this method is
relatively small. This approach only handles genus-zero objects and does not
perform topology-altering operations.

3   Geometric Carving

We propose to perform topological refinement by a top-down scheme. We consider the input model to be the objective of the process and begin with a coarse
initial approximation that bounds the input. By a series of operations we gradually introduce details into the initial approximation. Through each such operation we eliminate a part of the complement of the input with respect to the
initial approximation. Figure 2 illustrates an objective (f) carved from its convex
hull (a).

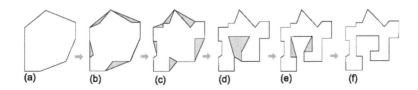

Fig. 2. The carving process

Carving can be applied to meshes in two or three dimensions. We describe the
method for 3-dimensional meshes and assume the application for 2-dimensional
meshes is clear. Given a mesh M, we first construct a coarse genus-zero mesh C_0
bounding M (for example, the convex hull of M). We then construct a tetrahedralization T of the complement of M with respect to C_0, C_0/M, in a fashion
constrained by the faces of C_0/M^1. Next, we iteratively remove the tetrahedra of
T from the current approximation until the objective M is reached. The tetrahedron elimination step is carried out by a carving operator, which takes the

---

[1] We will assume the complement is tetrahedralizable. Otherwise, we can introduce
Steiner vertices to tetrahedralize it [21].

mesh M and a tetrahedron $t \in T$ as input and performs the required update operations to turn the volume of t into volume external to the approximation. Through the process, we produce a chain of approximations, in which C_0 is the first and coarsest. Each application of the carving operator leads to a finer approximation of M. Let $C_i \xrightarrow{t} C_{i+1}$ denote that C_{i+1} is received by carving the tetrahedron t from C_i. A given order for the tetrahedra of T defines a chain of approximations for M,

$$C_0 \xrightarrow{t_0} C_1 \xrightarrow{t_1} \cdots \xrightarrow{t_{n-2}} C_{n-1} \xrightarrow{t_{n-1}} C_n = M, \tag{1}$$

where t_i is the ith tetrahedron carved and n is the number of tetrahedra in T. The key design decisions to be made in a carving algorithm are therefore the way of constructing the initial approximation, the manner of tetrahedralizing the complement and the order of removing the tetrahedra.

Tetrahedra that are on the surface of an approximation should be carved before internal tetrahedra. Otherwise, internal spaces will be formed, which increase the complexity of the model. For this purpose, let us first consider several definitions. A face of an approximation is said to be a constraining face if it belongs to M. The degree of a tetrahedron t with respect to an approximation C and an input mesh M, $d_{C,M}(t)$, is the number of faces of t that are non-constraining with respect to M and adjoin the exterior of C. We define a tetrahedron $t \in T$ to be $carvable$ from a mesh C_i if $d_{C_i,M}(t) > 0$, i.e., if at least one of its non-constraining faces adjoins the exterior domain of C_i. Equivalently, t is carvable from C_i if either t has a non-constraining face that adjoins the exterior domain of C_0 or one of the tetrahedra sharing a non-constraining face with t has been carved through the $i - 1$ first steps. We shall consider in each iteration only the carvable tetrahedra in T.

Notice that by the above definition of a carvable tetrahedron, internal isolated components of the complement are left unattained throughout the execution. These are caused by internal closed spaces of objects. We initially mark one inwards-facing face per isolated component to be a non-constraining face adjoining the exterior of the approximation. This causes a single tetrahedron per isolated component to be marked carvable and thus carved.

Using the above refinement scheme, topological features like tunnels are naturally formed, rather than eliminated, and can be more easily controlled. The process also naturally fuses multiple disconnected components together. The initial approximation is then a single object that splits throughout the process into separate components as the parts of the complement that connect them are eliminated. Figure 3 depicts two spheres splitting through the carving process. In addition, this process also reduces the number of vertices and faces required to represent the input, since the vertices and faces of the input mesh that are internal to an approximation and have not been introduced yet are not included in its representation.

If we consider the resulting multiresolution chain reversely, we begin with M and converge to the initial approximation C_0. The image-space area that is bounded by the silhouette of any approximation in the chain, from any viewpoint,

Fig. 3. Two spheres splitting through the carving process

contains the area bounded by the silhouette of the input. The approximation then monotonically expands within the boundaries of the initial approximation. This property prevents model parts from disappearing by reaching sub-pixel dimensions.

3.1 Initial Approximation

The initial mesh C_0 and the objective M form two opposite extremes for a range of approximations constructed through the process, since any approximation generated is contained by C_0 and contains M. C_0 is the coarsest approximation produced by the algorithm, which implies that in order to achieve maximal simplification of topology, C_0 should be a single component, genus-zero object.

The bounding-box of M seems to be, at first glance, a good initial approximation. The bounding-box of M is genus-zero, of constant complexity and can be calculated easily in linear time. Alas, the vertices of the bounding-box can lie far from any vertex of M, and can thus form features that hardly depend on M. Moreover, vertices of the bounding-box tend to receive relatively high degree in tetrahedralizations of the complement that do not introduce additional Steiner vertices into the faces of the bounding-box (see Figure 4(a)). This leads to unpleasant artifacts in produced approximations, such as sharp features.

The fairly opposite extreme for the choice of initial approximation is the convex hull of M, which consists solely of vertices of M and contains it (see Figure 4(b)). A problem of using the convex hull as initial approximation is that it prevents any simplification to the already fully detailed parts of the convex

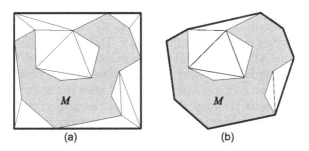

Fig. 4. Triangulation of the complements of M with respect to (a) the bounding box and (b) the convex hull of M

hull, which are the parts of the surface of M that lie on its convex hull. These parts start the process fully detailed and are not affected by any carving step. This problem can be handled in several ways and will be discussed shortly. For the results presented in this paper, we used the convex hull of M as initial approximation.

3.2   Complement Constrained Tetrahedralization

The complement can be tetrahedralized in advance to determine the pieces that ought to be removed in order to extract the input from the initial approximation. Although the tetrahedralization can be performed implicitly by carving operator applications, pretetrahedralization assists in clarifying the concept and still achieves quality results. The tetrahedralization method used is of high importance to the results of a carving algorithm, since it determines the characteristics of the tetrahedra to be carved and consequently the geometry and topology of generated approximations.

An important feature of a tetrahedron, which determines the visual change it introduces into a mesh when carved, is its circumradius-to-shortest edge ratio. Shewchuk [23] defines a d-simplex as *skinny* if its circumradius-to-shortest edge ratio is larger than some threshold constant to decide whether it should be eliminated from the triangulation. The effect of carving a skinny tetrahedron is expected to be relatively small and less perceptible, compared with the effect of carving a rounder tetrahedron of roughly the same size. Tetrahedralizations that consist of skinny tetrahedra thus usually effect in slow development of the approximation throughout the process, in which the carving is often tight with the surface of M. Carving round tetrahedra, on the other hand, yields rough and deep cavities, which can refine the approximation rapidly. Another feature that determines the effect of the carving operator on a mesh is the volume of the tetrahedron carved. Tetrahedra with larger volume introduce a larger effect into the approximation. As in any refinement scheme, we would like the process to converge rapidly toward the objective. These two features motivated our choice of the Constrained Delaunay Tetrahedralization (CDT) (see Section 4).

3.3   Carving Operator

The role of the carving operator is to perform the set of update operations to reflect a single carving operation to the surface. The effect of carving a triangle is illustrated in Figure 5 on a manifold two-dimensional mesh. Note that the surface of the approximation is denoted by bold lines before and after the operation. Carving a tetrahedron t from C is valid if and only if t is carvable. Taking a simple subset-placement strategy, this operation can be performed by removing a set of non-constraining faces from the mesh and introducing a set of new faces.

Let us formalize the operator for a mesh in R^d. Let t be a d-simplex, which is carvable from a mesh C, let F_{ext} be the set of hyperfaces of t that adjoin the exterior of C, and let F_{Mext} be the set of constraining hyperfaces in F_{ext},

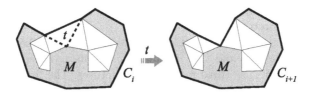

Fig. 5. The result of the carving operator applied by a triangle t to a mesh C_i, yielding C_{i+1}

i.e., $0 \leq |F_{Mext}| < |F_{ext}| \leq d + 1$. The carving operator eliminates the non-constraining hyperfaces of t in F_{ext} from C, and introduces the $d + 1 - |F_{ext}| + |F_{Mext}|$ remaining hyperfaces of t into C, as well as possibly a vertex of t if not already in the approximation. Care should be taken through the execution of the carving operator to avoid non-manifold configurations. Notice that while in a manifold object, and thus closed complement, constraining hyperfaces cannot simultaneously belong to a carvable d-simplex and adjoin the exterior domain, the situation is quite ordinary in non-manifold objects. In addition, by the above definition, for a non-manifold surface constraining hyperfaces are introduced along with their opposite-facing hyperfaces – their folds. Fold hyperfaces can easily be tracked and eliminated.

Let us consider the case in which $d_{C,M}(t) = 1$, i.e., a single non-constraining hyperface s of t adjoins the exterior of C, and assume carving t introduces a new vertex, $v \in t$ and $v \notin s$. The carving operator removes the hyperface s from the mesh and introduces the d remaining hyperfaces, which are adjacent to the new vertex v in t. Carving t from C in this case is equivalent to inserting a Steiner point inside s and contracting it to v. Note that only d-simplices of degree 1 may introduce a new vertex into the approximation, since in all other cases, the $d + 1$ vertices of the d-simplex already exist in the approximation.

If we reverse the basic carving operator to add d-simplices of degree 1 to the approximation, instead of carving them, we get a vertex-decimation operator, in which a single vertex of degree d is removed along with the hyperfaces in its neighborhood, and the resulting hole is retriangulated. Since the hole consists of exactly d vertices, the triangulated surface is a hyperface, consisting of the d vertices of the resulting hole.

4 A CDT Carving Algorithm

We propose to use the Constrained Delaunay Tetrahedralization (CDT) [6,22] to divide the complement. The CDT maintains several favorable properties, such as low circumradius-to-shortest edge ratio of tetrahedra. This allows reducing the number and severity of skinny tetrahedra to be carved, which introduce additional vertices and faces but contribute less to the refinement of the mesh through the multiresolution process. Skinny tetrahedra that do exist in the CDT are usually tight with constraining faces, which occlude most of the visibility within the circumsphere of the tetrahedron. Open spaces are regularly tetrahedralized by relatively round tetrahedra.

The CDT carving algorithm first constructs the CDT for the underlying domain of the complement and then begins to carve its tetrahedra until the objective is reached. We use three main data-structures. The first is a CDT data structure that maintains the connectivity between tetrahedra. We maintain the current approximation separately and update it in each step according to the operations performed on the CDT. In addition, we use a priority queue to manage the order of the tetrahedra, where at each stage tetrahedra that turn carvable are placed in the queue.

We would like to carve tetrahedra that fill wider spaces first. For instance, wide tunnels should be introduced in the refinement process before narrow tunnels. For this purpose, we experimented with various measure functions as queue keys, two of which yielded similar quality results - the volume of the tetrahedra and the longest-edge length of the tetrahedra, largest first in both cases. The longest-edge measure avoids artifacts met by the volume measure, such as sharp and small features. These result from the discontinuousness of the volume measure over neighbouring tetrahedra. For the resuls presented in this paper the longest-edge length measure was used.

Let us consider a heuristic that reduces the number of faces required for representing approximations, without changing the number of vertices. The purpose of the heuristic is generally to take advantage of each vertex introduced as much as possible for carving the complement. As mentioned above, only tetrahedra of degree 1 can introduce a new vertex into the approximation, while tetrahedra of higher degree consist solely of vertices that have already been introduced. Notice also that carving tetrahedra of degree higher than 1 does not increase the number of faces in the approximation. For instance, carving a tetrahedron of degree 2 is equivalent to flipping the exposed edge of the tetrahedron. Let us define a tetrahedron t to be a *dangling* tetrahedron of C with respect M, if $d_{C,M}(t) > 1$. Namely, a tetrahedron is dangling when more than one of its non-constraining faces adjoins the exterior of the current approximation. Dangling tetrahedra are first removed from the initial CDT. Then, through the carving process when a tetrahedron is carved its neighbors are checked to be dangling. If any of the neighbors is found to be dangling, it is carved as well and its neighbors are processed in the same manner.

Let us note a key feature of CDT carving that can be identified in the genus-reduction algorithm of El-Sana and Varshney [9]. To clarify the correlation, let us consider a refinement variation of the genus-reduction approach, in which the algorithm begins with an infinite-radius sphere, to which the area accessible on the surface of the mesh is exactly the area shared by the input and the initial approximation. To increase the level of detail one reduces the radius of the sphere, causing internal areas on the mesh that are less accessible from the exterior to be detailed. While the refinement version of the genus-reduction algorithm involves rolling a shrinking sphere over the mesh, CDT carving involves carving

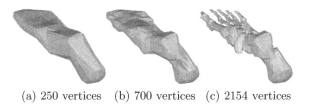

(a) 250 vertices   (b) 700 vertices   (c) 2154 vertices

Fig. 6. Topology simplified levels of detail for a foot bones model

shrinking tetrahedra inside the complement. The common effect of the methods is closing tighter spaces of the complement, while leaving open wider spaces.

The time complexity analysis of the algorithm for a 3D mesh is as follows. The extraction of the complement can be performed by checking for each face of C_0 whether it exists in M and vice-versa. This can be implemented using a hash table in time $O(c\log(c))$, where c is the maximum between the number of faces in M and the number of faces in C_0. The expected time of the convex hull construction step is $O(v\log(v))$, where v is the number of vertices in M. Using the sweep-plane algorithm by Shewchuk [22], one can construct the CDT of the complement in time $O(vs)$, where s is the number of tetrahedra in the CDT. Constructing the queue as a heap from the CDT is performed in $O(s\log(s))$ time. This is also the total time complexity of the carving steps through an algorithm execution. Since $v \le s$ and $c \le s$, the overall expected time complexity of the algorithm is $O(vs + s\log(s))$.

5 Results

Figure 8 demonstrates the effect of the CDT carving algorithm on a detailed torso model. The ribs of the torso form complex tunnels, which are impossible to simplify using topology preserving methods and intricate to simplify using existing topology simplifying operators, such as pair collapse (as demonstrated in Figure 1). The same difficulty holds for the sconce model in Figure 7. The tunnels of different size and shape are sealed by the CDT carving algorithm, due

Table 1. Sample running times

Model	Vertices	Hull Vertices	Faces	Tetra-hedra	Time (sec)
Yucca leave	801	87	1.2K	5.3K	0.3
Bones	2,154	124	4.2K	13.9K	1.3
Mushrooms	4,515	282	9.0K	30.0K	3.6
Elm	7,401	91	8.5K	47.4K	9.5
Horse	19,851	979	39.7K	137.9K	70.5
Scots Pine	90,125	151	61.7K	595.1K	847.9

Fig. 7. Topology simplified levels of detail of a sconce model

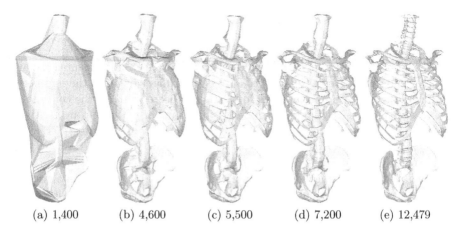

(a) 1,400   (b) 4,600   (c) 5,500   (d) 7,200   (e) 12,479

Fig. 8. Simplified levels of detail for a torso model and their corresponding vertex counts

to the relative size of tetrahedra in the area. Figure 8(d) presents a simplified version of the torso, which is visually almost identical to the original, due to the excessive detail of the spine in the original model. The genus of the torso remains low until the approximation acquires the desired shape of the input. Figure 9 illustrates a highly clustered foliage model, which consists of numerous parts (e.g. the leaves). The general shape of the branches becomes noticeable in Figure 9(d). Notice that throughout almost the whole refinement process, the foliage remains in the form of a single component. Figure 6 depicts the refinement process of a human foot bones model. The original model consists of numerous components for different bones. These remain connected throughout almost the whole process and the approximation remains bounding the bones.

The running-time of our current implementation for the CDT carving algorithm is depicted in Table 1. The machine used to achieve the results has 500Mhz Intel Pentium processor and 496MB of main memory. The running time of the CDT carving algorithm on each model consists of the running time required for constructing the constrained Delaunay tetrahedralization of the complement and for carving its tetrahedra. Our implementation makes use of the useful CDT library by Hang Si [24].

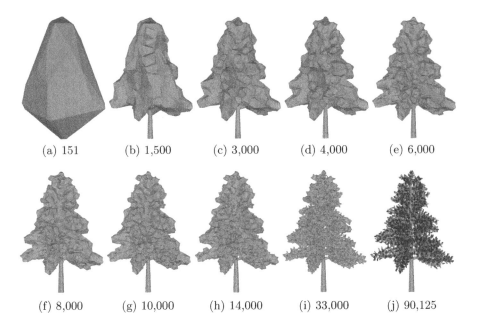

(a) 151	(b) 1,500	(c) 3,000	(d) 4,000	(e) 6,000

(f) 8,000	(g) 10,000	(h) 14,000	(i) 33,000	(j) 90,125

Fig. 9. Simplified levels of detail of a Scots Pine model and the number of vertices in each level

6   Conclusion and Future Work

We introduced a refinement method for the simplification of polygonal surfaces. The method aims at producing a chain of approximations morphing to an input mesh of arbitrary topology from a genus-zero, coarse, bounding mesh. The morphing is achieved by manipulating the input's complement with respect to the bounding mesh. We also described the CDT carving algorithm that follows the guidelines of the method to construct a multiresolution stream of quality approximations. The algorithm is useful for simplification of highly clustered meshes, a class of meshes that tend to lose their basic characteristics and shape by any general simplification method known to us so far. Viewing the approximation chain reversely as a decimation chain, the aggregation is monotonic, i.e., it constantly reduces the number of disjoint components. The approximations produced by the framework are silhouette-expanding, coarsened inside-out and converge to the convex-hull.

The proposed approach has several limitations that should be addressed. As explained in Section 3.1, a lower boundary on the complexity of approximations is derived from the initial approximation used, e.g. the convex hull. In some cases the convex hull may include much of the complexity of the mesh. A possible solution is to first perform a topology preserving, non-degenerating geometric simplification on the input (e.g., edge-collapse). The convex hull of the simplified result can then be calculated and used as initial approximation, and the simplified mesh as input to the carving process. This yields a decimation stream in which the model is first coarsened geometrically and then topologically. One of the advantages of this approach is that the carving calculations are performed on a smaller dataset. The CDT algorithm has relatively high time complexity. In fact, the carving process is performed offline at a preprocessing stage and generates approximations to be used online. The complexity of the algorithm is thus not necessarily a critical issue. In addition, in some special cases artifacts may be formed in the mesh through the carving process, such as sharp features. One of the ways to avoid these is by more sophisticated prioritization of the tetrahedra that measures the appropriate features.

We are working on extending the method in several main directions. Currently, inner vertices of the mesh naturally have lower priority than outer vertices. We would like to support semi-automatic prioritization of regions of the input mesh. This will allow preserving important geometric and topologic features, such as corners and tunnels. Another extension of the method is to combine it with geometric simplification, for instance by interleaving topology-preserving steps and carving steps. This will yield a general simplification method that correctly handles topology.

Acknowledgments. This work was supported by the Lynn and William Frankel Center for Computer Sciences. In addition, we would like to thank the reviewers for their constructive comments.

References

1. Amenta, N., Bern, M.W.: Surface reconstruction by voronoi filtering. In: Symposium on Computational Geometry, pp. 39–48 (1998)
2. Andujar, C., Ayala, D., Brunet, P., Joan-Arinyo, R., Sole, J.: Automatic generation of multi-resolution boundary representations. Computer-Graphics Forum 15(3), 87–96 (1996)
3. Attene, M., Spagnuolo, M.: Automatic surface reconstruction from point sets in space. Computer Graphics Forum 19(3), 457–465 (2000)
4. Bernardini, F., Bajaj, C.: Sampling and reconstructing manifolds using alpha-shapes. Technical Report CSD-97-013, Department of Computer Sciences, Purdue University (1997)
5. Bischoff, S., Kobbelt, L.P.: Isosurface reconstruction with topology control. Pacific Graphics 00(246) (2002)
6. Chew, L.P.: Constrained delaunay triangulations. Algorithmica 4(1), 97–108 (1989)
7. Edelsbrunner, H.: Weighted alpha shapes. Technical Report UIUCDCS-R-92-1760, Department of Computer Science, University of Illinois at Urbana-Champaign (1992)
8. Edelsbrunner, H., Mücke, E.: Three-dimensional alpha shapes. ACM transactions on Graphics 13(1), 43–72 (1994)
9. El-Sana, J., Varshney, A.: Topology simplification for polygonal virtual environments. IEEE Transactions on Visualization and Computer Graphics 4(2), 133–144 (1998)
10. Garland, M., Heckbert, P.: Surface simplification using quadric error metrics. In: Proceedings of SIGGRAPH 1997, pp. 209–216. ACM Press, New York (1997)

11. Gerstner, T., Pajarola, R.: Topology preserving and controlled topology simplifying multiresolution isosurface extraction. In: Proceedings of the IEEE Visualization 2000, pp. 259–266 (2000)
12. Giesen, J., John, M.: Surface reconstruction based on a dynamical system. Computer Graphics Forum 21(3) (2002)
13. Gopi, M., Krishnan, S., Silva, C.T.: Surface reconstruction based on lower dimensional localized delaunay triangulation. 19(3) (2000)
14. He, T., Hong, L., Varshney, A., Wang, S.: Controlled topology simplification. IEEE Transactions on Visualization and Computer Graphics 2(2), 171–184 (1996)
15. Ju, T., Zhou, Q.-Y., Hu, S.-M.: Editing the topology of 3d models by sketching. ACM Trans. Graph. 26(3), 42 (2007)
16. Kobbelt, L.P., Vorsatz, J., Labsik, U., Seidel, H.P.: A shrink wrapping approach to remeshing polygonal surfaces. Computer Graphics Forum 18(3), 119–130 (1999)
17. Luebke, D., Erikson, C.: View-dependent simplification of arbitrary polygonal environments. In: Proceedings of SIGGRAPH 1997, pp. 198–208. ACM Press, New York (1997)
18. Popović, J., Hoppe, H.: Progressive simplicial complexes. In: Proceedings of SIGGRAPH 1997, ACM SIGGRAPH, pp. 217–224. ACM Press, New York (1997)
19. Rossignac, J., Borrel, P.: Multi-resolution 3D approximations for rendering. In: Modeling in Computer Graphics, pp. 455–465. Springer, Heidelberg (1993)
20. Schroeder, W.J.: A topology modifying progressive decimation algorithm. In: IEEE Visualization 1997 Proceedings, pp. 205–212. SIGGRAPH Press (1997)
21. Shewchuk, J.R.: A condition guaranteeing the existence of higher-dimensional constrained delaunay triangulations. In: Proceedings of the Fourteenth Annual Symposium on Computational Geometry, Association for Computing Machinery, (Minneapolis, Minnesota), pp. 76–85 (1998)
22. Shewchuk, J.R.: Sweep algorithms for constructing higher-dimensional constrained delaunay triangulations. In: Proceedings of the Sixteenth Annual Symposium on Computational Geometry, pp. 350–208 (2000)
23. Shewchuk, J.R.: Delaunay refinement algorithms for triangular mesh generation. Computational Geometry: Theory and Applications 22(1), 86–95 (2002)
24. Si, H.: TetGen. A Quality Tetrahedral Mesh Generator and Three-Dimensional Delaunay Triangulator. Version 1.4. User's Manual (2006), `http://tetgen.berlios.de`
25. Wood, Z., Hoppe, H., Desbrun, M., Schröder, P.: Removing excess topology from isosurfaces. ACM Transactions on Graphics 23(2), 190–208 (2004)
26. Zhou, Q.-Y., Ju, T., Hu, S.-M.: Topology repair of solid models using skeletons. IEEE Transactions on Visualization and Computer Graphics 13(4), 675–685 (2007)

Comparing Small Visual Differences between Conforming Meshes[*]

Zhe Bian[1], Shi-Min Hu[1], and Ralph Martin[2]

[1] Department of Computer Science & Technology
Tsinghua University, Beijing, China, 100084
bz05@mails.tsinghua.edu.cn, shimin@tsinghua.edu.cn
[2] School of Computer Science, Cardiff University, Cardiff, UK
Ralph.Martin@cs.cardiff.ac.uk

Abstract. This paper gives a method of quantifying small visual differences between 3D mesh models with conforming topology, based on the theory of strain fields. Our experiments show that our difference estimates are well correlated with human perception of differences. This work has applications in the evaluation of 3D mesh watermarking, 3D mesh compression reconstruction, and 3D mesh filtering.

Keywords: 3D Conforming Meshes, Mesh Comparison, Perception, Strain Fields.

1 Introduction

3D surface triangle meshes are widely used in computer graphics and modelling; techniques such as watermarking, filtering and compression are often applied to meshes.

Watermarking is used to hide an 'invisible' digital signature into the mesh for information security and digital rights management [1,2,7,12,13,14]. The watermark information is encoded into small perturbations to the model's description, e.g. its vertex coordinates, changing the model's geometry by a small amount. Little work has been done on methods for evaluating the *quality* of watermarking schemes, which involves the *perceptibility* of the watermark (and other considerations such as how difficult it is to remove the watermark, etc). Mesh models constructed from 3D scanner data must be *denoised* before they are suitable for application purposes [18]. More generally, various *filters* may be applied to meshes to modify them in some way [19]. It is useful to be able to assess the *visual impact* of filtering algorithms. Both to economize use of bandwidth, and to save storage, *mesh compression* is useful [8,9,15,17]. Evaluation of *visual differences* between the reconstructed version and the original are again important.

Here, we give a new method for quantifying small visual differences between conforming meshes (i.e. meshes with the same number of triangles, connected in the same way). Conforming modification is often made in watermarking and filtering schemes, but perhaps less so in other applications.

[*] This work was supported by the National Basic Research Project of China (Project Number 2006CB303104), the Natural Science Foundation of China (Project Number 60673004) and by an EPSRC Travel Grant.

F. Chen and B. Jüttler (Eds.): GMP 2008, LNCS 4975, pp. 62–78, 2008.

Such a quality measure should agree with human perception, which is subjective and hard to quantify [20]. Nevertheless, our methodology, based on strain fields, provides a measure which is well correlated with perceptual results provided by human subjects.

2   Related Work

If a watermark becomes too obvious, distorting the model too much, a watermarking scheme is unacceptable. We may also wish to *compare* the perceptibility of the same watermarking information added by different schemes when deciding which scheme to adopt. A method adapted from image watermarking assessment is given by [10], while [6] provides two methods based on surface roughness. Few objective methodologies have been proposed, and even less attention has been paid to subjective evaluation [16]: as static 2D images of 3D models are inadequate for assessing the quality of a 3D model, there is a need for 3D quality metrics.

Other papers have discussed evaluation of the visual effects of mesh compression and filtering. Some are based on perceptual metrics taken from image processing [10], others consider geometric differences [3,5,6], and yet others combine both [22]. However, assessing perceptual degradation of images and 3D models are different tasks. [3,5] only consider geometric errors based on Hausdorff distance, which do not correlate well with human perception [6]. The two methodologies in [6] are better. One measures mesh distortion using roughness based on dihedral angles between faces. The other uses an equation based on variance of displacements (between the original and the smoothed meshes) along the normals of vertices in a 2-ring area of each vertex which performance is depended on smooth algorithms. One limitation is that these metrics use the roughness of the original mesh for normalization. If the roughness is large, distortion of smooth parts of the mesh is not captured, yet may be quite visible to an observer. These methods also take time more than linear in the number of mesh faces.

The most direct way of evaluating perceptibility of a change in a 3D mesh is to ask human observers. However, using enough 'typical' (whatever that means) human observers is time consuming and costly. Human beings do not give consistent and repeatable answers. Such tests can easily accidentally introduce bias . Thus, objective methodologies are preferable for assessing visual changes in meshes. This paper provides such a methodology, producing results well-correlated with those from human observers, in our limited testing.

Our approach analyzes shape and size changes of mesh triangles, ignoring any rigid-body motion. We use *strain energy* to quantify the deformation. Section 3 explains strain fields, and Section 4 shows how to compute strain in a mesh. Section 5 discusses how to compute a perceptibility distance based on strain energy. Section 6 and Section 7 describe an experiment to record human perception of visual changes in meshes, and experimental tests of our proposed methodology, including comparisons with other simpler approaches. Conclusions are drawn in Section 8.

3   Strain Fields

Strain and *stress* are used to describe pointwise mechanical behaviour inside a solid body [23]. At any point, strain represents the local deformation and stress the local

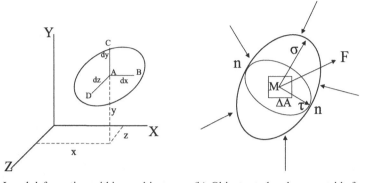

(a) Local deformation within an object   (b) Object acted on by an outside force

Fig. 1. *Strain* and *stress* in an object

force. Isotropic linear elasticity is the simplest mathematical model for continuum mechanics. It makes the following assumptions:

1. The material is continuous, and contains no voids.
2. The material is homogeneous.
3. The material is isotropic.
4. The material reverts to its original shape if external forces are removed; deformation is proportional to the applied external force.
5. Local deformation and rotation of the object are small.

The first four assumptions are plausible for most mesh models, so it is reasonable to treat them as elastic objects. The final assumption is also satisfied for the kind of application we consider. Thus, it is reasonable to use the concepts and methods of isotropic linear elasticity as a basis for measuring visual differences between conformal meshes.

The effects of the mesh processing algorithm under evaluation can be seen as equivalent to applying a fictitious external force which distorts the 3D mesh. We use the idea of a *strain field* to analyze the deformation of the mesh. Changes in position of the mesh vertices represent distortion of the mesh, but also incorporate a rigid-body motion which leaves the shape unchanged. To remove this rigid body motion, we use the concept of *strain*. In Figure 1(a), let A be some arbitrary point of the object. Consider three infinitesimal segments AB, AC, and AD, parallel to the coordinate axes and with lengths dx, dy and dz. When the object is acted upon by an external force, both the *lengths* of these lines and the *angles* between them change. The fractional changes in the lengths are called the *normal strain* at A, denoted by ϵ_x, ϵ_y and ϵ_z, respectively. The fractional changes in the angles are called the *shear strain*, denoted by γ_{xy}, γ_{yz} and γ_{zx}, respectively.

Given point A at (x, y, z), let $A' = (x + u, y + v, z + w)$ be its location after deformation and (u, v, w) be the corresponding displacement during the deformation. The strains describing the shape and volume change of the object can be expressed in terms of the displacements: if the local deformation and rotation are small, the relation between strain and displacement is linear:

$$
\begin{aligned}
\epsilon_x &= \partial u/\partial x, & \gamma_{xy} &= \partial v/\partial x + \partial u/\partial y \\
\epsilon_y &= \partial v/\partial y, & \gamma_{yz} &= \partial w/\partial y + \partial v/\partial z \\
\epsilon_z &= \partial w/\partial z, & \gamma_{zx} &= \partial u/\partial z + \partial w/\partial x
\end{aligned}
\tag{1}
$$

Deformation of a solid body requires *strain energy*. Associated with the local deformation, described by strain, there is a local force, *stress*, distributed throughout the object. The strain energy is the work done by the stress on the corresponding strain.

In Figure 1(b), let the plane n–n be the interface between the top and bottom parts of the object, M a fixed point on the plane, ΔA the differential area around M, and $\Delta \mathbf{P}$ the internal force acted on ΔA. The quantity $\Delta \mathbf{P}/\Delta A$ is called the average traction. At M, the stress component on the surface n–n is the limit of the average traction [21]:

$$
\mathbf{F} = \lim_{\Delta A \to 0} \Delta \mathbf{P}/\Delta A
\tag{2}
$$

The stress component $\mathbf{F}$ can be decomposed into two parts: the normal stress σ perpendicular to the n–n surface and the shear stress τ within the n–n surface.

4 Strain in a Mesh

We now discuss how strain and stress can be used to provide a measure of perceptibility of differences between two conforming meshes.

A 3D mesh is a shell composed of triangular faces of negligible thickness (see Figure 2(a)). If the mesh is distorted slightly, and we assume that its faces do not bend, to a first approximation the mesh triangles are unchanged in their normal direction: only their shapes and positions change [4]. Any elastic deformation only occurs within the plane of each triangle. The strain for each triangle can thus be computed in its own plane by ignoring any rigid body motion.

We interpolate the vertex displacements across each triangle to compute the strain for each triangle. See Figure 2(b). Let D, E, F be the vertices of a triangle in its initial position. We use a local (x, y) coordinate system for this triangle. After deformation, these vertices go to new positions. We project these in the z-direction back into the (x, y) plane, giving new points D', E', F'. The displacement of each vertex from its initial position is given by $x_{D'} = x_D + u_D$, $y_{D'} = y_D + v_D$. Finite element shape functions [24] are now used to interpolate the displacement functions u, v within the triangle. Consider u: v is computed similarly. We approximate u across each triangle by using: $u = a_1 + a_2 x + a_3 y$ where the a_i are to be determined. Putting u, x and y for each vertex of the triangle in turn into this expression and solving gives:

$$
\begin{aligned}
a_1 &= \frac{1}{2A}((x_E y_F - x_F y_E)u_D + (x_F y_D - x_D y_F)u_E + (x_D y_E - x_E y_D)u_F) \\
a_2 &= \frac{1}{2A}((y_E - y_F)u_D + (y_F - y_D)u_E + (y_D - y_E)u_F) \\
a_3 &= \frac{1}{2A}((x_F - x_E)u_D + (x_D - x_F)u_E + (x_E - x_D)u_F)
\end{aligned}
\tag{3}
$$

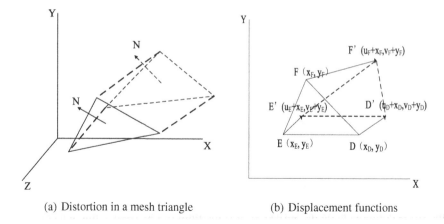

(a) Distortion in a mesh triangle   (b) Displacement functions

Fig. 2. Triangular face before and after distortion

where

$$2A = (y_E - y_F)x_D + (y_F - y_D)x_E + (y_D - y_E)x_F.$$

As we assume that each triangle deforms entirely within its own plane, there is no deformation or strain normal to each triangle. Thus, following the approach in [4], Equation (1) may be simplified in this case to give:

$$\epsilon_x = \frac{\partial u}{\partial x}, \quad \epsilon_y = \frac{\partial v}{\partial y}, \quad \epsilon_z = \frac{\nu}{\nu - 1}(\epsilon_x + \epsilon_y),$$

$$\gamma_{xy} = \frac{\partial v}{\partial x} + \frac{\partial u}{\partial y}, \quad \gamma_{yz} = 0, \quad \gamma_{zx} = 0, \tag{4}$$

where ν is Poisson's ratio (see later). We can now compute ϵ_x, ϵ_y and γ_{xy} from the displacement functions u, v obtained above.

5   Perceptibility and Strain Energy

5.1   Basic Approach

Following an approach used for images [7], we define a *perceptual distance* $P(m_0, m_p)$ between the original model m_0 and the processed model m_p: the larger the perceptual distance, the more visible. Our experiments show that we can define such a function using stress and strain which agrees well with human perception.

As ϵ_x, ϵ_y and γ_{xy} have different physical meanings and different directionality, it is not easy to construct a composite measure from them. The key is to use the concept of *strain energy*. *Strain energy density* is defined as

$$D = \frac{1}{2}(\sigma_x \epsilon_x + \sigma_y \epsilon_y + \sigma_z \epsilon_z + \tau_{xy}\gamma_{xy} + \tau_{yz}\gamma_{yz} + \tau_{zx}\gamma_{zx}).$$

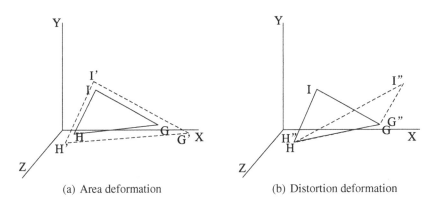

(a) Area deformation   (b) Distortion deformation

Fig. 3. Area and distortion deformation

Using the relations between stress and strain, the *strain energy* may be written as [23]:

$$W = DS_\triangle = \frac{1}{2}((\lambda + \frac{2}{3}G)\epsilon_{ii}^2 + (2G)\epsilon_{ij}'\epsilon_{ij}')S_\triangle \tag{5}$$

where $S_\triangle$ is the area of each triangle, and

$$\epsilon_{ii}^2 = (\epsilon_x + \epsilon_y + \epsilon_z)^2, \qquad \epsilon_{ij}'\epsilon_{ij}' = \epsilon_x'\epsilon_x' + \epsilon_y'\epsilon_y' + \epsilon_z'\epsilon_z' + \frac{1}{2}(\gamma_{xy}^2 + \gamma_{yz}^2 + \gamma_{zx}^2),$$

$$\epsilon_k' = \epsilon_k - \frac{1}{3}(\epsilon_x + \epsilon_y + \epsilon_z), \qquad k = x, y, z$$

E is *Young's modulus*, ν is *Poisson's ratio*, $\lambda = E\nu/((1 + \nu)(1 - 2\nu))$ is *Lamé's first parameter* and $G = E/(2(1 + \nu))$ is the *shear modulus*: these physical quantities determine the material's elastic properties. We simply fix them in our methodology to $E = 1$ and $\nu = 0$: these choices give good results in practice.

Elastic deformation of a planar triangle can be decomposed to into two parts, *area change* and *distortion* (shape) deformation: see Figure 3. As these two deformations may have different visual effects in a deformed mesh, we may divide the *strain energy* W into two corresponding parts: *area strain energy*, W_{Area} and *distortion strain energy*, $W_{\text{Distortion}}$ [23] which may be combined with different weights according to their respective effects. The *strain energy* for a single triangle can now be written as $W = W_{\text{Area}} + W_{\text{Distortion}}$, where

$$W_{\text{Area}} = \frac{1}{2}(\lambda + \frac{2}{3}G)\epsilon_{ii}^2 S_\triangle, \qquad W_{\text{Distortion}} = \frac{1}{2}(2G)\epsilon_{ij}'\epsilon_{ij}'S_\triangle.$$

5.2 Candidate Improvements

Distortions in 3D meshes may be of varying kinds, with varying perceptibility. We therefore considered how we might improve upon the basic ideas above. In particular, we considered two ideas:

Projection. Many applications of 3D meshes render them. The most important quantity determining the appearance of a triangle is its normal vector. Strain energy does not directly capture this idea. For example, when vertices are displaced within the local tangent plane of the mesh, the strain field can be large although a rendered image remains almost the same. Consider several adjacent triangles which lie in the same plane, and hence have the same normals. We may consider two types of distortion involving these triangles: ones in which the center vertex still lies in the plane, and ones in which the center vertex moves perpendicular to the tangent plane. The former distortions have no visual effect, and thus should be ignored. More generally, we can represent any local distortion as a combination of within-tangent-plane distortion and normal distortion. To remove the contribution of the in-tangent-plane distortion, we project the vertex after distortion back into the original tangent plane, and calculate the strain energy with respect to this adjusted position of this common vertex.

Edge triangles. Triangles in different locations generally have differing visual impact. For a model with smooth faces bounded by sharp edges, distortions in triangles at edges are likely to more noticeable. We thus accordingly apply a weight w_i to each triangle's strain as follows. We set $w_i = \pi - \alpha_i$, if α_i is less than $\pi/2$, and $w_i = 1$ otherwise, where α_i is the smallest dihedral angle between face i and its neighbours.

As we will see later, these two proposed improvements actually have little useful effect on our results.

5.3   Perceptual Distance

We now consider how to covert strain energy into a *perceptual distance*. The distortion must be measured *relative* to the size of the model, and should also be independent of the number of triangles. We thus define the *perceptual distance* $P(m_0, m_p)$ to be the weighted average strain energy (ASE) over all triangles (processing tangent triangular faces), normalized by S, the total area of the triangular faces:

$$P(m_0, m_p) = \frac{1}{S} \sum w_i W_i. \tag{6}$$

We may also split $P(m_0, m_p)$ into two separate components $P_{\text{Area}}(m_0, m_p)$ and $P_{\text{Distortion}}(m_0, m_p)$ corresponding to the area and distortion components of W_i. They may then be combined giving, $P(m_0, m_p) = P_{\text{Area}}(m_0, m_p) + P_{\text{Distortion}}(m_0, m_p)$. Alternatively, we might wish to ensure that *both* types of distortion are independently less than some threshold for a difference to be considered imperceptible, or that *both* quantities are less for one mesh change than another. Alternatively, we could try to determine the relative importance of these two strain energies, and produce an overall perceptual distance based on e.g. a weighted sum of these quantities. In practice, this does not seem to give any improvement.

6   Evaluation

We now give three experiments, and show that our measure of perceptibility produces results that correlate well with those assessed by human subjects.

Experiment I gives human opinions concerning the perceptibility of changes caused by making changes to models used in Experiments II and III. Experiment II considers two simple measures of perceptibility of changes, based on triangle areas, and triangle normals, and shows that we do better than using such simple measures. Experiment III tests the candidate improvements.

6.1 Experiment I: Human Perception of Differences Versus Strain Field Measure

To obtain some ground truth, i.e. subjective human results on the perceptibility of various changes in mesh models [20], we followed the approach in [6]. We embedded data into mesh models of a chess king and a horse, using the *Triangle Similarity Quadruple* (TSQ) watermarking method [13], and a noise embedding method.

To measure the subjective degree of mesh deformation perceived by human observers, we produced meshes with different amounts of deformation. We prepared variants of the horse, and of the chess king, resulting in 15 meshes for each: the undeformed model, 7 with watermarks embedded in different parts of the mesh, incorporating different amounts of data, and 7 with different amounts of noise embedded at different places. Subjects were asked to rate the differences between the original and processed models on a scale from from 0 to 10, where 0 meant identical and 10 meant very dissimilar, to give an *opinion score* (OS).

In order to help the subjects evaluate differences in the 3D mesh models, we paid careful attention to the rendering conditions, as suggested by [6]:

- *Color.* We used black for the background to help models stand out. Models were coloured grey, which makes edges more visible and deformations easier to see.
- *Light source.* All models used the same single white point light source: multiple lights can confuse observers.
- *Lighting.* Although a local illumination model can produce more realistic effects for textured models, it can hide parts with high distortion. Thus we also used a global illumination model.
- *Texture.* Models were untextured, as textures can hide any distortion.
- *Test subjects.* 30 test subjects (20 male, 10 female) were drawn from a pool of computer science students aged 22–25. For impartiality, some of the chosen test subjects had knowledge of computer graphics, and others did not.
- *Screen and Model Resolution.* The models were displayed on an 17-in LCD monitor, with resolution 1280×1024. The watermarked and original model were displayed together so as to fill the screen. The chess king model had 12170 triangles and horse model had 10024 triangles, allowing clear observation of detail. The screen was viewed from a distance of approximately 0.6m.
- *Interaction.* We allowed the subjects to rotate and zoom the models: [16] suggests that evaluation of alterations to 3D objects should permit interaction.

The experiment comprised three steps:

1. *Oral instructions and training.* First, we told the subjects about 3D mesh models, watermarking, compression and filters. We then gave examples of an unaltered mesh, to be scored as 0, and a worst-case altered mesh, to be scored as 10.

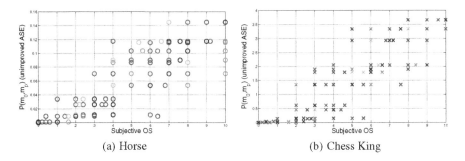

(a) Horse   (b) Chess King

Fig. 4. *Unimproved* ASE perceptual distance versus subjective opinion score (OS)

2. *Practice with a sample model.* Next, the subjects were allowed to interact with various processed models to familiarize themselves with the experiment.
3. *Experimental trials.* In this step, the subjects were asked to score the differences between the original models and altered models.

While human observers were rather variable in their opinions as to perceptibility of differences, there was general correlation between average strain energy (ASE) and opinion scores, as will be discussed in more detail in the next Section. Figures 4(a), 4(b) show plots of the individual opinion scores (OS) against strain energy. Each circle in Figure 4(a) and cross in Figure 4(b) corresponds to one model assessed by one subject.

In subsequent experiments, we used the *means* of these human opinion scores (MOS) for each altered model as being representative of the amount of visual differences perceived by human subjects.

6.2   Experiment II: Strain Field Measure Versus Other Simpler Measures

We next investigated the relationship between mean opinion score values (MOS) and two other simple perceptual distance measures which might plausibly be used for assessing mesh distortion: the *fractional change in the total area of the triangles* (P_{FTAR}) and the *fractional change in angle between normal vectors of adjacent faces* (P_{FNVANG}). The first of these measures is defined as follows:

$$P_{\mathrm{FTAR}}(m_0, m_p) = \sum_{i=1}^{n} |\Delta S_i| / \sum_{i=1}^{n} S_i,$$

where n is the number of faces of the mesh, ΔS_i is the change in area of face i and S_i is its area. The second is defined as

$$P_{\mathrm{FNVANG}}(m_0, m_p) = \sum_{i=1}^{m} |\Delta \alpha_i| / \sum_{i=1}^{m} \alpha_i,$$

where m is the number of the edges of the mesh, $\Delta \alpha_i$ is change between in angle of normal vectors between edge i and α_i is the angle of normal vectors between edge i.

Table 1. Correlation between various perceptual distance measures and human opinion done with watermarking and noisy separately (watermarking is noted by W and noisy is noted by N)

Perceptual distance based on	Correlation coefficient Horse	Correlation coefficient Chess King
FTAR (W)	0.9925	0.9854
FNVANG (W)	0.9893	0.9699
unimproved ASE (W)	0.9669	0.9827
FTAR (N)	0.9710	0.9552
FNVANG (N)	0.9622	0.9282
unimproved ASE (N)	0.9814	0.9776

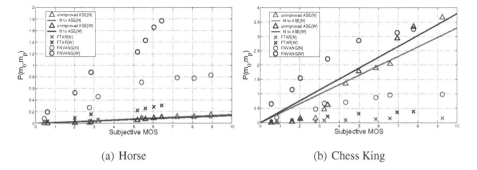

(a) Horse   (b) Chess King

Fig. 5. Various perceptual difference measures versus mean subjective opinion score (MOS) done with watermarking and noisy separately (watermarking is noted by W and noisy is noted by N)

Initially, we used three methodologies: unimproved ASE, FTAR and FNVANG, to separately evaluate the effects of watermarking and noise addition with different amounts of added information or noise. Figure 5 shows that *all* evaluation methodologies give results well correlated with the mean opinion scores from Experiment I. Table 1, shows the corresponding correlation coefficients. It is perhaps unsurprising that each assessment methodology gives a consistent result when the *same* processing method is used to distort the mesh more and more.

However, a good evaluation method should be independent of the way in which the mesh has been distorted. We thus put the distorted meshes produced by watermarking and adding noise into a single set, and did the experiment again. Figure 6 shows that the mean opinion scores from Experiment I are now *not* well correlated with the proposed P_{FTAR} or P_{FNVANG} perceptual distances, either for the horse or the chess king—we can see these two simpler perceptual distance measures produce much more scattered results than the average strain energy measure. They do not adequately predict human opinion of mesh differences—whereas the strain field perceptual distance measure produces results which lie much closer to a straight line.

To more precisely analyze this observation, we calculated the correlation coefficient between each perceptual distance measure and the human mean observation scores: see Table 2. The simpler FTAR and FNVANG perceptual distances have much lower

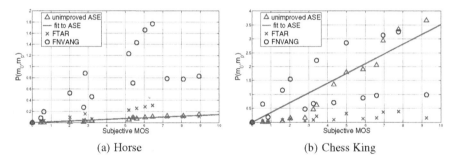

Fig. 6. Various perceptual difference measures versus mean subjective opinion score (MOS)

Table 2. Correlation between various perceptual distance measures and human opinion

Perceptual distance based on	Correlation coefficient Horse	Correlation coefficient Chess King
FTAR	0.56	0.71
FNVANG	0.67	0.54
unimproved ASE	0.97	0.97

correlation than our *unimproved* ASE strain energy perceptual distance, and hence are of less value for measuring human opinion of distortion. We further conclude, given the very high correlation coefficients observed, that the (unimproved) perceptual distance based on strain energy is a useful replacement for subjective mean human opinions of mesh differences.

From a mathematical point of view, we believe the better performance of the ASE perceptual distance arises because strain energy provides a well-defined L^2 measure in deformation space, unlike FTAR and FNVANG.

6.3   Experiment III: Variants of Strain Energy Measure

We next investigated the relationship between strain energy and mean opinion score values (MOS) when using the suggested improvements based on projection and edge weights (see Section 5.2). We compared 4 variants of our method: *unimproved* perceptual distance (without projection and face weights), perceptual distance using projection, perceptual distance using edge weights, and perceptual distances using both projection and edge weights.

Figure 7 shows a comparison of these four different measures with mean human opinion scores, as in Experiment II. In each case a similar close-to-linear relationship can be seen for both horse and chess king models. Using the projection idea has almost no effect on perceptual distances. Using edge weighting has a bigger effect (see Figure 7), although there is no obvious improvement in terms of linearity of relationship. The probable reason is that the models used (like many models) have relatively few flat regions or sharp triangles. We also note that the subjective scores are not very exact, making it difficult to distinguish whether the results before and after these modifications are really an improvement.

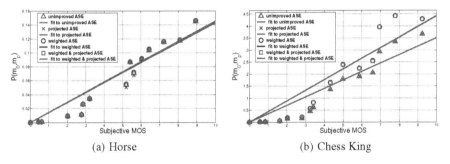

(a) Horse   (b) Chess King

Fig. 7. Kinds of ASE perceptual distance and mean opinion score

Table 3. Correlation between the four perceptual distance measures and human opinion

Perceptual distance based on	Correlation coefficient Horse	Correlation coefficient Chess King
unimproved ASE	0.9745	0.9660
projected ASE	0.9715	0.9661
weighted ASE	0.9751	0.9560
projected & weighted ASE	0.9752	0.9557

Again correlation coefficients were computed and are given in Table 3. The edge weights *do* make a small improvement for the horse model, but have the opposite effect for projection, while for the chess king model, the opposite is true, projection making a small improvement but edge weighting being worse. Ultimately, our experiments show no benefit to using either suggested improvement: there is almost no difference in correlation between the computed measure and human opinion. We thus recommend use of the *unimproved* ASE based perceptual distance, as it is simpler to compute. The other methods increase the computation time, without noticeable benefit.

Ultimately, the number and type of subjects used in these experiments was limited, as were the number and types of test models, and the range of methods used to add distortion to the models. We acknowledge that considerable further testing is necessary to fully validate these conclusions.

7 Further Demonstrations

We now provide two further demonstrations of our approach. The first is intended to give the reader a visual impression of how our measure works in practice, while the second demonstrates an application of our methodology by comparing the relative perceptibility of deformations produced by two different mesh deformation methods.

7.1 Demonstration I: Distortion in a Buddha Model

Here we simply present a series of Buddha models, with 62224 faces, which have been processed by watermarking (Figures 8(a–f)) and noise addition (Figures 8(g–l))

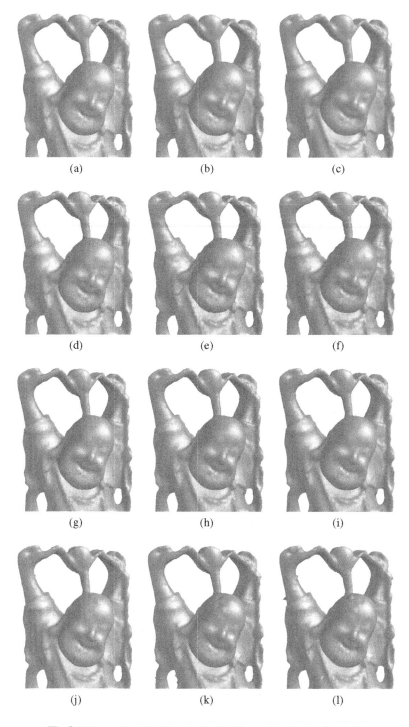

(a)   (b)   (c)

(d)   (e)   (f)

(g)   (h)   (i)

(j)   (k)   (l)

Fig. 8. Watermarked Buddhas (a–f), Buddhas with added noise (g–l)

Table 4. Perceptual distance based on strain energy $P \times 10^{-5}$ for Buddha models with varying deformations

Watermarked models	FTAR	P	Models with noise	FTAR	P
(a)	0	0	(g)	0	0
(b)	661	4	(h)	115	4
(c)	3335	114	(i)	568	114
(d)	6644	455	(j)	1074	523
(e)	10020	1042	(k)	1702	1778
(f)	12342	1533	(l)	2313	3988

algorithms, making changes controlled by successively increasing the FTAR measure as stated in Table 4, which also gives the perceptual distances of these models from the unperturbed model. These figures and numbers allow the reader to gain some impression of their own concerning our (unimproved) perceptual distance measure.

7.2 Demonstration II: Application to Comparison of Mesh Processing Methods

We next demonstrate a simple application of our methodology. We compared meshes processed in two different ways, and demonstrate that our method provides the same results as human opinion as to which change is less perceptible. Again we used the horse model with 10024 faces and the chess king with 12170 faces. The latter is smoother than the horse and its edges are more obvious, so we expected differences to be more visually apparent for this model. The meshes were distorted using the same two processing methods as in Experiment I—watermarking, and adding noise. We embedded an average 2 bits distortion into each triangular face of the meshes using each method. Figure 9(a) shows the original horse model, Figure 9(b) shows the watermarked horse model, and Figure 9(c) shows the noisy horse model. Differences in the distortions are

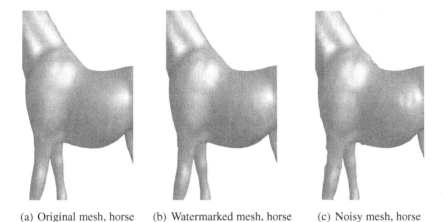

(a) Original mesh, horse   (b) Watermarked mesh, horse   (c) Noisy mesh, horse

Fig. 9. Original and processed models, horse

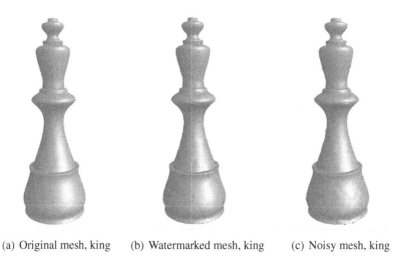

(a) Original mesh, king   (b) Watermarked mesh, king   (c) Noisy mesh, king

Fig. 10. Original and processed models, king

Table 5. Perceptibility distances for three measures and two models

Models	Projected & weighted ASE	FTAR	FNVANG
Watermarked Horse	0.0002	0.0623	0.3596
Noisy Horse	0.0017	0.0414	0.2712
Watermarked King	0.0140	0.0835	1.1492
Noisy King	0.0591	0.0499	0.4720

clearly perceptible in these close up views. Figures 10(a)–10(c) are analogous figures for the chess king model.

For both models, the reader will probably agree with other human observers that the original and watermarked meshes show quite small differences, whereas the differences between the original and noisy meshes are more obvious. The watermarks are embedded into the all parts of the mesh fairly evenly, whereas more noise is embedded into the mesh at some places than others. As a result distortion in the latter case is more obvious, and correspondingly our perceptual distance measure is also bigger.

Furthermore, as expected, because the king is smoother than the horse, evidence of deformation in the king is more obvious, as predicted in Table 5 by our strain energy perceptual distance. (here we used the projected and edge weighted measure). For both models, our measure of the difference is lower for the watermarked mesh than the noisy mesh, which corresponds with with visual inspection. Our own visual subjective results, and the measure, are in agreement that the watermarking method produces less visible differences than adding noise, both for the smoother king model and for the more textured horse model.

Note, however, that both of the simpler FTAR and FNVANG measures predict the *opposite*: that the watermarking changes should be more visible. This is further evidence that these particular, simple, perceptual distance measures are not satisfactory.

8   Conclusions and Future Work

This paper has proposed a new methodology for comparing small visual differences between conforming models. Our methodology is based on the use of strain field theory to quantify deformation produced by some process. Our experiments show that this objective method can produce results which are strongly correlated with the average of subjective human opinions. The same is *not* true of simpler measures based on changes in triangle areas, or surface normals.

This work has applications to assessing the perceptible effects of 3D mesh processing algorithms such as watermarking, compression and other filtering. For example, in watermarking we could use this approach to decide which part of a model is the most suitable place for embedding a watermark, to decide how much data can be hidden in a mesh model, or to choose between watermarking schemes.

Our methodology has certain limitations. One is that for certain objects, such as engineering components, sharp edges are very important, yet involve few triangles. We probably do not give a high enough weight to such triangles. A second is that the method is based on deformation of the triangular faces, yet a model may in principle deform without changes in the size and shape of faces. We assume that the strain in a mesh model occurs only in the surface of the model. Hence, the associated strain energy only measures intrinsic deformation in the surface, and deformation in the normal direction is ignored. For example, a cylindrical surface can be developed into a plane without changing the shape of any triangle; our proposed strain energy would report zero perceptual change in such a case. To what extent this is a problem in practice for real closed meshes is not clear: global considerations mean that some other part of the surface is likely to deform highly if such changes occur. Again, we emphasize that we in intend our method to be applied to deformations produced by applications in which triangles do not move far from their original positions.

Our experiments also clearly have some limitations. The number and type of persons, models, and processing algorithms used in evaluation would ideally be much higher.

One area we wish to explore in future is to try to statistically determine thresholds of perceptibility through subjective experiments, which can then be used with our method to decide whether a visual difference is perceptible.

References

1. Agarwal, P., Adi, K., Prabhakaran, B.: Robust Blind Watermarking Mechanism for Motion Data Streams. In: Proc. 8th Workshop on Multimedia and security, Geneva, Switzerland, pp. 230–235 (2006)
2. Arnold, M., Wolthusen, D., Schmucker, M.: Techniques and Applications of Digital Watermarking and Content Protection. Artech House Publishers (2003)
3. Aspert, N., Santa-Cruz, D., Ebrahimi, T.: Mesh: Measuring Error Between Surfaces Using the Hausdorff Distance. In: Proc. IEEE Int. Conf. Multimedia and Expo 2002 (ICME), vol. 1, pp. 705–708 (2002)
4. Chen, W.F., Salipu, A.F.: Elasticity and Plasticity. China Architechture & Building Press (2005)
5. Cignoni, R.S.P., Rocchini, C.: Metro: Measuring Error on Simplified Surfaces. Comput. Graph. Forum 17(2), 167–174 (1998)

6. Corsini, M., Gelasca, E.D., Ebrahimi, T., Barni, M.: Watermarked 3-D Mesh Quality Assessment. IEEE Trans. Multimedia 9(2), 247–256 (2007)

7. Cox, I., Miller, M., Bloom, J., Miller, M.: Digital Watermarking: Principles & Practice. Morgan Kaufmann, San Francisco (2002)

8. Isenburg, M., Gumhold, S.: Out-of-Core Compression for Gigantic Polygon Meshes. In: Proc. SIGGRAPH 2003, pp. 935–942 (2003)

9. Khodakovsky, A., Guskov, I.: Compression of Normal Mesh. Geometric Modeling for Scientific Visualization, pp. 189–206. Springer, Heidelberg (2004)

10. Lindstrom, P., Turk, G.: Image-Driven Simplification. ACM Trans. Graph. 19(3), 204–241 (2000)

11. Methodology for Subjective Assessment of the Quality of Television Pictures Recommendation BT. 500–11 Geneva, Switzerland (2002)

12. Ohbuchi, R., Masuda, H., Aono, M.: Watermarking Three-Dimensional Polygonal Models. In: Proc. ACM International Conference on Multimedia 1997, Seattle, pp. 261–272 (1997)

13. Ohbuchi, R., Masuda, H., Aono, M.: Watermarking Three-Dimensional Polygonal Models through Geometric and Topological Modifications. IEEE J. Selected Areas in Communication 16(4), 551–560 (1998)

14. Ohbuchi, R., Takahashi, S., Miyazawa, T., Mukaiyama, A.: Watermarking 3D Polygonal Meshes in the Mesh Spectral Domain. In: Proc. Graphics Interface 2001, pp. 9–17 (2001)

15. Peyre, G., Mallat, S.: Surface compression with geometric bandelets. In: Proc. SIGGRAPH 2005, pp. 601–608 (2005)

16. Rogowitz, B., Rushmeier, H.: Are Image Quality Metrics Adequate to Evaluate the Quality of Geometric Objects. In: Rogowitz, B.E., Pappas, T.N. (eds.) Proc. SPIE Human Vis. Electron. Imag. VI, vol. 4299, pp. 340–348 (2001)

17. Rossignac, J., Szymczak, A.: Wrap & Zip Decompression of the Connectivity of Triangle Meshes Compressed with Edgebreaker. Computational Geometry, Theory and Applications 14(1/3), 119–135 (1999)

18. Sun, X., Rosin, P.L., Martin, R.R., Langbein, F.C.: Fast and Effective Feature-Preserving Mesh Denoising. IEEE Trans. Visualization and Computer Graphics 13(5), 925–938 (2007)

19. Taubin, G.: A Signal Processing Approach to Fair Surface Design. In: SIGGRAPH 1995, pp. 351–358 (1995)

20. Watson, B., Friedman, A., McGaffey, A.: Measuring and Predicting Visual Fidelity. In: SIGGRAPH 2001, pp. 213–220 (2001)

21. Wilhelm, F.: Tensor Analysis and Continuum Mechanics. Springer, Berlin (1972)

22. Williams, N., Luebke, D., Cohen, J.D., Kelley, M., Schubert, B.: Perceptually Guided Simplification of Lit, Textured Meshes. In: Proc. 2003 Symp. Interactive 3D Graphics, Monterey, CA, pp. 113–121 (2003)

23. Xun, Z.: Elasticity Mechanics. Peoples Education Press (1979)

24. Zienkiewicz, O.C.: The Finite Element Method. McGraw-Hill, New York (2000)

Continuous Collision Detection between Two 2D Curved-Edge Polygons under Rational Motions

Wenjuan Gong and Changhe Tu

School of Computer Science and Technology,
Shandong University, Jinan, China
gongwenj@163.com, chtu@sdu.edu.cn

Abstract. This paper presents a novel approach which continuously detects the first collision between two curved-edge polygons moving under rational motions. Edges of the two polygons in this paper are planar curves, represented as conic splines, i.e. elliptic or parabolic sections. The curved-edge polygons are not confined to be convex and conic sections are only required to be GC^0 continuous. Motions of the polygons are modeled by interpolating between control points along motion trajectories. Our algorithm returns the first collision moment and collision position if there is a collision between the two moving polygons and returns no-collision otherwise. Collision condition of the two polygons moving under rational motions is represented as an univariate polynomial of time t. Bernstein form is used to improve the accuracy of solving the high degree polynomial. We also use bounding circles to improve the efficiency of our approach and compare our method with the PIVOT2D method and prove ours to be more accurate and faster.

Keywords: continuous collision detection, conic splines, rational motion.

1 Introduction

Collision detection is an important problem in animation, CAD and robotics. Collision detection is usually used to improve reality in virtual environment by avoiding penetrating between objects. Besides, it is used as path planning in robotics to calculate in advance paths of moving robots to avoid collision during motions. There are many applications like robot or vehicle path planning where the robots and vehicles are represented as $2D$ figures moving in $2D$ plane, and our concentration is collision detection of objects moving in $2D$ plane.

Currently, there are many collision detection approaches which are divided into two major categories: one is *discrete collision detection* [1,11], the other is *continuous collision detection* [2,3]. For discrete approach, collision detection is performed at discretely sampled time instants. In this approach, collision may be missed for fast moving or tiny objects; for example, a fast moving object may pass through small or thin objects without being detected if the collision occurs

F. Chen and B. Jüttler (Eds.): GMP 2008, LNCS 4975, pp. 79–91, 2008.
© Springer-Verlag Berlin Heidelberg 2008

between two consecutive sample instants. Shortening the time intervals produces higher precision, but the expense is reduced efficiency.

To overcome this drawback, the idea of *continuous collision detection* (CCD) has been explored, in which continuous motion between successive object positions is modeled and collision detection is performed without time sampling. The continuous motion interpolating key frames or sample positions of an object should be generic enough to cater for a wider spectrum of applications, while it should be simple enough to be computed efficiently. For motion planning in robotics applications the motion of an end-effector or a robot is usually pre-specified.

Present continuous collision approach like[3] deals only with two elliptic discs moving in $2D$ space but has no consideration of collision detection between polygons. Present collision detection methods between two polygons, like method based on graphics hardware[11], method based on Kinetic Data Structure(KDS) [2] and method like [1] process collision between two polygons with line segment edges. When dealing with curved-edge polygons, curves are approximated with line segments. Approach based on KDS solves problem of two polygons under simple motions, which needs to decompose motion in several directions when dealing with complex motion forms.

This paper solves the problem of continuously detecting the first collision between two moving curved-edge polygons. Conics are one of the most widely used curves, e.g. to design the bodies of aircraft[20], to express circular arcs, spheres or tori[21,22,23,24,25]. Conic splines could optimally approximate planar curves with respect to maximum norm[12]. Rational motions allow effective numerical processing by using various well-developed techniques of processing polynomials. According to[4,5,6,7,8], low degree rational motions are adequate to fulfill the need of motion design and representation in robotics and CAD/CAM, especially for planar motions. So the problem is simplified and our focus is continuously detecting the first collision between two curved-edge polygons represented as conic splines moving under rational motions.

Collision condition is represented as an univariate polynomial of time t. After solving this univariate polynomial of time t, the algorithm judges whether this collision point is valid. The minimal valid t is the first collision moment. Main procedures of this algorithm are: First, static collision condition of two curved-edge polygons is established; motion equations are computed by interpolating the dynamic images of control points defined on motion trajectory; then collision condition of two curved-edge polygons moving under rational motions is computed; Bernstein form is used to improve accuracy in solving the high degree univariate polynomial; bounding circles are used to improve the algorithm efficiency; finally, we compare our method with PIVOT2D[11] in accuracy and efficiency.

The remainder of this paper is organized as follows. Section 2 is preliminaries. In section 4, we give the collision condition of two moving curved-edge polygons under rational motions. In section 5, we introduce techniques to improve algorithm performance. In section 6, the algorithm skeleton of our approach is given. Section 7 is experimental results, including accuracy and efficiency comparisons with PIVOT2D. In section 8, we conclude this paper and discuss further research.

2   Preliminaries

2.1   2D Euclidean Rational Motions

Recent studies on rational motions[4,5,6,7] especially that on planar rational motions [8], have shown that low degree rational motions are adequate to fulfill the need of motion design and representation in robotics and CAD/CAM. Rational motions allow effective numerical processing by using various well-developed techniques of processing polynomials.

We consider only rational motions in our algorithm. The motion of an object is either given in the rational form or interpolated several given positions[8] of the object. In particular, we require that the motion is Euclidean so that the size and shape of a curved-edge polygon will not change during the motion. Now we would have a brief review of planar rational Euclidean motions.

According to [3], a Euclidean transformation in $\mathbb{E}^2$ is given by

$$X' = MX \tag{1}$$

where

$$M = \begin{pmatrix} R & V \\ 0^T & 1 \end{pmatrix} \tag{2}$$

is some constant matrix and X, X' are points in $\mathbb{E}^2$ in homogeneous coordinate forms. The rotational part of the transformation is described by a 2×2 orthogonal matrix R and the translational part by the vector V. If R and V are continuous functions of t, then M describes transformation over time and can therefore be denoted by $M(t)$. In particular, if the entries of $M(t)$ are rational functions and R is orthogonal for all t, then $M(t)$ is called a rational Euclidean motion whose degree is the maximal degree of its entries.

If we write

$$R = \begin{pmatrix} \cos\phi & -\sin\phi \\ \sin\phi & \cos\phi \end{pmatrix}, \qquad V = \begin{pmatrix} v_x \\ v_y \end{pmatrix},$$

then the kinematic image(see [8]) $d \in P^3$ of M is given by

$$d = \begin{pmatrix} d_0 \\ d_1 \\ d_2 \\ d_3 \end{pmatrix} = \begin{pmatrix} v_x \sin\phi/2 - v_y \cos\phi/2 \\ v_x \cos\phi/2 + v_y \sin\phi/2 \\ -2\cos\phi/2 \\ 2\sin\phi/2 \end{pmatrix}.$$

Conversely, any point d in P^3 with $d_2^2 + d_3^2 \neq 0$ corresponding to a Euclidean transformation M in $\mathbb{E}^2$ is given by

$$M = \begin{pmatrix} d_2^2 - d_3^2 & 2d_2d_3 & 2(d_0d_3 - d_1d_2) \\ -2d_2d_3 & d_2^2 - d_3^2 & 2(d_0d_2 + d_1d_3) \\ 0 & 0 & d_2^2 + d_3^2 \end{pmatrix} \tag{3}$$

After interpolating sample positions[8] using kinematic mapping, rational motion $M(t)$ is calculated. When applying a rational motion $M(t)$ to a conic section $\mathcal{A}_i : X^T A X = 0$, we get a moving conic section $\mathcal{A}_i(t) : X^T A(t) X = 0$ where

$$A(t) = (M^{-1}(t))^T A M^{-1}(t).$$

If M is represented as in equation(3), then

$$M^{-1} = \begin{pmatrix} d_2^2 - d_3^2 & -2d_2 d_3 & 2(d_1 d_2 + d_0 d_3) \\ 2d_2 d_3 & d_2^2 - d_3^2 & 2(d_1 d_3 - d_0 d_2) \\ 0 & 0 & d_2^2 + d_3^2 \end{pmatrix}.$$

The maximal degree of the entries in $A(t)$ is $2k$, if the degree of the motion $M(t)$ is k.(see [3])

2.2 Collision Condition of Two Conics under Rational Motions

Collision condition of two conics under rational motions can be characterized by root patterns of the characteristic polynomial $f(\lambda)$ according to [17].

Suppose the two moving conics represented by $X^T A_i(t) X = 0$ and $X^T B_j(t) X = 0$ are separate at $t = 0$. Characteristic polynomial of these two conics, is:

$$f(\lambda; t) = det(\lambda A_i(t) - B_j(t)).$$

Write:

$$f(\lambda; t) = g_3(t)\lambda^3 + g_2(t)\lambda^2 + g_1(t)\lambda + g_0(t).$$

The discriminant of $f(\lambda; t)$, as a function of t, is

$$\Delta(t) = 18g_3(t)g_2(t)g_1(t)g_0(t) - 4g_2^3(t)g_0(t) + g_2^2(t)g_1^2(t) - 4g_3(t)g_1^3(t) - 27g_3^2(t)g_0^2(t).$$

According to [17]: if $f(\lambda; t), t \in [0, 1]$ has real zeros, the two conics touch at $t_{min} \in [0, 1]$, where t_{min} is the smallest real zeros of $f(\lambda; t), t \in [0, 1]$, i.e. $t_{min} = min\{t | f(\lambda; t) = 0, t \in [0, 1]\}$.

3 Representation of Moving Conic Sections

Recent studies on planar curves[12] have shown that conic splines could optimally approximate planar curves with respect to maximum norm, so our curved-edge polygon is represented as conic splines, i.e. composite curve of conic sections. In addition, conics are one of the most widely used curves in CAD/CAGD and can be easily processed.

Considering the inconvenience of conic splines as input and the relationship between conics and rational quadratic Bézier curve segments, we use Bézier curves to input the two curved-edge polygons. We would give the correlation between rational quadratic Bézier curve segments and conics in the following part.

In our approach, curved-edge polygon is input as piecewise rational quadratic Bézier curve segments, defined by

$$R(r) = \frac{B_0^2(r)P_0 + \omega B_1^2(r)P_1 + B_2^2(r)P_2}{B_0^2(r) + \omega B_1^2(r) + B_2^2(r)}.$$

Here, $0 \le r \le 1, P_i \in R^2$ and $B_i^r = C_i^n r^i (1-r)^{i-1}$ are Bernstein polynomials. ω are called weights, and the P_i are control points. For $\omega < 1$, we obtain an ellipse; for $\omega = 1$, a parabola; and for $\omega > 1$, a hyperbola. (see [9])

A conic, to which a conic section converted from a rational quadratic Bézier curve segment belong, in the $2D$ Euclidean plane $\mathbb{E}^2$, can be written in quadratic form $X^T A X = 0$, where $A = [a_{i,j}]$ is a 3×3 real symmetric matrix, and X is a $3D$ column vector for the homogeneous coordinates of a point in $\mathbb{E}^2$.

The rational quadratic Bézier curve segment from input has its implicit form according to [19]. The quadratic form $X^T A X = 0$ represents one entire conic, while we need part of it to represent the input rational quadratic Bézier curve segment. Here we introduce control polygon of the Bézier curve to denote the part of the conic actually used in our approach. That is to say we denote one conic section of curved-edge polygon using the form:

$$\{X^T A_i X = 0, P_i, P_{i+2}, P_{i+1} | t \in [0,1]\},$$

where P_i, P_{i+2} and P_{i+1} are the three vertices of the control triangle[9] of this Bézier curve segment. P_i and P_{i+2} are two endpoints of this Bézier curve segment.

In the case of a moving conic section, its representation is:

$$\{X^T A_i(t) X = 0, P_i(t), P_{i+2}(t), P_{i+1}(t) | t \in [0,1]\},$$

where $A_i(t) = (M^{-1}(t))^T A_i M^{-1}(t)$, $P_i(t) = M(t)P_i$, $P_{i+2}(t) = M(t)P_{i+2}$ and $P_{i+1}(t) = M(t)P_{i+1}$ are three vertices of the control triangle under motion. $M(t)$ and $M^{-1}(t)$ is calculated according to the method described in former section.

4 Collision Condition of Two Moving Curved-Edge Polygons

In this section, we would give the collision condition of two moving curved-edge polygons represented by GC^0 continuous conic splines under rational motions. Two convex curved-edge polygons composed of GC^1 continuous conic splines collide when the conic section pair are tangent externally first. The collision condition of two curved-edge polygons composed of GC^0 continuous conic splines which are not required to be convex is more complicated. But they can be concluded into the following three conditions: two conic sections of the two curved-edge polygons touch first; an endpoint of one conic section and a conic section collide first; two endpoints of conic sections collide first. The third condition is included in the second condition, so we describe the first two in detail. The three different conditions could be found in Fig.1, where P_0, P_2, P_0' and P_2' are endpoints of the conic section, and P is the point where the two sections collide.

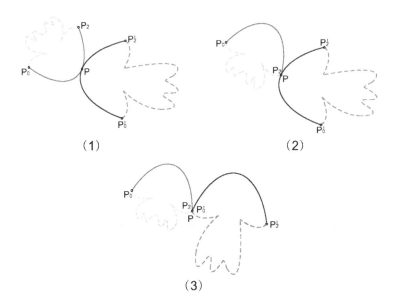

Fig. 1. The collision condition of (1)two conic sections; (2)one endpoint and one conic section; (3)two endpoints

4.1 Collision Condition of Two Conic Sections under Rational Motions

Given two conic sections $\mathcal{A}_i$ and $\mathcal{B}_j$ of the two moving polygons with motion matrix of $M_A(t)$ and $M_B(t), t \in [0, 1]$ respectively, according to former methods, their motion equations could be represented as: $\{X^T A_i(t)X = 0, P_i(t), P_{i+2}(t), P_{i+1}(t)|t \in [0, 1]\}$ and $\{X^T B_j(t)X = 0, P_j(t), P_{j+2}(t), P_{j+1}(t)|t \in [0, 1]\}$. The collision condition of two conic sections is when these two conics touch and the touching point belongs to both sections.

After calculating the first touching point of the two moving conics according to section2.2, we should judge whether the point belongs to both sections. Here we use barycentric coordinates $\{(s_1, s_2, s_3)|s_1+s_2+s_3 = 1\}$ to represent the touching point $V = s_1 P_0 + s_2 P_1 + s_3 P_2$, where P_0, P_1, P_2 are the three control points of one conic section. If the touching point belongs to this conic section(excluding two endpoints), it is also inside the control polygon and vice versa. In barycentric coordinates, a point inside the reference triangle which is the control polygon here is equivalent to three positive coordinates. So if s_1, s_2, s_3 are all positive, the colliding point belongs to the conic section (see fig.2).

When the three coordinates are not all positive, there might be two circumstances: first, the touching point doesn't belong to these two sections; second, the touching point is an endpoint of conic sections. In the first circumstance, these two moving conic sections still may collide first, which is the condition of a moving endpoint colliding with a moving conic section, so it is included in the following part. The second circumstance is also included in the following part.

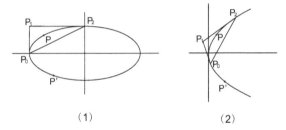

Fig. 2. Judging colliding point on (1)an elliptic section; (2)a parabolic section

4.2 Collision Condition of an Endpoint with a Conic Section under Rational Motions

Collision condition of an endpoint with a conic section of two moving curved-edge polygons is described in detail in this part. Suppose endpoint $\mathcal{A}_{Vi1}$ of the first polygon under rational motion is represented as

$$V_{i1}^A(t) = \begin{pmatrix} V_{i1x}^A(t) \\ V_{i1y}^A(t) \\ V_{i1z}^A(t) \end{pmatrix}.$$

Conic section $\mathcal{B}_j$ of the second polygon moving under rational motion is represented as $\{X^T B_j(t) X = 0, P_i(t), P_{i+2}(t), P_{i+1}(t) | t \in [0,1]\}$. Collision of a moving vertex with a moving conic section satisfies:

$$\left(V_{i1}^A(t)\right)^T B_j(t) \left(V_{i1}^A(t)\right) = 0 \tag{4}$$

Then judge whether the collision point of time t belongs to the conic section according to the method in the former part. If the collision point of time t does belong to the conic section, then the endpoint and the conic section collide at time t.

After calculating all the conic section pairs and all the endpoint-conic section pairs, we could find the least time t, which is the first collision time of the two curved-edge polygons. Curved-edge polygons in our methods are only required to be GC^0 continuous, so we could only calculate the first collision time.

5 Techniques to Improve Algorithm Performance

Bounding box is usually used in collision detection to improve efficiency by excluding collision before calculating exact collision points or allowing areas of interaction to be localized quickly. Many different geometric primitives have been used to construct bounding box, like Sphere, AABB(Axis-Aligned Bounding Box)[18], K-Dop(Discrete orientation polytopes), OBB(Oriented Bounding Box) and methods like [14,15,17].

In our approach, we use two layers of bounding circles to improve algorithm efficiency. The first layer is the circumcircle of the AABB of the curved-edge

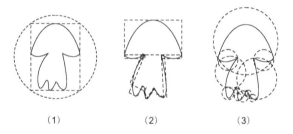

(1)   (2)   (3)

Fig. 3. (1)The first layer bounding circle of the first curved-edge polygon. (2)OBBs of conic sections of the first curved-edge polygon. (3)Bounding circles of the second layer.

polygon before any rotation. The second layer is the circumcircles of OBBs of conic sections before any rotation.(see Fig.3)

Our experiments also use Bernstein form to improve the robustness and accuracy of our collision detection procedure as described in [3].

6   Implementation

Based on collision conditions described in former sections, we outline the framework of our algorithm for collision detection of two moving curved-edge polygons under rational motions. The two polygons $\mathcal{A}$ and $\mathcal{B}$ are composed of conic splines(elliptic sections or parabolic sections).

Suppose $\mathcal{A}$ and $\mathcal{B}$ have m and n conic sections respectively. The number i conic section of curved-edge polygon $\mathcal{A}$:$\mathcal{A}_i$ under motion can be represented in a quadratic form $\{XA_i(t)X^T = 0, P_i(t), P_{i+2}(t), P_{i+1}(t)|t \in [0,1]\}$. The two endpoints of number j conic spline of curved-edge polygon $\mathcal{B}$: $\mathcal{B}_{Vj1}$ and $\mathcal{B}_{Vj2}$ under motion can be represented as $V_{j1}^B(t)$ and $V_{j2}^B(t)$ in homogeneous coordinates, where

$$V_{j1}^B(t) = \begin{pmatrix} V_{j1x}^B(t) \\ V_{j1y}^B(t) \\ V_{j1e}^B(t) \end{pmatrix}.$$

Also, the condition of intersecting polygons before motion is not considered in our approach, for there is no meaning in practical problems of motion planning.

Algorithm: *CD-Moving2DPolygons*

Input: Control points and weights of the conic splines that compose two polygons: $\mathcal{A}$ and $\mathcal{B}$. Control points of the trajectories of two moving curved-edge polygons.

Output: If the two moving curved-edge polygons collide: return when and where it occurs; otherwise return NO-COLLISION.

Step1: Compute motion matrices of $\mathcal{A}$ and $\mathcal{B}$. Compute whether the bounding circles of the two curved-edge polygons collide during motions. If there is a collision, go to step2. If there is no collision, Return NO-COLLISION.

Step2: Computer whether bounding circles of a conic section pair of two curved-edge polygons collide during motions. If there is no collision, repeat **Step2** until all mn pairs are finished. If there is a collision, go to setp3.

Step3:

(1)step1: Compute the characteristic polynomial of $XA_i(t)X^T = 0$ and $XB_j(t)X^T = 0$: $f(\lambda; t) = det(\lambda A_i(t) - B_j(t))$.

step2: Compute the discriminant $\Delta(t)$ of $f(\lambda; t)$.

step3: Determine whether $\Delta(t) = 0$ has any real roots $t \in [0, 1]$. If yes, compute the point positions and determine whether these points are valid. Insert valid collision moments into list T and go to (2).

(2)step1: Compute whether endpoints $\mathcal{A}_{Vi1}$, $\mathcal{A}_{Vi2}$ collide with conic section $\mathcal{B}_j$. Take endpoint $\mathcal{A}_{Vi1}$ for example: compute whether $\left(V_{i1}^A(t)\right)^T B_j(t) \left(V_{i1}^A(t)\right) = 0$ has roots $t \in [0, 1]$. If yes, compute the point positions and determine whether these points are valid. Insert valid collision moments into list T.

step2: Compute whether endpoints $\mathcal{B}_{Vj1}$, $\mathcal{B}_{Vj2}$ collide with conic section $\mathcal{A}_i$ using method as step2. Repeat **Step2** for the next conic section pair until all mn pairs are finished.

Step4: If T is null, return NO-COLLISION. Otherwise, return the minimum of all collision moments in T. That is the first collision moment of the two moving curved-edge polygons.

7   Experiments Results

In this section, we would compare our approach with PIVOT2D[11] in accuracy and efficiency. When solving collision detection of two curved-edge polygons under rational motions using PIVOT2D method, we need approximate conic splines using line segments. PIVOT2D uses graphics hardware and sampling method in collision detection, so its efficiency and accuracy are determined by resolution, approximation accuracy and sampling frequencies.

7.1   Accuracy Comparison

Here we give an example that proves our method to be more accurate than PIVOT2D.

Example: consider two curved-edge polygons moving under rational motions of degree 4. The first jellyfish is composed of 10 conic sections and the second jellyfish is composed of 12 conic sections. Fig.4 shows shapes of the two objects and position definition of the control points of trajectories.

Using our method, the first collision time is $t = 0.469891$. Using PIVOT2D method by setting resolution as 1024×1024, approximating each conic section using 25 line segments and sampling every 0.0001 second, it fails to detect the collision. Even using more dense sampling in this case, the collision point will not be detected, because accuracy is also determined by resolution and approximation accuracy of line segments.

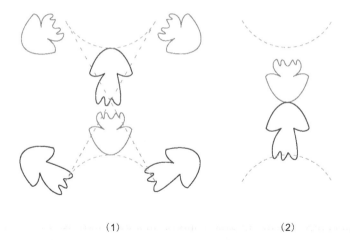

(1)   (2)

Fig. 4. (1)The position definition of two curved-edge polygons. (2)The first touching moment of the two moving curved-edge polygons.

7.2 Efficiency Comparison

We generated 600 test cases to compare the efficiency of our method with the PIVOT2D. There are three kinds of rational motions: linear translations and general degree 2, degree 4 rational motions. Among the cases 300 collide during motion, and the other 300 have no collision. The shapes are manually defined and varied from 2 conic sections to 8 conic sections. The positions of the control points of trajectory are generated randomly. Experiments were run on a PC with a $3GHz$ Intel CPU.

Fig.5-Fig.7 show comparison results between methods with bounding circles and without bounding circles. The horizontal axis represents product of segment numbers of two curved-edge polygons, and the vertical axis represents the time cost of detecting the first collision. The purple line represents conditions when there are collisions and the blue line represents conditions when there is no collision.

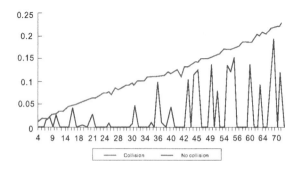

Fig. 5. collision and no collision instances under rational motion of degree 1

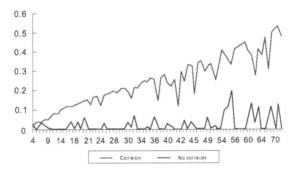

Fig. 6. collision and no collision instances under rational motion of degree 2

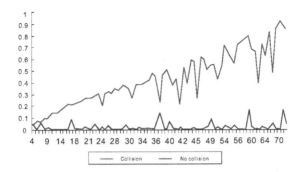

Fig. 7. collision and no collision instances under rational motion of degree 4

Table 1. Time costs comparison between our approach and PIVOT2D

Motion Degree		c1	c2	c3	n1	n2	n3
Our Approach		0.1174268	0.2498024	0.4287794	0.0268076	0.0260531	0.0224668
PIVOT2D	200	0.0173171	0.0170488	0.0159443	0.0121395	0.0143987	0.0145297
	500	0.0261333	0.0268481	0.0235952	0.0318770	0.0362878	0.0360371
	2000	0.0733630	0.0774930	0.0622216	0.1221851	0.1445268	0.1449021
	5000	0.1674449	0.1738885	0.1411358	0.3033577	0.3555660	0.3590165
	10000	0.3262970	0.3376892	0.2700015	0.6106828	0.7132211	0.7100643
	20000	0.6348025	0.6642116	0.5276802	1.2326582	1.4201348	1.4475557

According to experiment results of PIVOT2D method, we use 1024×1024 as resolution and use 25 line segments to approximate a conic section. Experiment results are shown in table1. The first row shows motion degree, e.g. c1 means curved-edge polygons under motion degree 1 collide during motion and n2 means curved-edge polygons under motion degree 2 have no collision during motion, c for collision and n for no-collision. The second row shows the average time cost calculating the first collision time and the first collision position using our method. The third to the eighth rows show the average time costs calculating the first collision time under 200 to 20000 sampling frequencies using PIVOT2D method.

From the table, we could see PIVOT2D method is indeed very fast when there is collision and low frequency. But as we former mentioned, this might result in omitting collision between sampling moments and time costs increase greatly as sampling frequencies increase. Our method performs better when there is no collision, for bounding circles reduce calculations greatly.

8   Conclusions and Future Work

We have presented an accurate and efficient collision detection algorithm for two curved-edge polygons moving under rational motions in 2D plane. The algorithm, called **CD-Moving2DPolygons**, is based on an algebraic collision condition between two conics under rational motions. Rational motions are considered because they are flexible enough for modeling general motions and their polynomial representation makes an algebraic treatment to collision detection possible. Our method uses exact representations for curves and therefore does not suffer from errors induced by curve approximation. Our algorithm determines collision by checking for the existence of real roots of a univariate function which is the discriminant of the characteristic equation of the two moving conic sections. Unlike many other collision detection algorithms, it does not need sampling on the motion path so that inaccuracy due to limited sampling frequency is avoided. Bernstein forms used for polynomial manipulation significantly increase numerical stability of the algorithm for high degree rational motions.

In our future work we hope to speed up the algorithm using GPU.The improvement of the bounding circle is not so efficient under conditions when the two curved-edge polygons collide and we hope that a better method could be found to improve that condition.

References

1. Jiménez, J.J., Segura, R.J., Feito, F.R.: Efficient Collision Detection between $2D$ Polygons. Journal of WSCG 12(1-3), 21–36 (2004)
2. Basch, J., Erickson, J., Guibas, L.J., Hershberger, J., Zhang, L.: Kinetic Collision Detection Between Two Simple Polygons. Computational Geometry: Theory and Applications 27(3), 211–235 (2004)
3. Choi, Y.-K., Wang, W., Liu, Y., Kim, M.-S.: Continuous Collision Detection for Elliptic Disks. IEEE Transactions on Robotics and Automation 22(2), 213–224 (2006)
4. Horsch, T., Jüttler, B.: Cartesian spline interpolation for industrial robots. Comput. Aided Design 30(3), 217–224 (1998)
5. Jüttler, B., Wagner, M.G.: Computer-aided design with spatial rational B-spline motions. ASME Journal of Mech. Design 118(2), 193–201 (1996)
6. Jüttler, B., Wagner, M.G.: Kinematics and Animation. In: Farin, G., Hoschek, J., Kim, M.-S. (eds.) Handbook of Computer Aided Geometric Design, pp. 723–748. North-Holland, Amsterdam (2002)
7. Röschel, O.: Rational motion design - a survey. Comput. Aided Design 30(3), 169–178 (1998)

8. Wagner, M.G.: Planar Rational B-Spline Motions. Comput. Aided Design 27(2), 129–137 (1995)
9. Farin, G.: Curvature Continuity and Offsets for Piecewise Conies. ACM Transactions on Graphics 10, 366–377 (1991)
10. Sederberg, T.W., Chen, F.: Implicitization using Moving Curves and Surfaces. In: Proceedings of the 22nd annual conference on Computer graphics and interactive techniques
11. Hoff III, K.E., Zaferakis, A., Lin, M., Manocha, D.: Fast and Simple 2D Geometric Proximity Queries Using Graphics Hardware. In: Proceedings of 2001 ACM Symposium on Interactive 3D Graphics, pp. 145–148 (2000)
12. Ahn, Y.J.: Conic approximation of planar curves. Comput. Aided Design 33(ER12), 867–872 (2001)
13. Hoff III, K.E., Culver, T., Keyser, J., Lin, M., Manocha, D.: Fast Computation of Generalized Voronoi Diagrams Using Graphics Hardware. In: Proc. of ACM SIGGRAPH, pp. 277–286 (1999)
14. Beckmann, N., Kriegel, H., Schneider, R., Seeger, B.: The r*-tree: An efficient and robust access method for points and rectangles. In: Proc.SIGMOD Conf. on Management of Data, pp. 322–331 (1990)
15. Cameron, S.: Approximation hierarchies and s-bounds. In: Proceedings. Symposium on Solid Modeling Foundations and CAD/CAM Applications, Austin, TX, pp. 129–137 (1991)
16. Floater, M.: An $O(h^{2n})$ Hermite approximation for conic sections. Computer Aided Geometric Design 14(2), 135–151 (1997)
17. Wang, W., Krasauskas, R.: Interference analysis of conics and quadrics. In: Goldma, R., Krasauska, R. (eds.) Topics in Algebraic Geometry and Geometric Modeling. AMS Contemporary Mathematics, vol. 334, pp. 25–36 (2003)
18. Van den Bergen, G.: Efficient Collision Detection of Complex Deformable Models using AABB Trees[J]. Journal of Graphics Tools 2(4), 1–13 (1997)
19. Farin, G.: Curves and Surfaces for CAGD. Academic Press, London (1993)
20. Adams, L.J.: The Use of Conics in Airplane Design. Mathematics Magazine 25(4), 195–202 (1952)
21. Piegl, L.: The sphere as a rational Bezier surface. Comput. Aided Geometric Design 3(1), 45–52 (1986)
22. Pratt, T.: Curve and surface constructions using rational B-splines. Comput. Aided Design 19(9), 485–498 (1987)
23. Sederberg, T., Anderson, D., Goldman, R.: Implicit representation of parametric curves and surfaces. Computer Vision, Graphics and Image Processing 28, 72–84 (1985)
24. Tiller, W.: Rational B-splines for curve and surface representation. IEEE Computer Graphics and its Application 3(6), 61–69 (1983)
25. Wilson, P.R.: Conic representations for sphere description. IEEE Computer Graphics and its Application 7(4), 1–31 (1987)
26. Barequet, G., Elber, G., Myung-Soo, K.: Computing the Minimum Enclosing Circle of a Set of Planar Curves. Computer Aided Design and Applications 2(1-4), 301–308 (2005)

Controlling Torsion Sign

E.I. Karousos, A.I. Ginnis, and P.D. Kaklis

National Technical University of Athens
School of Naval Architecture & Marine Engineering
{karousos,ginnis}@naval.ntua.gr, kaklis@deslab.ntua.gr

Abstract. We present a method for computing the domain, where a control point is free to move so that the corresponding spatial curve is regular and of constant sign of torsion along a subinterval of its parametric domain of definition. The approach encompasses all curve representations that adopt the *control-point paradigm* and is illustrated for a spatial Bézier curve.

Keywords: spatial curve, torsion, shape.

1 Introduction

Space curves, either as individual geometrical entities or as a means for compactifying surface or 3D-object representation, occur over a wide range of diverse application areas, for example: robotic path planning [4], simulation of one-dimensional flexible objects [11], [6], feature curves in automotive design scenarios [5], shape preservation [8] and shape matching for part inspection [14], just to name a few.

Provided that they can be robustly located, torsion-zeros are similarity invariant shape descriptors [9], that can be used for *3D curve matching*. E.g., the multiscale shape representation developed in [10] is based on the distributions of torsion-zero crossings. The shape influence of torsion sign can be further grounded by recalling results that link vanishing torsion with global convexity, namely the so-called spatial version of the four-vertex theorem, implying that simple closed nonvanishing torsion curves, lying on strictly convex surfaces in space, do not exist; see, e.g., [2], [12], [3].

The present paper addresses the problem of controlling the sign of torsion for the class of parametric curves $\mathbf{c}(t)$, that adopt the so-called *control-point paradigm*, namely,

$$\mathbf{c}(t) = \sum_{\rho} d_\rho N_\rho(t), \tag{1}$$

where d_ρ are the so-called *control points* and $\{N_\rho(t)\}$ is a finite set of weight functions with properties, e.g., non-negativity, partition of unity, variation diminishing, that render $\mathbf{d}_\rho$'s suitable for geometrically intuitive manipulation of the curve $\mathbf{c}(t)$. Note that most of the *continuous* curve representations, encountered in CAGD theory and CAD/CAM practice, are instances of the control-point paradigm: Bézier, *B*-splines, NURBS, trigonometric splines, exponential splines in tension, variable-degree splines and so on.

F. Chen and B. Jüttler (Eds.): GMP 2008, LNCS 4975, pp. 92–106, 2008.

The specific problem this work is dealing with, is to determine the so-called *domain of positive/negative torsion*, $D_{\tau\pm}(I)$, containing all possible loci of a user-specified control point $\mathbf{d}$, for which the resulting curves $\mathbf{c}(\mathbf{d};t)$ are regular and possess positive/negative torsion along the user-specified subinterval I of their parametric domain of definition. Note that, possessing a handy and easily computable representation of $D_{\tau\pm}(I)$, is clearly a prerequisite for handling the *constrained fairing problem for 3D curves*, namely optimizing curvature and/or torsion under shape (torsion-sign) constraints.

The rest of the paper is structured as follows. In §2, after introducing some necessary notation and terminology (Definition 2.1) and studying the auxiliary case when the interval I shrinks to a point (Proposition 2.1, Remark 2.1), we reach the conclusion that the problem in question can be formulated, under rather non-restrictive conditions, as a problem of intersecting a one-parameter family of half spaces; see in the last paragraph of § 2. Next, § 3 focuses on studying the properties of the envelope of the planes bounding the afore mentioned family of half spaces. In the most general and interesting case, the envelope is a tangential developable whose edge of regression is analyzed by Lemma 3.1 and Proposition 3.1. These results are subsequently exploited for quantifying the local properties of the envelope; see Proposition 3.2 and Remark 3.1.

Section 4 constitutes the kernel of the paper, eventually providing a three-step method for calculating the requested domain $D_{\tau\pm}(I)$ of positive/negative torsion; see in the last paragraph of § 4. The proposed method stems from combining the results of § 3 with those collected Proposition 4.1 and Corollary 4.1. The latter provide a precise and easily computable boundary representation of the required half-space intersection under a set of conditions, that can easily be met via Lemma 3.1 and Corollary 4.2. In this setting, the sought-for intersection is a convex domain bounded by a developable and two planar sets; see Fig. 1. The paper concludes with illustrating the developed method for a spatial quintic curve; see Figs 2-3.

2 Formulation of the Problem and Auxiliary Results

Suppose that a family of space parametric curves is given by

$$\mathbf{F}(\mathbf{d};t) := \mathbf{d}\, N(t) + \mathbf{s}(t), \quad t \in I \subset \mathbb{R}, \tag{2}$$

which maps each $\mathbf{d} \in \mathbb{R}^3$ to a spatial curve parameterized with respect to t, with $N: I \to \mathbb{R}$ and $\mathbf{s}: I \to \mathbb{R}^3$ being two sufficiently differentiable functions. The problem to be investigated is *which members of this family are regular curves of constant sign of torsion*. The same problem can be stated as follows: *Determine all admissible loci of a point $\mathbf{d}$ so that the curve $\mathbf{d}\, N(t) + \mathbf{s}(t)$ is regular and has constant sign of torsion.* Curve families as in (2) can naturally stem from curves whose parametric representation adopts the control point paradigm (1) with all control points kept fixed but one, say $\mathbf{d}$, that will be referred to as the free control point. Let us start dealing with this problem stating.

Definition 2.1. *The domain of positive torsion* $D_{\tau+}(t_0)$ *of a family of space parametric curves* $\mathbf{F}(\mathbf{d};t)$ *with respect to a fixed* $t_0 \in I$, *consists of all possible locations of* $\mathbf{d}$ *for which* $\mathbf{F}(\mathbf{d};t_0)$ *is an ordinary point of non-zero curvature and positive torsion.* $D_{\tau-}(t_0)$ *can be analogously defined.*

Let $\mathbf{d} = (x,y,z)$, $\mathbf{F} = (F_1, F_2, F_3)$ and $\mathbf{s} = (s_1, s_2, s_3)$. Recall that the sign of torsion at a regular and non-zero curvature point coincides with the sign of the quantity

$$h_t(x,y,z) := [\dot{\mathbf{F}}, \ddot{\mathbf{F}}, \dddot{\mathbf{F}}], \tag{3}$$

where dot signifies differentiation with respect to t and $[\mathbf{a}, \mathbf{b}, \mathbf{c}]$ denotes the standard triple scalar product of vectors $\mathbf{a}$, $\mathbf{b}$ and $\mathbf{c}$. Setting

$$a(t) := \begin{vmatrix} \dot{N} & \ddot{N} & \dddot{N} \\ \dot{s}_2 & \ddot{s}_2 & \dddot{s}_2 \\ \dot{s}_3 & \ddot{s}_3 & \dddot{s}_3 \end{vmatrix}, \quad b(t) := \begin{vmatrix} \dot{s}_1 & \ddot{s}_1 & \dddot{s}_1 \\ \dot{N} & \ddot{N} & \dddot{N} \\ \dot{s}_3 & \ddot{s}_3 & \dddot{s}_3 \end{vmatrix}, \quad c(t) := \begin{vmatrix} \dot{s}_1 & \ddot{s}_1 & \dddot{s}_1 \\ \dot{s}_2 & \ddot{s}_2 & \dddot{s}_2 \\ \dot{N} & \ddot{N} & \dddot{N} \end{vmatrix}, \quad d(t) := \begin{vmatrix} \dot{s}_1 & \ddot{s}_1 & \dddot{s}_1 \\ \dot{s}_2 & \ddot{s}_2 & \dddot{s}_2 \\ \dot{s}_3 & \ddot{s}_3 & \dddot{s}_3 \end{vmatrix}, \tag{4}$$

(3) can be written as

$$h_t(x,y,z) = a(t)x + b(t)y + c(t)z + d(t). \tag{5}$$

In order to state and prove the basic result of this section let us first introduce some notation. Let $\mathbf{a}\colon J \to \mathbb{R}^3$, $u\colon J \to \mathbb{R}$, $t \in J \subset \mathbb{R}$, be two arbitrary functions. For fixed $s \in J$ with $\mathbf{a}(s) \neq \mathbf{0}$, $H_s(\mathbf{a}, u) := \{\mathbf{x} \in \mathbb{R}^3 : \mathbf{a}(s)\mathbf{x} + u(s) = 0\}$ denotes a plane, while $H_s^{+(-)}(\mathbf{a}, u) := \{\mathbf{x} \in \mathbb{R}^3 : \mathbf{a}(s)\mathbf{x} + u(s) \geq (\leq)0\}$ denote the corresponding half-spaces.

Proposition 2.1. *Let* $t_0 \in I$ *be such that* $(a(t_0), b(t_0), c(t_0)) \neq \mathbf{0}$. *Then the domains of positive,* $D_{\tau+}(t_0)$, *and negative,* $D_{\tau-}(t_0)$, *torsion with respect to* t_0 *are the open half-spaces* $H_{t_0}^+(\mathbf{a}, u)\backslash H_{t_0}(\mathbf{a}, u)$ *and* $H_{t_0}^-(\mathbf{a}, u)\backslash H_{t_0}(\mathbf{a}, u)$, *respectively, where* $\mathbf{a} := (a, b, c)$, $u := d$.

Proof. Obviously, in view of (3), $H_{t_0}^{\pm}\backslash H_{t_0}$ is subset of $D_{\tau\pm}(t_0)$, respectively. Let us now take $\mathbf{d}$ on the plane H_{t_0}. If $\dot{\mathbf{F}}(\mathbf{d};t_0) \neq \mathbf{0}$ then $\mathbf{d}$ cannot belong to either $D_{\tau+}$ or $D_{\tau-}$, as a result of the fact that $\mathbf{F}(\mathbf{d};t_0)$ is either zero-curvature or zero-torsion point.

Now, we shall show that $\mathbf{F}(\mathbf{d};t_0)$ is a singular point if $\dot{\mathbf{F}}(\mathbf{d};t_0) = 0$. Differentiating (2), we get $\mathbf{d}^* = -\dot{\mathbf{s}}(t_0)/\dot{N}(t_0)$, where $\dot{N}(t_0) \neq 0$ as a consequence of the hypothesis $(a(t_0), b(t_0), c(t_0)) \neq \mathbf{0}$. Moreover, it can be checked that $\ddot{\mathbf{F}}(\mathbf{d}^*;t_0) \neq \mathbf{0}$. If, on the contrary, $\ddot{\mathbf{F}}(\mathbf{d}^*;t_0) = \mathbf{0}$, equivalently, $\ddot{\mathbf{s}}\dot{N} - \dot{\mathbf{s}}\ddot{N} = \mathbf{0}$, then $\ddot{\mathbf{s}} = (\ddot{N}\dot{\mathbf{s}})/\dot{N}$. Thus, $\ddot{\mathbf{F}} = \mathbf{d}\,\ddot{N} + \ddot{\mathbf{s}} = \mathbf{d}\,\ddot{N} + (\ddot{N}\dot{\mathbf{s}})/\dot{N} = (\ddot{N}/\dot{N})\dot{\mathbf{F}}$, which implies that each $\mathbf{d} \in \mathbb{R}^3$ lies on $H_{t_0}(\mathbf{a}, u)$, as $h_t = [\dot{\mathbf{F}}, \ddot{\mathbf{F}}, \dddot{\mathbf{F}}] = 0$. Therefore $(a, b, c) = \mathbf{0}$, which is a contradiction to the hypothesis. Summarizing, $\dot{\mathbf{F}}(\mathbf{d}^*;t_0) = \mathbf{0}$ while $\ddot{\mathbf{F}}(\mathbf{d}^*;t_0) \neq \mathbf{0}$, and thus, according to a standard result from singularity theory of curves ([13], Ch. VIII, §3), we deduce that $\mathbf{F}(\mathbf{d}^*;t_0)$ is a singular point. □

Remark 2.1. *If, in contrast with the assumption made in Proposition 2.1,* $(a(t), b(t), c(t)) = \mathbf{0}$, *it can be shown:*

- If $\dot{N}(t) \neq 0$, then $D_{\tau+}(t)$ is empty or contains exactly one element.
- If $\dot{N}(t) = 0$ and $\dot{\mathbf{s}}(t) \neq \mathbf{0}$, then $D_{\tau+}(t) = \varnothing$ or $D_{\tau+}(t) = \mathbb{R}^3$.
- If $\dot{N}(t) = 0$ and $\dot{\mathbf{s}}(t) = \mathbf{0}$, then to come into a conclusion about $D_{\tau+}(t)$, we need to find the first non-vanishing derivative of either $\dot{N}(t)$ or $\dot{\mathbf{s}}(t)$.

Analogous results hold true for $D_{\tau-}(t_0)$.

Unless stated otherwise and without loss of generality, we henceforth focus on the problem of maintaining positive torsion over an interval I. We denote by $D_{\tau+}(I)$ the domain of positive torsion of a family of space curves $\mathbf{F}(d;t)$, $t \in I$, i.e., the set of all possible locations of $\mathbf{d}$ for which $\mathbf{F}(\mathbf{d};t)$, defined as in (2), is ordinary with non-zero curvature and positive torsion over I. Note that

$$D_{\tau+}(I) = \bigcap_{t \in I} D_{\tau+}(t) = K \cap L, \tag{6}$$

where

$$K := \bigcap_{\substack{t \in I \\ (a,b,c) \neq \mathbf{0}}} D_{\tau+}(t), \qquad L := \bigcap_{\substack{t \in I \\ (a,b,c) = \mathbf{0}}} D_{\tau+}(t),$$

and a, b, c are the functions given in (4). Proposition 2.1 implies that the set

$$K = \bigcap_{\substack{t \in I \\ (a,b,c) \neq \mathbf{0}}} (H_t^+((a,b,c),d) \backslash H_t((a,b,c),d)) \tag{7}$$

is convex, being the intersection of a one-parameter family of half-spaces. In order to get a complete representation of $D_{\tau+}(I)$ we shall focus on K, aiming to express it as a finite intersection of easily computable convex sets. As for L, viewing Remark 2.1 in the context of the most popular representations of the control-point paradigm, it is easy to see that L can be expressed as the intersection of finite, possibly degenerate ($\varnothing$ or $\mathbb{R}^3$) sets, and in this connection its computation is left to be done on a case-by-case basis.

3 On the Envelope of a One-Parameter Family of Planes

Aiming to develop a method for computing K, the present and following (§ 4) sections deal with the more general problem of finding an analytical representation of the boundary of the closed convex set $G := \bigcap_{t \in I} H_t^+(\mathbf{a}, p)$, where $\mathbf{a} = (a_1, a_2, a_3) \colon I \to \mathbb{R}^3$ and $p \colon I \to \mathbb{R}$ are two general sufficiently differentiable functions of parameter t over an interval I. As it will be clearly seen in §4, G is fully characterized by the envelope of the associated one-parameter family of planes, $H_t(\mathbf{a}, p)$, $t \in I$, the local properties of which are studied in the current section. It is well known ([1], Ch. 5, §5.3) that the envelope (or *discriminant*) of this family is the (ruled) surface generated by the straight lines:

$$H_t(\mathbf{a}, p) \cap H_t(\dot{\mathbf{a}}, \dot{p}). \tag{8}$$

In many practically important cases, a characterization of the envelope is provided by the quantities

$$\delta(t) := \begin{vmatrix} a_1 & \dot{a}_1 & \ddot{a}_1 \\ a_2 & \dot{a}_2 & \ddot{a}_2 \\ a_3 & \dot{a}_3 & \ddot{a}_3 \end{vmatrix}, \qquad \omega(t) := \begin{vmatrix} a_1 & \dot{a}_1 & \ddot{a}_1 & \dddot{a}_1 \\ a_2 & \dot{a}_2 & \ddot{a}_2 & \dddot{a}_2 \\ a_3 & \dot{a}_3 & \ddot{a}_3 & \dddot{a}_3 \\ p & \dot{p} & \ddot{p} & \dddot{p} \end{vmatrix}, \tag{9}$$

as it can be seen by the following standard result : *a one-parameter family of planes, which are not all parallel and which do not form a pencil, has as its envelope a cylinder, if $\delta = 0$, a cone, if $\delta \neq 0$ and $\omega = 0$, or a tangential developable, if $\delta \neq 0$ and $\omega \neq 0$, for all $t \in I$*; see ([15], Ch. 2, §4).

The cylinder, the cone and the tangential developable are developable surfaces, since they share the property that they have a constant tangent plane along a generating line. We notice that, at regular points, the Gaussian curvature of a developable surface is identically zero.

Since the envelope being tangential developable is the most general and interesting case, we shall continue with examining the local properties of the envelope in this case. Let $\mathbf{r}(t)$ be the curve defined by:

$$\mathbf{r}(t) := H_t(\mathbf{a}, p) \cap H_t(\dot{\mathbf{a}}, \dot{p}) \cap H_t(\ddot{\mathbf{a}}, \ddot{p}). \tag{10}$$

Since $\delta \neq 0$ and $\omega \neq 0$, Lemma 3.1 below implies that $\mathbf{r}$ and its tangent lines are well defined. It is easy to check that the generating lines of the envelope and the planes $H_t(\mathbf{a}, p)$ coincide with the tangent lines and the osculating planes of $\mathbf{r}$, respectively. Hence, the envelope is a tangential developable with $\mathbf{r}(t)$ being its edge of regression, in other words, it is the tangent surface of $\mathbf{r}$, given by $\mathbf{r}(t) + v\dot{\mathbf{r}}(t)$, with $(t, v) \in I \times \mathbb{R}$. It is then natural to proceed with studying the local properties of $\mathbf{r}(t)$.

Lemma 3.1. *If $\delta\omega \neq 0$ then the edge of regression $\mathbf{r}$ is regular with nonzero curvature while its torsion τ is given by $\tau = -\delta^2/(\omega\|\mathbf{a}\|^2)$.*

Proof. One can easily see that δ is the determinant of the linear system in the right-hand side of (10). Since $\delta \neq 0$, (10) is uniquely solvable and thus $\mathbf{r}$ is well defined. Furthermore, it is straightforward to show that

$$\dot{\mathbf{r}} = (-\omega/\delta^2)\mathbf{a} \times \dot{\mathbf{a}} \qquad \text{and} \qquad \dot{\mathbf{r}} \times \ddot{\mathbf{r}} = (\omega^2/\delta^3)\mathbf{a}. \tag{11}$$

Since $\delta\omega \neq 0$, we have $\dot{\mathbf{r}} \neq 0$ and $\dot{\mathbf{r}} \times \ddot{\mathbf{r}} \neq 0$, which in its turn implies that $\mathbf{r}$ is regular with nonzero curvature. On the other hand, a simple computation shows that the torsion numerator $[\dot{\mathbf{r}}, \ddot{\mathbf{r}}, \dddot{\mathbf{r}}]$ equals to $-\omega^3/\delta^4$. As for the denominator, the second of (11) readily leads to $\|\dot{\mathbf{r}} \times \ddot{\mathbf{r}}\|^2 = |\omega^2/\delta^3|^2\|\mathbf{a}\|^2$ and, therefore, we arrive at the desired torsion formula. □

Let us now investigate how $\mathbf{r}$ behaves in the neighborhood of isolated zeros of $\delta\omega$.

Proposition 3.1. *Let* $\delta \neq 0$ *for* $t \in J \subset I$ *and* $t_0 \in J$ *be such that* $\omega(t_0) = 0$ *while* $\omega \neq 0$ *for* $t \in J\backslash\{t_0\}$. *If* ω *changes sign at* t_0 *then* $\mathbf{r}(t_0)$ *is a singular point, otherwise* $\mathbf{r}(t_0)$ *is an ordinary point.*

Proof. From the first of (11) we readily have that the unit tangent vector of $\mathbf{r}$ is given by

$$\frac{\dot{\mathbf{r}}}{\|\dot{\mathbf{r}}\|} = -\frac{\omega}{|\omega|} \frac{\mathbf{a} \times \dot{\mathbf{a}}}{\|\mathbf{a} \times \dot{\mathbf{a}}\|}.$$

Now, if ω changes sign at t_0 then $\mathbf{c}_0 = \lim_{t \to t_0^-} \frac{\dot{\mathbf{r}}}{\|\dot{\mathbf{r}}\|} = -\lim_{t \to t_0^+} \frac{\dot{\mathbf{r}}}{\|\dot{\mathbf{r}}\|}$, which implies that $\lim_{t \to t_0} \frac{\dot{\mathbf{r}}}{\|\dot{\mathbf{r}}\|}$ does not exist and thus $\mathbf{r}(t_0)$ is a singular point. Moreover, we observe that $\mathbf{c}_0 \mathbf{a}(t_0) = 0$ and therefore $H_{t_0}(\mathbf{a}, p)$ is the osculating plane of $\mathbf{r}$ at t_0. If ω does not change sign, then $\lim_{t \to t_0} \frac{\dot{\mathbf{r}}}{\|\dot{\mathbf{r}}\|}$ exists, implying that $\mathbf{r}(t_0)$ is an ordinary point. $\qquad\square$

When $\delta(t_0) = 0$ for some $t_0 \in J$, in which case the edge of regression is not defined, a variety of behaviors of $\mathbf{r}(t)$ can arise, depending on the properties of $\omega(t)$ and $\mathbf{a}(t) \times \dot{\mathbf{a}}(t)$ in the neighborhood of $t = t_0$. We proceed with a rough description of these behaviors:

(A) Let $\delta(t_0) = 0$ and $\mathbf{a}(t_0) \times \dot{\mathbf{a}}(t_0) \neq \mathbf{0}$. Then:
    (i) If $\omega(t_0) \neq 0$, then $\mathbf{r}(t)$ tends to infinity, approaching the plane $H_{t_0}(\mathbf{a}, p)$ while $\dot{\mathbf{r}}$ tends to be parallel to it.
    (ii) If $\omega(t_0) = 0$ and $\lim_{t \to t_0} \mathbf{r} = \mathbf{r}_0 \in \mathbb{R}^3$, we distinguish between the following two possibilities:
        1. ω retains sign at $t = t_0$, in which case $\mathbf{r}(t_0)$ is an ordinary point. Moreover, if δ changes sign at t_0, then $\mathbf{r}(t_0)$ is an inflection point.
        2. Otherwise, $\mathbf{r}(t_0)$ is a cusp point.
    (iii) If $\omega(t_0) = 0$ and $\lim_{t \to t_0} \mathbf{r} = \infty$, then $\mathbf{r}$ approaches the plane $H_{t_0}(\mathbf{a}, p)$ while $\dot{\mathbf{r}}$ tends to be perpendicular to $\mathbf{a}(t_0)$.
(B) If $\delta(t_0) = 0$ and $\mathbf{a}(t_0) \times \dot{\mathbf{a}}(t_0) = \mathbf{0}$ with $\mathbf{a_0} = \lim_{t \to t_0} \frac{\mathbf{a}}{\|\mathbf{a}\|}$, $p_0 = \lim_{t \to t_0} \frac{p}{\|\mathbf{a}\|}$, $\mathbf{a}_1 = \lim_{t \to t_0} \frac{\mathbf{a} \times \dot{\mathbf{a}}}{\|\mathbf{a} \times \dot{\mathbf{a}}\|}$, then the analysis in (A) can be repeated simply by using as tangent plane at $t = t_0$ the plane $H_{t_0}(\mathbf{a_0}, p_0)$. If, however, one of the above limits does not exist, one should examine the local behavior of curve $\mathbf{r}$ on a case-by-case basis.

We shall now proceed to exploit the previous results for studying the local properties of the envelope of the one parameter family of planes $H_t(\mathbf{a}, p)$, $t \in I$. For this purpose, we firstly state and prove

Lemma 3.2. *Let* $\mathbf{q}\colon I \to \mathbb{R}^3$ *be a regular parameterized curve of nonzero curvature* κ *and let* $\mathbf{x}(t, v)$ *be its tangent surface, i.e.,* $\mathbf{x}(t, v) := \mathbf{q}(t) + v\dot{\mathbf{q}}(t)$, $(t, v) \in I \times \mathbb{R}$. *The restriction* $\mathbf{x}\colon U \to \mathbb{R}^3$, *where* $U = \{(t, v) \in I \times \mathbb{R}\colon v \neq 0\}$, *is a regular parameterized surface with mean curvature*

$$H = -\frac{\|\dot{\mathbf{q}}\|^2 \|\dot{\mathbf{q}} \times \ddot{\mathbf{q}}\| \tau_*}{2|v|(\|\dot{\mathbf{q}}\|^2 \|\ddot{\mathbf{q}}\|^2 - (\dot{\mathbf{q}}\ddot{\mathbf{q}})^2)},$$

where τ_* *denoting the torsion of* $\mathbf{q}$.

Proof. Since $\kappa(t) = \|\dot{\mathbf{q}} \times \ddot{\mathbf{q}}\|/\|\dot{\mathbf{q}}^3\|$ is nonzero, we firstly observe that

$$\mathbf{x}_t \times \mathbf{x}_v = v\ddot{\mathbf{q}} \times \dot{\mathbf{q}} \neq 0, (t,v) \in U, \tag{12}$$

which implies that $\mathbf{x}\colon U \to \mathbb{R}^3$ is a regular parameterized surface. We now turn to calculate its mean curvature,

$$H = \frac{1}{2}\frac{eG - 2fF + gE}{EG - F^2}, \tag{13}$$

where E, F, G and e, f, g are the coefficients in the first and the second fundamental form, respectively. Since $\mathbf{x}_t = \dot{\mathbf{q}}(t) + v\ddot{\mathbf{q}}(t)$, $\mathbf{x}_v = \dot{\mathbf{q}}(t)$, $\mathbf{x}_{tt} = \ddot{\mathbf{q}}(t) + v\dddot{\mathbf{q}}(t)$, $\mathbf{x}_{vv} = 0$, $\mathbf{x}_{tv} = \ddot{\mathbf{q}}(t)$, the aforementioned coefficients have as follows:

$$\begin{array}{ll} E = \mathbf{x}_t\mathbf{x}_t = \|\dot{\mathbf{q}}\|^2 + v^2\|\ddot{\mathbf{q}}\|^2 + 2v\dot{\mathbf{q}}\ddot{\mathbf{q}}\ , & e = \mathbf{N}\mathbf{x}_{tt} = -|v|\frac{[\dot{\mathbf{q}},\ddot{\mathbf{q}},\dddot{\mathbf{q}}]}{\|\dot{\mathbf{q}}\times\ddot{\mathbf{q}}\|}, \\ F = \mathbf{x}_t\mathbf{x}_v = \|\dot{\mathbf{q}}\|^2 + v\dot{\mathbf{q}}\ddot{\mathbf{q}} & , f = \mathbf{N}\mathbf{x}_{tv} = 0, \\ G = \mathbf{x}_v\mathbf{x}_v = \|\dot{\mathbf{q}}\|^2 & , g = \mathbf{N}\mathbf{x}_{vv} = 0, \end{array} \tag{14}$$

where $\mathbf{N} := \frac{\mathbf{x}_t\times\mathbf{x}_v}{\|\mathbf{x}_t\times\mathbf{x}_v\|} = \frac{v}{|v|}\frac{\ddot{\mathbf{q}}\times\dot{\mathbf{q}}}{\|\ddot{\mathbf{q}}\times\dot{\mathbf{q}}\|}$. Substituting (14) into (13), we finally get the desired formula for H. $\qquad\square$

Let us now focus our attention on the restriction $\mathbf{s}_i$ of $\mathbf{x}(t,v)$ on $U_{I,i} := \{(t,v) \in I \times \mathbb{R}\colon (-1)^{i+1}v > 0\}$, $i = 1, 2$. Based on the previous pair of lemmata, one can prove

Proposition 3.2. *If $\delta\omega \neq 0$, $t \in J \subset I$, then the restrictions $\mathbf{s}_i$ of $\mathbf{x}(t,v)$ on $U_{J,i}$, $i = 1, 2$, are regular parameterized surfaces with mean curvature of constant sign and unit-normal vector field $\mathbf{N}(t,v)$ sharing the same or opposite direction with the normal $\mathbf{a}(t)$ of the associated one-parameter family of planes $H_t(\mathbf{a}, p), t\in J$.*

Proof. As $\mathbf{a}(t)$, $p(t)$ are sufficiently differentiable functions, we can say that $\delta(t)$, $\omega(t)$ are continuous which, in conjunction with the assumption that $\delta\omega \neq 0$ for $t \in J$, implies that δ and ω are of constant sign on J. From Lemma 3.1, we then have that the curvature of the edge of regression $\mathbf{r}$ is nonzero, while its torsion is of constant sign. Now, combining this result with Lemma 3.2 we readily see that the mean curvature of $\mathbf{s}_i$ on $U_{J,i}$, $i = 1, 2$, is of constant sign as well. Next, using (11) and (12) we obtain the following equalities

$$\mathbf{N}\mathbf{a} = \frac{\mathbf{x}_t \times \mathbf{x}_v}{\|\mathbf{x}_t \times \mathbf{x}_v\|}\mathbf{a} = -\frac{v}{|v|}\frac{\dot{\mathbf{r}} \times \ddot{\mathbf{r}}}{\|\dot{\mathbf{r}} \times \ddot{\mathbf{r}}\|}\mathbf{a} = (-1)^i\frac{|\delta|}{\delta}\|\mathbf{a}\|, \tag{15}$$

the last one guaranteeing that $\mathbf{N}(t,v)$ and $\mathbf{a}(t)$ share the same or the opposite direction on $U_{J,i}$. $\qquad\square$

Remark 3.1. *Given that the Gaussian curvature of $\mathbf{s}_i$ is zero, Proposition 3.2 implies that $\mathbf{s}_i$ is locally convex over $U_{J,i}$, $i = 1, 2$. Local convexity guarantees that it is possible to define a differential field of unit normal vectors $\mathbf{N}_i$ on $\mathbf{s}_i$*

such that, for each point $\mathbf{p} \in \mathbf{s}_i$, *there exists a surface neighborhood* V *of* $\mathbf{p}$ *which is contained in the side of the tangent plane towards which* $\mathbf{N}_i$ *at* $\mathbf{p}$ *is pointing. Moreover, it can easily be confirmed that if* $\delta\omega < 0$ *then* $\mathbf{a}$ *is in the same direction with* $\mathbf{N}_1$ *and in the opposite to* $\mathbf{N}_2$. *If* $\delta\omega > 0$, *the previous result holds after exchanging* $\mathbf{N}_1$ *with* $\mathbf{N}_2$.

4   On the Intersection of a One-Parameter Family of Half-Spaces

We now proceed to develop a method for computing the domain $G := \bigcap_{t \in I} H_t^+(\mathbf{a}, p)$ grounded on the assumption, supported by our study in §3, that the torsion of the edge of regression $\mathbf{r}$ of the envelope of the bounding planes of $H_t^+(\mathbf{a}, p)$ is of constant sign.

Proposition 4.1. *Let* $\mathbf{r}: (a, b) \to \mathbb{R}^3$ *be a sufficiently differentiable regular space curve with* $(\mathbf{t}, \mathbf{n}, \mathbf{b})$, κ *and* τ *denoting its Frénet trihedron, curvature and torsion, respectively; see helicoidal curve in Fig. 1. Assume that* $\kappa > 0$, $\tau > 0$ *and* $\mathbf{b}(v)\mathbf{t}(u) > 0, a < u, v < b, v \neq u$, *as well as the limits of* $\mathbf{r}$, $\mathbf{b}$ *and* $\mathbf{t}$ *at the endpoints of* (a, b) *exist. Furthermore, let* $\mathbf{c}^*(s)$ *be the intersection curve of the tangent surface* $\mathbf{x}(s, v) := \mathbf{r}(s) + v\dot{\mathbf{r}}(s)$, $(s, v) \in (a, b) \times \mathbb{R}$, *of* $\mathbf{r}$ *with the osculating plane* $H_b(\mathbf{b}, -\mathbf{rb})$ *of* $\mathbf{r}$ *at* $t = b$. *Then the boundary of* $\mathcal{G} := \bigcap_{s \in [a,b]} H_s^+(\mathbf{b}, -\mathbf{rb})$ *consists of three pieces, namely a developable and two planar sets (see Fig. 1) defined as below:*

(i) *The subset of the surface* $\mathbf{x}(s, v)$, $(s, v) \in (a, b) \times (0, \infty)$, *given by* $\mathbf{y}(s, v) := \mathbf{c}^*(s) + v\dot{\mathbf{r}}(s)$, $(s, v) \in (a, b) \times (0, \infty)$; *see the green (a) surface in Fig. 1.*

(ii) *The convex subset of plane* $H_a(\mathbf{b}, -\mathbf{rb})$, *obtained by intersecting the half-planes* $\mathfrak{p}_1 : H_a(\mathbf{b}, -\mathbf{rb}) \cap H_a^+(-\mathbf{n}, \mathbf{rn})$ *and* $\mathfrak{p}_2 : H_a(\mathbf{b}, -\mathbf{rb}) \cap H_b^+(\mathbf{b}, -\mathbf{rb})$, *where* $\mathbf{n}(a) := \mathbf{b}(a) \times \mathbf{t}(a)$; *see the red (b) planar surface in Fig. 1.*

(iii) *The convex subset of plane* $H_b(\mathbf{b}, -\mathbf{rb})$, *whose boundary consists of the trace of* $\mathbf{c}^*(s)$ *and, if they exist, two half-lines contained in the tangent lines at the endpoints of* $\mathbf{c}^*(s)$; *see the blue (c) planar surface in Fig. 1.*

Proof. Without loss of generality we assume that $\dot{\mathbf{r}}$ has unit length, that is $\dot{\mathbf{r}} = \mathbf{t}$. First we notice that the continuity of $\mathbf{b}$ and $\mathbf{r}$ in compact $[a, b]$ implies that every boundary point of $\mathcal{G}$ should lie on a plane $H_{s_1}(\mathbf{b}, -\mathbf{rb})$ [1] for some $s_1 \in [a, b]$. Thus, since $\mathcal{G}$ is closed, it is sufficient to compute the sets $\mathcal{G} \cap H_{s_0}$ and then take their union $\cup_{s_0 \in [a,b]}(\mathcal{G} \cap H_{s_0})$. The sets $\mathcal{G} \cap H_{s_0}$ can be alternatively considered as $\cap_{s \in [a,b]}(H_s^+ \cap H_{s_0})$, a point of view that is adopted from now on. For $s \neq s_0$, $H_s^+ \cap H_{s_0}$, $s \in [a, b]$, is a family of half planes with boundary lines $H_s \cap H_{s_0}$, whose envelope is the intersection of H_{s_0} with the tangent surface $\mathbf{r}$, represented as

$$\mathbf{c}(s) := \mathbf{r}(s) - u(s)\mathbf{t}(s), s \in (a, b), s \neq s_0,$$

---

[1] Within this proof and whenever no confusion is likely to arise, symbols of the form $H_\bullet^*(\mathbf{b}, -\mathbf{rb})$ will be replaced by $H_\bullet^*$.

with

$$u(s) = u_*(s_0, s), \quad u_*(s, w) = \frac{\mathbf{b}(s)(\mathbf{r}(w) - \mathbf{r}(s))}{\mathbf{b}(s)\mathbf{t}(w)}.$$

We now collect in the form of a lemma a set of properties of $\mathbf{c}$ that will be used in the sequel of the proof.

Lemma 4.1. *Let* $\mathbf{r}(s)$ *be a space curve parameterized by arc length. Let* $u(s)$ *be a real function such that the curve* $\mathbf{c}(s) := \mathbf{r}(s) - u(s)\mathbf{t}(s)$, $\mathbf{t} = \dot{\mathbf{r}}$, *is planar, that is,* $\mathbf{x_1}\dot{\mathbf{c}} = 0$ *for some fixed* $\mathbf{x_1} \in \mathbb{R}^3$. *Then:*

1. $\dot{\mathbf{c}}(s) = f(s)\mathbf{x_1} \times \mathbf{b}(s)$ *where* $f(s) := \frac{u(s)\kappa(s)}{\mathbf{x_1}\mathbf{t}(s)}$.
2. $\mathbf{x_1}(\dot{\mathbf{c}} \times \ddot{\mathbf{c}}) = \tau(s)f(s)^2\|\mathbf{x_1}\|^2\mathbf{x_1}\mathbf{t}(s)$.
3. $(\mathbf{x_1} \times \mathbf{x_2})\dot{\mathbf{c}} = f(s)\|\mathbf{x_1}\|^2\mathbf{x_2}\mathbf{b}(s)$ *for* $\mathbf{x_2} \in \mathbb{R}^3$ *and* $\mathbf{x_1}\mathbf{x_2} = 0$.
4. $(\mathbf{x_1} \times \dot{\mathbf{c}})\mathbf{b}(s) = -f(s)\|\mathbf{x_1} \times \mathbf{b}(s)\|^2$.

We can now easily see that $\mathbf{c}$ is indeed the envelope of the family of lines $H_s \cap H_{s_0}$, $s \neq s_0$, since the point $\mathbf{c}(s)$ lies on both H_{s_0} and H_s as well as $\mathbf{b}(s_0) \times \mathbf{b}(s)$ is collinear with $\dot{\mathbf{c}}(s)$, as it can readily be deduced by setting $\mathbf{x_1} = \mathbf{b}(s_0)$ in item *1* of Lemma 4.1.

To proceed we have to distinguish between the following three cases:

(i) $a < s_0 < b$. In this case, it can easily be seen that since $\mathbf{b}(v)\mathbf{t}(u) > 0$ for all $v, u \in (a, b)$, $v \neq u$, we have that $u(s) < 0$, $a < s < s_0$ and $u(s) > 0$, $s_0 < s < b$. In addition, one can prove that

$$\dot{\mathbf{c}} \neq \mathbf{0}, \quad \lim_{t \to s_0} \mathbf{c} = \mathbf{r}(s_0) \quad \text{and} \quad \lim_{t \to s_0} \frac{\dot{\mathbf{c}}}{\|\dot{\mathbf{c}}\|} = \mathbf{t}(s_0). \tag{16}$$

Setting $\mathbf{x_1} = \mathbf{b}(s_0)$ in item *4* of Lemma 4.1, we deduce that the half-plane $H_{s_0} \cap H_s^+$ is identical to the one determined by the line $H_{s_0} \cap H_s$ and its normal vector $\mathbf{b}(s_0) \times \dot{\mathbf{c}}(s)$, if $a < s < s_0$, or the opposite of it, if $s_0 < s < b$. Let us now focus on the limits $H_s^+ \cap H_{s_0}$ as $s \to s_0\pm$. As implied by (16), the intersection of the limits is the line

$$\ell_1 : \mathbf{r}(s_0) + v\mathbf{t}(s_0), v \in \mathbb{R},$$

that is, the generating line of the tangent surface $\mathbf{x}(s, v)$ at $s = s_0$. Consequently, $\cap_{s \in [a,b]}(H_s^+ \cap H_{s_0}) = \cap_{s \in [a,b]}(H_s^+ \cap \ell_1)$, which is obviously a subset of ℓ_1. Thus, it suffices to calculate $H_s^+ \cap \ell_1$ for $s \neq s_0$, which is but the half line ℓ_1 with $v \geq h(s) := -u_*(s, s_0)$, as $\mathbf{b}(v)\mathbf{t}(s_0) > 0$ for all $v \in (a, b)$, $v \neq s_0$. After simple calculations we obtain

$$h(s) = \frac{\tau(s)\mathbf{b}(s_0)\mathbf{t}(s)}{(\mathbf{b}(s)\mathbf{t}(s_0))^2}\mathbf{n}(s_0)(\mathbf{c}(s) - \mathbf{r}(s_0)).$$

Applying items *2* and *3* of Lemma 4.1 with $\mathbf{x_1} = \mathbf{b}(s_0)$ and $\mathbf{x_2} = \mathbf{t}(s_0)$, we have that $\mathbf{c}$ is of constant sign of curvature, while the tangent vector of each segment $\mathbf{c}|_{(a,s_0)}$ and $\mathbf{c}|_{(s_0,b)}$ turns at most π. Thus, the trace of $\mathbf{c}$

lies entirely on one of the closed half-planes determined by the tangent line at s_0. Combining this fact with the hypotheses $\tau > 0$ and $\mathbf{b}(v)\mathbf{t}(u) > 0$, we eventually conclude that $\dot{h}$ is positive and therefore h is an increasing function. Consequently, $\cap_{s\in[a,b],s\neq s_0}(H_s^+ \cap \ell_1)$ is the half-line $\mathbf{r}(s_0) + v\mathbf{t}(s_0)$, $v \geq h(b)$. Let us now s_0 vary in (a,b). Then, the sought-for boundary $\cup_{s_0\in(a,b)}(\cap_{s\in(a,b)}(H_s^+ \cap H_{s_0}))$ of $\mathcal{G}$ is the surface $\mathbf{y}(s,v) = \mathbf{c}^*(s) + v\dot{\mathbf{r}}(s)$, $(s,v) \in (a,b) \times (0,\infty)$.

(ii) $s_0 = a$. Since $\mathbf{b}(v)\mathbf{t}(u) > 0$, we have that $u(s) > 0, a < s < b$. Applying again Lemma 4.1 we have that the envelope $\mathbf{c}$ is now convex. Moreover, $\mathbf{b}(a) \times \dot{\mathbf{c}}(s)$ is the normal vector of $\mathbf{c}$ and it is in the opposite direction to that determined by the half-plane $H_s^+ \cap H_a$. Hence, $\cap_{s\in[a,b]}(H_s^+ \cap H_a)$ is the intersection of the half-planes which correspond to the endpoints of (a,b), namely $\mathfrak{p}_1$ and $\mathfrak{p}_2$.

(iii) $s_0 = b$. In this case, $u(s) < 0, a < s < b$. Analogously to case (ii), $\mathbf{c}$ is convex. Again, $\mathbf{b}(b) \times \dot{\mathbf{c}}(s)$ is the normal vector of $\mathbf{c}$, but now it is in the same direction with that determined by the half-plane $H_s^+ \cap H_a$. Then, $\cap_{s\in[a,b]}(H_s^+ \cap H_b)$ is the convex subset of H_b which is defined by the trace of $\mathbf{c}$ and, if they exist, its endpoint tangent lines. □

The remaining combinations regarding the sign of τ and $\mathbf{bt}$ can be analogously treated leading to:

Corollary 4.1. *(i) If $\tau < 0$, $\mathbf{b}(u)\mathbf{t}(v) < 0$ and $\mathcal{G} := \cap_{s\in[a,b]} H_s^-(\mathbf{b}, -r\mathbf{b})$, Proposition 4.1 remains valid simply by replacing H^+ with H^-.*
– [(ii)]If $\{\tau > 0$, $\mathbf{b}(u)\mathbf{t}(v) > 0$ and $\mathcal{G} := \cap_{s\in[a,b]} H_s^-(\mathbf{b}, -r\mathbf{b})\}$ or $\{\tau < 0$, $\mathbf{b}(u)\mathbf{t}(v) < 0$ and $\mathcal{G} := \cap_{s\in[a,b]} H_s^+(\mathbf{b}, -r\mathbf{b})\}$, one should reverse the orientation of the curve $\mathbf{r}$ which reduces the configuration to that of Proposition 4.1 or case (i) above, respectively.

The following result provides sufficient conditions for satisfying the assumption: $\mathbf{b}(v)\mathbf{t}(s) > 0$ for all $v, s \in (a,b), v \neq s$ of Proposition 4.1.

Proposition 4.2. *Let $\mathbf{r}(s) = (r_1(s), r_2(s), r_3(s))$ be a space curve with positive torsion over (a,b). Let $\mathbf{x_1}, \mathbf{x_2}$ be a pair of orthogonal to each other vectors. If $\mathbf{x_1}\mathbf{t}(s) \neq 0$ and $\mathbf{x_2}\mathbf{b}(s) \neq 0$ for all $s \in (a,b)$, then $\mathbf{b}(v)\mathbf{t}(s) > 0$ for all $v, s \in (a,b), v \neq s$.*

Proof. Without loss of generality, we assume that $\mathbf{x_1} = (0,0,1)$ and $\mathbf{x_2} = (1,0,0)$. The assumption $\mathbf{x_1}\mathbf{t}(s) \neq 0$ implies that $\dot{r}_3 \neq 0$ in (a,b) and therefore we can apply the inverse function theorem, guaranteeing that the inverse of r_3 exists and possesses the same order of differentiability with r_3. Hence, $\mathbf{r}(s)$ can be reparameterized as: $\mathbf{r}(s) = (f(s), g(s), s)$, where f, g are sufficiently differentiable functions. Computing $\mathbf{t}$ and $\mathbf{b}$, we have

$$\mathbf{t}(s) = a_1(s)(\dot{f}, \dot{g}, 1), \quad \mathbf{b}(s) = a_2(s)(-\ddot{g}, \ddot{f}, \dot{f}\ddot{g} - \ddot{f}\dot{g}), \quad \text{where} \quad a_1(s), a_2(s) > 0,$$

which yields:

$$\mathbf{b}(v)\mathbf{t}(s) = a_1(s)a_2(v)[\ddot{g}(v)(\dot{f}(v) - \dot{f}(s)) - \ddot{f}(v)(\dot{g}(v) - \dot{g}(s))]. \quad (17)$$

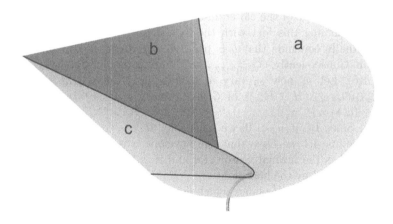

Fig. 1. Illustrating Proposition 4.1

The assumption $\mathbf{x_2}\mathbf{b}(s) \neq 0$ implies that $\ddot{g} \neq 0$. Assume further, without loss of generality, $\ddot{g} < 0$. As the torsion τ of $\mathbf{r}$ is positive and $\tau = a_3(s)(\dot{f}\ddot{g} - \ddot{f}\dot{g})$, with $a_3(s) > 0$ we have that $(\dot{f}\ddot{g} - \ddot{f}\dot{g}) > 0$. The last inequality implies that the planar curve $\mathbf{h} := (-\dot{g}, \dot{f})$ has positive curvature and moreover, since $-\dot{g}$ is strictly increasing function, it follows that $\mathbf{h}$ is a strictly convex curve. Therefore $\mathbf{n_h}(v)(\mathbf{h}(s) - \mathbf{h}(v)) > 0, s \neq v$, where $\mathbf{n_h}$ is the normal vector of $\mathbf{h}$. Noting that

$$\mathbf{n_h}(v)(\mathbf{h}(s) - \mathbf{h}(v)) = (-\ddot{f}(v), -\ddot{g}(v))(-\dot{g}(s) + \dot{g}(v), \dot{f}(s) - \dot{f}(v))$$

and recalling (17) as well as $a_1(s)a_2(v) > 0$, the validity of the proposition follows readily.   □

The corollary below guarantees that for a curve with positive torsion, its parametric domain of definition can be partitioned so that $\mathbf{b}(v)\mathbf{t}(s) > 0$ for all $v, s \in (a, b)$, $v \neq s$, holds true in each open subinterval of the partition. This result can be easily obtained by applying Proposition 4.2 with $\mathbf{x_1} = \mathbf{t}(s_0), \mathbf{x_2} = \mathbf{b}(s_0)$ for some $s_0 \in [a, b]$ and appealing to the Heine-Borel covering theorem.

Corollary 4.2. *Let* $\mathbf{r}(s)$, $s \in (a, b)$, *be a space curve for which* $\tau > 0$, *for all* $s \in (a, b)$, *and let the limits of* $\mathbf{t}, \mathbf{b}$ *at the endpoints of* (a, b) *exist. Then there exists a partition* $\mathcal{P} = \{a = t_0 < t_1 < \ldots < t_{n-1} < t_n = b\}$ *of* $[a, b]$, *such that* $\mathbf{b}(v)\mathbf{t}(s) > 0$ *for all* $v, s \in (t_{i-1}, t_i), v \neq s$, *and for each index* $1 \leq i \leq n$.

Based on the propositions and corollaries presented in §3, §4, we end this section with roughly describing a three-step method for computing the domain $G = \bigcap_{t \in [a,b]} H_t^+(\mathbf{a}, p)$, which is the aim of this section stated in its beginning paragraph. Firstly, we decompose the edge of regression $\mathbf{r}$ of the envelope of the associated family of planes $H_t(\mathbf{a}, p)$ into regular segments, where its torsion is of constant sign while the binormal $\mathbf{b}$ of $\mathbf{r}$ shares the same or opposite direction with $\mathbf{a}$. Appealing to Lemma 3.1, this decomposition can be achieved by isolating the roots t_i of equation $\delta(t)\omega(t) = 0$. Furthermore, according to Proposition 3.2, this decomposition segments the aforementioned envelope into locally convex pieces, where their unit normal vector field shares the same or opposite direction with $\mathbf{a}$; see also Remark 3.1. Next, based on Corollary 4.2, we further partition each interval $[t_i, t_{i+1}]$ by solving equations of the form $\mathbf{b}(t^*)\mathbf{t}(s) = 0$, $\mathbf{b}(s)\mathbf{t}(t^*) = 0$, t^*, $s \in [t_i, t_{i+1}]$, for a user-specified $t^* \in [t_i, t_{i+1}]$, $\mathbf{t}$ denoting the unit tangent of $\mathbf{r}$. In each interval of the so obtained finer partition, one can readily use Proposition 4.1 or Corollary 4.1, depending on the sign of $\mathbf{ba}$, to obtain the boundary representation of $G_i := \bigcap_{t \in [t_i, t_{i+1}]} H_t^+(\mathbf{a}, p)$. The third and final step consists in computing the intersection of the so resulting G_i's, that eventually reduces to finding intersections between developables.

5   Computing the Domain of Constant-Sign Torsion

The results of Sections 3 and 4 are readily applicable for computing the domain K (eq.(7)) where the control point $\mathbf{d}$ of a spatial curve, represented as in (2), is

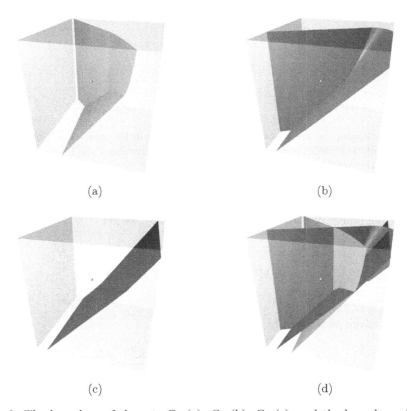

(a)   (b)

(c)   (d)

Fig. 2. The boundary of the sets G_1 (a), G_2 (b), G_3 (c), and the boundary of the negative torsion domain $D_{\tau-}([0,1]) = \bigcap_{i=1,2,3} G_i$, (d)

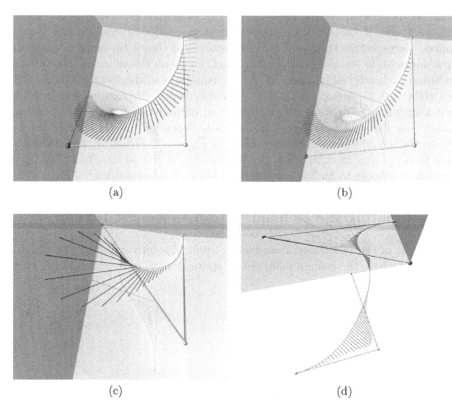

(a)   (b)

(c)   (d)

Fig. 3. Positioning the free control point (red sphere): in the interior (a), on a linear edge (b) and on a linear vertex of the boundary (c/d view from the interior/exterior) of the negative-torsion domain

free to move while its torsion retains constant sign over I. Applying (8), (9) and (10) for the family of planes defined by (4) and (5), we get

$$\mathbf{r} = -\frac{\dot{\mathbf{s}}}{\dot{N}}, \quad \delta = \dot{N}M^2, \quad \omega = M^3 \quad \text{with } M = \begin{vmatrix} \dot{N} & \ddot{N} & \dddot{N} & \ddddot{N} \\ \dot{s}_1 & \ddot{s}_1 & \dddot{s}_1 & \ddddot{s}_1 \\ \dot{s}_2 & \ddot{s}_2 & \dddot{s}_2 & \ddddot{s}_2 \\ \dot{s}_3 & \ddot{s}_3 & \dddot{s}_3 & \ddddot{s}_3 \end{vmatrix}, \qquad (18)$$

while the envelope is given by

$$\mathbf{x}(s,v) = \mathbf{r}(s) + v\dot{\mathbf{r}}(s) = -\frac{\dot{\mathbf{s}}}{\dot{N}} - v\frac{\ddot{\mathbf{s}}\dot{N} - \dot{\mathbf{s}}\ddot{N}}{\dot{N}^2}. \qquad (19)$$

Suppose that $\ddot{\mathbf{s}}\dot{N} - \dot{\mathbf{s}}\ddot{N} \neq \mathbf{0}$ for all $t \in I$. Then, as it is shown in [7], surface (19) is the locus of the free control point $\mathbf{d}$ that yields a curve with zero curvature point, while its regression curve $\mathbf{r}$ is the locus of $\mathbf{d}$ that guarantees a curve with a cusp.

This section ends with illustrating the methodology outlined in the last paragraph of §4 for a spatial quintic Bézier curve. In this example, we are interested in calculating the domain $D_{\tau-}(I)$, where the free control point, depicted as red sphere in Figs. 3.a-3.d, is free to move so that the sign of torsion is negative over $I = [0, 1]$. Employing the first step of the proposed methodology, we compute the roots $t_i = \{0, 0.4, 0.5, 1\}$ of the polynomial equation $\delta(t)\omega(t) = 0$ in $[0,1]$, which provides the required partition of the associated edge of regression. Going to the second step, in each of the three subintervals $[t_i, t_{i+1}]$ of the obtained partition, we check and find no zeros of the equations $\mathbf{b}(t_i)\mathbf{t}(s) = 0$, $\mathbf{b}(s)\mathbf{t}(t_i) = 0$, and thus we can proceed to compute the sets G_i, $i = 1, 2, 3$. The boundary of each one of these sets, clipped by a transparent box, is shown in Figs. 2.a-2.c with its interior being signified via a gold sphere. The outcome of the third and final step, namely the intersection $D_{\tau-}([0, 1]) = \cap_{i=1,2,3} G_i$, is given in Fig. 2.d.

In the remaining figures, we illustrate the effect of positioning the free control point in the interior (Fig. 3.a) as well as the boundary (Figs. 3.b-3.d) of the domain $D_{\tau-}([0, 1])$. In these figures, besides the curve and its control polygon, the vector $\tau\mathbf{n}$, appropriately scaled, is depicted in porcupine form. In Figs. 3.b-3.d the free control point is located on the boundary of $D_{\tau-}([0, 1])$, yielding points of null torsion. Analytically, in Fig. 3.b the free control point lies on a linear edge of the boundary, which vanishes torsion at the end points of the curve, while in Figs. 3.c and 3.d null torsion occurs also in the interior, as a result of positioning the free control point on a linear vertex of the boundary.

References

1. Bruce, J.W., Giblin, P.J.: Curves and Singularities. Cambridge University Press, Cambridge (1988)
2. Costa, S.I.R.: On closed twisted curves. Proceedings of the American Mathematical Society 109, 205–214 (1990)
3. Costa, S.I.R., Romero-Fuster, M.C.: Nowhere vanishing torsion closed curves always hide twice. Geometriae Dedicata 66, 1–17 (1997)
4. Dooner, D.B.: Introducing radius of torsure and cylindroid of torsure. Journal of Robotic Systems 20, 429–436 (2003)
5. Farin, G.: Class A Bézier Curves. CAGD 23, 573–581 (2006)
6. Grégoire, M., Schömer, E.: Interactive simulation of one dimensional flexible parts. In: SPM 2006, ACM Symposium on Solid and Physical Modeling, Cardiff, Wales, UK, June 2006, pp. 95–103 (2006)
7. Juhász, I.: On the singularity of a class of parametric curves. CAGD 23, 146–156 (2006)
8. Kong, V.P., Ong, B.H.: Shape preserving F^3 curve interpolation. CAGD 19, 239–256 (2002)
9. Li, S.Z.: Similarity invariants for 3D space curve matching. In: Proceedings of the First Asian Conference on Computer Vision, Osaka, Japan, November 1993, pp. 454–457 (1993)
10. Mokhtarian, F.: A theory of multiscale, torsion-based shape represenation for space curves. Computer Vision and Image Understanding 68, 1–17 (1997)

11. Moll, M., Kavraki, L.E.: Path planning for variable resolution minimal-energy curves of constant length. In: Proceedings of the 2005 IEEE International Conference on Robotics and Automation, Barcelona, Spain, April 2005, pp. 2130–2135 (2005)
12. Nuño-Balesteros, J.J., Romero-Fuster, M.C.: A four vertex theorem for strictly convex space curves. Journal of Geometry 46, 119–126 (1993)
13. Pogorelov, A.: Geometry. MIR Publishers, Moscow (1987)
14. Rao, P.V.M., Bodas, M., Dhande, S.G.: Shape matching of planar and spatial curves for part inspection. Computer-Aided Design & Applications 3, 289–296 (2006)
15. Struik, D.J.: Lectures on Classical Differential Geometry, 2nd edn. Addison-Wesley Publishing Company, INC., Reading (1961)

Cutting and Fracturing Models without Remeshing

Chao Song, Hongxin Zhang, Yuan Wu, and Hujun Bao

State Key Lab of CADCG, Zhejiang University,
Hangzhou 310027, P.R. China
{songchao,zhx,wuyuan,bao}@cad.zju.edu.cn

Abstract. A finite element simulation framework for cutting and fracturing model without remeshing is presented. The main idea of proposed method is adding a discontinuous function for the standard approximation to account for the crack. A feasible technique is adopted for dealing with multiple cracks and intersecting cracks. Several involved problems including extended freedoms of finite element nodes as well as mass matrix calculation are discussed. The presented approach is easy to simulate object deformation while changing topology. Moreover, previous methods developed in standard finite element framework, such as the stiffness warping method, can be extended and utilized.

Keywords: Physically based animation, finite element method, fracturing model, without remeshing.

1 Introduction

In industry design and digital entertainment, the simulation for cutting, fracturing models and their deformation has been widely applied. Hence relevant research has been an import area in computer graphics, virtual reality and computer aided design.

One of the key issue on this topic is how to deal with the changing of shape topology during simulating cutting and fracturing in which crack initialization and crack growth are included. Various methods based on dynamics and statics were proposed in recent research. There are mainly three kinds of simulation techniques on space discreting, which are mass-spring system, finite element method (FEM) and meshless method. Meanwhile, the research and applications of FEM and meshless methods are received increasing concern because of their high controllability and stability.

The standard FEM simulates fracturing and cutting problems by remeshing models around a growing crack. However, remeshing is computationally expensive and lots of physical parameters for new element nodes have to be calculated. On the other hand, it is increasingly difficult to guarantee the simulation stability. In this paper, we leverage an extended finite element method that adjusts the element approximation function to account for element discontinuous based on standard FEM framework. In our proposed approach, the remeshing procedure is not a necessary step. Moreover, many previous computation techniques, such as techniques of accessory calculation, can be easily utilized.

Our work has three main contributions:

- We propose an approach of attaching additional degree of freedoms (DOFs) on element nodes and make it pretty easily to be implemented.

F. Chen and B. Jüttler (Eds.): GMP 2008, LNCS 4975, pp. 107–118, 2008.

- We improve the approach for tackling multiple cracks and intersecting cracks to meet the demand in computer graphics and virtual reality.
- We adopt the stiffness warping technique to enhance our simulation framework for compensating the nonlinear factors of deformation.

Moreover, in order to optimize the calculation and stability, several relevant implementation issues, such as reasonable choices of the mass matrix, are discussed in this paper.

2 Related Work

In the literature of computer animation, Terzopoulos and Fleischer [1,2] used a distance threshold between two nodes to judge fracture during simulating viscoelastic and plastic deformation. They demonstrated this technique with sheets paper and cloth that could be torn apart. Later, the mass-spring system was applied to simulate the fracturing models, including work of Norton *et al.* [3], simulating the mud crack pattern by Hirota *et al.* [4]. The most important virtue of the mass-spring system in fracture simulation is relied on the simple data structure which leads to simple implementations. But two of its disadvantages are calculation instability and limited reality of resultant animations.

In 1999, O'Brien *et al.* did excellent work in simulating fracture models using standard FEM framework. They adopted separation tensor and remeshing finite element mesh of model to successfully simulate brittle fracture [5], ductile fracture [6] and surface cracks pattern [7]. Müller [8] did similar work to implement real-time simulation of brittle fracture.

Molino and Bao as well as their colleagues [9,10] originally presented a virtual node algorithm to deal with the troubles of remeshing in standard FEM. The algorithm duplicates cutting elements that meet specific conditions and the remeshing procedure is not necessary. The virtual node algorithm developed the simulation for topology changed by FEM. The drawback of the algorithm is in the complexity of geometry data structure and the strictly limitation in that the smallest possible unit would be individual nodes. Wicke *et al.* [11] presented a finite element on convex polyhedra to simulating cutting models. Their method also does not need additional remeshing, but expensive calculations are required.

Pauly *et al.* [12] applied the meshless method to fracture simulation in computer graphics. Although it has high cost in calculating approximation functions, meshless method has great advantages in dealing with point sample models and large strain deformation.

Fracture and cutting have been studied extensively in the mechanics literature. There are a lot of related work on the extended finite element method which has an initial form for small strain and statics application [13,14]. The method has been developing in mechanics and even the virtual node algorithm can also be included. In this paper, we adopt the foundational theory and develop it for application of movie industry and virtual reality.

The paper is organized as follows. Our method is mainly presented in Section 3. Section 4 discusses the solutions of multiple cracks and intersection cracks. In Section 5, a brief consideration on simulation control is given. Section 6 describes our implementation and several examples are provided. Paper summary and conclusions are presented in the last section.

3   Method Description

In general, a model in our computation framework is represented as a domain $\Omega \subset R^3$. For each point $p \in \Omega$, the movement is represented

$$p : \Omega \times R \rightarrow R^3 : (\mathbf{X}, 0) \mapsto \mathbf{x}(\mathbf{X}, t),$$

where $\mathbf{X}(p)$ is the material coordinates, $\mathbf{x}(\mathbf{X}, t)(p)$ is the location coordinates of point p at time t. Let $\mathbf{u}(p) = \mathbf{x}(p) - \mathbf{X}(p)$ be the displacement. If there are n cracks in the model, we represent them as $\Gamma_{ci}, i \in [1, n]$. The simulation goal is getting all points coordinates after the model movement or deformation. In addition, as for cutting simulation, the cutting face can be taken as a crack. It will not be distinguished in this paper.

In our simulation, a model is discreted into tetrahedron finite elements without regarding of the cracks. But our method can be utilized in other types of elements without any difficulty. The dynamics equation of simulation is a PDE based on time

$$\mathbf{M}\ddot{\mathbf{x}} + \mathbf{C}\dot{\mathbf{x}} + \mathbf{K}(\mathbf{x} - \mathbf{X}) = \mathbf{f}_{\text{ext}}, \tag{1}$$

where $\mathbf{M}$, $\mathbf{C}$, $\mathbf{K}$ and $\mathbf{f}_{\text{ext}}$ are mass matrix, damp matrix, stiffness matrix and extern force vector of nodes, respectively. In the above equation, symbol $\mathbf{x}$ and $\mathbf{X}$ are the column vectors which are composed by space coordinates and material coordinates of all nodes, respectively. All these coefficients are calculated as in standard FEM, that is, $\mathbf{M} = \sum \mathbf{m}^e$, $\mathbf{C} = \sum \mathbf{c}^e$, $\mathbf{K} = \sum \mathbf{k}^e$ and $\mathbf{f}_{\text{ext}} = \sum \mathbf{f}_{ext}^e$. The matrices with superscript e stand for the element distribution matrices.

3.1   Adding Extended Freedoms

In order to make the description clearer, we call the element with a crack as *crack element* (e.g., the element 1 to 6 in Figure 1). In this section, we will explain our simulation method by limiting one crack in one crack element without loss of generality. We will discuss more on the solutions for multiple cracks in one crack element in the next section.

Our crack propagation is per-element based one, i.e., a crack advances on a complete element within a time-step. Similar to add virtual nodes in the virtual node algorithm, we simply add extended freedoms for related nodes in our method instead. To achieve this goal, let K_{nT} denote the node set related to the crack n. Regarding for the crack n, we attach additional freedoms $\mathbf{a}_i = \{a_{ix}, a_{iy}, a_{iz}\}^T$ on node $i \in K_{nT}$ (Figure 1). The key problem is how to construct K_{nT} corresponding to crack n. Hence a two-step procedure is adopted in our framework:

(1) After mapping the crack to the initial configuration of a model, for each edge of the model, determine whether there exists intersection point between it and crack n. If it is true, add the intersection point into S_n. And then

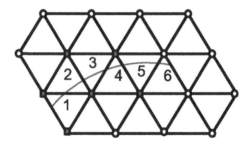

Fig. 1. Add extended freedoms for nodes. The red curve represents a crack. Nodes represented by red rectangle need to add extended freedoms. Nodes represented by circle without filling are not necessary to add extended freedoms.

    (a) if the intersection point is a vertex, add the vertex into K_{nT},
    (b) if the intersection is not a vertex, add two vertices of the edge into K_{nT};
(2) Find the crack fringe curve L_{nf}. For each point $q \in S_n$, if $q \in L_{nf}$ and meet two conditions: q is not in the model surface and is not in the other crack surface. Then,
    (a) if q is a vertex (element node), remove the vertex from K_{nT};
    (b) if q is not a vertex, remove two vertices of the edge including q from K_{nT}.

Compared with previous methods of adding extended finite element in mechanics literature, our above method is simpler in three dimension simulations.

3.2 Crack Element Approximation

After adding extended degree of freedoms, a crack element has at least one node with extended freedom. In our method, we assume that the displacement field can be decomposed into a continuous part and a discontinuous part, i.e., $\mathbf{u} = \mathbf{u}^{cont} + \mathbf{u}^{disc}$ in the crack element. Hence an approximating function of displacement field presented by [15] can be adopted:

$$\mathbf{u}(\mathbf{X}) = \sum_{I=1}^{4} N_I(\mathbf{X})\mathbf{u}_I + \sum_{I \in K_T} N_I(\mathbf{X})\left[H(f(\mathbf{X})) - H(f(\mathbf{X}_I))\right]\mathbf{a}_I, \tag{2}$$

where $f(\mathbf{X}) = 0$ is an implicit surface representation of the crack, $N_I(x)$ is tetrahedron shape function. In Equation 2, the first term $\mathbf{u}^{cont} = \sum_{I=1}^{4} N_I(\mathbf{X})\mathbf{u}_I$ is a continuous part and has the same shape as standard FEM approximation. And the second term $\mathbf{u}^{disc} = \sum_{I \in K_T} N_I(\mathbf{X})\left[H(f(\mathbf{X})) - H(f(\mathbf{X}_I))\right]\mathbf{a}_I$ is a discontinuous part that indicates the difference between the two sides of crack due to the *Heaviside function*

$$H(x) = \begin{cases} 1, x \geq 0, \\ 0, x < 0. \end{cases} \tag{3}$$

As for a normal element, the standard FEM approximation is performed actually. In fact, $\mathbf{u}^{disc}$ is a zero vector when the Heaviside function is constant in the whole element. It is worth noting that there always exists movement independence between the two

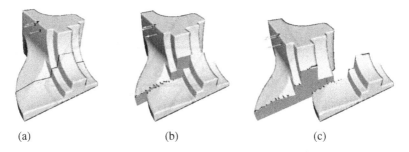

Fig. 2. Movement independence. (a) initial cutting model. (b) and (c) removing one part from the model.

sides of a crack. This is critical and is a primary criterion for judging the feasibility of a simulation method of cutting and fracturing. An example in Figure 2 shows the movement independence of our simulation.

3.3   Crack Element Calculating

In a crack element, the initial coordinates, initial velocity and initial accelerate of extended freedom is $(0, 0, 0)$. The calculation of the corresponding mass of extended freedom and the element mass matrix is vital and can affect the movement independence.

In standard FEM, there are mainly two choices for setting mass matrix. One is the *average mass matrix* which is setting matrix as a diagonal matrix, whose nonzero value is the model mass divided by the total number of the nodes. The other one is the *compatibility mass matrix* which is calculated by element approximation and density. It is straightforward to extend the average mass matrix by setting the mass value of corresponding the all extended freedoms be zero. Unfortunately, our experiences show it may make the coefficient matrix of solve system be singular when multiple cracks are occurred in a model and there are more than one crack in the same element.

We also tried to make the corresponding mass of extended freedom the same value as the mass of corresponding node as [10]. But it is evident to increase total mass of the model and simulation error. More seriously, the movement independence is failure in our simulation framework.

To sum up the above arguments, compatibility mass matrix have to be adopted. As for a crack element, the element mass matrix should be calculated as follows

$$\begin{cases} \mathbf{m}_{ij}^{uu} = \int_{\Omega^e} N_i N_j \mathrm{d}\omega^e \\ \mathbf{m}_{ij}^{ua} = \int_{\Omega^e} N_i N_j(\mathbf{X}) \left[H(f(\mathbf{X})) - H(f(\mathbf{X}_j)) \right] \mathrm{d}\omega^e \\ \mathbf{m}_{ij}^{au} = \int_{\Omega^e} N_i(\mathbf{X}) \left[H(f(\mathbf{X})) - H(f(\mathbf{X}_i)) \right] N_j \mathrm{d}\omega^e \end{cases} \quad (4)$$

where $\mathbf{m}_{ij}^{\xi\eta}(\zeta, \eta = u, a)$ denotes the corresponding mass between the ζ-th freedom of node i and the η-th freedom of node j, and Ω^e is the space domain of the element.

Strictly following Equation (4) can guarantee the movement independence. But it makes difficulty on applying Equation (4) in interactive application. That is, we do not know a model will fracture or be cut during deformation and choose average mass

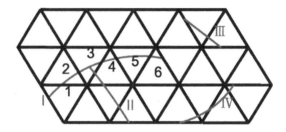

Fig. 3. Multiple cracks in a model. Red curves indicate cracks.

matrix at beginning of simulation. In order to avoid this problem, we calculate the mass matrix as follows. At beginning of simulation, we adopt the average mass matrix if there is no crack in the model. When the model has cracks, evaluate the mass matrix by calculating $\mathbf{m}_{ij}^{ua}$ and $\mathbf{m}_{ij}^{au}$ only if $i = j$, and keep the others $\mathbf{m}_{ij}^{\xi\eta}(\zeta, \eta = u, a)$ be zero. Our experience shows that our approach is feasible and decreases the cost of calculation.

Beside the evaluation of mass matrices, the calculation on crack elements is almost the same as in standard finite element method by applying the approximation function in Equation (2). The above treatment will be only invoked if the element includes cracks. Thus a small amount of additional calculation are required.

4   Multiple Cracks and Intersecting Cracks

Commonly there are multiple cracks and intersecting cracks in a model. We therefore improve the methods presented by mechanics literature to meet the demand of computer graphics applications.

4.1   Independent Cracks

In section 3, we have already known that an element is not affected by a crack if Heviside function is constant in it. So if there are multiple cracks in a model and any two cracks of them is far, not in the same element and not intersecting. We call these cracks be independent (independent cracks, as if III and IV in Figure 3). We can deal with these cracks separately. Certainly, corresponding crack elements crack set is different for every crack. The tackling can be expressed as

$$\mathbf{u}(\mathbf{X}) = \sum_{I=1}^{4} N_I(\mathbf{X})\mathbf{u}_I + \sum_{n=1}^{n_c} \sum_{I \in K_{nT}} N_I(\mathbf{X}) \left[H(f_n(\mathbf{X})) - H(f_n(\mathbf{X}_I)) \right] \mathbf{a}_I^n \qquad (5)$$

where n is the crack flag in the element, implicit function $f_n(\mathbf{X}) = 0$ represents the surface of crack n, $\mathbf{a}_I^n$ is the corresponding extended freedom of crack n at node I. And term K_{nT} is the node set in which every node need to add extended freedom of corresponding crack n.

4.2 Multiple Cracks and Intersecting Cracks

Independent cracks can grow to intersect in the same element during the simulation. Regarding for this problem, Equation (5) may be failed. Budyn *et al.* [16] proposed a method of intersecting cracks for engineering application. Our experiments proved applying their method for computer graphics application will be not able to guarantee the movement independence. Hence we have to extend and improve their methods.

For the reason of clarity but without loss of generality, we only explain the occurrence when there are two cracks in one element. Obviously, the two cracks can be intersecting or not. But we treat both cases unitedly.

In our simulation, it is necessary to assume that the two cracks cut the element of in order. As for intersecting crack, the later crack face c is stopped by the former crack face C (see Figure 4). But when adding corresponding extended freedom, the later crack should be assumed to whole cutting the element. Firstly, we must judge the two crack mutual location according to their implicitly representing. And then we apply the following equation [16]:

$$
\mathbf{u}(\mathbf{X}) = \sum_{I=1}^{4} N_I(\mathbf{X})\mathbf{u}_I + \sum_{I \in K_{cT}} N_I \varphi_{cI}(\mathbf{X})\mathbf{a}_I^c +
$$
$$
\sum_{I \in K_{CT}} N_I(\mathbf{X}) \left[H(f_C(\mathbf{X})) - H(f_C(\mathbf{X}_I)) \right] \mathbf{a}_I^C, \tag{6}
$$

where

$$
\varphi_{cI}(\mathbf{X}) = \begin{cases} H(f_C(\mathbf{X})) - H(f_C(\mathbf{X}_I)), & \mathbf{X} \in \text{A1}, \\ H(f_c(\mathbf{X})) - H(f_c(\mathbf{X}_I)), & \mathbf{X} \in \text{A2}. \end{cases}
$$

Here the extended freedoms $\mathbf{a}_I^c$, $\mathbf{a}_I^C$ are matching crack c and C.

It is worth noting that we apply Equation (6) for the elements not only including intersecting cracks as in literature [16]. In order to simulation for computer graphics and virtual reality, we improve the approach as following.

As for the case that two cracks are not intersecting in one element, Equation (6) is used directly. When two cracks intersect and coincide, they will fusion to one crack and stop growth. The element in which the fusion points of the two cracks may be tackled by taking it into intersecting crack element.

In addition, if other elements adjacent with the multiple cracks element meet: (i) be cut by crack C and (ii) have no less than one node with extended freedoms corresponding crack c. It also must be tackled with Equation (6) even though it is not cut by more

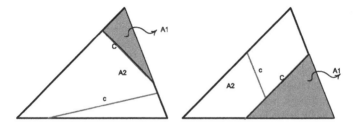

Fig. 4. Two cracks in one model. The filled region is A1, and the remaining part is A2.

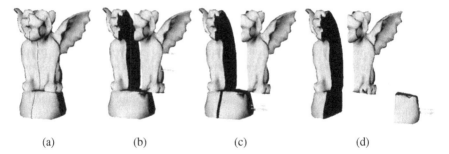

Fig. 5. The movement of intersecting cracking. (a) initial cutting model. (b) moving the right upper part. (c) moving the right lower part. (d) the finally movement.

than one crack. Element 1, 5 and 6 in Figure 3 must be tackled with equation (6) and the above method. Figure 6 shows the movement independent in a model with intersecting and cutting crack.

As for more than two cracks in an element, we deal with them according to Equation (6) similar with two cracks. In fact, the method is not increasing calculation other than judge and save the mutual location and easy to be implemented. Moreover, the restriction is not the same strictly as virtual nodes algorithm in that every fracture part must include no less than one node. In our simulation experiments, there is no trouble caused by the problem.

5   Simulation Control

In general, most of the previous research in mechanics is as for engineering application dealing with *small deformation*. We therefore want to enhance our simulation method for *large deformation*. In the method described in section 3, Cauchy strain is adopted to obtain stiffness matrix. The linear calculation will lead to non-realistic results. Therefore the stiffness warping technique for compensating the nonrealistic are applied in our simulation framework. As for fracture control, we mainly use the results from the previous research.

5.1   Stiffness Warping in Crack Element

By applying the stiffness warping technique proposed in [17] to our simulation framework, Equation (2) is rewritten as

$$\mathbf{M}\ddot{\mathbf{x}} + \mathbf{C}\dot{\mathbf{x}} + \sum_i \sum_j \mathbf{R}_{ij}^e \mathbf{k}_{ij}^e (\mathbf{R}_{ij}^{e-1} \mathbf{x}_j - \mathbf{X}_j) = \mathbf{f}_{ext}, \qquad (7)$$

where j is node index, i is index of the element which includes node j, $\mathbf{R}_{ij}^e$ is a corresponding sub-matrix of the rotation of element I matching node j, $\mathbf{k}_{ij}^e$ is a corresponding sub-matrix of stiffness matrix of element i. Please note that the rotation matrix is between current location and initial location.

As for a crack element, it is necessary to calculate rotation of every part taken into by cracks separately. This can be obtained by performing polar decomposition for the

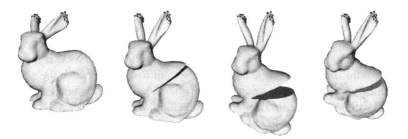

Fig. 6. Movement sequence of *Bunny* with cutting crack

conversion matrix of location. Obviously, in the crack element, the coefficient of matrix calculation in Equation (7) should be performed according to every part because of their rotation may be different. In the example showed in Figure 6, the *bunny* with a cutting crack moves under gravity while the two ears of the model are constrained. Our algorithm generally provides visual-pleasant during the simulation of model deformation.

5.2 Fracture Control

As for the model being cut, initial cracks are provided by users while the simulation procedure is performed automatically. When simulating fracturing model, the Rankine condition of maximal principal stress is used to control element fracture as [8,9]. The stress of an element is smoothed by volume weighted averaging with adjacent elements before the principal stress is obtained by doing eigenvalue decomposition to stress matrix. If the maximal principle stress is greater than a tensile threshold (positive), or the minimal principle stress is less than a compressive threshold (negative), the element is cut by a crack controlled along the principle direction.

  When the distance between one crack fringe and the other crack face is less than a threshold, the former directly propagates to intersect the latter. In addition, if the distance and the orientation angle between two crack fringes are less than threshold values, the two cracks fusion into one crack. After that, the tackle method is performed that has been discussed in Section 3 and 4.

6 Implementation and Examples

According to the description above, we implemented a prototype of model simulation in a compatible PC with Intel CPU. Similar to [18], we leverage a variant of implicit integration method for dynamics simulation in this paper. The main steps are:

1. Update velocities of all nodes according to $\mathbf{v}^{i+1} = \mathbf{v}^i + \Delta\mathbf{v}^i$.
2. Update displacements of all nodes according to $\mathbf{x}^{i+1} = \mathbf{x}^i + \mathbf{v}^i \cdot \Delta t$.
3. Processing collision and determining the boundary condition.
4. Solve difference of velocities $\Delta\mathbf{v}^{i+1}$ by

$$(\mathbf{M} + \Delta t\mathbf{C} + \Delta t^2\mathbf{K})\Delta\mathbf{v}^{i+1} = \Delta t(\mathbf{f}_{ext} - \mathbf{C}\mathbf{v}^i - \sum_i\sum_j \mathbf{R}_{ij}^e\mathbf{k}_{ij}^e(\mathbf{R}_{ij}^{e-1}\mathbf{x}_j - \mathbf{X}_j) - \Delta t\mathbf{v}^i).$$

In our simulation framework, the total number of elements is invariant while the number of freedoms of a model may be altered. Real displacement of every node are the displacement of initial freedoms according to Equation (2). The displacement of extended freedoms can be used for calculating point positions in cracks. It is worth mentioning that the dimension of the final linear system will rise up when the number of crack elements is rising. However, the only amended the approximation function of crack element, common element calculation is according with standard FEM. It follows that the calculating cost will not increase a lot. As our simulation framework can be tailored to emulate the virtual node algorithm (VNA), there is no significant advantage in computation performance compared with VNA. But regarding for the implementation of data structure and geometry processing operators, our method is simpler and easier to integrate with existing techniques of standard FEM.

In this paper, adjusting the approximation of crack element is based on the *theory of partition of unity* whose feasibility has been thoroughly discussed in [19,20]. In addition, we utilize a linear FEM framework with stiffness warping to simulate large deformations for the applications of computer graphics. These treatments ensure the stability of our simulation method. Compared with standard FEM, our method do not need any remeshing. Therefore the instability factor due to the occurrence of long and thin tetrahedra while cutting and fracturing models in standard FEM can be avoided.

Several computation examples (e.g., Figure 2, 5, 6, 8 and 9) are provided in this paper to demonstrate the performance of our proposed method. Figure 7 is a plane constrained on the left side while pulling on the right side. As expected, when the pulling force surpass a pre-set threshold, the plane is slit open into to two parts.

Figure 8 is an example of cutting model. We cut a plane into several pieces firstly, and then drop it on top of a sphere. During the period of falling and colliding with other objects, the movement of all of the parts is independent. In this example, the initial DOFs is 4671 and the DOFs is 7026 after cutting. The simulation refresh rate in this example can reach interactive speed, and it takes less than 1 second for every frame in average.

An example of a *flyman* model objecting to bump on the face is showed in Figure 9. The initial flyman, which is a complete model without any cracks, breaks into several pieces during the movement. The initial model has 450 nodes and 1320 elements. At the end of simulation, the model has total 2139 DOFs and every frame costs 700ms in average.

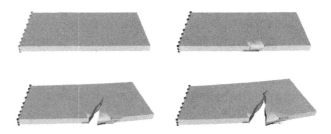

Fig. 7. The fracturing *plane* subject to tension. Red arrows represent the direction of pulling force.

Fig. 8. A *plane* being cut into several parts and falling down

Fig. 9. A *flyman* subjecting to bump on face

7 Conclusion and Future Work

In this paper, a simulation framework for cutting and fracturing is proposed. A novel approximation function for crack element is employed on the basis of standard FEM. Several key problems, such as tackling for the mass matrix, multiple cracks and intersecting cracks and compensating deformation, are dissolved. Compared with standard FEM, our proposed method does not need any remeshing, and can decrease computation cost as well as increase simulation stability.

However, there are still several limitations in our simulating. The first is that cutting or fracturing models in our method is element-based. When crack only cut an element half, our method in this paper cannot deal with it currently. The others, such as control fracture more efficiently and collision between difference fracturing parts, should be accomplished. We will explore more on those topics in near future.

Acknowledgements. This project is supported in partial by the Cultivation Fund of the Key Scientific and Technical Innovation Project, Ministry of Education of China (No. 705027), 973 Program of China (No.2002CB312102), and NSFC (No.60505001).

References

1. Terzopoulos, D., Fleischer, K.: Deformable models. The Visual Computer 4, 306–331 (1988)
2. Terzopoulos, D., Fleischer, K.: Modeling inelastic deformation:viscoelasticity, plasticity, fracture. In: Computer Graphics (SIGGRAPH 1988 Proceedings), pp. 269–278 (1988)
3. Norton, A., Turk, G., Bacon, B., Gerth, J., Sweeney, P.: Animation of fracture by physical modeling. The Visual Computer 7, 210–217 (1991)
4. Hirota, K., Tanoue, Y., Kaneko, T.: Generation of crack patterns with a physical model. The Visual Computer 14, 126–137 (1998)

5. O'Brien, J.F., Hodgins, J.K.: Graphical modeling and animation of brittle fracture. In: ACM SIGGRAPH 1999. Computer Graphics Proceedings, Annual Conference Series, pp. 137–146 (1999)
6. O'Brien, J.F., Bargteil, A., Hodgins, J.: Graphical modeling of ductile fracture. ACM Trans. Graph. 21, 291–294 (2002)
7. Iben, H.N., O'Brien, J.F.: Generating surface crack patterns. In: Symposium on Computer Animation, pp. 177–185 (2006)
8. Müller, M., McMillan, L., Dorsey, J., Jagnow, R.: Real-time simulation of deformation and fracture of stiff materials. In: Proc. Eurographics Workshop. Eurographics Asscociation. Compute. Anim. And Simu. 2001, pp. 99–111 (2001)
9. Molino, N., Bao, Z., Fedkiw, R.: A virtual node algorithm for changing mesh topology during simulation. ACM Trans. Graph. 23(3), 385–392 (2004)
10. Bao, Z., Hong, J.M., Teran, J., Fedkiw, R.: Fracturing rigid materials. IEEE Trans. Vis. Comput. Graph. 13(2), 370–378 (2007)
11. Wicke, M., Botsch, M., Gross, M.: A finite element method on convex polyhedra. In: Computer Graphics Forum (Proc. Eurographics 2007), vol. 26, pp. 355–364 (2007)
12. Pauly, M., Keiser, R., Adams, B., Dutré, P., Gross, M.H., Guibas, L.J.: Meshless animation of fracturing solids. ACM Trans. Graph. 24(3), 957–964 (2005)
13. Belytschko, T., Black, T.: Elastic crack growth in finite elements with minimal remeshing. International Journal for Numerical Methods in Engineering 45(5), 601–620 (1999)
14. Moes, N., Dolbow, J., Belytschko, T.: A finite element method for crack growth without remeshing. International Journal for Numerical Methods in Engineering 46(1), 131–150 (1999)
15. Zi, G., Belytschko, T.: New crack-tip elements for xfem and applications to cohesive cracks. International Journal for Numerical Methods in Engineering 57, 2221–2240 (2003)
16. Budyn, E., Zi, G., Moes, N., Belytschko, T.: A method for multiple crack growth in brittle materials without remeshing. International Journal for Numerical Methods in Engineering 62, 1741–1770 (2004)
17. Müller, M., Gross, M.H.: Interactive virtual materials. In: Graphics Interface (2004), pp. 239–246 (2004)
18. Baraff, D., Witki, A.P.: Large steps in cloth simulation. In: ACM SIGGRAPH (1998), pp. 43–54 (1998)
19. Duarte, C.A., Oden, J.T.: An h-p adaptive method using clouds. Computer Methods in Applied Mechanics and Engineering 139, 237–262 (1996)
20. Melenk, J.M., Bubska, I.: The partition of the unity finite element method: basic theory and applications. Computer Methods in Applied Mechanics and Engineering 139, 289–314 (1996)

Detection of Planar Regions in Volume Data for Topology Optimization

Ulrich Bauer and Konrad Polthier

FU Berlin,
Arnimallee 3, 14195 Berlin, Germany
{ubauer,polthier}@mi.fu-berlin.de
http://geom.mi.fu-berlin.de/

Abstract. We propose a method to identify planar regions in volume data using a specialized version of the discrete Radon transform operating on a structured or unstructured grid. The algorithm uses an efficient discretization scheme for the parameter space to obtain a running time of $\mathcal{O}(N(T \log T))$, where T is the number of cells and N is the number of plane normals in the discretized parameter space.

We apply our algorithm in an industrial setting and perform experiments with real-world data generated by topology optimization algorithms, where the planar regions represent portions of a mechanical part that can be built using steel plate.

Keywords: Discrete Radon transform, Hough transform, Plane detection, Topology optimization.

1 Introduction

The field of *topology optimization* studies the automatic generation of mechanical parts with an a priori unknown topological shape [1]. Prominent techniques include *continuous methods* [1,2,3], which optimize a 3D density function in the given work space, or truss methods [4,5], which optimize and rearrange a construction of stiff linear elements connected at junctions points. In the present paper we analyze data arising from continuous methods. Since these methods often produce output defined on unstructured grids such as tetrahedral meshes, our method is designed to work on this kind of data.

The density function obtained from continuous methods is a coarse model of the optimal structural design. Typically, post processing of the density function is needed in order to obtain a constructible shape. For example, planar metal plates are a simple and cheap building block to physically realize a mechanical part. We present an algorithm which automatically computes a set of planar regions which best approximates a given density function arising from topology optimization.

By a planar region, we denote a connected component of the intersection of some plane with the support of the given density function $\Omega \subset \mathbb{R}^3$, i.e. the domain on which the density function is defined. Our algorithm solves the problem

F. Chen and B. Jüttler (Eds.): GMP 2008, LNCS 4975, pp. 119–126, 2008.
© Springer-Verlag Berlin Heidelberg 2008

Fig. 1. A volume rendered density function and two planar regions, visualized by their corresponding planes. Volume data colored according to the closest planar region.

to find the planar region that covers most density. The process can be repeatedly applied after purging the density around the optimal planar region.

Our algorithm uses a discrete Radon transformation [6] and computes the integrated density of planar regions through the work space. Regions with highest integral density become candidates for the final construction (cf. Fig. 1). A similar technique for 2D images is known as Hough Transform [7,8]. The space of planes in $\mathbb{R}^3$ is discretized using a triangulation of the unit sphere together with an optimized traversal algorithm to reduce the number of sorting procedures.

The algorithm grew out of an industry collaboration with ThyssenKrupp Tallent Ltd., and has been tested on various datasets from automotive industry. We provide experimental evidence of the reliability and efficiency of our implementation.

1.1   Related Work

The proposed technique can be considered as a discretization of a variant of the Radon transform in $\mathbb{R}^3$ on piecewise constant functions defined on structured or unstructured grids. The most well-known discretization of a Radon transform is the Hough Transform [7,8], which is defined on a 2-dimensional grid representing a pixel image.

Other discretizations of the Radon transform have been proposed in [9] for 3-dimensional regular grids, and in [10,11] for unstructured point clouds in $\mathbb{R}^3$. A common problem in these schemes is that the configuration space, in particular the space of undirected plane normals, namely the projective plane $\mathbb{RP}^2$, is represented by a single-chart parametrization over a subset of $\mathbb{R}^2$, and contains singularities where a piece of the boundary of the parameter domain is mapped onto a single point. In the vicinity of these singularities, unbounded metric distortion is unavoidable. Uniform sampling in the parameter domain therefore leads to drastic oversampling near the singularities. We propose a discretization that aims at equal distribution of sampling points on $\mathbb{RP}^2$. Note that this is not an issue for the usual Hough transform in $\mathbb{R}^2$, since the space of unoriented normals $\mathbb{RP}^1$ can be parametrized without metric distortion.

A different scheme using clustering of local estimates for plane detection in point clouds has been proposed in [12]. This approach, however, has disadvantages: first, the local estimation of best-fitting planes is susceptible to noise and requires an additional parameter determining the estimation range. Moreover, finding clusters is a complex problem. In comparison, discrete variants of the Hough transform require no local estimates and no initial values, are very robust to noise, and deterministically find the optimum in the discrete parameter space by exhaustive search.

Other examples (with a different scope) of post-processing methods with the goal to extract production-relevant data from data generated by structural optimization tools can be found in [13,14].

1.2   Overview

In Section 2, we review the problem in the smooth setting. Section 3 describes the proposed discretization of the problem. In Section 4, we provide a complexity analysis of the algorithm, and Section 5 contains results of our algorithm on test data.

2   The Radon Transform and Generalizations

Consider a density function $\rho : \Omega \subset \mathbb{R}^3 \to \mathbb{R}$ with support on a compact domain $\Omega \subset \mathbb{R}^3$. The (generalized) Radon transform of ρ is defined as the integral of ρ over a hyperplane $H \in \mathbb{H}$ [6]:

$$\mathcal{R}[\rho](H) = \int_H \rho(x)\mathrm{d}x$$

In his original work, Radon's main interest was on the *inverse Radon transform*, which allows to reconstruct the function ρ from the integrals over hyperplanes; this transform found a particularly important application in the evaluation of computer tomography data. But also the Radon transform gained interest in a particular discretization for 2D images, called the Hough Transform [7,8], which is an important tool for line detection. One notable property of the Hough transform (and the Radon transform in general) is its high robustness to noise, because only quantities integrated over large domains are considered.

Maxima of this function over the set of hyperplanes $\mathbb{H}$ correspond to planar features. To incorporate geometric locality into the search for planar regions, we can restrict the integration to connected components of the intersection of H with the domain Ω and obtain the following optimization problem:

$$\max_{\substack{D \subset \Omega \cap (H \in \mathbb{H}) \\ D \text{ connected}}} \int_D \rho(x)\mathrm{d}x$$

To smooth the density function and to allow a certain tolerance for off-plane deviation, we can enlarge the integration domain to include all points having distance at most w from the hyperplane

$$H_w = \{x : d(H, x) \le w\} \,,$$

i.e. the intersection of two halfspaces, called *fattened plane* of width w, and consider the optimization problem

$$\max_{\substack{D \subset \Omega \cap H_w \\ D \text{ connected}}} \int_D \rho(x) \mathrm{d}x$$

This smoothing is of special interest in our application, since the density function computed by topology optimization algorithms is often concentrated in bone-like structures (cf. Figs. 1 and 3). To recognize planar regions formed by these "bones", one would like to set the width w to approximately match the width of such a bone. This is especially important when the optimization process is applied iteratively, because the density covered by the fattened plane has to be purged after each iteration. If the width w is chosen too small, some density in the vicinity of the optimal planar region is left and will be considered in the next iteration. As a consequence, several similar planar regions can be found, corresponding to the same feature in the input data. To avoid this problem, the parameter w must be chosen large enough that the planar features are completely covered by the fattened planes.

3   Detection of Planar Regions in Discrete Data

Assume that we are given a structured or unstructured grid with a piecewise constant nonnegative scalar density function assigned to each cell. We are now searching for the fattened plane that covers most density. To simplify computation, we are using a lumped mass model, i.e. we assume that the whole mass inside a cell is concentrated at its barycenter. The mass of a cell is computed as the density multiplied by the volume of the cell. If the width w of the fattened plane is considerably larger than the typical diameter of the cells, this simplification introduce only negligible artifacts.

3.1   Discretization of the Parameter Space

The described problem is a global nonlinear optimization problem, but it only has a 3-dimensional parameter space, which makes it feasible for exhaustive search in an appropriate discretization of the parameter space. We choose the following parametrization for the parameter space, which is the space of unoriented planes. We describe each plane by a normal vector $\boldsymbol{n}$ with $\|\boldsymbol{n}\| = 1$ and a distance to the origin d. Since we are only considering unoriented planes, we also assume $\langle \boldsymbol{n}, (1, 0, 0) \rangle \ge 0$. This means that the space of normals considered can be parametrized over a unit hemisphere. To obtain a discrete search space for the normals, we therefore are looking for an even distribution of points on the unit

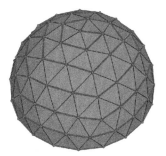

Fig. 2. A triangulated hemisphere, obtained by successive subdivision of the icosahedron, with a Hamiltonian path over the edge graph

(hemi-)sphere. This is achieved by repeated 1–4 subdivision of an icosahedron, each time followed by a projection of the new vertices onto the unit sphere (see Fig. 2). For a more in-depth discussion about the problem of distributing points evenly on the sphere, see [15].

To obtain a discretization for the range of distances d, we first move the origin to the centroid of the volume mesh. Now, for every direction n, we compute the smallest range of distances d containing all planes with normal n going through the data points, by sorting the data points with respect to their scalar product with the normal $d_P = \langle P, n \rangle$. This range is then equidistantly discretized.

3.2   Sweeping the Parameter Space

With the data points sorted in direction of the normal n, we can now easily slide a window $[d - w, d + w]$ over the range of distances. We use an accumulator variable to count the mass of all points in the range $[d - w, d + w]$ w.r.t. the normal n. Each time we proceed to the next discrete value of d, we add to the accumulator the mass of all newly covered points, and subtract the mass of all points no longer covered by the window. Since the points are sorted, this can be done in constant time for each point.

If we are only searching for the mass in connected components inside the window (where connectivity is induced by the grid), we also remember the first and the last index of the (sorted) points covered by the current window. We then do a traversal of the connectivity graph by breadth first search to find all connected components and to compute their respective masses.

When we iterate through the discrete set of normals, we make sure that subsequent normals are joined by an edge of the subdivided icosahedron, and therefore do not vary much. This is ensured by ordering the vertices of the triangulated hemisphere by a Hamiltonian path over the edge graph. Such a Hamiltonian path can easily be found by "spiraling" from the north pole to the equator (see Fig. 2). The benefit of using this ordering of the normals is that complexity of two subsequent sorting operations is low when the two normals are very similar, because the two resulting sorting sequences are also similar.

Algorithm 1. PlaneFinder

Input: A grid G with a piecewise constant density function ρ
    A width w of the fattened planes
Output: A planar connected component covering most mass

   discretize the unit hemisphere by iterated subdivision of the icosahedron
   order the vertices by a spiraling Hamiltonian path
   for each vertex n (i.e. normal) **do**
     sort the grid cells in direction n
     discretize the range in direction n equidistantly
     for each plane (n, d) **do**
       collect the cells closer than w to the plane
       compute the connected component having most mass
     end for
   end for
   return the connected component having most mass (and the corresponding plane)

3.3   Purging Dominant Planar Features

After a complete sweep through the discretized parameter space, we have found an optimal plane with a corresponding set of grid cells. We now purge the density values on this set of cells. This allows to find further planar regions, avoiding the possibility that another optimum covers essentially the same data.

4   Complexity Analysis

Let T denote the number of cells, N be the number of discrete normal directions for the planes (the number of vertices of the triangulated hemisphere), and D be the maximal number of planes checked in any direction. Assuming $D \in o(T)$ and $w \in \mathcal{O}(\frac{1}{D})$, the total running time of the algorithm is $\mathcal{O}(N(T \log T))$.

For each normal, the set of lumped mass points is sorted, which is done in time $\mathcal{O}(T \log T))$. Then the set of points covered by the current plane is collected, taking time $\mathcal{O}(T)$ in total. For each of these sets, a breadth-first search is performed to compute the connected components of the induced subgraph of the cell connectivity graph. Each vertex is covered by a constant number of fattened planes, since $w \in \mathcal{O}(\frac{1}{D})$. Moreover, since each vertex of this graph has degree at most 4, traversal of each of these graphs also takes $\mathcal{O}(T)$ in total.

5   Results

We tested our algorithm on real-world instances with between 150 000 and 300 000 tetrahedra. On a typical instance with 163 799 tetrahedra, 43 042 of which carried non-zero density, computation of the optimal plane took about 25 seconds on a standard laptop (2.16 GHz, 2GB RAM). The dimensions of the

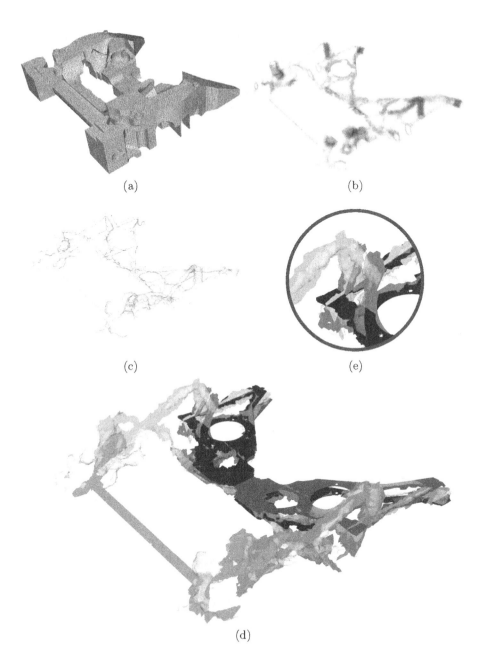

(a)

(b)

(c)

(e)

(d)

Fig. 3. Identifying planes on a typical density distribution generated with a topology optimization software. (a) The input domain. (b) A volume rendering of the density function. (c) An isosurface of the density function, shown transparent. (d) The 6 most dominant planar regions found using our algorithm. (e) A closeup of (d).

bounding box of the tetrahedral mesh are approximately $80 \times 90 \times 20$ cm; the parameter w was set to 2 cm. The results computed using 6 iterations of the algorithm are visualized in Figure 3.

Acknowledgments. The work of Ulrich Bauer was supported by ThyssenKrupp Tallent Ltd., UK. Konrad Polthier is supported by the DFG Research Center MATHEON "Mathematics for key technologies" in Berlin. Thanks to Adrian Chapple from ThyssenKrupp Tallent for providing volume data.

References

1. Bendsøe, M.: Topology Optimization: Theory, Methods, and Applications. Springer, Heidelberg (2003)
2. Stolpe, M.: On Models and Methods for Global Optimization of Structural Topology. KTH, Mathematics, Stockholm (2003)
3. Sigmund, O.: A 99 line topology optimization code written in Matlab. Structural and Multidisciplinary Optimization 21(2), 120–127 (2001)
4. Ben-Tal, A., Bendsøe, M.: A New Method for Optimal Truss Topology Design. SIAM Journal on Optimization 3, 322 (1993)
5. de Klerk, E., Roos, C.,T.T.: Semi-definite problems in truss topology optimization. Technical Report Report 95–128, Faculty of Technical Mathematics and Informatics, Delft, Netherlands (1995)
6. Radon, J.: Über die Bestimmung von Funktionen durch ihre Integralwerte längs gewisser Mannigfaltigkeiten. Berichte Sächsische Akademie der Wissenschaften, Leipzig, Mathematisch-Physikalische Klasse 69, 262–277 (1917)
7. Hough, P.: Method and Means for Recognizing Complex Patterns, United States Patent US 3,069,654, 18.12.2002 (1962)
8. Duda, R.O., Hart, P.E.: Use of the Hough transformation to detect lines and curves in pictures. Commun. ACM 15(1), 11–15 (1972)
9. Sarti, A., Tubaro, S.: Detection and characterisation of planar fractures using a 3d hough transform. Signal Processing 82(9), 1269–1282 (2002)
10. Vosselman, G., Dijkman, S.: 3D building model reconstruction from point clouds and ground plans. International Archives of Photogrammetry and Remote Sensing 34(3/W4), 37–43 (2001)
11. Kurdi, F.T., Landes, T., Grussenmeyer, P.: Hough-Transform and Extended RANSAC Algorithms for Automatic Detection of 3D Building Roof Planes from Lidar Data. In: Proceedings of the ISPRS Workshop on Laser Scanning, pp. 407–412 (2007)
12. Peternell, M., Steiner, T.: Reconstruction of piecewise planar objects from point clouds. Computer-Aided Design 36(4), 333–342 (2004)
13. Paralambros, P.Y., Chirehdast, M.: An integrated environment for structural configuration design. Journal of Engineering Design 1, 73–96 (1990)
14. Hornlein, H., Kocvara, M., Werner, R.: Material optimization: bridging the gap between conceptual and preliminary design. Aerospace Science and Technology 5(8), 541–554 (2001)
15. Katanforoush, A., Shahshahani, M.: Distributing Points on the Sphere, I. Experimental Mathematics 12(2), 199–209 (2003)

Determining Directional Contact Range of Two Convex Polyhedra

Yi-King Choi[1], Xueqing Li[2], Fengguang Rong[2], Wenping Wang[1],
and Stephen Cameron[3]

[1] The University of Hong Kong, Pokfulam Road, Hong Kong, China
{ykchoi,wenping}@cs.hku.hk
[2] Shandong University, Shandong, China
{xqli,rfguang}@sdu.edu.cn
[3] Oxford University Computing Laboratory, Parks Road, Oxford, OX1 3QD, U.K.
Stephen.Cameron@comlab.ox.ac.uk

Abstract. The *directional contact range* of two convex polyhedra is the range of positions that one of the polyhedron may locate along a given straight line so that the two polyhedra are in collision. Using the contact range, one can quickly classify the positions along a line for a polyhedron as "safe" for free of collision with another polyhedron, or "unsafe" for the otherwise. This kind of contact detection between two objects is important in CAD, computer graphics and robotics applications. In this paper we propose a robust and efficient computation scheme to determine the directional contact range of two polyhedra. We consider the problem in its dual equivalence by studying the Minkowski difference of the two polyhedra under a duality transformation. The algorithm requires the construction of only a subset of the faces of the Minkowski difference, and resolves the directional range efficiently. It also computes the contact configurations when the boundaries of the polyhedra are in contact.

Keywords: directional contact range, separating distance, penetrating distance, convex polyhedra, duality transformation, signed distance.

1 Introduction

The collision status of two objects, i.e., whether they are separate or intersecting, as well as their contact configurations, i.e., at which parts they are in contact, are important in many applications in CAD, computer graphics and robotics, or other areas that involve physical simulations, where responses are subsequently deduced based on these pieces of information. In this paper, we focus on the collision status and contact configurations of two convex polyhedra, assuming that they may only move along a given direction. The restriction regarding the direction is deemed reasonable, as there are a lot of applications in which object translations are only allowed in some specific directions. In industrial modeling or motion design, for example, the directions of movements that a mechanical part can take are limited by the constraints imposed by the degree of freedom of

F. Chen and B. Jüttler (Eds.): GMP 2008, LNCS 4975, pp. 127–142, 2008.

the part. The directional collision status of two objects is therefore useful, e.g., for object placements and motion design in a dynamic environment.

We define the *directional contact range* (DCR) of two convex polyhedra P and Q with respect to a direction $\mathbf{s}$ to be the range of positions that Q can locate along $\mathbf{s}$ so that P and Q are in contact or overlap, assuming that P is kept static. Equivalently, we say that

$$\mathrm{DCR}(P, Q, \mathbf{s}) = [\underline{\alpha}, \overline{\alpha}] \quad \text{iff} \quad P \cap Q^{t\hat{\mathbf{s}}} \neq \emptyset, \forall t \in [\underline{\alpha}, \overline{\alpha}]$$

where $\underline{\alpha}, \overline{\alpha} \in \mathbb{R}$, $\hat{\mathbf{s}} = \mathbf{s}/\|\mathbf{s}\|$ and $Q^{t\hat{\mathbf{s}}} = \{\mathbf{q} + t\hat{\mathbf{s}} \mid \mathbf{q} \in Q\}$ is the result of Q translated by $t\hat{\mathbf{s}}$. In particular, $Q^{\underline{\alpha}\hat{\mathbf{s}}}$ and $Q^{\overline{\alpha}\hat{\mathbf{s}}}$ are in external contact with P, i.e., they touch P only at some boundary points.

The DCR essentially gives the relative positions between the polyhedra at which they are in contact, and therefore can solve collision queries when Q is considered moving along $\mathbf{s}$. Since the polyhedra are convex, it is obvious that the DCR is either empty or is a single closed interval. The directional separating distance or penetration distance of P and Q, when they are separate or overlap, respectively, can also be computed from the DCR, so that if $\mathrm{DCR}(P, Q, \mathbf{s}) = [\underline{\alpha}, \overline{\alpha}]$, the required directional distance is given by $\min\{|\underline{\alpha}|, |\overline{\alpha}|\}$.

Object interference testing or collision detection has been intensively studied in the fields of computational geometry, computer graphics and robotics (see a survey in [1,2]). Given two convex polyhedra, Cameron and Culley considered their minimum translational distance [3]; and there are convex optimization methods [4] and feature-based algorithms [5,6] that determine their closest features. Kim et al. [7] estimated the penetration depth of two intersecting polyhedra using the Gaussian map of their Minkowski sum. In relation to directional contact, Dobkin et al. devised an $\mathcal{O}(\log^2 n)$ algorithm to compute the directional penetration depth of two intersecting convex polyhedra [8], and showed that the directional distance corresponds to the directional distance between the origin and the Minkowski difference polyhedron, M, of the polyhedra. Hence, a brute-force algorithm for finding the directional distance by intersecting a line from the origin and M has $\mathcal{O}(n^2)$ complexity, which is also the geometric complexity of M. There are efficient solutions for computing the intersection of line and a convex polyhedron, including linear programming approaches or geometrical methods such as [9,10] that transform the problem to locating a point in a convex plane partition in the dual space. Our algorithm differs by using another form of duality transformation, and most importantly, we exploits the fact that M is not a general convex polyhedron, but the Minkowski difference of two convex polyhedron with much simpler geometric complexity.

1.1   Major Contributions

In this paper, we present an algorithm to compute the directional contact range (DCR) of two convex polyhedra efficiently. The goal of the algorithm is to seek a face on the Minkowski difference of the two polyhedra which gives the contact features at their touching positions, given that one of them may move freely along a specific direction. We define a convex function which guarantees convergence and therefore guides the search in a robust manner. Moreover, we break down the search on the Minkowski difference into three different phases (corresponding to the three different types of faces), skipping most of the EE-type faces which is of $\mathcal{O}(n^2)$ where n is the number of faces of the polyhedra (Section 3.2), and thereby obtaining the target face efficiently.

The essence of our idea is to consider the DCR problem in its dual equivalence. We study the Minkowski difference under a duality transformation and a convex function is then defined as the signed distance of a vertex on the dual polyhedron to a plane. We also show that maximizing the convex function is essentially the same as to finding a face containing the intersection of a ray from the origin with the Minkowski difference in the primal space, hence solving the DCR problem. The convex nature of the search process is difficult to perceive in the primal space intuitively, but could be proved easily by its dual counterpart. Although our algorithm is based on the concept of duality transformation, its computation does not involve any explicit application of the transformation and therefore no overhead is incurred in this regard.

2   The Key Idea

In this section, we explain the fundamental idea of our algorithm, which relates the DCR problem of two polyhedra in the primal space to a search for a vertex on a Minkowski difference polyhedron in the dual space. Let P be a convex polyhedron in $\mathbb{E}^3$, then $\mathcal{V}_P$, $\mathcal{F}_P$, and $\mathcal{E}_P$ denote the set of vertices, faces, and edges of P, respectively.

2.1   Minkowski Difference of Two Polyhedra in Relation to DCR

Given two polyhedra P and Q, let $-Q = \{-q \mid q \in Q\}$. We consider the Minkowski difference of P and Q (or equivalently, the Minkowski sum of P and $-Q$) defined by $M \equiv P \oplus (-Q) = \{\mathbf{p} - \mathbf{q} \mid \mathbf{p} \in P, \mathbf{q} \in Q\}$. Since P and Q are both convex, M is also a convex polyhedron [11]. The origin $\mathbf{o}$ is in M if and only if there are some $\mathbf{p} \in P$ and $\mathbf{q} \in Q$ such that $\mathbf{p} = \mathbf{q}$, i.e., P and Q overlap and share a common point. Moreover, $\mathbf{o}$ is on the boundary of M if and only if P and Q share common boundary points only.

When Q moves in a direction $\mathbf{s}$, M moves in the opposite direction $-\mathbf{s}$. If P and Q do not intersect along $\mathbf{s}$, M does not contain the origin when moving along $\mathbf{s}$ and the DCR is empty. Otherwise, the DCR is the range of distances that M can travel along $\mathbf{s}$ with the origin remains in M. In other words, the DCR are bounded by the distances from the origin to the intersections of the line $\mathbf{os}$ and the boundary of M (Fig. 1). These two intersection points corresponds to when P and Q are in external touch. In the case where $\mathbf{os}$ has only one intersection with M, the DCR is a single value which is the distance from the origin to the intersection. The intersection points must lie on the boundary of M, and hence our primary task is to compute the intersection between the line **os** and the boundary M. Since intersections must always lie on some faces of M, faces of M (but not edges nor vertices) are only considered in our algorithm.

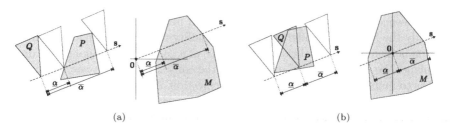

(a)   (b)

Fig. 1. The DCR $[\underline{\alpha}, \overline{\alpha}]$ of two convex polyhedra P and Q, and the corresponding distances from the origin to the boundary of the Minkowski difference M when P and Q are (a) separate; and (b) overlap

Proposition 1. *Let the line* **os** *be given by* $s(u) = u\hat{s}$, $u \in \mathbb{R}$. *Suppose* **os** *intersects* $M = P \oplus -Q$ *at the faces* $f_{\min}$ *and* $f_{\max}$ *with points of intersection* $s(u_{\min})$ *and* $s(u_{\max})$, *respectively, so that* $u_{\min} \leq u_{\max}$ *and* $s(u) \in M$ *if and only if* $u \in [u_{\min}, u_{\max}]$. *Then the DCR of* P *and* Q *is given by* $[u_{\min}, u_{\max}]$.

Note that $f_{\min}$ (or $f_{\max}$) could be more than one face which happens when **os** intersects M at a vertex or an edge.

2.2   The Dual of the Minkowski Difference

Given an arbitrary point $\mathbf{c} \in \mathbb{R}$, we may classify the faces of M into three groups depending on the positions of $\mathbf{c}$ to the plane $\mathcal{H}_f$ containing a face f of M: f is a supporting face, a convex face or a concave face if $\mathbf{c}$ lies on $\mathcal{H}_f$, in the inner half space (i.e., the half space in which M resides) of $\mathcal{H}_f$, or in the outer half space of $\mathcal{H}_f$, respectively.

Suppose M is transformed under a duality as described in Appendix A.1 using an interior point of M as the centre of duality. Then every faces of M are properly transformed to a vertex not at infinity and the dual M^* is a convex polyhedron, with vertices $\mathcal{F}_M^*$ and faces $\mathcal{V}_M^*$. The dual of an edge defined by two adjacent vertices $\mathbf{v}_0$, $\mathbf{v}_1$ in M is an edge common to two adjacent faces $\mathbf{v}_0^*$, $\mathbf{v}_1^*$ in M^*. In general, if an arbitrary point $\mathbf{c}$ not in the interior of M is used as the centre of duality, the above correspondence between the features of M and M^* still applies, but M^* is no longer compact in $\mathbb{E}^3$. Its boundary is defined by two disjoint shells that extends to infinity. In particular, the supporting faces (if any) of M with respect to $\mathbf{c}$ are transformed to points at infinity, while the dual of the convex faces are the convex faces w.r.t. the origin in the dual space which form one of the continuous convex shell of M^* (Fig. 2). As we shall see, our algorithm will work on the convex faces with respect to a given point.

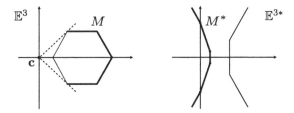

Fig. 2. A 2D illustration of M and its dual M^* with the centre of duality c not in M. The convex faces w.r.t. c on M (thick lines) corresponds to the black points lying on the convex shell (marked with thick lines) with respect to the origin in $\mathbb{E}^{3*}$.

2.3   Signed Distance of a Face from the Origin

Suppose that a plane is given by $\Pi : A^T\mathbf{x} = k$ where $A \in \mathbb{R}^3$, $\mathbf{x} = (x, y, z)^T$, $k \in \mathbb{R}$, and we assume that Π is normalized such that $\|A\| = 1$ and $k > 0$ is the shortest distance from the origin to Π. The *signed distance* of a point $\mathbf{x}_0$ to the plane Π is then given by $d_\Pi(\mathbf{x}_0) = A^T\mathbf{x}_0 - k$. Given a point $\mathbf{c} \notin M$, let $\hat{\mathcal{F}}_\mathbf{c}$ denote the set of convex faces of M with respect to $\mathbf{c}$. We define the *signed distance* of $f \in \hat{\mathcal{F}}_\mathbf{c}$, denoted by $d(f)$, to be the signed distance of f^* to the plane $\mathbf{o}^*$ in $\mathbb{E}^{3*}$, i.e., $d(f) = d_{\mathbf{o}^*}(f^*)$, where $\mathbf{c}$ is the centre of duality and $\mathbf{o}^*$ is the dual of the origin $\mathbf{o} \in \mathbb{E}^3$. Let $\mathcal{H}_f : N^T\mathbf{x} = k$, where $\|N\| = 1$ and $k > 0$, be the containing plane of f. Then $f^* = N/(k - N^T\mathbf{c}) \in \mathcal{V}_{M^*}$ in the dual space with $\mathbf{c}$ as the centre of duality. The origin $\mathbf{o}$ is first translated by $-\mathbf{c}$ and the plane equation of $\mathbf{o}^*$ is $-\mathbf{c}^T\mathbf{x} = 1$. Hence, we have

$$d(f) = d_{\mathbf{o}^*}(f^*) = \frac{-\mathbf{c}^T}{\|\mathbf{c}\|} \cdot \frac{N}{k - N^T\mathbf{c}} - \frac{1}{\|\mathbf{c}\|} = -\frac{k}{\|\mathbf{c}\|(k - N^T\mathbf{c})}. \tag{1}$$

The quantity $d_{\mathbf{o}^*}(f^*)$ uniquely determines a plane l^* in $\mathbb{E}^{3*}$ passing through f^* and parallel to $\mathbf{o}^*$ such that $d_{\mathbf{o}^*}(\mathbf{x}) = d_{\mathbf{o}^*}(f^*)$ for all points $\mathbf{x} \in l^*$ (Fig. 3). Since l^* and $\mathbf{o}^*$ have the same normal direction, it can also be shown that l, $\mathbf{c}$ and $\mathbf{o}$ are collinear. Moreover, l^* passes through f^* and hence l must lie on $\mathcal{H}_f$, the containing plane of f. This implies that l is the intersection of $\mathcal{H}_f$ and the line $\mathbf{co}$.

Suppose that the ray $\mathbf{co}$, given by $l(t) = -t\mathbf{c}, t > 0$, hits some faces in $\hat{\mathcal{F}}_\mathbf{c}$ (Fig. 3). Then the containing planes of all faces in $\hat{\mathcal{F}}_\mathbf{c}$ must intersect $\mathbf{co}$. The signed distance for a convex face $f_t \in \hat{\mathcal{F}}_\mathbf{c}$ whose containing plane intersects $\mathbf{co}$ at a point $l(t)$ is $d(f_t) = (1 - t)/(t\,\|\mathbf{c}\|)$. Since $d(f_t)$ is a decreasing function for $t > 0$, it means that the containing plane of the face with maximum signed distance over all faces in $\hat{\mathcal{F}}_\mathbf{c}$, has the closest intersection to $\mathbf{c}$ with the ray $\mathbf{co}$. Due to the convexity of M, this intersection must lie on a face of M and we have the following proposition:

Proposition 2. *Let $f_{\max} \in \hat{\mathcal{F}}_\mathbf{c}$ be the convex face with respect to $\mathbf{c}$ whose signed distance is the maximum over all faces in $\hat{\mathcal{F}}_\mathbf{c}$, i.e., $f_{\max} = \arg\max_f\{d(f) \mid f \in \hat{\mathcal{F}}_\mathbf{c}\}$. Then $f_{\max}$ contains an intersection of the ray $\mathbf{co}$ and M.*

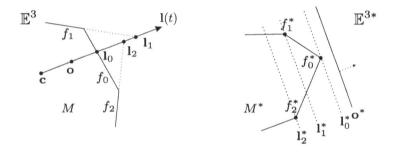

Fig. 3. The vertex f_0^* in $\mathbb{E}^{3*}$ attaining maximum signed distance to $\mathbf{o}^*$ is the dual of a face f_0 in $\mathbb{E}^3$ intersecting the ray $\mathbf{co}$

Since f^* lies on a convex shell, the signed distance function is convex over $\hat{\mathcal{F}}_c$. Starting from a face $f \in \hat{\mathcal{F}}_c$, we may therefore search for $f_{\max}$ at which the ray $\mathbf{co}$ intersects M. It is important to note that the intersection needs not be solved, as its distance from the origin can be computed directly from $\mathrm{d}(f_{\max})$ as follows. We established that $\mathrm{d}(f) = (1 - t)/(t\|\mathbf{c}\|)$ is the signed distance of a convex face $f \in \hat{\mathcal{F}}_c$ whose containing plane intersects $\mathbf{co}$ at $\mathbf{l}(t)$. Let $\alpha(f)$ be the signed distance of $\mathbf{l}(t)$ from $\mathbf{o}$ along the ray $\mathbf{co}$. Then,

$$\alpha(f) = (t - 1)\|\mathbf{c}\| = \left(\frac{1}{\mathrm{d}(f)\|\mathbf{c}\| + 1} - 1\right)\|\mathbf{c}\| = -\frac{\mathrm{d}(f)\|\mathbf{c}\|^2}{\mathrm{d}(f)\|\mathbf{c}\| + 1}.$$

Hence, the distance from the origin to the intersection of $f_{\max}$ and $\mathbf{co}$ is given by $\alpha(f_{\max})$.

3   The Algorithm

Given two convex polyhedra P and Q, and a direction $\mathbf{s} \in \mathbb{R}^3$, the following algorithm computes the DCR of P and Q with respect to $\mathbf{s}$:

Step 1:   Check whether the line $\mathbf{os}$ intersects $M = P \oplus (-Q)$. If not, we have $\mathrm{DCR}(P, Q, \mathbf{s}) = \emptyset$. Otherwise, choose a point $\mathbf{c}_{\min} = u\mathbf{s}$ for some $u > 0$, and that both $\mathbf{o}$ and M lie on the same side of $\mathbf{c}_{\min}$ on $\mathbf{os}$. Choose also $\mathbf{c}_{\max} = v\mathbf{s}$ for some $v < 0$, with both $\mathbf{o}$ and M lying on the same side of $\mathbf{c}_{\max}$ on $\mathbf{os}$.

Step 2:   Using $\mathbf{c}_{\min}$ as the centre of duality, search for $f_{\min}$ which attains the maximum signed distance among all convex faces with respect to $\mathbf{c}_{\min}$, i.e., $f_{\min} = \arg\max_f\{\mathrm{d}(f) \mid f \in \hat{\mathcal{F}}_{\mathbf{c}_{\min}}\}$. Then, use $\mathbf{c}_{\max}$ as the centre of duality and search for $f_{\max} = \arg\max_f\{\mathrm{d}(f) \mid f \in \hat{\mathcal{F}}_{\mathbf{c}_{\max}}\}$.

Step 3:   Report $\mathrm{DCR}(P, Q, \mathbf{s}) = [\underline{\alpha}, \overline{\alpha}]$ where $\underline{\alpha} = -\alpha(f_{\min})$ and $\overline{\alpha} = \alpha(f_{\max})$.

Our algorithm does not require the complete construction of the Minkowski difference M. Moreover, we devise a novel search scheme in step 2 which skips some faces in M in order to reach $f_{\min}$ and $f_{\max}$ efficiently. The details would be discussed in subsequent sections.

3.1 Determining the Center of Duality c

If two polyhedra P and Q do not meet no matter how far Q moves along a given direction $\mathbf{s}$, their DCR with respect to $\mathbf{s}$ is empty. In this case, the line $\mathbf{os}$ does not intersect the Minkowski difference M, which can be checked without constructing M as follows.

Let $\dot{P}$ and $\dot{Q}$ be the orthographic projection along $\mathbf{s}$ of P and Q to a plane $\mathcal{H}$ normal to $\mathbf{s}$. We construct the convex hull, $CH(\dot{P})$ and $CH(\dot{Q})$, of $\dot{P}$ and $\dot{Q}$, respectively. This can be done efficiently since the vertices of $CH(\dot{P})$ and $CH(\dot{Q})$ are the silhouette vertices of P and Q as viewed along $\mathbf{s}$. Then, we obtain $\dot{M} = CH(\dot{P}) \oplus (-CH(\dot{Q}))$. Now, $\mathbf{os}$ intersects M if and only if $CH(\dot{P})$ and $CH(\dot{Q})$ overlap, i.e., $\dot{M}$ contains the origin.

Suppose now that $\mathbf{os}$ intersects M. In general $\mathbf{os}$ has two intersections with M which is convex, and therefore we need to choose two points, each as the centre of duality to locate one intersection at one time. To locate the face $f_{\max}$ of M (see Proposition 1), the centre of duality $\mathbf{c}_{\max}$ should lie on $\mathbf{os}$ so that the ray $\mathbf{c}_{\max}\mathbf{o}$ hits $f_{\max}$ and that $f_{\max}$ is a convex face with respect to $\mathbf{c}_{\max}$. Hence, we require that $\mathbf{c}_{\max} = v\mathbf{s}$ for some $v < 0$, and $\mathbf{o}$ and M be on the same side of $\mathbf{c}_{\max}$ on $\mathbf{os}$. Now, $\mathbf{c}_{\max}$ can be computed easily by approximating M with its bounding box $M_{BB} = P_{BB} \oplus -Q_{BB}$, where P_{BB} and Q_{BB} are the bounding boxes of P and Q, respectively. The point $\mathbf{c}_{\min} = u\mathbf{s}$ for some $u > 0$ is then chosen similarly.

3.2 Searching the Face with Maximum Signed Distance

Step 2 of our algorithm involves searching the faces $f_{\max}$ and $f_{\min}$ at which $\mathbf{os}$ intersects M. We will only describe the search for $f_{\max}$, since $f_{\min}$ can be found in the same way using $\mathbf{c}_{\min}$ instead as the centre of duality.

A brute-force search for $f_{\max}$ is to first construct M, which is of $\mathcal{O}(n^2)$ complexity. Moreover, to locate $f_{\max}$ directly on M using its face adjacency information, is inefficient, as face traversal can only advance to an immediate neighbour at one step. We therefore break down the search for $f_{\max}$ in three successive phases, each within an independent face subset of M. This allows a quicker leap over the faces on M and hence a more rapid search of $f_{\max}$. Also, the number of faces on M that needs to be constructed are greatly reduced.

Let us define the supporting vertex of of a polyhedron P for a face f be $s_P(f) = \arg\max_{\mathbf{v}}\{\mathbf{n}(f) \cdot \mathbf{v} \mid \mathbf{v} \in \mathcal{V}_P\}$, where $\mathbf{n}(f)$ is the normal vector of f and $\cdot$ denote the vector dot-product. The Gaussian image of M, $G(M)$, is obtained by superimposing the Gaussian images $G(P)$ and $G(-Q)$ (Appendix A.2). For any face $f_p \in \mathcal{F}_P$, the point $G(f_p)$ must fall within the region $G(s_{-Q}(f_p))$. Similarly, the point $G(f_q)$ must fall within the region $G(s_P(f_q))$. Hence, each point in $G(P)$ and $G(-Q)$ corresponds to a face of FV- and VF-type, respectively, in M (Fig. 4). Furthermore, each arc-arc intersection on $\mathcal{S}^2$ corresponds to a pair of edges (one from P and one from $-Q$) sharing a common normal direction and

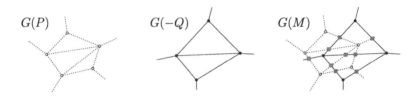

$G(P)$ $G(-Q)$ $G(M)$

Fig. 4. The planar representation of the Gaussian image $G(M)$ by superimposing $G(P)$ and $G(-Q)$. There are three types of vertices in $G(M)$: (i) (white point) a point of $G(P)$ falling within a region of $G(-Q)$, i.e., a face in $\mathcal{F}_{fv}$; (ii) (black point) a point of $G(-Q)$ falling within a region of $G(P)$, i.e., a face in $\mathcal{F}_{vf}$; and (iii) (shaded square) the intersection point of two arcs, each from $G(P)$ and $G(-Q)$, i.e., a face in $\mathcal{F}_{ee}$.

amounts to a EE-type face in M. The FV-, VF- and EE-type faces form three independent subsets $\mathcal{F}_{fv}$, $\mathcal{F}_{vf}$ and $\mathcal{F}_{ee}$, respectively, which are given as follows:

$\mathcal{F}_{fv}$: Each face $F(f_p, \mathbf{v}_q)$ is a point set $\{\mathbf{x} + \mathbf{v}_q \mid \mathbf{x} \in f_p\}$, where $f_p \in \mathcal{F}_P$ and $\mathbf{v}_q \in \mathcal{V}_{-Q}$. Also, $\mathbf{v}_q = s_{-Q}(f_p)$.

$\mathcal{F}_{vf}$: Each face $F(\mathbf{v}_p, f_q)$ is a point set $\{\mathbf{v}_p + \mathbf{x} \mid \mathbf{x} \in f_q\}$, where $f_q \in \mathcal{F}_{-Q}$ and $\mathbf{v}_p \in \mathcal{V}_P$. Also, $\mathbf{v}_p = s_P(f_q)$.

$\mathcal{F}_{ee}$: Each face $F(e_p, e_q)$ is a parallelogram with vertices $\mathbf{v}_0 = \mathbf{v}_{p_0} + \mathbf{v}_{q_0}, \mathbf{v}_1 = \mathbf{v}_{p_1} + \mathbf{v}_{q_0}, \mathbf{v}_2 = \mathbf{v}_{p_1} + \mathbf{v}_{q_1}, \mathbf{v}_3 = \mathbf{v}_{p_0} + \mathbf{v}_{q_1}$ where $\mathbf{v}_{p_0}, \mathbf{v}_{p_1} \in \mathcal{V}_P$, $\mathbf{v}_{q_0}, \mathbf{v}_{q_1} \in \mathcal{V}_{-Q}$, and $e_p = (\mathbf{v}_{p_0}, \mathbf{v}_{p_1}) \in \mathcal{E}_P$, $e_q = (\mathbf{v}_{q_0}, \mathbf{v}_{q_1}) \in \mathcal{E}_{-Q}$. Moreover, the Gaussian images of e_p and e_q intersect on $\mathcal{S}^2$ (Fig. 4).

The following pseudocode searches for f_{max} with the maximum signed distance d_{max} among all convex faces with respect to $\mathbf{c}$ on $M = P \oplus -Q$:

Procedure MaxSignedDistance($P, Q, \mathbf{c}$)
    $(f_{fv}, d_{fv}) \leftarrow$ Search-FV
    $(f_{vf}, d_{vf}) \leftarrow$ Search-VF
    $(f_{max}, d_{max}) \leftarrow$ Search-EE(d_{fv}, d_{vf})
    return (f_{max}, d_{max})

Search-FV. This procedure is to search for a face with the maximum signed distance among all convex faces with respect to $\mathbf{c}$ in $\mathcal{F}_{fv}$. The search is conducted according to face adjacency of P. It is important that we start from a convex face on M, which ensures that all subsequent faces in the search are convex faces, due to the convexity of the signed distance function. We choose $f_0 = \arg\max_f\{\mathbf{n}(f) \cdot \mathbf{co} \mid f \in \mathcal{F}_P\}$, where $\mathbf{n}(f)$ is the normal vector of a face f, as the initial face such that the corresponding face $F(f_0, s_{-Q}(f_0)) \in \mathcal{F}_{fv}$ is guaranteed to be a convex face with respect to $\mathbf{c}$. Starting from f_0, the search in Search-FV considers the neighbouring faces of the current face in P and advances to one whose corresponding face in M has the local maximum signed distance. Neighbouring (or adjacent) faces are those faces incident to the vertices of the current face in P. Two faces adjacent in P may not constitute adjacent faces in M, and therefore a gain (by skipping some faces in M) is obtained by advancing faces in the search based on their adjacency in P.

The procedure is described in the following pseudocode. The function Signed-Distance-FV(f) constructs a face $F(f, s_{-Q}(f)) \in \mathcal{F}_{\mathrm{fv}}$ and computes its signed distance using Eq. (1). The supporting vertex of $-Q$ for a face f is determined using the hierarchical representation of a polyhedron by Dobkin and Kirkpatrick [12].

Procedure Search-FV
$\quad d_{\mathrm{fv}} = $ SignedDistance-FV(f_0)
$\quad$ For each iteration i,
$\quad\quad$ For each of the n faces f_i^j, $j = 0, \ldots, n-1$, that are adjacent to f_i in P,
$\quad\quad\quad d_i^j \leftarrow $ SignedDistance-FV(f_i^j).
$\quad\quad$ If $d_{\mathrm{fv}} < d_i^k$, where $d_i^k = \max\{d_i^j\}$,
$\quad\quad\quad d_{\mathrm{fv}} \leftarrow d_i^k$, $f_{i+1} \leftarrow f_i^k$.
$\quad\quad$ Else,
$\quad\quad\quad$ Return (f_i, d_{fv}).

Theorem 1 states the correctness of Search-FV whose proof is omitted due to space limitation.

Theorem 1 (Correctness of Search-FV). *The face f_{fv} returned by Search-FV attains the maximum signed distance d_{fv} among all convex faces in $\mathcal{F}_{\mathrm{fv}}$ with respect to* $\mathbf{c}$, *i.e.,* $f_{\mathrm{fv}} = \arg\max_f\{\mathrm{d}(f) \mid f \in \mathcal{F}_{\mathrm{fv}} \cap \hat{\mathcal{F}}_{\mathbf{c}}\}$.

Search-VF. This procedure searches for a face with maximum signed distance among all convex faces with respect to $\mathbf{c}$ in $\mathcal{F}_{\mathrm{vf}}$. The face $f_{\mathrm{fv}} = F(f_p, s_{-Q}(f_p)) \in \mathcal{F}_{\mathrm{fv}}$ computed by Search-FV is supposed to be closest to $f_{\max}$ among all faces in $\mathcal{F}_{\mathrm{fv}}$, and it should give a good starting point for subsequent search. Hence, we choose the initial face for Search-VF as a face f_0 that is incident at $s_{-Q}(f_p)$ in $-Q$. The search then proceeds like Search-FV by interchanging the role of P and $-Q$. Similarly, we have the following theorem.

Theorem 2 (Correctness of Search-VF). *The face f_{vf} returned by Search-VF attains the maximum signed distance d_{vf} among all convex faces in $\mathcal{F}_{\mathrm{vf}}$ with respect to* $\mathbf{c}$, *i.e.,* $f_{\mathrm{vf}} = \arg\max_f\{\mathrm{d}(f) \mid f \in \mathcal{F}_{\mathrm{vf}} \cap \hat{\mathcal{F}}_{\mathbf{c}}\}$.

Search-EE. The next step is to search for the remaining convex faces in $\mathcal{F}_{\mathrm{ee}}$, starting from f_{fv} or f_{vf}, whichever attains the greater signed distance. Let e_p and e_q be edges in $\mathcal{E}_P$ and $\mathcal{E}_{-Q}$, respectively. A face $F(e_p, e_q) \in \mathcal{F}_{\mathrm{ee}}$ is formed only if the Gaussian images of e_p and e_q intersect on S^2 (Section A.2). The steps of Search-EE are given in the following pseudocode:

Procedure Search-EE

$d_{ee} \leftarrow \max\{d_{fv}, d_{vf}\}$.

$f_m \leftarrow$ the face f_{fv} or f_{vf} attains d_{ee}.

If $f_m = f_{fv} = F\big(f_p, s_{-Q}(f_p)\big)$, then

$\mathcal{FS}_0 \leftarrow$ all possible faces $F(e_p, e_q)$, where e_p is an edge incident to a vertex of f_p, and e_q is an edge incident with $s_{-Q}(f_p)$,

Else if $f_m = f_{vf} = F\big(s_P(f_q), f_q\big)$, then

$\mathcal{FS}_0 \leftarrow$ all possible faces $F(e_p, e_q)$, where e_p is an edge incident with $s_P(f_q)$, and e_q is an edge incident to a vertex of f_q.

For each iteration $i = 0, 1, 2, \ldots$

Let $\hat{f}_i = F(\hat{e}_p, \hat{e}_q)$ be the face in $\mathcal{FS}_i$ with the maximum signed distance.

If $d_{ee} < d(\hat{f}_i)$, then

$d_{ee} \leftarrow d(\hat{f}_i), \quad f_{ee} \leftarrow \hat{f}_i$

$\mathcal{FS}_{i+1} \leftarrow$ all possible faces $F(e_p, e_q)$, where e_p is an edge incident to an end vertex of $\hat{e}_p$, e_q is an edge incident to an end vertex of $\hat{e}_q$

Else

Return (f_{ee}, d_{ee}).

It can be shown that the initial face set $\mathcal{FS}_0$ contains all neighbouring EE-type faces of the initial face f_m, by considering all possible EE-type faces formed by an edge incident to the vertex that forms f_m and an edge incident with a face that forms f_m. Moreover, the subsequent face sets $\mathcal{FS}_i$ includes all the neighbouring EE-type faces of the current EE-type face $\hat{f}_i$ with the maximum signed distance, by considering all possible EE-type faces formed by two edges, each incident to an end vertex of an edge forming $\hat{f}_i$. Hence, we have the following theorem:

Theorem 3 (Correctness of Search-EE). *The face $f_{\max}$ returned by Search-EE attains the maximum signed distance among all convex faces in $\hat{\mathcal{F}}_c$ with respect to c, i.e., $f_{\max} = \arg\max_f \{d(f) \mid f \in \hat{\mathcal{F}}_c\}$.*

3.3   Computation Details

Contact configurations. The faces $f_{\min}$ and $f_{\max}$ indicate the contact configuration of P and Q when they are in external contact along the DCR direction. The contact features are given by the features on P and Q that form the faces $f_{\min}$ and $f_{\max}$. For example, if $f_{\max} = F(\mathbf{v}_p, f_q) \in \mathcal{F}_{vf}$, the contact features of P and $Q^{\bar{\alpha}s}$ are the vertex $\mathbf{v}_p \in P$ and the face $f_q \in Q$.

Supporting faces with respect to the centre of duality. We may encounter supporting faces with respect to the centre of duality in our algorithm, which are possible neighbours of a convex face. Supporting faces correspond to points at infinity in the dual space (Section 2.2) and can be identified if $k - N^T c = 0$ when evaluating the signed distance given by Eq. (1). Supporting faces are ignored in our algorithm without affecting its correctness.

To decide whether two arcs on S^2 intersect. In Search-EE, to decide whether two edges $e_p \in \mathcal{E}_P$ and $e_q \in \mathcal{E}_{-Q}$ form a face $F(e_p, e_q) \in \mathcal{F}_{ee}$, we check whether $G(e_p)$ and $G(e_q)$ intersect on the Gaussian sphere S^2. Let $\mathbf{a}, \mathbf{b}$ be the end points of $G(e_p)$, $\mathbf{c}, \mathbf{d}$ be the end points of $G(e_q)$ and $\mathbf{o}$ be the centre of S^2 (Fig. 5). The arcs $G(e_p)$ and $G(e_q)$ intersect if and only if (1) $\mathbf{c}, \mathbf{d}$ are on different sides of the plane $\mathbf{oba}$; (2) $\mathbf{a}, \mathbf{b}$ are on different sides of the plane $\mathbf{ocd}$; and (3) $\mathbf{a}, \mathbf{b}, \mathbf{c}, \mathbf{d}$ are on the same hemisphere. Consider the *signed volume*, $|\mathbf{cba}| = \det[\mathbf{c}\ \mathbf{b}\ \mathbf{a}]$, of a parallelepiped spanned by three vectors $\mathbf{a}, \mathbf{b}, \mathbf{c}$. The above three conditions can be formulated as (1) $|\mathbf{cba}| \times |\mathbf{dba}| < 0$; (2) $|\mathbf{adc}| \times |\mathbf{bdc}| < 0$; and (3) $|\mathbf{acb}| \times |\mathbf{dcb}| > 0$. We need to compute $|\mathbf{cba}|, |\mathbf{dba}|, |\mathbf{adc}|$ and $|\mathbf{bdc}|$ only, since $|\mathbf{acb}| = |\mathbf{cba}|$ and $|\mathbf{dcb}| = |\mathbf{bdc}|$.

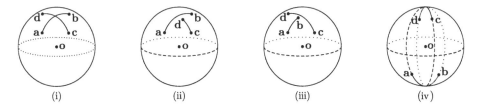

Fig. 5. Determining if two arcs intersect on S^2. Arcs intersect in (i). No intersection where (ii) condition (1); (iii) condition (2) and (iv) condition (3) is violated.

Span of faces with same normal direction. To simplify the preceding discussions, we assumed that all faces on M have distinct normal directions. However, this is not always true for convex polyhedra with arbitrary mesh structures. In this case, a face is augmented to include also its neighbouring span of faces with the same normal direction.

To avoid repetitive visits to a face. A hash table is used to record the visited faces in each procedure to avoid unnecessary repetitive computations for a face which is visited previously in the searching process.

4   Performance

The following experiments are designed to evaluate the performance of our algorithm. A set of six convex polyhedra (Fig. 6) are used (whose names and number of vertices are given in the brackets): a truncated elliptic cone (P_1 – 20), a truncated elliptic cylinder (P_2 – 50), two ellipsoids (P_3 – 200, P_4 – 500), the convex hull of a random point set in a cube (P_5 – 100), and the volume of revolution of a convex profile curve (P_6 – 200). The sizes of the polyhedra are all within a sphere of radius 5. The cone and the cylinder are with the aspect $a : b : h = 1 : 2 : 4$, where a, b are the sizes of the base ellipse and h is the height. The size of the ellipsoids are in $a : b : c = 2 : 2 : 5$ and $2 : 4 : 5$ respectively for P_3 and P_4, where a, b and c are the length of the three major axes.

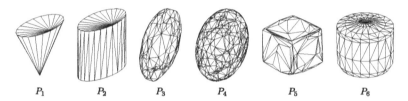

P_1 P_2 P_3 P_4 P_5 P_6

Fig. 6. The six objects used in the experiments

Five pairs of objects are chosen for DCR calculations: (P_2, P_2), (P_1, P_3), (P_6, P_6), (P_4, P_5) and (P_4, P_6), where the total number of vertices of the two objects are 100, 220, 400, 600, 700, respectively. We also generated another 10 pairs of objects which are ellipsoids of the same size aspect as P_4, but with different number of vertices. For each pair of objects P and Q, P is kept static and Q assumes 40 random orientations; for each orientation, Q is also translated so that the shortest distance between the two objects ranges from -1.5 to 1.5 in 11 samples, of which 5 samples correspond to the cases where P and Q intersect, 1 sample corresponds to touching, and 5 samples correspond to separation. Also, for each fixed shortest distance with a random orientation, we compute the DCR between P and Q with respect to 40 random directions. It means that, for each pair of convex polyhedra, we perform a total of $11 \times 40 \times 40$ different DCR computations and the average CPU time for each run is taken. Each reported non-empty DCR $[\underline{\alpha}, \overline{\alpha}]$ along a specified direction **s** is verified by translating Q along **s** by $\underline{\alpha}$ and $\overline{\alpha}$, and using the GJK algorithm to compute the shortest distance between P and the translated Q, which should be zero as the objects are then in external contact. We note that the average shortest distance is 1.9×10^{-6} with a standard deviation of 10^{-5}; the maximum of the absolute shortest distance is found to be 10^{-4}.

The experiments were carried out on a desktop computer with an Intel Core 2 Duo E6600 2.40 GHz CPU (single-threaded) and a 2GB main memory. The performance of our algorithm is shown in Fig. 7. It takes less than 0.25 milliseconds to compute the DCR of two convex polyhedra with a total of 1000 vertices. Although the cylinder pair (P_2, P_2) is of only 100 vertices in total, the running time in this case is disproportionally longer than expected. Not only that a face at the planar bases of cylinders needs to be augmented to include the span of all other coplanar faces (Section 3.3), the increase in running time is also due largely to the fact that a face at the base has a large number of adjacent faces—which are the 50 faces on the curved surface of a cylinder.

The result of ellipsoid pairs shows empirically an approximately linear growth in the running time with respect to the total number of vertices. Recall that not all faces on the Minkowski difference M are being constructed and visited. From the above experiments, it is found that in a search of a single intersection on M, on average 13.7% of the faces on M is visited. In particular, only 2.5% of the EE-type faces is visited on average, which means that most of the EE-type faces are skipped in our algorithm.

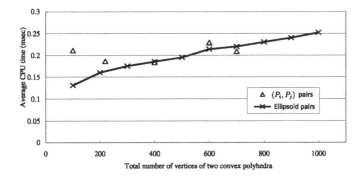

Fig. 7. The average CPU time for computing the DCR of five pairs (P_i, P_j) of specific convex polyhedra $(P_1 - P_6)$, and 10 pairs of ellipsoids with varying number of vertices

A worst case scenario is designed where the number of EE-type faces is of $\mathcal{O}(n^2)$, where n is the number of vertices on the polyhedra. Two cones are constructed, each having 21 vertices (20 on the circular rim and 1 at the apex), 38 faces (20 on the slanted surface) and 57 edges. Both cones are very flat with aspect ratio $a : b : h = 12 : 12 : 1$ (Fig. 8(a)). The Gaussian image of each cone has one point (corresponding to the 18 faces on the flat surface) that is antipodal to a set of 20 points (corresponding to the 10 faces on the slanted surface), and 20 great arcs that looks like great semicircles, which corresponds to the edges on the circular rim (Fig. 8(b)). One of the cones is rotated about the y-axis by 90 degrees, so that each of its 20 great semicircles in the Gaussian image intersects with half of the great semicircles in the Gaussian image of the other cone (Fig. 8(b)). Hence, there are in total 38 FV-type, 38 VF-type and 200 EE-type faces on the Minkowski difference of the cones. Our algorithm, on computing a single intersection on the Minkowski difference, requires a visit to only 21 FV-type, 21 VF-type and 20 EE-type faces, showing that most of the EE-type faces (90%) are skipped which renders an efficient computation.

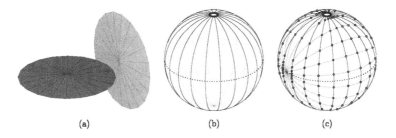

(a)   (b)   (c)

Fig. 8. (a) Two thin cones whose DCR is computed. (b) Gaussian image of one of the cones. (c) Gaussian image of the Minkowski difference of the cones; the white circles correspond to the EE-type faces (only features on the front-facing surface of the Gaussian sphere are shown.)

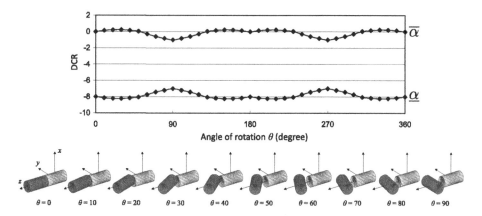

Fig. 9. The DCR between two cylinders with respect to the z-direction. The white cylinder is static and the grey cylinder rotates about its principle x-axis. The first 10 instants (with the angle of rotation for the grey cylinder $\theta = 0$ to 90) of the two cylinders are shown.

Next, we compute the DCR of two circular cylinders, both are of radius 1 and length 4. The height of both cylinders lie along the z-axis, and the first cylinder (in white) stays static at the origin, while the other cylinder (in grey), with centre at $z = 4$, assumes different rotations about its principal x-axis. The angle of rotation is from 0 to 360 degrees, with 10 degrees increments. The two cylinders have external contact at the planar faces when $\theta = 0, 180$ and 360 degrees. The DCR of the two cylinders with respect to the $+z$-direction computed by our algorithm is shown in Fig. 9.

5   Conclusion

We have presented a novel method for computing the directional contact range (DCR) between two convex polyhedra with respect to a given direction. The DCR of two convex polyhedra can be computed by finding the intersections of a line with the Minkowski difference M of the polyhedra. We consider the problem in the dual space where a face on M corresponds to a vertex on the dual polyhedron M^*, and formulate the DCR computation in forms of searching a vertex which attains the maximum signed distance from a plane. The search problem in the dual space is easily shown to be convex, and a search scheme is devised accordingly to locate the face that contains the required intersection on M efficiently. The search scheme is divided into three stages, each working only on a subset of the faces on M. This division allows the elimination of most EE-type faces whose worst case complexity is $\mathcal{O}(n^2)$. Our experimental results show that our algorithm exhibits efficient performance. Although our tests do not experience major robustness problems, we note here that the signed distance function $d(f)$ is non-linear with respect to the distance between the duality centre **c** and the intersection of f and **co**. Possible numerical issues thus induced will be further explored.

Acknowledgement

This work is supported in part by research grants from the Research Grant Council of the Hong Kong SAR (HKU 7031/01E), and in part by the Natural Science Foundation of Shandong Province (Y2005G03).

References

1. Jiménez, P., Thomas, F., Torras, C.: 3D collision detection: a survey. Computers & Graphics 25(2), 269–285 (2001)
2. Lin, M.C., Manocha, D.: Collision and proximity queries. In: Handbook. of Discrete and Computational Geometry (2003)
3. Cameron, S., Culley, R.: Determining the minimum translational distance between two convex polyhedra. In: Proc. IEEE Int. Conf. on Robotics and Automation, pp. 591–596 (1986)
4. Gilbert, E.G., Johnson, D.W., Keerthi, S.S.: A fast procedure for computing the distance between objects in three-dimensional space. IEEE J. Robot. Automat. (4), 193–203 (1988)
5. Lin, M.C., Canny, J.: A fast algorithm for incremental distance calculation. In: Proc. IEEE Int. Conf. on Robotics and Automation, Sacramento, pp. 1008–1014 (April 1991)
6. Mirtich, B.: V-Clip: Fast and robust polyhedral collision detection. ACM Trans. Graph. 17(3), 177–208 (1998)
7. Kim, Y.J., Lin, M.C., Manocha, D.: Incremental penetration depth estimation between convex polytopes using dual-space expansion. IEEE Trans. Visual. Comput. Graphics 10(2), 152–163 (2004)
8. Dobkin, D.P., Hershberger, J., Kirkpatrick, D.G., Suri, S.: Computing the intersection-depth of polyhedra. Algorithmica 9(6), 518–533 (1993)
9. Günther, O.: Efficient structures for geometric data management. Springer, New York (1988)
10. Kolingerová, I.: Convex polyhedron-line intersection detection using dual representation. The Visual Computer 13(1), 42–49 (1997)
11. Grünbaum, B.: Convex Polytopes. Wiley, Chichester (1967)
12. Dobkin, D.P., Kirkpatrick, D.G.: A linear algorithm for determining the separation of convex polyhedra. J. Algorithms 6(3), 381–392 (1985)
13. Levy, H.: Projective and Related Geometry. Macmillan, Basingstoke (1964)

A   Major Concepts

A.1   Duality Transformation

There are different formulations for duality transformation. A more general form is to consider the self-dual duality with respect to a given non-singular quadric surface $\mathcal{B} : X^T B X = 0$ where $X = (x, y, z, 1)^T$ is the homogeneous coordinates of a point, and B is a 4×4 real symmetric matrix. The dual of a point Y_0 is a plane $\mathcal{Y} : Y_0^T B X = 0$ (the *polar* of Y_0) and the dual of a plane $\mathcal{V} : V_0^T X = 0$ is a point $U_0 = B^{-1} V_0$ (the *pole* of $\mathcal{V}$ [13]). It is easy to verify that if $\mathcal{Y}$ is the dual of Y_0, then Y_0 is the dual of $\mathcal{Y}$. Also, if Y_0 is a point on $\mathcal{B}$, its dual is the tangent plane to $\mathcal{B}$ at Y_0. In this work, we consider duality transformation with respect to the unit sphere in $\mathbb{E}^3$. The dual relationship between a point and a plane in $\mathbb{E}^3$ in terms of affine coordinates $\mathbf{x} = (x, y, z)^T$ is as follows. Suppose a plane Π, not passing through the origin, is given by $A^T \mathbf{x} = k$ in the *primal space* $\mathbb{E}$, where $A \in \mathbb{R}^3$ and k is a nonzero real number. A *duality transformation* maps Π to a point $\mathbf{w} = A/k$ in the *dual space* $\mathbb{E}^{3*}$. A point $\mathbf{u} \neq \mathbf{0}$ in $\mathbb{E}^3$ is transformed to a plane $\mathcal{U} : \mathbf{u}^T \mathbf{x} = 1$ in $\mathbb{E}^{3*}$. If we extend $\mathbb{E}^3$ to include the plane at infinity (i.e., the extended Euclidean space), a plane passing through the origin in $\mathbb{E}^3$ is mapped to a point at infinity in $\mathbb{E}^{3*}$; whereas the origin in $\mathbb{E}^3$ is transformed to the plane at infinity in $\mathbb{E}^{3*}$. Note that $\mathbb{E}^3$ is the dual space of $\mathbb{E}^{3*}$. We use ψ^* to denote the dual counterpart of an entity ψ in $\mathbb{E}^3$. We may also consider a duality transformation centred at an arbitrary point $\mathbf{c} \in \mathbb{E}^3$. This can be done by first translating the origin in $\mathbb{E}^3$ to $\mathbf{c}$ before applying duality, and we call $\mathbf{c}$ the *centre of duality*.

A.2   Gaussian Image of a Polyhedron

The Gaussian image $G(P)$ of a convex polyhedron P is a planar graph embedded on the unit sphere $\mathcal{S}^2$ (Fig. 10). The Gaussian image of any feature (i.e., vertex, edge or face) ϕ of a polyhedron P is the set of normal directions of planes that may come into contact with P at ϕ. In other words, ϕ is the supporting feature of P in the directions represented by its Gaussian image. Hence, a face $f \in \mathcal{F}_P$ corresponds to a point $G(f) = \hat{\mathbf{n}}(f) \in \mathcal{S}^2$ where $\hat{\mathbf{n}}(f)$ is the unit normal vector of f; an edge $e \in \mathcal{E}_P$ common to two faces $f_0, f_1 \in P$ corresponds to a great arc $G(e)$ connecting two vertices $G(f_0)$ and $G(f_1)$; a vertex $\mathbf{v} \in \mathcal{V}_P$ common to the faces $f_0, \ldots, f_m$ corresponds to a convex spherical polygon $G(\mathbf{v})$ whose vertices are $G(f_0), \ldots, G(f_m)$.

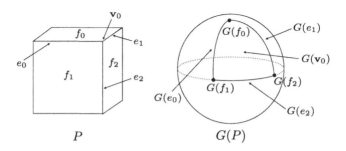

Fig. 10. A polyhedron P and its Gaussian image $G(P)$ on $\mathcal{S}^2$

Efficient Collision Detection Using a Dual Bounding Volume Hierarchy

Jung-Woo Chang[1], Wenping Wang[2], and Myung-Soo Kim[1]

[1] Seoul National University, Korea
[2] University of Hong Kong, Hong Kong

Abstract. We perform collision detection between static rigid objects using a bounding volume hierarchy which consists of an oriented bounding box (OBB) tree enhanced with bounding spheres. This approach combines the compactness of OBBs and the simplicity of spheres. The majority of distant objects are separated using the simpler sphere tests. The remaining objects are in close proximity, where some separation axes are significantly more effective than others. We select 5 from among the 15 potential separating axes for OBBs. Experimental results show that our algorithm achieved favorable speed up with respect to the existing OBB algorithms.

Keywords: Collision detection, Bounding volume hierarchy, OBB, Sphere.

1 Introduction

Collision detection is one of the most important geometric queries, with diverse engineering applications in areas such as robotics, computer graphics, animation, computer games, virtual reality, simulation and haptic rendering [1]. Because of their importance, collision detection algorithms have been studied for decades [12].

Among many different approaches, the bounding volume hierarchy has proved to be the most successful in contemporary systems [1]. The computation time of this approach can be formulated as follows [3]:

$$T = N_v \times C_v + N_p \times C_p, \tag{1}$$

where

T: total execution time
N_v: number of bounding volume pair overlap tests
C_v: time for testing a pair of bounding volumes
N_p: number of primitive pair overlap tests
C_p: time for testing a pair of primitives

This formula clearly shows that the performance mainly depends on two factors: the tightness of the bounding volumes and the simplicity of overlap test for a pair of bounding volumes. The first factor is related to N_v and N_p, whereas the second factor is related to C_v.

F. Chen and B. Jüttler (Eds.): GMP 2008, LNCS 4975, pp. 143–154, 2008.
© Springer-Verlag Berlin Heidelberg 2008

Spheres [6,14] and axially aligned bounding boxes (AABBs) [2] allow the simplest overlap tests. On the other hand, oriented bounding boxes (OBBs) [3] and discrete orientation polytopes (k-DOP) [8] fit volumes more tightly. In this paper, we propose a dual scheme that combines the simplicity of spheres and the compactness of OBBs to produce an efficient algorithm. The experiments show that the performance improvement over conventional algorithms is favorable.

Van den Bergen [19] suggested a simple but rough separation test for OBBs known as SAT lite. By using only 6 of the 15 potential separating axes, SAT lite demonstrates a better performance than the original OBB algorithm that uses all 15 axes [3]. In this paper, we use a sphere test followed by 5 separation axis tests. These 6 tests may look the same as the 6 tests of the SAT lite. However, the main difference is that two objects that have passed the sphere test can be expected to be in close proximity. In this stage, the choice of which 5 axes among the 15 possible axes becomes very important.

The main contribution of this work can be summarized as follows:

- A dual OBB-sphere tree is proposed, where each OBB node is enhanced with a bounding sphere.
- The more efficient sphere test is applied first to eliminate distant objects.
- We propose a selection of five separation axes that are effective in separating objects which are very close but not overlapping.
- For a wide range of experiments, the performance improvement has been observed over conventional OBB algorithms.

2 Related Work

In this section, we will briefly review related work on collision detection. The most basic type of collision detection algorithm deals with rigid bodies in static poses. But many recent studies have looked at more complicated problems, including detecting collisions between deformable models rather than rigid models [7,9,10,11,17,22], collision detection in the continuous time domain rather than static pose [15,16] and collision detection algorithms which can run on graphics hardware [4,22]. Even though these complicated problems deal with more general cases, the basic type collision detection algorithm is still quite important because the algorithms for the basic type problem are much more efficient than the algorithms for complicated problems. In this paper, we focus on the algorithm for the basic type problem.

The most widely used algorithms for detecting collisions between static, rigid bodies are based on a hierarchy of bounding volumes. As formulated in Equation (1), the performance of this approach is governed by the tightness of the bounding volumes and the simplicity of the overlap test for a pair of bounding volumes. Because the selection of the bounding volume is closely related to the performance of an algorithm, many different kinds of bounding volumes have been suggested.

Beckmann et al. [2] proposed the AABB tree and Palmer et al. [14] and Hubbard [6] put forward the sphere tree to solve the problem. By introducing OBBs, having additional rotational degrees of freedom, Gottschalk et al. [3] constructed a new and efficient collision detection algorithm. Klosowski et al. [8] suggested the k-DOP tree which tightly bounds the underlying mesh and efficiently tests the overlap of two bounding volumes. To preserve the geometric feature of k-DOP, the k-DOP-based algorithm should tumble and recompute the k-DOP for rotational motion. There are some efficient methods that solve this k-DOP tumbling problem [5,21]. Larsen et al. [13] proposed the rectangular swept sphere as a bounding volume to solve the distance measure problem, which is closely related to collision detection.

Two convex objects are disjoint if and only if there exists a separating axis such that the projections of the two objects on to that axis are separate. If the objects are both convex polytopes, it is possible to enumerate a finite set of axes that can guarantee their separation, if they are separate. Two convex polytopes are separate if and only if the projections of the two polytopes are separate on at least one of these potential separating axes. The number of such axes is determined by the geometry of the polytopes. For instance, AABBs have 3 potential separating axes, which correspond to the x, y and z axis of coordinate system. Two k-DOPs have k potential separating axes and two OBBs have 15 potential separating axes. The SAT lite algorithm of van den Bergen [19] uses 6 of the 15 axes. By this rough separation test, the overall performance of the SAT lite algorithm is better than the original algorithm based on the exact OBB test.

3   Collision Detection Algorithm

We will now present an efficient collision detection algorithm for static rigid bodies. We will first describe the construction of a dual OBB-sphere tree and show how the problem can be reduced to a situation of close proximity using a sphere test. Then we will go on to describe the selection of an effective set of axes to deal with the remaining problem.

3.1   Dual OBB-sphere Bounding Volume Tree

To combine the relative advantages of OBB and sphere trees, we construct a dual OBB-sphere tree which keeps both an OBB and a bounding sphere for each node of the tree. The basic structure of the dual OBB-sphere tree is the same as the OBB tree proposed by Gottschalk et al. [3]. For every node of an OBB tree, the dual OBB-sphere tree also contains a sphere which bounds the elements of the polygonal mesh that are at that node.

There are two ways commonly used for the construction of a bounding sphere. The first method is to construct the smallest enclosing sphere [20]. The second is to fix the center of the bounding sphere at the center of the OBB. In the latter case the centers of the two bounding spheres under a separation test can be reused for the subsequent OBB separation test, so the second method provides a simpler overlap test. But the first method naturally guarantees a tighter bounding sphere. From a series of experiments, we found that the tightness of the first method is more important than the simplicity of the second for the improvement of overall performance. The dual OBB-sphere bounding volume tree therefore keeps a smallest enclosing bounding sphere and an OBB at each node.

The test for separation of a pair of nodes in the dual OBB-sphere trees is as follows. The bounding spheres of two corresponding nodes are tested whether they overlap or not. If the bounding spheres are separate, the two nodes are separate too. If the bounding spheres overlap, then the OBBs are tested for overlap using the method to be described in the following subsection.

Testing the bounding spheres has two advantages. The first is enhanced tightness. Although OBBs are generally tighter than spheres, there are some cases in which the spheres are separate when the OBBs overlap. The second advantage is a more general reduction of the problem. OBBs are only tested when they are quite close because remote OBBs are eliminated by the sphere test. An overlap test can be designed especially for the case of OBBs in close proximity, as described in the following subsection.

3.2   The Selection of a Set of Separating Axes

Gottschalk et al. [3] proved that the separation test for two OBBs can be reduced to separation tests with respect to 15 potential separating axes. Van den Bergen [19] then suggested SAT lite, which is a looser but simpler separation test that uses a subset of 6 of the potential separating axes. However, any subset of the potential separating axes can provide a viable separation test. In this subsection, we suggest a near-optimal subset of the potential separating axes for OBBs which are already known to be in close proximity.

The 15 potential separating axes for two OBBs are $\{a_0, a_1, a_2, b_0, b_1, b_2, c_{00},$ $c_{01}, c_{02}, c_{10}, c_{11}, c_{12}, c_{20}, c_{21}, c_{22}\}$. The axes $\{a_0, a_1, a_2\}$ are defined by the orientation of the first OBB, and the subscripts are defined by the extents of the OBB, such that the extent corresponding to a_0 is smaller than the other extents, and the extent corresponding to a_2 is larger than the other extents. The axes $\{b_0, b_1, b_2\}$ are defined by the orientation of the second OBB and the subscripts are defined as before. The axis c_{ij} is the cross product of a_i and b_j. The exact separation test for OBBs uses all 15 axes but SAT lite uses $\{a_0, a_1, a_2, b_0, b_1, b_2\}$ only. On the other hand, we propose a sphere test followed by the separation test using 5 axes $\{a_0, b_0, c_{22}, c_{12}, c_{21}\}$.

Our selection of the separating axes is based on the fact that the extent corresponding to a potential separating axis is much more important for OBBs in close proximity than it is in the general case. Because this fact is hard to illustrate in 3D, we will consider the Minkowski sum of bounding volumes in 2D [1]. The separation test for two rectangles is closely related to the containment test for a reference point with respect to slabs, and it is also determined by the Minkowski sum of the rectangles. This relation is shown in Figures 1 and 2. The left images of Figures 1 and 2 denote the objects itself and the right images denote their configuration spaces defined by the Minkowski sum of two objects. In Figure 1, two rectangles are separated with respect to the y axis, and the relative center of the two rectangles is placed outside the slab, which is orthogonal to the y axis.

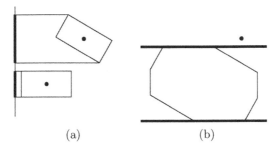

Fig. 1. The relation between the overlap test for an axis and the containment test for a slab - separation case

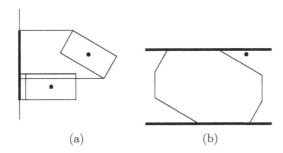

Fig. 2. The relation between the overlap test for an axis and the containment test for a slab - false positive case

Figure 2 shows a false positive case. Though the two rectangles are separated and their relative center is placed outside of the Minkowski sum, they are overlapping on the y axis and their relative center is located inside of the slab.

Figures 3 and 4 show the relation between the extent corresponding to an axis and the compactness achieved by that axis when the objects are in close proximity. In the case of 2D rectangles, the potential separating axes are defined by the orientation of each rectangle. The 4 potential separating axes can be categorized into 2 axes $\{a_0, b_0\}$ which correspond to smaller extents and 2 axes $\{a_1, b_1\}$ which correspond to larger extents. The axes $\{a_0, b_0\}$ can be called as minor axes and the axes $\{a_1, b_1\}$ can be called as major axes like the diameters of ellipse are commonly called. Diagram (c) in both Figures shows the importance of the extent corresponding to an axis when the objects are in close proximity. If the relative center of two objects is contained in the Minkowski sum of two bounding circles, the two bounding rectangles must be tested. Regarding the selection of separating axes, the Minkowski sum of two bounding circles can be subdivided to 4 regions - white region, light gray region, dark gray region and black region. The white region shows that the two rectangles overlap for all 4 potential separating axes. In the light gray region the two rectangles are separated with respect to minor axes and in the dark gray region the two rectangles are separated with respect to major axes. If the relative center of two objects is within the black

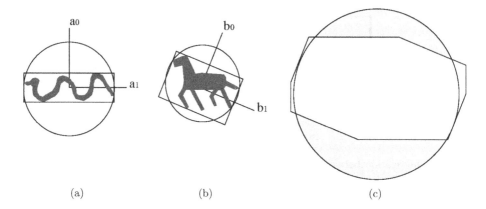

Fig. 3. Minkowski sum of bounding volumes (general case): (a) O_1 and its bounding rectangle and bounding circle; (b) O_2 and its bounding rectangle and bounding circle; (c) The Minkowski sum of the two bounding rectangles and that of the two bounding circles

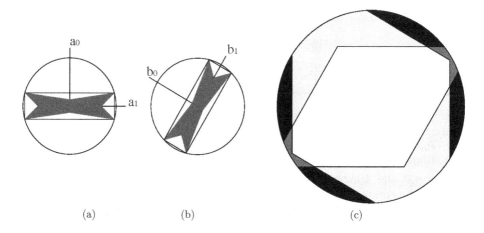

Fig. 4. Minkowski sum of bounding volumes (extreme case): (a) O_1 and its bounding rectangle and bounding circle; (b) O_2 and its bounding rectangle and bounding circle; (c) the Minkowski sum of the two bounding rectangles, that of the two bounding circles, and the clipped parallelograms defined by each set of separating axes

region, then either the separation test with respect to the minor axes or the separation test with respect to the major axes will detect the separation. In these figures, the reduction in discrimination that comes from eliminating the 2 major axes is denoted by the dark gray region, which is relatively small in Figure 4 and is absent in Figure 3. These examples support the elimination of the separation test for major axes. A more detailed discussion about the loss of discrimination region is presented in the Appendix.

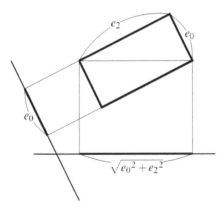

Fig. 5. The length of projection

The above suggests that an axis which corresponds to a smaller extent should be preferred. For three-dimensional OBBs, a_0 is likely to be a more discriminating axis than a_1 or a_2 and we would expect b_0 to be better than b_1 or b_2. In the case of axes defined as a cross product, we need to estimate the length of projection. This can be most easily shown by an example. The extents corresponding to each axis can be written $\{e_0, e_1, e_2\}$ and $\{f_0, f_1, f_2\}$, where e_0 corresponds to a_0 and f_0 corresponds to b_0. The axis c_{12} is orthogonal to a_1 and to b_2. Because of this orthogonality, the projection of the first OBB on to c_{12} can be reduced to the projection of a rectangle whose extents are e_0 and e_2. The length of the projection is bounded by the interval $[e_0, \sqrt{e_0^2 + e_2^2}]$ as shown in Figure 5. In the same way, the projection of the second OBB on to c_{12} is bounded by the interval $[f_0, \sqrt{f_0^2 + f_1^2}]$. Because the axis with the smallest length of projection is likely to be the most effective, c_{22} is chosen rather than c_{11} or c_{00}.

The foregoing discussion can be summarized as follows:

- In the case of axes defined by the orientation of OBBs, the axis with the smallest extent is preferred.
- An axis c_{ij} is preferred if the extents corresponding to a_i and b_j are relatively large.

By using these rules, the potential separating axes can be sorted. The remaining problem is to determine the size of the reduced set of axes. We used a greedy framework for a series of experiments. Each axis is added to the reduced set of axes in the order of preference and tested to see whether it improves the performance of collision detection. The experiments show that the optimal size of the reduced set of axes is 5 and the final set of axes is $\{a_0, b_0, c_{22}, c_{12}, c_{21}\}$.

4   Experimental Results

We have implemented our collision detection algorithm in C++ on an Intel Core Duo 1.66GHz PC with a 1.5GB main memory. To demonstrate the performance

Table 1. The execution time (in sec)

		RAPID	QuickCD	SAT lite	Dual
Scenario I	0%	27.2540	50.5709	25.3296	20.6053
	1%	14.0696	36.4444	13.2419	10.1924
	2%	8.6457	26.2898	8.1142	5.8939
	3%	6.2860	19.6893	5.8670	4.0741
	4%	4.9193	15.4700	4.5779	3.0381
	5%	4.0032	12.7353	3.7149	2.3816
Scenario II	10^{-1}	0.0012	0.0112	0.0010	0.0013
	10^{-2}	0.0138	0.4909	0.0113	0.0131
	10^{-3}	0.1754	0.8663	0.1418	0.1653
	10^{-4}	0.9418	0.9339	0.7621	0.7074
	10^{-5}	1.1858	0.9409	0.9682	0.8973
Scenario III		0.9640	0.5816	0.9935	0.8690

of our algorithm, we used three heterogeneous scenarios, each applied to different input meshes and position/orientation configurations.

In Scenario I, we employed two Happy Buddha models, each constructed with 15,536 triangles, and located the two models in 229,824 different configurations of relative position and orientation. The different configurations were generated by the sphere method of Trenkel et al. [18], which is a benchmark generation method for collision detection algorithms. We used 6 different relative distances: 0%, 1%, 2%, 3%, 4%, and 5% of the size of input models. Each distance is determined by the radius of a bounding sphere. A total of 266 different positions are generated on the sphere by sampling the spherical coordinates at every 15°. Moreover, a total of 144 different orientations are generated by sampling Euler rotation angles at every 60°. For each one of six distances, a total of 38,304 configurations are tested.

Scenario II considers two concentric spheres of radius 1 and $1+\epsilon$ respectively. (This test was also conducted in Gottschalk et al. [3] and Klosowski et al. [8].) Each sphere was approximated with 79,600 triangles, and five different values of ϵ were used: $10^{-1}, 10^{-2}, 10^{-3}, 10^{-4}$, and 10^{-5}.

In Scenario III, we repeat one test of Klosowski et al. [8], where a hand model with 404 triangles moves along a path consisting of 2,528 frames in the interior of an airplane constructed with 169,944 triangles.

For comparison purpose, in addition to ours we have also tested three other collision detection packages: RAPID, QuickCD, and SAT lite. RAPID is an OBB-based collision detection package which is available from http://www.cs.unc.edu/~geom/OBB/OBBT.html. QuickCD is a k-DOP-based package available from http://www.ams.sunysb.edu/~jklosow/quickcd/QuickCD.html. By slightly modifying the source code of RAPID, we implemented the SAT lite as well as our own algorithm called Dual.

Table 1 shows the execution times of collision detection tests applied to each scenario using four different packages. Dual is faster than RAPID about 24-40% for **Scenario I** and about 10% for **Scenario III**. It is also faster than RAPID

Table 2. The number of overlap tests

		RAPID		SAT lite		Dual		
		Box	Triangle	Box	Triangle	Sphere	Box	Triangle
Scenario I	0%	56,228,398	2,301,207	65,976,730	3,141,544	59,980,606	47,996,655	2,353,391
	1%	32,878,322	253,521	39,583,946	435,413	34,891,432	26,718,220	272,417
	2%	20,601,212	10,717	24,899,942	22,324	21,004,422	15,453,244	12,507
	3%	15,053,514	1,842	18,128,188	3,670	14,671,214	10,563,698	2,122
	4%	11,830,858	834	14,215,850	1,586	11,018,180	7,823,461	972
	5%	9,656,554	504	11,568,194	752	8,691,784	6,105,813	546
Scenario II	10^{-1}	2,735	0	2,879	0	3,778	3,081	0
	10^{-2}	34,195	0	35,283	0	44,522	35,895	0
	10^{-3}	445,727	0	460,279	0	591,726	477,699	0
	10^{-4}	2,224,243	89,284	2,299,687	102,236	2,373,534	1,981,233	89,286
	10^{-5}	2,780,453	136,796	2,874,967	184,672	2,972,484	2,498,341	146,129
Scenario III		1,760,646	168,962	2,208,346	264,964	2,055,120	1,922,216	170,038

about 5-25% for 4 cases of **Scenario II** and slower than RAPID about 8% only for one case (corresponding to the largest value of $\epsilon = 10^{-1}$). Moreover, it is faster than QuickCD about 60-80% for **Scenario I** and 5-97% for **Scenario II**. Nevertheless, it is slower than QuickCD about 50% for **Scenario III**. Dual is also faster than SAT lite about 18-36% for **Scenario I** and about 13% for **Scenario III**. It is slower about 16-30% for 3 cases of **Scenario II** and faster about 7% for 2 cases (corresponding to the smaller values of $\epsilon = 10^{-4}, 10^{-5}$).

By comparing the execution times, we realized that the k-DOP-based algorithm and the OBB-based algorithms show different patterns. This means that there are some applications which best fit to k-DOP-based algorithm and some to OBB-based algorithms. For our experiments, the outperformance of k-DOP in **Scenario III** is based on the fact that the orientation changes and the corresponding recomputation of k-DOPs are needed only for the moving hand model which is considerably smaller (404 triangles) than the whole environment (170,348 triangles). Since our algorithm is based on OBB, it showed a pattern similar to the other OBB-based algorithms. Thus we concentrate on the performance of three OBB-based algorithms below.

More detailed discussions on the performance of three OBB-based implementations are in Table 2. SAT lite uses only a subset of potential separating axes; thus the number of bounding volume overlap tests and triangle overlap tests for SAT lite is larger than that for RAPID. Though SAT lite needs more overlap tests, SAT lite is faster than RAPID for **Scenario I** and **Scenario II** since each bounding volume overlap test is considerably simpler. By comparing the number of bounding volume overlap tests and triangle overlap tests for SAT lite and that for Dual, we can show the advantage of enhanced tightness which was discussed in Section 3.1. The number of triangle overlap tests for Dual is smaller than that for SAT lite in all three scenarios. Moreover, the number of bounding volume overlap tests for Dual is smaller than for SAT lite in **Scenario I** and **Scenario III**. For 3 cases in **Scenario I**, the number of bounding volume overlap tests for Dual is even smaller than RAPID. This result implies that the reduced set of one sphere and five axes in our algorithm is more effective than the six axes of SAT lite.

Dual is faster than RAPID for all cases except the case of 10^{-1} in **Scenario II**. It is also faster than SAT lite except the three cases of $10^{-1}, 10^{-2}, 10^{-3}$ in **Scenario II**. Because the performance for the worst case is more important, the case of 0% is the most important for **Scenario I** and the case of 10^{-5} is the most important for **Scenario II**. In other words, the more difficult cases where Dual show better performance are more significant than the cases where RAPID or SAT lite show better performance.

The above experimental results show that our algorithm is a good choice when a collision detection package is needed for static rigid bodies.

5   Conclusions

We have presented a fast OBB-based collision detection algorithm that uses both OBBs and spherical bounding volumes. We have shown how to combine the compactness of OBBs and the efficient overlap test for spheres. Out of the 15 possible separation axes for two OBBs, we have selected 5 axes which detect separation most effectively. Experimental results show that our scheme makes a favorable speed up with respect to existing algorithms based on OBBs.

References

1. Akenine-Moller, T., Hains, E.: Real-Time Rendering. A K Peters (2002)
2. Beckmann, N., Kriegel, H.-P., Schneider, R., Seeger, B.: The R*-Tree: an efficient and robust access method for points and rectangles. In: ACM SIGMOD Conf. on the Management of Data, pp. 322–331 (1990)
3. Gottschalk, S., Lin, M.C., Manocha, D.: OBB-Tree: a hierarchical structure for rapid interference detection. In: ACM SIGGRAPH 1996, pp. 171–180 (1996)
4. Govindaraju, N.K., Redon, S., Lin, M.C., Manocha, D.: CULLIDE: interactive collision detection between complex models in large environments using graphics hardware. In: Proc. Eurographics/SIGGRAPH Graphics Hardware Workshop, pp. 25–32 (2003)
5. He, T.: Fast collision detection using QuOSPO trees. In: ACM Symp. on Interactive 3D Graphics, pp. 55–62 (1999)
6. Hubbard, P.M.: Collision detection for interactive graphics applications. IEEE Trans. on Visualization and Computer Graphics 1(3), 218–230 (1995)
7. James, D.L., Pai, D.K.: BD-Tree: output-sensitive collision detection for reduced deformable models. ACM Trans. on Graphics 23(3), 393–398 (2004)
8. Klosowski, J.T., Held, M., Mitchell, J.S.B., Sowizral, H., Zikan, K.: Efficient collision detection using bounding volume hierarchies of k-DOPs. IEEE Trans. on Visualization and Computer Graphics 4(1), 21–37 (1998)
9. Kavan, L., Zara, J.: Fast collision detection for skeletally deformable models. Computer Graphics Forum 24(3), 363–372 (2005)
10. Larsson, T., Akenine-Moller, T.: Collision detection for continuously deforming bodies. In: Proc. Eurographics, pp. 325–333 (2001)

11. Larsson, T., Akenine-Moller, T.: Efficient collision detection for models deformed by morphing. The Visual Computer 19(2-3), 164–174 (2003)
12. Lin, M.C., Gottschalk, S.: Collision detection between geometric models: a survey. In: Proc. IMA Conference on the Mathematics of Surfaces, pp. 37–56 (1998)
13. Larsen, E., Gottschalk, S., Lin, M.C.: Fast distance queries using rectangular swept sphere volumes. In: Proc. IEEE International Conf. on Robotics and Automation (ICRA), pp. 3719–3726 (2000)
14. Palmer, I., Grimsdale, R.: Collision detection for animation using sphere-trees. Computer Graphics Forum 14(2), 105–116 (1995)
15. Redon, S., Kheddar, A., Coquillart, S.: Fast continuous collision detection between rigid bodies. Computer Graphics Forum 21(3), 279–287 (2002)
16. Redon, S., Kim, Y.J., Lin, M.C., Manocha, D.: Fast continuous collision detection for articulated models. In: ACM Symp. on Solid Modeling and Applications, pp. 145–156 (2004)
17. Teschner, M., Kimmerle, S., Heidelberger, B., Zachmann, G., Raghupathi, L., Fuhrmann, A., Cani, M.-P., Faure, F., Magnenat-Thalmann, N., Strasser, W., Volino, P.: Collision detection for deformable objects. Computer Graphics Forum 24(1), 61–81 (2005)
18. Trenkel, S., Weller, R., Zachmann, G.: A Benchmarking Suite for Static Collision Detection Algorithms. In: International Conference in Central Europe on Computer Graphics, Visualization and Computer Vision (WSCG) (2007)
19. van den Bergen, G.: Efficient collision detection of complex deformable models using AABB trees. J. Graphics Tools 2(4), 1–14 (1997)
20. Welzl, E.: Smallest enclosing disks (balls and ellipsoids). New Results and New Trends in Computer Science 555, 359–370 (1991)
21. Zachmann, G.: Rapid collision detection by dynamically aligned dop-trees. In: Proc. of IEEE Virtual Reality Annual International Symposium (VRAIS), pp. 90–97 (1998)
22. Zhang, X., Kim, Y.J.: Interactive collision detection for deformable models using Streaming AABBs. IEEE Trans. on Visualization and Computer Graphics 13(2), 318–329 (2007)

Appendix

We present a statistical analysis of the loss of discrimination resulting from the elimination of the major axes in the two-dimensional case.

The overhead in discrimination corresponds to the dark gray region in Figure 4. A general formula for the area of this dark gray region would be very complicated; thus we have adopted a statistical analysis. By calculating the area for a number of samples, we provide an understanding of this overhead.

Because the space of all possible rectangles and circles is huge, we made the following assumptions:

- **Assumption 1:** The extents corresponding to a_1 and b_1 are fixed at 1.
- **Assumption 2:** To restrict the space of possible circles, each bounding circle surrounds the oriented bounding rectangle. The center of the bounding circle is fixed at the center of the bounding rectangle and its radius is determined by the size of the bounding rectangle. This assumption makes the calculation easy, and also makes the estimation very conservative.

- **Assumption 3:** Because the rectangle and the circle are both symmetrical, and the center of the bounding circle is fixed at the center of the bounding rectangle, the angle between a_0 and b_0 can be limited to the range of $0°\sim90°$ without loss of generality.

The extents corresponding to a_0 and b_0 are written h_1 and h_2, which are of course both less than or equal to 1. The angle is sampled at $10°$ increments and h_1 and h_2 are sampled with 0.1 increments. This gives 10 samples for each of 3 free variables, making 1,000 samples in total.

The ratio of the areas of two different regions is calculated as

$$\text{ratio} = \frac{\text{area of dark gray region}}{\text{area of white region} + \text{area of dark gray region}}.$$

The dark gray region denotes the overhead in discrimination. The white region and the dark gray region are the ones we test in our algorithm. This ratio represents the percentage of false positive case resulting from eliminating two major axes.

The average ratio for the 1,000 samples is 6.4%. For 80.7% of these samples, the ratio is less than 10%. In the worst case, the ratio is 22.2%. This worst case arises when the angle is $0°$, and h_1 and h_2 are equal to 1.0.

Fast and Local Fairing of B-Spline Curves and Surfaces

P. Salvi[1], H. Suzuki[1], and T. Várady[2]

[1] University of Tokyo
[2] Geomagic Hungary

Abstract. The paper proposes a fast fairing algorithm for curves and surfaces. It first defines a base algorithm for fairing curves, which is then extended to the surface case, where the isocurves of the surface are faired. The curve fairing process involves the discrete integration of a pseudo-arc-length parameterization of B-spline curves, with a blending and fitting phase concluding the algorithm. In the core of the fairing method, there is a fairness measure introduced in an earlier paper of the authors. This measure is based on the deviation from an ideal or target curvature. A target curvature is a series of smooth curvature values, generated from the original curve or surface. This curve and surface fairing technique is local and semi-automatic, but the user can also designate the region to be faired. The results are illustrated by a few examples on real-life models.

Keywords: Curves and Surfaces, Geometric Optimization, Reverse Engineering.

1 Introduction

Fairing curves and surfaces plays an important role in CAGD, especially in the automobile industry. Connected curve nets and surfaces have to be smoothed while preserving their original features and connectivity. Automating this process is a crucial problem of Digital Shape Reconstruction[1] [3].

Fairness does not have an exact mathematical definition, and it may have a different meaning depending on the context [9]. Still, there are some common properties of what we would call *fair*. It certainly includes some kind of mathematical continuity, e.g. C^2 or G^2. It also incorporates the requirement that a surface should have even reflections. Another important, although less intuitive, requirement is the smooth transition of curvature [6]. These properties can be tested using an arsenal of interrogation methods, e.g. isophote lines, curvature combs or curvature maps. But even with these, finding and mending small artifacts is a laborious manual process.

The concept of *target curvature* and an iterative algorithm based on it is reviewed in Section 2. A new, fast algorithm is explained in Section 3, accompanied by test results in Section 4.

---

[1] Formerly called Reverse Engineering.

F. Chen and B. Jüttler (Eds.): GMP 2008, LNCS 4975, pp. 155–163, 2008.

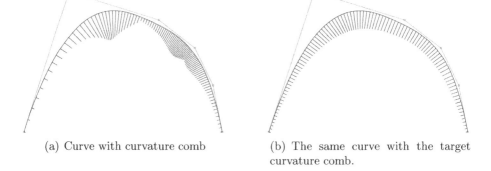

(a) Curve with curvature comb   (b) The same curve with the target curvature comb.

Fig. 1. Target curvature of a cubic curve

2 Related Work

There is an abundant literature on creating fair curves and surfaces, dealing with both the definition of fairness and smoothing algorithms. Here we will only cover a previous publication of the authors, defining a fairness measure based on a target curvature. The algorithm in Section 3 will use this measure to generate fair curves and surfaces. This section also presents an iterative algorithm based on the same measure. For a more comprehensive review on fairness measures and algorithms, look at [9,10].

2.1 Fairness Measure

Most fairing algorithms use some kind of fairness measure as a numerical representation of smoothness. In other words, we can say that a curve c is fair, if

$$\mathcal{F}(c) < \tau$$

applies, where τ is a user-defined tolerance. Salvi and Várady [10] propose a measure based on a *target curvature* — a sufficiently smoothed series of sampled curvature values. It can be thought of as a smoothed curvature comb (Fig. 1). From the target curvature, a new fairness measure naturally arises for curves:

$$E = \sum_i |\kappa(t_i) - g(t_i)|^2,$$

where g is the smoothed (target) curvature and t_i are sampled parameter points. The definition means that the less a curve's curvature deviates from the target curvature, the more fair it is. This measure is parameterization-independent, since it uses the curvature, that is a geometrical property of the curve. Most smoothness measures in use today are parameterization-dependent, and thus cannot be used safely when the parameterization considerably deviates from the arc-length parameterization [5].

Determining the Target Curvature. For creating the target curvature, simple, fast algorithms can be used. One way is averaging consecutive sampled values of the curvature. As global averaging may remove or flatten important parts of the curvature, the user should be allowed to modify the target curvature manually, or restrict its changes.

Another possibility is to fit a curve over sampled values of the curvature (fitting over curvature comb endpoints would be more intuitive, but it may have self-intersections). For smoother results, we may also minimize the curvature of the fitted curve. This leads to the minimization of the functional

$$F(g) = \sum_i |g(t_i) - \kappa(t_i)|^2 + \int \hat{\kappa},$$

where $\hat{\kappa}$ is the curvature of the fitted curve g.

For both of these methods, the number of samples may have a great impact on the resulting target curvature. Since smoothness is much more important than faithfulness to the original curvature, a loose sampling rate is recommended.

Extension to Surfaces. Most fairness measures for curves have their surface counterparts. This measure is no exception: it can be logically applied to surfaces using some combination of the principal curvatures, e.g.:

$$\Pi = \sum_i \sum_j (|\kappa_1(u_i, v_j) - g_1(u_i, v_j)|^2 + |\kappa_2(u_i, v_j) - g_2(u_i, v_j)|^2).$$

2.2   An Iterative Algorithm

The algorithm presented by the authors in [10] fixes all but one control point and minimizes the fairness measure defined above by moving the remaining one. It also constrains the deviation from the original curve using its control polygon. One may apply various heuristics to find the overall optimal shape, e.g. moving the control point with the largest error in every turn. Although not mentioned in the original paper, the downhill simplex method [8] can be used for minimization, and checking the deviation of a suitable neighbourhood instead of the whole domain can speed up the calculation of the fairness measure as well. Its extension to the surface case is straightforward; a sample result is presented in Fig. 2.

3   The Proposed Algorithm

The main idea is that if we reparameterize a curve to arc-length parameterization, the curvature will be equal to the norm of the second derivative [5], so we can create a target curvature and integrate it twice to find the faired

Fig. 2. Fiat body part faired by a previous algorithm of the authors [10]

curve directly. Discrete integration provides us with points we can use for fitting, eventually resulting in a faired curve or surface. The use of fitting harms locality of course, but in exchange we get a great increase in speed. In the following sections we first define the algorithm for curves and then extend it to surfaces.

3.1 Curve Case

The algorithm can be broken down to the following four phases:

1. Creating a pseudo arc-length parameterization for n points.
2. Integrating twice starting at the left end to the right and vice versa, with deviation control.
3. Blending the results.
4. Fitting a spline that approximates the blended points.

Pseudo arc-length parameterization means that the parameter domain is divided into $n-1$ small segments that have approximately equal length. At these points the curvature is the same as the norm of the second parametric derivative of the curve in arc-length parameterization, i.e. when the same distance in parameter space means the same distance in arc length. We will denote the parameter of the kth point with p_k.

The second phase is carried out both from the left and from the right side. Since these are symmetrical, we examine only the former. The numerical integration algorithm needs the direction of the second derivative, as well as the starting point. In arc-length parameterization the second derivative is in the normal direction, which can be easily calculated. In the first integration step we should approximate the first derivative at the starting point of the curve in an arc-length parametric sense, i.e. $\hat{f}'(0) = \frac{f(p_1)-f(p_0)}{h}$, where $h = \frac{\text{length}(p_0,\ p_{n-1})}{n-1}$, f denotes the actual, $\hat{f}$ the arc-length parametric representation of the curve, and the function $\text{length}(p_i,\ p_j)$ gives the length of the curve in the interval $[p_i,\ p_j]$. In the second integration step, we can use the coordinate of the original curve: $\hat{f}(0) = f(p_0)$.

The Euler method is the simplest numerical integration algorithm, and even this produces good results. The equation for the first step is

$$\hat{f}'(x + h) = \hat{f}'(x) + h\hat{f}''(x),$$

and for the second step:

$$\hat{f}(x + h) = \hat{f}(x) + h\hat{f}'(x).$$

The second-order Runge–Kutta method uses a midpoint for better approximation. Let $n = 4k$, then for the first step the equation is

$$\hat{f}'(x + 2h) = \hat{f}'(x) + 2h\hat{f}''(x + h),$$

and for the second step:

$$\hat{f}(x + 4h) = \hat{f}(x) + 4h\hat{f}'(x + 2h).$$

In both cases, we need to control the deviation from the original curve. This can be achieved by adding an extra term $c(x)$ to the equations, that pulls back the points when they get too close to a user-defined deviation tolerance. Imagine a band around the original curve with the tolerance as its breadth (Fig. 3). If the integrated curve gets close to the border of the band, it will be turned towards the middle.

$$c(x) = \min\left[1, \left(\frac{|d|}{m}\right)^2\right] \cdot d,$$

Fig. 3. Maximum deviation band around a curve

where $d = f(x) - \hat{f}(x)$ and m is the predefined tolerance.

In the third phase we apply a blending function to the two curves (integrated from the left and right side, respectively). We can use a suitable blending function[2], e.g. the Hermite blending function $\lambda(t) = 3t^2 - 2t^3$, or the 5th-degree polynomial $\lambda(t) = 6t^5 - 15t^4 + 10t^3$, so the blended points will be of the form

$$g(x) = (1 - \lambda(t))\hat{f}_{\text{left}}(x) + \lambda(t)\hat{f}_{\text{right}}(x),$$

where $t = \frac{x}{\text{length}(p_0, \ p_{n-1})}$.

Finally, an approximating spline is computed by fitting a least-squares B-spline over the points [7]. Usually the original knot vector can be passed to the fitting algorithm, and (e.g. curvature minimizing) smoothing terms can also be employed in the fitting process.

3.2   Surface Case

We can use the curve fairing method to fair individual isocurves of the surface. Let the B-spline surface be given in the form

$$S(u, v) = \sum_{i=0}^{n} \sum_{j=0}^{m} d_{ij} N_{i,k}^{U}(u) N_{j,l}^{V}(v),$$

$$(u, v) \in [u_{k-1}, u_{n+1}] \times [v_{l-1}, v_{m+1}],$$

where $U = (u_i)_{i=0}^{n+k}$ and $V = (v_j)_{j=0}^{m+l}$ represent the knot vectors. Then the u-isocurve for a fixed v parameter can be written as

$$C_v(u) = \sum_{i=0}^{n} \left(\sum_{j=0}^{m} N_{j,l}^{V}(v) d_{ij} \right) N_{i,k}^{U}(u),$$

i.e. a B-spline curve with the same knot vector U, and with control points $\sum_{j=0}^{m} N_{j,l}^{V}(v) d_{ij}$. We can apply steps 1–3 of the curve fairing method to get the faired points. These points can be collected to form two point clouds (one from the u-parametric and one from the v-parametric curves). Blending them and fitting a surface on the blended point set will result in a fair surface.

4   Test Results

Figure 4 shows a curve before and after fairing, using three different tolerances. We can see that fairness comes in exchange for a larger deviation. The change in fairness is more visible on the extrusion surfaces — the wiggles in the isophote lines are reduced in the tight and medium tolerance case, and totally smoothed out with loose tolerance.

Figure 5 shows local fairing on a Fiat body part. The region on the right was selected for fairing, and the curvature became much more smooth as a result. Other regions were not affected (substantially).

---

[2] A suitable blending function $\lambda(t)$ has the following properties: (i) $\lambda(0) = 0$ and $\lambda(1) = 1$ (ii) $\lambda^{(k)}(0) = 0$ and $\lambda^{(k)}(1) = 0$ for some $k = 1 \ldots$.

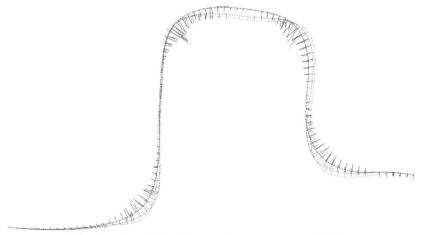

(a) Original: red; tight: blue; medium: green; loose: black.

(b) Isophotes of the extrusion surfaces.

Fig. 4. Fairing a curve with tight, medium and loose tolerance

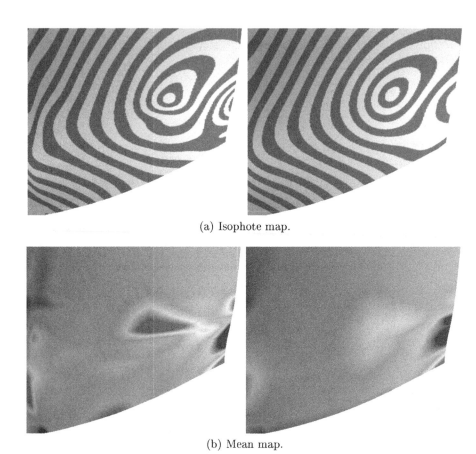

(a) Isophote map.

(b) Mean map.

Fig. 5. Local fairing of a Fiat body part

5 Conclusions

A fast and local algorithm was introduced for both curve and surface fairing. It uses a parameterization-independent fairness measure based on a target curvature. Effectiveness is achieved by integration and fitting instead of iteration. User interaction may be necessary, but minimal. Our future research will be the fairing of multiple curves / surfaces together, while preserving or enhancing the continuity between them.

References

1. Eck, M., Hadenfeld, J.: Local Energy Fairing of B-Spline Curves. Computing Supplement 10, 129–147 (1995)
2. Farin, G.: Curves and Surfaces for Computer Aided Geometric Design. A Practical Guide, 5th edn. Academic Press, London (2002)

3. Farin, G., Hoschek, J., Kim, M.-S. (eds.): Handbook of Computer Aided Geometric Design. North-Holland, Amsterdam (2002)
4. Hahmann, S., Konz, S.: Knot-Removal Surface Fairing Using Search Strategies. Computer Aided Design 30, 131–138 (1998)
5. Lee, E.T.Y.: Energy, fairness, and a counterexample. Computer Aided Design 22(1), 37–40 (1990)
6. Moreton, H.P., Séquin, C.H.: Minimum Variation Curves and Surfaces for Computer-Aided Geometric Design. In: Sapidis, N.S. (ed.) Designing Fair Curves and Surfaces, pp. 123–159. SIAM, Philadelphia (1994)
7. Piegl, L., Tiller, W.: The NURBS Book, 2nd edn. Springer, Heidelberg (1997)
8. Press, W.H., Teukolsky, S.A., Vetterling, W.T., Flannery, B.P.: Numerical Recipes in C, 2nd edn. Cambridge University Press, Cambridge (1992)
9. Roulier, J., Rando, T.: Measures of Fairness for Curves and Surfaces. In: Sapidis, N.S. (ed.) Designing Fair Curves and Surfaces, pp. 75–122. SIAM, Philadelphia (1994)
10. Salvi, P., Várady, T.: Local Fairing of Freeform Curves and Surfaces. In: Proceedings of the Third Hungarian Graphics and Geometry Conference (2005)

Finite Element Methods for Geometric Modeling and Processing Using General Fourth Order Geometric Flows

Guoliang Xu[*]

State Key Laboratory of Scientific and Engineering Computing
Institute of Computational Mathematics, Academy of Mathematics and
System Sciences, Chinese Academy of Sciences, Beijing 100080, China
xuguo@lsec.cc.ac.cn

Abstract. A variational formulation of a general form fourth order geometric partial differential equation is derived, and based on which a mixed finite element method is developed. Several surface modeling problems, including surface blending, hole filling and surface mesh refinement with the G^1 continuity, are taken into account. The used geometric partial differential equation is universal, containing several well-known geometric partial differential equations as its special cases. The proposed method is general which can be used to construct surfaces for geometric design as well as simulate the behaviors of various geometric PDEs. Experimental results show that it is simple, efficient and gives very desirable results.

Keywords: Geometric PDE, Surface blending, hole filling, Surface mesh refinement, Mixed finite element method.

1 Introduction

In this paper, we use a general form geometric partial differential equation (GPDE) to solve several surface modeling problems. The GPDE we use includes many well known equations as its special cases. It is nonlinear and geometrically intrinsic, which is solved using a mixed finite element method. The problems we solve include surface blending, hole filling and mesh refinement with the G^1 continuity. For the problems of surface blending and hole filling, we are given triangular surface meshes of the surrounding area. Triangular surface patches need to be constructed to fill the openings enclosed by the surrounding surface mesh and interpolate its boundary with the G^1 continuity (see Fig. 2 and 3). For mesh refinement problem, we are given an initial surface mesh. We construct a sequence of surface meshes which interpolates the vertices of the previous meshes with the specified G^1 continuity (see Fig. 4).

There are basically two approaches for solving a GPDE numerically on surface. One is the generalized finite divided difference method (see [25,28]), the

---

[*] Project support in part by NSFC grant 60773165 and National Key Basic Research Project of China (2004CB318000).

F. Chen and B. Jüttler (Eds.): GMP 2008, LNCS 4975, pp. 164–177, 2008.

other is the finite element method (see [1,10,16,18,26]). The approach we adopt in this paper is a mixed finite element method consisting of using the Loop's subdivision basis for representing the surfaces and the piecewise linear basis for representing the curvatures. Loop's basis have been used in finite element analysis and design for thin-shell structures (see [6,7]). The generalized finite divided difference method has been used in [25,28]. It is well known that the finite divided difference method is simpler and easier to implement, but it lacks the convergence analysis. The finite element method is not as simple as the finite divided difference method, but is based on a well developed mathematical foundation.

Previous Work. The well-known biharmonic equation and its various variations have been repeatedly used to solve the problems of geometry modeling, such as blending and the hole filling (see [3,4]), interactive surface design (see [15], [24]) and interactive sculpting (see [14]). A PDE method based on the biharmonic equation is proposed [19] with certain functional constraints, such as geometric constraints, physical criteria and engineering requirements, which can be incorporated into the shape design process. The main advantageous of using the biharmonic equation is that the equation is linear, hence it is easy to solve. However, the biharmonic equation is not geometrically intrinsic and its solution depends on the concrete surface parametrization used. Furthermore, the biharmonic equation is defined on a rectangular domain in general, hence it is inappropriate for modeling surfaces with arbitrary shaped boundaries. In contrast, the equations we use in this paper are geometrically intrinsic without the limitation of the shape of the surface boundaries. These equations are called geometric partial differential equations (GPDE).

The simplest GPDE may be the mean curvature flow (MCF). MCF and its variations have been intensively used to smooth or fair noisy surfaces (see [1,9,11] for references). These second order equations have been shown to be the most important and efficient flows for fairing or denoising surface meshes. However, for solving the surface modeling and design problems, MCF cannot achieve the G^1 continuity at the boundary. To achieve smooth joining of different surface patches in solving the surface blending problem (see [8,21,22,25]), the fourth order geometric flows, such as Willmore flow and surface diffusion flow, have been employed. For instance, the surface diffusion flow is used in [21] to fair surface meshes with the G^1 conditions in a special case where the meshes are assumed to have the subdivision connectivity. This equation is also used in [22] for smoothing meshes while satisfying the G^1 boundary conditions and the requirement of the outer fairness (the smoothness in the classical sense) and the inner fairness (the regularity of the vertex distribution). Willmore flow has been used for the surface restoration (see [8]) and surface modeling (see [26]) using the finite element method. Three fourth order flows (surface diffusion flow, Willmore flow and quasi surface flow) are treated separately using finite element method in [26].

Main Results. We derive a weak form formulation for a general form fourth order geometric flow. Based on this weak form a mixed form formulation is proposed. The equation is solved by a newly proposed mixed finite element method over surfaces. The proposed approach is simple, efficient and general as well. It handles a class of geometric PDEs in the same way and solves several surface modeling problems in the same manner. The approach is applied to solving each of the surface modeling problems and gives very desirable results for a range of the complicated surface models.

The rest of the paper is organized as follows: Section 2 introduces some background material from differential geometry and our earlier publications. Section 3 describes the used nonlinear GPDE and its variational form. In section 4, we give the details of the discretization of the variational form in both the spatial and the temporal directions. Comparative examples to illustrate the different effects achievable from the solutions of these GPDEs are given in section 5.

2   Differential Geometry Preliminaries

In this section, we introduce the used notations, curvatures and several geometric differential operators, including Laplace-Beltrami operator and Giaquinta-Hildebrandt operator etc. Some results used in this paper are also presented.

Let $\mathcal{S} = \{\mathbf{x}(u,v) \in \mathbb{R}^3; (u,v) \in \Omega \subset \mathbb{R}^2\}$ be a regular and sufficiently smooth parametric surface. To simplify the notation we sometimes write $w = (u,v) = (u^1, u^2)$. Let $g_{\alpha\beta} = \langle \mathbf{x}_{u^\alpha}, \mathbf{x}_{u^\beta} \rangle$ and $b_{\alpha\beta} = \langle \mathbf{n}, \mathbf{x}_{u^\alpha u^\beta} \rangle$ be the coefficients of the first and the second fundamental forms of $\mathcal{S}$ with

$$\mathbf{x}_{u^\alpha} = \frac{\partial \mathbf{x}}{\partial u^\alpha}, \quad \mathbf{x}_{u^\alpha u^\beta} = \frac{\partial^2 \mathbf{x}}{\partial u^\alpha \partial u^\beta}, \quad \alpha, \beta = 1, 2, \quad \mathbf{n} = (\mathbf{x}_u \times \mathbf{x}_v)/\|\mathbf{x}_u \times \mathbf{x}_v\|,$$

where $\langle \cdot, \cdot \rangle$ and $\times$ denote the usual scalar and cross products of two vectors, respectively, in Euclidean space $\mathbb{R}^3$. Set

$$g = \det[g_{\alpha\beta}], \quad [g^{\alpha\beta}] = [g_{\alpha\beta}]^{-1}, \quad b = \det[b_{\alpha\beta}], \quad [b^{\alpha\beta}] = [b_{\alpha\beta}]^{-1}.$$

Curvatures. To introduce the notions of mean curvature and Gaussian curvature, let us first introduce the concept of *Weingarten map* or *shape operator*. The shape operator of surface $\mathcal{S}$ is a self-adjoint linear map on the tangent space $T_{\mathbf{x}}\mathcal{S} := \mathrm{span}\{\mathbf{x}_u, \mathbf{x}_v\}$ defined by $\mathcal{W} : T_{\mathbf{x}}\mathcal{S} \to T_{\mathbf{x}}\mathcal{S}$, such that

$$\mathcal{W}(\mathbf{x}_u) = -\mathbf{n}_u, \quad \mathcal{W}(\mathbf{x}_v) = -\mathbf{n}_v. \tag{1}$$

We can easily represent this linear map by a 2×2 matrix $S = [b_{\alpha\beta}][g^{\alpha\beta}]$. In particular, $[\mathbf{n}_u, \mathbf{n}_v] = -[\mathbf{x}_u, \mathbf{x}_v]S^T$ is valid. The eigenvalues k_1 and k_2 of S are *principal curvatures* of $\mathcal{S}$ and their arithmetic average and product are the *mean curvature* and the *Gaussian curvature*, respectively. That is

$$H = \frac{k_1 + k_2}{2} = \frac{\mathrm{tr}(S)}{2}, \qquad K = k_1 k_2 = \det(S).$$

Tangential gradient operator. Let f be a C^1 smooth function on $\mathcal{S}$. Then the *tangential gradient operator* ∇ acting on f is given by (see [12], page 102)

$$\nabla f = [\mathbf{x}_u, \ \mathbf{x}_v][g^{\alpha\beta}][f_u, \ f_v]^T \in \mathbb{R}^3, \tag{2}$$

where f_u and f_v denote the first order partial derivatives of f with respect to arguments u and v. For a vector-valued function $\mathbf{f} = [f_1, \cdots, f_k]^T \in C^1(\mathcal{S})^k$, we define its gradient by $\nabla \mathbf{f} = [\nabla f_1, \cdots, \nabla f_k] \in \mathbb{R}^{3 \times k}$. It is easy to see that

$$\nabla \mathbf{x} = [\mathbf{x}_u, \ \mathbf{x}_v][g^{\alpha\beta}][\mathbf{x}_u, \ \mathbf{x}_v]^T, \quad \nabla \mathbf{n} = -[\mathbf{x}_u, \ \mathbf{x}_v][g^{\alpha\beta}]S[\mathbf{x}_u, \ \mathbf{x}_v]^T, \tag{3}$$

and both $\nabla \mathbf{x}$ and $\nabla \mathbf{n}$ are symmetric 3×3 matrices.

Second tangent operator. Let f be a C^1 smooth function on $\mathcal{S}$. Then we define the *second tangent operator* $\Diamond$ acting on f by (see [27])

$$\Diamond f = [\mathbf{x}_u, \ \mathbf{x}_v][h_{\alpha\beta}][f_u, \ f_v]^T \in \mathbb{R}^3, \quad \text{with} \quad [h_{\alpha\beta}] = \frac{1}{g} \begin{bmatrix} b_{22} & -b_{12} \\ -b_{12} & b_{11} \end{bmatrix}. \tag{4}$$

Divergence operator. Let $\mathbf{v}$ be a C^1 smooth vector field on $\mathcal{S}$. Then the *divergence* of $\mathbf{v}$ is defined by

$$\text{div}(\mathbf{v}) = \frac{1}{\sqrt{g}} \left[\frac{\partial}{\partial u}, \frac{\partial}{\partial v} \right] \left[\sqrt{g} \, [g^{\alpha\beta}] \, [\mathbf{x}_u, \ \mathbf{x}_v]^T \mathbf{v} \right]. \tag{5}$$

For a matrix-valued function $\mathbf{V} = [\mathbf{v_1}, \cdots, \mathbf{v_k}] \in C^1(\mathcal{S})^{3 \times k}$, we define $\text{div}(\mathbf{V}) = [\text{div}(\mathbf{v_1}), \cdots, \text{div}(\mathbf{v_k})]^T \in \mathbb{R}^k$.

Laplace-Beltrami operator. Let f be a C^2 smooth function on $\mathcal{S}$. Then ∇f is a smooth vector field on $\mathcal{S}$. The Laplace-Beltrami operator (LBO) Δ applying to f is defined by (see [13], page 83) $\Delta f = \text{div}(\nabla f)$.

Giaquinta-Hildebrandt operator. Let f be a C^2 smooth function on $\mathcal{S}$. Then the Giaquinta-Hildebrandt operator acting on f is given by (see [17], page 84) $\Box f = \text{div}(\Diamond f)$. For a vector-valued function $\mathbf{f} = [f_1, \cdots, f_k]^T \in C^1(\mathcal{S})^k$, we define

$$\Delta \mathbf{f} = \text{div}(\nabla \mathbf{f}), \quad \Box \mathbf{f} = \text{div}(\Diamond \mathbf{f}). \tag{6}$$

For the operators introduced above, we can prove the following theorems that are used in the sequel. Detailed proofs can be found in our technical report [29].

Theorem 1. *Let* $\mathbf{x} \in \mathcal{S}$. *Then*

$$\Delta \mathbf{x} = 2H\mathbf{n}, \quad \Delta \mathbf{n} = -2\nabla H - 2H\Delta \mathbf{x} + \Box \mathbf{x}. \tag{7}$$

Theorem 2. *Let* $\mathbf{x} \in \mathcal{S}$. *Then*

$$\Box \mathbf{x} = 2K\mathbf{n}, \quad \Box \mathbf{n} = -\nabla K - K\Delta \mathbf{x} = -\nabla K - H\Box \mathbf{x}. \tag{8}$$

To derive the weak form of the considered geometric flows, we introduce two Green's formulas.

Theorem 3 (Green's formula I for LB operator). ([5], page 142) *Let* $f \in C^2(\mathcal{S}), h \in C^1(\mathcal{S})$, *with at least one of them compactly supported. Then*

$$\int_{\mathcal{M}} \left(h\Delta f + \langle \nabla f, \nabla h \rangle \right) dA = 0.$$

Theorem 4 (Green's formula I for GH operator). *Let* $f \in C^2(\mathcal{S}), h \in C^1(\mathcal{S})$, *with at least one of them compactly supported. Then*

$$\int_{\mathcal{M}} \left(h\square f + \langle \nabla f, \Diamond h \rangle \right) dA = 0.$$

3   Weak and Mixed Forms of a General Fourth Order GPDE

Euler-Lagrange equation. Let $f(H, K) \in C^1(\mathbb{R} \times \mathbb{R})$ be a given Lagrange function. Define energy functional

$$\mathcal{E}(\mathcal{S}) = \int_{\mathcal{S}} f(K, H) dA. \tag{9}$$

In this paper we use a vector-valued Euler-Lagrange equation derived from the complete-variation

$$\underline{\mathbf{x}}(w, \varepsilon) = \mathbf{x}(w) + \varepsilon\Phi(w), \quad w \in \overline{\Omega}, \quad |\varepsilon| \ll 1, \quad \Phi \in C_c^\infty(\Omega)^3, \tag{10}$$

for the functional $\mathcal{E}(\mathcal{S})$. We refer a result from [28] as the following theorem:

Theorem 5. *Let* $f(H, K) \in C^1(\mathbb{R} \times \mathbb{R})$. *Then the Euler-Lagrange equation of the integral* $\mathcal{E}(\mathcal{S})$ *from the complete-variation* (10) *is*

$$\square(f_K \mathbf{n}) + \frac{1}{2}\Delta(f_H \mathbf{n}) - \operatorname{div}(f_H \nabla \mathbf{n}) - \operatorname{div}[(f - 2Kf_K)\nabla \mathbf{x}]) = \mathbf{0} \in \mathbb{R}^3. \tag{11}$$

Theorem 6. *The weak form of* (11) *can be written as*

$$\int_{\mathcal{S}} \left[f_H \nabla \mathbf{x} \Diamond \phi + (f - 2Hf_H - 2Kf_K)\nabla \mathbf{x} \nabla \phi + \frac{1}{2} f_H \mathbf{n}\Delta \phi + f_K \mathbf{n}\square \phi \right] dA = \mathbf{0},$$
$$\forall \phi \in C_c^\infty(\Omega). \tag{12}$$

Proof. By direct calculation, it is not difficult to derive that $[g^{\alpha\beta}]S = 2H[g^{\alpha\beta}] - K[b^{\alpha\beta}]$. Using this equality and (3), we have

$$\nabla \mathbf{n} = \Diamond \mathbf{x} - 2H\nabla \mathbf{x}. \tag{13}$$

Multiplying both sides of (11) with $\phi \in C_c^\infty(\Omega)$, then taking integral over $\mathcal{S}$, using Green formulas, and finally using (13), we have

$$\mathbf{0} = \int_{\mathcal{S}} \left[\square(f_K \mathbf{n}) + \frac{1}{2}\Delta(f_H \mathbf{n}) - \operatorname{div}(f_H \nabla \mathbf{n}) - \operatorname{div}[(f - 2Kf_K)\nabla \mathbf{x}]) \right] \phi \, dA$$

$$= \int_{\mathcal{S}} \left[f_K \mathbf{n}\square \phi + \frac{1}{2} f_H \mathbf{n}\Delta \phi + [f_H(\Diamond \mathbf{x} - 2H\nabla \mathbf{x})]\nabla \phi + [(f - 2Kf_K)\nabla \mathbf{x}]\nabla \phi) \right] dA.$$

Noticing that $\Diamond \mathbf{x} \nabla \phi = \nabla \mathbf{x} \Diamond \phi$, we obtain (12) in the end.

A General form Geometric Flow. For a given function $f(H, K)$, let $\mathcal{S}_0$ be a given initial surface with boundary Γ. Then the geometric flow consists of finding a family $\{\mathcal{S}(t); t \geq 0\}$ of smooth orientable surfaces in $\mathbb{R}^3$ which evolve according to the following equation

$$\begin{cases} \frac{\partial \mathbf{x}}{\partial t} + \Box(f_K \mathbf{n}) + \frac{1}{2}\Delta(f_H \mathbf{n}) - \operatorname{div}(f_H \nabla \mathbf{n}) - \operatorname{div}\left[(f - 2Kf_K)\nabla \mathbf{x}\right] = 0, \\ \mathcal{S}(0) = \mathcal{S}_0, \quad \partial \mathcal{S}(t) = \Gamma. \end{cases} \quad (14)$$

Weak form formulation. Using (12), we can obtain the weak form of equation (14) as follows:

$$\begin{cases} \displaystyle\int_{\mathcal{S}} \frac{\partial \mathbf{x}}{\partial t}\phi \, dA + \int_{\mathcal{S}} \Big[f_H \nabla \mathbf{x} \Diamond \phi + (f - 2Hf_H - 2Kf_K)\nabla \mathbf{x}\nabla \phi \\ \qquad\qquad + \frac{1}{2} f_H \mathbf{n}\Delta \phi + f_K \mathbf{n}\Box \phi \Big] \, dA = 0, \quad \forall \phi \in C_c^{\infty}(\Omega), \quad (15) \\ \mathcal{S}(0) = \mathcal{S}_0, \quad \partial \mathcal{S}(t) = \Gamma. \end{cases}$$

Assumptions. In order to make the generated flow meaningful and well-defined. We first require that f is a smooth function about its arguments. We also assume that f is an algebraic function, meaning it does not involve differential and integral operations about its variables. Suppose f_H and f_K could be represented as

$$f_H = 2\alpha H + 2\beta K + \mu, \quad f_K = 2\gamma H + 2\delta K + \nu. \quad (16)$$

Let $\mathbf{h} = 2H\mathbf{n}$, $\mathbf{k} = 2K\mathbf{n}$. Then using (15) and

$$\int_{\mathcal{S}} \mathbf{h}\psi \, dA = \int_{\mathcal{S}} \Delta \mathbf{x}\psi \, dA = -\int_{\mathcal{S}} \nabla \mathbf{x}\nabla \psi \, dA, \quad \forall \psi \in C_c^{\infty}(\Omega),$$

$$\int_{\mathcal{S}} \mathbf{k}\varphi \, dA = \int_{\mathcal{S}} \Box \mathbf{x}\varphi \, dA = -\int_{\mathcal{S}} \nabla \mathbf{x}\Diamond \varphi \, dA, \quad \forall \varphi \in C_c^{\infty}(\Omega),$$

we obtain a mixed form of equation (14) as follows: Find $[\mathbf{x}, \mathbf{h}, \mathbf{k}] \in \mathbb{R}^{3\times 3}$, such that

$$\begin{cases} \displaystyle\int_{\mathcal{S}} \frac{\partial \mathbf{x}}{\partial t}\phi \, dA + \int_{\mathcal{S}} \Big[f_H \nabla \mathbf{x}\Diamond \phi + (f - 2Hf_H - 2Kf_K)\nabla \mathbf{x}\nabla \phi \\ \qquad +\frac{1}{2}(\alpha \mathbf{h} + \beta \mathbf{k} + \nu \mathbf{n})\Delta \phi + (\gamma \mathbf{h} + \delta \mathbf{k} + \mu \mathbf{n})\Box \phi \Big] \, dA=0, \quad \forall \phi \in C_c^{\infty}(\Omega), \\ \displaystyle\int_{\mathcal{S}} \mathbf{h}\psi \, dA + \int_{\mathcal{S}} \nabla \mathbf{x}\nabla \psi \, dA = 0, \quad \forall \psi \in C_c^{\infty}(\Omega), \\ \displaystyle\int_{\mathcal{S}} \mathbf{k}\varphi \, dA + \int_{\mathcal{S}} \nabla \mathbf{x}\Diamond \varphi \, dA = 0, \quad \forall \varphi \in C_c^{\infty}(\Omega). \end{cases}$$

$$(17)$$

From the definitions of tangential operators, it is easy to derive that $\nabla \mathbf{x}\Diamond \phi = \Diamond \phi$, $\nabla \mathbf{x}\nabla \phi = \nabla \phi$. Further notice that $2H = \mathbf{n}^T \mathbf{h}$, $2K = \mathbf{n}^T \mathbf{k}$, the weak form (17) can be rewritten as: Find $[\mathbf{x}, \mathbf{h}, \mathbf{k}] \in \mathbb{R}^{3\times 3}$, such that

$$
\begin{cases}
\int_{\mathcal{S}} \frac{\partial \mathbf{x}}{\partial t} \phi \, dA + \int_{\mathcal{S}} \Big[\mu \nabla \mathbf{x} \Diamond \phi + \eta \nabla \mathbf{x} \nabla \phi + \Diamond \phi \mathbf{n}^T (\alpha \mathbf{h} + \beta \mathbf{k}) \\
\qquad + \nabla \phi \mathbf{n}^T [(\omega - f_H) \mathbf{h} + (\epsilon - f_K) \mathbf{k}] \\
\qquad + \frac{1}{2} (\alpha \mathbf{h} + \beta \mathbf{k} + \nu \mathbf{n}) \Delta \phi + (\gamma \mathbf{h} + \delta \mathbf{k} + \mu \mathbf{n}) \Box \phi \Big] \, dA = 0, \quad \forall \phi \in C_c^\infty(\Omega), \\
\int_{\mathcal{S}} \mathbf{h} \psi \, dA + \int_{\mathcal{S}} \nabla \mathbf{x} \nabla \psi \, dA = 0, \quad \forall \psi \in C_c^\infty(\Omega), \\
\int_{\mathcal{S}} \mathbf{k} \varphi \, dA + \int_{\mathcal{S}} \nabla \mathbf{x} \Diamond \varphi \, dA = 0, \quad \forall \varphi \in C_c^\infty(\Omega),
\end{cases}
\tag{18}
$$

where ω, ϵ and η are defined by $f = 2\omega H + 2\epsilon K + \eta$. System (18) is the starting point of the finite element discretization.

4   Solving the GPDE by Mixed Finite Element Methods

Consider a triangular surface mesh Ω with vertices $\mathbf{x}_1, \mathbf{x}_2, \cdots, \mathbf{x}_n$. For each vertex $\mathbf{x}_i$, we associate it with a basis function ϕ_i. Then the surface $\mathcal{S}$ is approximately represented as

$$
\mathbf{x} = \sum_{j=1}^{n} \phi_j(q) \mathbf{x}_j \in \mathbb{R}^3, \quad q \in \Omega,
\tag{19}
$$

where $\mathbf{x}_1, \mathbf{x}_2, \cdots, \mathbf{x}_n$ are regarded as the control vertices of $\mathcal{S}$.

Now we classify these control vertices into several categories. The first category consists of interior vertices, denoted as $\mathbf{x}_1, \mathbf{x}_2, \cdots, \mathbf{x}_{n_0}$. The positions of these vertices are subject to change (unknown). For the problems of hole filling and blending, the interior vertices are those in the openings (see Fig. 1(a), the solid dots). For the problem of mesh refinement, all the non-interpolating vertices are interior (see Fig. 1(b), the solid dots). Apart from the interior vertices, the remaining vertices are classified as 1-ring neighbors (see Fig. 1(a), the empty

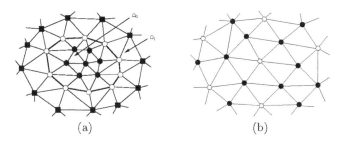

(a)   (b)

Fig. 1. Classification of vertices. (a) The solid dots are interior control vertices. The empty dots are 1-ring neighbor (boundary) vertices. The solid squares are 2-ring neighbor vertices. The shaded region is Ω_0. The region enclosed by the 2-ring neighbor vertices is Ω_1. (b) The solid dots are the interior control vertices. The empty ones are the interpolated vertices.

dots), 2-ring neighbors (see Fig. 1(a), the solid squares), $\cdots$, according to their distance to the interior vertices. For instance, the 1-ring neighbors are all the vertices adjacent to the interior ones, the 2-ring neighbors are all the remaining vertices adjacent to the 1-ring neighbors, $\cdots$. The k-ring neighbor control vertices are denoted as $\mathbf{x}_{n_{k-1}+1}, \cdots, \mathbf{x}_{n_k}$. All the 1-ring, 2-ring, $\cdots$, neighbor vertices are fixed.

In our mixed form formulations of GPDE, the mean and Gaussian curvature normals are also treated as unknown vector-valued functions. Hence, for each control vertex $\mathbf{x}_i$, we also associate it with a continuous basis function ψ_i for representing the curvatures. For instance, mean and Gaussian curvature normals are represented as

$$\mathbf{h}(\mathbf{x}) = \sum_{j=1}^{n} \psi_j(q)\mathbf{h}_j \in \mathbb{R}^3, \quad \mathbf{k}(\mathbf{x}) = \sum_{j=1}^{n} \psi_j(q)\mathbf{k}_j \in \mathbb{R}^3, \quad q \in \Omega, \qquad (20)$$

respectively. Here $\mathbf{h}_j$ and $\mathbf{k}_j$ denote the mean and Gaussian curvature normals at $\mathbf{x}_j$, respectively.

Suppose that $\phi_i(q)$ and $\psi_i(q)$ have compact supports and the sizes of their supports are 2 and 1, respectively. At a k-ring neighbor vertex, the surface point does not depend upon the interior vertices if $k \geq 2$, so the mean and Gaussian curvature normals there are well defined from the surface. However, at the interior and the 1-ring neighbor vertices, the corresponding surface points relate to unknown vertices. Then the mean and Gaussian curvatures there cannot be determined from the surface. Hence we treat these curvatures as unknowns. In a word, the unknown vertices are $\mathbf{x}_1, \mathbf{x}_2, \cdots, \mathbf{x}_{n_0}$, the unknown mean curvature normals are $\mathbf{h}_1, \mathbf{h}_2, \cdots, \mathbf{h}_{n_0}, \mathbf{h}_{n_0+1}, \cdots, \mathbf{h}_{n_1}$, and the unknown Gaussian curvature normals are $\mathbf{k}_1, \mathbf{k}_2, \cdots, \mathbf{k}_{n_0}, \mathbf{k}_{n_0+1}, \cdots, \mathbf{k}_{n_1}$.

In the next two sub-sections, we will obtain matrix forms of the used GPDE by substituting (19)–(20) into (18).

Note that the proposed scheme needs the values of the mean and Gaussian curvature normals at each vertex. For an interior or 1-ring neighbor vertex, the mean and Gaussian curvature normals are unknown in the equation to be solved, but initial values at $t = 0$ is required. For a k-ring neighbor vertex $(k \geq 2)$, these vectors need to be computed previously and keep fixed afterwards (see Remark 4.3 for how these vectors are computed). A default choice of these vectors is to set them as zeros.

4.1 Spatial Direction Discretizations

We construct one equation for each unknown vertex $\mathbf{x}_j$ $(j = 1, \cdots, n_0)$ and other two equations for the unknown vector-valued curvatures $\mathbf{h}_j$ and $\mathbf{k}_j$ $(j = 1, \cdots, n_1)$. This is achieved by the following operations: (i) Substituting (19)–(20) into (18). (ii) Taking the test functions $\phi = \phi_i$ $(i = 1, \cdots, n_0)$ and

$\psi = \varphi = \psi_i$ ($i = 1, \cdots, n_1$). (iii) Using the fact that $\frac{\partial \mathbf{x}_j(t)}{\partial t} = 0$ for $j > n_0$ (since $\mathbf{x}_j$ is fixed). We then obtain the following matrix form of (18):

$$\begin{cases} M_{n_0} \dfrac{\partial \mathcal{X}_{n_0}(t)}{\partial t} + L_{n_3}^{(p)} \mathcal{X}_{n_3}(t) + L_{n_2}^{(m)} \mathcal{H}_{n_2}(t) + L_{n_2}^{(g)} \mathcal{K}_{n_2}(t) + B_{n_0} = 0, \\ M_{n_2}^{(1)} \mathcal{H}_{n_2}(t) + L_{n_3}^{(1)} \mathcal{X}_{n_3}(t) = 0, \\ M_{n_2}^{(1)} \mathcal{K}_{n_2}(t) + L_{n_3}^{(2)} \mathcal{X}_{n_3}(t) = 0, \end{cases} \tag{21}$$

where

$$\mathcal{X}_j(t) = [\mathbf{x}_1(t), \cdots, \mathbf{x}_j(t)]^T \in \mathbb{R}^{j \times 3},$$
$$\mathcal{H}_j(t) = [\mathbf{h}_1(t), \cdots, \mathbf{h}_j(t)]^T \in \mathbb{R}^{j \times 3},$$
$$\mathcal{K}_j(t) = [\mathbf{k}_1(t), \cdots, \mathbf{k}_j(t)]^T \in \mathbb{R}^{j \times 3},$$

are unknowns and

$$M_K = (m_{ij})_{ij=1}^{n_0, K}, \qquad M_K^{(1)} = \left(m_{ij}^{(1)} \right)_{ij=1}^{n_1, K},$$
$$L_K^{(p)} = \left(l_{ij}^{(p)} \right)_{ij=1}^{n_0, K}, \quad L_K^{(m)} = \left(l_{ij}^{(m)} \right)_{ij=1}^{n_0, K}, \quad L_K^{(g)} = \left(l_{ij}^{(g)} \right)_{ij=1}^{n_0, K},$$
$$L_K^{(1)} = \left(l_{ij}^{(1)} \right)_{ij=1}^{n_1, K}, \quad L_K^{(2)} = \left(l_{ij}^{(2)} \right)_{ij=1}^{n_1, K}, \qquad B_{n_0} = (b_i^T)_{i=1}^{n_0},$$

are the coefficient matrices. The elements of these matrices are defined as follows:

$$m_{ij} = \int_S \phi_i \phi_j \, dA, \qquad\qquad m_{ij}^{(1)} = \int_S \psi_i \psi_j \, dA,$$

$$l_{ij}^{(p)} = \int_S \left[\mu (\nabla \psi_j)^T \Diamond \phi_i + \eta (\nabla \psi_j)^T \nabla \phi_i \right] \, dA \in \mathbb{R},$$

$$l_{ij}^{(m)} = \int_S \left[\psi_j [\alpha \Diamond \phi_i + (\omega - f_H) \nabla \phi_i] \mathbf{n}^T + (\tfrac{1}{2} \alpha \psi_j \Delta \phi_i + \gamma \psi_j \Box \phi_i) I_3 \right] \, dA \in \mathbb{R}^{3 \times 3},$$

$$l_{ij}^{(g)} = \int_S \left[\psi_j [\beta \Diamond \phi_i + (\epsilon - f_K) \nabla \phi_i] \mathbf{n}^T + (\tfrac{1}{2} \beta \psi_j \Delta \phi_i + \delta \psi_j \Box \phi_i) I_3 \right] \, dA \in \mathbb{R}^{3 \times 3}.$$

$$l_{ij}^{(1)} = \int_S (\nabla \phi_j)^T \nabla \psi_i \, dA, \qquad\qquad l_{ij}^{(2)} = \int_S (\nabla \phi_j)^T \Diamond \psi_i \, dA,$$

$$b_i = \int_S [\mathbf{n}(\tfrac{1}{2} \nu \Delta \phi_i + \mu \Box \phi_i)] dA.$$

The integrations in these matrix elements are computed by a 12-point Gaussian quadrature rule. That is, each domain triangle is subdivided into four sub-triangles and a three-point Gaussian quadrature rule is employed on each of the sub-triangles (see [2] for a set of Gaussian quadrature rules). In system (21), moving the terms relating to the known vertices $\mathbf{x}_{n_0+1}$, $\cdots$, $\mathbf{x}_{n_3}$, the known mean curvatures $\mathbf{h}_{n_1+1}$, $\cdots$, $\mathbf{h}_{n_2}$ and the known Gaussian curvatures $\mathbf{k}_{n_1+1}$, $\cdots$, $\mathbf{k}_{n_2}$ to the right-handed side, we can rewrite the system as follows

$$\begin{cases} M_{n_0} \dfrac{\partial \mathcal{X}_{n_0}(t)}{\partial t} + L_{n_0}^{(p)} \mathcal{X}_{n_0}(t) + L_{n_1}^{(m)} \mathcal{H}_{n_1}(t) + L_{n_1}^{(g)} \mathcal{K}_{n_1}(t) = B^{(0)}, \\ M_{n_1}^{(1)} \mathcal{H}_{n_1}(t) + L_{n_0}^{(1)} \mathcal{X}_{n_0}(t) = B^{(1)}, \\ M_{n_1}^{(1)} \mathcal{K}_{n_1}(t) + L_{n_0}^{(2)} \mathcal{X}_{n_0}(t) = B^{(2)}. \end{cases} \tag{22}$$

Note that $M_{n_0}^{(1)}$ and $M_{n_1}^{(2)}$ are symmetric and positive definite matrices.

4.2   Finite Element Spaces

Now let us introduce two finite element spaces E_h and F_h for representing the surfaces and the curvatures, respectively. Let

$$E_h = \operatorname{span}[\phi_1, \cdots, \phi_n], \quad F_h = \operatorname{span}[\psi_1, \cdots, \psi_n],$$

where the basis functions ϕ_i and ψ_i are defined for the vertex $\mathbf{x}_i$. In our implementation, ϕ_i is defined as the limit function of the Loop's subdivision scheme for the control values one at $\mathbf{x}_i$ and zero at other vertices. It is known that ϕ_i is a quartic box spline. If vertex $\mathbf{x}_i$ is irregular, local subdivision is needed around $\mathbf{x}_i$ until the parameter values of interest are interior to a regular patch (see [1] for details). The size of the support of ϕ_i is two. Our ψ_i is defined as a piecewise linear function which has the values one at $\mathbf{x}_i$ and zero at other vertices. Hence the size of the support of ψ_i is one. Since ψ_i is a piecewise linear function, its evaluation is straightforward. For evaluating ϕ_i, we employ the fast computation scheme proposed by Stam [23].

4.3   Temporal Direction Discretization

Suppose we have approximate solutions $\mathcal{X}_{n_0}^{(k)} = \mathcal{X}_{n_0}(t_k)$, $\mathcal{K}_{n_1}^{(k)} = \mathcal{K}_{n_1}(t_k)$ and $\mathcal{H}_{n_1}^{(k)} = \mathcal{H}_{n_1}(t_k)$ of system (22) at t_k. We want to construct approximate solutions $\mathcal{X}_{n_0}^{(k+1)}$, $\mathcal{K}_{n_1}^{(k+1)}$ and $\mathcal{H}_{n_1}^{(k+1)}$ for the next time step $t_{k+1} = t_k + \tau^{(k)}$ using a semi-implicit Euler scheme. That is, the derivative $\frac{\partial \mathcal{X}_{n_0}}{\partial t}$ is replaced by $[\mathcal{X}_{n_0}^{k+1} - \mathcal{X}_{n_0}^k]/\tau^{(k)}$, and all the M, L and B matrices in (22) are computed using the surface data at t_k. Such a treatment yields a linear system of the equations with $\mathcal{X}_{n_0}^{(k+1)}$, $\mathcal{H}_{n_1}^{(k+1)}$ and $\mathcal{K}_{n_1}^{(k+1)}$ as unknowns:

$$\begin{bmatrix} M_{n_0} + \tau^{(k)} L_{n_0}^{(p)} & \tau^{(k)} L_{n_1}^{(m)} & \tau^{(k)} L_{n_1}^{(g)} \\ L_{n_0}^{(1)} & M_{n_1}^{(1)} & 0 \\ L_{n_0}^{(2)} & 0 & M_{n_1}^{(1)} \end{bmatrix} \begin{bmatrix} \mathcal{X}_{n_0}^{(k+1)} \\ \mathcal{H}_{n_1}^{(k+1)} \\ \mathcal{K}_{n_1}^{(k+1)} \end{bmatrix} = \begin{bmatrix} \tau^{(k)} B^{(0)} + M_{n_0}^{(1)} \mathcal{X}_{n_0}^{(k)} \\ B^{(1)} \\ B^{(2)} \end{bmatrix}. \tag{23}$$

The coefficient matrix of this system is highly sparse, hence an iterative method for solving it is desirable. We use Saad's iterative method [20], named GMRES, to solve the system. The experiment shows that this iterative method works very well.

5   Examples

In this section, we give several examples to show the strength of the proposed approach. Considering the approach is general, there are infinitely many possibilities for choosing f as well as the geometric models. The examples provided here are just a few of them. In the illustrative figures in this section, τ and T stand for the temporal step-size and the number of iterations used. We present several application examples, including surface blending, hole filling and surface mesh refinement.

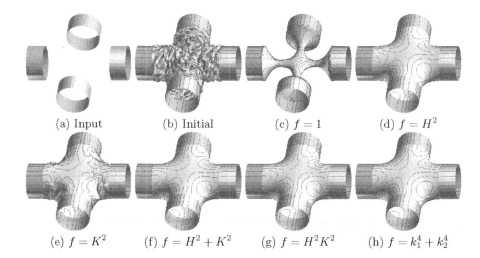

(a) Input   (b) Initial   (c) $f = 1$ (d) $f = H^2$

(e) $f = K^2$ (f) $f = H^2 + K^2$ (g) $f = H^2 K^2$ (h) $f = k_1^4 + k_2^4$

Fig. 2. Surface Blending: The input is four cylinders to be blended as shown in figure (a). The initial blending surface is shown in figure (b). We evolve the initial surface using the geometric flows with different Lagrange functions $f(H, K)$. Figures (c)-(h) are the evolution results of the corresponding flows, where figure (c) is the result after 135 iterations with temporal step-size 0.01 and figures (d)–(h) show the results after 1000 iterations with temporal step-size 0.00001.

1. Surface Blending

Given a collection of surface meshes with boundaries, we construct a fair surface to blend the meshes with the G^0 or G^1 continuity at the boundaries. The aim is to observe the smoothness at the blending boundaries. Fig. 2 shows some the blending results, where (a) is the input surface model to be blended. (b) shows the initial blending surface construction with noise added. (c)–(h) show the results using different f. It could be observed that all the used fourth order flows yield smooth joining bending surfaces at the boundaries. But the result yielded by $f = K^2$ is not good at all. The curves on these surfaces are isophotes. The smooth isophotes imply that the surfaces are at least G^1 smooth.

2. Hole Filling

Given a surface mesh with holes, we construct fair surfaces to fill the holes with the G^0 or G^1 continuity on the boundaries. Fig 3 shows such an example, where a bunny mesh with a lot of complex shaped holes is given (figure (a)). An initial G^0 minimal surface filler construction of the holes is shown in (b), which is generated by the geometric flow with Lagrange function $f = 1$. The fair blending surface (figure (c)) is generated by taking $f = H^2$. The mesh of (d) is the same as (c) but shaded in one color. It is easy to see that the fourth order flow used generate G^1 smooth filling surface, while the second order flow yields G^0 smooth surface.

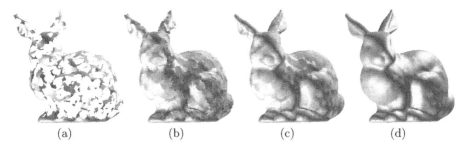

Fig. 3. Hole filing: (a) shows a bunny mesh riddled with all complex shaped holes. (b) shows a minimal surface filler construction. (c) (shaded in different colors) and (d) (shaded in one color) are the faired filler surfaces, after 50 iterations, generated by taking the Lagrange function $f = H^2$. The time step-length is chosen to be 10^{-6}.

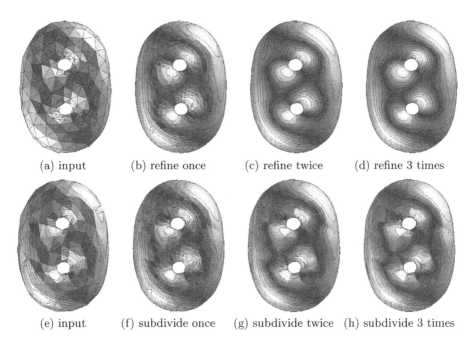

Fig. 4. Mesh refinement: (a) shows an input initial triangular mesh. (b)-(d) are the iterative refinements, where (b), (c) and (d) are generated using $\tau = 0.005, 0.0001, 0.000001$, respectively. The iteration numbers are $15, 50, 50$ respectively. (e) shows the same mesh as (a) but with isophotes. (f)–(h) are the repeatedly subdivision results using Butterfly subdivision scheme.

3. Surface Mesh Refinement

Mesh refinement is a process of subdividing each triangles into several subtriangles by inserting new vertices at certain places. For simplicity, we subdivide each triangle into 4 subtriangles by inserting a new vertex for each edge. The position of the newly inserted vertices are determined by the geometric PDEs. The original vertices are fixed.

The first row of Fig. 4 shows a sequence of meshes, where (a) shows the input triangular surface mesh. Figures (b)-(d) show the iteratively subdivision results using a geometric PDE by taking $f = H^2$. For comparing with the stationary interpolatory subdivision scheme, we present in the second row the subdivision results of Butterfly scheme. The isophotes on the surfaces illustrate that the geometric PDE approach yields high quality surfaces.

6   Conclusions

We have developed a mixed finite method for solving a general form fourth order geometric partial differential equation. We applied the proposed method to solve several surface modeling problems, including surface blending, hole filling and mesh refinement with the G^1 continuity. The used geometric partial differential equation is universal, containing several well-known geometric partial differential equations as its special cases. The proposed method is general which can be used to construct surfaces for geometric design as well as simulate the behaviors of various geometric PDEs. Experimental results have shown that the proposed approach can be used to handle a large number of geometric PDEs and the numerical algorithm is simple, efficient and gives very desirable results.

References

1. Bajaj, C., Xu, G.: Anisotropic diffusion of surface and functions on surfaces. ACM Transaction on Graphics 22(1), 4–32 (2003)
2. Bajaj, C., Xu, G., Warren, J.: Acoustics Scattering on Arbitrary Manifold Surfaces. In: Proceedings of Geometric Modeling and Processing, Theory and Application, Japan, pp. 73–82 (2002)
3. Bloor, M.I.G., Wilson, M.J.: Generating blend surfaces using partial differential equations. Computer Aided Design 21(3), 165–171 (1989)
4. Bloor, M.I.G., Wilson, M.J.: Using partial differential equations to generate free-form surfaces. Computer Aided Design 22(4), 221–234 (1990)
5. Chavel, I.: Riemannian Geometry – a Modern Introduction. Cambridge University Press, Cambridge (1993)
6. Cirak, F., Ortiz, M.: C^1-conforming subdivision elements for finite deformation thin-shell analysis. Internat. J. Numer. Methods Engrg. 51(7), 813–833 (2001)
7. Cirak, F., Ortiz, M., Schroder, P.: Subdivision Surfaces: A New Paradigm for Thin-Shell Finite-Element Analysis. Internat. J. Numer. Methods Engrg. 47, 2039–2072 (2000)
8. Clarenz, U., Diewald, U., Dziuk, G., Rumpf, M., Rusu, R.: A finite element method for surface restoration with boundary conditions. Computer Aided Geometric Design 21(5), 427–445 (2004)
9. Clarenz, U., Diewald, U., Rumpf, M.: Anisotropic geometric diffusion in surface processing. In: Proceedings of Viz2000, IEEE Visualization, Salt Lake City, Utah, pp. 397–405 (2000)

10. Deckelnick, K., Dziuk, G.: A fully discrete numerical scheme for weighted mean curvature flow. Numerische Mathematik 91, 423–452 (2002)
11. Desbrun, M., Meyer, M., Schröder, P., Barr, A.H.: Implicit fairing of irregular meshes using diffusion and curvature flow. In: SIGGRAPH 1999, Los Angeles, USA, pp. 317–324 (1999)
12. do Carmo, M.P.: Differential Geometry of Curves and Surfaces. Prentice-Hall, Englewood Cliffs (1976)
13. do Carmo, M.P.: Riemannian Geometry. Birkhäuser, Boston, Basel, Berlin (1992)
14. Du, H., Qin, H.: Direct manipulation and interactive sculpting of PDE surfaces. 19(3), 261–270 (2000)
15. Du, H., Qin, H.: Dynamic PDE-based surface design using geometric and physical constraint. Graphical Models 67(1), 43–71 (2005)
16. Dziuk, G.: An algorithm for evolutionary surfaces. Numerische Mathematik 58, 603–611 (1991)
17. Giaquinta, M., Hildebrandt, S.: Calculus of Variations. A Series of Comprehensive Studies in Mathematics, vol. I(310). Springer, Berlin (1996)
18. Kobbelt, L., Hesse, T., Prautzsch, H., Schweizerhof, K.: Iterative Mesh Generation for FE-computation on Free Form Surfaces. Engng. Comput. 14, 806–820 (1997)
19. Lowe, T., Bloor, M., Wilson, M.: Functionality in blend design. Computer-Aided Design 22(10), 655–665 (1990)
20. Saad, Y.: Iterative Methods for Sparse Linear Systems. Second Edition with corrections (2000)
21. Schneider, R., Kobbelt, L.: Generating Fair Meshes with G^1 Boundary conditions. In: Geometric Modeling and Processing, Hong Kong, China, pp. 251–261 (2000)
22. Schneider, R., Kobbelt, L.: Geometric fairing of irregular meshes for free-form surface design. Computer Aided Geometric Design 18(4), 359–379 (2001)
23. Stam, J.: Fast Evaluation of Loop Triangular Subdivision Surfaces at Arbitrary Parameter Values. In: SIGGRAPH 1998 Proceedings (1998), CD-ROM supplement
24. Ugail, H., Bloor, M., Wilson, M.: Techniques for interactive design using the PDE method. ACM Transaction on Graphics 18(2), 195–212 (1999)
25. Xu, G., Pan, Q., Bajaj, C.L.: Discrete surface modelling using partial differential equations. Computer Aided Geometric Design 23(2), 125–145 (2006)
26. Xu, G., Pan, Q.: G^1 Surface Modelling Using Fourth Order Geometric Flows. Computer-Aided Design 38(4), 392–403 (2006)
27. Xu, G., Zhang, Q.: Construction of Geometric Partial Differential Equations in Computational Geometry. Mathematica Numerica Sinica 28(4), 337–356 (2006)
28. Xu, G., Zhang, Q.: A General Framework for Surface Modeling Using Geometric Partial Differential Equations. In: Computer Aided Geometric Design (to appear, 2007)
29. Zhang, Q., Xu, G.: Geometric partial differential equations for minimal curvature variation surfaces. In: Research Report No. ICM-06-03. Institute of Computational Mathematics, Chinese Academy of Sciences (2006)

Geodesic as Limit of Geodesics on PL-Surfaces

André Lieutier[1,2] and Boris Thibert[2]

[1] Dassault Systemes, Aix-en-Provence, France
[2] Laboratoire Jean Kuntzmann, Grenoble, France
andre.lieutier@3ds.com,
Boris.Thibert@imag.fr

Abstract. We study the problem of convergence of geodesics on PL-surfaces and in particular on subdivision surfaces. More precisely, if a sequence $(T_n)_{n\in\mathbb{N}}$ of PL-surfaces converges in distance and in normals to a smooth surface S and if C_n is a geodesic of T_n (*i.e.* it is locally a shortest path) such that $(C_n)_{n\in\mathbb{N}}$ converges to a curve C, we wonder if C is a geodesic of S. Hildebrandt et al. [11] have already shown that if C_n is a shortest path, then C is a shortest path. In this paper, we provide a counter example showing that this result is false for geodesics. We give a result of convergence for geodesics with additional assumptions concerning the rate of convergence of the normals and of the lengths of the edges. Finally, we apply this result to different subdivisions surfaces (such as Catmull-Clark) assuming that geodesics avoid extraordinary vertices.

Keywords: subdivision surfaces, triangulations, PL-surfaces, geodesics, shortest paths, convergence.

1 Introduction

A geodesic is usually defined as a curve on a surface that is locally a shortest path. Geodesics appear naturally in several applications, among which we can mention: i) The modelling of the human heart: the heart left ventricle can be modelled by a family of embedded surfaces; a muscular fiber of the central region of the left ventricle has particular properties and can be considered as a geodesic of one of those surfaces [17, 20]. ii) In the fabrication of composite parts by filament winding, the filament must idealy wind along geodesics [1]. iii) Finally the computation of radar cross sections involves the simulation of creeping ray which follow geodesics of the object [2, 4]. In this context, and since piecewise linear 2-manifolds (denoted by PL-surfaces in the following) are widely used for surface modelling, it is natural to consider the modelling of geodesics on surfaces and in particular on PL-surfaces.

We distinguish the geodesics from the more restricted class of shortest paths. A shortest path is a curve on the surface that is connecting two points p and q such that any curve on the surface connecting p and q is longer. A geodesic is a curve on a surface whose length does not decrease if it is pertubed in a small neighborhood of any point. A shortest path is clearly a geodesic, but the

F. Chen and B. Jüttler (Eds.): GMP 2008, LNCS 4975, pp. 178–190, 2008.

converse is not true (for example, a great circle is a geodesic but not a shortest path of the sphere).

There exist several algorithms that build shortest paths on PL-surfaces [12, 13, 15, 18]. Concerning the geodesics, Pham-Trong et al. have also proposed an algorithm that builds geodesics on PL-surfaces [19]. In particular, they have also considered a sequence $(T_n)_{n \in \mathbb{N}}$ of PL-surfaces defined by the De Casteljau subdivision for Bezier surfaces that is converging to a Bezier surface S. On each T_n, they build a geodesic C_n whose sequence converges to a curve C. The natural question is then to wonder whether C is a geodesic of S.

The convergence of geodesics has already been studied in the case of shortest paths by Hildebrandt et al. [11] and Memoli et al. [16]. They show that if a sequence $(T_n)_{n \in \mathbb{N}}$ of PL-surfaces converges in Hausdorff distance to S, if the normals of T_n also converge to the normals of S, then the limit curve of a sequence of shortest paths is a shortest path of S. In this paper, we provide a counter-example showing that this result does not hold anymore for geodesics.

It is worth noting that the result of convergence of Hildebrandt et al. [11] cannot be used in some applications: for example, in the modelling of the human heart, the curves modelling the fibers are closed and are not shortest paths [17]. Furthermore, this result cannot be used to validate the algorithm given in [19]: indeed, Pham Trong and her coauthors build a sequence of geodesics that are not shortest paths in general.

The main result of this paper deals with convergence for geodesics. More precisely, we suppose, as for the result with shortest paths given in [11], that the sequence $(T_n)_{n \in \mathbb{N}}$ of PL-surfaces converges in Hausdorff distance and in normals to a smooth surface S. In addition, we also suppose that there exist two constants K_1 and K_2 independant on n such that the length of the edges of T_n is greater than $\frac{K_1}{2^n}$ and the maximal angle between the normals of T_n and the normals of S is less than $\frac{K_2}{2^n}$. In other words, the rate of convergence of the length of the edges cannot be faster than the rate of convergence of the normals.

This result is then applied to PL-surfaces T_n that follow subdivision schemes, such as for example schemes for B-spline surfaces, or Catmull-Clark schemes, or Bezier surfaces. In particular, our results validate the algorithm of Pham-Trong and her coauthors [19] that builds geodesics on subdivision surfaces.

It is interesting to note that shortest paths and geodesics do not deal with notions of the same order: the notion of shortest path relies on the notion of length, which is a quantity related to the first derivative. However, since a geodesic is defined locally, it depends on the infinitesimal variation of the length, which is a notion of second order.

In Section 3 we give a counter example showing that the results of convergence for shortest paths of [11] does not hold anymore for geodesics. We then give a result of convergence for geodesics with additional assumptions. We show in Section 4 that this result can be applied to several subdivision schemes (and in particular to the algorithm proposed in [19]). The last section gives an overview of the proof of the main result.

We acknowledge Cédric Gérot for his advises concerning subdivision surfaces.

2   Definitions

In the following, we will refer to triangulations instead of PL-surfaces.

2.1   Smooth Surfaces

In the following, a smooth surface means a C^2 surface which is regular, oriented, compact with or without boundary. We have the following proposition [10]

Proposition 1. *Let S be a smooth compact surface of $\mathbb{R}^3$. Then there exists an open set U_S of $\mathbb{R}^3$ containing S and a continuous map ξ from U_S onto S satisfying the following: if p belongs to U_S, then there exists a unique point $\xi(p)$ realizing the distance from p to S (ξ is nothing but the orthogonal projection onto S).*

This proposition allows to introduce the following notion introduced by H. Federer [10]: The *reach of a surface* S is the largest $r > 0$ for which ξ is defined on the open tubular neighborhood $U_r(S)$ of radius r of S.

2.2   Triangulations

A triangulation T is a connected topological 2-manifold made of a finite union of triangles of $\mathbb{R}^3$, such that the intersection of two triangles is either empty, or equal to a vertex, or equal to an edge.

2.3   Curves

In the following a curve C means a lipschitz parametrized curve $C : [0,1] \to \mathbb{R}^3$. Its length is denoted by $l(C)$. Similarly, for $0 \le t_a < t_b \le 1$ we denote by $l(C, t_a, t_b)$ the length of the curve C restricted to $[t_a, t_b]$. As a particular case of Rademacher theorem, we know that C is differentiable almost everywhere. Moreover, it is the integral of its derivative. Whenever it exists, we denote by $C'(t)$ the derivative of C at t.

• We say that C has a *uniform parametrization* if it satisfies for almost every $t \in [0,1]$ $\|C'(t)\| = l(C)$.
• A curve $C : [0,1] \to M \subset \mathbb{R}^3$ with uniform parametrization is said to be a geodesic of a lipschitz surface M (M can be a triangulation or a smooth surface) if it locally minimizes the length, *i.e.* if for every $t \in [0,1]$, there exists $0 \le t_a \le t \le t_b \le 1$ (where $t_a < t$ if $t > 0$ and $t_b > t$ if $t < 1$), such that any lipschitz curve $\widetilde{C} : [0,1] \to M$ such that $\widetilde{C}(0) = C(t_a)$ and $\widetilde{C}(1) = C(t_b)$ satisfies

$$l(C, t_a, t_b) \le l(\widetilde{C}).$$

The geodesic is said to be interior if for every $t \in [0,1]$, $C(t)$ is interior to the surface M. The geodesic C is a shortest path, if the length of any curve on M connecting $C(0)$ and $C(1)$ has a length greater than $l(C)$.

2.4   Characterisation of Geodesics on Triangulations

Let C be a polygonal curve of a triangulation T. Then C is a geodesic of T if and only if:

i) C it is a straight line on each triangle. (The vertices p of C then belongs to the edges of T or are vertices of T.)

ii) If p belongs to the interior of an edge of T, then the incident and refracted angles of C at p are equal (see Figure 1).

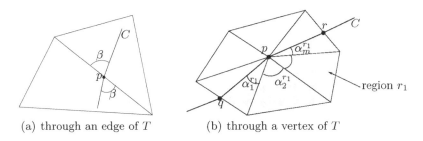

(a) through an edge of T (b) through a vertex of T

Fig. 1. A geodesic C of a $3D$ triangulation T

iii) If p is a vertex of T, then C separates the set of the triangles of T containing p into two connected regions r_1 and r_2 (see Figure 1). If one denotes by $\alpha_p^{r_1}$ the sum of the angles $\alpha_i^{r_1}$ of the triangles of region r_1 at p (*resp.* by $\alpha_p^{r_2}$ the angles $\alpha_i^{r_2}$ of the triangles of region r_2 at p), one has $\alpha_p^{r_1} \geq \pi$ and $\alpha_p^{r_2} \geq \pi$.

3   Convergence of Geodesics

The main result of this paper is given in this section. We first show , by building a counter-example, that the result of convergence for shortest paths of [11] does not hold anymore for geodesics in general. Then, we give a general result of convergence for geodesics that needs additional assumptions.

3.1   Counter-Example

Figure 2 shows a triangulation T_n for some n. The triangulation overlaps with the horizontal plane Π_H of equation $z = 0$ outside the large circle and inside the small one. In the ring between the two circles, it is made of 4^n identical small "roof shaped" bumps detailed on the right of Figure 2. First remark that if the angles β_i (see Figure 2) satisfy the equation $\beta_1 + \beta_2 + \beta_3 + \beta_4 > \pi$, then the polygonal curve wrapped around the outer circle is a geodesic (see the characterisation of geodesics in Section 2).

Detail of the construction
The points d_n^1, d_n^2, p_n and m_n are on the plane Π_H while the points t_n^1 and t_n^2 stand at some height above the plane. The faces $d_n^1 t_n^1 d_n^2$, $p_n t_n^2 m_n$, $d_n^1 t_n^1 t_n^2 m_n$ and

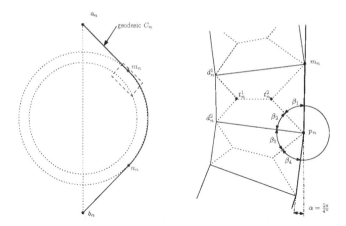

Fig. 2. Triangulation T_n and geodesic C_n seen from above: on the left we see the whole surface; the region in the dashed rectangle is depicted on the right

$d_n^2 t_n^1 t_n^2 p_n$ are planar and all make a slope s_n with Π_H. If we take $s_n = \frac{3}{2^n}$, one has, for each $n \in \mathbb{N}$, $\beta_1 + \beta_2 + \beta_3 + \beta_4 > \pi$.

Observe that the sequence $(T_n)_{n \in \mathbb{N}}$ of triangulations converges toward the plane in the Hausdorff sense. Furthermore the normals of T_n tend to the normals of the plane. The shortest path on T_n between the point a_n and b_n is the straight line (dotted line on Figure 2). However, the line $a_n m_n n_n b_n$ (denoted by C_n), wraped around the circle between m_n and n_n is a local minimum, that is a geodesic, between a_n and b_n (see the characterisation of geodesics in Section 2). These geodesics converge in the Hausdorff sense toward a curve C composed of two lines and an arc of circle.

We can notice that the triangulations T_n and the curves C_n satisfy all the assumptions of the result of [11], except that C_n is a geodesic (and not a shortest path). However the limit curve C is not a geodesic. This counter example implies that the convergence in distance and in normals of T_n to S is not sufficient to expect a result of convergence for geodesic.

3.2   Convergence Toward a Geodesic

The following theorem states that if a sequence $(T_n)_{n \in \mathbb{N}}$ of triangulations is converging to a smooth surface S, then, under reasonable assumptions, a sequence of geodesics of T_n is converging to a geodesic of S.

Theorem 1. *Let S be a smooth surface of $\mathbb{R}^3$, r denote the reach of S, and (T_n) be a sequence of triangulations. Let K, $\tilde{K}$, θ_{min} be positive constants and (d_n) a sequence of real numbers converging to 0. Suppose that for every n:*

a) T_n belongs to the tubular neighborhood $U_r(S)$ of radius r of S;
b) for every $m \in T_n$, $\|m - \xi(m)\| \le d_n$;

c) for every $m \in T_n$, the angle between any triangle Δ containing m and the tangent plane $\Pi^S_{\xi(m)}$ of S at $\xi(m)$ is smaller than $\frac{K}{2^n}$;

d) the lengths of the edges of T_n are greater than $\frac{\tilde{K}}{2^n}$;

e) all the angles of T_n are greater than θ_{min};

Let $(C_n)_{n \in \mathbb{N}}$ be a sequence of polygonal curves $C_n : [0,1] \to \mathbb{R}^3$ with uniform parametrization such that C_n is an interior geodesic of T_n and $\xi(C_n)$ does not intersect the boundary of S. If $(C_n)_{n \in \mathbb{N}}$ converges toward a curve C in the sup norm sense, then C is of class $\mathcal{C}^2$ and is a geodesic of S.

The sketch of the proof of this theorem is given in Section 5. For a complete proof, we can refer to the research report [14].

Remark 1. The result of Theorem 1 still holds if the sequence $(C_n)_{n \in \mathbb{N}}$ does not converge to a curve C, but if we suppose that the sequence $(l(C_n))_{n \in \mathbb{N}}$ is uniformly bounded. Indeed, in that case, we can show that the family $\{C_n, \ n \in \mathbb{N}\}$ is equicontinuous. The Arzela-Ascoli theorem (see [5] for example) then implies that a subsequence $(C_{n_k})_{k \in \mathbb{N}}$ of $(C_n)_{n \in \mathbb{N}}$ uniformly converges to a curve C. Theorem 1 can then be applied to $(C_{n_k})_{k \in \mathbb{N}}$.

4   Application to Subdivision Surfaces

The previous theorem can be easily applied to subdivision surfaces. In this section, we first give a general corollary, Corollary 1, that can be easily applied to several subdivision schemes. We then show that the result of convergence for geodesics holds when the triangulations follow a subdivision scheme for either B-splines surfaces, or Catmull-Clark schemes, or Bezier surfaces. We first need to give a few definitions.

Let $(P_n)_{n \in \mathbb{N}}$ be a sequence of parametrized triangulations $P_n : [0,1]^2 \to \mathbb{R}^3$ that is converging to a paramatrized smooth surface $f : [0,1]^2 \to \mathbb{R}^3$. The parameter domain $[0,1]^2$ of each P_n can be triangulated so that P_n is linear on each triangle of $[0,1]^2$.

– We say that the parameter domain of P_n is a triangular grid if its vertices are $p^{i,j}_n = \left(\frac{i}{2^n}, \frac{j}{2^n}\right)$ (where $i,j \in \{0,..2^n\}$) and the edges are $p^{i,j}_n p^{i+1,j}_n$, $p^{i,j}_n p^{i,j+1}_n$ and $p^{i,j}_n p^{i+1,j+1}_n$.

– We say that $(P_n)_{n \in \mathbb{N}}$ uniformly converges to a function f with rate of convergence $\frac{1}{2^n}$ if there exists $K \in \mathbb{R}$ such that:

$$\sup_{(u,v) \in [0,1]^2} \|P_n(u,v) - f(u,v)\| \leq \frac{K}{2^n}.$$

– We say that $(P_n)_{n \in \mathbb{N}}$ uniformly converges in derivative to f with rate of convergence $\frac{1}{2^n}$ if there exists $K > 0$, such that for any $n \in \mathbb{N}$:

$$\sup_{\substack{i \in \{0,..,2^n-1\} \\ j \in \{0,...,2^n\}}} \left\| 2^n \left[P_n\left(\frac{i+1}{2^n}, \frac{j}{2^n}\right) - P_n\left(\frac{i}{2^n}, \frac{j}{2^n}\right) \right] - \frac{\partial f}{\partial u}\left(\frac{i}{2^n}, \frac{j}{2^n}\right) \right\| \leq \frac{K}{2^n},$$

and

$$\sup_{\substack{i\in\{0,..,2^n\}\\j\in\{0,..,2^n-1\}}} \left\| 2^n \left[P_n\left(\frac{i}{2^n},\frac{j+1}{2^n}\right) - P_n\left(\frac{i}{2^n},\frac{j}{2^n}\right)\right] - \frac{\partial f}{\partial v}\left(\frac{i}{2^n},\frac{j}{2^n}\right) \right\| \le \frac{K}{2^n}.$$

Corollary 1. *Let $(P_n)_{n\in\mathbb{N}}$ be a sequence of parametrized triangulations $P_n : [0,1]^2 \to \mathbb{R}^3$ and $f : [0,1]^2 \to \mathbb{R}^3$ be a parametrized surface of class $\mathcal{C}^2$, such that:*

a) the parameter domain of each P_n is a triangular grid,
b) $(P_n)_{n\in\mathbb{N}}$ uniformly converges to f with rate of convergence $\frac{1}{2^n}$,
c) $(P_n)_{n\in\mathbb{N}}$ uniformly converges in derivative to f with rate of convergence $\frac{1}{2^n}$,
d) f is regular, i.e. $\forall (u,v)\in[0,1]^2 \; \frac{\partial f}{\partial u}(u,v)\wedge\frac{\partial f}{\partial v}(u,v)\ne 0$.

Let $(C_n)_{n\in\mathbb{N}}$ be a sequence of polygonal curves $C_n : [0,1] \to \mathbb{R}^3$ with uniform parametrization such that C_n is an interior geodesic of P_n. If $(C_n)_{n\in\mathbb{N}}$ converges in the sup norm sense toward a curve C which is interior to S, then C is of class $\mathcal{C}^2$ and is a geodesic of S.

The proof of this corollary can be found in [14].

The previous corollary can be easily applied to some subdivision schemes. As an example, we give the following corollary concerning subdivision schemes for B-spline surfaces of arbitrary degree [9] (we call B-spline surface a smooth surface expressed in the B-spline basis). First remark that the subdivision scheme for a B-spline surface generates a sequence of quadrangulations. Each quadrangulation can be considered as a triangulation by dividing each quadrangle into two triangles.

Corollary 2. *Let $(P_n)_{n\in\mathbb{N}}$ be a sequence of triangulations (or quadrangulations) defined by the subdivision scheme for a B-spline surface of arbitrary degree that is converging to a regular $\mathcal{C}^2$ B-spline f (i.e. satisfies assumption d) of Corollary 1).*

Let $(C_n)_{n\in\mathbb{N}}$ be a sequence of polygonal curves $C_n : [0,1] \to \mathbb{R}^3$ with uniform parametrization such that C_n is an interior geodesic of P_n. If $(C_n)_{n\in\mathbb{N}}$ converges in the sup norm sense toward a curve C which is interior to S, then C is of class $\mathcal{C}^2$ and is a geodesic of S.

Proof. The assumptions a) and d) of Corollary 1 are clearly satisfied. Let $u \in [0,1]$. The polygonal curve $v \to P_n(u,v)$ follows the subdivision scheme for a B-spline curve and is uniformly converging to the function $f_u : v \to f(u,v)$. More precisely, there exists $K_u \in \mathbb{R}$ such that for every $n \in \mathbb{N}$ one has (see Theorem 4.12 of [8] or Corollary 3.3 of [7]):

$$\sup_{v\in[0,1]} \|P_n(u,v) - f(u,v)\| \le \frac{K_u}{2^n}.$$

In fact, one can show that K_u does not depend on u: let us denote by $P_{i,j} = P_0(u_i,v_j)$ the poles of the initial control mesh (where $i,j \in \{0,..M\}$). The kth pole $P_{u,k}$ of f_u is the evaluation at u of the B-spline whose control net is $P_{0,k},...P_{M,k}$. That implies that $P_{u,k}$ belongs to the convex hull of $P_{0,k},...P_{M,k}$ and then to the convex hull of the $P_{i,j}$.

Since K_u only depends on the maximum distance between two poles of the control net of f_u (see the proof of Theorem 4.11 of [8]), it only depends on the diameter of the convex hull of the $P_{i,j}$ and does not depend on u.

This implies that $(P_n)_{n \in \mathbb{N}}$ uniformly converges toward f. Assumption b) is then proved. Similarly, assumption c) is also proved on the divided difference scheme (see for example [21] for details on divided difference schemes).

Remark 2. Corollary 2 directly implies that the result of convergence still holds for Catmull-Clark schemes if the limit curve C does not contain extraordinary points of S (see [3] for details on the Catmull-Clark scheme). By extraordinary point on the limit surface, we precisely mean the limit of the sequence of vertices corresponding to an extraordinary point of the triangulation through successive subdivisions. Indeed, after a sufficient number of iterations, the curve traverses only a finite number of bicubic B-splines patches where Corollary 2 can be applied.

Remark 3. A proof similar to the one of Corollary 2 shows that this result also holds for Bezier surfaces and their successive control nets defined by the De Casteljau algorithm.

5   Overview of the Proof of Theorem 1

We give in this section an overview of the proof. We first need to introduce definitions. We then give intermediate results in sections 5.2, 5.3 and 5.4. We finally give the sketch of the proof in section 5.5. The complete proof of Theorem 1 and of these intermediate results can be founded in the research's report [14].

5.1   Preliminary Definitions

Let ϵ be smaller than the reach of S. Let T be a triangulation such that ξ induces an injection from T to S. Let denote by $R(T)$ the set of polygonal curves C of T that are linear on each triangle of T, to be more precise, if τ is a triangle (a *triangle* is defined here as a closed simplex, i.e. containing its boundary edges and vertices) of T, the image of each connected component of $\{t \in [0,1], C(t) \in \tau\}$ is a line segment: geodesics on T trivially satisfy this condition.

Let $C \in R(T)$ be a polygonal curve that belongs to the tubular neighborhood $V_\epsilon(S)$ of radius ϵ of S. In the following, if $m \in S$, we denote by P_m^S the orthogonal projection onto the tangent plane Π_m^S of S at the point m.

- The *total curvature of C* is given by:

$$TC_{3D}(C) = \sum_{p \; vertex \; of \; C} \beta_{dev}^{3D}(p),$$

where $\beta_{dev}^{3D}(p)$ is the deviation angle of C at the vertex p (see Figure 3). We call here deviation angle of a polygonal curve $P_1, .., P_n$ at the vertex P_i the angle $\angle \overrightarrow{P_{i-1}P_i}, \overrightarrow{P_iP_{i+1}}$.

Similarly, for $0 \leq t_a < t_b \leq 1$ we denote by $TC_{3D}(C, t_a, t_b)$ the total curvature of the curve C restricted to $[t_a, t_b]$.

• The *tangent total curvature of C with respect to S* is defined by

$$TC_{Tan}^{TS}(C) = \sum_{p \ vertex \ of \ C} \beta_{dev}^{TS}(p),$$

where $\beta_{dev}^{TS}(p)$ is the deviation angle of $P_{\xi(p)}^{S}(C)$ at the vertex $\xi(p)$ (see Figure 3).

Similarly, for $0 \leq t_a < t_b \leq 1$ we denote by $TC_{Tan}^{S}(C, t_a, t_b)$ the tangent total curvature of the curve C restricted to $[t_a, t_b]$.

• Let $\sharp C$ be the number of intersections between C and the edges of T. More precisely, if one denotes by E the set of edges of T and $N_{CC}(X)$ the number of connected components of a set X:

$$\sharp C = \sum_{e \in E} N_{CC}(C([0,1]) \cap e).$$

Similarly, for $0 \leq t_a < t_b \leq 1$ we denote by $\sharp(C, t_a, t_b)$ the number of intersections between the curve C restricted to $[t_a, t_b]$ and the edges of T.

5.2   Majoration of the Deviation Angles of a Geodesic

Let T_n be a triangulation close enough to a smooth surface S (so that the map ξ of Proposition 1 is well defined on T). The following proposition gives a bound on the deviation angle at a vertex of a geodesic of a triangulation that depends on the angles between all the adjacent triangles. Furthermore, it also gives a bound on the deviation angle of the orthogonal projection of the geodesic on a plane tangent to the surface S.

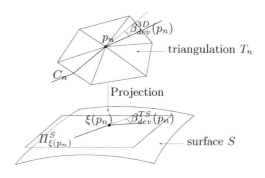

Fig. 3. Deviation angle of a geodesic and of its orthogonal projection (Proposition 2)

Proposition 2. *There exists K_1, such that for every n: if C_n is a geodesic of T_n and p_n is a vertex of C_n, then we have:*

$$\beta_{dev}^{3D}(p_n) \leq K_1 \, \alpha_n, \quad and \quad \beta_{dev}^{TS}(p_n) \leq K_1 \, \alpha_n^2,$$

where α_n is the maximal angle between all the triangles of T_n containing p_n and $\Pi_{\xi(p_n)}^{S}$.

Remark 4. We can note that $\beta_{dev}^{TS}(p_n)$ is much smaller than $\beta_{dev}^{3D}(p_n)$. Roughly speaking, the normal component of $\beta_{dev}^{3D}(p_n)$ is in $O(\alpha_n)$ and the tangential component is in $O(\alpha_n^2)$ (we consider here a normal to a triangle of T_n). The calculus of $\beta_{dev}^{TS}(p_n)$ then involves the projection of a unitary normal to a triangle of T_n onto $\Pi_{\xi(p_n)}^S$, whose norm is less than $\sin(\alpha_n) \le \alpha_n$.

5.3 Number of Intersections of a Polygonal Curve with Edges of a Triangulation

The following proposition indicates that the number of times a polygonal curve C_n intersects the edges of a triangulation T_n depends on its length and on its total curvature. In other words, if C_n is not too long and if it does not "turn too much" then it does not intersect too many edges (see Figure 4).

Proposition 3. *There exists a constant K_2, such that for any curve $C_n \in R(T_n)$, one has:*

$$\sharp C_n \le K_2 \left[1 + TC_{Tan}^{TS}(C_n) + 2^n\, l(C_n) \right].$$

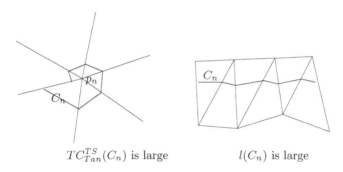

$$TC_{Tan}^{TS}(C_n) \text{ is large} \qquad\qquad l(C_n) \text{ is large}$$

Fig. 4. $\sharp C_n$ depends on $TC_{Tan}^{TS}(C_n)$ and $l(C_n)$(Proposition 3)

5.4 A Sufficient Condition for the Regularity of the Limit Curve

The following proposition gives a sufficient condition on a sequence of polygonal curves so that its limit curve is of class $C^{1,1}$ (*i.e.* the function is differentiable, and its derivation is lipschitz).

Proposition 4. *Let $(C_n)_{n\in\mathbb{N}}$ be a sequence of polygonal curves $C_n : [0,1] \to \mathbb{R}^3$, with uniform parametrization that converges toward a non constant curve C in the sup norm sense. If*

$$\exists k_1, k_2, \; \forall t_a, t_b \in [0,1] \quad TC_{3D}(C_n, t_a, t_b) \le \frac{k_1}{2^n} + k_2\, l(C_n, t_a, t_b),$$

then the curve C has uniform parametrization, is of class $C^{1,1}$ and has curvature bounded by k_2. Moreover $l(C) = \lim_{n\to\infty} l(C_n)$ and for any $t_0 \in (0,1)$:

$$\lim_{n\to\infty} \frac{dC_n}{dt^+}(t_0) = \frac{dC}{dt}(t_0).$$

5.5   Sketch of Proof of Theorem 1

We take the notations of Theorem 1. Let α_n be the smallest real number such that for every $m \in T_n$, the angle between any triangle Δ containing m and the tangent plane $\Pi^S_{\xi(m)}$ of S at $\xi(m)$ is smaller than α_n. By assumption, we have $\alpha_n \le K/2^n$.

Step 1: By combining Propositions 2 and 3, we have the following circular in-equality:

$$TC^{TS}_{Tan}(C_n, t_a, t_b) \le K_1 \, \alpha_n^2 \, \sharp(C_n, t_a, t_b)$$
$$\le K_1 \, K_2 \, \alpha_n^2 \left[1 + TC^{TS}_{Tan}(C_n, t_a, t_b) + 2^n \, l(C_n, t_a, t_b) \right].$$

Therefore, for some constants K_3 and K_4 independant of t_a and t_b, one has:

$$TC^{TS}_{Tan}(C_n, t_a, t_b) \le \frac{K_3}{2^n} l(C_n, t_a, t_b) + \frac{K_4}{4^n}. \tag{1}$$

Step 2: By using again Propositions 2 and 3, one can show that for some constant K and K':

$$TC_{3D}(C_n, t_a, t_b) \le K \, l(C_n, t_a, t_b) + K' \, \frac{1}{2^n}. \tag{2}$$

Proposition 4 then implies that the limit curve C is of class $\mathcal{C}^{1,1}$.

Step 3 We consider a point $p_0 = C(t_0)$. In this step, by using equations (1) and (2), we prove that the projection $P^S_{C(t_0)} \circ C$ of C on the plane tangent to S at $C(t_0)$ is twice differentiable at t_0 and that $\left(P^S_{C(t_0)} \circ C \right)''(t_0) = 0$. One can show that this implies that C is of class $\mathcal{C}^2$, then that it has zero geodesic curvature and then that it is a geodesic of S [6].

6   Conclusion and Future Works

The main result of this work gives sufficient conditions for a sequence of geodesics on PL-surfaces to converge toward a geodesic on a smooth limit surface. We believe this is a significant step toward an effective notion of geodesic: indeed, the usual definition of geodesic is not effective because it relies on the notions of smooth curves and surfaces and on the pointwise curvature which can not be exactly represented on computers. Our main theorem states that the usual notion of geodesic coincides with the limit of a sequence of PL-curves. Therefore, by using our result, a realistic algorithm can output a sequence of curves whose limit is a geodesic of a smooth surface. Notice that, given a smooth surface with bounded curvature, there exists a sequence of PL-surfaces converging to it (and that matches the conditions of our theorem). However, in order to completely get the effective notion of geodesic, one still has to quantify the rate of convergence of this sequence of curves.

We also believe that our result could be improved by relaxing the condition on the edge lengths: indeed, in the counter-example the lengths decrease with the order $\frac{1}{4^n}$ with respect to a decrease rate of $\frac{1}{2^n}$ of the angular convergence. We believe that it is possible to improve the theorem between the $\frac{K}{2^n}$ condition of the theorem and the $\frac{1}{4^n}$ of the counter-example.

Another possible improvement of the result is to suppose that the limit surface is of class $\mathcal{C}^{1,1}$ (instead of $\mathcal{C}^2$). Notice that such a generalisation (and also a weaker condition on the edge lengths in the main theorem) would be very useful for some subdivision surfaces with extraordinary points. Indeed, at extraordinary points, the limit surface of some subdivision surfaces is only of class $\mathcal{C}^{1,1}$. We proved (for example for the Catmull-Clark scheme) that if the limit curve of a sequence of geodesics does not contain extraordinary points, then it is a geodesic. We believe that the result still holds if the limit curve contains extraordinary points.

References

1. Blais, F., Chazal, F.: Modélisation de déviateurs CRT: empilement du fil et tassage, research report for Thomson-Multimdia, 46 pages, ref. 102.02.CC-YKE, 01/07/02
2. Bouche, D., Molinet, F.: Méthodes asymptotiques en électromagnétisme (French) Mathématiques & Applications (Berlin), 16. Springer, Paris, xviii+416 (1994)
3. Catmull, E., Clark, J.: Recursively generated B-spline surfaces on arbitrary topological surfaces. Computer-Aided Design 10(6), 350–355 (1978)
4. Dessarce, R.: Calculs par lancer de rayons, PhD thesis, université Joseph Fourier (1996)
5. Dieudonné, J.: Éléments d'analyse. Tome I: Fondements de l'analyse moderne(French). Cahiers Scientifiques, Fasc. XXVIII Gauthier-Villars, Éditeur, Paris, pp. xxi+390 (1968)
6. Do Carmo, M.P.: Differential geometry of curves and surfaces (Translated from the Portuguese. Prentice-Hall Inc., Englewood Cliffs (1976)
7. Dyn, N.: Subdivision schemes in CAGD. In: Light, W.A. (ed.) Advances in Numerical Analysis, Wavelets, Subdivision Algorithms and Radial Basis Functions, pp. 36–104. Oxford University Press, Oxford (1992)
8. Dyn, N., Levin, D.: Subdivision schemes in geometric modelling. Acta Numerica 11, 73–144 (2002)
9. Farin, G.: Curves and surfaces for computer aided geometric design, a practical guide. Academic Press, Boston, London, Sydney (1988)
10. Federer, H.: Curvature measures. Trans. Amer. Math. Soc. 93, 418–491 (1959)
11. Hildebrandt, K., Polthier, K., Wardetzky, M.: On the Convergence of Metric and Geometric Properties of Polyedral Surfaces. Geometria Dedicata (to appear, 2007)
12. Kimmel, R., Amir, A., Bruckstein, A.M.: Finding shortest paths on surfaces. In: laurent, P.J., LeMéhauté, A., Schumaker, L.L. (eds.) Curves and Surfaces in Geometric Design, A.K. Peters, Wellesley, MA, pp. 259–268 (1994)
13. Kimmel, R., Sethian, J.: Computing geodesic paths on manifolds. Proc. Natl. Acad. Sci. 95(15), 8431–8435
14. Lieutier, A., Thibert, B.: Geodesics as limits of geodesics on PL-surfaces. IMAG's research report n 1086-M (July 2007), https://hal.archives-ouvertes.fr/hal-00160820
15. Maekawa, T.: Computation of shortest paths on free-form parametric surfaces. J. Mechanical Design 118, 499–508 (1996)

16. Memoli, F., Sapiro, G.: Distance functions and geodesics on submanifolds of $\mathbb{R}^d$ and point clouds. SIAM J. APPL. MATH. 65(4), 1227–1260

17. Mourad, A.: Description topologique de l'architecture fibreuse et modélisation mécanique du myocarde. PhD thesis, Institut National Polytechnique de Grenoble (2003)

18. Peyré, G., Cohen, L.: Geodesic Computations for Fast and Accurate Suface Remeshing and Parametrization. Progess in Nonlinear Differential Equations and Their Applications 63, 157–171

19. Pham-Trong, V., Biard, L., Szafran, N.: Pseudo-geodesics on three-dimensional surfaces and pseudo-geodesic meshes. Numer. Algorithms 26(4), 305–315 (2001)

20. Streeter, D.: Gross morphology and fiber geometry of the heart. In: Berne, R.M., Sperelakis, N. (eds.) Handbook of physiology section 2. The Heart (American Physiology Society), pp. 61–112. Williams and Wilkins, Baltimore (1979)

21. Warren, J., Weimer, H.: Subdivision Methods for Geometric Design: A constructive Approach. Morgan Kaufmann, San Francisco (2001)

Hausdorff and Minimal Distances between Parametric Freeforms in $\mathbb{R}^2$ and $\mathbb{R}^3$

Gershon Elber[1] and Tom Grandine[2]

[1] Dept. of Computer Science, Technion – IIT, Haifa 32000, Israel
[2] The Boeing Company, Seattle, USA

Abstract. We present algorithms to derive the precise Hausdorff distance and/or the minimal distance between two freeform shapes, either curves or surfaces, in $\mathbb{R}^2$ or $\mathbb{R}^3$. The events at which the Hausdorff/minimal distance can occur are identified and means to efficiently compute these events are presented. Examples are also shown and the extension to arbitrary dimensions is briefly discussed.

Keywords: Bisectors, Antipodal points, Algebraic constraints, Spline geometry, Collision detection.

1 Introduction and Previous Work

The need to compute the maximal or minimal distance between two entities in $\mathbb{R}^2$ or $\mathbb{R}^3$ emerges in a whole variety of applications. Both collision detection calculations and Haptic interaction can greatly benefit from such black boxes [7]. Force feedback for Haptic devices is typically applied once the interaction tool gets closer to the surface of the approached object and is typically applied in the direction from the closest point on that surface. Similarly, the Hausdorff distance computation plays a major role in any approximation method of a curve or a surface by lower degree curves or surfaces or even piecewise linear approximations, as it provides L_∞ bounds over the approximation.

Given two objects, $\mathcal{O}_1, \mathcal{O}_2 \in \mathbb{R}^n$, the Hausdorff distance between them is defined as:

$$D_H(\mathcal{O}_1, \mathcal{O}_2) = \max\left(\max_{P \in \mathcal{O}_1} \min_{Q \in \mathcal{O}_2} ||P - Q||, \max_{Q \in \mathcal{O}_2} \min_{P \in \mathcal{O}_1} ||P - Q||\right).$$

Figure 1 illustrates this definition using the geometric insight that the Hausdorff distance can sometimes (but not always!) be captured as the last contact point of the offset of one shape with the other shape, and vice versa.

That said, not much can be found on the problem and its solution for freeform polynomial geometry or even for piecewise linear polygonal geometry. In [2], point sampling over two polygonal meshes is proposed as an approximation for the Hausdorff distance between the two polygonal meshes. [2] offered the

F. Chen and B. Jüttler (Eds.): GMP 2008, LNCS 4975, pp. 191–204, 2008.

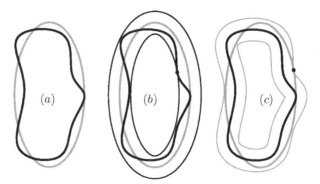

Fig. 1. The Hausdorff distance between two planar curves in (a) could be computed as the maximum offset amount between the last contact point of the offset front of one shape with the second shape (b) and vice versa (c)

'Metro' point sampling tool that is the acceptable tool in the computer graphics community to estimate Hausdorff distances over polygonal meshes.

In [8], bounds on the Hausdorff distance of two given curves are presented. The cases of two implicits, an implicit and a parametric form, and two parametric form are considered but only to give an upper bound. The bound between two parametric forms is the worst offered of the three and is derived by examining the difference vectors of the corresponding control points, bringing both curves into a comparable representation. This approach further assumes the two curves are fairly close to each other, an assumption that might yield poor answers if not.

It is not common that, in principle, a problem is simpler in the piecewise polynomial domain than in the piecewise linear case. The Hausdorff distance computation between two objects is one such problem. The fact that the polygonal mesh is not tangent plane continuous, makes it very difficult to track the exact position when an event of an extreme distance can occur. In [1], the exact Hausdorff distance between points and freeform planar parametric curves is investigated, taking advantage of the fact that the input is C^1 continuous. The events where the Hausdorff distance can occur are then identified and reduced to a set of differential algebraic constraints. The finite solution set of these constraints is then examined for the actual Hausdorff distance. Our work here builds upon [1] and while we follow a similar approach, we will lay out the differences and also go beyond to consider freeform geometry in $\mathbb{R}^3$ and $\mathbb{R}^n$ as well.

This paper is organized as follows. In Section 2, we consider the problem of the Hausdorff distances in the plane. Extensions of the result to $\mathbb{R}^3$ and $\mathbb{R}^n$ are discussed in Section 3. Minimal distances between freeforms are discussed in Section 4 and examples for distance computations in $\mathbb{R}^2$ and $\mathbb{R}^3$ are presented in Section 5. Possible extensions, and computational considerations are discussed in Section 6 and finally, we conclude in Section 7.

2   Hausdorff Distance in the Plane

We following [1], who presents the necessary algebraic constraints for Hausdorff distances in the plane, and express all the events at which the Hausdorff distance could occur at, between two planar C^1 parametric curves.

Let $C_1(r)$, $r \in [0,1]$ and $C_2(t)$, $t \in [0,1]$ be two regular [1] C^1 continuous planar parametric curves. Then,

Definition 1. *A **normal-line** to $C_1(r)$ at the parameter $r = r_0$ is a line through $C_1(r_0)$ that is parallel to the curve's normal, $N_{C_1}(r_0)$.*

The Hausdorff distance could clearly occur at the end (or C^1 discontinuity) points of one of the curves, if the curves are open (or only piecewise C^1). See Figure 2 (a) and (b). This amounts to examining the distance between the end points of the two curves but also to looking for the normal-lines of $C_2(t)$ that go through $C_1(r_0), r_0 = 0, 1$, if any, or vice versa. These normal-lines' locations could be identified by resolving the following algebraic constraint:

$$\langle C_1(r_0) - C_2(t), C_2'(t) \rangle = 0, \tag{1}$$

having one non-linear equation in one unknown, t, to solve for. The Hausdorff distance between a point and and curve in the plane is now a simple problem that could be reduced to examining end-point vs. end-point events as well as the events satisfied by Equation (1). To consider more events at which the Hausdorff distance could occur between two planar curves, we also need the following:

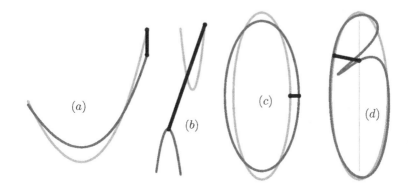

Fig. 2. The Hausdorff distance events (in black) between two curves (in two different gray colors) can occur at either the end points (a), end point of one curve along a normal-line of the other curve (b), antipodal locations (c), or when one (dark-gray) curve intersects with the (think line) self-bisector of the other (light-gray) curve (d)

---

[1] A parametric form is considered regular if its derivatives span the form's tangent space, at every point in the domain. For a curve C, this amounts to the constraints $||C'|| > 0$.

Definition 2. *The line through $C_1(r)$ and $C_2(t)$ is denoted as a* **curves' binormal-line** *at parameters $r = r_0$ and $t = t_0$, if it is a curve normal-line at both $C_1(r_0)$ and $C_2(t_0)$. Points $C_1(r_0)$ and $C_2(t_0)$ are then denoted* **antipodal points**.

The Hausdorff distance between $C_1(r)$ and $C_2(t)$ could also occur at antipodal points of the two curves. See Figure 2 (c) for an example. These antipodal location could be identified using the following set of constraints,

$$\langle C_1(r) - C_2(t), C_1'(r) \rangle = 0,$$
$$\langle C_1(r) - C_2(t), C_2'(t) \rangle = 0, \tag{2}$$

having two equations and two unknowns, t and r, to solve for.

Interestingly enough, these are not the only events at which the Hausdorff distance (event) can occur between two planar curves, and [1] identifies a third case. Let B_i be the self bisector of planar curve C_i. I.e. the locus of points that are equidistant from two different locations on C_i. Then, the Hausdorff distance between $C_1(r)$ and $C_2(t)$ could occur at the locations where $C_1(r)$ intersects B_2 or when $C_2(t)$ intersects B_1 (See Figure 2 (d)). Algebraically speaking, the constraint for $C_1(r)$ to intersect B_2 means that $C_1(r)$ is on the intersection of two independent normal-lines of C_2, at $C_2(t)$ and $C_2(s)$, and further, this intersection is at equal distance from the two *foot points* of these normal-lines, as $C_1(r)$ is on B_2:

$$\langle C_1(r) - C_2(t), C_1(r) - C_2(t) \rangle - \langle C_1(r) - C_2(s), C_1(r) - C_2(s) \rangle = 0,$$
$$\langle C_1(r) - C_2(s), C_2'(s) \rangle = 0,$$
$$\langle C_1(r) - C_2(t), C_2'(t) \rangle = 0, \tag{3}$$

where the first constraint makes sure the distances to the two bisector's foot points are the same and the last two constraints ensure the foot directions are orthogonal to the tangents of the curve. In all, Equation (3) presents three constraints in three unknowns, r, s, and t. The first constraint in Equation (3) could be rewritten as

$$\langle C_2(t) - C_2(s), C_2(t) + C_2(s) - 2C_1(r) \rangle = 0, \tag{4}$$

hinting to the fact that the term $(t - s)$ exists in this constraint. Hence, for $t = s$, the first constraint is always satisfied. Further, the last two constraints coalesce so the solver is likely to return the entire domain as a valid solution to Equations (3). In [1], a partial remedy that alleviates the problem is offered by adding a fourth constraint (and a fourth variable u) in the form of $1 - u(t - s) = 0$ to ensure that $t \neq s$, having u within some finite parametric domain. The solution of [1] is not only expensive due to the expansion of the formulation into four equations and four unknowns but will also miss any valid answer where $(t - s)$ is below the $1/u$ selected resolution.

A simpler yet more efficient and more robust alternative approach that one can employ in this specific case, is to divide all input curves at all locations where the curvature, κ, achieves an extremum. I.e. solve first for the locations where C_i satisfy $\kappa_i' = 0$, $i = 1, 2$, and split the two curves at those extrema. One should note that while κ is not rational in general, κ^2 is. Then, and since s and t must be on the opposite sides of some curvature extrema parameter value, one only needs to deal with three different curves, two of which are segments of C_2.

An even better and more general solution would aim at eliminating the $(t - s)$ term from Equations (3) before attempting to solve Equations (3), an approach we are taking in this work. In [9], we present an algorithm to algebraically decompose and remove a $(t - s)$ term from a function known to hold such a term, when the function is in either a Bézier or a B-spline form.

In [1], the subdivision solver of [10] is employed to solve these algebraic constraint. While [10] supports only Bernstein polynomials, as part of this work we use a similar solver that is capable of handling piecewise polynomials B-spline constraints as well [3,5]. All examples presented in this work employ the solver [3,5] over the B-spline domain, that is implemented using the IRIT [6] solid modeling environment.

3   Hausdorff Distances in $\mathbb{R}^3/\mathbb{R}^n$

Interesting enough, Equations (1), (2) and (3) holds for $\mathbb{R}^n$, and specifically, for $\mathbb{R}^3$. The direct extensions of Definitions 1 and 2 to $\mathbb{R}^n$ paves the way to the rest of the necessary extensions:

Definition 3. *A **normal-line** in $\mathbb{R}^n$ to a parametric form $F(\mathbf{u})$, $\mathbf{u} = (u_1 \cdots u_m)$ at the parametric location $\mathbf{u} = \mathbf{u_0}$ is a line through $F(\mathbf{u_0})$ that is also in the normal space of F at $\mathbf{u_0}$.*

Definition 4. *The line in $\mathbb{R}^n$ through $F(\mathbf{u})$, $\mathbf{u} = (u_1 \cdots u_m)$ and $G(\mathbf{v})$, $\mathbf{v} = (v_1 \cdots v_n)$, is denoted as F and G's **bi-normal-line** at parameters $\mathbf{u} = \mathbf{u_0}$ and $\mathbf{v} = \mathbf{v_0}$, if it is a normal-line at both $F(\mathbf{u_0})$ and $G(\mathbf{v_0})$. Points $F(\mathbf{u_0})$ and $G(\mathbf{v_0})$ are then denoted **antipodal points**.*

We now consider the more involved cases of a curve and a surface (in Section 3.1) and two surfaces in space (in Section 3.2), in $\mathbb{R}^3$, while we also portray the necessary steps for these constraints in $\mathbb{R}^n$.

3.1   Hausdorff Distance between a Curve and a Surface

In order to further extend the ability to compute the Hausdorff distance and support it between a curve C and a surface S, similar events to those presented in Section 2 should first be extended to $\mathbb{R}^3$. If S is open, all its boundary corner points and boundary curves should be examined against C as space point-point, point-curve, and curve-curve Hausdorff distances cases. However, we also need to consider a new type of a Hausdorff event between a space curve, $C(t)$, and a freeform surface, $S(u, v)$, in $\mathbb{R}^3$.

An equivalent condition to the antipodal curve-curve event, following Definition 4, can be expressed by requiring that the line through $C(t)$ and $S(u,v)$ be indeed a bi-normal-line and reside in the normal space of C and the normal space of S. Algebraically, we have,

$$\langle S(u,v) - C(t), C'(t) \rangle = 0,$$
$$\left\langle S(u,v) - C(t), \frac{\partial S(u,v)}{\partial u} \right\rangle = 0,$$
$$\left\langle S(u,v) - C(t), \frac{\partial S(u,v)}{\partial v} \right\rangle = 0, \tag{5}$$

having three constraints in three unknowns.

Extending Constraint (5) to $\mathbb{R}^n$ between parametric manifolds $F(\mathbf{u})$, $\mathbf{u} = (u_1 \cdots u_m)$ and $G(\mathbf{v})$, $\mathbf{v} = (v_1 \cdots v_n)$ is fairly straight forward having m orthogonality constraints of the form $\frac{\partial F}{\partial u_i} = 0$ and n orthogonality constraints of the form $\frac{\partial G}{\partial v_j} = 0$, in $m + n$ degrees of freedom, to solve for.

Similarly, we are required to extend the events resulting from intersecting one shape with the self-bisector of the other. Considering the self-bisector of C (parametrized twice by independent parameters t and s) yields,

$$\langle S(u,v) - C(t), S(u,v) - C(t) \rangle$$
$$- \langle S(u,v) - C(s), S(u,v) - C(s) \rangle = 0,$$
$$\langle S(u,v) - C(s), C'(s) \rangle = 0,$$
$$\langle S(u,v) - C(t), C'(t) \rangle = 0, \tag{6}$$

having three equations and four unknowns. Indeed, this should not come as a surprise as the self-bisector sheet of C in $\mathbb{R}^3$ is a bivariate surface and its intersection with S yields the univariate solution space that Equation (6) seeks. Let $N_S(u,v)$ be a normal field of $S(u,v)$. Interested in the extreme distances only along this univariate, Equation (6) could, for example, be augmented with the extreme distance condition, that occurs when the three vectors of $C(t) - C(s)$, $\frac{C(t)+C(s)}{2} - S(u,v) = \frac{1}{2}(C(t) + C(s) - 2S(u,v))$, and $N_S(u,v)$, are all coplanar, or,

$$\langle (C(t) - C(s)) \times (C(t) + C(s) - 2S(u,v)), N_S(u,v) \rangle = 0. \tag{7}$$

Figure 3 shows this special case, with this augmented constraint.

The first constraint in Equation (6) could be rewritten as

$$\langle C(t) - C(s), C(t) + C(s) - 2S(u,v) \rangle = 0, \tag{8}$$

clearly hinting once more to the fact that the term $(t-s)$ exists in this constraint as well. By subdividing C at the locations of maximum curvature, or better yet, algebraically eliminating the $(t-s)$ term from Equations (6) altogether, we avoid the need to introduce an additional parameter, as in [1].

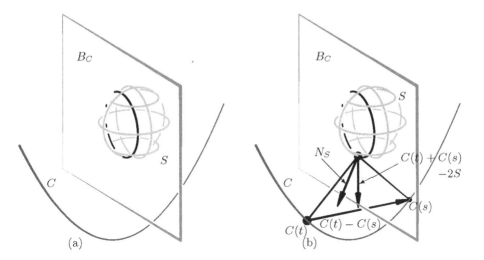

Fig. 3. The Hausdorff distance event between a surface S and a curve C can occur at the intersection of S with the self-bisector of C, B_C. While this intersection is a (black) curve in (a), we only seek the extreme distances along this curve, using, for example, a condition that occurs when the three vectors of $C(t) - C(s)$, $C(t) + C(s) - 2S$, and N_S, the normal of S, are all coplanar (b).

Now consider the intersection of the self-bisector of S (parametrized independently twice as $S(u, v)$ and $S(r, s)$) with $C(t)$, to yield,

$$
\langle S(u, v) - C(t), S(u, v) - C(t) \rangle
$$
$$
\langle S(r, s) - C(t), S(r, s) - C(t) \rangle = 0,
$$
$$
\left\langle S(u, v) - C(t), \frac{\partial S(u, v)}{\partial u} \right\rangle = 0,
$$
$$
\left\langle S(u, v) - C(t), \frac{\partial S(u, v)}{\partial v} \right\rangle = 0,
$$
$$
\left\langle S(r, s) - C(t), \frac{\partial S(r, s)}{\partial r} \right\rangle = 0,
$$
$$
\left\langle S(r, s) - C(t), \frac{\partial S(r, s)}{\partial s} \right\rangle = 0, \tag{9}
$$

having five equations and five unknowns.

Unfortunately, Equations (9) form, again, a singular set of constraints as every location for which $u = r$ and $v = s$ is identically satisfying the first constraint in Equations (9). The solution is again to eliminate the terms $(u = r)$ and $(v = s)$, a process that is beyond the scope of this paper. See [4] for more on this algebraic decomposition.

3.2   Hausdorff Distance between Two Surfaces

Continuing to the most general case of the Hausdorff distance in $\mathbb{R}^3$ between two different surfaces, $S(u, v)$ and $R(r, s)$, we now need to consider the computation of bi-normal-lines and detect all antipodal locations between these surfaces,

$$\left\langle S(u, v) - R(r, s), \frac{\partial S(u, v)}{\partial u} \right\rangle = 0,$$

$$\left\langle S(u, v) - R(r, s), \frac{\partial S(u, v)}{\partial v} \right\rangle = 0,$$

$$\left\langle S(r, s) - R(r, s), \frac{\partial R(r, s)}{\partial r} \right\rangle = 0,$$

$$\left\langle S(r, s) - R(r, s), \frac{\partial R(r, s)}{\partial s} \right\rangle = 0, \tag{10}$$

having four equations and four unknowns.

Considering the self bisector of one surface, say R (parametrized as $R(r, s)$ and $R(a, b)$), against the other surface S would again yield a univariate solution as the intersection of one (self-bisector of R) surface with another (S). Interested in the extreme distance only, we once more augment this set of constraints with an extreme distance constraint, having in all,

$$\langle S(u, v) - R(r, s), S(u, v) - R(r, s) \rangle \rangle$$
$$\langle S(u, v) - R(a, b), S(u, v) - R(a, b) \rangle = 0,$$

$$\left\langle S(u, v) - R(a, b), \frac{\partial R(a, b)}{\partial a} \right\rangle = 0,$$

$$\left\langle S(u, v) - R(a, b), \frac{\partial R(a, b)}{\partial b} \right\rangle = 0,$$

$$\left\langle S(u, v) - R(r, s), \frac{\partial R(r, s)}{\partial r} \right\rangle = 0,$$

$$\left\langle S(u, v) - R(r, s), \frac{\partial R(r, s)}{\partial s} \right\rangle = 0, \tag{11}$$

and one possible co-planarity extreme distance constraint to fully constraint the system of equations, following Equation (7), of

$$\langle (R(a, b) - R(r, s)) \times (R(a, b) + R(r, s) - 2S(u, v)), N_S(u, v) \rangle = 0,$$

having six equations and six unknowns, in all.

4   Minimal Distance between Curves and Surfaces

Having all this machinery we developed so far, it can also be used to determine the minimal distance between two curves or surfaces in $\mathbb{R}^2$ or $\mathbb{R}^3$. The minimal distance events could occur at either the boundaries (end points for curves, boundary curves and corner points for surfaces) or at the interior of the domain at antipodal locations. Since we have already seeing how to compute these events, we can deduce the minimal distances as well. Note that the self-bisector event is not relevant here.

In the next section, we presents some examples of both the Hausdorff distance computation and the minimal distance testing.

5   Examples

In this section, we present a few examples of the implemented-so-far portion of the computation of distances portrayed in Sections 2, 3 and 4. We present results of deriving the Hausdorff distance and minimal distance between curves in $\mathbb{R}^2$ and $\mathbb{R}^3$.

Figures 4 to 7 presents four examples of increasing complexity starting from an approximation of a sine function in the plane (Figure 4), a circular function in $\mathbb{R}^3$ (Figure 5), a helical function in $\mathbb{R}^3$ (Figure 6) and a general space curve (Figure 7). All curves are B-spline curves of degrees 3 or 4. The Hausdorff distance computation times are between a few seconds to several dozens seconds for the most complex example of Figure 7, on a modern PC workstation. The minimal distance computation took a small fraction of that.

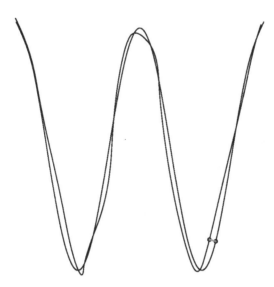

Fig. 4. The Hausdorff distance between a polynomial sine function approximation and a perturbed sine function, in the plane. Both curves are quadratic with 21 control points.

Fig. 5. The Hausdorff distance between a polynomial circular function approximation and a perturbed circular function, in $\mathbb{R}^3$. Both curves are cubic with 10 and 24 control points, respectively.

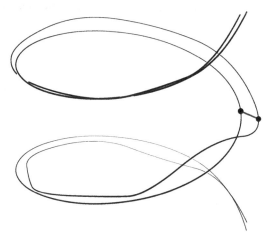

Fig. 6. The Hausdorff distance between a polynomial helical function approximation and a perturbed helical function. Both curves are quadratic with 21 control points.

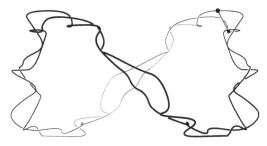

Fig. 7. The Hausdorff distance between two similar yet general space curves. Both curves are quadratic with 53 control points.

Fig. 8. The minimal distance between a polynomial circular function approximation and a perturbed circular function, in $\mathbb{R}^3$. Both curves are cubic with 10 and 24 control points, respectively. See also Figure 5.

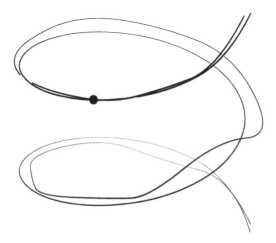

Fig. 9. The minimal distance between a polynomial helical function approximation and a perturbed helical function. Both curves are quadratic with 21 control points. See also Figure 6.

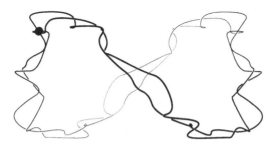

Fig. 10. The minimal distance between two similar yet general space curves. Both curves are quadratic with 53 control points. See also Figure 7.

Figures 8 to 10 presents computations of minimal distances between two shapes, for the same pairs of curves as in Figures 5 to 7, for completion. In all cases, the minimal distance is *not* zero.

6   Extensions and Computational Comments

We have derived conditions for the computations of the events where the Hausdorff distance and/or minimal distance between two regular C^1 freeform parametric shapes in $\mathbb{R}^2$ and $\mathbb{R}^3$ can take place. So far, we have implemented and tested all the cases for curves in $\mathbb{R}^2$ and $\mathbb{R}^3$, as was demonstrated in Section 5.

The presented constraints, even for curves, impose major computational burdens, when attempting to solve them. Consider a pair of curves, each with $O(n)$ coefficients. The addition/subtraction and/or product operations between this pair of curves, in all presented constraints, are typically derived as outer (tensor) products. Hence, any constraint that holds only a pair of independent curves (i.e. Constraint (2)) will possess $O(n^2)$ coefficients whereas a constraint involving three independent curves (i.e. Constraint (4)) will contain $O(n^3)$ coefficients.

In general, having k independent parametric forms, would yield constraints with $O(n^k)$ coefficients. This exponential growth renders this tensor product representation futile, when more than a few independent variables are involved. While beyond this writeup, we are working on an approach that reduces this exponential complexity from $O(n^k)$ to $O(np)$, where p is the number of operators (i.e. addition, subtraction, or product) in the constraint. p is typically small and in the order of k. To get a hint at the expected benefit from using this approach, both in speedup and in memory consumption reduction, the set of Equations (3) was solved in both the traditional, tensor product, way and using the new approach, for a few curves of different sizes. Table 1 summarizes the result. Using up to 1470 Mbytes when solving for the Hausdorff distance between two curves of 50 coefficients, the traditional, tensor product, approach converts the first constraint in Equations (3) into a tensor product trivariate of $O(50^3)$ coefficients. Due to the fact that the constraint also involves (inner) products, it ends up with around one million coefficients. With 8 bytes per double, a single constraint will consume around 10 Mbytes of memory!

Table 1. A comparison of solving Equations (3) for different pairs of curves, of different sizes, using the traditional, tensor product, approach and the new approach. Benefits are clearly significant in both speedup and memory consumption, as the complexity is increased.

Num. of Coeffs.	Traditional		New Approach	
	Time (Secs.)	Size (MB.)	Time (Secs.)	Size (MB.)
10	2.1	14.5	1.8	7
20	8.4	77	2.9	7.5
50	326	1470	23	11.5

7   Conclusions

In this paper, we have presented algorithms to derive the precise Hausdorff and minimal distance between regular C^1 freeform shapes in $\mathbb{R}^2$ and $\mathbb{R}^3$. We hope to continue and completely implement all cases in $\mathbb{R}^3$, including surfaces, in the future.

Clearly some of the posed constraints, like in Equations (1), (2) and (3), could be extended with ease to $\mathbb{R}^n$. Others, like (6) and (7), are more difficult to extend. One can expect that the way Equations (6) (Equations (11)) could be augmented with an extreme condition (7) (Equations (12)) to yield a zero dimensional solution space could also be applied to higher dimensions as well.

In all the above constraints, the implicit assumption was that the shapes do not (self-) interest. Many of the presented constraints vanish at an intersection, and hence, an implicit preprocessing step to all the above computation should examine for intersections first and preclude these intersection locations from the computation.

The overall computation is not fast. The need to solve for the simultaneous zeros of several piecewise polynomials of many coefficients, makes the computations consumes seconds of processing, even for curves. Further methods to make this computation more efficient are to be sought.

Acknowledgment

This research was partly supported by the New York metropolitan research fund, Technion, in part by the Israel Science Foundation (grants no. 346/07) and in part by the Israeli Ministry of Science Grant No. 3-4642.

References

1. Alt, H., Scharf, L.: Computing the Hausdorff distance between sets of curves. In: Proceedings of the 20th European Workshop on Computational Geometry (EWCG), Seville, Spain, pp. 233–236 (2004)
2. Cignoni, P., Rocchini, C., Scopigno, R.: Metro: Measuring error on simplified surfaces. Computer Graphics Forum 17(2), 167–174 (1998)
3. Elber, G., Kim, M.S.: Geometric constraint solver using multivariate rational spline functions. In: Proc. of the ACM symposium on Solid Modeling and Applications, pp. 1–10 (2001)
4. Elber, G., Grandine, T., Kim, M.S.: Surface Self-Intersection Computation via Algebraic Decomposition. functions. In: SPM 2007 (submitted for publication, 2007)
5. Hanniel, I., Elber, G.: Subdivision Termination Criteria in Subdivision Multivariate Solvers. Computer Aided Design 39, 369–378 (2007)
6. IRIT 9.5 User's Manual, Technion (2005), http://www.cs.technion.ac.il/~irit
7. Johnson, D.: Minimum distance queries for haptic rendering. PhD thesis, Computer Science Department, University of Utah (2005)

8. Jüttler, B.: Bounding the Hausdorff Distance of Implicitely Defined and/or parametric Curves. In: Lyche, T., Schumaker, L.L. (eds.) Matematical Methods in
   CAGD, Oslo, pp. 1–10 (2000)
9. Pekerman, D., Elber, G., Kim, M.-S.: Self-Intersection Detection and Elimination
   in freeform Curves. In: Computer Aided Design (accepted for publication)
10. SYNAPS (SYmbolic Numeric ApplicationS),
    http://www-sop.inria.fr/galaad/software/synaps

On Interpolation by Spline Curves with Shape Parameters

Miklós Hoffmann[1] and Imre Juhász[2]

[1] Institute of Mathematics and Computer Science, Károly Eszterházy College,
Leányka str. 4., H-3300 Eger, Hungary
hofi@ektf.hu
http://www.ektf.hu/tanszek/matematika/hofi.html
[2] Department of Descriptive Geometry, University of Miskolc,
H-3515 Miskolc-Egyetemváros, Hungary
agtji@uni-miskolc.hu
http://abrpc09.abrg.uni-miskolc.hu/Juhasz/index.html

Abstract. Interpolation of a sequence of points by spline curves generally requires the solution of a large system of equations. In this paper we provide a method which requires only local computation instead of a global system of equations and works for a large class of curves. This is a generalization of a method which previously developed for B-spline, NURBS and trigonometric CB-spline curves. Moreover, instead of numerical shape parameters we provide intuitive, user-friendly, control point based modification of the interpolating curve and the possibility of optimization as well.

Keywords: interpolation, spline curve, shape parameter.

1 Introduction

Interpolation of an ordered sequence of points is one of the most widely used methods in practical curve modeling, hence there is a vast number of papers and book chapters dealing with this topic (c.f. the books [2], [11]). Designers generally prefer spline curves, where most of the methods work globally. Even in the case of B-spline or NURBS curves, which are standard description methods in curve design and have local control properties, the process of finding control points of an interpolating curve is global and the resulting curve cannot be locally controlled (see e.g. [1], [3], [4], [7]). To overcome this problem some methods have been developed by means of which the shape of the interpolating curve can be adjusted by numerical techniques (see e.g. [8], [9], [12]). Shape parameters and other numerical techniques, however, do not provide intuitive shape control methods such as control point repositioning in approximation. Moreover, in these global methods a large system of equations has to be solved with relatively high computational cost. Especially for large set of data points, local methods have the advantage of solving smaller systems, since the computation of each curve segment is based on only a subset of data. Unfortunately, these local methods typically attain only C^1 continuity [10].

F. Chen and B. Jüttler (Eds.): GMP 2008, LNCS 4975, pp. 205–214, 2008.

In the last couple of years a new local method has been developed for some types of spline curves that requires only local computation and yields C^2 continuous spline curves. This technique - which is based on linear blending - has been implemented for NURBS in [13], for B-spline curve in [14] and for trigonometric CB-spline curve in [15]. In this method the shape of the interpolating curve can also be adjusted numerically by some shape parameters.

In this paper we will focus on this latter method. In Section 2 we generalize the linear blending interpolation method for a large class of curves. On the other hand, since designers generally prefer geometric entities instead of numerical values, we provide intuitive, control point based modification of the interpolating curve. We let the designer alter the shape of the curve similarly to the approximating curves, meanwhile the interpolation property is continuously preserved. This method will be discussed in Section 3. Finally in Section 4 further generalization in terms of parametrization and the possibility of optimization is provided.

2   Interpolation by Linear Blending

At first we describe the curve generating method which we refer to as linear blending. Consider the points $\mathbf{p}_i$, $(i = 0, ..., n)$ and the piecewisely defined curve

$$\mathbf{b}(u) = \sum_{i=-1}^{n+1} F_i(u)\mathbf{p}_i, \quad u \in [u_0, u_n], \tag{1}$$

where $\mathbf{p}_i$ are called control points and $F_i(u)$ are basis functions of some space. To ensure that the number of arcs equals the number of control legs, we define two artificial control points $\mathbf{p}_{-1} := \mathbf{p}_0$ and $\mathbf{p}_{n+1} := \mathbf{p}_n$. The only restriction is that each arc of the curve has to be defined by four consecutive control points, that is the arcs of this curve can be written as

$$\mathbf{b}_j(u) = \sum_{l=j-1}^{j+2} F_l(u)\mathbf{p}_l, \quad u \in [u_j, u_{j+1}], \quad j = 0, ..., n-1,$$

where the values u_j are called knots.

Now, consider the jth control leg $\mathbf{p}_j, \mathbf{p}_{j+1}, (j = 0, ..., n-1)$ and parameterize this line segment by some function $s(u)$ as

$$\mathbf{l}_j(u) = (1 - s(u))\mathbf{p}_j + s(u)\mathbf{p}_{j+1}, \quad u \in [u_j, u_{j+1}].$$

At this moment the function $s(u)$ can be any function which fulfills two conditions

$$s(u_j) = 0, \quad s(u_{j+1}) = 1. \tag{2}$$

Possible choices of the function will be discussed in detail in Section 4.

Linearly blending the arcs $\mathbf{b}_j(u)$ of the spline curve with the corresponding line segments $\mathbf{l}_j(u)$ we obtain the linear blending curve consisting of the arcs

$$\mathbf{c}_j(u, \alpha) = (1 - \alpha)\mathbf{b}_j(u) + \alpha\mathbf{l}_j(u),$$

where α is a global shape parameter of the curve. To achieve more flexibility in shape modification, the shape parameter α can also be the function of u. A straightforward way is to define it piecewisely by local shape parameters α_i^* associated to each point $\mathbf{p}_i$, and to use the same blending function $s(u)$ as for the line segment, i.e.,

$$\alpha_j(u) = (1 - s(u))\alpha_j^* + s(u)\alpha_{j+1}^*.$$

This way each arc of the linear blending curve will have three parameters, the running parameter u and the two local shape parameters $\alpha_j^*, \alpha_{j+1}^*$:

$$\mathbf{c}_j(u, \alpha_j^*, \alpha_{j+1}^*) = (1 - \alpha_j(u))\mathbf{b}_j(u) + \alpha_j(u)\mathbf{l}_j(u),$$

that is
$$\begin{aligned}\mathbf{c}_j(u, \alpha_j^*, \alpha_{j+1}^*) = &(1 - (1 - s(u))\alpha_j^* - s(u)\alpha_{j+1}^*)\mathbf{b}_j(u) + \\ &((1 - s(u))\alpha_j^* + s(u)\alpha_{j+1}^*)\mathbf{l}_j(u).\end{aligned} \tag{3}$$

In order to obtain C^m continuity of the consecutive arcs of the curve, the kth derivatives of the function $s(u)$ have to satisfy the conditions

$$s^{(k)}(u_j) = 0, s^{(k)}(u_{j+1}) = 0 \quad 1 \le k \le m, \quad j = 0, ..., n - 1. \tag{4}$$

Now, based on the idea described in [13], we modify the linear blending method to interpolate a given sequence of points. As we have seen the curve arc $\mathbf{c}_j(u, \alpha_j^*, \alpha_{j+1}^*)$ is "between" the arc $\mathbf{b}_j(u)$ and the control leg $\mathbf{l}_j(u)$, that is the curve approximates the given points (except the case when all the shape parameters are set to be 1). But there is no specific reason to choose the control leg as the blended line segment, so we can define the linear blending of the spline arc $\mathbf{b}_j(u)$ with some other line segment as well. In this way we can obtain an appropriate line segment by means of which the resulted curve will interpolate the given points.

Let the points $\mathbf{p}_i$, associated parameter values u_i and shape parameters α_i^*, $(i = 0, ..., n)$ be given. If the parameter values u_i are unknown, then choose them by some of the well-known methods, like centripetal or chord length method - these are only initial values which will be allowed to be changed during the interactive modification of the interpolating curve. Now, the problem is to find a linear blending curve having the given shape parameters α_i^*, which interpolates the given points $\mathbf{p}_i$ at the given parameter values u_i.

Consider the approximating spline curve defined by the given points as control points

$$\mathbf{b}_j(u) = \sum_{l=j-1}^{j+2} F_l(u)\mathbf{p}_l, \quad u \in [u_j, u_{j+1}], \quad j = 0, ..., n - 1.$$

Also consider the parameterized line segments

$$\mathbf{l}_j(u) = (1 - s(u))\mathbf{v}_j + s(u)\mathbf{v}_{j+1}, \quad j = 0, ..., n - 1,$$

where the points $\mathbf{v}_j$ are unknown. Analogously to (3), arcs of the linear blending curve can be written as

$$
\begin{aligned}
\mathbf{c}_j(u, \alpha_j^*, \alpha_{j+1}^*) = &(1 - (1 - s(u))\alpha_j^* - s(u)\alpha_{j+1}^*)\mathbf{b}_j(u) + \\
&((1 - s(u))\alpha_j^* + s(u)\alpha_{j+1}^*)((1 - s(u))\mathbf{v}_j + s(u)\mathbf{v}_{j+1}).
\end{aligned}
\tag{5}
$$

After some calculations, the interpolation assumptions yield equalities

$$
\mathbf{v}_j(u_j, \alpha_j^*) = \mathbf{p}_j + \frac{(1 - \alpha_j^*)}{\alpha_j^*}(\mathbf{p}_j - \mathbf{b}(u_j)) \quad j = 0, ..., n.
\tag{6}
$$

Using these points, the curve composed of the arcs (5) will interpolate the given points at the given parameter values.

Since we had very little assumptions about the initial base curve $\mathbf{b}(u)$, especially no assumptions about its basis functions $F_i(u)$ and their space, this general method includes the methods described in [13], [14] and [15] as special cases.

As one can observe, there are infinitely many possibilities to choose the points $\mathbf{v}_j$. In the following Section we will discuss the geometric consequences of the alteration of these points, which allows us to modify the interpolating curve in an intuitive way.

3   Control Point Based Modification

Each of the points $\mathbf{v}_i$ depends on two parameters: the corresponding parameter value u_i and the local shape parameter α_i^*. Instead of manipulating these values numerically and calculate the points $\mathbf{v}_i$ and finally the interpolating curve arcs $\mathbf{c}_j(u, \alpha_j^*, \alpha_{j+1}^*)$, our main idea is to start to think in the opposite direction: what happens to these values and what happens to the curve if we change the position of the points $\mathbf{v}_i$?

3.1   Altering the Shape Parameter α_i^*

The modification of the point $\mathbf{v}_i$ affects only two consecutive arcs of the interpolating curve. If the parameter value u_i is fixed, then due to Equation (6) the point $\mathbf{v}_i$ can move along a straight line defined by the points $\mathbf{p}_i$ and $\mathbf{b}(u_i)$.

What is the geometric effect of this modification to the points of the curve? We describe the movement of a fixed point of the curve arc $\mathbf{c}_j(u, \alpha_j^*, \alpha_{j+1}^*)$ as a function of α_j^*. Fixing $u = \bar{u}$ and $\alpha_{j+1}^* = \bar{\alpha}_{j+1}^*$ and substituting Equation (6) into (5), after some calculation we get

$$
\mathbf{c}_j(\bar{u}, \alpha_j^*, \bar{\alpha}_{j+1}^*) = \mathbf{A}\alpha_j^* + \frac{\mathbf{B}}{\alpha_j^*} + \mathbf{C}, \quad \alpha_j^* \in \mathbb{R} \backslash \{0\},
$$

where $\mathbf{A}, \mathbf{B}$ and $\mathbf{C}$ are constants, depending on the knot values u_j, u_{j+1}, the fixed parameters $\bar{u}, \bar{\alpha}_{j+1}^*$ and the positions of the points $\mathbf{p}_j, \mathbf{p}_{j+1}$

$$\mathbf{A} = (s\,(\bar{u}) - 1)\mathbf{b}(\bar{u}) + (1 - s\,(\bar{u}))((1 - s\,(\bar{u}))\mathbf{b}(u_j) + s\,(\bar{u})\,\mathbf{v}_{j+1}),$$
$$\mathbf{B} = s\,(\bar{u})\,\bar{\alpha}^*_{j+1}(1 - s\,(\bar{u}))(\mathbf{p}_j - \mathbf{b}(u_j)),$$
$$\mathbf{C} = (1 - s\,(\bar{u})\,\bar{\alpha}^*_{j+1})\mathbf{b}(\bar{u}) + s\,(\bar{u})\,\bar{\alpha}^*_{j+1}((1 - s\,(\bar{u}))\mathbf{b}(u_j) + s\,(\bar{u})\,\mathbf{v}_{j+1}) +$$
$$(1 - s\,(\bar{u}))^2(\mathbf{p}_j - \mathbf{b}(u_j)).$$

This is the equation of a hyperbola with center $\mathbf{C}$ and asymptotic directions $\mathbf{A}$ and $\mathbf{B}$. This means, that modifying the position of the point $\mathbf{v}_j$ along the straight line defined by Equation (6), points of the interpolating curve will move along a well-defined hyperbola (see Fig.1).

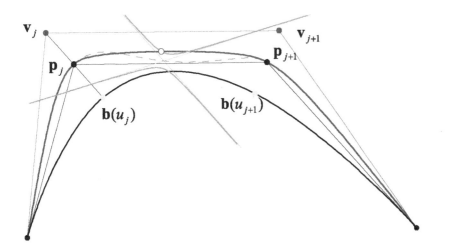

Fig. 1. The interpolating curve (thick red), the base curve (black) and the line segments (red). Modifying the shape parameter α^*_j, the vertex $\mathbf{v}_j$ moves along the line $\mathbf{p}_j\mathbf{b}\,(u_j)$, while the points of the interpolating curve arc $\mathbf{p}_j\mathbf{p}_{j+1}$ move along hyperbolas. One of the asymptotes of these hyperbolas is always parallel to the direction $\mathbf{p}_j - \mathbf{b}\,(u_j)$. The green dashed line is the set of the centers of these hyperbolas. Figure shows a cubic B-spline implementation.

Moreover, one of the asymptotes of these hyperbolas are parallel to the direction $\mathbf{p}_j - \mathbf{b}\,(u_j)\ \forall u \in [u_j, u_{j+1}]$, and the movement of the points of the interpolating curve flows typically along this asymptote. This hyperbola is degenerated to a straight line when $j = 0$.

3.2   Altering the Parameter Value u_i

As we have seen, each of the points $\mathbf{v}_i$ depends on two parameters: the corresponding parameter value u_i and the local shape parameter α^*_i. If the shape parameter value α^*_i is fixed, and we let the parameter value u_i vary between u_{i-1} and u_{i+1} then the movement of the point $\mathbf{v}_i$ and finally the change of the shape of the curve will be a little more complicated. This is a consequence of the fact that the value u_i also serves as a knot value of the original base curve $\mathbf{b}(u)$. Therefore, the alteration of the knot value of a curve changes the shape of this base curve as well. This alteration has been studied for B-spline and NURBS curves in [5], [6]. In general, altering the value $u_i \in [u_{i-1}, u_{i+1}]$ we get a one-parameter family of base curves $\mathbf{b}(u, u_i)$ with family parameter u_i. Let us consider those points of these curves that correspond to $u = u_i$, that is the points $\mathbf{b}(u_i, u_i)$. These points form a curve which will be denoted by

$$\mathbf{e}(u_i) = \mathbf{b}(u_i, u_i), \quad u_i \in [u_{i-1}, u_{i+1}]. \tag{7}$$

It is easy to see from Equation (6), that altering the value u_i, the point $\mathbf{v}_i$ will move along a path curve that can be obtained by a central similarity from the curve (7), where the center of similitude is the given point $\mathbf{p}_i$, and the ratio is $(1 - \alpha_i^*)/\alpha_i^*$

$$\mathbf{v}_i(u_i) = \mathbf{p}_i + \frac{(1 - \alpha_i^*)}{\alpha_i^*} \left(\mathbf{p}_i - \mathbf{e}(u_i)\right). \tag{8}$$

3.3   Repositioning of $\mathbf{v}_i$

As we have learned from the previous subsections, the position of the point $\mathbf{v}_i$ depends on u_i and α_i^* in such a way, that modifying α_i^* the point will move along straight lines passing through $\mathbf{p}_i$, while modifying u_i the point will move along similar path curves (8), where the center of similitude is $\mathbf{p}_i$. Changing the direction of this deduction simply means that repositioning the point $\mathbf{v}_i$ one can calculate the actual values of u_i and α_i^* by finding the corresponding line and a path curve.

The permissible positions of $\mathbf{v}_i$ are on the surface $\mathbf{v}_i(u_i, \alpha_i^*)$ (or a planar region in degenerate cases). The isoparametric lines of this surface are straight lines passing through $\mathbf{p}_i$ and the path curves (8), which means that the permissible positions of $\mathbf{v}_i$ are on the surface of a cone, the apex of which is $\mathbf{p}_i$, and the two limiting generators are the lines connecting $\mathbf{e}(u_{i-1})$, $\mathbf{p}_i$, and $\mathbf{e}(u_{i+1})$, $\mathbf{p}_i$, respectively. If the cone degenerates to a planar region these two lines define the boundary of the permissible region (see Fig.2).

These results enable us to modify the shape of the interpolating curve by an intuitive way, that is similar to control point repositioning in the case of approximating curves. Given a sequence of points $\mathbf{p}_i$, associated parameter values u_i and shape parameters α_i^*, $(i = 0, ..., n)$, we can calculate the initial interpolating curve piecewisely by Equation (5). Then we can change the position of any point $\mathbf{v}_i$ inside its region of permissible positions, without dealing with numerical values and parameters. The effect of this repositioning is described above, the actual values of the parameters can automatically be calculated, thus the new curve will remain interpolating curve. Although, these points are not "real" control points of the curve, the effect of their modifications is intuitive and the designers do not have to bother with numerical parameters.

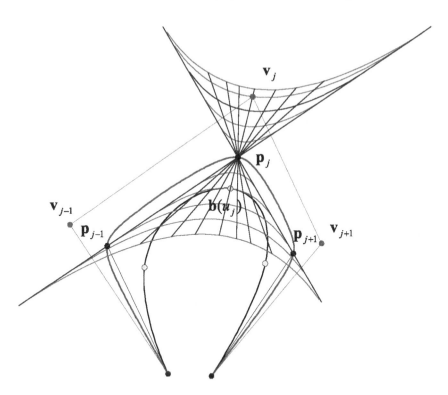

Fig. 2. The interpolating curve (thick red) and its control points $\mathbf{v}_i$. Permissible positions of the point $\mathbf{v}_j$ are on a cone (or planar region). The α_j^* isoparametric curves are straight lines (blue), while the u_i isoparametric curves are path curves (8) (grey). Figure shows a cubic B-spline implementation.

4   Parametrization of the Line Segment

One of the key steps of constructing the interpolating curve (5) is the parametrization of the line segment $\mathbf{l}_j(u)$ by means of the function $s(u)$. This function is also used for the definition of the shape parameter function $\alpha_j(u)$ from the local shape parameters α_j^* and α_{j+1}^*. Beside the obvious endpoint conditions (2) some continuity conditions (4) also have to be fulfilled by this function. Certainly, the choice of this function depends on the base curve $\mathbf{b}(u)$, more precisely the type of its basis functions $F_i(u)$. If, for example, the base curve is a cubic polynomial curve, the function $s(u)$ should also be cubic polynomial in order to achieve the interpolating curve to be cubic polynomial curve as well. However, in this specific case the function $s(u)$ is bound to be higher degree polynomial by the continuity conditions mentioned above. In [13] the minimum degree polynomial is found to be the quintic Hermite polynomial $6u^5 - 15u^4 + 10u^3$.

Tai and Wang [14] suggested piecewisely defined (i.e. spline) function for $s(u)$ to decrease the degree meanwhile the desired C^2 continuity is preserved.

Instead of the quintic polynomial, they use a function defined piecewisely over the interval $[u_j, u_{j+1}]$ as

$$s(u) = \frac{9}{2}\left(\frac{u - u_j}{u_{j+1} - u_j}\right)^3, \quad u \in \left[u_j, \frac{2}{3}u_j + \frac{1}{3}u_{j+1}\right]$$

$$s(u) = \frac{9}{2}\left[\left(\frac{u - u_j}{u_{j+1} - u_j}\right)^3 - 3\left(\frac{u - u_j}{u_{j+1} - u_j} - \frac{1}{3}\right)^3\right],$$

$$u \in \left[\frac{2}{3}u_j + \frac{1}{3}u_{j+1}, \frac{1}{3}u_j + \frac{2}{3}u_{j+1}\right]$$

$$s(u) = 1 - \frac{9}{2}\left(\frac{u_{j+1} - u}{u_{j+1} - u_j}\right)^3, \quad u \in \left[\frac{1}{3}u_j + \frac{2}{3}u_{j+1}, u_{j+1}\right].$$

A similarly structured trigonometric spline function $s(u)$ has been applied in the case of trigonometric spline base curve in [15].

But these methods can be considered as special cases of the following method defining the appropriate spline function $s(u)$ for parametrization of the line segment. If the base curve is a cubic polynomial curve, then let us consider a general spline function over the interval $[u_j, u_{j+1}]$. We divide the domain of definition into three subintervals with boundaries $u_j < t_1 < t_2 < u_{j+1}$ and we specify cubic polynomials $s_0(u), s_1(u), s_2(u)$ on these smaller ranges. Coefficients of polynomials

$$s_j(u) = a_{j0} + a_{j1}u + a_{j2}u^2 + a_{j3}u^3, \quad (j = 0, 1, 2)$$

can be computed by means of the continuity constraints. The coefficients of the resulted polynomials are

$$a_{00} = 1, \; a_{01} = 0, \; a_{02} = 0, \; a_{03} = \frac{-1}{t_1 t_2}$$

$$a_{10} = \frac{t_2 - t_1 - t_2 t_1}{(t_1 - 1)(t_1 - t_2)},$$

$$a_{11} = \frac{3t_1}{(t_1 - 1)(t_1 - t_2)},$$

$$a_{12} = \frac{-3}{(t_1 - 1)(t_1 - t_2)},$$

$$a_{13} = \frac{1 + t_2 - t_1}{(t_1 - 1)(t_1 - t_2)t_2}$$

$$a_{20} = \frac{1}{(1 - t_1)(1 - t_2)},$$

$$a_{21} = \frac{-3}{(1 - t_1)(1 - t_2)},$$

$$a_{22} = \frac{3}{(1 - t_1)(1 - t_2)},$$

$$a_{23} = \frac{-1}{(1 - t_1)(1 - t_2)}.$$

Thus we have two more free parameters t_1, t_2 that heavily affect the overall parametrization of the interpolating curve, which can be used for optimization, especially because they have only minor effects on the shape of the curve. Moreover, for other types of base curves this technique can also be applied, even for higher order C^m continuity, when the range has to be divided into $m + 1$ parts and the spline function has to be defined over these intervals.

5   Conclusion

An interpolation method has been provided in this paper, that linearly blends the arcs of a base curve and the line segments of a calculated polygon. The method works for a large class of spline curves, actually the only restriction is that each arc has to be determined by four consecutive control points. The method is a generalization of formerly published techniques.

A user-friendly, control point based shape control of the interpolating curve is also proposed, where the vertices of the calculated polygon play the role of control points. The effect of repositioning of these points on the shape is very intuitive, the permissible area of the points are large and easily defined. Therefore, the user does not have to bother with numerical parameters in the fine tuning of the interpolating curve.

Acknowledgements

The authors wish to thank the Hungarian National Scientific Fund (OTKA No. T048523) and the National Office of Research and Technology (Project CHN-37/2005) for their financial support of this research.

References

1. Barsky, B.A., Greenberg, D.P.: Determining a set of B-spline control vertices to generate an interpolating surface. Computer Graphics and Image Processing 14, 203–226 (1980)
2. Hoschek, J., Lasser, D.: Fundamentals of CAGD, AK Peters, Wellesley, MA (1993)
3. Lavery, J.E.: Univariate cubic Lp splines and shape-preserving multiscale interpolation by univariate cubic L1 splines. Computer Aided Geometric Design 17, 319–336 (2000)
4. Ma, W., Kruth, J.P.: NURBS curve and surface fitting and interpolation. In: Daehlen, M., Lyche, T., Schumaker, L. (eds.) Mathematical Methods for Curves and Surfaces, Vanderbilt University Press, Nashville & London (1995)
5. Juhász, I., Hoffmann, M.: The effect of knot modifications on the shape of B-spline curves. Journal for Geometry and Graphics 5, 111–119 (2001)
6. Juhász, I., Hoffmann, M.: Modifying a knot of B-spline curves. Computer Aided Geometric Design 20, 243–245 (2003)
7. Juhász, I., Hoffmann, M.: On parametrization of interpolating curves. Journal of Computational and Applied Mathematics (2007), doi:10.1016/j.cam.2007.05.019
8. Kaklis, P.D., Sapidis, N.S.: Convexity-preserving interpolatory parametric splines of non-uniform polynomial degree. Computer Aided Geometric Design 12, 1–26 (1995)

9. Kong, V.P., Ong, B.H.: Shape preserving F3 curve interpolation. Computer Aided Geometric Design 19, 239–256 (2002)
10. Li, A.: Convexity preserving interpolation. Computer Aided Geometric Design 16, 127–147 (1999)
11. Piegl, L., Tiller, W.: The NURBS Book. Springer, Berlin (1995)
12. Sapidis, N., Farin, G.: Automatic fairing algorithm for B-spline curves. Computer-Aided Design 22, 121–129 (1990)
13. Tai, C.-L., Barsky, B.A., Loe, K.-F.: An interpolation method with weight and relaxation parameters. In: Cohen, A., Rabut, C., Schumaker, L.L. (eds.) Curve and Surface Fitting: Saint-Malo 1999, pp. 393–402. Vanderbilt Univ. Press, Nashville (2000)
14. Tai, C.-L., Wang, G.-J.: Interpolation with slackness and continuity control and convexity-preservation using singular blending. J. Comp. Appl. Math. 172, 337–361 (2004)
15. Pan, Y.-J., Wang, G.-J.: Convexity-preserving interpolation of trigonometric polynomial curves with shape parameter. Journal of Zhejiang University Science A 8, 1199–1209 (2007)

Lepp Terminal Centroid Method for Quality Triangulation: A Study on a New Algorithm

Maria-Cecilia Rivara and Carlo Calderon

Department of Computer Science, University of Chile
{mcrivara,carcalde}@dcc.uchile.cl

Abstract. We introduce a new Lepp-Delaunay algorithm for quality triangulation. For every bad triangle t with smallest angle less than a threshold angle θ, a Lepp-search is used to find an associated convex terminal quadrilateral formed by the union of two terminal triangles which share a local longest edge (terminal edge) in the mesh. The centroid of this terminal quad is computed and Delaunay inserted in the mesh. The algorithm improves the behavior of a previous Lepp-Delaunay terminal edge midpoint algorithm. The centroid method computes significantly smaller triangulation than the terminal edge midpoint variant, produces globally better triangulations, and terminates for higher threshold angle θ (up to 36°). We present geometrical results which explain the better performance of the centroid method. Also the centroid method behaves better than the off-center algorithm for θ bigger than 25°.

Keywords: mesh generation, triangulations, Lepp Delaunay algorithms, Lepp centroid.

1 Introduction

In the last decade, methods that produce a sequence of improved constrained Delaunay triangulations (CDT) have been developed to deal with the quality triangulation of a planar straight line graph D. The combination of edge refinement and Delaunay insertion has been described by George and Borouchaki [3,2] and Rivara and collaborators, [8,9,10,11]. Mesh improvement properties for iterative Delaunay refinement based on inserting the circumcentre of triangles to be refined have been established by Chew, [1], Ruppert [12], and Shewchuk [15]. Applications of this form of refinement have been described by Weatherill et al [17] and Baker [13]. Baker also published a comparison of edge based and circumcenter based refinement [14]. Algorithms based on off-center insertions have been recently presented by Üngor and collaborators [19,24]. Algorithms for uniform triangular meshes are discussed in [16]. For a theoretical review on mesh generation see the monograph of Edelsbrunner [4].

Longest edge refinement algorithms. The longest edge bisection of any triangle t is the bisection of t by the midpoint of its longest edge and the opposite

F. Chen and B. Jüttler (Eds.): GMP 2008, LNCS 4975, pp. 215–230, 2008.

vertex (Figure 3). Longest edge based algorithms [5,6,7,8,9,11] were designed to take advantage of the following mathematical properties on the quality of triangles generated by iterative longest edge bisection of triangles [18,21,22], and require of a resonably good quality input triangulation to start with.

Theorem 1. *For any triangle t_0 of smallest angle α_0.*
(i) The iterative longest edge bisection of t_0 assures that for any longest edge son t of smallest angle α_t, it holds that $\alpha_t \geq \alpha_0/2$, the equality holding only for equilateral triangles. (ii) A finite number of similarly distint triangles is generated. (iii) The area of t_0 tends to be covered by quasi equilateral triangles (for which at most 4 similarly different, good quality triangles are obtained by longest edge bisection of t_0).

Lepp (Delaunay terminal edge) midpoint method. The algorithm was designed to improve the smallest angles in a Delaunay triangulation. This proceeds by iterative selection of a point M which is midpoint of a Delaunay terminal edge (a longest edge for both triangles that share this edge) which is then Delaunay inserted in the mesh. This method uses the longest edge propagating path associated to a bad quality processing triangle to determine a terminal edge in the current mesh. The algorithm was introduced in a rather intuitive basis as a generalization of previous longest edge algorithms in [9,10,11]. This was supported by the improvement properties of both the longest edge bisection of triangles (Theorem 1) and the Delaunay algorithm, and by the result presented in Theorem 2 in next section. Later in [10] we discussed some geometrical properties including a (rare) loop case for angle tolerance greater than 22° and its management. However, while empirical studies show that the method behaves analogously to the circumcircle method in 2-dimensions [9,10,11], formal proofs on algorithm termination and on optimal size property have not been fully established due to the difficulty of the analysis. Recently in [20] we have presented some geometrical improvement properties of an isolated insertion of a terminal edge midpoint M in the mesh. In [23] a first termination proof is presented and several geometric aspects of the algorithm are studied.

Lepp-centroid algorithm. In order to improve the performance of the previous Lepp midpoint algorithm, in this paper we introduce a new Lepp-centroid algorithm for quality triangulation. For any general (planar straight line graph) input data, and a quality threshold angle θ, the algorithm constructs constrained Delaunay triangulations that have all angles at least θ as follows: for every bad triangle t with smallest angle less than θ, a Lepp-search is used to find an associated convex terminal quadrilateral formed by the union of two terminal triangles which share a local longest edge (terminal edge) in the mesh. The centroid of this terminal quad is computed and Delaunay inserted in the mesh. The process is repeated until the triangle t is destroyed in the mesh.

In section 2 we introduce the basic concepts of longest edge propagating path (Lepp), terminal edges and terminal triangles, and a relevant constraint on the largest angle of Delaunay terminal triangles. In section 3 we describe the Lepp-midpoint algorithm, discuss a special loop case that rarely occurs for angles greater than 22°, and a geometric characterization on Delaunay terminal triangles. We use this characterization to state improved angle bounds on the smallest angles, and to prove that the new points are not inserted too close to previous vertices in the mesh. In section 4 we formulate the new Lepp centroid algorithm and state geometrical results which explain the better performance of the Lepp centroid method, and guarantee that the loop case is avoided. In section 5 we present an empirical study that compares the behavior of Lepp-centroid and Lepp-midpoint methods. The centroid method computes significantly smaller triangulation than the terminal edge midpoint variant, produces globally better triangulations, and terminates for higher threshold angle θ (up to 36°). We also show that the Lepp centroid method behaves better than the off-center algorithm for $\theta > 25°$.

2   Concepts and Preliminary Results

An edge E is called a terminal edge in triangulation τ if E is the longest edge of every triangle that shares E, while the triangles that share E are called terminal triangles [9,10,11]. Note that in 2-dimensions either E is shared by two terminal triangles t_1, t_2 if E is an interior edge, or E is shared by a single terminal triangle t_1 if E is a boundary (constrained) edge. See Figure 1 where edge AB is an interior terminal edge shared by two terminal triangles t_2, t_3, while edge CD is a boundary terminal edge with associated terminal triangle t_3.

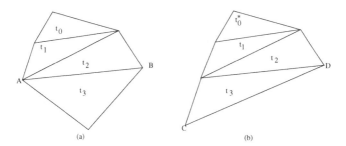

Fig. 1. (a) AB is an interior terminal edge shared by terminal triangles (t_2, t_3) associated to Lepp(t_0) $=\{t_0, t_1, t_2, t_3\}$; (b) CD is a boundary terminal edge with unique terminal triangle t_3 associated to Lepp(t_0^*) $= \{t_0^*, t_1, t_2, t_3\}$

For any triangle t_0 in τ, the longest edge propagating path of t_0, called *Lepp(t_0)*, is the ordered sequence $\{t_j\}_0^{N+1}$, where t_j is the neighbor triangle on a longest edge of t_{j-1}, and longest-edge $(t_j) >$ longest-edge (t_{j-1}), for $j=1,...$ N. Edge $E =$ longest-edge$(t_{N+1}) =$ longest-edge(t_N) is a terminal edge in τ and

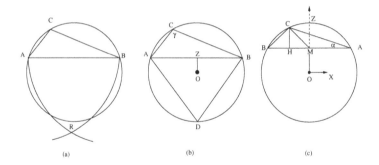

Fig. 2. R is the geometrical place of the fourth vertex D for Delaunay terminal triangles ABC, ABD; (b) R reduces to one point when $\gamma = 2\pi/3$ (triangle ADB equilateral); (c) For a bad terminal triangle BAC, Lepp-midpoint method inserts midpoint M

this condition determines N. Consequently either E is shared by a couple of terminal triangles (t_N, t_{N+1}) if E is an interior edge in τ, or E is shared by a unique terminal triangle t_N with boundary (constrained) longest edge. See Figure 1.

For a Delaunay mesh, an unconstrained terminal edge imposes the following constraint on the largest angles of the associated terminal triangles [8,9,11]:

Theorem 2. *For any pair of Delaunay terminal-triangles t_1, t_2 sharing a non-constrained terminal edge, largest angle$(t_i) \leq 2\pi/3$ for $i = 1, 2$.*

Proof. For any Delaunay terminal triangles BAC of longest edge AB (see Figure 2(a)), the third vertex D of the neighbor terminal triangle ABD must be situated in the exterior of circumcircle CC(BAC) and inside the circles of center A, B and radius $\overline{AB}$. This defines a geometrical place R for D which reduces to one point when $\angle BCA = 2\pi/3$ where OZ = r/2 (see Figure 2(b)), implying that $R = \phi$ when angle $BCA > 2\pi/3_\square$

For a single longest edge bisection of any triangle t, into two triangles t_A, t_B, the following result holds:

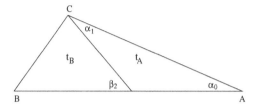

Fig. 3. Notation for longest edge bisection

Proposition 1. *For the longest edge bisection of any triangle t (see Figure 3), where $BC \leq CA \leq BA$, its holds that: a) $\alpha_1 \geq \alpha_0/2$ which implies $\beta_2 \geq 3\alpha_0/2$; b) If t is obtuse, $\alpha_1 \geq \alpha_0$ which implies $\beta_2 \geq 2\alpha_0$*

Lemma 1. *The longest edge bisection of any bad triangle BAC produces an improved triangle t_B and a bad quality triangle t_A. Usually t_A has largest angle greater than $2\pi/3$ and it is consequentely eliminated by edge swapping.*

3   Lepp (Terminal Edge) Midpoint Method

Given an angle tolerance θ_{tol}, the algorithm can be simply described as follows: iteratively, each bad triangle t_{bad} with smallest angle less than θ_{tol} in the current triangulation is eliminated by finding Lepp(t_{bad}), a pair of terminal triangles t_1, t_2, and associated terminal edge l. If non-constrained edges are involved, then the midpoint M of l is Delaunay inserted in the mesh. Otherwise the constrained point insertion criterion described below is used. The process is repeated until t_{bad} is destroyed in the mesh, and the algorithm finishes when the minimum angle in the mesh is greater than or equal to an angle tolerance θ_{tol}.

When the second longest edge CA is a constrained edge, the swapping of this edge is forbidden. In such a case, the insertion of point M would imply that the later processing of bad quality triangle MAC would introduce triangle MAM_1 (see Figure 4(a)) similar to triangle ABC implying an infinite loop situation. To avoid this behavior we introduce the following additional operation, which guarantees that M is not inserted in the mesh by processing triangle M_1BA.

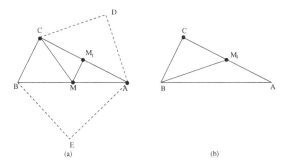

(a)   (b)

Fig. 4. (a) Over constrained edge CA, the insertion of M and M_1 produces triangle MAM_1 similar to triangle BAC; (b) Insertion of M_1 avoids this situation

Constrained edge point insertion: If CA is a constrained edge and BA is not a constrained edge, then insert midpoint M_1 of edge CA.

Special loop case. For Lepp-midpoint method, there is a rare special loop case discussed in [10], where a triangle MAM_1 similar to a bad-quality triangle

ABC can be obtained for a non-constrained edge CA. This happens when quadri-talerals BEAC and ADCM (see Figure 4(a)) are terminal quadrilaterals (where edges BA and CA are terminal edges respectively) together with some non-frequent conditions on neighbor constrained items. A necessary but not sufficient condition on the triangle ABC for this to happen is that angle $BMC \geq \pi/3$ which implies $\alpha_0 \geq \alpha_{limit} = arctan \frac{\sqrt{15}-\sqrt{3}}{3+\sqrt{5}} > 22°$ for obtuse triangle BAC [10]. This loop case can be avoided by adding some extra conditions to the algorithm. To simplify the analysis we restrict the angle tolerance to α_{limit} which is slightly bigger than the limit tolerance, equal to 20.7°, used to study both for the circuncenter and the off-center methods. The algorithm is given below:

Lepp Midpoint Algorithm
Input = a CDT, τ, and angle tolerance θ_{tol}
Find S_{bad} = the set of bad triangles with respect to θ_{tol}
for each t in S_{bad} **do**
    while t remains in τ **do**
        Find Lepp (t_{bad}), terminal triangles t_1, t_2 and terminal edge l. Triangle t_2 can be null for boundary l.
        Select Point (P, t_1, t_2, l)
        Perform constrained Delaunay insertion of P into τ
        Update S_{bad}
    end while
end for
Select Point (P, t_{term1}, t_{term2}, l_{term})
if (second longest edge of t_{term1} is not constrained and second longest edge of t_{term2} is not constrained) or l_{term} is constrained **then**
    Select P = midpoint of l_{term} and return
else
    for j = 1,2 **do**
        if t_{termj} is not null and has constrained second longest edge l^* **then**
            Select P = midpoint of l^* and return
        end if
    end for
end if

Angle and edge size bounds for Lepp midpoint method. The results of this section improve results discussed in [23]. Firstly we present a characteriza-tion of Delaunay terminal triangles based on fixing the second longest edge CA and choosing the smallest angle at vertex A. The diagram of Figure 5 (a) shows the possible locations for vertex B and the midpoint M. The diagram is defined as follows: (1) Since CB is a shortest edge, B lies inside the circular arc EFA of centre C and radius $|\overline{CA}|$. Consequently, M lies inside the circular arc $E'F'A$

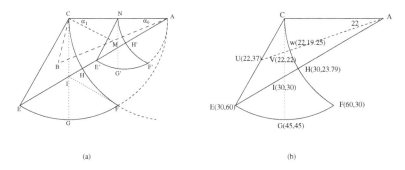

(a)   (b)

Fig. 5. (a) EFC and E'F'N' are geometrical places for vertex B and midpoint M for a terminal triangle BAC with respective smallest and largest angle of vertices A and C. (b) Distribution of angles (α_0, α_1).

of centre $N = (C + A)/2$ and radius $|\overline{CA}|/2$; (2) Since BA is a longest edge, B lies outside the circular arc CF of centre A and radius $|C - A|$, and so M lies outside circular arc NF' of centre A, radius $|\overline{CA}|/2$; (3) According to Theorem 2, the line CE makes an angle of 120° with CA.

Now we use the diagram of Figure 5(a) both to improve the bounds on α_1 (Figure 3) and to bound the minimum distance from M to the previous neighbor vertices in the mesh. To this end consider the distribution of the ordered pair of angles (α_0, α_1) illustrated in Figure 5(b). As expected for the right triangles (B over CG) $\alpha_1 = \alpha_0$. Also note that the ratio α_1/α_0 decreases from 2 to 1 along line EC (obtuse triangles with largest angle equal to $2\pi/3$), while the ratio α_1/α_0 increases from 1/2 to 1 along arc F to C (isosceles acute triangles with two longest edges). Note that segment lines UW and EH correspond to fixed smallest angle equal to α_{limit} and 30° respectively.

Lemma 2. *For acute Delaunay terminal triangles of smallest angle α, there exist constants C_1, C_2 such that:*

a) $\alpha_0 \leq 30°$ (B in region CIH) implies $\alpha_1 \geq C_1\alpha_0$ with $C_1 \approx 0.79$.
b) $\alpha_0 \leq \alpha_{limit}$ (B in region CVW) implies $\alpha_1 \geq C_2\alpha_0$ $C_2 \approx 0.866$
c) The ratio α_1/α_0 approaches 1.0 both when α_0 decreases, and when BAC becomes a right triangle.
d) Using the notation of Figure 3, $\beta_2 \geq (1 + C_1)\alpha_0$ for $\alpha_0 \leq 30°$.

Proof for (d) note that $\beta_2 = \alpha_1 + \alpha_0$ (Figure 3)   □

In order to bound the minimum distance from M to previous vertices in the mesh, we use both the properties of the longest edge bisection of a Delaunay terminal triangle BAC and the constraint on the empty circumcircle. Note that the circumcenter O of an obtuse (acute) triangle is situated in the exterior (the interior) of the triangle. Furthermore for any non constrained Delaunay obtuse triangle t, the distance $d = MO$ from the circumcenter O to the longest edge

BA (see Figure 2(c)) satisfies that $0 < d < r/2$, where r is the cicunradius. We will consider the limit cases $d = r/2$ and $d = 0$, which respectively correspond to largest angles equal to $2\pi/3$ and $\pi/2$, as well as the cases $\alpha = \alpha_{limit}$ and $\alpha = 14°$ to state bounds for obtuse and acute triangles.

Lemma 3. *Consider any Delaunay terminal triangle t of smallest angle α and terminal edge AB of midpoint M (Figure 2(c)), and let d(M) be the minimum distance from M to any vertex of the mesh. Then there exists constants C_1, C_2, C_3, C_4 such that*

a) For acute t
   (i) $\alpha \leq \alpha_{limit}$ implies $d(M) \geq C_1 \mid BC \mid$ with $C_1 > 1.3$
   (ii) $\alpha \leq 30°$ implies $d(M) \geq C_2 \mid BC \mid$ with $C_2 = 1$
b) For obtuse t
   (i) $\alpha \leq 14°$ implies $d(M) \geq C_3 \mid BC \mid$ with $C_3 > 1$
   (ii) $\alpha \leq \alpha_{limit}$ implies $d(M) \geq C_4 \mid BC \mid$ wiht $C_4 > 0.66$
   (iii) $\alpha \leq 30°$ implies $d(M) \geq C_5 \mid BC \mid$ with $C_4 = 0.5$
c) obtuse t with $\alpha > 14°$ implies $\beta_2 > 28° > \alpha_{limit}$

The following theorem based on Proposition 1 and Lemmas 2 and 3 assure that bad quality terminal triangles are quickly improved by introducing a sequence of better triangles of edge CB, but not introducing points too close to the previous vertices in the mesh. This improves the results presented in [23].

Theorem 3. *Consider $\theta_{tol} \leq \alpha_{limit}$. Then for any bad quality terminal triangle of smallest angle α, a finite sequence of improved triangles t_B (Figure 3), is obtained until t_B is good such that: (a) For $\alpha \leq 14°$, none edge smaller than the existing neighbor edges is inserted in the mesh. (b) Only at the last improvement step (when $\alpha > 14°$) a small smallest edge, at least 0.66 times the size of a previous neighbor smallest edge, can be occassionaly introduced in the mesh for obtuse triangle t.*

Remark: Note that the worst $d(M)$ value is obtained for obtuse triangles, for which in turn the angles are most improved.

4   Lepp Centroid Algorithm

The Lepp centroid method was designed both to avoid the loop situation discussed in section 3 and to improve the slower convergence reported in [11] for $\theta_{tol} > 25°$ due to the fact that good quality acute terminal triangles can produce a slightly bad triangle t_A (Figure 3). Instead of selecting an edge aligned midpoint M, we select the centroid of a terminal quad defined as the quadrilateral $ACBD$ formed by a couple of terminal triangles BAC and BDA (Figure 6) sharing an unsconstrained terminal edge. The algorithm is given below:

Lepp–Terminal-Centroid Algorithm
Input = a CDT, τ, and angle tolerance θ_{tol}
Find S_{bad} = the set of bad triangles with respect to θ_{tol}
for each t in S_{bad} **do**
   while t remains in τ **do**
      Find Lepp (t_{bad}), terminal triangles t_1, t_2 and terminal edge l. Triangle
      t_2 can be null for boundary l.
      Select Point (P, t_1, t_2, l)
      Perform constrained Delaunay insertion of P into τ
      Update S_{bad}
   end while
end for

New Select Point (P, t_{term1}, t_{term2}, l_{term})
if l_{term} is constrained **then**
   Select P = midpoint of l_{term} and return
end if
if (second longest edge of t_{term1} is not constrained and second longest edge
of t_{term2} is not constrained) **then**
   Select P = centroid of quad (t_{term1}, t_{term2}) and return
else
   for j = 1,2 **do**
      if t_{termj} is not null and has constrained second longest edge l^* **then**
         if t_{termj} does not contain the centroid **then**
            Select P = midpoint of l^* and return
         else
            Select P = centroid of quad $((t_{term1}, t_{term2})$
         end if
      end if
   end for
end if

Geometrical properties of the centroid selection. Consider a terminal quadrilateral $ACBD$ formed by the union of a pair of terminal triangles ABC, ABD sharing the terminal edge AB (Figure 6). In what follows we will prove that inserting the terminal centroid Q, defined as the centroid of the terminal quadrilateral $ACBD$, always produce a better new triangle t_B (Figure 3). Consider a coordinate system based on a bad quality triangle ABC of longest edge BA, with center in B and x-axis over BA as shown in Figure 6. This is the longest edge coordinate system introduced by Simpson [25]. As discussed in the proof of Theorem 2, D must lie in the 'triangle' UVW defined by circumcircle $CC(ABC)$ and the lens VA and VB which are arcs of the circles of radious BA and respective centers A and B. The coordinates of the vertices are $B(0,0)$, $A(l_{max}, 0)$, $C(a_c, b_c)$ and $D(a_D, b_D)$.

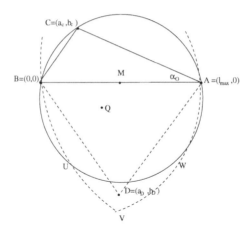

Fig. 6. Estimating location of centroid Q

Let $U(a_{min}, b_{min})$ [$W(a_{max}, b_{min})$] be the leftmost [*rightmost*] intersection point of the lens and $CC(ABC)$ as shown. The following lemma bounds the location of Q by parameters determined by triangle ABC.

Lemma 4. *Let the centroid Q have longest edge coordinates (a_Q, b_Q) with respect to triangle ABC. Then*

$$a_{min}/4 + l_{max}/4 < a_Q < a_{max}/4 + l_{max}/2, \text{ and } b_V/4 < b_Q < (b_c + b_{min})/4$$

Proof. *The proof follows from computing the centroid coordinates*

$$a_Q = (a_C + a_D + l_{max})/4 \text{ and } b_Q = (b_c + b_D)/4$$

Corollary 1. *For any terminal quad involving a pair of (obtuse acute) terminal triangles, the quad centroid is situated in the interior of the acute triangle.*

Corollary 2. *Let t_a, t_b be terminal triangles forming a terminal quadrilateral, and sharing a terminal edge, E. The shortest edge that results from the insertion of the terminal centroid into the mesh is longer than the shortest edge that results from inserting the midpoint of E.*

Theorem 4. *The algorithm does not suffer of the special loop case (section 3).*

Proof. The centroid Q is not aligned with the vertices of the terminal edge excepting the case of right isosceles triangles sharing a longest edge.

Remark 1. The constraint on θ_{tol} is not necessary and the analysis of the algorithm can be extended until $\theta_{tol} = 30°$.

5   Empirical Study and Concluding Remarks

We consider the 3 test problems of Figure 7 whose (bad quality) initial triangulations are shown in this figure. They correspond to a square with 400

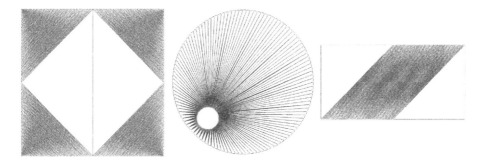

Fig. 7. Initial triangulations for the test problems

equidistributed boundary points (Square400) and discretized circle having a discretized circular hole close to the boundary with 240 boundary points (Circle 240), a rectangle having 162 boundary points distributed as shown in Figure 7 (Rectangle 162). We also consider an A test case having 21 boundary points, provided at the Triangle site. The initial triangulations are Delaunay excepting the triangulation of the Rectangle 162 case, which is an example proposed by Edelsbrunner [4] for the point triangulation problem.

Lepp centroid versus Lepp midpoint. We used the test cases of Figure 7 to compare the algorithms. We ran every case for different θ_{tol} values, with the input meshes of Figure 7, until reaching the maximum practical tolerance angle θ_{tol}. We studied the mesh quality both with respect to the smallest angle and with respect to the area quality measure defined as $q(t) = CA(t)/l^2$, where $A(t)$ is the area of triangle t, l is its longest edge and C is a constant such that $q(t) = 1$ for the equilateral triangle. For both algorithms we studied the evolution of the minimum smallest angle, the minimum area quality measure, the average smallest angle, the normalized minimum edge size (wrt the minimum edge size in the input mesh) and the average Lepp size. As expected the smallest angle and the area quality measure show analogous behavior.

The empirical study shows that for every test problem, the Lepp centroid method computes significantly smaller triangulations than the Lepp midpoint variant and terminates for higher threshold angle θ_{tol}. The Lepp centroid works for θ_{tol} up to $36°$ while the previous midpoint method works for $\theta_{tol} \leq 30°$ for these examples. If is worth noting that the Lepp centroid method produces globally better triangulations having both significantly higher average smallest angle and a smaller percentage of bad quality triangles than the Lepp midpoint variant for the same θ_{tol} value. To illustrate this see Figure 8 and Figure 9 which respectively show the behavior of the minimum area quality measure and of the average smallest angles for both algorithms. This is also illustrated in Figure 11 which shows the evolution of the area quality distribution for the Square 400 test problem for θ_{tol} equal to $10°, 25°$.

The normalized minimum edge size in the mesh (wrt the smallest edge in the initial mesh) is shown in Figure 10. Note that this parameter behaves better

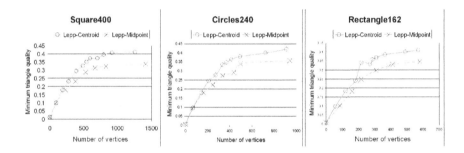

Fig. 8. Evolution of the minimum area quality measure as a function of the number of vertices for Lepp-centroid and Lepp-midpoint algorithms

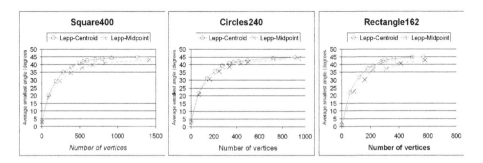

Fig. 9. Evolution of the average smallest angle (degrees) as a function of the number of vertices for Lepp-centroid and Lepp-midpoint algorithms

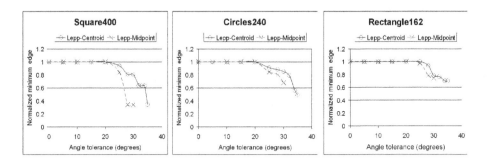

Fig. 10. Normalized minimum edge size (wrt the smallest edge in the initial mesh) as a function of the angle tolerance θ_{tol} for centroid and midpoint algorithms

than predicted by the theoretical results of section 3, even for the Lepp midpoint method. For all these problems the average Lepp size remains between 3 to 5, being this value slightly higher for the Lepp centroid method. Finally the triangulation obtained with the Lepp-centroid method (and for the Triangle method) for the Circles 240 test problem for $\theta_{tol} = 32°$ is shown in Figure 12.

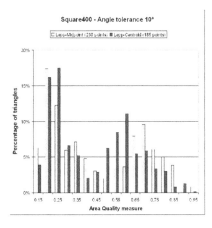
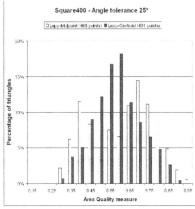

Fig. 11. Area quality distribution for Square 400 for $\theta_{tol} = 10°$ (midpoint 230 points, centroid 185 points) and $\theta_{tol} = 25°$ (midpoint 668 points, centroid 491 points)

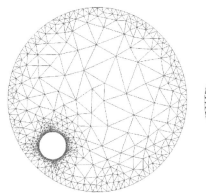
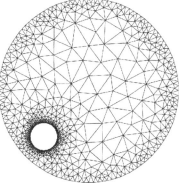

Centroid 32 degrees, 493 points added   Triangle 32 degrees, 639 points added

Fig. 12. Triangulations obtained with Lepp-centroid and Triangle methods, $\theta_{tol} = 32°$

A comparison with Triangle. As reported previously in [11] the Lepp midpoint method showed a behavior analogous to the circumcenter algorithm implemented in a previous Triangle version [15]. Here we use the current Triangle version based on the off-center point selection [19,24], to perform a comparison with the Lepp centroid method for the test problems of Figure 7 and the A test case. The evolution of the minimum angle in the mesh as a function of the number of vertices is shown in Figure 13 for the same set of θ_{tol} values used in the preceding subsection. Note that for all these cases the Lepp-centroid method worked well (with reasonable number of vertices inserted) for θ_{tol} up to 36°, while that the Triangle method worked for θ_{tol} up to 35° but increasing highly the number of points inserted for θ_{tol} bigger than 25°. Only for the small A-test-case, where basically boundary points are inserted, both algorithms have more

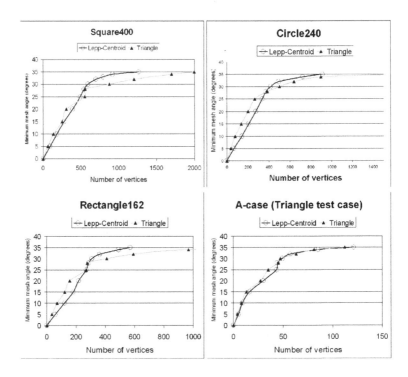

Fig. 13. Evolution of the minimum angle in the mesh as a function of the number of vertices for Lepp-centroid and Triangle methods

Number of points added							
Minimum mesh angle (degrees)	Square400		Circles240		Rectangle162		A-case	
	Centroid	Triangle	Centroid	Triangle	Centroid	Triangle	Centroid	Triangle
20	391	310	274	202	217	156	30	27
25	491	553	335	266	271	270	43	35
28	543	557	371	386	286	277	44	45
30	595	880	416	501	310	411	49	47
32	703	1202	493	639	363	592	57	63
34	919	1700	722	897	490	968	85	81
35	1264	1997	911	1811	570	4853	121	112
36	1843	-	1630	-	5473	-	239	-

similar behavior. A quantitative view of the behavior of both algorithms is given in the Table below which shows the number of points added to the mesh for all the test problems and different values of θ_{tol}. The triangulations obtained for the Circles240 case with the Lepp-centroid and Triangle methods for $\theta_{tol} = 32°$ are shown in Figure 12.

Concluding remarks. The results of this paper suggest that for Lepp-centroid method the algorithm analysis can be extended until 30°. In effect, in section 3 we prove that for the Lepp-midpoint method, under the constraint $\theta_{tol} \leq \alpha_{limit} \approx 22°$, and for improving a bad quality triangle ABC and its t_B sons (Figure 3), a small number of points is introduced which are not situated too close to the previous vertices in the mesh. In exchange, the results of section 4 guarantee both that the centroid method does not suffer of special looping conditions and that smallest edges bigger than those introduced by the midpoint method are introduced, which suggest that for the Lepp-centroid method the analysis can be extended for $\theta_{tol} \leq 30°$. In this paper we also provide empirical evidence that show that the Lepp-centroid method behaves in practice better than the off-center algorithm for $\theta_{tol} > 25°$.

Recently Erten and Ungor [24] have introduced algorithms that improve the off-center performance with respect to the mesh size and the angle θ_{tol} by using point selections depending on some triangle cases. We plan to improve the Lepp-based algorithms also in this direction.

Acknowledgements. Research supported by DI ENL 07/03. We are grateful with Bruce Simpson who contributed to an early formulation of this paper.

References

1. Chew, L.P.: Guaranteed-quality triangular meshes. Technical report TR-98-983, Computer Science Department, Cornell University, Ithaca, NY (1989)
2. George, P.L., Borouchaki, H.: Delaunay Triangulation and Meshing. In: Hermes (1998)
3. Borouchaki, H., George, P.L.: Aspects of 2-D Delaunay Mesh Generation. International Journal for Numerical Methods in Engineering 40, 1997 (1975)
4. Edelsbrunner, H.: Geometry and Topology for Mesh Generation. In: Cambridge Monographs on Applied and Computational Mathematics, Cambridge Univ. Press, Cambridge (2001)
5. Rivara, M.C.: Algorithms for refining triangular grids suitable for adaptive and multigrid techniques. International Journal for Numerical Methods in Engineering 20, 745–756 (1984)
6. Rivara, M.C.: Selective refinement/derefinement algorithms for sequences of nested triangulations. International Journal for Numerical Methods in Engineering 28, 2889–2906 (1989)
7. Rivara, M.C., Levin, C.: A 3d Refinement Algorithm for adaptive and multigrid Techniques. Communications in Applied Numerical Methods 8, 281–290 (1992)
8. Rivara, M.C.: New mathematical tools and techniques for the refinement and / or improvement of unstructured triangulations. In: Proceedings 5th International Meshing Roundtable, Pittsburgh, pp. 77–86 (1996)
9. Rivara, M.C.: New longest-edge algorithms for the refinement and/or improvement of unstructured triangulations. International Journal for Numerical Methods in Engineering 40, 3313–3324 (1997)
10. Rivara, M.C., Hitschfeld, N., Simpson, R.B.: Terminal edges Delaunay (small angle based) algorithm for the quality triangulation problem. Computer-Aided Design 33, 263–277 (2001)

11. Rivara, M.C., Palma, M.: New LEPP Algorithms for Quality Polygon and Volume Triangulation: Implementation Issues and Practical Behavior. In: Cannan, S.A. (ed.) Trends unstructured mesh generation, Saigal, AMD, vol. 220, pp. 1–8 (1997)
12. Ruppert, J.: A Delaunay refinement algorithm for quality 2-dimensional mesh generation. J. of Algorithms 18, 548–585 (1995)
13. Baker, T.J.: Automatic mesh generation for complex three dimensional regions using a constrained Delaunay triangulation. Engineering with Computers 5, 161–175 (1989)
14. Baker, T.J.: Triangulations, Mesh Generation and Point Placement Strategies. In: Caughey, D. (ed.) Computing the Future, pp. 61–75. John Wiley, Chichester (1995)
15. Shewchuk, J.R.: Triangle: Engineering a 2D Quality Mesh Generator and Delaunay Triangulator. In: First Workshop on Applied Computational Geometry, pp. 124–133. ACM Press, New York (1996)
16. Aurehammer, F., Katoh, N., Kokima, H., Ohsaki, M., Xu, Y.-F.: Approximating uniform triangular meshes in polygons. Theoretical Computer Science, 879–895 (2002)
17. Weatherill, N.P., Hassan, O.: Efficient three-dimensional Delaunay triangulation with automatic point creation and imposed boundary constraints. International Journal for Numerical Methods in Engineering (2039)
18. Rosenberg, I.G., Stenger, F.: A lower bound on the angles of triangles constructed by bisecting the longest side. Mathematics of Computation 29, 390–395 (1975)
19. Üngor, A.: Off-centers: a new type of Steiner points for computing size-optimal quality-guaranteed Delaunay triangulations. In: Orejas, F., Spirakis, P.G., van Leeuwen, J. (eds.) ICALP 2001. LNCS, vol. 2076, pp. 152–161. Springer, Heidelberg (2001)
20. Simpson, B., Rivara, M.C.: Geometrical mesh improvement properties of Delaunay terminal edge refinement. In: Kim, M.-S., Shimada, K. (eds.) GMP 2006. LNCS, vol. 4077, pp. 536–544. Springer, Heidelberg (2006)
21. Stynes, M.: On Faster Convergence of the Bisection Method for certain Triangles. Mathematics of Computation 33, 1195–1202 (1979)
22. Stynes, M.: On Faster Convergence of the Bisection Method for all Triangles. Mathematics of Computation 35, 1195–1202 (1980)
23. Rivara, M.C.: A study on Delaunay terminal edge method. In: Pébay, P.P. (ed.) Proceedings of the 15th International Roundtable, pp. 543–562. Springer, Heidelberg (2006)
24. Erten, H., Üngor, A.: Triangulations with locally optimal Steiner points. In: Belyaev, A., Garland, M. (eds.) Eurographics Symposium on Geometry Processing (2007)
25. Simpson, R.B.: Gometry Independence for a Meshing Engine for 2D Manifolds. International Journal for Numerical Methods in Engineering, 675–694 (2004)

Mean Value Bézier Maps

Torsten Langer[1], Alexander Belyaev[2], and Hans-Peter Seidel[1]

[1] MPI Informatik
Campus E1 4, 66123 Saarbrücken, Germany
{langer,hpseidel}@mpi-inf.mpg.de
http://www.mpi-inf.mpg.de
[2] Heriot-Watt University
Riccarton, Edinburgh EH14 4AS, Scotland, United Kingdom
a.belyaev@hw.ac.uk
http://www.eps.hw.ac.uk/~belyaev

Abstract. Bernstein polynomials are a classical tool in Computer Aided Design to create smooth maps with a high degree of local control. They are used for the construction of Bézier surfaces, free-form deformations, and many other applications. However, classical Bernstein polynomials are only defined for simplices and parallelepipeds. These can in general not directly capture the shape of arbitrary objects. Instead, a tessellation of the desired domain has to be done first.

We construct smooth maps on arbitrary sets of polytopes such that the restriction to each of the polytopes is a Bernstein polynomial in mean value coordinates (or any other generalized barycentric coordinates). In particular, we show how smooth transitions between different domain polytopes can be ensured.

Keywords: Mathematical foundations of CAGD, Shape representation, Curves and Surfaces, Computer Graphics, Mean value coordinates, Bézier surfaces.

1 Introduction

Bernstein polynomials are at the core of classical Computer Aided Design. In the 1960s, they were used for the construction of Bézier surfaces [1, 2, 3], which remain an important tool until today. Later, Bernstein polynomials were applied to define free-form deformations of 3D space [4, 5]. More general, they can be used to construct any kind of smooth map that requires local control.

In this paper, we use the notion of *Bézier maps* to denote polynomial functions $f : \mathbb{R}^d \to \mathbb{R}^e$ in the form of simplicial Bézier maps

$$f(\lambda) = \sum_{|\alpha|=n} b_\alpha B_\alpha^n(\lambda) \tag{1}$$

or tensor product Bézier maps

$$f(\mathbf{x}) = \sum_{i_1,\dots,i_d=0}^{n} b_{i_1 \dots i_d} \prod_{j=1}^{d} B_{i_j}^n(x_j) \tag{2}$$

F. Chen and B. Jüttler (Eds.): GMP 2008, LNCS 4975, pp. 231–243, 2008.
© Springer-Verlag Berlin Heidelberg 2008

where $\lambda := \lambda(\mathbf{x})$ are barycentric coordinates of $\mathbf{x} := (x_1, \dots x_d)$ with respect to a domain simplex (or polytope) $P \subset \mathbb{R}^d$ with vertices $\{\mathbf{v}_1, \dots \mathbf{v}_k\}$ ($k = d + 1$ if P is a simplex) while (2) is defined over the domain $[0, 1]^d$. n is the polynomial degree, $b_\alpha \in \mathbb{R}^e$ and $b_{i_1 \dots i_d} \in \mathbb{R}^e$ are the control points, and B_α^n and B_i^n are the Bernstein polynomials defined by

$$B_\alpha^n(\lambda) = \frac{n!}{\alpha!}\lambda^\alpha \,, \qquad B_i^n(x) = \binom{n}{i}(1 - x)^{n-i}x^i \qquad (3)$$

where we use the standard multi-index notation $\alpha := (\alpha_1, \dots \alpha_k) \in \mathbb{N}^k$ with $|\alpha| := \sum_i \alpha_i$, $\alpha! := \prod_i \alpha_i!$, and $\lambda^\alpha := \prod_i \lambda_i^{\alpha_i}$.

Important special cases of Bézier maps are on the one hand Bézier curves and (hyper-) surfaces where $e > d$ and usually $d = 1$ or $d = 2$. On the other hand, if $d = e$, we obtain space deformations. Sederberg and Parry [5] used tensor product Bernstein polynomials defined on parallelepipeds in $\mathbb{R}^3$ to specify such free-form deformations. In this case, the control points b_{ijk} indicate the position and shape of the deformed parallelepiped. However, the restriction on the shape of the domain makes it sometimes difficult to adapt the deformation to complex real objects. This restriction can be overcome by generalizing the barycentric coordinates λ_i in (1) from simplices to more general polytopes. A first step in this direction was done by Loop and DeRose [6] who introduced coordinate functions l_i in order to define Bézier surfaces over regular k-gons. These coordinates are a special case of the Wachspress coordinates [7] that are defined inside of arbitrary convex polygons and were introduced to computer graphics by Meyer et al. [8]. A further generalization led to the definition of Wachspress coordinates for convex polytopes of higher dimensions [9, 10].

Another generalization of barycentric coordinates are the mean value coordinates [11, 12], which were extended to higher dimensions later on [13, 14, 15] (see also references therein). They have the advantage of being defined for arbitrary, convex and non-convex, polytopes. Unfortunately, mean value coordinates are only C^0-continuous at vertices [16]. Langer and Seidel addressed the latter problem and showed that the higher order discontinuities at the vertices vanish in the context of Bézier maps if the control points b_α satisfy certain continuity constraints [17]. They pointed out that mean value Bézier maps have a greater number of control points, and hence greater flexibility, than traditional Bézier maps. Their solution, however, is only valid for Bézier maps defined on a square. Thus, the mean value coordinates lost their greatest strength: to be defined with respect to arbitrary polytopes. Finally, it should be remarked that also a class of coordinates exists that is defined on point clouds instead of polytopes. These coordinates are not dependent on the connectivity of a particular polytope but they can not take advantage of such a structure either. The probably best known representatives for these kind of coordinates are Sibson coordinates [18]. Their use in defining generalized Bézier surfaces was discussed by Farin [19].

When constructing a smooth map consisting of several polynomials that are defined on adjoining polytopes, we have to ensure that the respective polynomials connect smoothly. For connecting simplicial and tensor product polynomials, a well developed theory is available. In [20], a smooth joint for a regular pentagon is constructed. Loop and DeRose [6, Sections 6 and 7] show how regular k-gons and triangles can be smoothly connected if Bernstein polynomials in Wachspress coordinates are used. This approach is extended in [21] where the control net for a complete surface is constructed. Unfortunately, their polynomial representation algorithm requires coordinates that are rational polynomial functions. The reason, in short, is that their proof uses the polarization of a polynomial. Hence, it cannot be carried over to mean value Bézier maps (Bézier maps based on mean value coordinates). Furthermore, their method does not cover the case of general domain polygons but only regular k-gons.

In this paper, we derive constraints on the control points of Bézier maps in arbitrary generalized barycentric coordinates to obtain smooth transitions between arbitrary domain polytopes. One essential requirement, as noted in [22], is to adopt an indexing scheme that is adapted to the given polytopes. We chose to use multi-indices (as has been done before in [6]). They correspond to the Minkowski sum approach in [22]. (The set of multi-indices of degree n for a polytope with vertex set V corresponds to $\bigoplus_{i=1}^{n} V$.)

2 Theoretical Foundation

Classical barycentric coordinates specify local coordinates $\lambda_i(\mathbf{x})$ for a point $\mathbf{x}$ with respect to a simplex. When generalizing this concept from simplices to arbitrary polytopes[1] P with vertices $\{\mathbf{v}_1 \ldots \mathbf{v}_k\}$, we require that the λ_i satisfy

$$\sum_i \lambda_i(\mathbf{x}) = 1 \qquad \text{partition of unity,} \tag{4}$$

$$\sum_i \lambda_i(\mathbf{x})\mathbf{v}_i = \mathbf{x} \qquad \text{linear precision.} \tag{5}$$

We call a set of continuous functions $\lambda_i(\mathbf{x})$ that satisfies (4) and (5) *barycentric coordinates*. They are *positive* if additionally

$$\forall i \; \lambda_i(\mathbf{x}) > 0 \qquad \text{positivity} \tag{6}$$

holds for all points $\mathbf{x}$ within convex polytopes.

Barycentric coordinates for polytopes can be inserted in (1) to obtain (generalized) Bézier maps. Wachspress coordinates and mean value coordinates are the most prominent positive barycentric coordinates. Further types of coordinates can be found in [23, 24, 15, 25, 26, 27]. It has been observed [6] that Bézier maps based on Wachspress coordinates defined on a square lead to the well-known tensor product Bézier maps. Mean value Bézier maps have the advantage that their domain is not restricted to convex polygons. For all kinds of Bézier maps the following properties are satisfied.

Proposition 2.1. *Let λ_i be barycentric coordinates with respect to a polytope P, and let the Bernstein polynomials B_α^n and a Bézier map f be defined as in (3) and (1). Then the following properties hold:*

1. $B_\alpha^n(\lambda) = \sum_{i=1}^k \lambda_i B_{\alpha-\mathbf{e}_i}^{n-1}(\lambda)$ *(for any Bernstein polynomial B_β^m, $|\beta| = m$, we use the convention $B_\beta^m(\lambda) := 0$ if one of the $\beta_i < 0$),*

---

[1] A n-dimensional polytope is bounded by $(n-1)$-dimensional hyper-planes. It is a generalization of polygons and polyhedra to higher dimensions.

2. *let $(\mathbf{v}_{i_0}, \mathbf{v}_{i_1})$ be an edge of P, then the boundary curve $f(\lambda((1 - t)\mathbf{v}_{i_0} + t\mathbf{v}_{i_1}))$ is a Bézier curve with control points $(b_{(n-j)\mathbf{e}_{i_0}+j\mathbf{e}_{i_1}})_{j=0}^{n}$,*

3. *$\{B_\alpha^n\}$ forms a partition of unity; if P is convex and the λ_i are positive coordinates, the partition of unity is positive within P,*

4. *if P is convex and the λ_i are positive coordinates, the image of P under $f(\lambda(\mathbf{x}))$ is contained in the convex hull of the b_α,*

5. *the de Casteljau algorithm works: let $f(\lambda) = \sum_{|\alpha|=n} b_\alpha B_\alpha^n(\lambda)$ be a Bézier map with coefficients b_α. For $m \in \mathbb{N}$ and a given β with $|\beta| = n - m$, let $b_\beta^m(\lambda) := \sum_{|\alpha|=m} b_{\beta+\alpha} B_\alpha^m(\lambda)$. Then $P(\lambda) = b_0^n(\lambda)$ can be computed from the $b_\beta^0(\lambda) = b_\beta$ via the recursive relation $b_\beta^m(\lambda) = \sum_{i=1}^{k} \lambda_i b_{\beta+\mathbf{e}_i}^{m-1}(\lambda)$.*

($\mathbf{e}_i$ *denotes the multi-index with components* $(\mathbf{e}_i)_j = \delta_{ij}$, *and* $\mathbf{0}$ *denotes the multi-index with components* $\mathbf{0}_j = 0$.)

To join several Bézier maps smoothly, it is important to know their derivatives. In the remainder of the paper, we will assume that the λ_i are differentiable everywhere apart from the vertices $\mathbf{v}_i$. This is in particular true for Wachspress and mean value coordinates. Using the chain rule, it is straightforward to obtain

Lemma 2.2. *Let*

$$f(\lambda) = \sum_{|\alpha|=n} b_\alpha B_\alpha^n(\lambda) \ . \tag{7}$$

Then the first derivatives of f are given by

$$\frac{\partial}{\partial x_i} f(\lambda(\mathbf{x})) = n \sum_{|\alpha|=n-1} \sum_{j=1}^{k} \frac{\partial}{\partial x_i} \lambda_j(\mathbf{x}) b_{\alpha+\mathbf{e}_j} B_\alpha^{n-1}(\lambda(\mathbf{x})) \ . \tag{8}$$

However, the derivatives $\frac{\partial}{\partial x_i} \lambda_j$ are in general not easy to compute. Nevertheless, in [17], constraints on the control points b_α to achieve smooth derivatives across common (hyper-)faces of polytopes have been derived without exact knowledge of the derivatives of λ_j. Unfortunately, the proof is based on properties of barycentric coordinates that are specific to coordinates defined in a square. Since we want to have smooth transitions of Bézier maps defined on arbitrary polytopes, we need a more general approach. In the following, we give sufficient conditions for the control points b_α to join arbitrary polytopes smoothly.

Basically, the control points at the common (hyper-)faces and adjacent to it must be determined by affine functions A_β and these functions must coincide across these faces. This is visualized in Figure 1. The figure shows a Bézier surface and its control net from several viewpoints. The domain consists of a pentagon and an L-shaped hexagon that share two common edges (shown in black below the surface). On the right, the control net is shaded to indicate the smoothness conditions. The parts of the control net that correspond to the three common vertices of the two polygons are affine images of the domain polygons. They are drawn in different shades of gray.

We make this idea more precise in the following theorems. We begin by expressing the derivatives of a Bézier map with respect to the control points.

Theorem 2.3. *Let* $f(\lambda) = \sum_{|\alpha|=n} b_\alpha B_\alpha^n(\lambda)$ *be a Bézier map defined with respect to a polytope P with vertices* $\mathbf{v}_i$. *Assume that for every multi-index* β *with* $|\beta| = n - 1$ *an affine function* A_β *exists such that* $b_{\beta+\mathbf{e}_i} = A_\beta(\mathbf{v}_i)$ *for all* $i = 1..k$.

Then, the derivative of f with respect to a differential operator $\partial \in \left\{ \frac{\partial}{\partial x_i} \right\}$ *is*

$$\partial f(\lambda(\mathbf{x})) = n \sum_{|\beta|=n-1} \partial A_\beta \cdot B_\beta^{n-1}(\lambda(\mathbf{x})) \ .$$

Proof. The proof consists of a direct calculation.

$$\partial f(\lambda(\mathbf{x})) = n \sum_{|\beta|=n-1} \sum_{i=1}^{k} \partial \lambda_i(\mathbf{x}) b_{\beta+\mathbf{e}_i} B_\beta^{n-1}(\lambda(\mathbf{x})) \qquad \text{Lemma 2.2}$$

$$= n \sum_{|\beta|=n-1} \sum_{i=1}^{k} \partial \lambda_i(\mathbf{x}) A_\beta(\mathbf{v}_i) B_\beta^{n-1}(\lambda(\mathbf{x})) \qquad \text{definition of } A_\beta$$

$$= n \sum_{|\beta|=n-1} B_\beta^{n-1}(\lambda(\mathbf{x})) \partial \sum_{i=1}^{k} \lambda_i(\mathbf{x}) A_\beta(\mathbf{v}_i) \qquad \text{linearity of } \partial$$

$$= n \sum_{|\beta|=n-1} B_\beta^{n-1}(\lambda(\mathbf{x})) \partial A_\beta\Big(\sum_{i=1}^{k} \lambda_i(\mathbf{x}) \mathbf{v}_i \Big) \qquad \text{affine linearity of } A_\beta$$

$$= n \sum_{|\beta|=n-1} B_\beta^{n-1}(\lambda(\mathbf{x})) \partial A_\beta(\mathbf{x}) \qquad \text{linear precision (5) for } \lambda(\mathbf{x})$$

$$= n \sum_{|\beta|=n-1} B_\beta^{n-1}(\lambda(\mathbf{x})) \partial A_\beta \qquad A_\beta(\mathbf{x}) \text{ has constant derivative.} \quad (9)$$

$\square$

In the same way, we can compute higher derivatives:

Corollary 2.4. *In the situation of Theorem 2.3, assume that for every multi-index* γ *with* $|\gamma| = n - 2$ *an affine function* A_γ' *exists such that* $\partial A_{\gamma+\mathbf{e}_i} = A_\gamma'(\mathbf{v}_i)$ *for all* $i = 1..k$.

Then, the derivative $\partial' \partial f$ *of f with respect to a differential operator* $\partial' \in \left\{ \frac{\partial}{\partial x_i} \right\}$ *is*

$$\partial' \partial f(\lambda(\mathbf{x})) = n(n-1) \sum_{|\gamma|=n-2} \partial' A_\gamma' \cdot B_\gamma^{n-2}(\lambda(\mathbf{x})) \ .$$

Respective statements hold for the higher derivatives of f.

Proof. Since $\frac{1}{n}\partial f(\lambda) = \sum_{|\beta|=n-1} \partial A_\beta B_\beta^{n-1}(\lambda)$ is a Bézier map with coefficients ∂A_β, the claim follows immediately from Theorem 2.3 $\square$

Corollary 2.5 (Smooth mean value Bézier maps). *Let* $f(\lambda) = \sum_{|\alpha|=n} b_\alpha B_\alpha^n(\lambda)$ *be a Bézier map where the* λ_i *are the mean value coordinates with respect to a polytope P with vertices* $\mathbf{v}_i$. *Assume that an affine function* A_i *exists such that* $b_{(n-1)\mathbf{e}_i+\mathbf{e}_j} = A_i(\mathbf{v}_j)$ *for all* $j = 1..k$.

Then, the derivative of f with respect to any differential operator $\partial \in \left\{ \frac{\partial}{\partial x_i} \right\}$ has a continuous extension to $\mathbf{v}_i$ and

$$\lim_{\mathbf{x} \to \mathbf{v}_i} \partial f(\mathbf{x}) = n\, \partial A_i \ .$$

Respective statements hold for the higher derivatives of f.

Proof. We observe that the outer sum in (9) collapses to a single summand if the limit $\mathbf{x} \to \mathbf{v}_i$ is considered. We obtain the claim from the remaining term. $\square$

Finally, we obtain sufficient constraints on the b_α to achieve smooth Bézier maps across common (hyper-)faces of polytopes (common vertices or edges in the case of polylines or polyhedra). Note that the extent, to which these constraints are necessary as well, is not yet known.

Corollary 2.6 (Continuity across polytope boundaries). *Let $f(\lambda) = \sum_{|\alpha|=n} b_\alpha B_\alpha^n(\lambda)$ and $f'(\lambda') = \sum_{|\alpha|=n} b'_\alpha B_\alpha^n(\lambda')$ be Bézier maps defined with respect to polytopes P and P' that share a common (hyper-)face $\mathbf{f}$ (without loss of generality, let corresponding vertices have the same indices; that implies $\lambda_i(\mathbf{x}) = \lambda'_i(\mathbf{x})$ for all i and $\mathbf{x} \in \mathbf{f}$). Let $V := \{\mathbf{v}_{i_j}\}_{j=1}^l = \{\mathbf{v}'_{i_j}\}_{j=1}^l$ be the vertex set of $\mathbf{f}$, and let I_V be the set of all multi-indices β with $|\beta| = n - 1$ such that all non-zero entries β_i of β correspond to vertices in V, $i \notin V \Rightarrow \beta_i = 0$. Assume that, for every multi-index $\beta \in I_V$, an affine function A_β exists such that $b_{\beta+\mathbf{e}_i} = A_\beta(\mathbf{v}_i)$ for all $i = 1..k$ and $b'_{\beta+\mathbf{e}_i} = A_\beta(\mathbf{v}'_i)$ for all $i = 1..k'$.*
Then, the derivative of f and f' at points $\mathbf{x} \in \mathbf{f}$ with respect to a differential operator $\partial \in \left\{ \frac{\partial}{\partial x_i} \right\}$ is

$$\partial f(\lambda(\mathbf{x})) = \partial f'(\lambda'(\mathbf{x})) = n \sum_{\beta \in I_V} \partial A_\beta \cdot B_\beta^{n-1}(\lambda(\mathbf{x})) \ .$$

Respective statements hold for the higher derivatives of f and f'.

Proof. Observe that (9) is still valid if we substitute the sum over all $\beta \in I_V$ for the sum over all β with $|\beta| = n - 1$ (for $\mathbf{x} \in \mathbf{f}$). This implies the claim. $\square$

For Bézier surfaces, it is often sufficient if the tangent plane varies smoothly without requiring smoothness of the parameterization. In this case, slightly weaker constraints on the control points are sufficient.

Corollary 2.7 (Geometric continuity across polytope boundaries). *In the situation of Corollary 2.6, let Q be any affine transformation of the domain $\mathbb{R}^d$ that keeps $\mathbf{f}$ fixed such that $b_{\beta+\mathbf{e}_i} = A_\beta(\mathbf{v}_i)$ for all $i = 1..k$ and $b'_{\beta+\mathbf{e}_i} = A_\beta(Q\mathbf{v}'_i)$ for all $i = 1..k'$.*
Then $\partial f(\lambda) \cdot \partial Q = \partial f'(\lambda')$.

Proof. Factoring out Q in (9) yields the claim. $\square$

3   Applications

In this section, we present several applications of mean value Bézier maps. Although the results obtained in the previous chapter are general and hold for any barycentric coordinates, Wachspress and mean value coordinates are the only known positive three-point coordinates [23]. Wachspress coordinates, however, have already been used to some extent in the past in the form of tensor product Bézier maps (with parallelepipeds as domain) and S-patches [6] (with regular k-gons as domain). Therefore, it seemed more appropriate to us to use mean value Bézier maps to demonstrate our results.

In all our applications, we begin by specifying several domain polytopes and their respective control points to achieve a smooth Bézier map $f : \mathbb{R}^d \to \mathbb{R}^e$. To determine the polytope in which a point $\mathbf{x} \in \mathbb{R}^d$ lies, we can use another property of mean value coordinates: the mean value coordinates with respect to a polytope P are defined in the whole space $\mathbb{R}^d$ and the denominator (for normalization) in the construction is positive if and only if $\mathbf{x}$ lies within P [16,28]. Thus, we can automatically determine the polytope P containing $\mathbf{x}$ when computing the mean value coordinates of $\mathbf{x}$ with respect to P.

3.1 Bézier Curves and Surfaces

If we choose $d = 1$ or $d = 2$ and $e > d$, Bézier maps specialize to Bézier curves and surfaces. In the case $d = 1$, however, barycentric coordinates on the unique 1-dimensional polytope, which is the 1-simplex or line segment, are uniquely determined (t and $1-t$ on $[0, 1]$). Our results coincide with the well-known theory for Bézier curves.

Therefore, we present an example of a mean value Bézier surface, that is a mean value Bézier map $f : \mathbb{R}^2 \to \mathbb{R}^3$. Figure 1 shows a C^1-continuous Bézier surface from several viewpoints. It consists of two patches of degree 2. The domain is the union of a pentagon and an L-shaped hexagon, which share two common edges. Since the hexagon is not convex, some of the coordinates can become negative, and we cannot guarantee that the Bézier surface is contained in the convex hull of its control points. However, in practice, this posed no problem in our experiments. Nevertheless, it is probably possible to construct cases in which a mean value Bézier surface leaves the convex hull of its control points.

Note that the highlights vary smoothly across the common edges. The two domain polygons are shown in black below the surface. The control nets, which determine the shape of the surface, are also depicted. We followed the suggestion in [6] and drew all polygons $(b_{\beta+e_i})_{i=1}^k$ with $|\beta| = n - 1 = 1$. (For drawing purposes, we shifted the control net belonging to the pentagon slightly to make sure that it does not overlap with the other one.) On the left and in the middle, we shaded the control net for the pentagon in dark gray and the control net for the hexagon in light gray. On the right, we chose common shades of gray for the parts of the control net that belong to a common vertex of both polygons. They can be discerned as affine images of the domain.

3.2 Space Deformations

A Bézier map with $d = e$ is a space deformation of $\mathbb{R}^d$. While geometric continuity is often sufficient for Bézier curves and surfaces, we need "real" analytic continuity to obtain a smooth space deformation. Even a discontinuity of the absolute value of the derivative in a single direction may be clearly visible if a textured object is deformed.

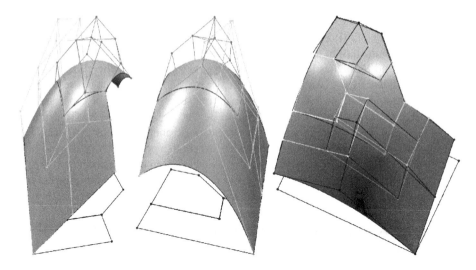

Fig. 1. Our method makes it possible to use non-convex polygons in the construction of Bézier surfaces. We present three views of a Bézier surface consisting of a pentagonal and an L-shaped hexagonal patch. Note that the highlights vary smoothly across the common edges.

To display a control polyhedron P, we note that each set $(b_{\beta+e_i})_{i=1}^k$ with $|\beta| = n - 1$ corresponds naturally to the polyhedron with vertices $(\mathbf{v}_i)_{i=1}^k$. Therefore, we connect control points $b_{\beta+e_i}$ and $b_{\beta+e_j}$ if and only if $(\mathbf{v}_i, \mathbf{v}_j)$ is an edge in P.

Figure 2 demonstrates a space deformation of $\mathbb{R}^3$. In (a), We show the cuboid that we want to twist by 180°. We align the control polyhedron with the edges of the cuboid. (b) depicts the result if the twist is done directly with 3D mean value coordinates (that is Bernstein polynomials of degree one). The lack of local control leads to a singularity. In (c), we include four additional vertices in the middle of the long edges without changing the total shape of the control polyhedron. This allows us better local control, but C^1-discontinuities are introduced in the middle and at the vertices. (The bead shaped reflection at the top left corner of the cuboid indicates the C^1-discontinuity of mean value coordinates at the vertices.) In (d), we split the control net into two identical, adjoining control polyhedra and deform them independently of each other. This gives us the desired local control but we still have the C^1-discontinuities. In (e), we use a Bézier map of degree 3 to join the two control polyhedra smoothly. It allows us to enforce C^1-continuity while maintaining local control. Observe that the C^1-discontinuities at the vertices have vanished as well. The control net shows how the continuity conditions are satisfied here. The left-most and right-most part is an affine image of the domain cuboids to make the deformation smooth at the respective vertices. (The left part is identically mapped, and the right part is rotated by 180° degree.) The two middle "columns" are mapped by a common affine map (both are rotated by 90° degree) to ensure a smooth transition between the adjoining control polyhedra.

As an example for stretching, we consider the cuboid once more and stretch its right half by a factor of two (Figure 3). To do this, we use the control net from Figure 2(e) again (shown in light gray). This time, we depict the underlying triangle mesh of the

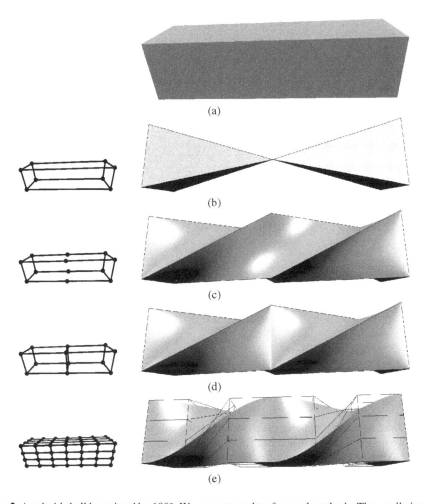

Fig. 2. A cuboid shall be twisted by 180°. We present results of several methods. The small picture on the left shows the corresponding control net. (*a*) The undeformed cuboid. (*b*) Interpolation of the twist with 3D mean value coordinates. (*c*) Interpolation of the twist with 3D mean value coordinates using additional control points. (*d*) We split the cuboid into two halves and interpolate both halves with 3D mean value coordinates. (*e*) Our method. Although we use the same two halves as interpolation domains as in (*d*), the use of third order polynomials allows us to control the smoothness. If we had increased the number of control points without using higher order polynomials, we would have introduced new discontinuities as in (*c*) and (*d*).

cuboid as well (in black) to demonstrate that no discontinuity is introduced in the middle of the cuboid but the stretching increases gradually. Hence, a texture could be mapped to the stretched cuboid without artifacts due to the stretching.

Figure 4 shows how a complex model can be handled by specifying a control net that is adapted to the shape of the model. It also shows that Bézier maps of different degrees can be mixed under certain circumstances. (Here, the body is mapped

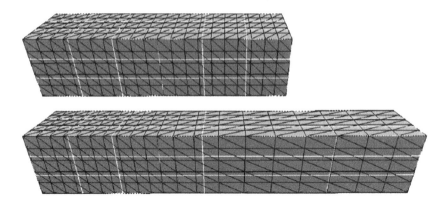

Fig. 3. A stretched cuboid

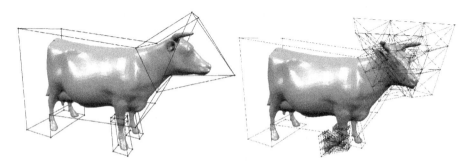

Fig. 4. The control net containing the cow consists of 6 polyhedra. One for the body, one for the head, two for the front knees, and two for the front legs (*left*). It demonstrates the ability of our method to handle complex control nets that are adapted to the shape of the object. We specified the deformation, which is C^1-continuous, by moving the vertices of the control polyhedra. The intermediate control points were computed automatically (*right*).

identically.) While the body and left front leg is mapped by a degree one map, the Bézier maps for the head and the right leg have degree three. Note that we didn't need to specify all the control points manually. We explain this in the case of the head: The seven vertices of the head polyhedron can be classified into two groups: the four vertices at the neck, which are connected to the body of the cow, and the three remaining, "exterior" vertices. Since we consider a Bézier map of order three, all control points can be classified as neck control points (if two or three entries of the multi-index belong to neck vertices) or as exterior control points (if two or three entries belong to exterior vertices). All neck control points are mapped identically to ensure a smooth connection to the body of the cow (Corollary 2.6), and all exterior control points are mapped by the linear function defined by the deformation of the head polyhedron to capture the total deformation and to make sure that the deformation is smooth at the three exterior vertices as well (Corollary 2.5). However, the task of providing a general and convenient method for the placement of the control points (for space deformations and for Bézier surfaces) remains a topic for future research.

4 Conclusions and Future Work

We developed criteria for the construction of smooth Bézier maps. A Bézier map is a map that is piecewise (on a given polytope) a homogeneous polynomial in generalized barycentric coordinates. We showed how the coefficients of the Bernstein polynomials can be chosen to enforce smoothness of any desired order across common (hyper-)faces of the polytopes. We chose to develop the theory in full generality although we mainly aim at Bézier maps in mean value coordinates. This allows the use of our results for any other barycentric coordinates that might come to the focus of attention in the future. Moreover, it shows that many results from the well developed field of simplicial and tensor product Bézier theory can be considered as a special case of our findings if Wachspress coordinates are used. Our indexing scheme, however, does not coincide with the traditional indexing scheme for tensor product Bézier maps. This sheds new light on the classical theory, which will hopefully lead to an better understanding of the tensor product Bézier maps as well.

Probably the most important examples of Bézier maps are Bézier curves and surfaces and space deformations. We presented examples of mean value Bézier surfaces and free-form deformations based on Bernstein polynomials in mean value coordinates as possible applications. Nearly without additional effort, we can ensure that our Bézier maps exhibit the desired smoothness even at the polytope vertices, although the mean value coordinates themselves are only C^0-continuous at these points. Thus, it is now possible to construct smooth mean value Bézier maps with arbitrary polytopes as domains.

Nevertheless, a number of open questions remain, which we intend to address in future work. Foremost, some kind of spline representation of Bézier maps has to be found that takes care of any continuity issues fully automatically. These splines should allow to place meaningful control points directly during the design of surfaces and deformations without the necessity to spend much time on the cumbersome process of satisfying the continuity constraints manually. Another issue that we did not discuss in the current paper are rational Bézier maps. The use of rational Bézier maps greatly expanded the capabilities of classical Bézier theory. The same should be done for generalized Bézier maps.

Acknowledgements. The research of the authors has been supported in part by the EU-Project "AIM@SHAPE" FP6 IST Network of Excellence 506766.

References

1. Bézier, P.: How Renault uses numerical control for car body design and tooling. In: Paper SAE 680010, Society of Automotive Engineers Congress, Detroit, Mich. (1968)
2. de Casteljau, P.: Outillage méthodes calcul, Paris: André Citroën Automobiles S.A (1959)
3. Forrest, A.R.: Interactive interpolation and approximation by Bézier polynomials. Comput. J. 15(1), 71–79 (1972)

4. Bézier, P.: General distortion of an ensemble of biparametric surfaces. Computer-Aided Design 10(2), 116–120 (1978)
5. Sederberg, T.W., Parry, S.R.: Free-form deformation of solid geometric models. In: SIGGRAPH 1986: Proceedings of the 13th annual conference on Computer graphics and interactive techniques, pp. 151–160. ACM Press, New York (1986)
6. Loop, C.T., DeRose, T.D.: A multisided generalization of Bézier surfaces. ACM Transactions on Graphics 8(3), 204–234 (1989)
7. Wachspress, E.L.: A Rational Finite Element Basis. Mathematics in Science and Engineering, vol. 114. Academic Press, New York (1975)
8. Meyer, M., Barr, A., Lee, H., Desbrun, M.: Generalized barycentric coordinates on irregular polygons. Journal of Graphics Tools 7(1), 13–22 (2002)
9. Warren, J.: Barycentric coordinates for convex polytopes. Advances in Computational Mathematics 6(2), 97–108 (1996)
10. Ju, T., Schaefer, S., Warren, J., Desbrun, M.: A geometric construction of coordinates for convex polyhedra using polar duals. In: Desbrun, M., Pottmann, H. (eds.) Third Eurographics Symposium on Geometry Processing, Eurographics Association, pp. 181–186 (2005)
11. Huiskamp, G.: Difference formulas for the surface Laplacian on a triangulated surface. J. Comput. Phys. 95(2), 477–496 (1991)
12. Floater, M.S.: Mean value coordinates. Computer Aided Geometric Design 20(1), 19–27 (2003)
13. Floater, M.S., Kós, G., Reimers, M.: Mean value coordinates in 3D. Computer Aided Geometric Design, Special Issue on Geometric Modelling and Differential Geometry 22(7), 623–631 (2005)
14. Ju, T., Schaefer, S., Warren, J.: Mean value coordinates for closed triangular meshes. ACM Transactions on Graphics 24(3), 561–566 (2005)
15. Langer, T., Belyaev, A., Seidel, H.-P.: Spherical barycentric coordinates. In: Sheffer, A., Polthier, K. (eds.) Fourth Eurographics Symposium on Geometry Processing, Eurographics Association, pp. 81–88 (June 2006)
16. Hormann, K., Floater, M.S.: Mean value coordinates for arbitrary planar polygons. ACM Transactions on Graphics 25(4), 1424–1441 (2006)
17. Langer, T., Seidel, H.-P.: Mean value Bézier surfaces. In: Martin, R., Sabin, M.A., Winkler, J.R. (eds.) Mathematics of Surfaces XII. LNCS, vol. 4647, pp. 263–274. Springer, Heidelberg (2007)
18. Sibson, R.: A vector identity for the Dirichlet tessellation. Math. Proc. of Cambridge Philosophical Society 87, 151–155 (1980)
19. Farin, G.: Surfaces over Dirichlet tessellations. Computer Aided Geometric Design 7(1–4), 281–292 (1990)
20. Charrot, P., Gregory, J.A.: A pentagonal surface patch for computer aided geometric design. Computer Aided Geometric Design 1(1), 87–94 (1984)
21. Loop, C., DeRose, T.D.: Generalized B-spline surfaces of arbitrary topology. In: SIGGRAPH 1990: Proceedings of the 17th annual conference on Computer graphics and interactive techniques, pp. 347–356. ACM, New York (1990)
22. Goldman, R.: Multisided arrays of control points for multisided Bézier patches. Computer Aided Geometric Design 21(3), 243–261 (2004)
23. Floater, M.S., Hormann, K., Kós, G.: A general construction of barycentric coordinates over convex polygons. Advances in Computational Mathematics 24(1–4), 311–331 (2006)
24. Ju, T., Liepa, P., Warren, J.: A general geometric construction of coordinates in a convex simplicial polytope. Computer Aided Geometric Design 24(3), 161–178 (2007)

25. Lipman, Y., Kopf, J., Cohen-Or, D., Levin, D.: GPU-assisted positive mean value coordinates for mesh deformations. In: SGP 2007: Proceedings of the fifth Eurographics symposium on Geometry processing, Aire-la-Ville, Switzerland, Switzerland, Eurographics Association, pp. 117–123 (2007)
26. Joshi, P., Meyer, M., DeRose, T., Green, B., Sanocki, T.: Harmonic coordinates for character articulation. ACM Transactions on Graphics 26(3), 71 (2007)
27. Langer, T., Seidel, H.-P.: Higher order barycentric coordinates. In: Eurographics (accepted, 2008)
28. Langer, T., Belyaev, A., Seidel, H.-P.: Mean value coordinates for arbitrary spherical polygons and polyhedra in $\mathbb{R}^3$. In: Chenin, P., Lyche, T., Schumaker, L.L. (eds.) Curve and Surface Design: Avignon 2006. Modern Methods in Mathematics, pp. 193–202. Nashboro Press, Brentwood (2007)

Meaningful Mesh Segmentation Guided by the 3D Short-Cut Rule

Zhi-Quan Cheng, Bao Li, Gang Dang, and Shi-Yao Jin

PDL Laboratory, National University of Defense Technology, P.R. China
Cheng.zhiquan@gmail.com

Abstract. Extended from the 2D silhouette-parsing short-cut rule [25], a 3D short-cut rule, which states " as long as a cutting path mainly crosses local skeleton and lies in concave regions, the shorter path is (other things being equal) the better " , is defined in the paper. Guided by the 3D short-cut rule, we propose a hierarchical model decomposition paradigm, which integrates the advantages of the skeleton-driven and minima-rule-based meaningful segmentation. Our method defines geometrical and topological functions of skeleton to locate initial critical cutting points, and then employs salient contours with negative minimal principal curvature values to determine natural boundary curves among parts. Sufficient experiments have been carried out on many meshes, and have shown that our framework could provide more perceptual results than pure skeleton-driven or minima-rule-based algorithm.

Keywords: Meaningful mesh segmentation, 3D short-cut rule, minimal-rule, skeleton.

1 Introduction

Mesh segmentation [1][2] refers to partitioning a mesh into a series of disjoint elements. It has become a key ingredient in many mesh operations, such as texture mapping [3][4], shape manipulation [5][6], simplification [7], mesh editing [8], mesh deformation [9], collision detection [10], shape analysis and matching [11][12][13]. Especially, the process that decomposes a model into visually meaningful components is called part-type segmentation [1] (or meaningful segmentation). Actually, meaningful segmentation is a challenging work and still in its infancy, e.g. compared to image segmentation (hundreds of papers). So, more researches are seeking to find more effective procedures, which can produce natural results that are in keeping with human recognition. Especially, more advanced coherency issues should be addressed, such as pose invariance [14], handling more complex models, e.g. David and Armadillo, extracting similar shapes over similar objects.

1.1 Related Work

The basic mesh segmentation problem can be viewed as clustering primitive elements (vertices, edges and faces) into sub-meshes, and the techniques finishing the partition include hierarchical clustering [3][11], iterative clustering [4][5][7], spectral analysis

F. Chen and B. Jüttler (Eds.): GMP 2008, LNCS 4975, pp. 244–257, 2008.

[15], and region growing [16]. We prefer readers to get recent surveys from [1] and [2]. Different from former explicit decomposition algorithms, two type implicit methods can also produce reasonable results.

One type, including [16][17][18][19], is guided by the minima rule [20], which states that human perception usually divides an object into parts along the concave discontinuity and negative minima of curvature. Enlightened by the minima rule, the mesh's concave features, identified as natural boundaries, are used for segmentation in the region growing watershed algorithm [16]. Due to the limitation of region growing, the technique can not cut a part if the part boundary contains non-negative minimum curvatures. And then, Page et al. [17] have used the factors proposed in [21] to compute the salience of parts by indirect super-quadric model. In order to directly compute the boundary of a part and avoid complex super quadric, Lee et al. [18] have experientially combined four functions (distance, normal, centricity and feature) to guide the cutting path in a reasonable way. And our previous work [19] could achieve similar results by principal component analysis technique. However, the segmentation results of these algorithms are dependent on the structures of underlying manifold surface, so they are sensitive to surface noises and tending to incur over-segmentation (one instance is shown in Fig. 9.a).

The other, including [6][10][22][23], is driven by curve-skeleton/skeleton [24] that is 1D topologically equivalent to the mesh. The skeleton-driven algorithms can guarantee perceptually salient decomposition, however, boundaries among parts do not always smoothing and follow natural visual perception. Consequently, it would be a better choice to combine the skeleton-driven approach with the minima rule.

1.2 Overview

Traditional 2D short-cut rule [25] states that human vision divides a silhouette into parts based on two important geometric factors: cut length and local symmetry (Fig. 1.a). For a 3D mesh, its local symmetry axis is its skeleton [24], and the cut length can be computed as the length of cutting loop curve that goes over the surface. Consequently, on the basis of the skeleton and cutting path, we extend the short-cut rule from 2D parsing to 3D segmentation. The extended 3D short-cut rule (Fig. 1.b) can be defined as " as long as a cutting path mainly crosses local skeleton and lies in concave regions, the shorter path is (other things being equal) the better ".

Aiming at integrating the minima-rule-based and skeleton-driven approaches, we would develop a robust meaningful segmentation paradigm guided by the extended

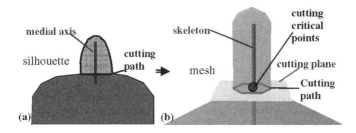

Fig. 1. Structural diagram of short-cut rule. (a)2D short-cut rule (b) 3D short-cut rule.

3D short-cut rule. And the new approach would also guarantee to divide an arbitrary genus mesh into a collection of parts isomorphic to disk / half-disk (base of planar parameterization). Especially, the algorithm would be resolution-independent, which means that the same segmentation is achieved at different levels of detail (e.g., in Fig. 2, two Armadillo meshes in different fidelity are decomposed into similar components, although segmented separately).

In a nutshell, the partitioning algorithm is carried out in two stages. Firstly, the hierarchical skeleton (Fig. 2.e) of a mesh is computed by a repulsive force field (Fig. 2.d) over the discretization (Fig. 2.c). And for every skeleton level, the critical cutting points (the larger points in Fig. 2.e) are preliminarily identified by geometrical and topological functions. Secondly, for each critical point, corresponding final boundary is obtained using local feature contours in valley regions. As a result, our algorithm can automatically partition a mesh into meaningful components with natural boundary (Fig. 2.f and 2.g).

In general, the paper makes the following contributions:

- We first extend the short-cut rule from 2D parsing to 3D mesh segmentation, and address the extension mechanism in detail.
- Guided by the extended 3D short-cut rule, an automatic meaningful segmentation paradigm is presented, which can successfully integrate two type segmentation algorithms in theory and practice. The decomposition is robust due to using new excellent geometrical and topological properties of skeleton, and the final borders of the parts are natural on the ground of the minima rule.

The rest of the paper is structured as follows. Critical cutting points are located in section 3 based on the core skeleton extracted in section 2, and then the cutting path completion mechanism is illustrated in section 4. Section 5 describes the hierarchical decomposition of meshes. Section 6 demonstrates some results and compares them with related works. Finally, section 7 makes a conclusion and gives some future researching directions.

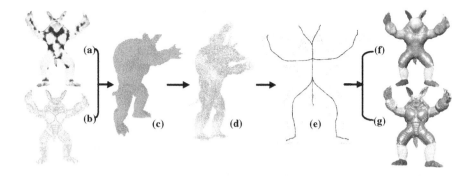

Fig. 2. Stages of Armadillo in our segmentation algorithm. (a)low-resolution Armadillo with 2,704 vertices (b)high-resolution Armadillo with 172,974 vertices (c)voxelized volume representation in 96^3 grids (d)corresponding repulsive force field (e)core skeleton with critical cutting points (the larger red points) (f)segmentation of low-resolution Armadillo (g) segmentation of high-resolution Armadillo.

2 Core Skeleton Extraction and Branch Priority

2.1 Core Skeleton Extraction

The skeleton of a mesh is generated by directly adapting a generalized potential field method [22], which works on discrete volumetric representation [26] of the model. Core skeleton, starting at the identified seed point, is discovered by common force following method on underlying vector field. At seed point, where the force vanishes, the initial directions are determined by evaluating eigen-values and eigen-vectors of corresponding Jacobian matrix. The force following process evaluates vector (force) value at current point and moves in the direction of the vector with a small pre-defined step (value σ, set as 0.2). Consequently, the obtained core skeleton, e.g. Fig 2.e, consists of a set of points sampled by above process.

Once the core point skeleton is extracted, similar smoothing procedure [27], is applied to alleviate the problem of noise. Basically, this procedure expands the fluctuant skeleton to a narrow sleeve, defined by each point's bounding sphere with radius of given threshold value σ, and then it fines the shortest polygonal path lying in the 2σ wide sleeve. The procedure gives a polygonal approximation of the skeleton, and can be imagined as stretching a loose rubber band within a narrow sleeve. Subsequently, the position of each point in the original skeleton is fine-tuned by sliding to the nearest point on the polygonal path.

2.2 Skeleton Branch Selection

According to the number of neighbouring points, skeleton points are classified into three kinds: terminal nodes (one neighbor), common nodes (two neighbors) and branch nodes (more than two neighbors). In the paper, terminal points and branch points would be viewed as feature points. Any subset of the skeleton, bounded by the feature points, is called a skeleton branch. It is important to determine the order of the branches, since our approach would like to detect initial critical cutting points by measuring related geometrical and topological properties and the separated parts would not be taken account into subsequent computation of critical points. Basically, the ordering should allow small but significant components to be extracted first, so that they are not absorbed by larger components in an improper way. We use three criteria to find the best branch: the type, length and centricity.

For each branch b, which is a set of points, we define its priority $P(b)$ as its type value adding the product of the reciprocal length and sum of all normalized centricity of its points, since we treat the total number of points as its length.

$$P(b)=Type(b)+\frac{1}{Length(b)}\sum_{t\in b}C(t) \qquad \text{(Equation 1)}$$

The type of the branch b is determined by the category of its two end points. The type weight of the branch with two branch nodes is low (value is 0), that with one terminal and one branch node is medium (value is 1), and that with two terminal nodes is high (value is 2).

The centricity of a point t is defined as the average hops from t to all the points of the skeleton. In a mesh, let $maxH$ represent the maximum average hopping numbers among all points, i.e. $maxH=max_t(avgH(t))$. We normalize the centricity value of vertex t as $C(t)=avgH(t)/maxH$. After a mesh has been partitioned based on the critical cutting points in the selected branch, the current centricity values of points are no longer valid in the following segmentation. Hence, we should re-compute the centricity values after each partitioning when we want to select another branch.

3   Locating Critical Cutting Points

Just as the principle, observed by Li et al. [10], geometrical and topological properties are two important characteristics, which distinguish one part from the others in mesh segmentation. We adapt the space sweeping method, used in computational geometry, to sweep a given mesh perpendicular to the skeleton branches.

Let b be a selected branch. If b is a medium type, we sweep it from its terminal point P_{start} to the other node P_{end}. And some points nearby P_{end} are excluded from the sweeping, since no effective critical points lie in the region. In the paper, the nearby region is a sphere, whose centre is P_{end} and its radius is the minimal distance from the point P_{end} to the nearest vertex on the surface. In other ways, b would be a high or low type, and be swept from the point with smaller cross-section area.

3.1   Geometrical Function

For the selected branch b, we compute the area of cross-section at each point p on sweeping path from P_{start} to P_{end-1}, and then define new geometrical function $G(p)$.

$$G(p)=\frac{AreaCS(p+1)-AreaCS(p)}{AreaCS(p)}, p \in [\,p_{start}, p_{end-1}\,] \qquad \text{(Equation 2)}$$

To accelerate the computation of cross-section area at each point, we approximately calculate it by summing up the number of the voxels that are intersected by the perpendicular sweeping plane of the current point.

By lining the dots of $G(p)$, we get a connected polygonal curve, which fluctuates in the way that there are a few outburst pulses on almost straight line close to zero. Three kinds of dots would be filtered out, and Fig. 3 shows the three kind sample profile of $AreaCS(p)$ and $G(p)$. On the viewpoint of $AreaCS(p)$, it's obvious that Fig. 5.a denotes a salient concavity, Fig. 5.b and 5.c respectively mean the fact how a thin part connects with a thick part. In the $G(p)$ curve, if the rising edge of one pulse goes through the p axis and its trough is less than -0.15 (Fig. 5.a or 5.c), the cross dot t would be selected. In addition, the peak of positive pulse (Fig. 5.b), whose value is more than a threshold 0.4, would also be identified. The points located in the skeleton, corresponding to dot t indicated by dashed lines, are marked as candidate critical points that would be used to divide the original model.

By using the $G(p)$ function, our method can segment L-shaped object (Fig 4.a), which is just the ambiguity of the minima rule theory [16]. However, it is not right to directly treat all selected points (e.g., Fig. 4.b) as real critical points, since straight absorption would lead to over-segmentation, as shown in Fig. 4.c. Hence, we avoid some over-parts by excluding some candidate points from the set of critical points. The exclusion is performed by checking whether three nearby candidates are located in a same space, defined as in the second paragraph of section 3. And if the nearby phenomenon happens, only the first critical point is preserved. Therefore, the

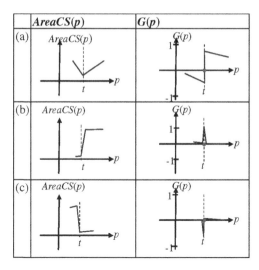

Fig. 3. Three kind sample profiles of $AreaCS(p)$ and $G(p)$

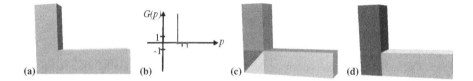

Fig. 4. The segmentation of L-shaped object. (a)the object can't be partitioned by watershed algorithm [16] (b)corresponding profile of $G(p)$, calculated by equation 2, with candidate critical dots in red color (c)over-segmentation problem (d)final result after critical point exclusion.

over-segmentation problem would be effectively resolved, and the natural result is gotten as Fig. 4.d. Besides these, if the nearby instance, between two adjacent critical points, existed, the point with smaller peak value would be discarded.

3.2 Topological Constraint

The topology of a mesh is characterized by its genus, its orientation, the number of its connected components, and the number of its boundary components [28]. In the paper, we use the genus property to achieve topological guarantee. This means that any genus mesh can be eventually divided into a collection of parts isomorphic to disk / half-disk. We equivalently define the genus of a connected oriented manifold by counting the number of rings in the skeleton of a mesh, since the number of rings is equal to that of handles. A ring is a closed path, on which one point can reach itself again when leaving away in a direction. For example, there is no ring in a sphere, one cycle in a torus skeleton, and two rings in an eight-shaped surface. Thus, their genera are zero, one and two, respectively. The rings are employed to constrain the segmentation process as follows: once a ring existed in the skeleton and only one cutting critical point was detected, the partitioning process should be in progress until the ring has been divided into at least two parts by inserting another critical point. Our topological constraint capacity is determined by the accuracy of skeleton. If no ring detected, the function would not do its work, e.g. Dragon in Fig. 7.

4 Cutting Path Completion

Direct segmentation only based on critical cutting points may not have exact and smooth boundaries among parts. To resolve the limitation, the ultimate boundaries between parts are completed at the aid of the feature contours of the underlying valley surface. Therefore, the skeleton-driven and minima-rule-based algorithms are consistently embodied in the 3D short-cut rule. The operational principle of boundary extraction can be sketched out as Fig. 5, and its carryout together with locating critical points (integrated by the 3D short-cut rule) is given by a precise algorithm framework in pseudocode (Algorithm 1).

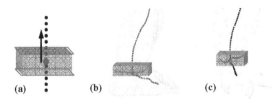

Fig. 5. Operational principle sketch of the boundary extraction guided by the extended 3D short-cut rule. (a)the restricting zone, built on a larger red critical cutting point, is formed by two parallel planes, which are perpendicular to the direction of the critical point and distance to the point with a threshold value. (b)(c)the feature contour, located at the armadillo's ankle in high and low resolution meshes, is used to define the partitioning boundary.

On one hand, the 3D short-cut rule requires to employ local medial axis, when partitioning a mesh. Until now, we have computed its skeleton, and found initial cutting positions marked by the critical points. For each critical point, we build corresponding segmentation region, enclosed by a restricting zone. For example, the ankle of the armadillo is enclosed by a restricting zone, shown as Fig. 5.b and 5.c. One zone (illustrated in Fig 5.a) is sliced by two parallel planes, whose normal is identical to the direction of the corresponding critical point, and both planes keep a same distance to the critical point with a threshold d value.

$$d = \begin{cases} 2*\sigma, & if \ \sigma > LNG_{edge} \\ 2*LNG_{edge}, & else \end{cases} \qquad \text{(Equation 3)}$$

where, value σ, defined in section 2, is the distance between the adjacent skeleton points, and LNG_{edge} is the average edge length of the mesh.

On the other hand, the 3D short-cut rule implies that it prefers to divide a given mesh into disjoint parts along the concave regions. Therefore, if one concave region lies in the defined restricting zone, we are inclined to extract the cutting boundary from the region. For instances, the Fig 5.b and 5.c demonstrate that the dark blue contour, located in the restricting pink zone, is used to get natural perceptual boundary between foot and leg of the Armadillo model. Similar to [19], we use proper normalization to unify the minima curvature value of each vertex, obtain the concave feature regions by filtering out the vertices with higher normalized value, extract contour curves from the graph structures of the regions (e.g. the blue regions in Fig. 1.a and 1.b) and complete best curve path going over the mesh in the shortest way. We refer readers to [19] for details regarding the feature contour extraction and path completeness. For every feature contour, we compute its main direction based on principle component analysis of its vertices. But only the feature contour, whose main direction is approximate to the orientation of the corresponding critical point, is treated as one boundary curve. The approximation is measured by the angle between them. And if the separation angle is less than $\pi/4$ in radian, we say that the points are approximate. Note that, once there is none concaving contour locating in a restricting zone, corresponding critical point can be removed and no partitioning action happens.

In order to get the accurate segmented parts, the part salience theory [21] is employed. The theory is implemented by the testing criterion method [18], and is used to check whether a part is significant enough or not. The criterion combines three factors, which are area, protrusion, and feature. Since our approach is heuristic, it is possible that there is manual rejection for some model. Fortunately, all models, used in the paper, require no manual interference. Therefore, we can generally address that our approach may be viewed as automatic meaningful segmentation.

Algorithm 1. 3D short-cut rule implementation framework in pseudocode

```
1 : void 3D_short-cut_rule (PolygonMesh& mInputMesh){
2 :    compute core skeleton s of mInputMesh;
3 :    construct branches set bs by breaking s;
4 :    while(sizeof(bs)!= 0){
5 :       select b with the maximal priority from bs;
6 :       build critical points set ps by sweeping b;
7 :       for (each point p in ps) {
8 :          create corresponding restricting zone rz of p ;
9 :          calculate the minimal principal curvature value pc of
                       each vertex located in zone rz;
10:          feature contours extraction in zone rz;
11:          complete the longest feature contour and form a
                       closed boundary;
12:       } //end for
13:    remove b from bs;
14:    }//end while
15:    part salience testing;
16: }
```

5   Hierarchical Segmentation

In addition to seed points, which generate the core skeleton for a given mesh, the divergence of the vector field of the mesh can also be used to compute level 1 skeleton. The divergence in a given region of space is a scalar that characterizes the rate of flow leaving that region, and a negative divergence value at a point measures the "sinkness" of the point [29]. In other words, one low divergence point implies that there is a protrusion in the corresponding area, e.g. horns and ears of Cow in Fig. 6.

Consequently, by sweeping the level 1 skeleton and repeating the previous steps (described in algorithm 1) in sequence except that core skeleton is replaced by level 1 skeleton, it is easy to hierarchically segment a given mesh. Fig. 6 has shown one instance, a cow mesh is hierarchically decomposed. Besides these, the voxels and skeletons of each obtained part can also be re-computed, and further divided by the former procedure. A part in the hierarchy can be recursively decomposed until at least one of the following conditions is met: (a) there is no critical cutting point detected in the part; (b) there is no ring in the skeleton, so that the topological constraint is satisfied; (c) the value of hierarchical segmentation level does not exceed a given threshold, which can be manually pre-set by user.

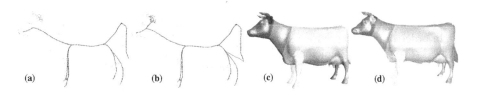

Fig. 6. Hierarchical segmentation of a cow with 2,903 vertices in 128^3 voxelization. (a)core skeleton (b)level 1 skeleton with 30% divergence (c)segmentation corresponding to (a) (d)segmentation corresponding to (b).

6 Results and Discussion

6.1 Results

Guided by the 3D short-cut rule, we can not only reasonably locate the meaningful segmentation positions by using the skeleton-driven space-sweeping method, but also

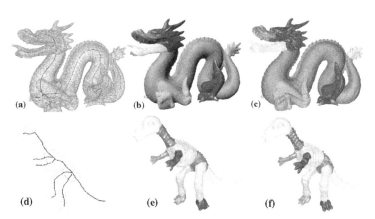

Fig. 7. Segmentation instances in 96^3 grids. (a)core skeleton of Dragon with critical cutting points (b)low level segmented Dragon with 5,000 vertices (c)high level segmented Dragon with 50,000 vertices (d)core skeleton of Disonaur with critical cutting points (e)low level segmented Disonaur with 3,514 vertices (f)high level segmented Disonaur with 56,194 vertices.

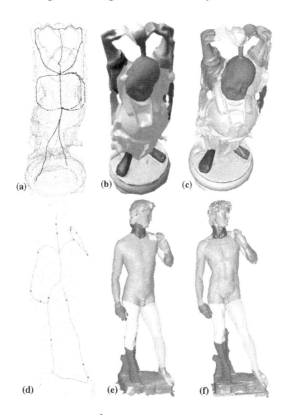

Fig. 8. Segmentation instances in 256^3 grids. (a)core skeleton of Buddha with critical cutting points (b)low level segmented Buddha with 10,000 vertices (c)high level segmented Buddha with 100,000 vertices (d)core skeleton of David with critical cutting points (e)low level segmented David with 4,068 vertices (f)high level David with 127,465 vertices.

get natural cutting boundary between different parts based on the minima-rule. Fig. 7 demonstrates the decomposition of two different resolution Dragon and Disonaur models and their core skeleton in 96^3 voxelized resolution. And Fig. 8 deals with more complex Buddha and David in 256^3 grids. As shown, our algorithm can obtain a more likely segmentation that is consistent with the human perceptual results. The figure also implies that our algorithm is resolution-independent.

6.2 Comparison and Discussion

Fig. 9 gives a comparison of the visual effects between the state-of-the-art minima-rule-based segmentation [18][19] and our algorithm using the core skeleton. For the Disonaur mesh, the over-segmentation would definitely happen in [19] and reproduced algorithm [18] (Fig. 9.a), while the problem does not happen in the paper (Fig. 9.b). As shown in the Fig. 9.c, [18] has inexact a thumb of hand. By contrast, our approach (Fig. 9.d) can divide the Hand in natural way.

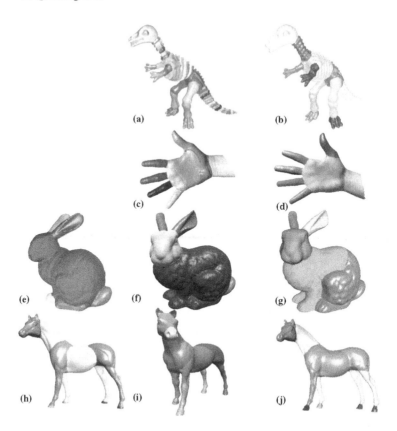

Fig. 9. Comparison with minima-rule-based algorithms [18][19]. (a) over-segmented Disonaur with 56,194 vertices generated by [19] and reproduced algorithm [18] (b)Disonaur partitioned by the paper (c)Hand with 10,070 vertices decomposed by [18] (d)Hand by the paper (e)Bunny with 34,834 vertices partitioned by [19] (f)Bunny by [18] (g)Bunny by the paper (h)Horse with 48,485 vertices segmented by [19] (i)Horse by [18] (j)Horse by the paper.

And then, we compare our results with the typical skeleton-driven algorithms [6] [10][23] in Fig. 10. On the one hand, our approach uses the more robust skeleton of the mesh and is not sensitive to surface noises any more, on the other hand, it is obvi- ous that the cutting boundaries of all parts have been improved by our approach and locate in concaving regions for all compared models.

The voxelized resolution is an important external factor affecting the skeleton, since it defines the precision of the repulsive force fields and determines the comput- ing time and storing memory requirement. It is evident that a 10^3 grid will yield a less accurate result than a 100^3 grid. However, it does not mean that the finer resolution is the better, in view of the application request and algorithm complexity. Especially, for different multi-resolutions of a mesh, the volume representation is almost similar, if the voxelized resolution is less than 512^3 grids.

It must be note that the computation of skeletonization occupies most (almost 99.5%) of the segmentation time, which ranges from ten minutes for common objects to more than hours for complex models, such as happy Buddha and David model.

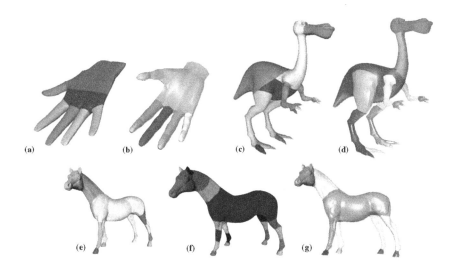

Fig. 10. Comparison with skeleton-driven algorithms [6][10][23]. (a)Hand in [10] (b)likely Hand with 1,572 vertices segmented by the paper (c)Dinopet in [10] (d)Dinopet with 4,388 vertices partitioned by the paper (e)Horse with 48,485 vertices segmented by [6] (f)Horse decomposed by [23] (g)Horse partitioned by the paper.

Therefore, the urgentest affair is to find a real-time skeletonizing approach, which can finish its work in seconds.

7 Conclusion

In this paper, we have proposed a 3D short-cut rule, which is extended from 2D silhouettes parsing domain, to guide model decomposition. The 3D short-cut rule, which states that " as long as a cutting path mainly crosses local skeleton and lies in concave regions, the shorter path is (other things being equal) the better " , can effectively integrate the advantages of existed skeleton-driven and minima-rule-based segmentation method in theory and practice. In a nutshell, our approach initially finds out all possible partitioning positions, which are marked by the critical cutting points. These points are identified by new functions of topological and geometrical properties, which are defined by the area of sweeping cross-section along skeleton branches sorted by priority. And then, we extract feature contours on the local surface, which lie in restricting zones defined by the critical points, and close the longest contour to produce natural boundaries between different parts. Consequently, our meaningful segmentation algorithm can achieve proper results fitting with human visual perception, such as, the boundary curves among parts would be in concaving regions and robust to surface noises.

Several enhancements should be added to our algorithm. Firstly, the skeletonizing procedure spends too much time that is additional to the mesh segmentation, and it must be speeded up by new skeleton algorithm. Furthermore, the skeleton may not be the best abstraction for all types of objects. For example, thin disk-like objects or a mushroom shape is not well represented by a curve-skeleton, and the skeleton extraction used in the paper may not be topology-preserving to some special models, e.g. Dragon and Buddha. So it is necessary to develop more accurate skeleton computation approaches. Finally, more proper geometrical and topological functions may exist and should be further investigated.

Acknowledgements

Special thanks to Dachille for the voxelization source code, Cornea for the demonstrating skeleton code, and the following groups for the 3D models they provided: the Stanford Computer Graphics Laboratory, the Caltech Multi-resolution Modeling Group and 3D Meshes Research Database by INRIA GAMMA Group. Finally, we wish to thank the anonymous reviewers of Eurographics 2007 for their valuable comments and suggestions. The work is funded by National Natural Science Foundation of China under Grant No. 60773020.

References

1. Shamir, A.: Segmentation Algorithms for 3D Boundary Meshes. In: EuroGraphics 2006, STAR, pp. 1–26. Blackwell Publishers, Malden (2006)
2. Attene, M., Katz, S., Mortara, M., Patane, G., Spagnuolo, M., Tal, A.: Mesh Segmentation - a Comparative Study. In: International Conference on. Shape Modeling and Applications 2006, pp. 14–25. IEEE Press, New York (2006)
3. Sander, P.V., Snyder, J., Gortler, S.J., Hoppe, H.: Texture Mapping Progressive Meshes. In: SIGGRAPH 2001, pp. 409–416. ACM Press, New York (2001)
4. Zhang, E., Mischaikow, K., Turk, G.: Feature-Based Surface Parameterization and Texture Mapping. ACM Transactions on Graphics 24(1), 1–27 (2005)
5. Katz, S., Tal, A.: Hierarchical Mesh Decomposition Using Fuzzy Clustering and Cuts. ACM Transactions on Graphics 22(3), 954–961 (2003)
6. Lien, J.M., Keyser, J., Amato, N.M.: Simultaneous Shape Decomposition and Skeletonization. In: ACM symposium on Solid and physical modeling 2006, pp. 219–228. ACM Press, New York (2006)
7. Cohen-Steiner, D., Alliez, P., Desbrun, M.: Variational Shape Approximation. ACM Transactions on Graphics 23(3), 981–990 (2004)
8. Mitani, J., Suzuki, H.: Making Papercraft Toys from Meshes Using Strip-based Approximate Unfolding. ACM Transactions on Graphics 23(3), 259–263 (2004)
9. Huang, J., Shi, X., Liu, X., Zhou, K., Wei, L.-Y., Teng, S., Bao, H., Guo, B., Shum, H.-Y.: Subspace Gradient Domain Mesh Deformation. ACM Transactions on Graphics 25(3), 1126–1134 (2006)
10. Li, X., Toon, T.W., Huang, Z.: Decomposing Polygon Meshes for Interactive Applications. In: ACM Symposium on Interactive 3D Graphics 2001, pp. 35–42. ACM Press, New York (2001)
11. Attene, M., Falcidieno, B., Spagnuolo, M.: Hierarchical Mesh Segmentation Based-on Fitting Primitives. The Visual Computer 22(3), 181–193 (2006)
12. Podolak, J., Shilane, P., Golovinskiy, A., Rusinkiewicz, S., Funkhouser, T.: A Planar-Reflective Symmetry Transform for 3D Shapes. ACM Transactions on Graphics 25(3), 549–559 (2006)

13. Mortaram, M., Patan, G., Spagnuolo, M., Falcidieno, B., Rossignac, J.: Plumber – A Method for a Multi-scale Decomposition of 3D Shape into Tubular Primitives and Bodies. In: ACM Symposium on Solid Modeling and Applications 2004, pp. 139–158. ACM Press, New York (2004)
14. Katz, S., Leifman, G., Tal, A.: Mesh Segmentation Using Feature Point and Core Extraction. The Visual Computer 21(8-10), 649–658 (2005)
15. Liu, R., Zhang, H.: Segmentation of 3D Meshes Through Spectral Clustering. In: 12th Pacific Conference on Computer Graphics and Applications, pp. 298–305. IEEE Press, New York (2004)
16. Page, D.L., Koschan, A.F., Abidi, M.A.: Perception-based 3D Triangle Mesh Segmentation Using Fast Marching Watersheds. In: IEEE Computer Vision and Pattern Recognition, vol. II, pp. 27–32. IEEE Press, New York (2003)
17. Page, D.L., Abidi, M.A., Koschan, A.F., Zhang, Y.: Object Representation Using the Minima Rule and Superquadrics for Under Vehicle Inspection. In: The IEEE Latin American Conference on Robotics and Automation, pp. 91–97. IEEE Press, New York (2003)
18. Lee, Y., Lee, S., Shamir, A., Cohen-Or, D., Seidel, H.-P.: Mesh Scissoring with Minima Rule and Part Salience. Computer Aided Geometric Design 22, 444–465 (2005)
19. Cheng, Z.-Q., Liu, H.-F., Jin, S.-Y.: The Progressive Mesh Compression Based on Meaningful Segmentation. The Visual Computer 23(9-11), 651–660 (2007)
20. Hoffman, D., Richards, W.A.: Parts of Recognition. Cognition 18, 65–96 (1984)
21. Hoffman, D., Signh, M.: Salience of Visual Parts. Cognition 63, 29–78 (1997)
22. Cornea, D.N., Silver, D., Yuan, X.S., Balasubramanian, R.: Computing Hierarchical Curve-Skeletons of 3D Objects. The Visual Computer 21(11), 945–955 (2005)
23. Tierny, J., Vandeborre, J.-P., Daoudi, M.: Topology Driven 3D Mesh Hierarchical Segmentation. In: The IEEE International Conference on Shape Modeling and Applications 2007, pp. 215–220. IEEE Press, New York (2007)
24. Cornea, D.N., Silver, D., Min, P.: Curve-Skeleton Applications. In: The IEEE Visualization 2005, pp. 23–28. IEEE Press, New York (2005)
25. Singh, M., Seyranian, G.D., Hoffman, D.D.: Parsing Silhouettes, the Short-Cut Rule. Perception and Psychophysics 61(4), 636–660 (1999)
26. Dachille, F., Kaufman, A.: Incremental Triangle Voxelization. In: Graphics Interface 2000, pp. 205–212. IEEE Press, New York (2000)
27. Kalvin, A., Schonberg, E., Schwartz, J.T., Sharir, M.: Two Dimensional Model Based Boundary Matching Using Footprints. International Journal of Robotics Research 5(4), 38–55 (1986)
28. Stahl, S.: Introduction to Topology and Geometry. John Wiley & Sons, Hoboken (2004)
29. Bouix, S., Siddiqi, K.: Divergence-Based Medial Surfaces. In: Vernon, D. (ed.) ECCV 2000. LNCS, vol. 1842, pp. 603–618. Springer, Heidelberg (2000)

Mesh Simplification with Vertex Color

Hyun Soo Kim, Han Kyun Choi, and Kwan H. Lee

Gwangju Institute of Science and Technology
1 Oryong-dong, Buk-gu, Gwangju 500-712, Korea
{hskim,korwairs,khlee}@gist.ac.kr
http://ideg.gist.ac.kr

Abstract. In a resource-constrained computing environment, it is essential to simplify complex meshes of a huge 3D model for visualization, storing and transmission. Over the past few decades, quadric error metric(QEM) has been the most popular error evaluation method for mesh simplification because of its fast computation time and good quality of approximation. However, quadric based simplification often suffers from its large memory consumption. Since recent 3D scanning systems can acquire both geometry and color data simultaneously, the size of model and memory overhead of quadric increases rapidly due to the additional color attribute. This paper proposes a new error estimation method based on QEM and half-edge collapse for simplifying a triangular mesh model which includes vertex color. Our method calculates geometric error by the original QEM, but reduces the required memory for maintaining color attributes by a new memory-efficient color error evaluation method.

Keywords: mesh simplification, level of detail, vertex color, multi-resolution.

1 Introduction

Representing complex 3-dimensional models as triangular mesh has been very popular across a broad spectrum of applications such as reconstructing surfaces, creating objects in virtual reality, and rendering of 3D objects. The triangular mesh of a 3D object can be acquired by a 3D scanning system. With the advance of non-contact optical scanners, it is possible to generate a high-resolution model, which contains millions of points with speed and accuracy. Due to the occlusion problem of the optical scanner, multiple scans are generally required for reconstructing a complete 3D model. The acquisition by scanning operation often leads to the size of model unnecessarily larger. Although the capability of graphics hardware has grown rapidly, the size of models has become too big to be managed by the computing hardware and network. Simplification of a model is necessary to visualize, store and transmit large size models in a resource constrained environment. Over the past few decades, preservation of the geometry of a model has been the main subject of simplification as reviewed in [1]. But nowadays many commercial scanners can obtain both geometry and RGB color

F. Chen and B. Jüttler (Eds.): GMP 2008, LNCS 4975, pp. 258–271, 2008.

attribute of a model simultaneously by using the digital camera techniques. Furthermore, the color of a model becomes more and more important in many application areas such that there has been a renewal of interest in simplifying the model with respect to both geometry and color.

Simplification algorithms have been evaluated in terms of memory consumption, computation time and the quality of approximation. Many algorithms have been published and each one has both advantages and disadvantages in view of the above factors[4]. Some algorithms[3,5] are fast but generate poor results, on the other hands, some methods[6,7] produce nice surfaces but take a long computation time. The QEM based algorithm proposed by Garland[2] has been considered as one of the most popular simplification methods providing a good combination of speed, robustness, and fidelity. But it suffers from large memory overhead. Garland also extended his algorithm in [9] to a color model. However, this caused the use of additional memory for having vertices of the model placed in a 6-dimensional space. This paper proposes a new error estimation method based on QEM and half-edge collapse for simplifying a triangular mesh model which also includes vertex color. Our method calculates geometric error by using the original QEM, but reduces memory consumption for maintaining color attributes.

The rest of the paper is organized as follows. We reviewed some of previous work related to simplification in section 2. Among simplification error metrics, we made an overview of quadric based error metrics in section 3. Section 4 presents the details of our error estimation method. In section 5, we applied our method to various models and discussed the performance of our method by comparing with the original QEM algorithm. Section 6 concludes the paper.

2   Previous Work

The simplification process of meshes generally consists of two parts; one to apply a simplification operator and the other to decide on the priority. Simplification operators reduce the complexity of a mesh by removing geometric primitives of a mesh such as a vertex, edge and triangle, and modifying the connectivity of primitives respectively. The choice of simplification operator depends on the conditions imposed by the computing environment, the ease of implementation, the target application and the data structure of mesh. In the simplification process, it is necessary to decide the priority of mesh primitives for applying the simplification operator in order. The determination of the priority is performed by measuring the error between the original model and the simplified model, where the error measure largely affects the quality of the simplification result.

Among various algorithms that have been used for simplification, some recent simplification methodologies are briefly reviewed below according to the types of simplification operator and the methods of measuring error.

2.1   Simplification Operator

Many simplification operators have been published and used over the past few decades. Researchers[1,10] have surveyed simplification algorithms with respect to simplification operators. We discussed some well-known simplification opera-tors below.

Polygon merging is one of the earliest simplification operators[15,16]. This operator merges nearly coplanar and adjacent polygons into bigger ones. For triangular mesh, re-triangulation is performed after merging the polygons. This gives the equivalent result as using the vertex removal operator[3,17]. Merg-ing criteria are often based on distance measure or surface curvature. Polygon merging has been applied under different names, such as superfaces[15] and face clustering[16]. The polygon elimination operator simplifies a mesh by collapsing a polygon to a single vertex[19]. The size of polygons and surface curvature can be used as the priority for elimination. We can replace a polygon elimination operator by a combination of edge collapses.

The edge collapse operator proposed by Hoppe et al.[19] is the most popu-lar operator for mesh simplification. This operator collapses an edge to a new vertex, preventing the need of re-triangulation. Additionally, faces that include the target edge are also removed after collapse. Fig. 1 depicts an edge collapse operation. When an edge collapses from v_i to v_j, the new position of the vertex can be chosen either by calculating the optimal position or by moving to a sur-viving vertex of the collapsed edge. The edge collapse using the first strategy for determining the new vertex position is called the full-edge collapse, and using the latter strategy is called the half-edge collapse. While the full-edge collapse in Fig. 1(a) requires recalculating the optimal position of the destination vertex additionally, the half-edge collapse just needs to choose one vertex on the col-lapsed edge in Fig. 1(b). Since the half-edge collapse is easier and more efficient in constructing progressive mesh than the full edge collapse, we used half-edge collapse as our simplification operator in this paper. Progressive mesh can be easily constructed by storing inverse transformation of edge collapse[7].

(a)   (b)

Fig. 1. (a) full-edge collapse (b) half-edge collapse

2.2   Error Measure of Simplification

The determination of the priority in simplification process is made by measuring the error during and after the simplification. The output quality of simplifica-tion algorithm generally depends on the error measure. Since most algorithms simplify meshes iteratively, the incremental error in each step by the simplifi-cation operator is very important for optimization and implementation. On the other hand, the total error between the original mesh and the simplified mesh is more useful for a good approximation. However, measuring the total error often increases the computing time as the model gets simplified. A variety of errors can be used as simplification criteria. The types of errors used in the previous studies are discussed below.

For a geometric error measure, simplification algorithms incorporated several conceptual elements such as the distance in Euclidean geometry, the size of polygons and the surface curvature. Lindstrom and Turk[8] defined their geometric error as the sum of squared tetrahedral volume. Hussain et al.[4] measured the geometric error by calculating the change of the face normals around a vertex. Other algorithms[3, 13, 14] were based on the distance error metric. The distance based algorithms usually promised speed and ease of implementation. Kobbelt et al.[11] constrained their algorithm by placing a limit on the approximated mesh. Garland and Heckbert proposed the quadric error metric[12] which offers a combination of speed, robustness and good approximation of the surface. The quadric error metric basically computes the sum of squared distances between a point and a set of planes. This algorithm begins by pairs of vertices and collapses those pairs by the priority until it reaches the desired level of simplification. Quadric based simplifications have been widely used in recent years because of its advantages such as the quality of the result mesh, fast computation time and robustness. As such, quadric based methods have created many variations. For example, Kim et al.[20] combined the tangent and curvature error to the original quadric error. Their error metric could preserve the feature region with small geometric error. Recent papers address the problem of representing a multiresolution model from a data structure point of view. [21,22] propose a new data structure for non-manifold simplical meshes in nD Euclidean space. Their data structure is compact and efficient to representing n-dimensional mesh and can be integrated with a multiresolution framework. In [23], Cignoni et al. also try to manage external memory for huge-size meshes by octree based data structure.

Although many researchers have proposed their own geometric error measures, few of them have addressed the simplification of the 3D color model. Garland and Heckbert[9] generalized their earlier QEM scheme[2] to deal with color attributes. They used a 6-dimensional space which consists of 3-dimensional coordinate (x, y, z) of vertices for shape information and additional color triples (r, g, b) of corresponding vertices for color representation. However, this expansion of dimension caused QEM to increase the number of required variables per a vertex from 10 to 28. Another problem of their work was that their algorithm could not handle color discontinuity. Hoppe proposed a more intuitive error metric and reduced the number of variables to 23 in [13], but the memory consumption still remained heavy compared to using only the 3-dimensional coordinate values. To solve the discontinuity problem, he introduced wedges into the data structure. Rigiorli et al.[24] evaluate the simplification error of color mesh in CIE-Luv color space. Certain et al.[8] propose simplification with texture color by using the wavelet transform and Fan et al.[25] take face color into account. These studies are appropriate to handle meshes with texture or face color. However, the raw data from 3D scanners usually consists of geometry and vertex color. Accordingly, our concern is to simplify large-size mesh with vertex color.

3   Quadric Error Metrics

In this section, we briefly review the concept of quadric error metric before introducing our proposed method.

3.1   Quadric for Geometry

Garland's original QEM[2] defines the geometric error as the sum of squared distances between an arbitrary point and a set of planes which contains neighboring faces of a vertex. In a 3-dimensional space, the standard representation of a plane is,

$$n^T v + d = 0 \tag{1}$$

where $n = [a \quad b \quad c]^T$ is a unit normal and d is a scalar constant.

For a vertex $v = [x \quad y \quad z]^T$ with a set of planes, the geometric error can be written as follows

$$E_{planes}(v) = \sum_i D_i^2(v) = \sum_i (n_i^T v + d_i)^2 \tag{2}$$

where i = neighbor triangles

Garland rewrites the squared distance into a new form as follows:

$$\sum D^2(v) = \sum (n_i^T v + d_i)^2 \tag{3}$$
$$= \sum \left(v^T (n_i n_i^T) v + 2 d_i n_i^T + d_i^2 \right)$$
$$= v^T \left(\sum (n_i n_i^T) \right) v + 2 \left(\sum d_i n_i^T \right) v + \sum d_i^2$$
$$= v^T A v + 2 b^T v + c$$

where A is a symmetric 3×3 matrix, b is a column vector of size 3 and c is a scalar[2].

From [2], a quadric is defined as a triple as follows.

$$Q = (A, b, c) \tag{4}$$

In a homogeneous matrix form, the quadric Q can be given by

$$Q = \begin{pmatrix} A & b \\ b^T & c \end{pmatrix} \tag{5}$$

Given this matrix, we can compute the geometric error $Q(v)$ using the quadric form below.

$$Q(v) = \tilde{v}^T Q \tilde{v} \quad \text{where} \quad \tilde{v} = [v \quad 1]^T = [x \quad y \quad z \quad 1]^T \tag{6}$$

The addition of quadrics can be defined component-wise: $Q_i(v) + Q_j(v) = (Q_i + Q_j)(v)$ where $(Q_i + Q_j) = (A_i + A_j, b_i + b_j, c_i + c_j)$.

Thus, the geometric error in (2) is computed by the sum of the quadrics Q_i which is determined by a set of planes.

$$E_{planes}(v) = \sum D_i^2(v) = \sum Q_i(v) = Q(v) \quad \text{where } Q = \sum Q_i \quad (7)$$

In equation (7), each quadric of a plane is given an equal weight. However, the area-weighted quadric is used in practice as follows:

$$Q = \sum area(f_i) \cdot Q_i \quad \text{where } f_i \text{ is a neighboring face of the given vertex} \quad (8)$$

Since a quadric matrix is symmetric, 10 coefficients are required to store a quadric Q. Let us now consider the quadric error when contracting an edge (v_i, v_j) to v. If the sum of quadrics is assigned to each vertex on the original mesh, the quadric of an edge is given by

$$Q_v = Q_{v_i}(v) + Q_{v_j}(v) \quad (9)$$

After the collapse of each edge, it is necessary to determine the position of a new vertex v. The simplest way is to choose one of the end points as the target point. Another solution is to find the optimal position which minimizes the quadric error. The minimum of the quadric error occurs where the gradient of the quadric $(\nabla Q(v) = 2Av + 2b)$ equals to zero.

Solving for $\nabla Q(v_{min}) = 0$, the optimal position becomes $v_{min} = -A^{-1}b$.

3.2   Quadric for Geometry and Attributes

Garland and Heckbert generalized the quadric error metric which includes vertex attributes such as normal, vertex color and texture in [9]. The original quadric error metric lies in R^3. They placed their framework into a n-dimensional space where n equals to $3+m$, and m is the number of vertex properties. For example, m equals to 3 in the case of the colored surface, because color consists of RGB triple. This makes the quadric of a vertex formulated as the distance-to-hyperplane in R^n space. They assumed that all attributes are linearly interpolated over triangular faces and can be measured by the Euclidean distance between them. For scale-invariance, the model is scaled within the unit cube in R^n.

Let us consider the triangle $T = (v_1, v_2, v_3)$ as depicted in Fig. 2. Given a colored mesh, each vertex can be represented by $v_i = (x_i \quad y_i \quad z_i \quad r_i \quad g_i \quad b_i)^T$.

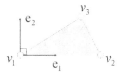

Fig. 2. Triangle in R^n and its corresponding orthonormal vectors[9]

To construct the quadric in the n-dimensional space, Garland and Heckbert computed two orthonormal unit vectors e_1 and e_2. These vectors are given by Gram-Schmidt orthogonalization

$$e_1 = \frac{v_2 - v_1}{\|v_2 - v_1\|} \tag{10}$$

$$e_2 = \frac{v_3 - v_1 - (e_1 \cdot (v_3 - v_1))e_1}{\|v_3 - v_1 - (e_1 \cdot (v_3 - v_1))e_1\|} \tag{11}$$

Given these orthonormal unit vectors, we can formulate the quadric as $D^2(v) = v^T A v + 2b^T v + c$ where

$$
\begin{aligned}
A &= I - e_1 e_1^T - e_2 e_2^T \\
b &= (v_1 \cdot e_1)e_1 + (v_1 \cdot e_2)e_2 - v_1 \\
c &= v_1 \cdot v_1 - (v_1 \cdot e_i)^2 - (v_2 \cdot e_2)^2
\end{aligned} \tag{12}
$$

In this generalized quadric, A is a symmetric $n \times n$ matrix, b is a column vector of size n and the total number of coefficients becomes $\big((n+1)(n+2)\big)/2$. For the vertex color, therefore, $28((7 \times 8)/2 = 28)$ coefficients are required to store a quadric.

4   Evaluation of Edge Cost

Our simplification method is similar to other algorithms and consists of two parts. First, we calculate the half-edge cost and place all half-edge collapses in a priority heap. Then we apply the simplification operator to the half-edge $(v_0 \rightarrow v_1)$ having the lowest cost and update the costs of all half-edges involving v_0 and v_1. In order to reduce memory overhead, we split the simplification error into two types of error term: the geometric error and the color error. We use the original QEM for our simplification of geometry because of its quality of approximation. On the other hand, our color error is classified into two terms: visual importance and collapse color error. We first define the visual importance, because we adopt half-edge collapse as our simplification operator. In the case of half-edge collapse, one of the vertices on an edge is removed while the other remains after collapse. Thus, it is desired to decide on the value that helps the final model to keep visually important vertices with respect to the global distribution of their color difference. On the other hand, the collapse color error represents the incremental color change as a result of each half-edge collapse. These two terms are described in section 4.1 and 4.2 and the details of the overall procedure are presented in section 4.3.

4.1   Visual Importance for Vertex Color

Visual importance $I(v_0)$ determines the color difference, that is, RGB distance between a vertex and its neighboring vertices in the original mesh. In other words, this value reflects the initial color difference. Since our color error does not store any history of color changes, we propose to use visual importance to represent the color change between the initial mesh and the result mesh. The colors of a triangular mesh are typically stored as (r, g, b) triples with each value in the range $[0, 1]$. The most straightforward way to measure the color difference is to treat the color space as a Euclidean space and compute RGB distance between two vertices as the equation below.

$$\|c_0 - c_i\| = \sqrt{(r_0 - r_i)^2 + (g_0 - g_i)^2 + (b_0 - b_i)^2} \tag{13}$$

where $c_0(r_0, g_0, b_0)$ is the color triple of vertex v_0 and c_i is that of its neighboring vertex v_i.

We define the visual importance of vertex v_0 as follows:

$$I(v_0) = \begin{cases} \dfrac{\sum_{i=\text{neighbor vertices}} \|c_0 - c_i\|}{\max\|c_0 - c_i\|} & \text{,when } \max(c_0 - c_i) \neq 0 \\ 0 & \text{,when } \max(c_0 - c_i) = 0 \end{cases} \tag{14}$$

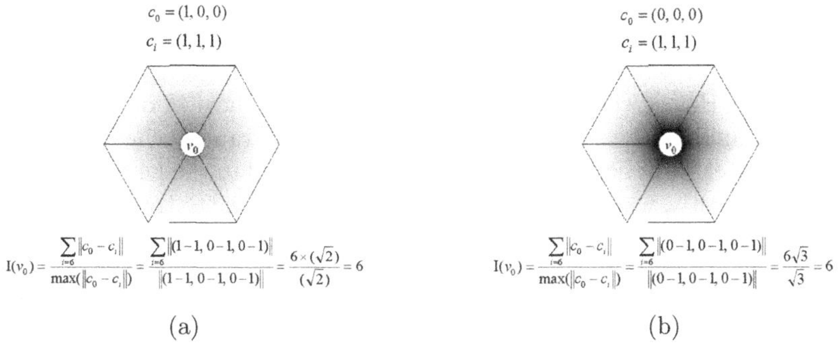

$c_0 = (1, 0, 0)$
$c_i = (1, 1, 1)$

$c_0 = (0, 0, 0)$
$c_i = (1, 1, 1)$

$$I(v_0) = \frac{\sum \|c_0 - c_i\|}{\max(\|c_0 - c_i\|)} = \frac{\sum \|1 - 1, 0 - 1, 0 - 1\|}{\|1 - 1, 0 - 1, 0 - 1\|} = \frac{6 \times (\sqrt{2})}{(\sqrt{2})} = 6$$

$$I(v_0) = \frac{\sum \|c_0 - c_i\|}{\max(\|c_0 - c_i\|)} = \frac{\sum \|0 - 1, 0 - 1, 0 - 1\|}{\|0 - 1, 0 - 1, 0 - 1\|} = \frac{6\sqrt{3}}{\sqrt{3}} = 6$$

(a)   (b)

Fig. 3. Examples of visual importance

The above definition means that the visual importance is zero if the color of vertex v_0 is equal to its neighbor vertices, otherwise it is a real value determined by the maximum color difference and the color difference with the neighboring vertices. We divide the sum of color difference by the maximum color difference to measure the relative color difference. Fig. 3 shows the visual comparison of two vertices having different color distance. The color triples of vertex v_0 in Fig. 3(a) and (b) are different, but it is obvious that vertex v_0 has different color from neighbors in both cases. Consequently, the visual importance values in Fig. 3(a) and (b) are equal. We cannot represent the human perception of color just using the Euclidean distance in RGB space unlike the geometric error. For example, the distance between white and red is much shorter than that between white and black in the RGB space, but human does not perceive the color difference proportional to the distance. The visual importance metric, therefore, is added to emphasize the local color difference between vertices.

After the half-edge collapse ($\boldsymbol{v}_0 \rightarrow \boldsymbol{v}_1$), the visual importance of remaining vertex $\boldsymbol{v}_1$ is updated as

$$\mathrm{I}^{\mathrm{after}}(\boldsymbol{v}_1) = \mathrm{I}^{\mathrm{before}}(\boldsymbol{v}_0) + \mathrm{I}^{\mathrm{before}}(\boldsymbol{v}_1) \tag{15}$$

In order to store and update visual importance, we need two floats per vertex, that is, the visual importance itself and the maximum color distance.

4.2   Collapse Color Error

The priority for simplification is stored in a heap with the lowest cost at the top. All half-edges have their own collapse error. The priority of a vertex $\boldsymbol{v}_0$ is determined by the minimum cost out of its out-going half-edges. Since the collapse color error should estimate the color change before and after the collapse, we define the collapse color error of half-edge collapse $\mathbf{C}(\boldsymbol{v}_0 \rightarrow \boldsymbol{v}_1)$ as

$$\mathbf{C}(\boldsymbol{v}_0 \rightarrow \boldsymbol{v}_1) = \sum_{i=\text{neighbor vertices of } \boldsymbol{v}_0} \left| \, \|c_0 - c_i\| \cdot \|v_0 - v_i\| - \|c_1 - c_i\| \cdot \|c_1 - c_i\| \, \right|$$

$$\tag{16}$$

When using vertex color, we interpolate the color of a triangle by colors of three vertices. Hence, the length of an edge should be short in order to preserve colors of triangles in the semantic area where the color rapidly changes. It means that the color error of half-edge collapse is proportional to the length of the half-edge. Consequently, we multiply the length of the half-edge to the color difference for each half-edge in [15]. In the case of half-edge collapse, the color change of collapse always occurs on out-going half-edges of vertex $\boldsymbol{v}_0$ which is inside the gray area in Fig. 4. The connectivity in the white area is not changed by collapse.

Fig. 4. The area of color change by a half-edge collapse

4.3   Simplification Algorithm

In the previous sections, we described two types of color error: visual importance and collapse color error. The final expression of our simplification algorithm combines the geometric error with these two color error terms. The total cost of half-edge collapse is now defined as

$$\text{error}(\boldsymbol{v}_0 \rightarrow \boldsymbol{v}_1) = (1 + \text{quadric}) \cdot (1 + \text{visual importance}) \cdot (1 + \text{color error})$$

$$= (1 + Q(\boldsymbol{v}_0 \rightarrow \boldsymbol{v}_1)) \cdot (1 + \mathrm{I}(\boldsymbol{v}_0)) \cdot (1 + \mathbf{C}(\boldsymbol{v}_0 \rightarrow \boldsymbol{v}_1)) \tag{17}$$

Since each of the above three error terms can be zero, we add one to each term to prevent the total error from becoming zero. Upon completion of computing the cost of every half-edge, the minimum cost among the out-going half-edge of each vertex is chosen as the vertex cost.

In summary, our simplification algorithm performs the following to simplify a 3D color mesh.

1. Compute the quadrics, visual importance and collapse color error.
2. Evaluate the cost of half-edge using (2)
3. Place all the half-edge collapses in a priority heap and collapse the half-edge which has the lowest cost.
4. Update the cost of all half-edge collapses involving vertices on the half-edge collapsed in step 3.
5. Repeat step 3 and 4 until there is no half-edge in the priority heap.

5   Results and Discussion

We compared the proposed method and the color QEM by Garland and Heckbert [11] using several 3D color models. They were compared in terms of computation time, memory consumption, geometric distance and color error. All models used were scanned by a commercial 3D scanner which can acquire both geometry and color. We ran these models on an Intel Pentium4-3.5GHz machine with 2G RAM. Fig. 5 shows the results of simplified models using the color QEM and our method, respectively. The reduction ratio of each model is described in Table 1. From the results, it is observed that our method can keep the geometry better and reduce the memory usage down to about a half of what is needed for the color QEM. However, the color QEM is faster and maintains color details better than our algorithm.

5.1   Computation Time

Table 1 lists the running time of the color QEM and our method in simplifying the three models at the reduction ratio presented in the table. Our method takes more time due to recalculating the color error that considers the change of connectivity after half-edge collapse, while the color QEM simply adds two quadrics after collapse. Although it depends on the size and connectivity of each model, our method is about 1.6 times slower than that of the color QEM.

5.2   Memory Consumption

The color QEM requires 28 floats per vertex. It means that the memory overhead is $112n$ bytes, where n is the number of vertices. The new quadric proposed by Hoppe[15] reduces the memory consumption to 23 floats per vertex. However, it is still heavy to construct a quadric for a huge size mesh. We use the original QEM which needs 10 floats for geometry. Additionally, we consume 2 floats for color error; one for the visual importance and the other to store the maximum color difference. Consequently, our method only spends $48n$ bytes of memory in total, which gives a significant advantage compared to the color QEM.

Table 1. Reduction ratio and computation times with respect to color QEM and our method

Model	Model Size (# of vertices/ # of faces)	Reduction Ratio(%) (# of vertices/# of faces)		Computation time(sec)	
		Color QEM	Our Method	Color QEM	Our Method
Monkey	50,503/ 101,002	90% (5,050/ 10,096)	90% (5,050/ 10,096)	3.13	5.46
Wood Duck	50,052/ 99,997	90% (5,005/ 9,925)	90% (5,005/ 9,926)	3.13	5.34
GIST Ari	1,175,653/ 2,361,302	97% (35,269/ 70,534)	97% (35,269/ 70,534)	83.78	143.20

5.3   Geometric Comparison

For geometric comparison between the original model and the simplified model, we used the well-known surface comparison tool - Metro[16] which is used by many researchers. In this research, we measured the mean distance and the RMS distance between the simplified mesh and the original mesh. Table 2 shows the results of the geometric comparison. The Mean and RMS distances of our method are always smaller than those of the color QEM for all tested models.

Table 2. Mean and RMS geometric error by Metro[14]

Model	Color QEM		Our Method	
	Mean error	RMS error	Mean error	RMS error
Monkey	3.95	5.30	3.44	4.69
Wood Duck	7.60	11.32	5.92	10.14
GIST Ari	1.45	2.35	0.82	1.41

5.4   Similarity of Color Appearance

For a given view, the appearance of a model is determined by the corresponding raster image which a renderer would produce[1]. For this, we implemented our own triangular mesh viewer which renders a mesh by OpenGL. To compare the

Table 3. Color error with comparison of rendered images

Model	Color QEM	Our Method
	RMS error	RMS error
Monkey	3.60	4.44
Wood Duck	9.07	10.41
GIST Ari	7.73	9.62

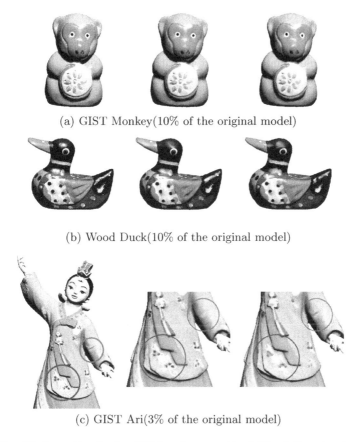

(a) GIST Monkey(10% of the original model)

(b) Wood Duck(10% of the original model)

(c) GIST Ari(3% of the original model)

Fig. 5. Simplified results of color QEM and our method. The leftmost column is the original model. The middle column is the color QEM result and the right column is ours.

original color of vertex, we disabled the lighting condition in OpenGL. After rendering a model, we obtained two 24-bit bitmap images of the original model and its simplified model separately. We compare color differences between two images pixel by pixel and calculate the RMS error of Euclidean distances between corresponding pixels of two images. Table 3 shows the RMS error between the rendered images of the original and the simplified models. As shown from Table 3, the color QEM can maintain the color distribution better than our algorithm.

6   Conclusion and Future Work

In this paper, we have presented a novel and memory-efficient half-edge cost evaluation method to simplify 3D mesh with vertex color. The main benefit of our method is to reduce memory consumption less than half of the original color QEM. Additionally, our method can keep the geometric shape better and the running time is also comparable to that of the color QEM.

However, our proposed method shows a weakness in maintaining accurate color details in an area with fine shapes. We will try to solve this problem by integrating hierarchical multiresolution framework. Our future work will focus on solving this problem and making our algorithm faster. In the proposed algorithm, it is also ignored that the RGB space is not perceptually linear. In our future work, a uniform color space[24]]such as CIE-Lab or XYZ space will be used in replacement of RGB space. The discontinuity problem of the attribute value will also be addressed for preserving the color attribute.

Acknowledgments. This research is supported by the NRL (National Research Laboratory) funded by the Ministry of Science and Technology of Korea and the CTI development project supported by the MCT and KOCCA in S. Korea.

References

1. Garland, M.: Multiresolution Modeling: Survey & Future Opportunities. In: Eurographics 1999, State of the Art Report (1999)
2. Garland, M., Heckbert, P.: Surface Simplification Using Quadric Error Metric. In: Computer Graphics(SIGGRAPH 1997 Proceeding), pp. 209–216 (1997)
3. Schröeder, W.J., Zarge, A., Loresen, W.E.: Decimation of Triangle Mesh. Computer Graphics(SIGGRAPH 1992 Proc.) 26(2), 65–70 (1992)
4. Hussain, M., Okada, Y., Niijima, K.: Efficient and Feature-Preserving Triangular Mesh. Journal of WSCG 12(1-3), 167–174 (2004)
5. Brodsky, D., Watson, B.: Model Simplification through Refinement. In: Graphics Interface 2000, pp. 221–228 (2000)
6. Ciampalini, A., Cignoni, P., Montani, C., Scopigno, R.: Multiresolution Decimation based on Global Error. The Visual Computer 13, 223–246 (1997)
7. Hoppe, H.: Progressive Mesh. In: Proc. SIGGRAPH 1996, pp. 99–108 (1996)
8. Certain, A., Popović, J., DeRose, T., Duchamp, T., Salesin, D., Stuetzle, W.: Interactive Multiresolution Surface Viewing. In: Proceedings of ACM SIGGRAPH 1996, pp. 91–98 (1996)
9. Garland, M., Heckbert, P.: Simplifying Surfaces with Color and Texture Using Quadric Error Metrics. In: Proc. Of IEEE Visualization 1998, pp. 263–270 (1998)
10. Wünsche, B.: A Survey and Evaluation of Mesh Reduction Techniques. In: Proc. IVCNZ 1998, pp. 393–398 (1998)
11. Kobbelt, L., Campagna, S., Seidel, H.P.: A General Framework for Mesh Decimation. In: Graphics Interface 1998 proceedings, pp. 43–50 (1998)
12. Garland, M., Heckbert, P.: Surface Simplification Using Quadric Error Metric. In: Visualization 98 proceedings, IEEE, pp. 263–269 (1998)
13. Hoppe, H.: New Quadric Metric for Simplifying Meshes with Appearance Attributes. In: IEEE Visualization 1999, pp. 59–66 (1999)
14. Cignoni, P., Rocchini, C., Scopigno, R.: Measuring Error on Simplified Surfaces. Computer Graphics Forum 17(2), 167–174 (1998)
15. Kalvin, A.D., Taylor, R.H.: Superfaces: Polygonal Mesh Simplification with Bounded Error. IEEE Computer Graphics and Applications 16(3), 64–77 (1996)
16. Garland, M., Willmott, A., Heckbert, P.S.: Hierarchical Face Clustering on Polygonal Surfaces. In: Proceedings of 2001 ACM Symposium on Interactive 3D Graphics, pp. 49–58 (2001)

17. Klein, R., Krämer, J.: Multiresolution Representations for Surface Meshes. In: Proceedings of Spring Conference on Computer Graphics 1997, pp. 57–66 (1997)

18. Hamann, B.: A Data Reduction Scheme for Triangulated Surfaces. Computer Aided Geometric Design 11, 197–214 (1994)

19. Hoppe, H., DeRose, T., Duchamp, T., McDonald, J., Stuetzle, W.: Mesh Optimization. In: Proceddings of SIGGRAPH 1993, pp. 19–26 (1993)

20. Kim, S.J., Kim, S.K., Kim, C.H.: Discrete Differential Error Metric for Surface Simplification. In: Proceedings of the 10th Pacific Conference on Computer Graphics and Applications, pp. 276–283 (2002)

21. De Floriani, L., Magillo, P., Puppo, E,, Sobrero, D.: A multi-resolution topological representation for non-manifold meshes. Computer Aided Design 36(2), 141–159 (2004)

22. De Floriani, L., Hui, A.: A Dimension-Independent Representation for Multiresolution Nonmanifold Meshes. Journal of Computing and Information Science in Engineering 6(4), 397–404 (2006)

23. Cignoni, P., Montani, C., Rocchini, C., Scopigno, R.: Interactive Multiresolution Surface Viewing. IEEE Transactions on Visualization and Computer Graphics 9(4), 525–537 (2003)

24. Rigiroli, P., Campadelli, P., Pedotti, A., Borghese, N.A.: Mesh refinement with color attributes. Computers & Graphics 25, 449–461 (2001)

25. Fahn, C.-.S., Chen, H.-.K., Shiau, Y.-.H.: Polygonal Mesh Simplification with Face Color and Boundary Edge Preservation Using Quadric Error Metric. In: Proceedings of IEEE Fourth International Symposium on Multimedia Software Engineering, pp. 174–181 (2002)

A Multistep Approach to Restoration of Locally Undersampled Meshes

Alexandra Bac, Nam-Van Tran, and Marc Daniel

LSIS
Ecole Suprieure d'Ingnieurs de Luminy
Case 925 - 13288 Marseille Cedex 09 - France
{alexandra.bac,van.tran-nam,marc.daniel}@esil.univmed.fr
http://www.lsis.org

Abstract. The paper deals with the problem of remeshing and fairing of undersampled areas (called "holes") in triangular meshes. In this work, we are particularly interested in meshes constructed with geological data but the method can however be applied to any kind of data. With such input data, the point density is often drastically lesser in some regions than in others: this leads to what we call "holes". Once these holes identified, they are filled using a multistep approach. We iteratively: insert vertices in the hole in order to progressively converge towards the density of its neighbourhood, then deform this patch mesh (by minimizing a discrete thin-plate energy) in order to restore the local curvature and guarantee the smoothness of the hole boundary. The main goal of our method is to control both time and space complexity in order to handle huge models while producing quality meshes.

Keywords: Resampling, holes, meshes, thin plate energy, surface fairing, multistep.

1 Introduction

Petrol industry has acquired in the last years, an enormous amount of seismic data which is, up to now, neither interpreted nor transformed into some "Shared Earth Model". In order to satisfy the needs of petrol companies, methodologies were recently proposed to reconstruct updatable, modifiable and extendable 3D geological models constructed with such data [1].

The construction process starts by the acquisition of point clouds defining the geologic surfaces (called "horizons"). These clouds are dense and unorganized. Triangular or parametric meshes (with BSpline surfaces) can be used to mesh these clouds [2], [3]. However, triangular meshes are preferable due to their simplicity and accuracy for data representation. According to the complexity of the geologic surfaces and the data acquisition technology (e.g. geoseismic or geomagnetic), data can miss in some regions of the surfaces. Figure 1 presents such cases. Such undersampled (if not empty) regions are called "holes". Such holes are not acceptable either for geologists or computer scientists and can induce

F. Chen and B. Jüttler (Eds.): GMP 2008, LNCS 4975, pp. 272–289, 2008.

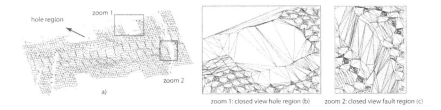

zoom 1: closed view hole region (b)   zoom 2: closed view fault region (c)

Fig. 1. Example of "holes" in a geological surface: (a): a horizon is represented by a point cloud (22,000 vertices), (b): a closer view of a hole region and (c): a fault region of the triangular mesh. Notice that in the regions where data is lacking, the triangle geometries are totally different to those of the surrounding triangles.

Fig. 2. The steps of our filling process for a simple object: (a) Hole identification, followed by a succession of (b_i) hole refinement, (c_i) fairing of the inserted patch

unexpected results when reconstructing 3D models. Other important regions of interest are geological faults. A fault consists in a brutal discontinuity in the horizon which often corresponds to a change in the point density of the data (fig. 1c). But the treatment of faults is of different nature and does not overlap with our hole filling process.

Our approach is directed by the nature of the problem: geological surfaces can be very large (we expect to handle billion triangle surfaces) and as surface improvement is only a preliminary step, its running time must remain low. Therefore, a global approach is not possible (for both space and time reasons). Moreover, as our input data is issued from seismic techniques, the "holes" we have to fill are localized - the input data is generally rather uniformly sampled, except in the holes where it is absent.

For efficiency, our method consists in a multistep approach (see figure 2). First, hole areas are identified (section 3). Then, we iteratively: (1) refine (or fill) holes (section 4) - vertices are inserted in order to progressively approach the density of the hole neighbourhood - and (2) fair the inserted patches (section 5) in order to satisfy a blending condition and restore the local shape of the surface. At step (2), we use a partial differential equation scheme (PDE) issued from a discretization of the bending energy (the thin plate energy) to restore the inserted patch shape. Results of this local resampling method and comparisons with other existing approaches, briefly recalled in section 2, are presented in section 7.

2   Related Works

Various methods have been proposed to fill undesirable holes in triangular meshes. The most popular approach for surface reconstruction and repairing is based on the use of implicit or parametric functions. In [4], Carr and *al* use Radial Basis Functions to represent scattered data. Other techniques based on the Moving Least Square projection [5], [6] are also proposed. The holes are then filled by extracting iso-surfaces from the implicit representation. Like any other implicit methods, these hole filling algorithms construct their implicit representations from the neighbourhood of the holes. Therefore, they tend to poorly perform if the hole topology is complex or if the hole geometry is twisted. Nevertheless, these algorithms guarantee that the hole filling patches smoothly blend with the original model.

In [7], Davis and *al* apply a volumetric diffusion technique to fill holes in range scans. The technique aims at reconstructing densely sampled incomplete closed surfaces. The process consists in converting a surface into a voxel-based representation with a clamped signed distance function; the final surface is extracted using marching cubes. This method is relevant to fill complex holes but does not guarantee the boundary continuity.

In [8], T. Ju, starting from a similar octree representation, partitions the space into disjoint inside and outside volumes (by means of a sign function consistently generated over the voxels). A closed surface is then extracted using a contouring algorithm such as a the marching cubes algorithm of dual contouring (see [9]). In [10] obtains such a closed surface by generating, from the voxelization of the space, a closed membrane (the algorithm si similar to the α-shape algorithm). This membrane is approximated to produce a closed discrete or smooth surface. Let us also mention the algorithm proposed by A. Hornung and L. Kobbelt in [11]; instead of using a reconstructing the surface as the zero level-set of a sign function (as in [8]), they compute a dilated crust around the cloud of points together with a distance function (local surface confidence). The closed surface maximizing the global confidence is obtained as a max cut over a weighted graph structure. Last, this surface is smoothed using an iterative bilaplacian operator. Such approaches efficiently produce closed smooth surfaces starting from irregular noisy data. However, in our setting, such global approaches are not appropriate. First, because undersampling is a local phenomenon in very dense and large clouds of points (millions to billions of points) and second because they produce closed surfaces whereas our surfaces are clearly not closed.

In [12] holes are first identified and filled by minimizing the area triangulation and the dihedral angle of its 3D contour. The triangulation is then refined with respect to the point density of neighbouring areas. Inserted patches are finally smoothed using an umbrella operator [13]. A method relatively similar to [12] is proposed by Pernot et *als* [14] for hole filling in reverse engineering. Pernot and *al* fill a hole with a regular grid, after having removed distorted triangles around the hole contour. The patch mesh is then deformed with the minimum of external force variation. Two other fairing processes ([12], [14]) are obtained by solving linear systems. Let us mention that both [12] and [14] focus on hole filling inside surfaces (and do not consider the problem of boundary holes which often occurs in the geological setting).

Schneider and *al* [15] propose a smoothing technique satisfying a G^1 boundary condition (actually a pseudo-G^1 condition adapted to the discrete case) based on solving non-linear fourth order partial differential equations. They introduce a notion of outer fairness (corresponding to the smoothness in the classical sense) and an inner fairness (the regularity of the vertex distribution). These criteria are used in their fairing process. In [16], the authors exploit these results in their freeform designing system from a collection of 3D curves. Recently, Xu et *al* [17] used surface diffusion flows of high order to solve different problems of smooth surface reconstruction. This approach guarantees a G^1 (fourth order PDE) and a G^2 (sixth order PDE) continuity along the boundary curves. For a similar surface restoration problem, Clarenz and *al*, in [18], use a finite element method to solve Willmore equations (let us also mention the work of Bobenko and Schrder [19] using Discrete Willmore Flow). In [13], L. Kobbelt and al. present a fast multi-resolution smoothing algorithm based on incremental fairing (by means of the "umbrella" operator (see [15])) of an improved hierarchy of nested spaces. Let us finally mention that in [18], the fairing techniques require an iterative resolution process.

The approaches by Liepa and Pernot are more relevant in our setting as their behaviour is purely local. Our approach, based on Liepa, intends on the one hand to improve the quality of the patching meshes (both for boundary and internal holes) and on the other hand to enhance the efficiency of the process. Our goal is reached by developping first a more appropriate filling procedure for open surfaces (adequate for boundary holes), second a new discretization of the thin plate energy in order to improve the quality of fairing and third a multistep process in order to enhance the efficiency of our algorithm in terms of both space and time.

3   Identification of Hole Regions

As explained in the introduction, our method has been designed to be as local as possible. Therefore, the first step consists in identifying the so-called hole regions: areas where the data is sparser than elsewhere (or absent) in the point cloud. In the geological setting, such areas are actually patches corresponding to an absence of seismic response. Many techniques are available to solve such problems (such as the powerful algorithm of α-shapes for instance), but they are computationally complex whereas this detection is only a preliminary step for us. As for the methods proposed in [10], [11] and [8], they are based on a voxelization of the space and produce a global smooth approximation of the cloud of point, not a local one.

Actually, once the data is triangulated (fig.1) using a Delaunay-like triangulation, hole regions can be properly detected by means of an elementary geometric parameter: comparison between the perimeter of triangles and the average perimeter over the mesh. This criterion performs particularly well on the Delaunay-like triangulations.

As pointed out in the first section, there also exist fault areas in geological surfaces inside which triangles present similar singular perimeter characteristics. In the present work, these areas are identified by a simple procedure based on the variation of the local normal with respect to the average normal of the surface. Such areas are discarded in the hole detection step (and therefore in the whole subsequent filling process). Indeed in the 3D reservoir modelling process, faults receive an appropriate and dedicated treatment.

Let us mention that the detection of faults can be improved through an efficient two steps algorithm. First, the rough detection previously mentionned, based on the variation of the local normal, is performed; second, the identified regions are more precisely analysed with techniques based on ridge and ravine detection (see for instance [20]). But this is not the purpose of the present article and still work in progress.

4   Filling Hole

Once undersampled regions have been detected, we wish to fill each one with a patch mesh approximating the density of the surrounding mesh. We inspire on the idea mentioned in [21], [12]. The principle consists in distributing the average length of the edges around the hole by subdividing the triangles and by exchanging the interior edges to improve the mesh regularity.

In order to create a regular patch mesh, vertices are inserted in singular triangles according to their geometry. We consider two cases: for sharp triangles, as in [12], a new point is inserted at the barycenter of the triangle which is therefore replaced by three new triangles. However, for obtuse triangles, the resulting new triangles would be even stretcher than the original one (fig. 3a); in particular the triangles along the boundary (which are most often obtuse) would become stretcher and stretcher eventually compromising the convergence of the filling process. Therefore in the case of obtuse triangles (which mostly occur on the boundary), we inspire on Rivara bisection: the new point is inserted in the middle of the obtuse edge (fig. 3b).

In order to maintain the Delaunay-like criterion, newly inserted edges are optimized by checking whether the condition "$\alpha + \beta > \Pi$" ($\sharp$) is true of false (fig. 4). If it is true, the newly inserted point C lies inside the circum-circle of the triangle opposite to this edge, in this case, in order to maintain the Delaunay-like criterion (see [2]), such an edge is flipped. This conditional edge flip will be called relaxation.

Given the final edge length ℓ, the density control factor λ and $\|\cdot\|$ the Euclidean distance, the hole refinement algorithm can be summarized as follows:

1. For each triangle $\mathcal{T}(P_i, P_j, P_k)$ in the hole regions:
   - If $\mathcal{T}$ is sharp,
       if for any $m = i, j, k$,
       $\|CP_m\| > \lambda * \ell$,
       with C the barycenter of $\mathcal{T}$

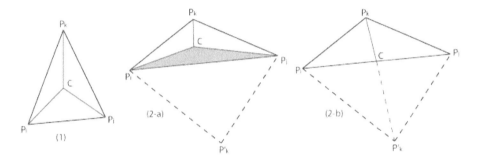

Fig. 3. (1) Sharp triangle - (2) Obtuse triangle (a) new point C inserted at the barycenter of triangles, one of the three new triangles is stretcher than the original one, (b) new point C inserted on the longest edge, both new triangles are more regular

   tri-subdivide $\mathcal{T}$ (Fig. 31)
    and then relax edges (P_iP_j, P_jP_k, P_iP_k)
  – If $\mathcal{T}$ is obtuse,
    let e is the edge opposite to the obtuse angle (ie. P_iP_j)
    let $\mathcal{T}'$ be the triangle sharing edge e with $\mathcal{T}$
    if $||e|| > \lambda * \ell$
    where
     bi-subdivide $\mathcal{T}$ and $\mathcal{T}'$ (Fig. 31-b)
     and relax edges $(P_iP_k, P_jP_k, P_iP'_k, P_jP'_k)$
2. If no new points were inserted in step 1, the hole filling process is complete.
3. Do
  relax all edges in the patch mesh
 While some edges are exchanged
4. Go to step 1

We have also tested other edge optimization criteria such as: triangle quality factor [22] or curvature variation [23]. However, the Delaunay-like condition ($\sharp$) experimentally led to better patch mesh regularity quality as well as shorter running time.

Even if the method is stable, the result can slightly change with the choice of parameters λ and ℓ. This choice is based on experimentations of different types of input models. In our implementation, choosing the parameter ℓ to be slightly

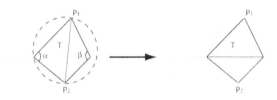

Fig. 4. Flip of the interior edges P_1P_2

less than the average edge length of the 2-neighbour ring and the density control factor λ in the range $[0.7 \ldots 1.4]$ yields to good experimental results (ie. to patch meshes that visually match the density of surrounding mesh).

5   Fairing Hole

After the previous hole filling process, the vertex density on the resulting surface is homogeneous but the surface is not smooth enough as we have used so far only C^0-like continuity conditions. Restoring the shape of hole area in order to obtain a smoother or more visually pleasant result is necessary.

For continuous surfaces, it is popular to minimize a fairness functional which is usually chosen according to either physical analogy (e.g. minimizing a membrane or thin plate energy functional) or geometric reasoning (e.g. minimizing area, curvature or curvature variations). In our case, we will focus on the thin plate energy functional:

$$S \mapsto \int_\Omega a(\kappa_1^2 + \kappa_2^2) + 2(1 - b)\kappa_1\kappa_2 \mathrm{d}\omega_S$$

where $S : \Omega \to \mathbb{R}^3$ is a parameterized surface, $\mathrm{d}\omega_S$ denotes the surface element, κ_1 and κ_2 are the principal curvatures of the surface and a and b are respectively constants describing resistance to bending and sheering. In the special case ($a = b = 1$), this formula becomes:

$$E_{TP}(S) = \int_\Omega (\kappa_1^2 + \kappa_2^2)d\omega_S = \int_\Omega \left((k_1 + k_2)^2 - 2k_1k_2\right)d\omega_S \tag{1}$$
$$= \int_\Omega 4H^2 - 2K\mathrm{d}\omega_S \tag{2}$$

where $H = \frac{\kappa_1 + \kappa_2}{2}$ and $K = \kappa_1\kappa_2$ are respectively the mean and Gaussian curvature of the surface.

5.1   Minimization of the Thin Plate Energy

Naturally, this thin plate energy functional can not be directly applied to the discrete surfaces. Let us first derive a discrete equivalent to this formula.

Approaches such as [15] overcome the difficulty of the high order of E_{TP} by transforming it into the following simplified version (which is quadratic and hence easily minimized):

$$\int F_{uu}^2 + 2F_{uv}^2 + F_{vv}^2 \tag{3}$$

where F_{uu}, F_{vv} and F_{uv} denote the second order partial derivatives.

In our setting, starting from equation (2), by the Gauss-Bonnet theorem, the integral of Gaussian curvature can be expressed as the integral of the geodesic curvature over the boundary of the hole surface. This boundary and the tangent planes along it will be preserved throughout the fairing process. Therefore, minimization of the thin plate energy reduces to the minimization of the following energy: $\int H^2 \mathrm{d}\omega$.

In [24], Greiner shows that a good approximation (better than equation (3)) is given by:

$$S \mapsto \int_{\Omega} \langle \Delta_{S_0} S | \Delta_{S_0} S \rangle d\omega_S \tag{4}$$

where S_0 is a parametrized surface close to the desired surface, Δ_{S_0} is the Laplace-Beltrami operator related to S_0 and $\langle \cdot | \cdot \rangle$ denotes the euclidean scalar product. This functional is a positive definite, quadratic function. Therefore, the variational calculus leads to a simple characterization of the minimum of its energy:

$$\Delta_{S_0}^2 S = 0. \tag{5}$$

There exist several discretization schemes for the Laplace-Beltrami operator. We adopt the approach mentioned in [25] and [26]. For a vertex P_i with valence m, denote by $N = \{P_{i_1}, P_{i_2}, ..., P_{i_m}\}$ the set of the P_i one-ring neighbours.

$$\triangle_S P_i = \sum_{P_j \in N} \frac{\cot \alpha_{ij} + \cot \beta_{ij}}{2 \times A(P_i)} (P_j - P_i) \tag{6}$$

$$\triangle_S^2 P_i = \sum_{P_j \in N} \frac{\cot \alpha_{ij} + \cot \beta_{ij}}{2 \times A(P_i)} (\triangle_S P_j - \triangle_S P_i) \tag{7}$$

where $\alpha_{ij} = \angle(P_i, P_{j-1}, P_j)$, $\beta_j = \angle(P_i, P_{j+1}, P_j)$ and the factor $A(P_i)$ denotes the area of the Voronoi cell around vertex P_i (fig. 5). The evaluation of $\Delta_S^2(P) =$ brings the two-ring neighbourhood of P into play. Therefore, a hole detected in section 3 is extended with its two-ring neighbourhood (we will speak about "extended hole"); vertices of the hole are tagged outer if they belong to this neighbourhood and inner otherwise. Outer vertices are fixed, therefore, the boundary of the extended hole and the tangent planes along it are constant during the fairing process. Thus, we are in position to use equation (5) to minimize the thin plate energy.

Let $P_1, ...P_n$ be the set of inner vertices of the hole, $P_{n+1}, ..., P_m$ be the outer vertices, let us set $P = [P_1, ..., P_n]^T \in \mathbb{R}^{n \times 3}$ and $\triangle_S P = [\triangle_S P_1, ..., \triangle_S P_n]^T \in \mathbb{R}^{n \times 3}$, where $\triangle_S P_i$ is the discrete Laplace-Beltrami operator at vertex P_i and $N_1(P_i) = \{i_1, i_2, .., i_m\}$ the set of the vertex indices of one-ring neighbourhood of P_i.

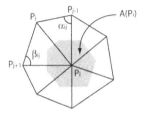

Fig. 5. The definition of α_{ij} and β_{ij} and the Voronoi cell $A(P_i)$

Equation(6) can be written in matrix form (the inner vertices are treated as unknowns in the discretized equations and the outer vertices are incorporated into its left-hand side):

$$\triangle_S P = M^{(1)}P - B^{(1)} \tag{8}$$

where $M^{(1)} = \{\omega_{ij}^{(1)}\}_{i,j=1}^{n}$ with

$$\omega_{ij}^{(1)} = \begin{cases} -\sum_{k \in N_1(P_i)} \frac{\cot \alpha_{ik} + \cot \beta_{ik}}{2 \times A(P_i)} & \text{if } i = j \\ \frac{\cot \alpha_{ij} + \cot \beta_{ij}}{2 \times A(P_i)} & j \in N_1(P_i), j \text{ inner} \\ 0 & \text{otherwhise} \end{cases} \tag{9}$$

(j inner means that P_j is an inner vertex) and $B^{(1)} = [\alpha_1^{(1)} \dots \alpha_n^{(1)}]^T \in \mathbb{R}^{n \times 3}$ with

$$\alpha_i^{(1)} = -\sum_{\substack{k \in N_1(P_i) \\ k \text{ outer}}} \frac{\cot \alpha_{ik} + \cot \beta_{ik}}{2 \times A(P_i)} P_k \tag{10}$$

(k outer means that P_k is an outer vertex). It follows from equations (7) and (8) that:

$$\triangle_S^2 P = M^{(1)}M^{(1)}P - (M^{(1)}B^{(1)} + B^{(2)}) \tag{11}$$

where $B^{(2)} = [\alpha_1^{(2)} \dots \alpha_n^{(2)}]^T \in \mathbb{R}^{n \times 3}$ with

$$\alpha_i^{(2)} = \sum_{\substack{j \in N_1(P_i) \\ j \text{ outer}}} \omega_{ij}^{(1)} \times \triangle_S P_j \tag{12}$$

Then, similarly, $B^{(2)}$ can be written as:

$$B^{(2)} = M^{(2)}P - B^{(3)} \tag{13}$$

Now substituting $B^{(2)}$ by (13) in (11), we obtain:

$$\triangle^2 P = MP - B \tag{14}$$

with $M = M^{(1)}M^{(1)} + M^{(2)}$ and $B = M^{(1)}B^{(1)} + B^{(3)}$

The matrix M is sparse and non symmetric, using the "compact memory technique" storage of array and the preconditioned bi-conjugate gradient stabilized method [27] to solve equation 14 provides good results (fig. 6).

5.2   Boundary Condition

Notice that the previous discretization of the second order Laplace-Beltrami operator involves the 2-neighbour ring of the considered vertex. Therefore, it is inapplicable to the boundary of the surface.

As a consequence, in our implementation, we discretize the quadric Laplace-Beltrami operator only for vertices (inner and outer) having a topological distance to the boundary greater than 1, and we only perform a first order discretization for

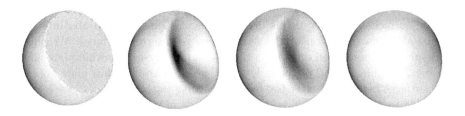

Fig. 6. Shape evolution at different iteration steps of the fairing process by BiGCS (from left to right)

vertices at distance 1 of the boundary. For boundary inner vertices for which the neighbouring information exists only on one side, applying such a discretization, would eventually shrink the surface. Such vertices usually exist in our setting. It would be tempting either to fix these vertices and treat them as outer vertices or to add some further constraints for boundary vertices displacement (see [28]). However, the first solution leads to visually unpleasant results whereas the second one yields to non linear systems or to large linear systems (due to the addition of Lagrange factors).

In order to overcome these limits, we want to control the displacement of boundary vertices with respect to their neighbours on the boundary. The discretizations (6) and (7) become:

$$\triangle P_i = \sum_{P_j \in N_1(P_i)} \omega_{ij}(P_j - P_i) \qquad \text{with } \omega_{ij} = \frac{1}{||P_i P_j||} \qquad (15)$$

$$\triangle^2 P_i = \sum_{P_j \in N_1(P_i)} \omega_{ij}(\triangle P_j - \triangle P_i) \qquad (16)$$

where $\{P_j\}_{j \in \{1...N_1(P_i)\}}$ are the neighbours of the boundary vertex P_i.

Developing (16)(15) as previously, we obtain again equation (14). Therefore, we solve two linear systems: one for boundary inner vertices and the other for other inner vertices (after having updated the new position of boundary inner vertices). Technically, it is however possible to regroup both of these systems in order to solve them simultaneously.

5.3 Fairing Method Efficiency

In order to estimate and compare the efficiency of our method (especially of the fairing step), it is necessary to define quality measures for the results. An important criterion is the restoration quality: starting from regular mesh surfaces in which holes have been created, we estimate the Hausdorff distance between the original and the restored surface. More precisely, we will call D_{max} this Hausdorff distance and D_{avg} the average symmetric distance between both surfaces [29].

Another interesting quality criterion is the smoothness of the restored surface. Many characterizations of smoothness exist in terms of: curvature and curvature variations, parametric and geometric continuity behaviour of the reflection lines of the surface... Because of this diversity, their is no clear and absolute mathematical characterization of the smoothness of discrete surfaces. We chose a measure of smoothness issued from the domain of physics: the thin plate energy. In the 2-neighbourhood of the hole areas, we compare the thin plate energy of the surfaces before and after the fairing process.

Table 1 and the figure 8 present a comparison of two fairing approaches: after identifying and filling holes (as described in sections 3 and 4), in (1) the mesh is faired by the Kobbelt "umbrella" operator weighted by a cotangent factor whereas in (2) we minimize the thin plate energy as described in section 5.

In conclusion, fairing by minimizing the thin plate energy leads to better results (in terms of restauration quality as well as in terms of smoothness which is well illustrated by a lower thin plate energy of faired patches and visually by a uniform distribution of errors (see figure 8d)). However, as the size of the faired area becomes larger, thin plate energy minimization becomes slower and slower. Our goal in section 6, is to introduce a multistep fairing approach (based on our thin plate energy minimization) combining quality and speed.

6   A Multiphase Approach to Fairing

Whatever the fairing method, restoration of large holes or strongly bent surfaces is expensive since many new vertices must be inserted during the filling step, leading to a large and expensive fairing step.

In order to propose a new approach, we started from the following observation: for large holes, fairing by minimizing the thin plate energy is computationally expensive. Therefore, instead of first filling and second fairing the patch, we only partially fill the hole (see figure 9), fair it, then fill again the resulting patch and fair the inserted vertices. Thus, each fairing step is now pretty quick whereas the quality of the final mesh is similar to the previous simple step approach. The first step actually restores the global shape of the hole whereas the second step adjusts its density and its smoothness. The global process is presented in figure 7. Observe that our multistep approach is not a multi-scale approach: only newly inserted vertices are considered as inner - and thus faired.

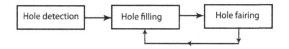

Fig. 7. Multistep hole filling process

In order to produce a quality smooth surface, one main parameter is the choice of the filling rate of each algorithm step which is basically controlled by the density factor λ_i (defined in section 4). We start with a factor λ_1 larger than λ in the first step and take λ_2 equals the user's defined value λ. In our implementation, we use two filling-fairing steps and the first filling parameter λ_1

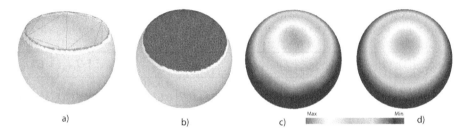

Fig. 8. Comparison between different fairing techniques: (a) initial model - (b) filled and faired model (minimization of the thin plate energy) - (c) error map for Liepa's algorithm - (d) error map for the thin plate energy minimization

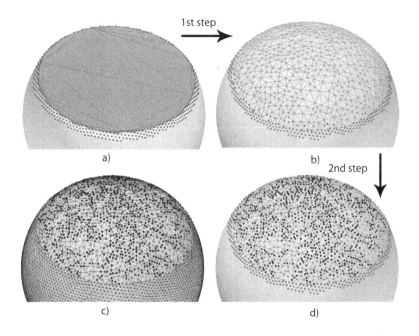

Fig. 9. Multistep hole filling process: top, first step - bottom, second step

is chosen in $[1.5\ldots 2.5]$ (whereas $\lambda_2 \in [0.7\ldots 1.4]$). However, more steps can be used, especially for large holes. An optimal strategy defining the best number of step could be developed.

7   Results and Conclusions

We implemented the hole-filling algorithm previously described using C/C++. In order to estimate the efficiency of our method, we have tested it over different

Table 1. Comparison of two fairing methods: (1) Kobbelt umbrella operator weighted by a cotangent factor (Liepa)-(2) our method: minimization of the thin plate energy

No.	(1)/(2)	N	Time (s)	E_0	E_1	E_1 gain (1)/(2)	D_{max}	D_{avg}	G_{max}	G_{avg}
1	(1)	276	0.81	41918.7	7944.28		0.02491	0.21834		
	(2)	276	0.73	41918.7	2175.92	72.61%	0.01171	0.12277	53.00%	43.77%
2	(1)	576	2.3	65213.5	7048.2		0.01779	0.17484		
	(2)	572	4.4	65213.5	3764.32	46.59%	0.01090	0.11734	38.74%	32.88%
3	(1)	994	8.5	86574	7749.57		0.014195	0.13887		
	(2)	994	14.1	86574	5761.98	25.65%	0.010731	0.11610	24.40%	16.40%
4	(1)	1564	21.1	143608	11890.3		0.01290	0.13296		
(Fig. 8)	(2)	1564	61.3	143608	8629.52	27.42%	0.01103	0.11949	14.55%	10.13%

where:

- N is the number of inserted vertices in the filling step
- E_0 (resp. E_1) is the thin plate energy of the surface before (resp. after) the fairing process
- E_1 gain is the difference between the thin plate energy of the surfaces obtained with both fairing approaches
- D_{max} and D_{avg} are the Hausdorff and average Hausdorff distance between the original and the restored surfaces
- G_{max} and G_{avg} are the gains between both approaches for D_{max} and D_{avg}.

Table 2. Comparison between (1) Liepa's algorithm - (2) our multistep algorithm. For the surface 1 and 2, D_{max}, D_{min} and G_{avg} are not computable because of inexistence of the reference surfaces.

Model	Model size	(1)/(2)	Inserted vertices	Time (s)	D_{max}	D_{avg}	G_{avg}
Sphere	7787	(1)	1872	33	0.12094	0.01123	
		(2)	1861	9	0.11686	0.01077	4.12%
		(1)	1402	17	0.12143	0.01160	
		(2)	1406	4	0.11852	0.01095	5.59%
		(1)	1089	10	0.14645	0.01384	
		(2)	1077	3	0.11718	0.01084	21.67%
		(1)	878	6.7	0.14559	0.01417	
		(2)	879	2.2	0.11628	0.01081	23.72%
Horse (Fig. 10)	44276	(1)	4788	131	0.18913	0.00240	
		(2)	4827	20	0.18269	0.00230	4.32%
		(1)	3702	76	0.18487	0.00302	
		(2)	3816	11	0.18340	0.00237	21.45%
		(1)	2888	48	0.19388	0.00342	
		(2)	2812	6	0.18422	0.00240	29.79%
		(1)	1910	21	0.19749	0.00416	
		(2)	1933	3	0.18607	0.00239	42.57%
Surface 1 (Fig. 11)	20628	(1)	3615	31.2	#	#	#
		(2)	3647	8.19	#	#	#
Surface 1 simplified (Fig. 12)	6500	(2)	573	0.3	#	#	#
Surface 2 (Fig. 13)	6080	(2)	2179	3.5	#	#	#

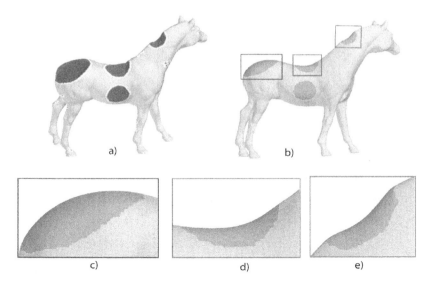

Fig. 10. Hole filling with our multistep method on a horse model: (a) initial model (44276 vertices) - (b) filled model (4827 vertices have been inserted) - (c), (d), (e) closer view of filled areas

models (a simple sphere model, a horse model and two geological surfaces). We have also run Liepa's algorithm on the same models. The results of these tests are presented in table 2 and illustrated in figure 10.

For a same number of inserted vertices, our method produces better results (the average error is between 4% to 40% lesser) with shorter running times (for large holes, our running times are from 3 to 7 times shorter). Therefore, the combination of the thin plate minimization and of our multistep approach fairing proves efficient.

We have then applied our algorithm to geological surfaces (our initial motivation). Each surface describes a geological horizon. A set of such surfaces tells about the evolution of the basement across geological ages. In order to create the initial triangulation of the point cloud, we use an incremental Delaunay triangulation algorithm (see [2]). The resulting meshes are quite voluminous and noisy (they contain thousands to hundreds of thousands of vertices). In order to simultaneously reduce the mesh complexity and homogenize them, the surfaces are pre-treated by a mesh simplification algorithm. We then apply our hole-filling algorithm to these simpler surfaces. Figure 11 and 13 present the results of our hole-filling process for both geological surfaces mentioned in table 2.

In conclusion, we propose a complete hole filling process for triangular meshes based on a multistep filling/fairing approach. Hole fairing is performed by minimizing a discretization of the thin-plate energy, which avoids the estimation of normals, tangents and curvatures on the hole neighbourhood. This method produces meshes of good quality (reconstructed surfaces are smooth and very close from the initial model) and our multistep approach leads to low running times.

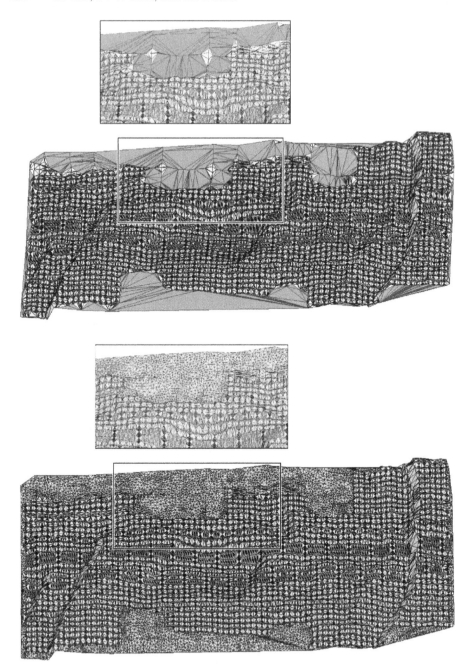

Fig. 11. Results of our hole-filling method for the geological surfaces 1. Top figures represent the initial model whereas bottom figures present the filled models.

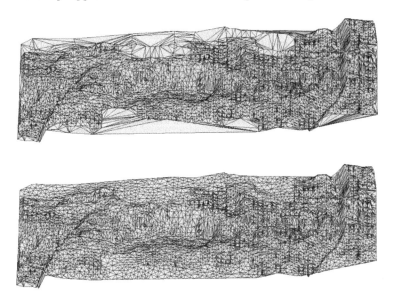

Fig. 12. Results of our hole-filling method for the simplified geological surfaces 1. Top figures represent the initial model whereas bottom figures present the filled models.

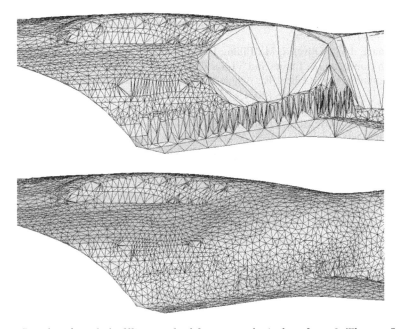

Fig. 13. Results of our hole-filling method for two geological surfaces 2. The top figures represent the initial model whereas the bottom figures present the filled models.

Also, because we do not use intermediate implicit surfaces, our approach is very fast and light. Let us mention that it can also apply to the reconstruction of meshes from curve nets (e.g. reconstruction of fault surfaces (fig. 1) from ridge and ravine curves).

Acknowledgements

The authors would like to thank the French Institute of Petrol for supporting this research.

Also we would like to thank the referees for their constructive remarks; it helped a lot in finalizing this article.

References

1. Brandel, S., Schneider, S., Guiard, N., Rainaud, J.F., Perrin, M.: Automatic building of structured geological models. Computing, Information Science and Engineering 5(2), 138–148 (2005)
2. Hjelle, O.: A triangulation template library (ttl): Generic design of triangulation software. Technical Report STF42 A00015, SINTEF (November 2000)
3. Hjelle, O.: Approximation of scattered data with multilevel b-splines. Technical Report STF42 A01011, SINTEF (November 2001)
4. Carr, J.C., Fright, W.R., Beatson, R.K.: Surface interpolation with radial basis functions for medical imaging. IEEE Transactions on Medical Imaging 16(1), 96–107 (1997)
5. Manuel, J.W., Oliveira, M.: Filling holes on locally smooth surfaces reconstructed from point clouds. Image and Vision Computing 25, 103–113 (2007)
6. Tekumalla, L.S., Cohen, E.: A hole-filling algorithm for triangular mesh. Uucs-04-019, University of Utah. (2004)
7. Davis, J., Marschner, S.R., Garr, M., Levoy, M.: Filling holes in complex surface using volumetric diffusion. In: Proceedings of the First international Symposium on 3D Data Processing, Visualization, Transmission, pp. 428–438 (2002)
8. Ju, T.: Robust repair of polygonal models. In: ACM SIGGRAPH, vol. 23(3) (2004)
9. Ju, T., Losasso, F., Schaffer, S., Warren, J.: Dual contouring of hermite data. ACM transactions on Graphics 21(3), 339–346 (2002)
10. Esteve, J., Brunet, P.A.V.: Approximation of variable density cloud of points by shrinking a discrete membrane. Computer Graphics Forum 24(4), 791–808 (2005)
11. Hornung, A.L.K.: Robust reconstruction of watertight 3d models from non-uniformly sampled clouds without normal information. In: Eurographics Symposium on Geometry Processing, pp. 41–50 (2006)
12. Liepa, P.: Filling holes in meshes. In: Eurographics Symposium on Geometry Processing, pp. 200–205 (2003)
13. Kobbelt, L., Campagna, S., Vorsatz, J., Seidel, H.: Interactive multi-resolution modeling on arbitray meshes. In: Computer Graphics (SIGGRAPH 1998 Proceedings), pp. 105–114 (1998)
14. Pernot, J.P., Verron, P.: Filling holes in meshes using a mechanical model to simulate the curvature variation minimization. Computer and Graphics 30(6) (2006)
15. Schneider, R., Kobbelt, L.: Geometric fairing of irregular meshes for free-form surface design. Computer Aided Geometric Design 18, 359–379 (2001)

16. Nealen, A., Igarashi, T., Sorkine, O., Alexa, M.: Fibermesh: designing freeform surfaces with 3d curves. ACM Trans. Graph - Siggraph 26(3), 41 (2007)
17. Xu, G., Pan, Q.: Discrete surface modeling using partial differential equations. CAGD 23, 125–145 (2006)
18. Clarenz, U., Diewald, U., Dziuk, G., Rumpf, M., Rusu, R.: A finite element method for surface restoration with smooth boundary conditions. CAGD 21(5), 427–445 (2004)
19. Bobenko, A., Schröder, P.: Discrete willmore flow. In: Eurographics Symposium on Geometric Processing (2005)
20. Yoshizawa, S., Belyaev, A., Seidel, H.P.: Fast and robust detection of crest lines on meshes. Technical report, ACM Symposium on Solid and Physical Modeling (2005)
21. Pfeifle, R., Seidel, H.P.: Triangular b-splines for blending and filling of polygonal holes. In: Graphics Interface, pp. 186–193 (1996)
22. Frey, P.J., Borouchaki, H.: Surface mesh quality evaluation. International Journal for Numerical Methods in Engineering 45(45), 101–108 (1999)
23. Dyn, N., Hormann, K., Kim, S., Levin, D.: Optimizing 3d triangulations using discrete curvature analysis. In: Mathematical Methods for Curves and Surfaces, Oslo, pp. 135–146 (2000)
24. Greiner, G.: Variational design and fairing of spline surfaces. Computer Graphics Forum 13(3), 143–154 (1994)
25. Desbrun, M., Meyer, M., Schroder, P.: Implicit fairing of irregular meshes using diffusion and curvature flow. In: Computer Graphics Proceedings, Annual Conference Series, (August 1999), vol. 7, pp. 317–324 (1999)
26. Meyer, M., Desbrun, M., Schroder, P., Barr, H.: Discrete differential-geometry operators for triangulated 2-manifolds. In: Proceedings VisMath 2002, Berlin (2002)
27. Rienen, U.V.: Numerical methods in computational electrodynamics. linear systems in practical application. Lecture Notes in Computational Science and Engineering 12 (2001)
28. Desbrun, M., Mayer, M., Alliez, P.: Intrinsic parameterizations of surface meshes. In: Computer Graphics Forum (Proceedings of Eurographics 2002), vol. 21(3), pp. 209–218 (2002)
29. Aspert, N., Santa-Cruz, D., Ebrahimi, T.: Mesh: Measuring errors between surfaces using the hausdorff distance. In: Proceedings of the IEEE International Conference on Multimedia and Expo, vol. 1, pp. 705–708 (2002)

Noise Removal Based on the Variation of Digitized Energy⋆

Qin Zhang[1],[⋆⋆], Jie Sun[2], and Guoliang Xu[3]

[1] School of Sciences, Beijing Information Science
and Technology University, Beijing 100192, China
[2] School of Mathematical Sciences,
Capital Normal University, Beijing 100037, China
[3] Computational Mathematics Institute,
Academy of Mathematics and System Sciences,
Chinese Academy of Sciences, Beijing 100080, China
{zqyork,xuguo}@lsec.cc.ac.cn, id_sunjie@yahoo.com

Abstract. A general formulation based on the variation of digitized energy to denoise image is proposed in this paper. This method is different from classical variational method employed in image processing. For a digitized energy functional, we first compute the variation, then design algorithms leading to digital filters. Numerical experiments and comparative examples are thus carried out to verify the effectiveness of the proposed method, which is efficient, adaptive and easily implemented. Higher quality images can be obtained with characteristic singular features preserved. The method can be easily expanded to multichannel image denoising.

Keywords: Image denoising, Digitized energy, Variational method, Graph.

1 Introduction

Image denoising is historically one of the oldest concerns in image processing and is still a necessary preprocessing step for many applications. What makes denoising so challenging is that a successful approach also must preserve characteristic singular features of images such as edges. Preservation of important singularities is absolutely necessary in image analysis and computer vision since digital "objects" are very much defined (or detected) via edges (i. e. segmentation) and other singularities (e. g., the corners of eyes and lips). In the past decades, deterministic and stochastic models are proposed to solve this problem. The topic of image denoising has occupied a large part of many monographs on image processing and analysis, such as in [1,2].

⋆ Project is supported by Beijing Educational Committee Foundation (KM200811232009) and NSFC grant 60773165 and National Key Basic Research Project of China (2004CB318000).
⋆⋆ Corresponding author.

F. Chen and B. Jüttler (Eds.): GMP 2008, LNCS 4975, pp. 290–303, 2008.

As two tightly linked methods, variational methods and Partial Differential Equation (PDE) methods have been two successful tools in image processing in recent years. Given an energy functional, we are able to obtain an Euler-Lagrange equation by variational calculus and a steady state equation or an evolution equation can be solved for the minimizer of the energy functional. On the other hand, many PDEs used in image processing can be deduced from energy functionals. This has been surveyed in [3]. Suppose that $u : \Omega \subset \mathbb{R}^2 \to \mathbb{R}$ is an original image describing a real scene, and u^0 is the observed image of the same scene (i. e., a degradation of u). We always use $u^0 = Ru + n$ to model the process, where n stands for white additive Gauss noise and R is a linear operator representing the blur (usually a convolution). In this paper, $R = I$. The image denoising problem is to recover u from u^0. As an inverse problem, image denoising, according to the well-posedness theory of Hadamard, is ill-posed. A classical way to overcome this ill-posedness, based on the theory of Tikhonov regularization, is to minimize the following regularized minimization problem,

$$E(u) = \int_{\Omega} \phi(\|\nabla u\|)d\mathbf{x} + \frac{\lambda}{2} \int_{\Omega} (u - u^0)^2 d\mathbf{x}, \tag{1}$$

where the first term is a regular term to describe the smoothness of the image and the second term measures the fidelity to the data. Parameter λ is a positive penalty constant to weigh these two parts. Function ϕ is chosen to depict the strength of the smoothness. The common choices for ϕ are $\phi(x) = x^2$ (the corresponding energy is Dirichlet energy) and $\phi(x) = x$ (the corresponding energy is usually named as total variation).

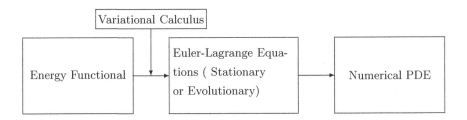

Fig. 1. Classical variational methods

Classical variational methods compute the Euler-Lagrange equation for (1) or construct an evolution equation by adding a temporal parameter (see Fig. 1). Then, the resulting differential equations (either the equilibrium equation or the evolution equation) are discretized numerically on a rectangular grid. The advantage of this method is that one can easily adapt many existing numerical methods for PDEs to this problem and establish efficient algorithms. Large amount of publications have appeared along this line (see e.g., [4,5,6]), and more papers utilized the variants of the PDEs obtained to denoising images. Since the Euler-Lagrange equation associated with (1) is usually a nonlinear PDE, when applying it to a digital image, one has to carefully choose numerical schemes to take care of the nonlinearity. Therefore, Osher and Shen [7] established a self-contained "digital" theory for the PDE method, in which knowledge of PDEs and numerical approximations is not required. The "digital" method does not mean the discrete numerical implementation of the existing differential equations. Instead, we digitalize the methodology. That is, we start directly with the discrete variational problem, from which algebraic equilibrium equations analogous to the PDEs are established. This method is thus self-contained. It is unnecssary to search for the meaning of these algebraic equations from their PDE counterparts. Remarkably, the underlying domain can be fairly irregular. Similar digitizing work on the evolution equation can also be found in Weickert [8]. However, the mathematical foundation of the latter is still the numerical discretization of PDEs on rectangular grids.

Motivated by the theory of [7], we propose to digitalize the energy (1) to obtain a general formulation for image denoising. An ideal graph model is considered and regularization operation for the local variation is avoided. Since this method avoids solving a continuous PDE on a rectangular grid, a linear system of equations is therefore unnecessarily solved. This formulation is implemented by a local iterative scheme which is more efficient than PDE methods. Numerical examples are provided to show this fact.

The method of computing the variation for discrete energy other than continuous energy also appears in other arenas. For instance, in [9], discrete minimal surfaces and their conjugates are calculated by the direct variation of discrete Dirichlet integral other than classical method of discretizing the PDE obtained by the variation of continuous energy. In [10], discrete Willmore flow is considered from the variation of discrete Willmore energy.

The remainder of the paper is organized as follows. In Section 2, the graph model, energy functional and denoising equation are given. We provide the detailed algorithm in Section 3. Numerical experiments and comparative examples are presented in Section 4. Section 5 concludes the paper.

2   Graph Model, Energy Functional and Denoising Equation

2.1   Basic Definitions for a Graph

For a digital image u is a function $u : \Omega \to \mathbb{R}$, let $[\Omega, G]$ denote an undirected graph with a finite set Ω of vertices (or nodes, pixels) and a dictionary G of edges. The graph is assumed to have no self-loops and general vertices are denoted by $\alpha, \beta, \ldots$(cf. Fig. 2). If α and β are linked by an edge with length r, then we write $\alpha \overset{r}{\sim} \beta$. We use

$$N_1 = \{\beta \in \Omega; \beta \overset{1}{\sim} \alpha\} \quad \text{and} \quad N_{\sqrt{2}} = \{\gamma \in \Omega; \gamma \overset{\sqrt{2}}{\sim} \alpha\}$$

to denote the 1-neighbors and $\sqrt{2}$-neighbors of α, respectively. The value at vertex α is denoted by u_α. Similar to the definition of length of gradient ∇u,

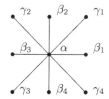

Fig. 2. Vertex α and its neighborhood

we can define the *local variation* or *strength* $\|\nabla_g u_\alpha\|$ of its digital version at any vertex α as

$$\|\nabla_g u_\alpha\| := \sqrt{\sum_{\beta \overset{1}{\sim} \alpha} (u_\beta - u_\alpha)^2 + \sum_{\gamma \overset{\sqrt{2}}{\sim} \alpha} \frac{(u_\gamma - u_\alpha)^2}{2}}.$$

If we only consider the N_1 neighborhood of α, then the definition of local variation coincides with the definition in [7].

Expanding the notations in [11], we define the *edge derivative*. Let e and f be the edges $\alpha \overset{1}{\sim} \beta$ and $\alpha \overset{\sqrt{2}}{\sim} \gamma$, respectively. Then the edge derivatives of u along e and f are correspondingly defined to be

$$\left.\frac{\partial u}{\partial e}\right|_\alpha := u_\beta - u_\alpha \quad \text{and} \quad \left.\frac{\partial u}{\partial f}\right|_\alpha := \frac{u_\gamma - u_\alpha}{\sqrt{2}}.$$

Apparently,

$$\left.\frac{\partial u}{\partial e}\right|_\alpha = -\left.\frac{\partial u}{\partial e}\right|_\beta \quad \text{and} \quad \left.\frac{\partial u}{\partial f}\right|_\alpha = -\left.\frac{\partial u}{\partial f}\right|_\gamma$$

hold. We can also write

$$\|\nabla_g u_\alpha\| = \sqrt{\sum_{e \vdash \alpha} \left[\left.\frac{\partial u}{\partial e}\right|_\alpha\right]^2 + \sum_{f \vdash \alpha} \left[\left.\frac{\partial u}{\partial f}\right|_\alpha\right]^2},$$

where $e \vdash \alpha$ means that α is a 1-node of e and $f \vdash \alpha$ means that α is a $\sqrt{2}$-node of f.

2.2   Digitized Energy and Equations

For energy functional (1), the fitted digitized version considered in this paper is as follows:

$$\mathcal{E}(u) := \sum_{\alpha \in \Omega} \phi(\|\nabla_g u_\alpha\|) + \frac{\lambda}{2} \sum_{\alpha \in \Omega} (u_\alpha - u_\alpha^0)^2, \tag{2}$$

where function $\phi(x)$ is simply chosen as S. Kim in [12] as $\phi(x) = x^{2-q}, 0 \le q < 2$. Obviously, $\phi(x)$ can be selected as other forms, such as listed in [1, p. 83]

$\phi(x) = \frac{x^2}{1+x^2}, \log(1 + x^2), 2\sqrt{1 + x^2} - 2$ etc. In practice, the penalty parameter (or Lagrange multiplier) λ is of great importance, and we will address this in Section 4.

Theorem 1. *With respect to digitized energy (2), the denoising equation is*

$$0 = \sum_{\beta \in N_1} (u_\alpha - u_\beta) \left(\frac{\phi'(\|\nabla_g u_\alpha\|)}{\|\nabla_g u_\alpha\|} + \frac{\phi'(\|\nabla_g u_\beta\|)}{\|\nabla_g u_\beta\|} \right)$$

$$+ \sum_{\gamma \in N_{\sqrt{2}}} \frac{(u_\alpha - u_\gamma)}{2} \left(\frac{\phi'(\|\nabla_g u_\alpha\|)}{\|\nabla_g u_\alpha\|} + \frac{\phi'(\|\nabla_g u_\gamma\|)}{\|\nabla_g u_\gamma\|} \right) + \lambda(u_\alpha - u_\alpha^0), \ \alpha \in \Omega(3)$$

Proof: To compute the variation of digitized energy (2), we take derivative for $\mathcal{E}(u)$ with respect to u_α.

Since only $1 + N_1 + N_{\sqrt{2}}$ terms in $\mathcal{E}(u)$ contain u_α (cf. Fig. 2), we thus compute the variation as

$$\frac{\partial \mathcal{E}(u)}{\partial u_\alpha}$$

$$= \frac{\partial \phi(\|\nabla_g u_\alpha\|)}{\partial u_\alpha} + \frac{\partial \left(\sum_{\beta \in N_1} \phi(\|\nabla_g u_\beta\|) \right)}{\partial u_\alpha} + \frac{\partial \left(\sum_{\gamma \in N_{\sqrt{2}}} \phi(\|\nabla_g u_\gamma\|) \right)}{\partial u_\alpha} + \lambda(u_\alpha - u_\alpha^0)$$

$$= \phi'(\|\nabla_g u_\alpha\|) \frac{\partial \left(\left(\sum_{\beta \in N_1} (u_\beta - u_\alpha)^2 + \sum_{\gamma \in N_{\sqrt{2}}} \frac{(u_\gamma - u_\alpha)^2}{2} \right)^{1/2} \right)}{\partial u_\alpha}$$

$$+ \sum_{\beta \in N_1} \phi'(\|\nabla_g u_\beta\|) \frac{\partial(\|\nabla_g u_\beta\|)}{\partial u_\alpha} + \sum_{\gamma \in N_{\sqrt{2}}} \phi'(\|\nabla_g u_\gamma\|) \frac{\partial(\|\nabla_g u_\gamma\|)}{\partial u_\alpha}$$

$$+ \lambda(u_\alpha - u_\alpha^0)$$

$$= \phi'(\|\nabla_g u_\alpha\|) \frac{1}{\|\nabla_g u_\alpha\|} \left(\sum_{\beta \in N_1} (u_\alpha - u_\beta) + \sum_{\gamma \in N_{\sqrt{2}}} \frac{u_\alpha - u_\gamma}{2} \right)$$

$$+ \sum_{\beta \in N_1} \phi'(\|\nabla_g u_\beta\|) \frac{1}{\|\nabla_g u_\beta\|} (u_\alpha - u_\beta) + \sum_{\gamma \in N_{\sqrt{2}}} \phi'(\|\nabla_g u_\gamma\|) \frac{1}{2\|\nabla_g u_\gamma\|} (u_\alpha - u_\gamma)$$

$$+ \lambda(u_\alpha - u_\alpha^0)$$

$$= \sum_{\beta \in N_1} (u_\alpha - u_\beta) \left(\frac{\phi'(\|\nabla_g u_\alpha\|)}{\|\nabla_g u_\alpha\|} + \frac{\phi'(\|\nabla_g u_\beta\|)}{\|\nabla_g u_\beta\|} \right)$$

$$+ \sum_{\gamma \in N_{\sqrt{2}}} \frac{(u_\alpha - u_\gamma)}{2} \left(\frac{\phi'(\|\nabla_g u_\alpha\|)}{\|\nabla_g u_\alpha\|} + \frac{\phi'(\|\nabla_g u_\gamma\|)}{\|\nabla_g u_\gamma\|} \right) + \lambda(u_\alpha - u_\alpha^0).$$

Consequently, the necessary condition for u_α being the minimizer of (2) is (3).

Corollary 1. *If* $\phi(x) = x^{2-q}, 0 \le q < 2$, *then (3) turns out to be*

$$0 = (2 - q)\left(\sum_{\beta \in N_1} (u_\alpha - u_\beta)(\|\nabla_g u_\alpha\|^{-q} + \|\nabla_g u_\beta\|^{-q}) \right.$$

$$\left. + \sum_{\gamma \in N_{\sqrt{2}}} \frac{(u_\alpha - u_\gamma)}{2}(\|\nabla_g u_\alpha\|^{-q} + \|\nabla_g u_\gamma\|^{-q}) \right) + \lambda(u_\alpha - u_\alpha^0), \quad \alpha \in \Omega. \tag{4}$$

In particular, if we select $q = 0$ and $q = 1$, then the equations are associated with the fitted digitized Dirichlet energy and total variation, respectively.

If $0 \le q \le 1$, as proved in [7], the minimizer of the fitted digitized energy is exist and unique since the energy functionals are strictly convex functionals of u. If $1 < q < 2$, according to the theory of the calculus of variations, we could not expect the existence and uniqueness of minimizer. But in our numerical test, the results of this case are better than the former. This phenomenon is quiet similar to the results obtained in [13]. In that paper, an analogous PDE model has been utilized to perform simultaneous image denoising and edge enhancement and is called *convex-concave anisotropic diffusion* (CCAD) model

$$\frac{\partial u}{\partial t} - \text{div}\left(\frac{\nabla u}{\|\nabla u\|^q} \right)\|\nabla u\|^q = \lambda(u^0 - u)\frac{\|\nabla u\|^q}{2 - q}.$$

When $q = 1$, the model reduces to Improved Total Variation (ITV) model ([14]), that is

$$\frac{\partial u}{\partial t} - \text{div}\left(\frac{\nabla u}{\|\nabla u\|} \right)\|\nabla u\| = \lambda(u^0 - u)\|\nabla u\|.$$

It has been numerically verified that for $1 < q < 2$, the CCAD model is superior to the ITV model. In fact, this phenomenon also supports the result of Perona-Malik (PM) model ([15]),

$$\frac{\partial u}{\partial t} = \text{div}(c(\|\nabla u\|)\nabla u),$$

where $c(x) = (1 + x^2/K^2)^{-1}$ for a threshold K and div is the classical divergence operator. Note that if we select $\phi(x) = \frac{1}{2}K^2 \ln(1 + x^2/K^2)$, then $c(x) = \phi'(x)/x$. The function $\phi(x)$ is strictly convex for $x < K$ and strictly concave for $x > K$. PM model has been regarded as a revolution in the field of PDE models in image denoising.

Corollary 2. *Equation (3) is the digital version of*

$$\text{div}\left(\phi'(\|\nabla u\|)\frac{\nabla u}{\|\nabla u\|} \right) + \lambda(u^0 - u) = 0, \tag{5}$$

which is the Euler-Lagrange equation of (1).

Proof: According to the definitions of edge derivative, we can rewrite

$$\sum_{\beta \in N_1} (u_\alpha - u_\beta) \left(\frac{\phi'(\|\nabla_g u_\alpha\|)}{\|\nabla_g u_\alpha\|} + \frac{\phi'(\|\nabla_g u_\beta\|)}{\|\nabla_g u_\beta\|} \right) = \sum_{e \vdash \alpha} \frac{\partial}{\partial e} \left[\frac{-\phi'(\|\nabla_g u\|)}{\|\nabla_g u\|} \frac{\partial u}{\partial e} \right] \bigg|_\alpha \quad (6)$$

and

$$\sum_{\gamma \in N_{\sqrt{2}}} \frac{(u_\alpha - u_\gamma)}{2} \left(\frac{\phi'(\|\nabla_g u_\alpha\|)}{\|\nabla_g u_\alpha\|} + \frac{\phi'(\|\nabla_g u_\gamma\|)}{\|\nabla_g u_\gamma\|} \right) = \sum_{f \vdash \alpha} \frac{\partial}{\partial f} \left[\frac{-\phi'(\|\nabla_g u\|)}{\|\nabla_g u\|} \frac{\partial u}{\partial f} \right] \bigg|_\alpha, \quad (7)$$

where as before, $e \vdash \alpha$ and $f \vdash \alpha$ mean that α is a 1-node of e and $\sqrt{2}$-node of f, respectively. Substituting (6) and (7) into (3) yields

$$\sum_{e \vdash \alpha} \frac{\partial}{\partial e} \left[\frac{-\phi'(\|\nabla_g u\|)}{\|\nabla_g u\|} \frac{\partial u}{\partial e} \right] \bigg|_\alpha + \sum_{f \vdash \alpha} \frac{\partial}{\partial f} \left[\frac{-\phi'(\|\nabla_g u\|)}{\|\nabla_g u\|} \frac{\partial u}{\partial f} \right] \bigg|_\alpha + \lambda(u_\alpha^0 - u_\alpha) = 0, \quad (8)$$

which is the digital version of (5).

Notice there is a sign difference between the first terms in (5) and (8). If $\phi(x) = x$, the result is consistent with the result in [11].

3   Detailed Denoising Algorithm

For a vertex α being dealt with, if $\|\nabla_g u_\alpha\| \neq 0, \|\nabla_g u_\beta\| \neq 0$ and $\|\nabla_g u_\gamma\| \neq 0$, we define weighted function

$$\omega_{\alpha\beta}(u) = \frac{\phi'(\|\nabla_g u_\alpha\|)}{\|\nabla_g u_\alpha\|} + \frac{\phi'(\|\nabla_g u_\beta\|)}{\|\nabla_g u_\beta\|}, \quad (9)$$

$$\omega_{\alpha\gamma}(u) = \frac{\phi'(\|\nabla_g u_\alpha\|)}{2\|\nabla_g u_\alpha\|} + \frac{\phi'(\|\nabla_g u_\gamma\|)}{2\|\nabla_g u_\gamma\|}. \quad (10)$$

If $\|\nabla_g u_\alpha\| = 0$, then we define $\omega_{\alpha\beta}(u) = \omega_{\alpha\gamma}(u) = 0$. If $\|\nabla_g u_\alpha\| \neq 0$ and $\|\nabla_g u_\beta\| = 0$, we first prescribe $\|\nabla_g u_\beta\| = a$, where a is a small positive number (e. g., $a = 10^{-4}$), and then define weight function according to (9).

Then equation (3) becomes

$$\left(\lambda + \sum_{\beta \in N_1} \omega_{\alpha\beta}(u) + \sum_{\gamma \in N_{\sqrt{2}}} \omega_{\alpha\gamma}(u) \right) u_\alpha - \sum_{\beta \in N_1} \omega_{\alpha\beta}(u) u_\beta - \sum_{\gamma \in N_{\sqrt{2}}} \omega_{\alpha\gamma}(u) u_\gamma$$

$$= \lambda u_\alpha^0, \quad (11)$$

for all $\alpha \in \Omega$. This is usually a system of nonlinear equations. For simplicity, we further define

$$h_{\alpha\beta}(u) = \frac{\omega_{\alpha\beta}(u)}{\lambda + \sum_{\beta \in N_1} \omega_{\alpha\beta}(u) + \sum_{\gamma \in N_{\sqrt{2}}} \omega_{\alpha\gamma}(u)}, \quad (12a)$$

$$h_{\alpha\gamma}(u) = \frac{\omega_{\alpha\gamma}(u)}{\lambda + \sum\limits_{\beta \in N_1} \omega_{\alpha\beta}(u) + \sum\limits_{\gamma \in N_{\sqrt{2}}} \omega_{\alpha\gamma}(u)}, \tag{12a}$$

$$h_{\alpha\alpha}(u) = \frac{\lambda}{\lambda + \sum\limits_{\beta \in N_1} \omega_{\alpha\beta}(u) + \sum\limits_{\gamma \in N_{\sqrt{2}}} \omega_{\alpha\gamma}(u)}. \tag{12b}$$

It is easy to see that $h_{\alpha\alpha} + h_{\alpha\beta} + h_{\alpha\gamma} = 1$.

To solve the system of equations (11), the simplest local iteration is the Gauss-Jacobi method

$$u_\alpha^{k+1} = \sum_{\beta \in N_1} h_{\alpha\beta}(u^k)u_\beta^k + \sum_{\gamma \in N_{\sqrt{2}}} h_{\alpha\gamma}(u^k)u_\gamma^k + h_{\alpha\alpha}(u^k)u_\alpha^0, \tag{13}$$

for all $\alpha \in \Omega$, where k denotes the iteration step. This process can be independently explained as a forced local low-pass digital filter. The update u_α is a weighted average of the existing u_β on its 1-neighbors and u_γ on its $\sqrt{2}$-neighbors and the raw data at α. The raw data serves as an attracting force preventing u from wandering far away.

In fact, image denoising is a process of weighted average. The key point is how to choose the weight for each pixel. For classical linear filters, a solid template (e. g., 3×3 or 5×5) with fixed coefficients is used to scan the whole image, therefore it is difficult to distinguish between features and homogeneous regions. Although the weights vary with different pixels in the nonlinear median filter (e. g., [16]), this filter is an exclusive filter in the sense that the filter coefficients are 0 or 1. In most cases, features can be preserved while smoothness of images could not be expected unless large filter window. However, large filter windows in general collide with the local characters of images.

For (13), the locality of images information can be preserved since the iteration is carried out merely in a neighborhood of each pixel. If suitable weighted functions $\omega_{\alpha\beta}(u)$ and $\omega_{\alpha\gamma}(u)$ can be selected, to discern the features and homogeneous regions for images is automatically performed. This adaptivity is easy to understand qualitatively. The key is the competition between the Lagrange multiplier λ and the local weights $\omega_{\alpha\beta}(u)$ and $\omega_{\alpha\gamma}(u)$. The local weights dominate λ if the current data u^k are very flat near a pixel α. Then the fitting term becomes less important and the filter acts like low-pass filtering purely on u^k, which makes the output u_α^{k+1} even flatter at the spot. On the other hand, if the current data u^k undergo an abrupt change of large amplitude at α, then the local weights $\omega_{\alpha\beta}$ and $\omega_{\alpha\gamma}(u)$ are insignificant compared with λ. If this is the case, the filter intelligently sacrifices smoothness for faithfulness. This mechanism is obviously important for faithful denoising of edges in image processing.

Another possible scheme for (11) is Gauss-Seidel method. We omit this here for saving space. The interested reader can write it out according to the schemes in [7].

From the iterative scheme (13), the algorithm can be given as follows.

Filtering algorithm:

1) Assign a linear order to all pixels: $\alpha_1 < \alpha_2 < \cdots < \alpha_{|\Omega|}$. Set $k = 0$.
2) $k = k + 1$. For each pixel α, calculate local variation $\|\nabla_g u_\alpha\|$.
3) For each pixel α and all its 1-neighbors β and $\sqrt{2}$-neighbors γ, calculate weighted function $\omega_{\alpha\beta}(u^k)$ and $\omega_{\alpha\gamma}(u^k)$ according to (9) and (10), respectively.
4) For each pixel α, compute $h_{\alpha\beta}(u), h_{\alpha\gamma}(u), h_{\alpha\alpha}(u)$ according to (12).
5) For each pixel α, compute u_α^{k+1} according to (13).
6) If $k > N$(a prescribed iteration step) or other termination condition is satisfied, stop. Otherwise, go to 2).

4   Numerical Experiments and Comparative Results

Numerical experiments and comparative results are presented in this section. All the experiments are carried out in a laptop (Intel© Core(TM) 2 CPU T7200 2.00GHz, 1GB RAM) with Matlab© software.

First, we address on some issues regarding implementation ([11]).

1) One attribute of this digitized energy variation model is that it does not require any artificial "boundary" condition. In the classical literature, the continuous diffusion equation is usually accomplished by the adiabatic Neumann condition. In the digital model, the boundary condition has been encoded into the structure of the graph and the definition of the local variation $\|\nabla_g u_\alpha\|$. For instance, each of the four corner pixels has only two neighbors and a typical boundary pixel has three neighbors. A simple checking on the definition of the local variation $\|\nabla_g u_\alpha\|$ verifies that the above boundary structure of the graph indeed corresponds to a flat outward extension of u, or the discrete outward Neumann condition.

2) For the most existing digital filters, the raw noisy image u^0 is "abandoned" right after the first iteration. That is, the filtering process can be described as

$$u^0 \longrightarrow u^1 \longrightarrow u^2 \longrightarrow \cdots .$$

At step k, u^k depends solely on u^{k-1}. The ignorance of u^0 at later steps causes intrinsic singularities to be smeared step by step (the discrete scale-space filter is an extreme example [8].) Therefore, a stopping time is required for such filtering processes. It is often a difficult task to determine an optimal stopping time. This is avoided by the digital filter since it recycles u^0 at each step:

$$u^0 \longrightarrow u^1 \xrightarrow{u^0} u^2 \xrightarrow{u^0} u^3 \xrightarrow{u^0} \cdots$$

The presence of u^0 at each step constantly reminds the filter not to forget the noisy image, which has information about the original singular features such as jumps and edges. We should point out, however, although the stopping time is unnecessary for the digital filter, the difficulty goes into the determination of the degree of influence that u^0 should impose at each step. Next item addresses this problem.

3) The Lagrange multiplier λ is important for the denoising effect. Practical concerns and estimates are discussed in Rudin and Osher [4], Blomgren and Chan [17]. In terms of the digital model, an estimation of the optimal λ is by

$$\lambda \approx \frac{1}{\sigma^2} \frac{1}{|\Omega|} \sum_{\alpha \in \Omega} \left[\sum_{\beta \sim \alpha} \omega_{\alpha\beta}(u_\beta - u_\alpha)(u_\alpha - u_\alpha^0) + \sum_{\beta \overset{\sqrt{2}}{\sim} \gamma} \omega_{\alpha\gamma}(u_\alpha - u_\alpha^0) \frac{u_\gamma - u_\alpha}{\sqrt{2}} \right],$$

where σ^2 is the variance of the noise, which is known or can be estimated from homogeneous regions in the image. $\omega_{\alpha\beta}$ and $\omega_{\alpha\gamma}$ are weights defined by (9) and (10), respectively. $|\Omega|$ is the size of the image, i. e., the total number of pixels. The formula suggests that λ is comparable to $1/\sigma^2$. For more details on the choice of λ, see [11].

4.1 The Effect of the Convex and Concave Properties of Function ϕ

As we have narrated under Corollary 1, the convex and concave properties of function ϕ influence the effect of denoising and features preservation. Since we merely select $\phi(x) = x^{2-q}$ in this paper, we select $q = 0, 0.5, 1, 1.02, 1.2, 1.5$ for testing. When $0 \leq q \leq 1$, the density function is convex. If $1 < q < 2$, the density function is concave, therefore the existence and uniqueness of the minimizer for energy functional can not be guaranteed. Similarly phenomena happened in PM model and in [13]. We use 1-neighborhood in this experiment. In Fig. 3, figure (a) is the original image. (b) is the contaminated image with Gauss noise ($\sigma = 1/7$) added. (c)–(h) are the six cases for q, respectively. We can find that figure (g) is the best among all these figures. This fact is verified in Table 1. Thus we can draw a conclusion that the concave functionals can preserve more features in some sense than convex functionals.

4.2 Pepper and Salt Noise Removal

For different choices of q, the model can not only reduce Gauss noise but also be able to eliminate pepper & salt noise efficiently. In Fig. 4, such an example is provided. Image (b) is the degraded version of original image (a) with 25% pepper & salt noise added. Figures (c)-(h) are the denoised results for different q. We can find that g is the best and this fact can be verified by Table 1. This time the function $\phi(x)$ is also concave.

4.3 Color Image Denoising by This Model

This model can be easily expanded to multichannel images. The easiest method is channel by channel scheme. Fig. 5 is such an example. Image in figure (b) is the polluted version of image (a) after 25% pepper & salt noise added. (c)–(h)

Fig. 3. The effect of convex and concave properties of function ϕ. (a) is the original image. (b) is the contaminated image with Gauss noise $\sigma = 1/7$. (c) is the result of L^2 energy, i. e., $q = 0$. (d) $q = 0.5$. (e) is the result of TV energy, i. e., $q = 1$. (f) $q = 1.02$. (g) $q = 1.2$. (h) $q = 1.5$.

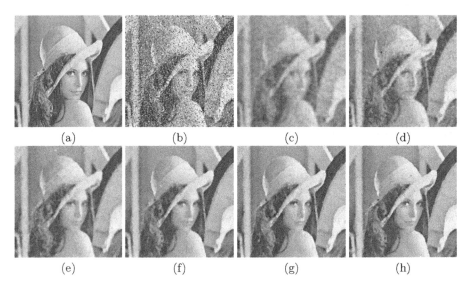

Fig. 4. The effect of pepper & salt noise removal for different q. (a) is the original image. (b) is the degraded image with 25% pepper & salt noise. (c) is the result of L^2 energy, i. e., $q = 0$. (d) $q = 0.5$. (e) is the result of TV energy, i. e., $q = 1$. (f) $q = 1.5$. (g) $q = 1.8$. (h) $q = 1.9$.

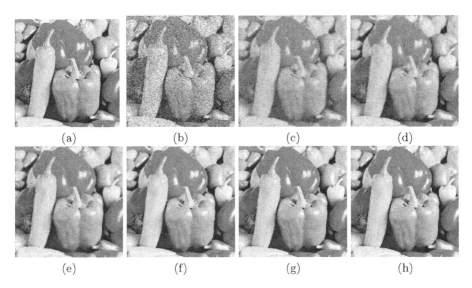

Fig. 5. The effect of pepper & salt noise removal for color image. (a) is the original image. (b) is the contaminated image with 25% pepper & slat noise. (c) is the result of L^2 energy, i. e., $q = 0$. (d) $q = 0.5$. (e) is the result of TV energy, i. e., $q = 1$. (f) $q = 1.5$. (g) $q = 1.8$. (h) $q = 1.9$.

are the results for different q. We can find that (g) is the best and this fact is consistent with the results in Table 1.

Before we go to another example, let us stop for one problem of denoising performance. To give an objective standard to detect which method is superior is of significant importance in image denoising field, but this is not a trivial task. In [18], the denoising performance are compared in four ways, that is: mathematical: asymptotic order of magnitude of the method noise under regularity assumptions; perceptual-mathematical: the algorithms artifacts and their explanation as a violation of the image model; quantitative experimental: by table of L^2 distances of the denoised version to the original image. The most powerful evaluation method seems, however, to be the visualization of the method noise on natural images. In the above figures, although we can discern which is the best, Table 1 is more persuasive. That is the L^2 distance from the original image to its estimate, which can be calculated by

$$d(\text{original}, \text{estimate}) = \left(\sum_{\alpha \in \Omega} (u_\alpha^{\text{orig}} - u_\alpha^{\text{est}})^2 \right)^{1/2}.$$

We also call it mean square error in this paper, noticing there is a square discrepancy between this definition with the one in the paper [18]. Obviously, for the same original image, the smaller is mean square error, the more faithful of the estimate image is. The table corroborates our visualizing. Another frequently used criteria is the residue of the estimate image and original one (e. g., in [19]).

Table 1. Mean square error

figure Fig.	(c)	(d)	(e)	(f)	(g)	(h)
Fig. 3	135.54	124.48	113.95	115.77	109.62	156.10
Fig. 4	142.75	122.35	96.48	80.64	76.47	77.08
Fig. 5	557.65	453.67	345.13	276.41	270.77	272.76

Since it depends still upon the visualizing, we do not adapt this comparison method here.

4.4 Time Consuming of This Method Compared with PDE Methods

Since PDE methods always involve the solution of a linear system of equation, they must consume more than this method, which utilizes an explicitly iterative scheme. Fig. 6 is a comparative example of Gauss noise removal with this model and ROF model [5]. Figure (b) is the polluted image of figure (a) with $\sigma = 1/7$. Figure (c) is the result of this model with $q = 1.2$ consuming 3.38 seconds. (d) is the result of ROF model consuming 26.80 seconds. The efficiency of this method is obviously higher than the PDE methods.

| (a) | (b) | (c) | (d) |

Fig. 6. Time consuming of this model campared with ROF model. (a) is the original image. (b) is the contaminated image with Gauss noise ($\sigma = 1/7$) added. (c) is the result of this model when $q = 1.2$. (d) is the result of ROF model.

5   Conclusion

A general formulation of image denoising method based on the variation of digitized energy functional is presented. We investigate a special case of density function which can produce higher quality restored image when it is concave. An explicitly iterative scheme is presented to solve the system of equations. Large variety of numerical experiments are performed to verify the effectiveness of the proposed model and comparative experiments are also carried out.

References

1. Aubert, G., Kornprobst, P.: Mathematical Problems in Image Processing: Partial Differential Equations and the Calculus of Variations, Applied Mathematical Sciences, 2nd edn, vol. 147. Springer, Heidelberg (2006)
2. Chan, T.F., Shen, J.H.: Image Processing and Analysis–Variational, PDE, Wavelet, and Stochastic Methods. SIAM, Philadelphia (2005)
3. Sun, J., Zhang, Q., Xu, G.: A survey of variational models in image denoising (submitted)
4. Rudin, L., Osher, S.: Total variation based image restoration with free local constraints. In: Proceedings of the IEEE International Conference on Image Processing, vol. 1, pp. 31–35 (1994)
5. Rudin, L., Osher, S., Fatemi, E.: Nonlinear total variation based noise removal algorithms. Physica D 60, 259–268 (1992)
6. Lysaker, M., Lundervold, A., Tai, X.C.: Noise removal using fourth-order partial differential equation with applications to magnetic resonance images in space and time. IEEE Trans. Image Processing 12(12), 1579–1590 (2003)
7. Osher, S., Shen, J.: Digitized PDE method for data restoration. In: Anastassiou, G. (ed.) Handbook of Analytic-Computational Methods in Applied Mathematics, Chapman & Hall/CRC, pp. 751–771 (2000)
8. Weickert, J.: Anisotropic Diffusion in Image Processing. ECMI. Teubner-Verlag, Stuttgart, Germany (1998)
9. Pinkall, U., Polthier, K.: Computing discrete minimal surfaces and their conjugates. Experim. Math. 2(1), 15–36 (1993)
10. Bobenko, A.I., Schröder, P.: Discrete Willmore flow. In: Desbrun, M., Pottmann, H. (eds.) Eurographics Symposium on Geometry Processing, pp. 101–110 (2005)
11. Chan, T.F., Osher, S., Shen, J.H.: The digital TV filter and nonlinear denoising. IEEE Trans. Image Processing 10(2), 231–241 (2001)
12. Kim, S., Lim, H.: A non-conver diffusion model for simultaneous image denoising and edge enhancement. Electronic Journal of Differential Equation (Conference 15), 175–192 (2007); Six Mississippi State Conference on Differential Equations and Computational Simulations.
13. Kim, S.: Image denoising via diffusion modulation. International Journal of Pure and Applied Mathmatics 30(1), 72–91 (2006)
14. Marquina, A., Osher, S.: Explicit algorithms for a new time dependent model based on level set motion for nonlinear deblurring and noise removal. SIAM J. Sci. Comput. 22(2), 387–405 (2000)
15. Perona, P., Malik, J.: Scale-space and edge detection using anisotropic diffusion. IEEE Trans. Pattern Anal. Mach. Intell. 12(7), 629–639 (1990)
16. Gonzalez, R.C., Woods, R.E.: Digital Imgae Processing. Addison–Wesley, New York (1992)
17. Bloor, M.I.G., Wilson, M.J.: Modular solvers for image restoration problems using the discrepancy principle. Numerical Linear Algebra with Applications 9(5), 347–358 (2002)
18. Buades, A., Coll, B., Morel, J.M.: On image denoising methods. Technical Report 2004-15, Centre de Mathématiques et de Leurs Applications(CMLA) (2004)
19. Joo, K., Kim, S.: PDE-based image restoration I: Anti-staircasing and anti-diffusion. Technical Report 2003-07, Department of Mathematics, University of Kentucky (2003)

Note on Industrial Applications of Hu's Surface Extension Algorithm

Yu Zang, Yong-Jin Liu*, and Yu-Kun Lai

Tsinghua National Laboratory for Information Science and Technology,
Department of Computer Science and Technology,
Tsinghua University, Beijing, P.R. China
liuyongjin@tsinghua.edu.cn

Abstract. An important surface modeling problem in CAD is to connect two disjoint B-spline patches with the second-order geometric continuity. In this paper we present a study to solve this problem based on the surface extension algorithm [Computer-Aided Design 2002; 34:415–419]. Nice properties of this extension algorithm are exploited in depth and thus make our solution very simple and efficient. Various practical examples are presented to demonstrate the usefulness and efficiency of our presented solution.

Keywords: Differential geometry, skinning, partial differential equations, splines.

1 Introduction

B-spline surfaces are widely used in most industrial CAD systems. Diverse practical algorithms have been proposed for various operations on B-spline surfaces, such as knot insertion and removal, degree elevation and reduction, etc. See [3] for an overview. One operation — extending a B-spline surface to a target curve — is recently proposed in [2].

The Hu's algorithm in [2] extends a given B-spline surface S to a target curve and represents the extended surface S' in B-spline form. A nice property of this algorithm is that the shape and the parameterization of the original surface S are preserved. In this paper, we present a note to show that by using the surface extension algorithm in a novel way, an important industrial problem presented below can be efficiently solved.

In industrial practice, usually large surfaces of complex physical object are designed by many small patches smoothly joined together. In many design activities (below we list two main cases), the smooth joins are frequently broken and small gaps appear among the surface patches:

- The designer may not be satisfied with one patch on the surface and will delete it to design a new patch with various constraints. However, the new patch may not fit the boundary of old patches very well;

---

* Corresponding author.

F. Chen and B. Jüttler (Eds.): GMP 2008, LNCS 4975, pp. 304–314, 2008.

– When the model transfers between different CAD system, due to the different data precision, small gaps may occur among patches.

So an important engineering problem is that, given two surface patches with a small gap in between (usually the gap is about 1mm-5mm), join two patches with G^0, G^1 and G^2 continuity, respectively. In this paper we present a simple and efficient solution with the surface extension algorithm to solve the G^2-joint problem.

At first glance, this problem can be solved by using the general surface blending algorithms [5]. However, it does not work in practice because: (1) due to the small gap, the blending surface, behaving as the smooth transitional surface among geometric objects, is heavily wrinkled; (2) adding an additional blending surface for each small gap will increase the number of fragments in the modeling surface.

Our solution is efficient in the fashion that we extend one surface patch to smoothly join the other one, such that no additional blending surface is created. Another nice property is that the shape and the parameterization of the two original patches will not be changed.

This paper is organized as follows. The Hu's surface extension algorithm is briefly reviewed in Section 2. Section 3 presents our solution to the G^2-joint problem with the Hu's surface extension algorithm. Section 4 shows some examples in industrial applications. Conclusion is given in Section 5.

2   Hu's Surface Extension Algorithm

A pth-degree B-spline curve is defined by

$$C(u) = \sum_{i=0}^{n} N_{i,p}(u)\mathbf{P}_i, \quad 0 \le u \le 1$$

where the knot vector is $U = \{\underbrace{0,\cdots,0}_{p+1}, u_{p+1}, \cdots, u_n, \underbrace{1,\cdots,1}_{p+1}\}$. Using the algorithm on page 577 of Ref. [3], the vector U can be unclamped at either end. The same curve after unclamping can be defined by

$$C(u) = \sum_{i=0}^{n} N_{i,p}(u)\widetilde{\mathbf{P}}_i, \quad 0 \le u \le 1$$

over the knot vector $U = \{\underbrace{0,\cdots,0}_{p+1}, u_{p+1}, \cdots, u_n, 1, \underbrace{s,\cdots,s}_{p}\}$, $s \ge 1$. To extend the curve C to a target point $\mathbf{R}$, the parameter s is determined by the arc-length estimation

$$s = 1 + \frac{\|\widetilde{\mathbf{P}}_n - \mathbf{R}\|}{\sum_{i=0}^{n-p} \|C(u_{i+p+1}) - C(u_{i+p})\|}$$

and the extended curve is represented as

$$C_1(u) = \sum_{i=0}^{n+1} N_{i,p}(u)\tilde{\mathbf{Q}}_i, \quad 0 \le u \le 1$$

over the knot vector $U = \{\underbrace{0, \cdots, 0}_{p+1}, \frac{u_{p+1}}{s}, \cdots, \frac{u_n}{s}, \frac{1}{s}, \underbrace{1, \cdots, 1}_{p+1}\}$, where

$$\tilde{\mathbf{Q}}_i = \begin{cases} \tilde{\mathbf{P}}_i, & i = 0, 1, \cdots, n; \\ \mathbf{R}, & i = n+1 \end{cases}$$

A degree $p \times q$ B-spline surface is defined by

$$S(u, v) = \sum_{i=0}^{n} \sum_{j=0}^{m} N_{i,p}(u)N_{j,q}(v)\mathbf{P}_{i,j}, \quad 0 \le u \le 1, 0 \le v \le 1$$

with knot vectors

$$U = \{\underbrace{0, \cdots, 0}_{p+1}, u_{p+1}, \cdots, u_n, \underbrace{1, \cdots, 1}_{p+1}\}$$
$$V = \{\underbrace{0, \cdots, 0}_{q+1}, u_{q+1}, \cdots, u_m, \underbrace{1, \cdots, 1}_{q+1}\}$$

Given a target curve $C(v)$, by knot insertion and degree elevation algorithms [3], $C(v)$ can have the same degree q as the knot vector V of the surface $S(u, v)$. Denote the control points of $C(v)$ by $\mathbf{Q}_i$, $i = 0, \cdots, m$. To extend $S(u, v)$ in the direction of u to $C(v)$, it suffices to extend the $m + 1$ B-spline curves

$$C_j(u) = \sum_{i=0}^{n} N_{i,p}(u)\mathbf{P}_{i,j}, \quad j = 0, \cdots, m$$

to the target points $\mathbf{Q}_i$ with the same parameterization.

3   Surface Blending Based on Extension

Refer to Fig. 1. The goal is to smoothly connect two disjoint patches with the second-order geometric continuity (G^2 for short), and without creating any additional patch fragments. Our basic idea is simple as follows.

Let two patches A and B be connected between boundary curves C_1 and C_4. By knot insertion and degree elevation, these two curves can have the same knot vector. First, two intermediate curves

$$C_2(v) = \tfrac{2}{3}C_1(v) + \tfrac{1}{3}C_4(v)$$
$$C_3(v) = \tfrac{1}{3}C_1(v) + \tfrac{2}{3}C_4(v) \tag{1}$$

are created. The positions of C_2 and C_3 will be further optimized by surface fairness criterion as stated in Section 3.3. Then the surface A is extended three times in turn: first to C_2, then to C_3 and finally to C_4.

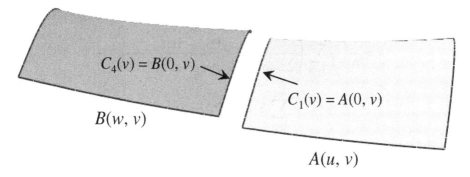

$C_4(v) = B(0, v)$

$C_1(v) = A(0, v)$

$B(w, v)$

$A(u, v)$

Fig. 1. Two disjoint patches that need to be G^2 smoothly joined along the boundaries C_1 and C_4

A nice property of Hu's surface extension algorithm is that the control points of C_2 and C_3 can be adjusted freely without changing the shape and parameterization of the original surface A. This property offers us sufficient degrees of freedom to make a G^2-join between A and B.

Let surfaces A and B be described by

$$A(u, v) = \sum_{i=0}^{m_1} \sum_{j=0}^{n} N_{i,p_1}(u) N_{j,q}(v) \mathbf{P}_{i,j},$$
$$B(w, v) = \sum_{i=0}^{m_2} \sum_{j=0}^{n} N_{i,p_2}(w) N_{j,q}(v) \mathbf{Q}_{i,j},$$

$$U = \{\underbrace{0, \cdots, 0}_{p_1+1}, u_{p_1+1}, \cdots, u_{m_1}, \underbrace{1, \cdots, 1}_{p_1+1}\}$$

$$V = \{\underbrace{0, \cdots, 0}_{q+1}, v_{q+1}, \cdots, v_n, \underbrace{1, \cdots, 1}_{q+1}\}$$

$$W = \{\underbrace{0, \cdots, 0}_{p_2+1}, u_{p_2+1}, \cdots, u_{m_2}, \underbrace{1, \cdots, 1}_{p_2+1}\}$$

Let the two surfaces be connected between boundary curves $C_1(v) = A(0, v)$ and $C_4(v) = B(0, v)$. The following is the algorithmic detail.

3.1 Surface Extension to a Target Curve

Refer to Fig. 1. By unclamping the knot vector U at the left end, the surface A can be extended left to a curve $C(v) = \sum_{j=0}^{n} N_{j,q}(v) \mathbf{R}_j$ with the knot vector V. The extended surface is represented as

$$\widetilde{A}(u, v) = \sum_{i=0}^{m_1+1} \sum_{j=0}^{n} N_{i,p_1}(u) N_{j,q}(v) \widetilde{\mathbf{P}}_{i,j} \qquad (2)$$

where the new knot vector is

$$\widetilde{U} = \{\underbrace{-a, \cdots, -a}_{p_1+1}, \widetilde{u}_{p+1} = 0, \widetilde{u}_{p+2} = u_{p_1+1}, \cdots, u_{m_1}, \underbrace{1, \cdots, 1}_{p_1+1}\} \qquad (3)$$

$a > 0$ and is estimated by

$$a = \frac{1}{n+1} \sum_{j=0}^{n} \left(\frac{\|\mathbf{P}_{0,j} - \mathbf{R}_j\|}{\sum_{r=0}^{m_1-p_1} \|A_j(u_{p_1+r+1}) - A_j(u_{p_1+r})\|} \right)$$

$$A_j(u) = \sum_{i=0}^{m_1} N_{i,p_1}(u)\mathbf{P}_{i,j}, \quad 0 \le j \le n$$

By the surface extension algorithm [2] and the knot vector unclamping algorithm [3], the new control points in eq.(2) can be obtained by:

(i) $\widetilde{\mathbf{P}}_{i,j}^0 = \mathbf{P}_{i,j}, \quad i = 0, 1, \cdots, p-1; \quad j = 0, 1, \cdots, m_1$

(ii) $\begin{cases} \widetilde{\mathbf{P}}_{i,j}^r = \widetilde{\mathbf{P}}_{i,j}^{r-1}, \quad i = r, \cdots, p-1 \\ \widetilde{\mathbf{P}}_{i,j}^r = \frac{\widetilde{\mathbf{P}}_{i,j}^{r-1} - \gamma \widetilde{\mathbf{P}}_{i+1,j}^r}{1-\alpha}, \quad i = r-1, r-2, \cdots, 0 \\ \text{where } \gamma = \frac{u_{p_1} - u_{p-r+i}}{u_{p+i+1} - u_{p-r+i}}, \quad r = 1, 2, \cdots, p-1, \quad j = 0, 1, \cdots, n \end{cases}$

(iii) $\widetilde{\mathbf{P}}_{i,j} = \begin{cases} \mathbf{P}_{i-1,j}, \quad i = p+1, p+2, \cdots, m_1+1 \\ \widetilde{\mathbf{P}}_{i-1,j}^{p-1}, \quad i = 1, 2, \cdots, p \\ \mathbf{R}_j, \quad i = 0 \\ j = 0, 1, \cdots, n \end{cases}$

The knot vector (3) can be rewritten as

$$\widetilde{U} = \{\underbrace{0, \cdots, 0}_{p_1+1}, \frac{a}{1+a}, \frac{u_{p_1+1}+a}{1+a}, \cdots, \frac{u_{m_1}+a}{1+a}, \underbrace{1, \cdots, 1}_{p_1+1}\}$$

Refer to Fig. 2 (a-c). By extending the surface A from the boundary curve C_1 to the curves C_2, C_3 in eq.(1) and C_4 in turn, the two patches A and B are connected without changing the shape and parameterization of the original patches. To this end, the extended patch A is represented as

$$\widehat{A}(u,v) = \sum_{i=0}^{m_1+3} \sum_{j=0}^{n} N_{i,p_1}(u)N_{j,q}(v)\widehat{\mathbf{P}}_{i,j}$$

$$\widehat{U} = \{\underbrace{0, \cdots, 0}_{p_1+1}, \widehat{u}_{p_1}, \cdots, \widehat{u}_{m_1+3}, \underbrace{1, \cdots, 1}_{p_1+1}\}$$

3.2   Modify the Extended Surface to Meet G^2 Continuity

Refer to Fig. 2(c-d). By surface extension, we can adjust the second and the third rows (i.e., $\widehat{\mathbf{P}}_{1,j}$ and $\widehat{\mathbf{P}}_{2,j}$) of control points of the extended surface A to make a G^2 connection with surface B, while the shape of the original patches A and B does not change.

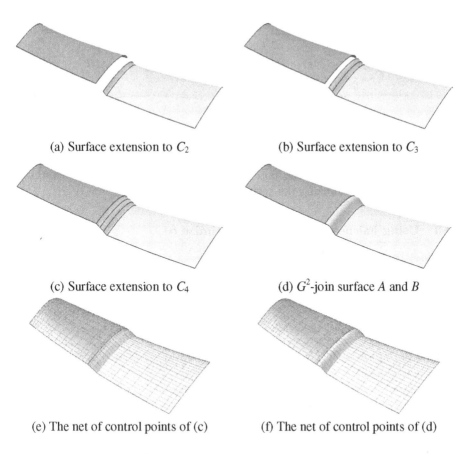

(a) Surface extension to C_2

(b) Surface extension to C_3

(c) Surface extension to C_4

(d) G^2-join surface A and B

(e) The net of control points of (c)

(f) The net of control points of (d)

Fig. 2. G^2-connect two disjoint patches in four steps

To connect surfaces A and B with G^2 continuity, the control points $\widehat{\mathbf{P}}_{1,j}$ and $\widehat{\mathbf{P}}_{2,j}$, $j = 0, 1, \cdots, n$ of A should be adjusted such that the following equation is satisfied:

$$\begin{cases} \widehat{A}_u(0, v) = \alpha B_w(0, v) \\ \widehat{A}_{uu}(0, v) = \alpha^2 B_{ww}(0, v) + \beta B_w(0, v) \end{cases}$$

where α, β are some constants.

For easy programming, we set $\beta = 0$ and the new positions of control points $\widehat{\mathbf{P}}_{1,j}$ and $\widehat{\mathbf{P}}_{2,j}$ of A are determined by

$$\begin{cases} \widehat{\mathbf{P}}_{0,j} = \mathbf{Q}_{0,j} \\ N'_{0,p_1}(0)\widehat{\mathbf{P}}_{0,j} + N'_{1,p_1}(0)\widehat{\mathbf{P}}_{1,j} = \alpha\left(N'_{0,p_2}(0)\mathbf{Q}_{0,j} + N'_{1,p_2}(0)\mathbf{Q}_{1,j}\right) \\ N''_{0,p_1}(0)\widehat{\mathbf{P}}_{0,j} + N''_{1,p_1}(0)\widehat{\mathbf{P}}_{1,j} + N''_{2,p_1}(0)\widehat{\mathbf{P}}_{2,j} = \\ \alpha^2\left(N''_{0,p_2}(0)\mathbf{Q}_{0,j} + N''_{1,p_2}(0)\mathbf{Q}_{1,j} + N''_{2,p_2}(0)\mathbf{Q}_{2,j}\right) \\ j = 0, 1, \cdots, n \end{cases} \qquad (4)$$

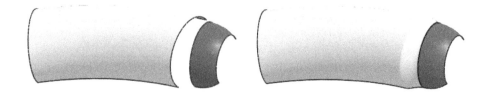

Fig. 3. G^2-connect two disjoint patches and minimize the surface fairness energy

The constant α offers us one more degree of freedom such that we can optimize the blending by minimizing the surface fairness energy.

3.3   Using Fairness Criterion to Optimize α

We use the standard second order energy

$$E = \iint A_{uu}^2 + 2A_{uv}^2 + A_{vv}^2 \, dudv \qquad (5)$$

to measure the fairness of surface A [6]. By solving eq.(4), we have

$$\widehat{\mathbf{P}}_{1,j} = \frac{1}{N'_{1,p_1}(0)} \left[\alpha \left(N'_{0,p_2}(0)\mathbf{Q}_{0,j} + N'_{1,p_2}(0)\mathbf{Q}_{1,j} \right) - N'_{0,p_1}(0)\widehat{\mathbf{P}}_{0,j} \right]$$
$$= \alpha M_{1,j} + N_{1,j}$$
$$\widehat{\mathbf{P}}_{2,j} = \frac{1}{N''_{2,p_1}(0)} \left[\alpha^2 \left(N''_{0,p_2}(0)\mathbf{Q}_{0,j} + N''_{1,p_2}(0)\mathbf{Q}_{1,j} + N''_{2,p_2}(0)\mathbf{Q}_{2,j} \right) \right.$$
$$\left. - N''_{0,p_1}(0)\widehat{\mathbf{P}}_{0,j} - N''_{1,p_1}(0)\widehat{\mathbf{P}}_{1,j} \right]$$
$$= \alpha^2 C_{2,j} + \alpha D_{2,j} + E_{2,j}$$

where $M_{1,j}, N_{1,j}, C_{2,j}, D_{2,j}, E_{2,j}$ are some constants. It can be calculated that energy $E(\alpha)$ in eq.(5) is a polynomial of degree four. Minimizing energy $E(\alpha)$ leads to solving a cubic equation, whose close-form formula can be found analytically [7].

4   Examples

By using formula (4) and solving eq.(5), our solution is very simple. The major advantage is that by extending one patch to G^2-connect the other patch, no additional patch fragment is created inbetween to blend the two disjoint patches; this nice property makes our solution particularly suitable for sewing two patches with small gaps. Two examples are shown in Figs. 2 and 3. An extreme case is illustrated in Fig. 4 in which two plane patches are G^2-connected.

In our solution, the boundaries of two patches to be connected do not need to be open. In Fig. 5, two tubes with closed profiles, generated by swept surfaces,

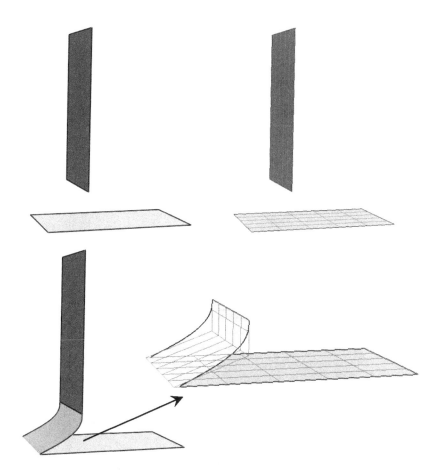

Fig. 4. G^2-connect two disjoint plane patches. The yellow patch is extended to meet the red plane patch; the right columns show the net of control points. By solving eq.(6), the optimal α is 7.63 to minimize the surface fairness energy.

are connected with G^2 continuity. To better illustrate the effect of surface fairness control by the simple parameter α, in these figures we show the faired surfaces with two different α values. Both tools of highlight lines [1] and isophotes [4] are used to inspect the surface quality (ref. Figs.5-6). The results clearly show the quality improvement in terms of surface fairness by minimizing energy (5) with one simple parameter α. Our method is simple and efficient because the exact minimization can be easily found by solving a cubic equation [7].

An application of our solution to the surfaces' G^2-connection problem is shown in Fig. 6, in which a scoop model is generated by G^2-connecting the handle part and the container part. Both tools of highlight lines and isophotes are shown to demonstrate the fairness and smoothness of the generated scoop model.

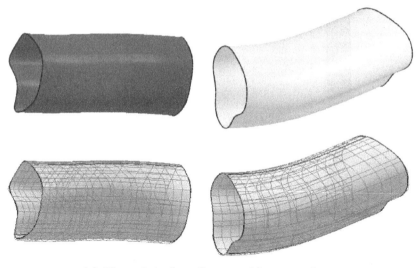

(a) The original surfaces and its control net

The extended yellow patch joining with red one

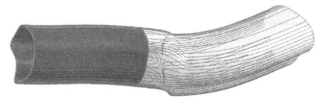

Fairness shown with highlight lines

(b) G^2-connection with $\alpha = 15.94$ where $E(\alpha) = 560.514$

(c) G^2-connection with $\alpha = 11.39$ where $E(\alpha) = 541.347$

Fig. 5. G^2-connect two disjoint tubes. Surface fairness of different α values are illustrated and, highlight lines are used to inspect the surface quality.

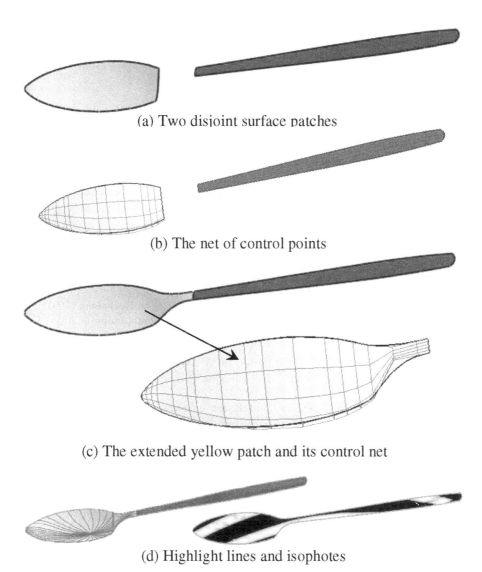

(a) Two disjoint surface patches

(b) The net of control points

(c) The extended yellow patch and its control net

(d) Highlight lines and isophotes

Fig. 6. Scoop modeling by surface extension with G^2 continuity

5   Conclusions

In this paper, we propose a simple and efficient method to connect two disjoint surface patches with G^2 continuity. Two major advantages are:

- the shape and parameterization of two original patches do not changes;
- no additional fragments are created between two disjoint patches; instead, the user can select one patch to extend and G^2-connect the other one.

We have successfully incorporated this method into a plugin module for a commercial CAD software. This module has been intensively tested and satisfied by industrial users.

Acknowledgments

The work was supported by the National Natural Science Foundation of China (Project Number 60736019, 60603085), and the National High Technology Research and Development Program of China (Project Number 2007AA01Z336, 2006AA01Z304).

References

1. Beier, K.P., Chen, Y.: Hightlight-line algorithm for realtime surface-quality assessment. Computer-Aided Design 26(4), 268–277 (1994)
2. Hu, S.M., Tai, C.L., Zhang, S.: An extension algorithm for B-splines by curve unclamping. Computer-Aided Design 34, 415–419 (2002)
3. Piegl, L., Tiller, W.: The NURBS Book, 2nd edn. Springer, Heidelberg (1997)
4. Poeschl, T.: Detecting surface irregularities using isophotes. Computer Aided Geometric Design 1(2), 163–168 (1984)
5. Vida, J., Martin, R.R., Varady, T.: A survey of blending methods that use parametric surfaces. Computer-Aided Design 26(5), 341–365 (1994)
6. Wallner, J.: Note on curve and surface energies. Computer Aided Geometric Design (in press, 2007), doi:10.1016/j.cagd.2007.05.007
7. http://mathworld.wolfram.com/CubicEquation.html

Parameterizing Marching Cubes Isosurfaces with Natural Neighbor Coordinates

Gregory M. Nielson[1], Liyan Zhang[2], Kun Lee[3], and Adam Huang[4]

[1] Arizona State University, Tempe, AZ, USA
[2] Nanjing University of Aeronautics and Astronautics, Nanjing, China
[3] Handong Global University, Pohang, Kyungbuk, South Korea
[4] National Taiwan University Hospital, Taipei, Taiwan

Abstract. The triangular mesh surfaces (TMS) which result form the Marching Cubes (MC) algorithm have some unique and special properties not shared by general TMS. We exploit some of these properties in the development of some new, effective and efficient methods for parameterizing these surfaces. The parameterization consists of a planar triangulation which is isomorphic (maps one-to-one) to the triangular mesh. The parameterization is computed as the solution of a sparse linear system of equations which is based upon the fact that locally the MC surfaces are functions (height-fields). The coefficients of the linear system utilize natural neighbor coordinates (NNC) which depend upon Dirchlet tessellations. While the use of NNC for general TMS can be somewhat computationally expensive and is often done procedurally, for the present case of MC surfaces, we are able to obtain simple and explicit formulas which lead to efficient computational algorithms.

Keywords: Marching Cubes, Parameterization, Natural Neighbor Coordinates.

1 Introduction

Triangle mesh surfaces (TMS) have become widely used as a convenient method of representing surfaces. The need for parameterizations of these surfaces has been well established in the literature. Methods which are nearly isometric (see [12]) and which approximately preserve area (see [2]), angles (see [9], [10], [20]), or shape (see [3]) have recently been discussed. Additional interest is based upon a strong connection to generalized barycentric coordinates (see [2], [24], [5], [6], [15]) and planar graphs (see [7], [22], [23]).

The particular triangle mesh surfaces which are produced by the Marching Cubes (MC) algorithm represent a growing and important special subset of TMS. These special TMS have some interesting and special properties not shared by general TMS. We exploit some of these properties in order to develop some efficient and effective methods for parameterizing these surfaces. In a nutshell:

1. We first note that with the proper edge choices (see [17]), the surface segment of a MC surface consisting of all triangles containing a vertex and all its neighbors (the *1-ring* or *star*) is a single-valued function (height-field) relative to a domain perpendicular to the edge (of the 3D rectilinear lattice) containing the vertex. See Fig. 2.

F. Chen and B. Jüttler (Eds.): GMP 2008, LNCS 4975, pp. 315–328, 2008.

2. Next, we represented the domain point of the vertex in terms of the domain points of its neighbors using natural neighbor coordinates (NNC). This yields the domain point of a vertex as a convex combination of the domain points of its neighbors. See Fig. 4. One of the unique contributions here is the fact that we are able to obtain simple, explicit representations (see Eq. (2)) which can be contrasted with the usual procedural approach used for NNC.
3. Next, this same linear constraint (convex combination) is imposed on the corresponding points in the parameter domain. The solution of this linear system, along with boundary constraints, yields a parameterization.
4. We further enhance the efficiency of the technique by noting that is sufficient to parameterize a subset (tiling) of the surface with edges on the 4*-network. The 4*-network is a subset of the edges of the MC surface lying on a collection of mutually orthogonal planes. See Fig. 1.

2 Parameterization Using Nearest Neighbor Coordinates

Prior to the specifics of our particular method, we include some general background and notation that will be used. In general, the *geometry* of a triangular mesh surface S, consists of a list of vertices $P_n = (x_n, y_n, z_n)$, $n = 1, \cdots, N$. The *topology* consists of a list of triple indices specifying the triangles of S, $N_T = \{(i, j, k): V_i, V_j, V_k \text{ are vertices of triangle of } S\}$. It is also useful to specify the topology with an edge list, $N_e = \{(i, j): V_i, V_j \text{ is an edge of } S\}$. For the most part, we are interested in triangular mesh, isosurfaces with boundaries which are topologically equivalent to a disc. We use ∂S to denote the polygonal boundary and the collection of indices of the vertices on the boundary is denoted by N_∂.

A planar parameterization of S consists of a polygon bounded, planar domain Q containing the points $Q_n = (s_n, t_n)$, $n = 1, \cdots, N$ which correspond one-to-one to the points P_n of S. The boundary of Q is denoted by ∂Q with boundary indices, $N_{\partial Q}$ exactly the same as N_∂ and the topology of S inherited by Q must be a valid 2D triangulation. This means that the triangles $Q_i, Q_j, Q_k, (i, j, k) \in N_T$ collectively cover P and only intersect at the common edges. The parameterization is the piecewise linear map, Φ, which maps each triangle of Q to the corresponding triangle of S. The map is one-to-one and onto.

Input to marching cubes algorithms consists of a three dimensional array of values $F_{i,j,k} = F(i\Delta x, j\Delta y, k\Delta z)$ representing samples of a function over a rectilinear grid consisting of lattice points $\{(i\Delta x, j\Delta y, k\Delta z), i = 1, \cdots, N_x; j = 1, \cdots, N_y; k = 1, \cdots, N_z\}$. For a given threshold value, α, the MC algorithm produces a triangular mesh surface which separates the lattice points with $F_{i,j,k} > \alpha$ from those with $F_{i,j,k} \leq \alpha$. The basic strategy of the MC algorithm consists of producing fragments of the surface one voxel (lattice cell) at a time. Details for the implementation of the MC algorithm including how to obtain the topology information by only one pass through the voxels can be found in [16].

We now make some observations about the special structure of the triangular mesh surfaces produced by the MC algorithm. Let S be the triangular mesh surface produced by the MC algorithm and consider the planar slices of S:

$X_i = S \cap \{(x,y,z): x = i\Delta x\}, \ i = 1, \cdots, N_x,$

$Y_j = S \cap \{(x,y,z): y = j\Delta y\}, \ j = 1, \cdots, N_y,$

$Z_k = S \cap \{(x,y,z): y = k\Delta z\}, \ k = 1, \cdots, N_z.$ These three collections form a mutually orthogonal network of planar, polygon curves lying on the surface S. Each vertex $V_i \in S$ is at the intersection of exactly two of these planar polygons; one from one of the collection of planar curves and the other from a collection that is orthogonal to it. An example of the 4*-network is shown in Fig. 1. We note that the valence of each vertex in the 4*-network is always four (except on the boundary) and the polygon patches are those provided by the various cases of the MC algorithm and will have 3, 4, 5 or 6 vertices. For a general P_i we denote the four neighbor points as P_{N_i}, P_{S_i}, P_{E_i} and P_{W_i}. For example, if $P_i = (j\Delta x, k\Delta y, z)$ then $P_{N_i} = (j\Delta x, y, z), y > k\Delta y$, $P_{S_i} = (j\Delta x, y, z), y \le k\Delta y$ $P_{W_i} = (x, k\Delta y, z), x > j\Delta x$, $P_{E_i} = (x, k\Delta y, z), x \le j\Delta x$. This is illustrated in Fig. 2.

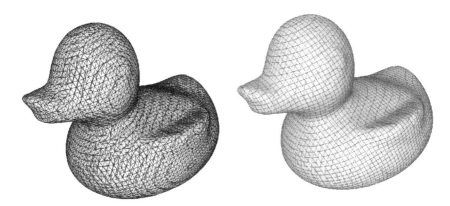

Fig. 1. The Marching Cubes triangular mesh surface is shown on the left and the 4*-network is shown on the right

It has previously been established (see [4] & [18]) that it is sufficient to only utilize the topology (edge connectivity) of the 4* network for a parameterization of a MC isosurface. This means that we only have to solve a linear system of the form

$$Q_i = C_{N_i} Q_{N_i} + C_{S_i} Q_{S_i} + C_{E_i} Q_{E_i} + C_{W_i} Q_{W_i} \ , i \notin N_\partial$$

$$Q_i \in \partial, \ i \in N_\partial \tag{1}$$

where

$$C_{N_i} \ge 0, C_{S_i} \ge 0, C_{E_i} \ge 0, C_{W_i} \ge 0, C_{N_i} + C_{S_i} + C_{E_i} + C_{W_i} = 1.$$

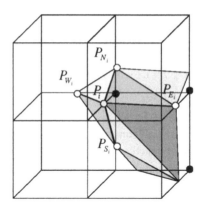

Fig. 2. The notation for the vertices of the 4*-network is illustrated. Also it is noted that the full star (1-ring) is a single-valued function relative to the plane perpendicular to the edge containing $P_i = (j\Delta x, k\Delta y, z)$.

The solution of (1) will lead to a tiling which can be subsequently triangulated leading to a parameterization of the MC surface S. See Fig. 3. We now set out to define the coefficients, $C_{N_i}, C_{S_i}, C_{E_i}, C_{W_i}$. With Fig. 2 in mind, we project the points P_{N_i}, P_{S_i}, P_{E_i} and P_{W_i} onto the domain plane perpendicular to the edge of the 3D lattice containing P_i and move to a local coordinate system where we introduce the values N_i, S_i, E_i and W_i so that domain point $(0, N_i)$ corresponds with P_{N_i}; $(0, S_i)$ corresponds with P_{S_i}; $(W_i, 0)$ corresponds with P_{W_i} and $(E_i, 0)$ corresponds with P_{E_i} and the domain point $(0,0)$ corresponds with the point P_i. This is illustrated in Fig. 4.

We now need to compute the natural neighbor coordinates (NNC) of domain point $(0,0)$ in terms of the domain points of the 4* neighbors. That is, we need the coefficients $C_{N_i}, C_{S_i}, C_{E_i}, C_{W_i}$ so that

$$(0,0) = C_{N_i}(0, N_i) + C_{S_i}(0, S_i) + C_{E_i}(E_i, 0) + C_{W_i}(W_i, 0) \quad i = 1, \cdots, N; i \notin N_\partial.$$

Before we proceed, we regress with a short discussion on natural neighbor coordinates (NNC) (see [14]). Given a collection of planar points (x_i, y_i), the Dirichlet tile $\tau(x_i, y_i)$ for the point (x_i, y_i) is the region consisting of all those points which are closer to (x_i, y_i) than any other point in the collection. Two existing points are called neighbors if their tiles share a common edge. Let (x, y) be an arbitrary point in the convex hull of the collection of points. If this point is inserted into the tessellation, it then acquires it own tile $\tau(x, y)$, made of portions of tiles of existing points, called neighbors of (x, y). This is illustrated in Fig. 5. Assume that (x, y) has m neighbors $(x_1, y_1), \cdots, (x_m, y_m)$. Now let $u(x, y)$ be the area of that portion of the

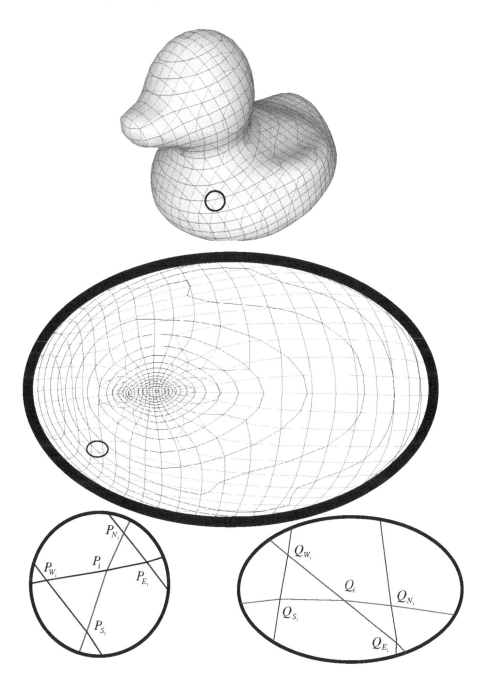

Fig. 3. Center image shows the tiling of the 4*-Network which will lead to a parameterization of the MC triangular mesh surface

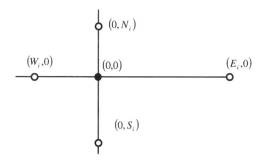

Fig. 4. The domain points of the 4*-network points

Dirichlet tile $\tau(x,y)$. Moreover, let $u_i(x,y)$ be the area of that portion of the Dirichlet tile $\tau(x,y)$ that intersects the tile $\tau(x_i, y_i)$, the NNC is then defined to be the ratio of these two areas, namely $\dfrac{u_i(x,y)}{u(x,y)}$ for $i=1,\cdots,m$. In the case where the collection of points consist of only three points, then NNC are the barycentric coordinates. In the same manner that barycentric coordinates give rise to piecewise linear interpolation over triangulated domains. The NNC can be used to define a function which interpolates to dependent values F_i at (x_i, y_i). This interpolant has been discussed by Sibson [21], Watson [26] and Boissonnat et al. [1].

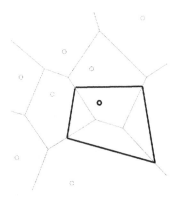

Fig. 5. The Dirichlet tiles of existing points are shown and the Dirichlet tile of the inserted point is boldly outlined

From Fig. 6, we can easily see that

$$C_{X_i} = \frac{A_{X_i}}{A_{N_i} + A_{E_i} + A_{W_i} + A_{S_i}}, \quad X = N, S, E, W$$

and so we need only compute the areas A_{N_i}, A_{E_i}, A_{W_i} and A_{S_i}. Two of these regions are triangles and two are trapezoids and so we have

$$A_{N_i} = \frac{-2E_i^2 W_i S_i + 2E_i W_i^2 S_i - N_i W_i S_i^2 + N_i E_i S_i^2}{8E_i W_i}$$

$$A_{E_i} = \frac{N_i W_i S_i^2 - N_i^2 W_i S_i}{8E_i W_i}$$

$$A_{W_i} = \frac{N_i^2 E_i S_i - N_i E_i S_i^2}{8E_i W_i}$$ (2)

$$A_{S_i} = \frac{2N_i E_i^2 W_i - 2N_i E_i W_i^2 + N_i^2 W_i S_i - N_i^2 E_i S_i}{8E_i W_i}$$

$$A_{N_i} + A_{E_i} + A_{W_i} + A_{S_i} = \frac{(E_i - W_i)(N_i - S_i)}{4}$$

We have left the above expressions unsimplified in order to more fully reveal the symmetry involved. The formulas of (2) are based upon Fig. 6 and therefore assume that $N_i S_i \geq E_i W_i$. For the other case, symmetric formulas based upon switching the roles of N_i, S_i and E_i, W_i must be used and subsequently simplified for efficient implementation.

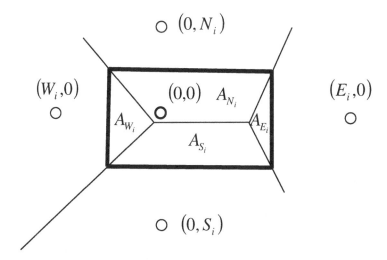

Fig. 6. The Domain points of the 4*-network vertices and the Dirichlet tile of the domain points of P_i is boldly outlined. The four intersection areas are labeled.

3 Examples

We show several examples that illustrate the use of this new method of parameterizing MC isosurfaces. The new method does not apply to general TMS and only applies to TMS within the context of the MC algorithm. Consequently, all of the examples here have an underlying field function from which the TMS is extracted by means of the MC algorithm (see [16]). If an extracted TMS is manipulated (e. g. an affine map is applied), then the embedding of the TMS in its rectilinear lattice must be reestablished prior to the application of the present method.

3.1 Geometric Representation of a Casting of a Mechanical Part

The need to parameterize the geometry of mechanical parts arises often in CAD applications. An example, using the techniques described here is shown in Fig. 7 and Fig. 8.

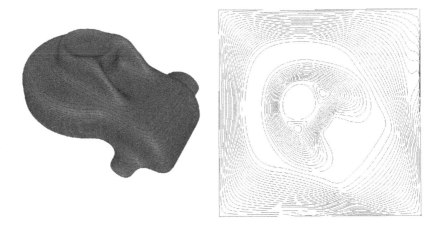

Fig. 7. The left image shows the MC isosurface of the geometry of a casting for the fabrication of a mechanical part. The right image illustrates the parameterization with one of the collection of 4*-network curves.

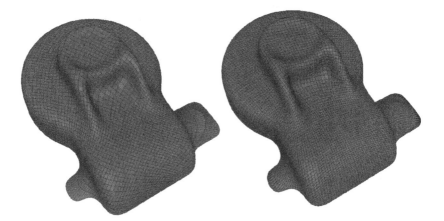

Fig. 8. The images illustrate the use of the parameterization for re-sampling the surface. This not only illustrates a practical use (tool path control) of the parameterization, but also a method for graphically assessing the quality of a particular parameterization.

3.2 Brain Morphology and Flattening by Parameterization

The term "flattening" has been used in the medical imaging field to convey the concept of mapping the surface of an object of interest to a planar domain (see [8], [11]). For example, with functional magnetic resonance (fMR) data it is possible to observe neural activity deep within the folds or sulci of the brain cortical surface using brain flattening techniques. There is also the potential use of flattening techniques in virtual colonoscopy where "the approach favored by pathologists, which involves cutting open the tube represented by the colon, and laying it out flat for comprehensive inspection" is simulated (Haker et al. [9], [10]). One approach to this flattening process is to use the parameterization techniques described here. We illustrate this possibility with an example that utilizes a volume data set that is available from the Surgical Planning Laboratory at Brigham and Women's Hospital, www.spl.harvard.edu. It is data that has been segmented from a MRI scan representing a brain which has had a tumor removed. (See Warfield et al. [27]). It consists of a binary array of size $124 \times 256 \times 256$ with a value of $F_{ijk} = 1$ indicating the presence of brain matter and a value $F_{ijk} = 0$ representing anything other than brain material. The MC algorithm can be used to compute a triangular mesh surface at threshold (say) 0.5 that will separate all lattice points in the volume from those outside the volume. This surface utilizes the midpoints of edges and it rather blocky and difficult to perceive. Using some recently developed techniques (see [19]), we

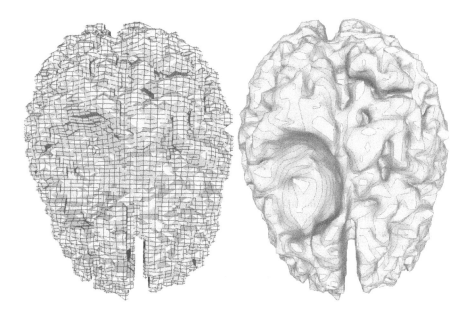

Fig. 9. The left image show the midpoint separating surface for the segmented data. The right image is the smooth shroud (see [19]). This brain has had a tumor removed (see [27] & [13]).

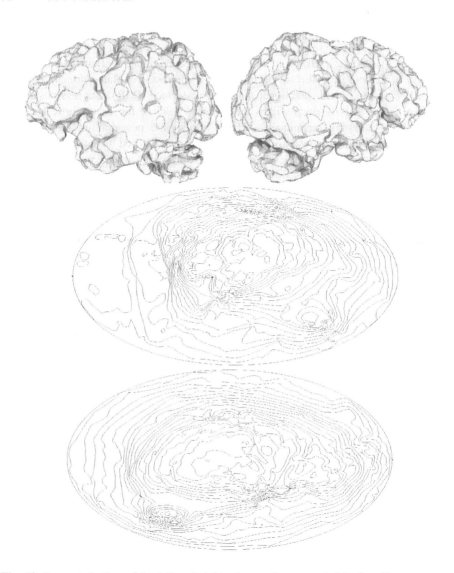

Fig. 10. Parameterization of the left and right spheres of a segmented brain with a common boundary. This contour plot allows for the correspondence of "features" in the object and the parameterization.

can obtain a smoother surface that has the same topology (edges) as the midpoint surface. This surface has vertices on the same edges as the midpoint surface, but the actual positions are chosen so as to optimize (subject to constraints) an energy/cost functional used to measure global smoothness. The left image of Fig. 9 shows the midpoint surface for a down-sampled volume data set of size $62 \times 128 \times 128$. Even though our algorithms have no problems at the higher resolution, the graphs illustrating the parameterization are too cluttered for this context and so a lower

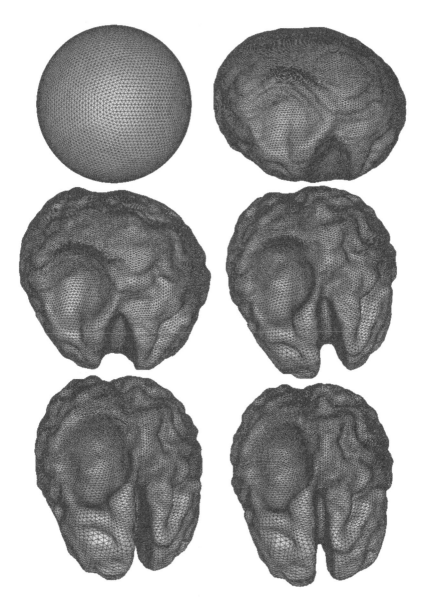

Fig. 11. A sequence of TMS morphing from a sphere to the brain revealing more fully the relationship of morphological features of the brain. The sphere TMS is "evaluated" over the brain surface using the spherical parameterization.

resolution model is used. Fig. 10 has views of the left and right hemispheres along with the respective parameterizations

Parameterizations conveniently lead to methods for mapping (morphing) one triangular mesh to another. If $Q_n = (s_n, t_n)$ are the vertices of the parameter domain

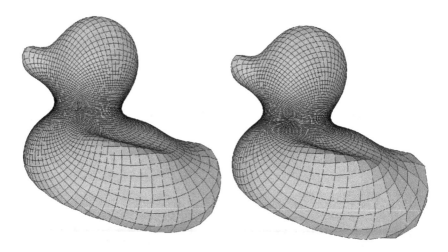

Fig. 12. A parameterization is used to re-sample a surface. On the left, the new NNC parameterization is used and on the right the mean value coordinates (see [5]) are used.

of a one surface, S, then we can sample a second surface, B, over the same triangulation and compute the vertices of the "in-between" surface as follows:

$$V(S, B, \alpha)_n = \alpha \varphi_S(s_n, t_n) + (1 - \alpha)\phi_B(s_n, t_n), \ 0 \le \alpha \le 1$$

This process is illustrated in Fig. 11 where the sequence of mappings more fully reveals the relationship of geometric features of the surface of the brain. The parameter domain consists of a unit sphere. The common ellipse domain of the left and right hemispheres of Fig. 10 is mapped to a unit circle followed by the conformal stereographic projection to the sphere.

Illustrated in Fig. 12 is a comparison of the new NNC method and the method of [5] based upon mean value coordinates. The two parameterizations (one for the head and the other for the body) have a common boundary consisting of a 4*-network curve in the vicinity of the neck. The parameterization has been sampled over a uniform 64x64 grid. The mean value coordinate approach leads to larger quadrilaterals near the center and most deeply embedded portion of the parameterization.

Acknowledgments. We wish to acknowledge the support of the Army Research Office under contract W911NF-05-1-0301. We wish to thank Aim@Shape for the data that led to the objects in Example 3.1. We wish to thank Drs. S. Warfield, M. Kaus, R. Kikinis, P. Black and F. Jolesz for sharing the tumor database. We wish to thank R. Holmes for his help and contributions.

References

1. Boissonnat, J.D., Cazals, F.: Smooth surface reconstruction via natural neighor interpolation of distance functions. In: Proc. 16th Annu. Sympos. Comput. Geom., pp. 223–232 (2000)
2. Desbrun, M., Meyer, M., Alliez, P.: Intrinsic parameterizations of surface meshes. Eurographics, Computer Graphics Forum 21(2), 666–911 (2002)

3. Floater, M.S.: Parametrization and smooth approximation of surface triangulations. Computer Aided Geoemtric Design 14(3), 231–250 (1997)
4. Floater, M.S.: Parametric tilings and scattered data approximation, Intern. J. Shape Modeling 4, 165–182 (1998)
5. Floater, M.S.: Mean Value Coordinates. Computer Aided Geometric Design 20(1), 19–27 (2003)
6. Floater, M.S., Hormann, K.: Parameterization of triangulations and unorganized points. In: Iske, A., Quak, E., Floater, M.S. (eds.) Tutorial on Multiresolution in Geometric Modelling, pp. 287–314. Springer, Heidelberg (2002)
7. Floater, M.S., Hormann, K.: A Tutorial and Survey. In: Dodgson, N.A., Floater, M.S., Sabin, M.A. (eds.) Advances in Multiresolution for Geometric Modellling, Springer, Heidelberg (2006)
8. Gu, X., Wang, Y., Chan, T.F., Thompson, P.M., Yau, S.-T.: Genus zero surface conformal mapping and its applications to brain surface mapping. IEEE Trans. Med. Imaging 23(8), 949–958 (2004)
9. Haker, S., Angenent, S., Tannenbaum, A., Kikinis, R., Sapiro, G.: Conformal surface parameterization for texture mapping. Transactions on Visualization and Computer Graphics 6, 181–189 (2000)
10. Haker, S., Angenent, S., Tannenbaum, A., Kikinis, R.: Non-distorting Flattening for Virtual Colonoscopy. In: Delp, S.L., DiGoia, A.M., Jaramaz, B. (eds.) MICCAI 2000. LNCS, vol. 1935, pp. 358–366. Springer, Heidelberg (2000)
11. Hong, W., Gu, X., Qiu, F., Jin, M., Kaufman, A.E.: Conformal virtual colon flattening. In: Symposium on Solid and Physical Modeling, vol. 85, pp. 85–93 (2006)
12. Hormann, K., Greiner, G.: MIPS: an efficient global parametrizaton method. In: Laurent, P.-J., Sablonniere, P., Schumaker, L.L. (eds.) Curve and Surface Design: Saint-Malo 1999, pp. 153–162. Vanderbilt University Press, Nashville (1999)
13. Kaus, M., Warfield, S.K., Nabavi, A., Black, P.M., Jolesz, F.A., Kikinis, R.: Segmentation of MRI of brain tumors. Radiology 218, 586–591 (2001)
14. Lodha, S., Franke, R.: Scattered Data Techniques for Surfaces. In: Hagen, H., Nielson, G., Post, F. (eds.) Scientific Visualization Dagstuhl 1997, pp. 181–222 (1999)
15. Meyer, M., Barr, A., Lee, H., Desbrun, M.: Generalized barycentric coordinates or irregular polygons. Journal of Graphical Tools 17, 13–22 (2002)
16. Nielson, G.M.: On Marching Cubes. IEEE Transactions on Visualization and Computer Graphics 9, 283–297 (2003)
17. Nielson, G.M.: MC*: Star functions for marching cubes. In: Proceedings of Visualization 2003, pp. 59–66 (2003b)
18. Nielson, G.M., Zhang, L., Lee, K.: Lifting Curve Parameterization Methods to Isosurfaces. Computer Aided Geometric Design 21, 751–756 (2004)
19. Nielson, G.M., Graf, G., Huang, A., Phliepp, M., Holmes, R.: Shrouds: Optimal separating surface for enumerated volumes. In: VisSym 2003, Eurographics Association, pp. 75–84 (2003)
20. Pinkall, U., Polthier, K.: Computing discrete minimal surfaces and their conjugates. Exp. Math. 2, 15–36 (1993)
21. Sibson, R.: A Brief Description of Natural Neighbor Interpolation. In: Barnett, D.V. (ed.) Interpreting Multivariate Data, pp. 21–36. Wiley, New York (1981)
22. Tutte, W.T.: Convex representation of graphs. Proc. London Math. Soc. 10, 304–320 (1960)
23. Tutte, W.T.: How to draw a graph. Proc. London Math. Soc. 13, 743–768 (1963)
24. Wachpress, E.: A Rational Finite Element Basis. Academic Press, London (1975)

25. Warfield, S.K., Kaus, M., Jolesz, F.A., Kikinis, R.: Adaptive, Template Moderated, Spatially Varying Statistical Classification. Med. Image Anal. 4(1), 43–55 (2000)
26. Watson, D.F.: nngridr: an implementation of natural neighbor interpolation. David Watson, Claremont, Australia (1994)
27. Warfield, S.K., Kaus, M., Jolesz, F.A., Kikinis, R.: Template Moderated, Spatially Varying Statistical Classification. Med. Image Anal. 4(1), 43–55 (2000)

Parametric Polynomial Minimal Surfaces of Degree Six with Isothermal Parameter

Gang Xu and Guozhao Wang

Institute of Computer Graphics and Image Processing, Department of Mathematics,
Zhejiang University, 310027 Hangzhou, P.R. China
xugangzju@yahoo.com.cn,
wanggz@zju.edu.cn

Abstract. In this paper, parametric polynomial minimal surfaces of degree six with isothermal parameter are discussed. We firstly propose the sufficient and necessary condition of a harmonic polynomial parametric surface of degree six being a minimal surface. Then we obtain two kinds of new minimal surfaces from the condition. The new minimal surfaces have similar properties as Enneper's minimal surface, such as symmetry, self-intersection and containing straight lines. A new pair of conjugate minimal surfaces is also discovered in this paper. The new minimal surfaces can be represented by tensor product Bézier surface and triangular Bézier surface, and have several shape parameters. We also employ the new minimal surfaces for form-finding problem in membrane structure and present several modeling examples.

Keywords: minimal surface, harmonic surfaces, isothermal parametric surface, parametric polynomial minimal surface of degree six, membrane structure.

1 Introduction

Minimal surface is an important class of surfaces in differential geometry. Since Lagrange derived the minimal surface equation in $\mathbf{R}^3$ in 1762, minimal surfaces have a long history of over 200 years. Because of their attractive properties, the minimal surfaces have been extensively employed in many areas such as architecture, material science, aviation, ship manufacture, biology, crystallogeny and so on. For instance, the shape of the membrane structure, which has appeared frequently in modern architecture, is mainly based on the minimal surfaces [1]. Furthermore, triply periodic minimal surfaces naturally arise in a variety of systems, including block copolymers, nanocomposites , micellar materials, lipid-water systems and certain cell membranes[11]. So it is meaningful to introduce the minimal surfaces into CAGD/CAD systems.

However, most of the classic minimal surfaces, such as helicoid and catenoid, can not be represented by Bézier surface or B-spline surface, which are the basic modeling tools in CAGD/CAD systems. In order to introduce the minimal surfaces into CAGD/CAD systems, we must find some minimal surfaces in the parametric polynomial form. In practice, the highest degree of parametric surface used in CAD systems is six, hence, polynomial minimal surface of degree six with

F. Chen and B. Jüttler (Eds.): GMP 2008, LNCS 4975, pp. 329–343, 2008.

isothermal parameter is discussed in this paper. The new minimal surfaces have elegant properties and are valuable for architecture design.

1.1 Related Work

There has been many literatures on the minimal surface in the field of classical differential geometry [19,20]. The discrete minimal surface has been introduced in recent years in [2,4,12,21,22,27,30]. As the topics which are related with the minimal surface, the computational algorithms for conformal structure on discrete surface are presented in [7,8,10]; and some discrete approximation of smooth differential operators are proposed in[31,32]. Cosín and Monterde proved that Enneper's surface is the unique cubic parametric polynomial minimal surface [3]. Based on the nonlinear programming and the FEM(finite element method), the approximation to the solution of the minimal surface equation bounded by Bézier or B-spline curves is investigated in [14]. Monterde obtained the approximation solution of the Plateau-Bézier problem by replacing the area functional with the Dirichlet functional in [15,16]. The modeling schemes of harmonic and biharmonic Bézier surfaces to approximate the minimal surface are presented in [3,17,18,29]. The applications of minimal surface in aesthetic design, aviation and nano structures modeling have been presented in [6,25,26,28].

1.2 Contributions and Overview

In this paper, we employ the classical theory of minimal surfaces to obtain parametric polynomial minimal surfaces of degree six. Our main contribution are:

- We propose the sufficient and necessary condition of a harmonic polynomial parametric surface of degree six being a minimal surface. The coefficient relations are derived from the isothermal condition.
- Based on the sufficient and necessary condition, two kinds of new minimal surfaces with several shape parameters are presented. We analyze the properties of the new minimal surfaces, such as symmetry, self-intersection, containing straight lines and conjugate minimal surfaces.
- Using surface trimming method, we employ the new minimal surfaces for form-finding problems in membrane structure.

The remainder of this paper is organized as follows. Some preliminaries and notations are presented in Section 2. Section 3 presents the sufficient and necessary condition of a harmonic polynomial parametric surface of degree six being a minimal surface. From the condition, two kinds of new minimal surface and their properties are treated in Section 4. The topic of Section 5 is trimming of the new minimal surfaces and its application in membrane structure. Finally, we conclude and list some future works in Section 6.

2 Preliminary

In this section, we shall review some concepts and results related to minimal surfaces [5,19].

If the parametric form of a regular patch in $\mathbf{R}^3$ is given by

$$r(u, v) = (x(u, v), y(u, v), z(u, v)), u \in (-\infty, +\infty), v \in (-\infty, +\infty),$$

Then the coefficients of the first fundamental form of $r(u, v)$ are

$$E = \langle r_u, r_u \rangle, F = \langle r_u, r_v \rangle, G = \langle r_v, r_v \rangle,$$

where r_u, r_v are the first-order partial derivatives of $r(u, v)$ with respect to u and v respectively and $\langle, \rangle$ defines the dot product of the vectors. The coefficients of the second fundamental form of $r(u, v)$ are

$$L = (r_u, r_v, r_{uu}), M = (r_u, r_v, r_{uv}), N = (r_u, r_v, r_{vv}),$$

where r_{uu}, r_{vv} and r_{uv} are the second-order partial derivatives of $r(u, v)$ and $(,,)$ defines the mixed product of the vectors. Then the mean curvature H and the Gaussian curvature K of $r(u, v)$ are

$$H = \frac{EN - 2FM + LG}{2(EG - F^2)}, \quad K = \frac{LN - M^2}{EG - F^2}.$$

Definition 1. If parametric surface $r(u, v)$ satisfies $E = G, F = 0$, then $r(u, v)$ is called *surface with isothermal parameter.*

Definition 2. If parametric surface $r(u, v)$ satisfies $r_{uu} + r_{vv} = 0$, then $r(u, v)$ is called *harmonic surface.*

Definition 3. If $r(u, v)$ satisfies $H = 0$, then $r(u, v)$ is called *minimal surface.*

Lemma 1. *The surface with isothermal parameter is minimal surface if and only if it is harmonic surface.*

Definition 4. If two differentiable functions $p(u, v), q(u, v) : U \mapsto R$ satisfy the Cauchy-Riemann equations

$$\frac{\partial p}{\partial u} = \frac{\partial q}{\partial v}, \frac{\partial p}{\partial v} = -\frac{\partial q}{\partial u}, \tag{1}$$

and both are harmonic. Then the functions are said to be *harmonic conjugate.*

Definition 5. If $P = (p_1, p_2, p_3)$ and $Q = (q_1, q_2, q_3)$ are isothermal parametrizations such that p_k and q_k are harmonic conjugate for $k = 1, 2, 3$, then P and Q are said to be *parametric conjugate minimal surfaces.*

For example, the helicoid and catenoid are conjugate minimal surface. Two conjugate minimal surfaces satisfy the following lemma.

Lemma 2. *Given two conjugate minimal surface P and Q and a real number t, all surfaces of the one-parameter family*

$$P_t = (\cos t)P + (\sin t)Q \tag{2}$$

satisfy

(a) *P_t are minimal surfaces for all $t \in \mathbf{R}$;*
(b) *P_t have the same first fundamental forms for $t \in \mathbf{R}$.*

Thus, from above lemma, any two conjugate minimal surfaces can be joined through a one-parameter family of minimal surfaces, and the first fundamental form of this family is independent of t. In other words, these minimal surfaces are isometric and have the same Gaussian curvatures at corresponding points.

3   Sufficient and Necessary Condition

The main idea of construction of new minimal surfaces is based on Lemma 1. We firstly consider the harmonic parametric polynomial surface of degree six.

Lemma 3. *Harmonic polynomial surface of degree six* $r(u, v)$ *must have the following form*

$$
\begin{aligned}
r(u, v) = {} & a(u^6 - 15u^4v^2 + 15u^2v^4 - v^6) + b(3u^5v - 10u^3v^3 + 3uv^5) + c(u^5 \\
& -10u^3v^2 + 5uv^4) + d(v^5 - 10u^2v^3 + 5u^4v) + e(u^4 - 6u^2v^2 + v^4) \\
& +f\, uv(u^2 - v^2) + gu(u^2 - 3v^2) + hv(v^2 - 3u^2) + i(u^2 - v^2) + \\
& juv + ku + lv + m,
\end{aligned}
$$

where $a, b, c, d, e, f, g, h, i, j, k, l, m$ *are coefficient vectors.*

Theorem 1. *Harmonic polynomial surface of degree six* $r(u, v)$ *is a minimal surface if and only if its coefficient vectors satisfy the following system of equations*

$$
\begin{cases}
4a^2 = b^2 \\
a \cdot b = 0 \\
2a \cdot c - b \cdot d = 0 \\
2a \cdot d + b \cdot c = 0 \\
25c^2 - 25d^2 + 48a \cdot e - 6b \cdot f = 0 \\
25d \cdot c + 12b \cdot e + 6a \cdot f = 0 \\
16e^2 - f^2 + 30c \cdot g - 30d \cdot h + 24a \cdot i - 6b \cdot j = 0 \\
4e \cdot f - 15c \cdot h + 15d \cdot g + 6b \cdot i + 6a \cdot j = 0 \\
9g^2 - 9h^2 + 16e \cdot i - 2f \cdot j + 10c \cdot k - 10l \cdot d = 0 \\
9g \cdot h - 2f \cdot i - 4e \cdot j - 5d \cdot k - 5c \cdot l = 0 \\
4i^2 - j^2 + 6g \cdot k + 6h \cdot l = 0 \\
2i \cdot j - 3g \cdot l - 3h \cdot k = 0 \\
18a \cdot g + 9b \cdot h + 20e \cdot c - 5f \cdot d = 0 \\
18a \cdot h - 9b \cdot g - 20e \cdot d - 5f \cdot c = 0 \\
6a \cdot k - 3b \cdot l + 10c \cdot i - 5d \cdot j + 12e \cdot g + 3f \cdot\ h = 0 \\
6a \cdot l + 3b \cdot k + 5c \cdot j + 10d \cdot i + 3f \cdot g - 12e \cdot h = 0 \\
4e \cdot k - f \cdot l + 3h \cdot j + 6g \cdot i = 0 \\
4e \cdot l + f \cdot k + 3g \cdot j - 6h \cdot i = 0 \\
2l \cdot i + k \cdot j = 0 \\
2k \cdot i - l \cdot j = 0 \\
k^2 = l^2 \\
k \cdot l = 0
\end{cases}
\tag{3}
$$

Remark. The proof of this theorem will be given in the Appendix.

4   Examples and Properties

Obviously, it is difficult to find the general solution for the system (3). But we can construct some special solutions from the condition. In order to simplify the system (3), we firstly make some assumptions about the coefficient vectors,

$a = (a_1, -a_2, 0), b = (2a_2, 2a_1, 0), c = (c_1, c_2, c_3), d = (d_1, d_2, d_3), e = (e_1, e_2, e_3), f = (f_1, f_2, f_3),$
$g = (g_1, g_2, g_3), h = (h_1, h_2, h_3), i = (i_1, i_2, i_3), j = (j_1, j_2, j_3), k = (k_1, k_2, k_3), l = (l_1, l_2, l_3),$

From $2a \cdot c - b \cdot d = 0$ and $2a \cdot d + b \cdot c = 0$, we have

$$a_1(c_1 - d_2) - a_2(c_2 + d_1) = 0, \tag{4}$$

$$a_2(c_1 - d_2) + a_1(c_2 + d_1) = 0, \tag{5}$$

From $(4) \times a_1 + (5) \times a_2$ and $(5) \times a_1 - (4) \times a_2$, we have

$$(a_1^2 + a_2^2)(c_1 - d_2) = 0, (a_1^2 + a_2^2)(c_2 + d_1) = 0.$$

Hence, we obtain

$$d_2 = c_1, d_1 = -c_2. \tag{6}$$

In the following subsections, we shall use this method to obtain the solutions.

4.1   Example 1

Supposing $c = d = g = h = k = l = 0, j_1 = 2i_2, j_2 = -2i_1$ in (3), we obtain

$$\begin{cases} 8a \cdot e - b \cdot f = 0 \\ 2b \cdot e + a \cdot f = 0 \\ 16e^2 - f^2 + 24a \cdot i - 6b \cdot j = 0 \\ 4e \cdot f + 6b \cdot i + 6a \cdot j = 0 \\ 8e \cdot i - f \cdot j = 0 \\ f \cdot i + 2e \cdot j = 0 \\ 4i_3^2 - j_3^2 = 0 \\ i_3 \cdot j_3 = 0 \end{cases} \tag{7}$$

From (7), we have

$$f_1 = f_2 = e_1 = e_2 = i_3 = j_3 = 0,$$
$$e_3 = \frac{\sqrt{6}}{2}\sqrt{\sqrt{(a_1^2 + a_2^2)(i_1^2 + i_2^2)} + (a_2 i_2 - a_1 i_1)},$$
$$f_3 = -2\sqrt{6}\sqrt{\sqrt{(a_1^2 + a_2^2)(i_1^2 + i_2^2)} - (a_2 i_2 - a_1 i_1)}.$$

Then we obtain a class of minimal surface with four shape parameters a_1, a_2, i_1 and i_2:

$$r(u, v) = (X(u, v), Y(u, v), Z(u, v)) \tag{8}$$

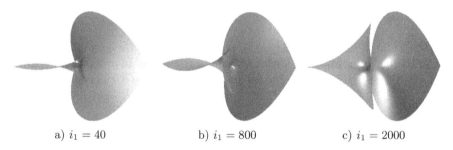

a) $i_1 = 40$ b) $i_1 = 800$ c) $i_1 = 2000$

Fig. 1. The effect of shape parameter i_1 in $\boldsymbol{r}_1(u,v)$. Here $a_1 = 4, u, v \in [-4, 4]$.

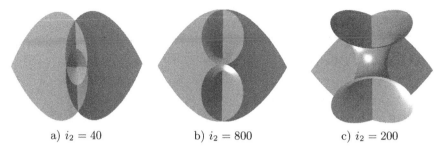

a) $i_2 = 40$ b) $i_2 = 800$ c) $i_2 = 200$

Fig. 2. The effect of shape parameter i_2 in $\boldsymbol{r}_2(u,v)$. Here $a_2 = 4, u, v \in [-4, 4]$.

where

$$X(u,v) = a_1(u^6 - 15u^4v^2 + 15u^2v^4 - v^6) + 2a_2(3u^5v - 10u^3v^3 + 3uv^5)$$
$$+ i_1(u^2 - v^2) + 2i_2uv,$$
$$Y(u,v) = -a_2(u^6 - 15u^4v^2 + 15u^2v^4 - v^6) + 2a_1(3u^5v - 10u^3v^3 + 3uv^5)$$
$$+ i_2(u^2 - v^2) - 2i_1uv,$$
$$Z(u,v) = \frac{\sqrt{6}}{2}\sqrt{\sqrt{(a_1^2 + a_2^2)(i_1^2 + i_2^2)} + (a_2i_2 - a_1i_1)(u^4 - 6u^2v^2 + v^4) -}$$
$$2\sqrt{6}\sqrt{\sqrt{(a_1^2 + a_2^2)(i_1^2 + i_2^2)} - (a_2i_2 - a_1i_1)}uv(u^2 - v^2).$$

When $a_2 = i_2 = 0$, we denote the minimal surface in (8) by $\boldsymbol{r}_1(u,v)$. The Gaussian curvature of $\boldsymbol{r}_1(u,v)$ is

$$K = -192a_1i_1(u^2 + v^2)^2. \tag{9}$$

Fig 1 shows the effect of the shape parameter i_1.

In the case $a_1 = i_1 = 0$, the minimal surface in (8) is denoted by $\boldsymbol{r}_2(u,v)$. The Gaussian curvature of $\boldsymbol{r}_2(u,v)$ is

$$K = -192a_2i_2(u^2 + v^2)^2. \tag{10}$$

Fig 2 illustrates the effect of the shape parameter i_2.

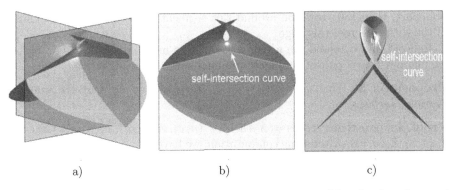

Fig. 3. The minimal surface $r_2(u, v)$ and its symmetric planes: (a) $r_2(u, v)$ with $a_2 = 4$ and $i_2 = 500$, $u, v \in [-4, 4]$, its symmetric planes $X = 0$ and $Y = 0$; (b) self-intersection curve on the plane $X = 0$;(c) self-intersection curve on the plane $Y = 0$

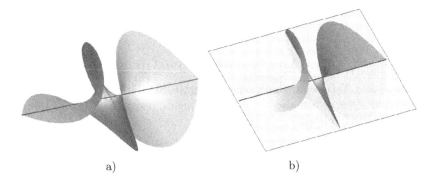

Fig. 4. The minimal surface $r_1(u, v)$ and the straight lines on it. Here $a_1 = 4, i_1 = 2000$ $u, v \in [-4, 4]$.

Enneper surface is the unique cubic isothermal parametric polynomial minimal surface, and it has several interesting properties, such as symmetry, self-intersection, and containing straight lines on it. For $r(u, v)$, we have the following propositions.

Proposition 1. *The minimal surface* $r_2(u, v)$ *is symmetric about the plane* $X = 0$ *and the plane* $Y = 0$.

Furthermore, there exists two self-intersection curves of $r_2(u, v)$ on the plane $X = 0$ and the plane $Y = 0$; besides the two self-intersection curves, there are no other self-intersection points on $r_2(u, v)$. Fig 3 shows the symmetric planes and self-intersection curves when $a_2 = 4$ and $i_2 = 500$.

Proposition 2. *The minimal surface* $r_1(u, v)$ *contains two orthogonal straight lines* $x = \pm y$ *on the plane* $Z = 0$.

Fig 4 shows the minimal surface and the straight lines on it. It is consistent with the fact that if a piece of a minimal surface has a straight line segment on its boundary, then 180° rotation around this segment is the analytic continuation of the surface across this edge.

Helicoid and catenoid are a pair of conjugate minimal surfaces. For $r(u,v)$, we find out a new pair of conjugate minimal surfaces as follows.

Proposition 3. *When* $a_1 = a_2, i_1 = i_2$, $r_1(u,v)$ *and* $r_2(u,v)$ *are conjugate minimal surfaces.*

Proof. Suppose that $r_1(u,v) = (X_1(u,v), Y_1(u,v), Z_1(u,v))$, $r_2(u,v) = (X_2(u,v), Y_2(u,v), Z_2(u,v))$. After some computation, we have

$$\frac{\partial X_1(u,v)}{\partial u} = a_1(6u^5 - 60u^3v^2 + 30uv^4) + 2i_1 u,$$

$$\frac{\partial X_1(u,v)}{\partial v} = a_1(60u^2v^3 - 30u^4v - 6v^5) - 2i_1 v$$

$$\frac{\partial X_2(u,v)}{\partial u} = a_2(30u^4v - 60u^2v^3 + 6v^5) + 2i_2 v,$$

$$\frac{\partial X_2(u,v)}{\partial v} = a_2(6u^5 - 60u^3v^2 + 30uv^4) + 2i_2 u.$$

When $a_1 = a_2, i_1 = i_2$, $\dfrac{\partial X_1(u,v)}{\partial u} = \dfrac{\partial X_2(u,v)}{\partial v}$, $\dfrac{\partial X_1(u,v)}{\partial v} = -\dfrac{\partial X_2(u,v)}{\partial u}$. That is, $X_1(u,v)$ and $X_2(u,v)$ are harmonic conjugate. Similarly, $Y_1(u,v)$ and $Y_2(u,v)$, $Z_1(u,v)$ and $Z_2(u,v)$ are also harmonic conjugate respectively. From Definition 5, the proof is completed.   □

From Lemma 2, when $a_1 = a_2, i_1 = i_2$, the surfaces of one-parametric family

$$r_t(u,v) = (\cos t)r_1(u,v) + (\sin t)r_2(u,v) \tag{11}$$

are minimal surfaces with the same first fundamental form. These minimal surfaces are isometric and have the same Gaussian curvature at corresponding points. It is consistent with (9)and (10).

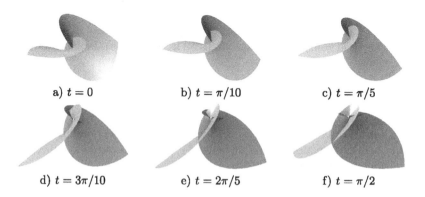

a) $t = 0$ b) $t = \pi/10$ c) $t = \pi/5$

d) $t = 3\pi/10$ e) $t = 2\pi/5$ f) $t = \pi/2$

Fig. 5. Dynamic deformation between $r_1(u,v)$ and $r_2(u,v)$. Here $a_1 = a_2 = 500$, $i_1 = i_2 = 5, u, v \in [-4, 4]$.

Let $t \in [0, \pi/2]$. When $a_1 = a_2$ and $i_1 = i_2$, for $t = 0$, the minimal surface r_t reduces to $r_1(u, v)$; for $t = \pi/2$, it reduces to $r_2(u, v)$. Then when t varies from 0 to $\pi/2$, $r_1(u, v)$ can be continuously deformed into $r_2(u, v)$, and each intermediate surface is also minimal surface. Fig 5 illustrates the dynamic deformation when $a_1 = a_2 = 500, i_1 = i_2 = 5$. It is similar with the dynamic deformation between helicoid and catenoid.

4.2   Example 2

Supposing $k = l = i = j = 0, c_3 = d_3 = g_3 = h_3 = 0$ in (3), we have

$$
\begin{cases}
8a \cdot e - b \cdot f = 0 \\
2b \cdot e + a \cdot f = 0 \\
16e^2 - f^2 + 30c \cdot g - 30d \cdot h = 0 \\
4e \cdot f - 15c \cdot h + 15d \cdot g = 0 \\
g^2 - h^2 = 0 \\
g \cdot h = 0 \\
18a \cdot g + 9b \cdot h + 20e \cdot c - 5f \cdot d = 0 \\
18a \cdot h - 9b \cdot g - 20e \cdot d - 5f \cdot c = 0 \\
4e \cdot g + f \cdot h = 0 \\
f \cdot g - 4e \cdot h = 0
\end{cases}
\tag{12}
$$

From (6) and (12), two solutions can be obtained: if $c_1 g_1 + c_2 g_2 > 0$, then

$$
f_2 = 4e_1, f_1 = -4e_2, h_1 = g_2, h_2 = -g_1,
$$
$$
e_3 = 0, f_3 = 2\sqrt{15}\sqrt{c_1 g_1 + c_2 g_2};
$$

if $c_1 g_1 + c_2 g_2 < 0$, then

$$
f_2 = 4e_1, f_1 = -4e_2, h_1 = g_2, h_2 = -g_1,
$$
$$
f_3 = 0, e_3 = \frac{\sqrt{15}}{2}\sqrt{-c_1 g_1 - c_2 g_2};
$$

Then we obtain two classes of minimal surface with eight shape parameters $a_1, a_2, c_1, c_2, e_1, e_2, g_1$ and g_2:

$$
r(u, v) = (X(u, v), Y(u, v), Z(u, v))
\tag{13}
$$

where

$$
\begin{aligned}
X(u, v) = {}& a_1(u^6 - 15u^4v^2 + 15u^2v^4 - v^6) + 2a_2(3u^5v - 10u^3v^3 + 3uv^5) \\
& + c_1(u^5 - 10u^3v^2 + 5uv^4) - c_2(v^5 - 10u^2v^3 + 5u^4v) + e_1(u^4 - 6u^2v^2 + v^4) \\
& - 4e_2 uv(u^2 - v^2) + g_1 u(u^2 - 3v^2) + g_2 v(v^2 - 3u^2), \\
Y(u, v) = {}& -a_2(u^6 - 15u^4v^2 + 15u^2v^4 - v^6) + 2a_1(3u^5v - 10u^3v^3 + 3uv^5) \\
& + c_2(u^5 - 10u^3v^2 + 5uv^4) + c_1(v^5 - 10u^2v^3 + 5u^4v) + e_2(u^4 - 6u^2v^2 + v^4) \\
& + 4e_1 uv(u^2 - v^2) + g_2 u(u^2 - 3v^2) - g_1 v(v^2 - 3u^2), \\
Z(u, v) = {}& 2\sqrt{15}\sqrt{c_1 g_1 + c_2 g_2}\, uv(u^2 - v^2),
\end{aligned}
$$

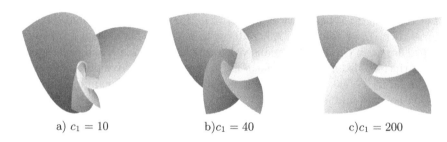

a) $c_1 = 10$ b)$c_1 = 40$ c)$c_1 = 200$

Fig. 6. The effect of shape parameter c_1 in $r_3(u,v)$. Here $a_1 = e_1 = g_1 = 4, u, v \in [-2, 2]$.

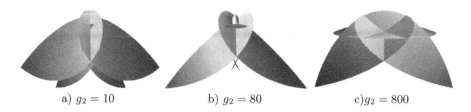

a) $g_2 = 10$ b) $g_2 = 80$ c)$g_2 = 800$

Fig. 7. The effect of shape parameter g_2 in $r_6(u,v)$. Here $a_2 = e_2 = 4, c_2 = -4$, $u, v \in [-2, 2]$.

or

$$\bar{r}(u,v) = (\bar{X}(u,v), \bar{Y}(u,v), \bar{Z}(u,v)) \tag{14}$$

where

$$
\begin{aligned}
\bar{X}(u,v) = & \, a_1(u^6 - 15u^4v^2 + 15u^2v^4 - v^6) + 2a_2(3u^5v - 10u^3v^3 + 3uv^5) \\
& + c_1(u^5 - 10u^3v^2 + 5uv^4) - c_2(v^5 - 10u^2v^3 + 5u^4v) + e_1(u^4 - 6u^2v^2 + v^4) \\
& - 4e_2uv(u^2 - v^2) + g_1u(u^2 - 3v^2) + g_2v(v^2 - 3u^2), \\
\bar{Y}(u,v) = & -a_2(u^6 - 15u^4v^2 + 15u^2v^4 - v^6) + 2a_1(3u^5v - 10u^3v^3 + 3uv^5) \\
& + c_2(u^5 - 10u^3v^2 + 5uv^4) + c_1(v^5 - 10u^2v^3 + 5u^4v) + e_2(u^4 - 6u^2v^2 + v^4) \\
& + 4e_1uv(u^2 - v^2) + g_2u(u^2 - 3v^2) - g_1v(v^2 - 3u^2), \\
\bar{Z}(u,v) = & \, \frac{\sqrt{15}}{2}\sqrt{-c_1g_1 - c_2g_2}(u^4 - 6u^2v^2 + v^4).
\end{aligned}
$$

When $a_2 = c_2 = e_2 = g_2 = 0$, we denote the minimal surface $r(u,v)$ in (13) by $r_3(u,v)$; similarly, in the case $a_1 = c_1 = e_1 = g_1 = 0$, $r(u,v)$ in (13) is denoted by $r_4(u,v)$. When $a_1 = a_2, c_1 = c_2, e_1 = e_2, g_1 = g_2$, we can obtain $r_4(u,v)$ from $r_3(u,v)$ by rotation transformation. Fig 6 illustrates the effect of c_1 of $r_3(u,v)$.

In the case $a_2 = c_2 = e_2 = g_2 = 0$, the minimal surface $\bar{r}(u,v)$ in (14) is denoted by $r_5(u,v)$; similarly, when $a_1 = c_1 = e_1 = g_1 = 0$, we denote $\bar{r}(u,v)$ in (14) by $r_6(u,v)$. In the case $a_1 = a_2, c_1 = c_2, e_1 = e_2, g_1 = g_2$, $r_6(u,v)$ can be obtained from $r_5(u,v)$ by rotation transformation. The effect of g_2 of $r_6(u,v)$ is shown in Fig 7.

For $r_5(u,v)$ and $r_6(u,v)$, we have the following proposition.

Fig. 8. Two different views of the minimal surface $r_6(u, v)$ and its symmetric plane. Here $a_2 = e_2 = g_2 = 4, c_2 = -4, u, v \in [-2, 2]$.

a)   b)

Fig. 9. Tensor product Bézier surface representation of $r(u, v)$ in (8) and (13):(a)$r(u, v)$ in (8) and its control mesh, $a_1 = 3, a_2 = 500, i_1 = i_2 = 1, u, v \in [0, 1]$ (b)$r(u, v)$ in (13) and its control mesh, $a_1 = c_1 = e_1 = g_1 = a_2 = c_2 = e_2 = 4, g_2 = 400, u, v \in [0, 1]$

Proposition 4. *The minimal surface* $r_5(u, v)$ *is symmetric about the plane* $Y = 0$; $r_6(u, v)$ *is symmetric about the plane* $X = 0$.

Fig 8 presents the symmetric plane of $r_6(u, v)$ with $a_2 = e_2 = g_2 = 4$ and $c_2 = -4$.

5   Application in Architecture

Obviously, the proposed minimal surfaces can be represented by tensor product Bézier surface or triangular Bézier surface. Fig 9 shows the tensor product Bézier surface representation of $r(u, v)$ in (8) and (13).

Geometric design and computation in architecture has been a hotspot in recent years [13,23,24]. In the surface of membrane structure, we need that the resultant nodal forces (i.e. residual forces) must be reduced to zero, so that there is no pressure difference across the surface. Hence, minimal surface is the ideal shape of the membrane structure. In particular, the minimal surfaces proposed in the current paper can be used for the form-finding problem, which is the first stage in the construction process of membrane structure.

From the classical theory of minimal surface, any trimmed surfaces on minimal surface are also minimal surfaces. Hence, the traditional surface trimming

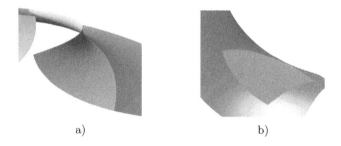

Fig. 10. Trimmed surfaces (yellow) on minimal surfaces $r(u,v)$(green) in (8) with $i_1 = a_1 = 1, i_2 = a_2 = 5$

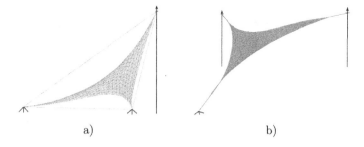

Fig. 11. Two modeling examples of membrane structures by using the triangular trimmed surfaces on minimal surface $r_1(u,v)$ with $a_1 = i_1 = 50$

methods in [9] are employed for form finding problems. Fig 10 illustrates two trimmed minimal surface. Modeling examples of membrane structure are shown in Fig 11.

6   Conclusion and Future Work

In order to introduce minimal surfaces into CAGD/CAD systems, the parametric polynomial minimal surface of degree six is studied in this paper. We propose the sufficient and necessary condition of a harmonic polynomial parametric surface of degree six being a minimal surface. Two kinds of new minimal surface with several shape parameters are obtained from the condition. We analyze the properties of the new minimal surface, such as symmetry, self-intersection, containing straights lines and conjugate property. Hence, the new minimal surfaces have the similar properties as the classical Enneper surface. In particular, the conjugate property is similar with the catenoid and the helicoid.

The Weierstrass representation is another method to obtain new minimal surfaces. However, it is difficult to choose the proper initial functions to obtain the parametric polynomial minimal surfaces. The method presented in this paper can directly derive the parametric polynomial minimal surface.

In the future, we will investigate the other applications of the new minimal surfaces. How to give the general formula of the parametric polynomial minimal surfaces, is also our future work.

Acknowledgments. The work described in this article is supported by the National Nature Science Foundation of China (No. 60773179, 60473130) and Foundation of State Key Basic Research 973 Development Programming Item of China (No. G2004CB318000).

References

1. Bletzinger, K.-W.: Form finding of membrane structures and minimal surfaces by numerical continuation. In: Proceeding of the IASS Congress on Structural Morphology: Towards the New Millennium, Nottingham, pp. 68–75 (1997) ISBN 0-85358-064-2
2. Bobenko, A.I., Hoffmann, T., Springborn, B.A.: Minimal surfaces from circle patterns: Geometry from combinatorics. Annals of Mathematics 164, 231–264 (2006)
3. Cosín, C., Monterde, J.: Bézier Surfaces of Minimal Area. In: Sloot, P.M.A., Tan, C.J.K., Dongarra, J., Hoekstra, A.G. (eds.) ICCS 2002. LNCS, vol. 2330, pp. 72–81. Springer, Heidelberg (2002)
4. Desbrun, M., Grinspun, E., Schroder, P.: Discrete differential geometry: an applied introduction. In: SIGGRAPH Course Notes (2005)
5. Do Carmo, M.: Differential geometry of curves and surfaces. Prentice-Hall, Englewood Cliffs (1976)
6. Grandine, S., Del Valle, T., Moeller, S.K., Natarajan, G., Pencheva, J., Sherman, S.: Wise. Designing airplane struts using minimal surfaces IMA Preprint 1866 (2002)
7. Gu, X., Yau, S.: Surface classification using conformal structures. ICCV, 701-708 (2003)
8. Gu, X., Yau, S.: Computing conformal structure of surfaces CoRR cs.GR/0212043 (2002)
9. Hoscheck, J., Schneider, F.: Spline conversion for trimmed rational Bezier and B-spline surfaces. Computer Aided Design 9, 580–590 (1990)
10. Jin, M., Luo, F., Gu, X.F.: Computing general geometric structures on surfaces using Ricci flow. Computer-Aided Design 8, 663–675 (2007)
11. Jung, K., Chu, K.T., Torquato, S.: A variational level set approach for surface area minimization of triply-periodic surfaces. Journal of Computational Physics 2, 711–730 (2007)
12. Li, X., Guo, X.H., Wang, H.Y., He, Y., Gu, X.F., Qin, H.: Harmonic volumetric mapping for solid modeling applications. In: Symposium on Solid and Physical Modeling, pp. 109–120 (2007)
13. Liu, Y., Pottmann, H., Wallner, J., Yang, Y.L., Wang, W.P.: Geometric modeling with conical meshes and developable surfaces. ACM Trans. Graphics 3, 681–689 (2006)
14. Man, J.J., Wang, G.Z.: Approximating to nonparameterzied minimal surface with B-spline surface. Journal of Software 4, 824–829 (2003)
15. Monterde, J.: The Plateau-Bézier Problem. In: Wilson, M.J., Martin, R.R. (eds.) Mathematics of Surfaces. LNCS, vol. 2768, pp. 262–273. Springer, Heidelberg (2003)

16. Monterde, J.: Bézier surfaces of minimal area: The Dirichlet approach. Computer Aided Geometric Design 1, 117–136 (2004)
17. Monterde, J., Ugail., H.: On harmonic and biharmonic Bézier surfaces. Computer Aided Geometric Design 7, 697–715 (2004)
18. Monterde, J., Ugail., H.: A general 4th-order PDE method to generate Bézier surfaces from boundary. Computer Aided Geometric Design 2, 208–225 (2006)
19. Nitsche, J.C.C.: Lectures on minimal surfaces, vol. 1. Cambridge Univ. Press, Cambridge (1989)
20. Osserman, R.: A survey of minimal surfaces, 2nd edn. Dover publ., New York((1986)
21. Pinkall, U., Polthier, K.: Computing discrete minimal surface and their conjugates. Experiment Mathematics 1, 15–36 (1993)
22. Polthier, K.: Polyhedral surface of constant mean curvature. Habilitationsschrift TU Berlin (2002)
23. Pottmann, H., Liu, Y.: Discrete surfaces in isotropic geometry. In: Mathematics of Surfaces, vol. XII, pp. 341-363 (2007)
24. Pottmann, H., Liu, Y., Wallner, J., Bobenko, A., Wang, W.P.: Geometry of multilayer freeform structures for architecture. ACM Transactions on Graphics 3, 1–11 (2007)
25. Séquin, C.H.: CAD Tools for Aesthetic Engineering. Computer Aided Design 7, 737–750 (2005)
26. Sullivan, J.: The aesthetic value of optimal geometry. In: Emmer, M. (ed.) The Visual Mind II, pp. 547–563. MIT Press, Cambridge (2005)
27. Wallner, J., Pottmann, H.: Infinitesimally flexible meshes and discrete minimal surfaces. Monatsh. Math (to appear, 2006)
28. Wang, Y.: Periodic surface modeling for computer aided nano design. Computer Aided Design 3, 179–189 (2007)
29. Xu, G., Wang, G.Z.: Harmonic B-B surfaces over triangular domain. Journal of Computers 12, 2180–2185 (2006)
30. Xu, G.L., Zhang, Q.: G^2 surface modeling using minimal mean-curvature-variation flow. Computer-Aided Design 5, 342–351 (2007)
31. Xu, G.L.: Discrete Laplace-Beltrami operators and their convergence. Computer Aided Geometric Design 10, 767–784 (2004)
32. Xu, G.L.: Convergence analysis of a discretization scheme for Gaussian curvature over triangular surfaces. Computer Aided Geometric Design 2, 193–207 (2006)

Appendix: The Proof of Theorem 1

Theorem 1 can be proved from the isothermal condition and the linear independence of the power basis. The partial derivatives of the harmonic surface $r(u,v)$ in Lemma 3 has the following forms:

$$r_u(u,v) = 6aA_5^o + 3bA_5^e + 5cA_4^e + 10dA_4^o + 4eA_3^o + fA_3^e + 3gA_2^e - 6hA_2^o + 2iA_1^o + jA_1^e + k$$
$$r_v(u,v) = -6aA_5^e + 3bA_5^o + 5dA_4^e - 10cA_4^o + fA_3^o - 4eA_3^e - 3hA_2^e - 6gA_2^o + jA_1^o - 2iA_1^e + l$$

where $A_5^o = u^5 - 10u^3v^2 + 5uv^4, A_5^e = 5vu^4 - 10v^3u^2 + v^5, A_4^e = u^4 - 6u^2v^2 + v^4, A_4^o = 2u^3v - 2uv^3, A_3^o = u^3 - 3uv^2, A_3^e = 3u^2v - v^3, A_2^e = u^2 - v^2, A_2^o = uv, A_1^o = u, A_1^e = v$.

Hence, from $F = \langle \boldsymbol{r}_u, \boldsymbol{r}_v \rangle$, the term u^{10} in F is related with A_5^o, then we obtain $\boldsymbol{a} \cdot \boldsymbol{b} = 0$ from $F = 0$. The term $u^9 v$ is related with A_5^e and A_5^o, then we get $4\boldsymbol{a}^2 = \boldsymbol{b}^2$. Similarly, the other equations in (3) can be obtain from $F = 0$ and $E = G$.

It is noted that we obtain only two equation for the terms $u^i v^j$, $i + j = k$, $k = 0, 1, 2, \cdots 9, 10$. One is for the case of i is even, and the other one is for the case of i is odd. The equations derived from $F = 0$ are the same as the case of $E = G$ except for the equations $\boldsymbol{k}^2 = \boldsymbol{l}^2$ and $\boldsymbol{k} \cdot \boldsymbol{l} = 0$. Hence, the number of equations in system (3) is $2 \times 2 \times (6 - 1) + 2 = 22$. Thus, the proof is completed. □

Physically-Based Surface Texture Synthesis Using a Coupled Finite Element System

Chandrajit Bajaj[1], Yongjie Zhang[2], and Guoliang Xu[3]

[1] Department of Computer Sciences and Institute for Computational Engineering and Sciences, The University of Texas at Austin, USA
bajaj@cs.utexas.edu
[2] Department of Mechanical Engineering, Carnegie Mellon University, USA
jessicaz@andrew.cmu.edu
[3] Institute of Computational Mathematics, Academy of Mathematics and System Sciences, Chinese Academy of Sciences, China
xuguo@lsec.cc.ac.cn

Abstract. This paper describes a stable and robust finite element solver for physically-based texture synthesis over arbitrary manifold surfaces. Our approach solves the reaction-diffusion equation coupled with an anisotropic diffusion equation over surfaces, using a Galerkin based finite element method (FEM). This method avoids distortions and discontinuities often caused by traditional texture mapping techniques, especially for arbitrary manifold surfaces. Several varieties of textures are obtained by selecting different values of control parameters in the governing differential equations, and furthermore enhanced quality textures are generated by fairing out noise in input surface meshes.

Keywords: texture synthesis, finite element method, reaction-diffusion equation, anisotropic diffusion.

1 Introduction

Texture mapping is an essential technique to map textures onto computer models, but it usually introduces distortions or discontinuities. An alternate approach is to synthesize a texture directly on the surface of objects. The finite difference method (FDM) has been extended to generate patterns over surfaces, but the problem of distortions or discontinuities still exist, and the results can be often divergent if non-suitable time steps or incorrect initial conditions are chosen. Here we use the finite element method (FEM) for the stable and robust solution of reaction-diffusion partial differential equations (PDEs). These FEM solutions allow for smooth and distortion free textures to be directly synthesized and visualized on arbitrary surface manifolds.

However when the input surface meshes are noisy, the textures synthesized by FEM solutions are often equally noisy, and as shown in Figure 1. To correct this, we couple an anisotropic diffusion PDE with the reaction-diffusion PDE's to remove noise on the surface while preserving geometric features, and synthesizing

F. Chen and B. Jüttler (Eds.): GMP 2008, LNCS 4975, pp. 344–357, 2008.
© Springer-Verlag Berlin Heidelberg 2008

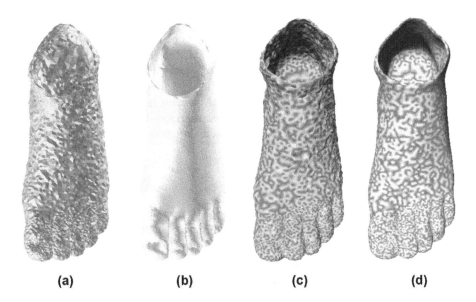

Fig. 1. Physically-based textures synthesized on the manifold foot surface. (a) - the noisy surface; (b) - anisotropic diffusion is applied on (a) to remove noise while preserving geometric features; (c) - reaction-diffusion is applied on (a) to synthesize textures; (d) - coupled reaction-diffusion and anisotropic diffusion are used to synthesize enhanced quality textures.

smooth textures at the same time. Enhanced and distortion-free textures are generated by using this coupled PDE system.

The main steps of our finite element texture synthesis approach are as follows:

– Variational formulation
– Discretize in the time domain
– Discretize in the spatial domain
– Refine finite elements and basis functions
– Construct and solve the resulting linear system

The finite element solution based on Galerkin discretization, use a variational approach to first generate a weak integral formulation, and then a discretization into a linear system.. For time-varying systems, the time domain is discretized using a semi-implicit backward Euler method. Recursive subdivision techniques are adopted to refine finite elements and basis functions. The mass matrix, the stiffness matrix, and the force vectors are constructed after evaluating each finite element. Finally textures are generated by constructing and solving a linear system.

Various textures can be synthesized by choosing different values of the parameters in the governing equations. In this paper, reaction-diffusion is used to form stable patterns such as spots and stripes when two or more chemicals diffuse at different rates and react with one another, and different coefficients in the

PDEs will influence the formation of textures. Anisotropic diffusion is coupled with reaction-diffusion equations to smooth the noise of the input surface mesh, and thereby toncrease the accuracy/quality of texture synthesis.

The remainder of this paper is organized as follows: Section 2 summarizes previous work related to texture synthesis, and finite element simulations; Section 3 explains the algorithm of Galerkin based finite element method solutions; Section 4 discusses the two coupled physical systems, reaction-diffusion and anisotropic diffusion; The final section presents our conclusion and future work.

2   Previous Work

Previous work on texture synthesis with high visual or numerical accuracy focuses on pattern generation and texture mapping, including statistical, non-parametric, as well as optimization-based techniques. In recent years, many texture synthesis techniques have been developed.

Reaction-diffusion: The reaction-diffusion equations were proposed as a model of biological pattern formation for texture synthesis. The traditional reaction-diffusion systems are extended by allowing anisotropic and spatially non-uniform diffusion, as well as multiple competing directions of diffusion [4]. There have been some attempts searching for different ways to generate textures on arbitrary surfaces based on FDM. Reaction-diffusion textures are generated to match the geometry of an arbitrary polyhedral surface by creating a mesh over a given surface and simulating the reaction-diffusion process directly on this mesh [5].

Texture Synthesis: During the past decade, many example-based texture synthesis methods have been proposed, including parametric methods [12,19,6,7], non-parametric methods [13,14,20,27,28], optimization-based methods [15,10], and appearance-space texture synthesis [16]. In order to synthesize textures over surfaces based on a given texture example, parametric methods attempt to construct a parametric model of the texture. Differently, non-parametric methods grows the texture one pixel/patch at a time. Optimization-based methods evolve the texture as a whole and improve the quality of the results. Besides texture synthesis on surfaces, various techniques have also been developed for solid texture synthesis [17,18].

Anisotropic Diffusion: The isotropic diffusion method can remove noise, but blurs features such as edges and corners. In order to preserve features during the process of noise smoothing, anisotropic diffusion [22] was proposed by introducing a diffusion tensor. Generally, a Gaussian filter is used to calculate the anisotropic diffusion tensor before smoothing, but it also blurs features. Bilateral filtering [23], a nonlinear filter combining domain and range filtering, was introduced to solve this problem. Anisotropic diffusion can be used for fairing out noise both in surface meshes and functions defined on the surface [1,24].

Simulation Using FEM: FEM has been used extensively in solving physically based problems. A finite element solver, CHARMS (conforming, hierarchical, adaptive refinement methods), constructs a framework for adaptive simulation by refining basis functions instead of refining elements [2]. An automated procedure [3] to generate a 3D finite element model of an individual patient's mandible with dental implants inserted was presented. Various methods of image processing, geometric modeling and finite element analysis were combined and extended. The deformation field between 3D images was computed by locally minimizing the sum of the squared differences between the images to be matched [11]. Nonlinear FEM using mass lumping was applied to produce a diagonal mass matrix that allows real time computation, and dynamic progressive meshes were generated to provide detailed information while minimizing computation [21].

3   Galerkin Based Finite Element Method

Textures can be generated by simulating physical phenomena, which are simplified into mathematical models represented by PDEs. Although FDM and its variants have been used extensively to solve PDEs, it is still challenging to synthesize textures directly over surfaces. FEM is a more stable and robust method in solving PDEs over arbitrary manifold surfaces.

Given PDEs with the required property parameters and the corresponding boundary and initial conditions, FEM tends to solve the weak form of these governing equations. The trial space and the test space are introduced. After the spatial and temporal discretization, each element is analyzed, and the recursive subdivision is used to refine the elements and basis functions. In the Galerkin method, the same format is adopted to construct the trial function and the test function. The variational formulations are rewritten by plugging the trial and test functions into the weak form, and are modified with the boundary conditions. In the end, a simplified linear system is built by uniting all the finite elements,

$$Kx = b \tag{1}$$

where x is the unknown vector. The matrix K and the vector b are calculated over the surface domain, which is discretized into small elements such as triangles. The trial and test functions are defined in the same format as a linear combination of basis functions, whose weights are elements of the unknown vector x. As a result, textures are generated by solving the linear system.

3.1   Variational Formulation

A generalized PDE over surface $\Omega \subset \mathbb{R}^3$ is shown in Equation (2), which can represent different physical phenomena by choosing corresponding variables and coefficients, such as the reaction-diffusion and the anisotropic diffusion.

$$\frac{\partial u}{\partial t} = C_0 \mathrm{div}(C_1 \nabla u) + C_2 u + C_3 \tag{2}$$

where div and ∇ are the divergence and the gradient operator over surface Ω (see [26] for their definitions), u represents different variables for various physical problems. u can be a scalar, for example, u is the concentration of a chemical in the reaction-diffusion equations. u can also be a vector in $\mathbb{R}^3$ such as the geometric position or function vectors at each vertex on Ω in the anisotropic diffusion equation. C_0, C_1, C_2, and C_3 are coefficients which can be functions of u, or just constants. C_1 could be a scalar or a 3×3 matrix.

Suppose u, ν are selected from the *trail space* and the *test space* respectively, the inner product of u and ν is defined as follows,

$$(u, \nu) = \int_\Omega u\nu dx. \tag{3}$$

By using the Divergence Theorem and the Partial Integration Rule, Equation (2) can be written in a variational form as in Equation (4),

$$\left(C_0^{-1}\frac{\partial u}{\partial t}, \nu\right) = -(C_1\nabla u, \nabla \nu) + (C_0^{-1}C_2 u, \nu)$$
$$+ (C_0^{-1}C_3, \nu). \tag{4}$$

The variational form is the starting point for the following spatial and temporal discretizations.

3.2   Discretization

For time-varying systems, the variational form needs to be discretized in the temporal space as well as in the spatial space. A *semi-implicit backward Euler method* is used for temporal discretization. For the spatial discretization, functions over the integration domain can be refined using the recursive subdivision schemes [1].

Spatial Discretization: The variational problem in Equation (4) is discretized in a function space which is defined by the limit of Loop's recursive subdivision. The function is locally parameterized in our finite element space, which is spanned by C^1 smooth quartic box spline basis functions.

Temporal Discretization: For time-varying systems, we have to discretize Equation (4) in the temporal space. Two issues need to be addressed for the temporal discretization: the choice of the time step, and the decision of which term needs to analyzed implicitly and which term needs to be handled explicitly. Here a *semi-implicit backward Euler discretization* is chosen,

$$(\frac{u^{n+1} - u^n}{C_0\tau}, \nu) = -(C_1\nabla u^{n+1}, \nabla \nu)$$
$$+ (C_0^{-1}C_2 u^{n+1}, \nu) + (C_0^{-1}C_3, \nu), \tag{5}$$

where τ is the time step, u at $t = (n+1)\tau$ is derived from u at $t = n\tau$. Equation (5) is rewritten as follows,

$$(C_0^{-1}u^{n+1}, \nu) + [(C_1\nabla u^{n+1}, \nabla \nu) - (C_0^{-1}C_2 u^{n+1}, \nu)]\tau$$
$$= (C_0^{-1}u^n, \nu) + \tau(C_0^{-1}C_3, \nu). \tag{6}$$

3.3   Element and Basis Refinement

For each vertex of a control mesh, its basis function is defined by the limit of the Loop's subdivision for the zero control values everywhere except at itself where it is one [1]. The recursive subdivision schemes are used to refine elements and basis functions, and smooth surfaces are generated via a limit procedure of an iterative refinement. Here, an approximating algorithm proposed by Loop [8] is adopted. C^2 limit surfaces are generated except for some extraordinary points at where C^1 continuity is achieved.

If all the vertices of a triangle have a valence of 6, then the triangle is called regular, otherwise it is irregular. A regular patch is controlled by twelve basis functions with explicit polynomial representations. For an irregular patch, the mesh needs to be subdivided repeatedly until the parameter values of interest are inside a regular one [9].

The basis functions are defined by the same recursive scheme. For each vertex of a control mesh, we associate it with a basis function, which is defined as the limit of the Loop's subdivision from control value one at the vertex and control value zero at every other vertices. Each triangle patch is defined locally by only a few related basis functions. Triangles can be grouped according to their vertex valences. All triangles with the same vertex valences have the same set of related basis functions, which only need to be calculated once. Therefore, the computation costs in the numerical integration can be reduced.

3.4   Linear System Construction

The Galerkin approximation is applied to our variational formulation. The trial and test functions are defined in the same format - a linear combination of basis functions,

$$u = \sum_{i=1}^{N} u_i \phi_i, \qquad \nu = \sum_{i=1}^{N} \nu_i \phi_i. \qquad (7)$$

Substitute Equation (7) into Equation (6), and rewrite it for each ν_j, we obtain

$$(M + \tau L)u^{n+1} = Mu^n + \tau F, \qquad (8)$$

where the mass matrix M, the stiffness matrix L, and the force vector F can be calculated as follows,

$$M_{ij} = (C_0^{-1}\phi_i, \phi_j), \qquad (9)$$
$$L_{ij} = (C_1\nabla\phi_i, \nabla\phi_j) - (C_0^{-1}C_2\phi_i, \phi_j), \qquad (10)$$
$$F_j = (C_0^{-1}C_3, \phi_j). \qquad (11)$$

The resulting linear system arising in each time step τ can be solved by a *preconditioned conjugate gradient method*. After each u_i ($i = 1, ..., N$) is calculated, we can obtain u by substituting them back into Equation (7).

4 A Coupled System of Reaction-Diffusion and Anisotropic Diffusion

As shown in Figure 2, our finite element solver can be applied to solve multiply finite element problems over arbitrary manifold surfaces when we choose different coefficients C_0, C_1, C_2, C_3 and u in Equation (2). Different systems are simulated using our solver, such as texture synthesis directly on surfaces by solving reaction-diffusion equations. The anisotropic diffusion is coupled to the reaction-diffusion equations in order to generate better quality textures by smoothing the integral domain.

Physical Problems	Equations	Coefficients in the generalized PDEs, Equation (2)			
		(C_0)	(C_1)	(C_2)	(C_3)
Reaction-Diffusion	Eqn.(12)	D_a	1	$-sb$	$16s$
Reaction-Diffusion	Eqn.(13)	D_b	1	$s(a-1)$	$-s\beta$
Reaction-Diffusion	Eqn.(16)	1	1	$(1+\Gamma cos\omega_f t)uv - d$	$c+\beta$
Reaction-Diffusion	Eqn.(17)	δ	1	$-u^2$	du
Anisotropic Diffusion	Eqn.(18)	$a(x)^\mu$	$a(x)^{1-\mu}D(x)$	0	0

Fig. 2. Coefficients in the generalized PDEs for different physical problems, reaction-diffusion and anisotropic diffusion

When we construct the linear system for each application system, we do not need to calculate each term since some coefficients are zero. There are a common inner product term (ϕ_i, ϕ_j) in the mass and stiffness matrices if $C_2 = constant$. This term only needs to be calculated once for each element. For our time-varying systems, we do not update each entry at each time step. In order to speed up the whole process, we take the value from the previous time step for some entries instead of recalculating all of them.

4.1 Reaction-Diffusion

As a biologically motivated method of texture synthesis, the reaction-diffusion is a process in which two or more chemicals diffuse and react with each other to form stable patterns. For a two chemical (a and b) reaction-diffusion system [5],

$$\frac{\partial a}{\partial t} = F(a,b) + D_a \Delta a, \tag{12}$$

$$\frac{\partial b}{\partial t} = G(a,b) + D_b \Delta b, \tag{13}$$

where $F(a,b) = s(16-ab)$, $G(a,b) = s(ab-b-\beta)$, a and b are concentrations, D_a and D_b are diffusion constants, β is the simulation random seed, s is a constant, and t is time.

In this paper, FEM is adopted to solve the PDEs. System (12-13) is weakened by using the divergence theorem and the partial integration, then both trial functions (a and b) and the test function (ν) are discretized in the same format. The reformulated variational form can be written as,

$$(a^{n+1}, \nu) + (s\tau b^n a^{n+1}, \nu) + (D_a \tau \nabla a^{n+1}, \nabla \nu)$$
$$= (a^n, \nu) + (16s\tau, \nu), \tag{14}$$
$$(b^{n+1}, \nu) - (s\tau a^n b^{n+1}, \nu) + (D_b \tau \nabla b^{n+1}, \nabla \nu)$$
$$+ (s\tau b^{n+1}, \nu) = (b^n, \nu) - (s\beta\tau, \nu), \tag{15}$$

where τ is the time step, and a^n is the concentration at time $t = n\tau$. Equation (14-15) can be rewritten into a linear system in Equation (8). The mass matrix, the stiffness matrix and the force vector of Equation (14) are as follows,

$$M_{ij} = (D_a^{-1}\phi_i, \phi_j),$$
$$L_{ij} = (\nabla\phi_i, \nabla\phi_j) - (sbD_a^{-1}\phi_i, \phi_j),$$
$$F_j = (16sD_a^{-1}, \phi_j).$$

Similarly, the mass matrix, the stiffness matrix and the force vector of Equation (15) are as follows,

$$M_{ij} = (D_b^{-1}\phi_i, \phi_j),$$
$$L_{ij} = (\nabla\phi_i, \nabla\phi_j) + (saD_b^{-1}\phi_i, \phi_j) - (sD_b^{-1}\phi_i, \phi_j),$$
$$F_j = -(s\beta D_b^{-1}, \phi_j).$$

Given the initial values $a = b = 4.0$, the concentrations of chemical a and b can be obtained at each time step iteratively. Specifically, for each temporal step, the system consisting of equations (14)-(15) is solved iteratively, until a and b achieve their stable states. Figure 3 and 4 show different patterns generated over a sphere surface. Spot and stripe patterns can be controlled by selecting different values of the parameters.

Additionally as shown in Figure 5, we used the same variational algorithm to solve another reaction-diffusion equations,

$$\frac{\partial u}{\partial t} = c + \beta - du + (1 + \Gamma \cos\omega_f t)u^2 v + \Delta u, \tag{16}$$

$$\frac{\partial v}{\partial t} = du - u^2 v + \delta \Delta v. \tag{17}$$

Figure 6 shows more textures generated by choosing different parameters in the physical phenomena of reaction-diffusion, Equation (12-13) or (16-17).

4.2 Anisotropic Diffusion

If $F(a, b) = 0$, then Equations (12) turns into an isotropic diffusion equation. Another kind of diffusion problem is anisotropic diffusion, which usually couples with the above reaction-diffusion system to generate better textures by smoothing noticeable size variances existing in the surface meshes. In order to remove those artifacts, an anisotropic diffusion tensor is introduced in the governing

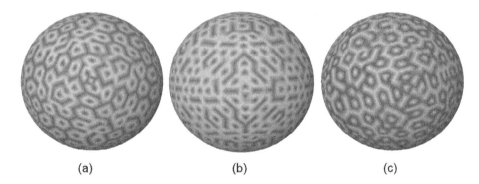

Fig. 3. Textures generated over sphere surfaces using Equation (12-13), where $F(a,b) = s(16 - ab)$, and $G(a,b) = s(ab - b - \beta)$. $\beta = 12.0+/-0.1$, $D_a = 0.125$, $D_b = 0.03125$. $s = 0.025$ in (a), $s = 0.0375$ in (b), and $s = 0.05$ in (c).

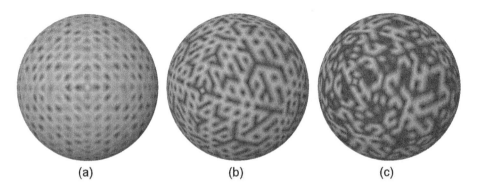

Fig. 4. Textures generated over sphere surfaces using Equation (12-13), where $F(a,b) = a - a^3 - b$, and $G(a,b) = s(a - s_1 b - s_0)$. $\beta = 12.0+/-0.1$, $D_a = 0.125$, $D_b = 0.03125$. $s = 0.025$ in (a), $s = 0.0375$ in (b), and $s = 0.05$ in (c).

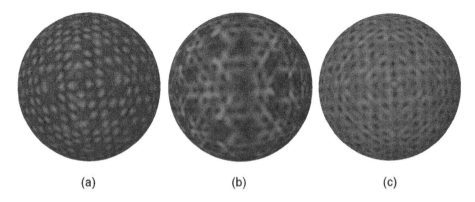

Fig. 5. Textures generated over sphere surfaces using Equations (16-17). $\beta = +/-0.1$, $c = 0.5$, $d = 1.5$, $\omega_f = 1.69$. $\delta = 0.0$ in (a), $\delta = 1.0$ in (b), and $\delta = 5.0$ in (c).

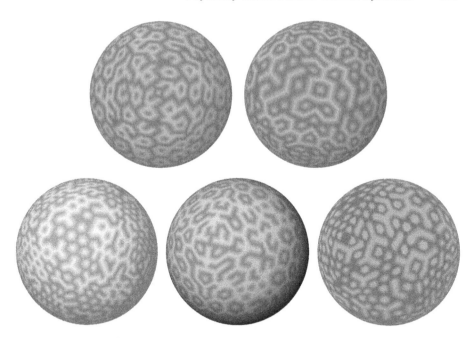

Fig. 6. More textures are generated over sphere surfaces by simulating the physical phenomia of reaction-diffusion

equation, and the Loop's subdivision techniques are combined with the diffusion model,

$$\frac{\partial x(t)}{\partial t} - a(x)^{\mu} \operatorname{div}[(a(x)^{1-\mu} D(x)\nabla x(t))] = 0, \tag{18}$$

where x is the geometric position vector of a point on the surface, t is time, $D(x)$ is a diffusion tensor to enhance sharp features, $a(x)$ is a smooth function which is adaptive to the mesh density, and $\mu \in [0, 1]$ is a parameter which changes the smoothing behavior of the equation. $a(x)$ and $D(x)$ are defined in [1]. Let k_1, k_2 be the two principal curvatures, and $e_1(x)$, $e_2(x)$ be the principal directions of the surface at point $x(t)$. Then the diffusion tensor is defined as

$$Dz = \alpha g(k_1)e_1(x) + \beta g(k)2)e_2(x) + N(x), \tag{19}$$

where $N(x)$ is the normal component of z, and $g(s)$ is

$$g(s) = \begin{cases} 1, & |s| \le \lambda, \\ (1 + \frac{(s-\lambda)^2}{\lambda^2})^{-1}, & |s| > \lambda, \end{cases} \tag{20}$$

λ is a given parameter which detects sharp features. After the spatial and time discretization of the weak form, Equation (18) can be rewritten as follows,

$$(a(x^n)^{-\mu}x^{n+1}, \nu) + \tau(a(x^n)^{1-\mu}D(x^n)\nabla x^{n+1}, \nabla \nu)$$
$$= (a(x^n)^{-\mu}x^n, \nu), \tag{21}$$

where ν is the test function, τ is the time step, and n is the step number. The mass matrix, the stiffness matrix and the force vector are as follow,

$$M_{ij} = (a(x)^{-\mu}\phi_i, \phi_j),$$
$$L_{ij} = (a(x)^{1-\mu}D(x)\nabla\phi_i, \nabla\phi_j),$$
$$F_j = 0.$$

Our finite element solver can smooth both geometric positions and function values of each vertex on the surface. Anisotropic diffusion is used to improve the quality of the synthesized textures by denoising the surface meshes while preserving geometric features. Figure 1(a-b) show one example.

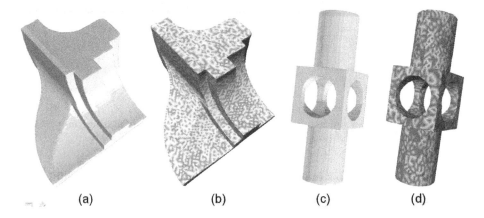

(a)   (b)   (c)   (d)

Fig. 7. Textures generated on isosurfaces with sharp feature preservation. (a) - the input surface model of the fandisk; (b) - the texture generated on (a); (c) - the input surface of the mechanical part; (d) - the texture generated on (c).

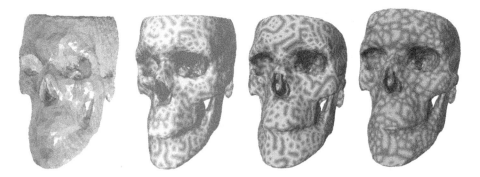

Fig. 8. Textures generated over the skull model by solving the reaction-diffusion governing equations, and the anisotropic diffusion equation is used to smooth noise in the surface mesh (the leftmost picture)

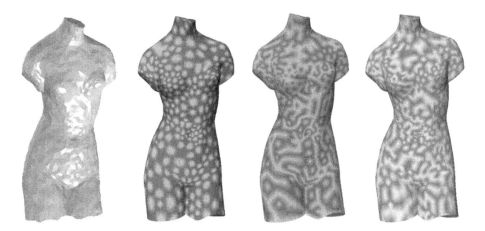

Fig. 9. Textures generated over the venus model by solving the reaction-diffusion governing equations, and the anisotropic diffusion equation is used to smooth noise in the surface mesh (the leftmost picture)

Fig. 10. Textures generated over the rabbit model by solving the reaction-diffusion governing equations, and the anisotropic diffusion equation is used to smooth noise in the surface mesh (the leftmost picture)

4.3   A Coupled System

Sometimes noise exists on the geometric surface, therefore in order to generate good quality textures on it, we need to remove the noise. Anisotropic diffusion is a good method for surface fairing because it enhances curve features on the surface by the careful choice of an anisotropic diffusion tensor. In our system, we couple the anisotropic diffusion with the reaction-diffusion together. In other words, the surface is smoothed at the same time when the texture is synthesized.

Figure 7 shows two examples with sharp features, the fandisk model and the mechanical part. The input mesh has some noises on the surface. After applying our coupled system, the surface is smoothed with sharp features preserved and good quality textures are generated as well. Figure 8, 9, and 10 show more results.

It is obvious that the noise on the input surface is removed with the preservation of geometric features, and good quality textures are synthesized over the skull, venus and bunny surfaces.

5   Conclusion and Future Work

We have described a stable and robust finite element solver over arbitrary manifold surfaces for a generalized PDE in the format of equation (2). This is used to simulate reaction-diffusion equations coupled with anisotropic diffusion for the synthesis of stable and continuous surface textures without distortions. Different control coefficients of the reaction-diffusion equations are used for the formation of different textures. Additionally, the anisotropic diffusion helps to generate better textures on surfaces by reducing noise in input surface meshes while preserving the surface's geometric features.

There are several directions for future work. Our finite element solvers also work for 3D domain, therefore in the near future we would like to solve the coupled texture synthesis system over a volumetric domain to generate 3D textures directly. Another problem is to tradeoff efficiency vs complexity of pattern generation. In other words, we would like to study if we can solve the PDEs approximately by not updating the matrix for a large number of time steps. Similarly we may not need a semi-implicit time discretization, but perhaps an approximate explicit time discretization to see if we can generate a wide range of patterns or not.

Acknowledgments. C. Bajaj was supported in part by NSF grants EIA-0325550, CNS-0540033, and NIH grants P20-RR020647, R01-GM074258, R01-GM073087 and R01-EB04873. G. Xu was supported in part by National Science Foundation of China grant (60773165), and National Key Basic Research Project of China (2004CB318000).

References

1. Bajaj, C., Xu, G.: Anisotropic Diffusion of Subdivision Surfaces and Functions on Surfaces. ACM Transactions on Graphics 22(1), 4–32 (2003)
2. Grinspun, E., Krysl, P., Schroder, P.: CHARMS: a simple framework for adaptive simulation. ACM SIGGRAPH, 281–290 (2002)
3. Futterling, S., Klein, R., Strasser, W., Weber, H.: Automated Finite Element of Human Mandible with Dental Implants. In: The Sixth International Conference in Central Europe on Computer Graphics and Visualization (1998)
4. Witkin, A., Kass, M.: Reaction-Diffusion Texture. ACM SIGGRAPH, 299–308 (1991)
5. Turk, G.: Generating Textures on Arbitrary Surfaces Using Reaction-Diffusion. ACM SIGGRAPH, 289–298 (1991)
6. Wei L., Levoy M.: Texture Synthesis Over Arbitrary Manifold Surfaces. ACM SIGGRAPH, 355–360 (2001)
7. Turk, G.: Texture Synthesis on Surfaces. ACM SIGGRAPH, 347–354 (2001)

8. Loop, C.: Smooth Subdivision Surfaces Based on Triangles. Master thesis. Technical Report, Department of Mathematics, University of Utah (1978)
9. Stam, J.: Evaluation Of Loop Subdivision Surfaces. ACM SIGGRAPH (1998)
10. Balmelli, L., Taubin, G., Bernardini, F.: Space-Optimized Texture Maps. Proceedings of Eurographics 21(3) (2002)
11. Ferrant, M., Warfield, S., Guttmann, C., Mulkern, R., Jolesz, F., Kikinis, R.: 3D Image Matching Using a Finite Element Based Elastic Deformation Model. In: Taylor, C., Colchester, A. (eds.) MICCAI 1999. LNCS, vol. 1679, pp. 202–209. Springer, Heidelberg (1999)
12. Heeger, D.J., Begen, J.R.: Pyramid-based Texture Analysis/Synthesis. ACM SIGGRAPH, 229–238 (1995)
13. De Bonet, J.S.: Multiresolution Sampling Procedure for Analysis and Synthesis of Texture Images. ACM SIGGRAPH, 361–368 (1997)
14. Wei, L.-Y., Levy, B.: Fast Texture Synthesis using Three-Structured Vector Quantization. ACM SIGGRAPH, 479–488 (2000)
15. Kwatra, V., Essa, I., Bobick, A., Kwatra, N.: Texture Optimization for Example-based Synthesis. ACM SIGGRAPH, 795–802 (2005)
16. Lefebvre, S., Hoppe, H.: Appearance-space Texture Synthesis. ACM SIGGRAPH, 541–548 (2006)
17. Qin, X., Yang, Y.-H.: Aura 3D Textures. IEEE Transactions on Visualization and Computer Graphics 13(2), 379–389 (2007)
18. Kopf, J., Fu, C.-W., Cohen-Or, D., Deussen, O., Lischinski, D., Wong, T.-T.: Solid Texture Synthesis from 2D Exemplars. ACM SIGGRAPH 26(3) (2007)
19. Levy, B., Mallet, J.-L.: Non-distorted Texture Mapping for Sheared Triangluated Meshes. ACM SIGGRAPH, 343–352 (1998)
20. Soler, C., Cani, M.: Angelidis A.: Hierarchical Pattern Mapping. ACM SIGGRAPH, 673–680 (2002)
21. Wu, X., Downes, M., Goktekin, T., Tendick, F.: Adaptive Nonlinear Finite Elements for Deformable Body Simulation Using Dynamic Progressive Meshes. In: Proceedings of Eurographics, vol. 20(3) (2001)
22. Weickert, J.: Anisotropic Diffusion in Image Processing. B. G. Teubner Stuttagart (1998)
23. Tomasi, C., Manduchi, R.: Bilateral Filtering for Gray and Color Images. IEEE Computer Vision, 839–846 (1998)
24. Clarenz, U., Diewald, U., Rumpf, M.: Nonlinear anisotropic diffusion in surface processing. IEEE Visualization Conference, 397–405 (2000)
25. Lin, A.L., Hagberg, A., Ardelea, A., Bertram, M., Swinney, H.L., Meron, E.: Four-Phase patterns in forced oscillatory systems. Phys. Rev. E. 62, 3790–3798 (2000)
26. Xu, G., Zhang, Q.: Construction of Geometric Partial Differential Equations in Computational Geometry. Mathematica Numerica Sinica 28(4), 337–356 (2006)
27. Tong, X., Zhang, J., Liu, L., Wang, X., Guo, B., Shum, H.-Y.: Synthesis of Bidirectional Texture Functions on Arbitrary Surfaces. ACM SIGGRAPH, 665–672 (2002)
28. Zhang, J., Zhou, K., Velho, L., Guo, B., Shum, H.-Y.: Synthesis of Progressively-Variant Textures on Arbitrary Surfaces. ACM SIGGRAPH, 295–302 (2003)

Planar Shape Matching and Feature Extraction Using Shape Profile

Yong-Jin Liu[1], Tao Chen[1],
Xiao-Yu Chen[2], Terry K. Chang[2], and Matthew M.F. Yuen[2]

[1] Tsinghua National Laboratory for Information Science and Technology,
Department of Computer Science and Technology,
Tsinghua University, Beijing, P.R. China
[2] The Hong Kong University of Science and Technology,
Hong Kong, P.R. China

Abstract. In this paper a novel cross correlation technique is proposed for shape matching between two similar objects. The proposed technique can not only evaluate the similarity between any two objects, but also has two distinct advantages compared to previous work: (1) the deformed articulated objects such as human being with different poses, can be matched very well; (2) the local feature extraction and correspondence can be established at the same time. The basic tool we used is the shape profile driven from the curvature map of the object profile. The cross correlation technique is applied to the shape profile of the two objects to evaluate their similarity. Filtering scheme is used to enhance the quality of both shape matching and extracted features. The invariant property, the robustness and the efficiency of the shape profile in shape matching and feature extraction are discussed.

Keywords: Feature extraction, shape matching, cross correlation.

1 Introduction

Shape matching is an important problem in image understanding and vision computing. The goal of this paper is to provide a robust technique to match two similar 2D profiles and simultaneously establish the feature point correspondence within a single unified framework.

Traditionally the shape matching process is classified into two levels: global and local shape matching. Global shape matching refers to the comparison of the overall shape of an object which deals with shape similarity. Local shape matching refers the matching of a pair of points in two similar shapes which deals with feature correspondence.

Common shape matching methods include shape distribution [4,12], the shape context method [2], and the curvature scale space method [9]. Ip et al. [4] applied the shape distribution method to meshed and faceted representations and used the membership classification distribution as the basis for a shape classification histogram. Osada et al. [12] used sampled distributions of simple shape functions, such as the distance between two random points on a surface, to check

F. Chen and B. Jüttler (Eds.): GMP 2008, LNCS 4975, pp. 358–369, 2008.
© Springer-Verlag Berlin Heidelberg 2008

shape similarity. These methods fail to cater for feature extraction from the object. Belgonie et al. [2] present the shape context method to represent each point on a object relative to all remaining points. This resulted in characteristics histograms used for shape comparison. This method gives reasonable good performance in shape matching and feature correspondence, but it suffers from high computational cost. The curvature scale space method of Mokhtatian [7,8] took advantage of connectivity between contour points and the shape of the object is represented by a graph of the parametric positions of the points of extreme curvature on the contour, however only curvature extreme points can be extracted.

The approach presented in this paper combines the conventional Fourier descriptor technique first developed by Zahn and Roskies [14] (see also [5,10]) and the advantage of the curvature scale space method [8,13]. The cross correlation technique is used to improve the shape matching procedure and is extended to cover the feature extraction process. An interesting related work in [6] for shape matching proposed to use integral invariant as the tool while in the proposed approach the differential invariants with filtering is applied. Both approaches can handle the noise robustly. The proposed approach is competitive by achieving both global and local shape matching in a single unified framework.

2   Shape Profile

In the shape profile approach, the shape ξ of the object segmented from the image is first represented by a finite set of its discrete profile points in the closed silhouette. This is achieved using common region-based segmentation followed by a simple edge linking algorithm [11] giving the silhouette profile $P = \{p(i) = (x(i), y(i))|i = 1, \cdots, n\}$, $p_i \in \mathbb{R}^2$ of n points. Normalized arc length parameterization on the profile P is adopted and the profile can be represented as $P = \{p(u)|u \in [0, 1]\}$, $p(u) \in \mathbb{R}^2$.

Given the object shape ξ, the shape profile is represented by the curvature $\kappa(u)$ along the contour:

$$\kappa(u) = \frac{\dot{x}(u) \times \ddot{y}(u) - \ddot{x}(u) \times \dot{y}(u)}{[\dot{x}(u)^2 + \dot{y}(u)^2]^{\frac{3}{2}}} \tag{1}$$

The curvature profile ψ along the contour of a particular object shape ξ is represented as $\psi(\xi) = \{\kappa(u)|u \in [0, 1]\}$. Due to the quantization error in the silhouette extraction process, the window-based method [1] is adopted to smooth out the noise; that is, the value of $\dot{x}(u)$ and $\ddot{x}(u)$ is each calculated by the neighboring points in the parametric space $x(u + w) - x(u - w)$ respectively where $2w$ is the width of the window.

To label the curvature of the contour, a signed curvature is adopted with positive value indicating convexity and vice versa. This eliminates the ambiguity of the curvature representation in view of its scalar property. The peaks in the curvature profile indicate the curvature extremities on the contour. Convex peaks are designated by the positive curvature value while the concave peaks

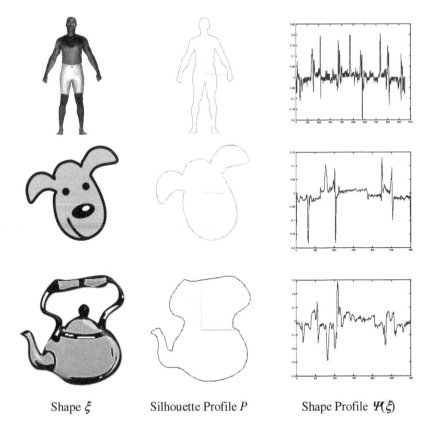

Shape ξ Silhouette Profile P Shape Profile $\Psi(\xi)$

Fig. 1. Object shape and curvature profile

are represented by negative peaks. Examples of the shape representation of the shape descriptor are shown in Fig. 1.

2.1   Shape Matching by Cross Correlation

An object, represented by the curvature profile or shape profile, is evaluated on shape similarity with the other object using the cross correlation technique. Let the two shapes each represented by the shape profile $\psi(\xi_1) = \{\kappa_1(u) | u \in [0,1]\}$ and $\psi(\xi_2) = \{\kappa_2(u) | u \in [0,1]\}$. The similarity between ξ_1 and ξ_2 is measured using the cross correlation coefficient defined by

$$C(\xi_1, \xi_2) = \frac{\sum_{u \in [0,1]}(\kappa_1(u) - \bar{\kappa}_1)(\kappa_2(u+v) - \bar{\kappa}_2)}{\sqrt{\sum_{u \in [0,1]}(\kappa_1(u) - \bar{\kappa}_1)^2}\sqrt{\sum_{u \in [0,1]}(\kappa_2(u+v) - \bar{\kappa}_2)^2}} \quad (2)$$

where $\bar{\kappa}$ is the average of the curvature κ, and v the phase shift between two shape profiles in the curvature parametric space representing the profile of the

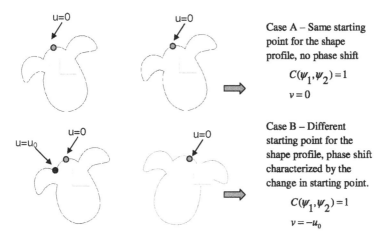

Fig. 2. Shape matching of the same profile – different starting point does not affect the cross correlation value between the shape profiles

two shapes. The higher the correlation, the higher the similarity is between the two shapes. The coefficient $C(\xi_1, \xi_2)$ is dependent on the phase shift v which is determined by searching the whole phase range to achieve the minimum of $C(\xi_1, \xi_2)$.

Given two shapes ξ_1 and ξ_2, ξ_1 is similar to ξ_2 if and only if ξ_1 can achieve a strong cross correlation with ξ_2 by a sequence of translation, scaling and rotation operations. Fig. 2 uses the same dog head shape of Fig. 1 to demonstrate that, the change in the location of the starting point for evaluating the shape profile of the same object does not affect the result of cross correlation value between the shape profiles. The maximum correlation value remains unity with the addition of a phase shift in the cross correlation output. As the profile is normalized to a range of 0 to 1, the phase shift simply reflects the relative shift of the starting point on the profile. This will apply to objects of the same class with a strong shape similarity.

In the example shown in Fig. 7, the shape profiles on two pairs of four similar shapes ξ_a, ξ_b, ξ_c and ξ_d, are adopted as samples for shape matching and the correlation coefficients are computed. Despite being male and female figures, the cross correlation between the front view pair and side view pair, $C(\xi_a, \xi_b)$ and $C(\xi_c, \xi_d)$, both show a distinct peak with a maximum cross correlation value of 0.6 at zero phase shift. The cross correlation value between the front view and side view, $C(\xi_b, \xi_c)$, does not exhibit a distinct peak and the maximum cross correlation has a much lower value of 0.15 and a phase shift of $v = -0.44$. This example illustrates the fundamental attractiveness of the method which utilizes a parametric space for shape matching eliminating the effect of size, position, and orientation.

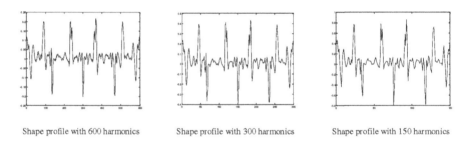

Shape profile with 600 harmonics   Shape profile with 300 harmonics   Shape profile with 150 harmonics

Fig. 3. Shape Profile with harmonics filtering

2.2   Noise Filtering in Shape Profile

Curvature is a second order quantity which is strongly affected by the outcome in profile segmentation from the images. There is likely to be noise in the profile. To improve on the result of the shape matching scheme, filtering of higher harmonics of the shape profile is introduced. Equations 3 and 4 respectively shows the discrete Fourier transform (DFT) and the inverse Fourier transform formulation.

$$X(\kappa) = \sum_{n=0}^{N-1} x(n)e^{-j2n\kappa\pi/N} \tag{3}$$

$$x(n) = \frac{1}{N} \sum_{\kappa=0}^{N-1} X(\kappa)e^{j2n\kappa\pi/N} \tag{4}$$

By using inverse DFT in equation (4), the shape profiles with different number of harmonics are constructed. Fig.3 shows the results for the filtered results of a typical shape profile for the front view human model. The result with 600 harmonics is the raw data after DFT, and the shape profiles with 300 and 150 harmonics are presented. It is observed that the overall shape of the shape profile remain relatively unaffected by the filtering. The shape profile retains the critical peaks after the filtering. The choice of the number of harmonics will affect the detection in curvature change but is not critical if it is above 150. This again depends on the class of objects under evaluation and the quality of the shape profile extraction program. The size or the length of the perimeter of the object is not a significant factor as the shape profile is always normalized to a range of 0 to 1.

3   Feature Extraction Using Feature Correspondence

In the current approach, the feature extraction process is formulated as a feature correspondence problem: Given two similar shapes ξ_1 and ξ_2, determine the "best" corresponding feature point $q_v \in \xi_2$ that matches a feature point $p_u \in \xi_1$. The feature extraction process is divided into two steps:

1. Shape profile for feature point extraction;
2. Feature point correspondence and extraction.

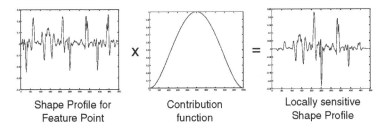

| Shape Profile for Feature Point | Contribution function | Locally sensitive Shape Profile |

Fig. 4. The effect of contribution function on shape profile

3.1 Shape Profile for Feature Point Extraction

Using the filtered template shape consisting of point $p_{u_0} \in \xi_1$ with parametric value u_0, the shape profile for feature point extraction is defined as:

$$\psi(p_{u_0}) = \{c(u)\kappa(u)|u \in [u_0, -w, u_0 + w], \kappa(u) \in \xi_1\} \tag{5}$$

where w is the half-width of the filtering window defined in section 2 and $c(u)$ is a contribution function. The feature point extraction from the shape profile ξ_1 depends on the curvature sequence within the filtering window. This is consistent with the observation that, curvature is influenced more by neighboring points than by remote points in the profile [2]. The contribution function is added to the shape profile in equation (5) to address the provide local sensitivity. The contribution function should satisfy the following properties. Firstly, the function $c(u)$ should act as a weighting function across the parametric domain $u_0 \pm w$ attaining a maximum value at the parametric value u_0. Secondly, $c(u)$ should gradually decrease to zero at the limits of the parametric domain $u_0 \pm w$, i.e. $c(u) = 0$ at $u = u_0 \pm w$. This makes the shape profile more sensitive to the region near u_0. Lastly the function should be symmetric about u_0, i.e. $c(u_0 + w) = c(u_0 - w)$ to provide an unbiased weighting before and after u_0.

Combining the requirements for the properties of the contribution function, we adopted the contribution function as follow:

$$c(u) = \begin{cases} \left\{ \left[\frac{u-u_0}{w}\right]^2 - 1 \right\}^2, & |u - u_0| < w \\ 0 & \text{otherwise} \end{cases} \tag{6}$$

By applying this contribution function, the shape profile becomes more sensitive to the neighboring points. The effect of the contribution function in improving the local sensitivity can be seen in Fig.4. In the examples, shown in Fig.5, the effect on the feature extraction with the contribution function is illustrated. The first column contains the feature point on the image template. The second column shows the extracted feature point without using the contribution function. The third column shows the extracted feature point using the contribution function. The improvement in the feature point extraction is obvious. It is also interesting to note that the contribution function is more effective when dealing with features located in a less distinct position with low curvature values.

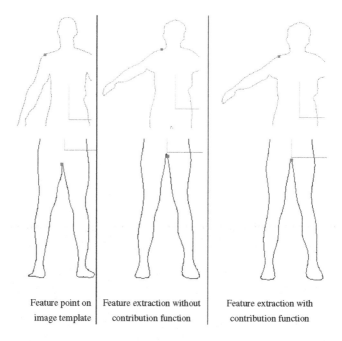

Feature point on | Feature extraction without | Feature extraction with
image template | contribution function | contribution function

Fig. 5. The effect of contribution function on feature extraction

3.2   Feature Point Correspondence and Extraction

In extracting the feature point q_v of the target shape ξ_2 corresponding to the feature point p_u on the template object, ξ_1, the shape profile $\psi(p_{u_0})$ of each feature point $p_u \in \xi_1$ is constructed by equation 5. Note that the shape profile is specific to the feature point due to the contribution function. The parametric value $v_0 = u_0 + \gamma$, where γ is the phase shift between the two shape profiles, ξ_1 and ξ_2, is adopted as the initial trial value. The shape profile of the initial trial point $q_{v_0} \in \xi_2$ is given as $\psi(q_{v_0}) = \{c(v)\kappa(v)|v \in [v_0 - w, v_0 + w], \kappa(v) \in \xi_2\}$.

To estimate the degree of similarity between two points $p_{u_0} \in \xi_1$ and $q_{v_0} \in \xi_2$, the cross correlation coefficient is computed. Similar to the shape similarity, the higher the value of the coefficient means better correspondence between the two points. As the contribution function is sample dependent, a neighborhood search is performed after the initial trial to obtain the best matched point. The neighborhood search evaluates the cross correlation of the feature point shape profile against the profile of the template. The best matched point is labeled as the matching feature point in target ξ_2 for the template feature point $p_{u_0} \in \xi_1$.

4   Properties of Shape Profile

Using the signed curvature-based representation and cross-correlation technique, the following properties are achieved.

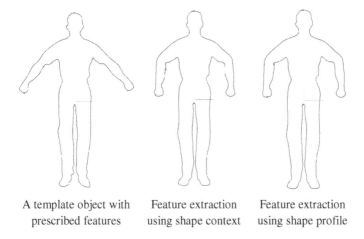

A template object with   Feature extraction   Feature extraction
prescribed features   using shape context   using shape profile

Fig. 6. Comparison of feature extraction using shape context and shape profile

4.1   Invariance

The shape profile represents the shape of an object as curvature parameterized along the outline contour. When considering the size effect, curvature is a local property which will not be affected by translation and rotation of the object, the descriptor is completely translation and rotation invariant and therefore provides size invariance. This property is important for the feature extraction from an image sequence in which translation and rotation of the object may be involved.

When considering the size effect, the curvature ratio is independent of the size when the scaling is uniform in all directions. In the formulation of cross correlation coefficient in equation (2), the value has already been normalized and therefore giving the size invariant property of the descriptor.

4.2   Robustness

Compared with curvature scale space, the shape profile additionally matches non-curvature-extreme points on the contour. Compared to the shape context method, the proposed descriptor is more robust in feature extraction for models exhibiting joint rotations. This is very important in matching articulated objects such as human beings in which the rotational joint movement always takes place.

To illustrate the robustness of our descriptor, feature extraction using both shape context and shape profile is performed on a human model exhibiting rotational joint movement. In the example shown in Fig.6, it is shown that both descriptors give the same performance in feature extraction for non-rotated parts such as the head point and crotch point. However, the shape context fails to extract feature points on the hand in which rotation occurs. Using the proposed shape profile, the problem can be solved. Therefore, the shape profile is more robust in feature extraction.

366   Y.-J. Liu et al.

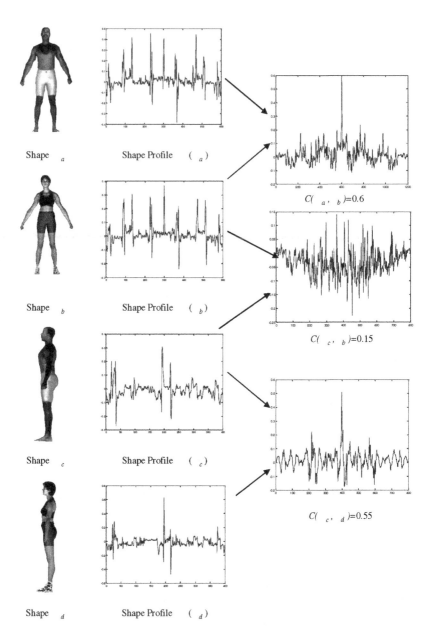

Fig. 7. Cross correlation of different shapes

4.3   Efficiency

When considering the efficiency of the descriptor, the shape matching could be carried out in $O(n^2)$ while the shape context performs in $O(n^3)$. To further speed up the matching process, one could reduce the resolution of the shape profile by using a smaller number of harmonics in the shape profile. In the above test, the silhouette consists of two thousand points. The matching is performed on a personal computer with 1.5GHz CPU. The time required for shape context is 5 seconds, while our descriptor completes the matching process in 1 second. This shows the speed improvement of our descriptor.

5   Experiments

The distinct feature of the proposed technique is to unify the global shape match and local feature correspondence in a single and efficient framework. More testing examples are shown in Fig. 8. The coefficient of the cross correlation (see

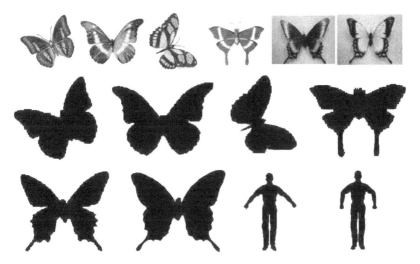

Fig. 8. More test examples of different shapes

	1.000000	0.935894	0.883248	0.570996	0.226203	0.385866	0.174578	0.218106
	0.935894	1.000000	0.879017	0.590056	0.285909	0.399657	0.265880	0.330852
	0.883248	0.879017	1.000000	0.595074	0.317536	0.514079	0.285271	0.348188
	0.570996	0.590056	0.595074	1.000000	0.868313	0.915651	0.716290	0.741140
	0.226203	0.285909	0.317536	0.868313	1.000000	0.957345	0.800244	0.790267
	0.385866	0.399657	0.514079	0.915651	0.957345	1.000000	0.773791	0.787230
	0.174578	0.265880	0.285271	0.716290	0.800244	0.773791	1.000000	0.985190
	0.218106	0.330852	0.348188	0.741140	0.790267	0.787230	0.985190	1.000000

Fig. 9. The similarity ranking between the examples in Fig. 8

Fig. 10. The feature correspondence by indices of two pairs of objects: one is for butterfly and the other is for human with different pose

eq.(2)) gives an overall ranking of the global shape matching as shown in Fig. 9. By applying the same framework with feature point correspondence extraction, the local features between similar objects are related by the same indices as demonstrated in Fig. 10. By using curvature profile, our technique is insensitive to the rigid motion between different subparts of the same object; this makes our method particularly suitable for shape matching and local feature correspondence of articulated objects.

6   Conclusions

The paper has presented a novel method for 2D shape matching and feature extraction based on the concept of curvature profile and the cross correlation technique. The method is similar to the curvature space method but utilizes the cross correlation technique to provide a unified method to enable global and local shape matching. This enables an efficient mapping of the feature points from the template to the segmented object in the target image. In shape matching, the phase shift in the cross correlation method can be used to rotationally aligned the object against the template and the cross correlation coefficient can be used to evaluate the shape similarity. By applying the method locally, the shape profile was shown to be able to map and extract the feature points from the profile at location where the curvature is not high. The invariance property, robustness, and efficiency of the method were discussed. In both methods noises can be effectively handled. Future work includes extension of the proposed cross correlation approach to hierarchical shape matching in the framework of [3].

Acknowledgments

The work was supported by the NSFC (Project Number 60603085), the 863 Program of China (Project Number 2007AA01Z336) and the 973 Program of China (Project Number 2006CB303102).

References

1. Baroni, M., Barletta, G., Fantini, A., Toso, A., Fantini, F.: Assessing LV wall motion by frame-to-frame curvature matching and optical flow estimation. In: IEEE Proceedings Computers in Cardiology, pp. 477–480 (1991)
2. Belongie, S., Malik, J., Puzicha, J.: Shape matching and object recognition using shape contexts. IEEE Transactions Pattern Analysis and Machine Intelligence 24(4), 509–522 (2002)
3. Gavrila, D.M.: A bayesian, exemplar-based approach to hierachical shape matching. IEEE Transactions Pattern Analysis and Machine Intelligence 29(8), 1408–1421 (2007)
4. Ip, C.Y., Lapadat, D., Sieger, L., Regli, W.C.: Using shape distributions to compare solid models. In: Proc. 7th ACM Symp. on Solid Modeling and Applications, pp. 273–280 (2002)
5. Lee, S.M., Abbott, A.L., Clark, N.A., Araman, P.A.: A shape representation for planar curves by shape signature harmonic embedding. In: Proc. CVPR 2006, pp. 1940–1947 (2006)
6. Manay, S., Hong, B.W., Yezzi, A., Soatto, S.: Integral invariant signatures. In: Proc. ECCV 2004, pp. 87–99 (2004)
7. Mohanna, F., Mokhtarian, F.: An Efficient Active Contour Model Through Curvature Scale Space Filterin. Multimedia Tools and Applications 21(3), 225–242 (2003)
8. Mokhtarian, F., Mackworth, A.K.: A theory of Mutliscale, Curvature-Based Shape Representation for Planar Curves. IEEE Trans. Pattern Anal and Machine Intell. 14(8), 789–804 (1992)
9. Mokhtarian, F., Suomela, R.: Curvature Scale Space for Image Point Feature Detection. In: Proc. International Conference on Image Processing and its Applications, Manchester, UK, pp. 206–210 (1999)
10. Persoon, E., Fu, K.S.: Shape discrimination using Fourier descriptors. IEEE Transactions on Systems, Man and Cybernetics 7(3), 170–179 (1977)
11. Pitas, I.: Digital Image Processing algorithms and applications. Wiley, Chichester (2000)
12. Osada, R., Funkhouser, T., Chazelle, B., Dobkin, D.: Shape Distributions. ACM Transactions on Graphics 21(4), 807–832 (2002)
13. Xiao, P., Barnes, N., Caetano, T., Lieby, P.: An mrf and gaussian curvature based shape representation for shape matching. In: Proc. CVPR 2007, (BMG Workshop) (2007)
14. Zahn, G.T., Roskies, R.Z.: Fourier descriptors for plane closed curves. IEEE Transactions on Computers C-21(3), 269–281 (1972)

Reconstructing a Mesh from a Point Cloud by Using a Moving Parabolic Approximation

Zhouwang Yang[1], Yeong-Hwa Seo[2], and Tae-Wan Kim[3]

[1] Department of Naval Architecture and Ocean Engineering,
Seoul National University, Seoul 151-744, Korea
[2] CAD Information Technology Laboratory, Seoul National University,
Seoul 151-744, Korea
[3] Department of Naval Architecture and Ocean Engineering, and Research Institute
of Marine Systems Engineering, Seoul National University, Seoul 151-744, Korea
taewan@snu.ac.kr
http://plaza.snu.ac.kr/~caditlab

Abstract. We use a moving parabolic approximation (MPA) to reconstruct a triangular mesh approximating the underlying surface of a point cloud. We suggest an efficient procedure to generate an initial mesh from a point cloud of closed shape. Then we refine this mesh selectively by comparing estimates of curvature from the point cloud with curvatures computed from the current mesh. We present several examples which demonstrate robustness of our method in the presence of noise, and show that the resulting reconstructions preserve geometric detail.

Keywords: point cloud, triangular mesh, refinement, mean curvature, subdivision, projection.

1 Introduction

A three-dimensional mesh made out of pieces of simple surface patches is a representation that is often used in engineering applications and computer graphics. The simplest example is a triangular mesh consisting of a set of triangles that meet at their shared edges. Many graphics software packages and hardware devices can operate efficiently on triangular meshes.

A problem that naturally arises in reverse engineering is this: given a finite sample $\mathcal{P} = \{\mathbf{p}_j\}_{j=1}^n \subset \mathbb{R}^3$ of an unknown surface $\mathcal{S}$, compute a mesh model $\mathcal{M}$. This model is referred to as the mesh reconstruction of $\mathcal{S}$ from $\mathcal{P}$. Typically, the desired form of $\mathcal{M}$ is a triangular mesh that can be directly used by downstream programs for further processing. In any case, the mesh reconstruction should well match the original surface in terms of geometrical and topological criteria [1,2]. The difficulty of achieving this purpose depends on the properties of the sample, as well as those of the sampled surface. Mesh reconstruction is an ill-posed problem since it has no unique solution: several triangular meshes may fulfill a particular set of criteria. A common hazard is the nonuniformity and noisiness of typical input data, which is usually the result of some physical measurement process such as scanning.

F. Chen and B. Jüttler (Eds.): GMP 2008, LNCS 4975, pp. 370–383, 2008.

The problem of mesh reconstruction from unorganized point clouds has received significant attention in the computational geometry and computer graphics communities. Various algorithms have been proposed for this task, which can be roughly classified into approaches based on Delaunay triangulation, implicit function and region-growing. Some surveys are available in the literature [3,4,2].

The Delaunay triangulation (tetrahedralization) of a set of scattered points is the dual of the Voronoi diagram [1] that decomposes $\mathbb{R}^3$ into convex polyhedra. Delaunay-based approaches establish topological connections between sample points and then filter out a subset of the resulting simplices to become the reconstructed mesh. Boissonnat [5] described a sculpting technique to remove tetrahedra from the Delaunay tetrahedralization so as to create a solid whose boundary is a surface mesh. Edelsbrunner and Mücke [6] introduced the notion of alpha-shapes which are polytopes derived from the Delaunay tetrahedralization of a given point-set with a controlling parameter alpha, and they then extracted the desired surface from these alpha-shapes. The power crust algorithm of Amenta et al. [7] produces a surface mesh as the interface between the inner and outer cells of a power diagram, which is defined by the union of balls located at the Voronoi poles. All these techniques come with theoretical guarantees of correct reconstruction if the sample meets certain conditions. However, they also have a high cost in terms of computation time and memory.

In essence, implicit function approaches reconstruct a mesh by computing an implicit function from the input data and extracting its isosurface using the Marching Cubes algorithm [8]. For instance, a signed distance to the point set or to the estimated tangent plane [9] was proposed as possible implicit functions. More recently, Poisson reconstruction and variants [10,11] are introduced for oriented or unoriented point sets where an implicit function f is derived by matching the gradients ∇f to the input or estimated normals $\mathbf{n}$. A disadvantage of these implicit approaches is that the mesh resolution cannot be adapted to local details of the surface, so that the resulting mesh must be further refined and optimized. This is especially necessary when the original point cloud is highly nonuniform.

Region-growing approaches start with a seed triangle (or edge) and incrementally extend the current mesh until the whole of the input data is covered. Bernardini et al. [12] designed the ball pivoting algorithm to generate an interpolating triangular mesh from a given unstructured point cloud. In this algorithm, a ball of user-specified radius is pivoted around an arbitrary edge of the current boundary until it touches another point in the point cloud, and then a new triangle is created from the edge and the point that was touched. Some modifications to the methods of forming new triangles during the growing process have been suggested [13,14]. Unfortunately, these region-growing algorithms are sensitive to noise, which is inevitably introduced in capturing point clouds from real objects.

Approaches to mesh reconstruction can be divided into two groups, depending on whether they interpolate the input data so that the vertices in the resulting mesh are selected from the original sample points, or whether the mesh is constructed from new points which approximate the input data. If the original data points are noisy, then an approximating mesh usually gives a more desired result than an interpolating mesh.

While a number of algorithms can now reconstruct meshes from point clouds, fewer methods are able to adapt given models well in curvature properties. In this paper, we propose a new method of reconstructing a triangular mesh to approximate a given point cloud. Starting with an initial mesh that roughly approximates the point cloud, we locate new mesh vertices closer to the underlying surface using the projection of moving parabolic approximation [15] which is robust in the presence of noise, and progressively refine the mesh on the basis of the mean curvature vector. Our algorithm produces a mesh with a resolution that is adapted to local details of the target shape.

The rest of the paper is organized as follows. In Section 2 we give an overview of our method of mesh construction. In Section 3, an efficient procedure is suggested to construct initial meshes from given point clouds. In Section 4, we present the curvature-based refinement of triangular meshes in detail. Examples and experimental results are reported in Section 5. Finally, we conclude this paper in Section 6.

2   Overview

Our method of mesh reconstruction from point clouds by the moving parabolic approximation has the following stages:

0. A rough initial mesh $\mathcal{M}^{(0)} = (V^{(0)}, E^{(0)})$ is specified, or constructed from a given point cloud $\mathcal{P} = \{\mathbf{p}_j\}_{j=1}^n \subset \mathbb{R}^3$, as described in Section 3. Let $V^{\text{New}} := V^{(0)}$ be the initial set of newly inserted vertices.
1. The following three steps of curvature-based refinement are repeatedly applied until the approximation error is within a predefined tolerance or the iteration counter reaches a maximal value:
   (a) Project each $\mathbf{v}^{\text{N}} \in V^{\text{New}}$ on to the underlying surface of the point cloud $\mathcal{P}$ using the MPA algorithm, and simultaneously obtain an estimate of the mean curvature vector $\mathbf{K}_{\mathcal{P}}(\mathbf{v})$ at $\mathbf{v} = \text{MPA}(\mathbf{v}^{\text{N}})$. After projection, the set of potential vertices is denoted by $V^{\text{Potential}} = \{\mathbf{v} = \text{MPA}(\mathbf{v}^{\text{N}}) \mid \forall \, \mathbf{v}^{\text{N}} \in V^{\text{New}}$ and $\|\mathbf{K}_{\mathcal{P}}(\mathbf{v})\| > \sigma\}$.
   (b) Calculate the mean curvature normal $\mathbf{K}_{\mathcal{M}}(\mathbf{v})$ using the differential geometry operator, and define $V^{\text{Active}} = \{\mathbf{v} \in V^{\text{Potential}} \mid \|\mathbf{K}_{\mathcal{M}}(\mathbf{v}) - \mathbf{K}_{\mathcal{P}}(\mathbf{v})\| > \varepsilon \|\mathbf{K}_{\mathcal{P}}(\mathbf{v})\|\}$ to be the collection of active vertices.
   (c) Insert a new vertex at the midpoint of every edge adjacent to any active vertex $\mathbf{v} \in V^{\text{Active}}$, and then recompute $V^{\text{New}} = \{\mathbf{v}^{\text{N}} = \frac{\mathbf{v}+\mathbf{v}_i}{2} \mid \forall \, \mathbf{v} \in V^{\text{Active}}$ and $(\mathbf{v}, \mathbf{v}_i) \in E\}$. The approximating mesh is updated by adding topological connections to the newly inserted vertices.
2. Output the resulting mesh $\mathcal{M} = (V, E)$ as the final approximation to the input point cloud $\mathcal{P}$.

3   Initial Mesh

In this section, we present an efficient procedure to construct rough meshes from point clouds which represent closed shapes. If the shape is open, the initial mesh must be provided with the point cloud as input.

We will denote a point cloud as $\mathcal{P} = \{\mathbf{p}_j = (x_{\mathbf{p}_j}, y_{\mathbf{p}_j}, z_{\mathbf{p}_j})^T\}_{j=1}^n$, and let

$$
\begin{aligned}
x_{\min} &= \min_{1 \le j \le n} x_{\mathbf{p}_j}, \; x_{\max} = \max_{1 \le j \le n} x_{\mathbf{p}_j}, \; x_{\text{cen}} = 0.5(x_{\min} + x_{\max}), \\
y_{\min} &= \min_{1 \le j \le n} y_{\mathbf{p}_j}, \; y_{\max} = \max_{1 \le j \le n} y_{\mathbf{p}_j}, \; y_{\text{cen}} = 0.5(y_{\min} + y_{\max}), \\
z_{\min} &= \min_{1 \le j \le n} z_{\mathbf{p}_j}, \; z_{\max} = \max_{1 \le j \le n} z_{\mathbf{p}_j}, \; z_{\text{cen}} = 0.5(z_{\min} + z_{\max}).
\end{aligned} \tag{1}
$$

We create a cube Ω that is slightly bigger than needed to enclose the whole point cloud. This cube is centered at $(x_{\text{cen}}, y_{\text{cen}}, z_{\text{cen}})^T$, and the length of its side is set to

$$
L = \frac{m}{m-3} \max\{(x_{\max} - x_{\min}), (y_{\max} - y_{\min}), (z_{\max} - z_{\min})\}, \tag{2}
$$

where $m > 3$ is an appropriate integer. By partitioning Ω uniformly along the x, y and z axes, we obtain a tensor-product grid $\mathcal{G} = \{\mathbf{g}_{i_x, i_y, i_z}\}_{i_x, i_y, i_z=0}^m$ and $m \times m \times m$ cubic cells $\{\mathcal{C}_{i_x, i_y, i_z}\}_{i_x, i_y, i_z=0}^{m-1}$ of size $\delta_L = L/m$. The intersection of the underlying surface of the point cloud with the partitioning of the cube is the interface that separates the exterior and interior cells. Combining this idea with the connectedness of the cells, we can propose an efficient strategy to identify the exterior cells, and hence to produce an initial mesh. For clarity of exposition, we define the following three sets:

- The set of exterior cells is denoted by
  $\Omega^E = \{\mathcal{C}_{i_x, i_y, i_z} \mid$ cell contains no data point and can be connected to infinity$\}$.
- The set of data cells is denoted by
  $\Omega^D = \{\mathcal{C}_{i_x, i_y, i_z} \mid$ cell contains at least one data point$\}$.
- The set of interior cells is denoted by Ω^I, and is the complement of $\Omega^E \cup \Omega^D$.

For a given point cloud of closed shape, it is apparent that all cells can be classified into one of these three sets. First we identify the set of data cells, Ω^D, by calculating the index

$$
\left(i_x = \left\lfloor \frac{x_{\mathbf{p}_j} - x_{\text{cen}} + 0.5L}{\delta_L} \right\rfloor, i_y = \left\lfloor \frac{y_{\mathbf{p}_j} - y_{\text{cen}} + 0.5L}{\delta_L} \right\rfloor, i_z = \left\lfloor \frac{z_{\mathbf{p}_j} - z_{\text{cen}} + 0.5L}{\delta_L} \right\rfloor \right) \tag{3}
$$

of the cell which contains a data point $\mathbf{p}_j$. Then we can identify the set of exterior cells, Ω^E, by an expansion algorithm that traces the direct neighbors of a cell $\mathcal{C}_{i_x, i_y, i_z}$, which are $\{\mathcal{C}_{i_x \pm 1, i_y, i_z}, \mathcal{C}_{i_x, i_y \pm 1, i_z}, \mathcal{C}_{i_x, i_y, i_z \pm 1}\}$. This expansion algorithm

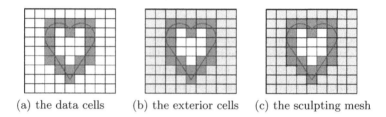

(a) the data cells   (b) the exterior cells   (c) the sculpting mesh

Fig. 1. Extraction of a sculpting mesh from a point cloud in 2D

starts with a seed in Ω^E. For instance, the corner cell $\mathcal{C}_{0,0,0}$ is known to be an exterior cell from Equation (2). As each new cell is added to Ω^E, we mark all its direct neighbors that contain no data point, and add them to Ω^E. The expansion stops when all unmarked direct neighbors of the expanding boundary are data cells. Ultimately, the remaining cells in the complement of $\Omega^E \cup \Omega^D$ will form the set of interior cells, Ω^I.

After having identified Ω^E, Ω^D and Ω^I, we can easily extract a sculpting mesh, which is the interface between Ω^E (the exterior cells) and Ω^D (the data cells). Figure 1 illustrates the analogous procedure in 2D. Let $\mathcal{M}^{(S)} = (V^{(S)}, E^{(S)})$ be a sculpting mesh in 3D. While preserving the connectivity information, we can tighten the mesh using a hybrid energy model

$$\min h(V^{(0)}) = \sum_i \frac{\|\mathbf{v}_i^{(0)} - \mathbf{P}_{j(i)}\|^2}{1 + \|\mathbf{v}_i^{(S)} - \mathbf{P}_{j(i)}\|^2} + \sum_{(i,l) \in E^{(S)}} \|\mathbf{v}_i^{(0)} - \mathbf{v}_l^{(0)}\|^2, \qquad (4)$$

where the position of each vertex $\mathbf{v}_i^{(0)} \in V^{(0)}$ is unknown, and $\mathbf{p}_{j(i)} \in \mathcal{P}$ is the nearest point to $\mathbf{v}_i^{(S)} \in V^{(S)}$ of the sculpting mesh. This minimization leads to a system of linear equations, which we can solve to obtain an initial mesh $\mathcal{M}^{(0)} = (V^{(0)}, E^{(0)})$ whose topological connections are the same as those of the sculpting mesh, i.e., $E^{(0)} = E^{(S)}$. Two examples of the construction of initial meshes in 3D are given in Figures 2 and 3.

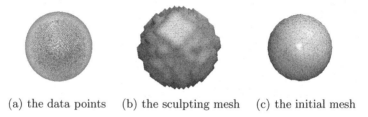

(a) the data points   (b) the sculpting mesh   (c) the initial mesh

Fig. 2. Construction of an initial mesh for the Sphere model

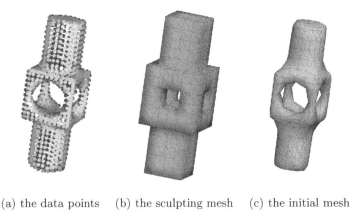

(a) the data points   (b) the sculpting mesh   (c) the initial mesh

Fig. 3. Construction of an initial mesh for the Mechpart model

4   Curvature-Based Refinement

The mean curvature is adopted in our refinement since its distribution of high magnitude indicates the region where sharp features appear. After obtaining an initial mesh with correct topology, we aim to refine the approximation based on mean curvature quantities. We will now give more detail about the curvature-based refinement of an approximating mesh to fit a given point cloud. This involves the steps of MPA projection, mean curvature normal calculation and adaptive subdivision.

4.1   MPA Projection

The main idea of moving parabolic approximation [15] is to locally fit a point cloud by an osculating paraboloid, and then recover the differential properties of its underlying surface. Suppose that $\mathcal{S}$ is the underlying surface of a point cloud $\mathcal{P} = \{\mathbf{p}_j\}_{j=1}^{n}$ in 3D. We project each newly inserted vertex on to the underlying surface $\mathcal{S}$ using the MPA algorithm, and simultaneously obtain the mean curvature vector at the projected point.

Let V^{New} be the set of newly inserted vertices in the current approximating mesh. Generally, $\mathbf{v}^{\mathrm{N}} \in V^{\mathrm{New}}$ will not lie on the underlying surface $\mathcal{S}$. Let the foot-point of $\mathbf{v}^{\mathrm{N}}$ on the underlying surface be denoted as

$$\mathbf{v} = \mathbf{v}^{\mathrm{N}} + \zeta \mathbf{n}, \tag{5}$$

where $\mathbf{n}$ is the unit normal to $\mathcal{S}$, and ζ is the signed distance from $\mathbf{v}^{\mathrm{N}}$ to $\mathbf{v}$ along $\mathbf{n}$. We aim to compute the foot-point $\mathbf{v}$ and the differential quantities (especially the mean curvature vector) at the foot-point. Let $\{\mathbf{t}_1(\mathbf{n}), \mathbf{t}_2(\mathbf{n})\}$ be the perpendicular unit basis vectors of the tangent plane, so that $\{\mathbf{v}; \mathbf{t}_1, \mathbf{t}_2, \mathbf{n}\}$

forms a local orthogonal coordinate system. Writing $\mathbf{q}_j = \mathbf{p}_j - \mathbf{v}^N$, we can formulate the MPA model as a constrained optimization:

$$\min f(\mathbf{n}, \zeta, a, b, c) =$$
$$\sum_j \left[\mathbf{q}_j^T \mathbf{n} - \zeta - \tfrac{1}{2} \left(a(\mathbf{q}_j^T \mathbf{t}_1)^2 + 2b(\mathbf{q}_j^T \mathbf{t}_1)(\mathbf{q}_j^T \mathbf{t}_2) + c(\mathbf{q}_j^T \mathbf{t}_2)^2 \right) \right]^2 e^{-\frac{\|\mathbf{q}_j - \zeta \mathbf{n}\|^2}{\rho^2}} \qquad (6)$$
$$\text{s.t. } \mathbf{n}^T \mathbf{n} - 1 = 0,$$

where $(\mathbf{n}, \zeta, a, b, c)$ are decision variables and ρ is a scale parameter.

From the optimum solution $(\mathbf{n}^*, \zeta^*, a^*, b^*, c^*)$ of the MPA model, we get the mean curvature vector of the underlying surface

$$\mathbf{K}_{\mathcal{P}}(\mathbf{v}) = \frac{1}{2}(a^* + c^*)\mathbf{n}^*, \qquad (7)$$

at the foot-point

$$\mathbf{v} = \text{MPA}(\mathbf{v}^N) = \mathbf{v}^N + \zeta^* \mathbf{n}^*. \qquad (8)$$

After projection of all the newly inserted vertices, we denote

$$V^{\text{Potential}} = \{ \mathbf{v} = \text{MPA}(\mathbf{v}^N) \mid \forall\, \mathbf{v}^N \in V^{\text{New}} \text{ and } \|\mathbf{K}_{\mathcal{P}}(\mathbf{v})\| > \sigma \}, \qquad (9)$$

as the set of potential vertices where the underlying shape should visibly curve. For this reason, we set $\sigma = 0.5/L$, so that it depends on the size of the space occupied by the whole cloud.

4.2   Mean Curvature Normal

We now use discrete differential geometry operators [16] to calculate the mean curvature normal on the triangular mesh. Let $N_1(\mathbf{v})$ be the index set of 1-ring neighbors of a vertex $\mathbf{v}$. The mean curvature normal is expressed as

$$\mathbf{K}_{\mathcal{M}}(\mathbf{v}) = \frac{1}{2\mathcal{A}_{\text{Mixed}}} \sum_{i \in N_1(\mathbf{v})} \left[\cot \beta_i (\mathbf{v} - \mathbf{v}_{i+1}) + \cot \gamma_i (\mathbf{v} - \mathbf{v}_i) \right], \qquad (10)$$

where β_i and γ_i are the angles subtended by two sides of the triangle $\triangle_i = \triangle \mathbf{v}\mathbf{v}_i\mathbf{v}_{i+1}$, as depicted in Figure 4, and $\mathcal{A}_{\text{Mixed}}$ denotes a mixed area around the vertex $\mathbf{v}$.

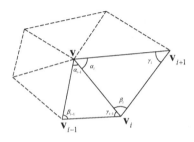

Fig. 4. The 1-ring neighbors of a vertex

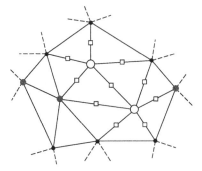

Fig. 5. New vertices inserted on the edges adjacent to active vertices

At each $\mathbf{v} \in V^{\text{Potential}}$, we use Equation (10) to compute the mean curvature normal $\mathbf{K}_{\mathcal{M}}(\mathbf{v})$ from the approximating mesh, and then compare it with an estimate of the mean curvature vector $\mathbf{K}_{\mathcal{P}}(\mathbf{v})$ obtained from the point cloud. For a specified ε, we define

$$V^{\text{Active}} = \{\mathbf{v} \in V^{\text{Potential}} \mid \|\mathbf{K}_{\mathcal{M}}(\mathbf{v}) - \mathbf{K}_{\mathcal{P}}(\mathbf{v})\| > \varepsilon\|\mathbf{K}_{\mathcal{P}}(\mathbf{v})\|\} \qquad (11)$$

to be a collection of active vertices in whose vicinity the current mesh needs further refinement.

4.3   Adaptive Subdivision

Using the curvature information, the collection of active vertices, V^{Active}, can be obtained from Equation (11). To achieve a high-quality approximation of the point cloud which preserves geometric details, we adaptively subdivide the current mesh by inserting a new vertex at the midpoint of every edge adjacent to each $\mathbf{v} \in V^{\text{Active}}$. This process is depicted in Figure 5, where a hollow circle '◯' denotes an active vertex and a hollow box '□' denotes a newly inserted vertex. Subsequently, we recreate the set of newly inserted vertices:

$$V^{\text{New}} = \{\mathbf{v}^{\text{N}} = \frac{\mathbf{v} + \mathbf{v}_i}{2} \mid \forall \, \mathbf{v} \in V^{\text{Active}} \text{ and } (\mathbf{v}, \mathbf{v}_i) \in E\}. \qquad (12)$$

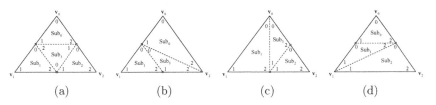

Fig. 6. Topological connections of triangles created by adaptive subdivision

Before going to the next iteration of curvature-based refinement, we update the approximating mesh by adding the topological connections to these new vertices. Figure 6 shows how the new triangles are created by our adaptive subdivision.

5   Implementation and Examples

We will now describe the implementation of our algorithm and test its performance on different models of point clouds.

Table 1. Performance of our algorithm on different point cloud models

Model	Number of Points	Resolutions of Sculpting Mesh	Iterations of Refinement	Number of Triangles	Computation Time (secs)
Knot	28659	38	1	9700	14.2
Bunny	35947	48	4	18266	39.1
Horse	48485	43	2	14566	37.1
Venus	134345	33	4	15848	25.4
Armadillo	172974	78	2	32038	87.1
Dragon	437645	118	1	73798	143.8

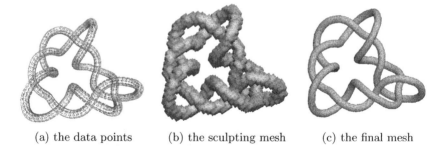

(a) the data points   (b) the sculpting mesh   (c) the final mesh

Fig. 7. Reconstruction of the Knot model

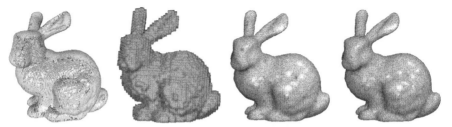

(a) the data points (b) the sculpting mesh (c) the initial mesh (d) the final mesh

Fig. 8. Reconstruction of the Bunny model

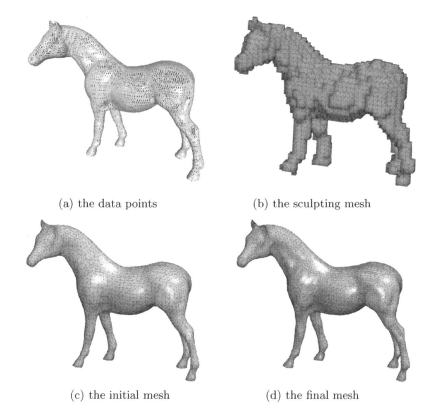

(a) the data points

(b) the sculpting mesh

(c) the initial mesh

(d) the final mesh

Fig. 9. Reconstruction of the Horse model

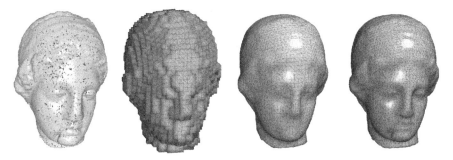

(a) the data points (b) the sculpting mesh (c) the initial mesh (d) the final mesh

Fig. 10. Reconstruction of the Venus model

We implemented our mesh reconstruction algorithm using Visual C++ 6.0 under Microsoft Windows, running on a PC with an Intel Pentium 3.0GHz processor and 1.0Gb of memory. For MPA projection, we gave an efficient implementation with the ANN library, which supports algorithms and data structures

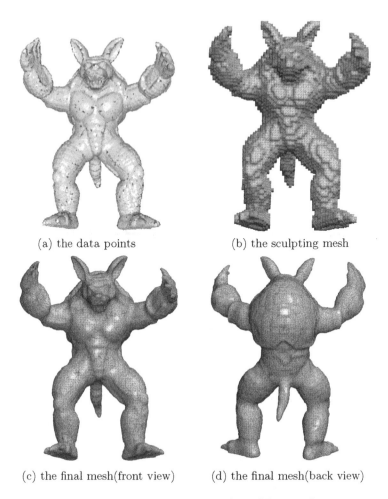

(a) the data points

(b) the sculpting mesh

(c) the final mesh(front view)

(d) the final mesh(back view)

Fig. 11. Reconstruction of the Armadillo model

based on kd-trees for searching nearest neighbors in the point cloud. We used a half-edge data structure [17] to maintain the mesh entities (vertices, edges, faces) and their connection information. Circulating around a vertex to obtain its 1-ring neighbors is a frequent operation in our curvature-based mesh refinement, and the half-edge structure allows this functionality to be provided in constant time without conditional branching [18].

We dynamically choose the minimum radius of a sphere that contains k nearest points to $\mathbf{v}^N$ as the scale parameter ρ in Equation (6), and set $k = 20$ to ensure that only points within the k-neighborhood contribute noticeably to the MPA model. In our experiments, we set $\varepsilon = 0.2$ for the relative tolerance of mean curvatures in Equation (11), which is used to detect the active vertices in each iteration.

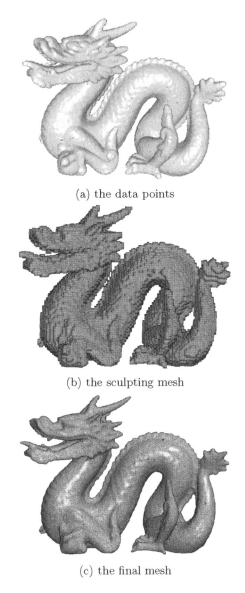

(a) the data points

(b) the sculpting mesh

(c) the final mesh

Fig. 12. Reconstruction of the Dragon model

We tested our algorithm on models with different geometric features, containing between 28659 to 437645 points, as set out in the second column of Table 1. The models consist of unorganized point clouds and exhibit noise, nonuniform density, sharp features and visible holes. As shown in the third column of Table 1, the different grid resolutions corresponding to m in Equation (2) are used for creating the sculpting meshes that are sufficiently coarse to fill holes in the point clouds. The fourth and fifth columns of Table 1 give the number of iterations required for mesh refinement and the number of triangles in the final meshes. The total computation time for each model is shown in the last column of Table 1. Figures 7 to 12 show the meshes reconstructed from given point clouds using our algorithm.

6   Conclusions

We have shown how to construct a triangular mesh approximating the underlying surface of a given point cloud. Our algorithm is based on the MPA projection and works robustly in the presence of noise, while yielding detailed mesh reconstructions. The effectiveness of the algorithm has been demonstrated in the reconstruction of meshes for a range of models with different complexities and features.

There are several areas for future work. We would like to look at the use of an adaptive grid in constructing the initial meshes. It is also clear that the performance of our algorithm could be improved by further optimization of aspects such as the strategy for choosing active vertices and the criterion used in adaptive subdivision. Finally, we would like to develop a method of constructing initial meshes from point clouds of open shape, which would allow us to mesh all clouds without providing initial meshes.

Acknowledgement. This work was supported by grant No. R01-2005-000-11257-0 from the Basic Research Program of the Korea Science and Engineering Foundation, and in part by Seoul R&BD Program.

References

1. Edelsbrunner, H.: Geometry and Topology for Mesh Generation. Cambridge University Press, Cambridge (2001)
2. Boissonnat, J.-D., Teillaud, M. (eds.): Effective Computational Geometry for Curves and Surfaces. Mathematics and Visualization Series. Springer, Heidelberg (2006)
3. Mencl, R., Müller, H.: Interpolation and approximation of surfaces from three-dimensional scattered data points. In: Proceedings of Eurographics (1998) State of the Art Report
4. Kobbelt, L., Bischoff, S., Botsch, M., Kähler, K., Rössl, C., Schneider, R., Vorsatz, J.: Geometric modeling based on polygonal meshes. In: Proceedings of Eurographics 2000, Course Notes (2000)
5. Boissonnat, J.-D.: Geometric structures for three-dimensional shape representation. ACM Transactions on Graphics 3(4), 266–286 (1984)
6. Edelsbrunner, H., Mücke, E.P.: Three-dimensional alpha shapes. ACM Transactions on Graphics 13(1), 43–72 (1994)
7. Amenta, N., Choi, S., Kolluri, R.K.: The power crust. In: Proceedings of the Sixth ACM Symposium on Solid Modeling and Applications, pp. 249–260 (2001)
8. Lorensen, W.E., Cline, H.E.: Marching cubes: A high resolution 3D surface construction algorithm. Computer Graphics 21(4), 163–169 (1987)

9. Hoppe, H., DeRose, T., Duchamp, T., McDonald, J., Stuetzle, W.: Surface reconstruction from unorganized points. In: Proceedings of ACM Siggraph 1992, pp. 71–78 (1992)

10. Kazhdan, M., Bolitho, M., Hoppe, H.: Poisson surface reconstruction. In: SGP 2006: Proceedings of the Fourth Eurographics Symposium on Geometry Processing, pp. 61–70 (2006)

11. Alliez, P., Cohen-Steiner, D., Tong, Y., Desbrun, M.: Voronoi-based variational reconstruction of unoriented point sets. In: SGP 2007: Proceedings of the Fifth Eurographics Symposium on Geometry Processing, Aire-la-Ville, Switzerland, Eurographics Association, pp. 39–48 (2007)

12. Bernardini, F., Mittleman, J., Rushmeier, H., Silva, C., Taubin, G.: The ball-pivoting algorithm for surface reconstruction. IEEE Transactions on Visualization and Computer Graphics 5(4), 349–359 (1999)

13. Huang, J., Menq, C.H.: Combinatorial manifold mesh reconstruction and optimization from unorganized points with arbitrary topology. Computer-Aided Design 34(2), 149–165 (2002)

14. Lin, H.W., Tai, C.L., Wang, G.J.: A mesh reconstruction algorithm driven by an intrinsic property of a point cloud. Computer-Aided Design 36(1), 1–9 (2004)

15. Yang, Z., Kim, T.-W.: Moving parabolic approximation of point clouds. Computer-Aided Design 39(12), 1091–1112 (2007)

16. Meyer, M., Desbrun, M., Schroder, P., Barr, A.H.: Discrete differential-geometry operators for triangulated 2-manifolds. In: Hege, H.C., Polthier, K. (eds.) Visualization and Mathematics, vol. III, pp. 35–57. Springer, Heidelberg (2003)

17. Cgal: Computational Geometry Algorithms Library (2007), http://www.cgal.org

18. Kettner, L.: Using generic programming for designing a data structure for polyhedral surfaces. Computational Geometry: Theory and Applications 13(1), 65–90 (1999)

A Revisit to Least Squares Orthogonal Distance Fitting of Parametric Curves and Surfaces

Yang Liu and Wenping Wang

Dept. of Computer Science, The University of Hong Kong,
Pokfulam Road, Hong Kong SAR, P.R. China
{yliu,wenping}@cs.hku.hk

Abstract. Fitting of data points by parametric curves and surfaces is demanded in many scientific fields. In this paper we review and analyze existing least squares orthogonal distance fitting techniques in a general numerical optimization framework. Two new geometric variant methods (GTDM and CDM) are proposed. The geometric meanings of existing and modified optimization methods are also revealed.

Keywords: orthogonal distance fitting, parametric curve and surface fitting, non-linear least squares, numerical optimization.

1 Introduction

Effective and accurate curve/surface fitting plays an important role and serves as a basic module in CAGD, computer graphics, computer vision and other scientific and engineering fields. We consider a common problem which occurs often in practical applications: fit a parametric curve/surface $\mathbf{C}(\mathbf{P};\mathbf{t})$ (whose parametric form is known but the parameter values are to be determined) to a set of given data points $\{\mathbf{X}_j\}_{j=1}^n \subset \mathbb{R}^s$. Here $\mathbf{P}$ are the shape parameters and $\mathbf{t} = (\tau_1, \ldots, \tau_m)$ are location parameters (For instance, $\mathbf{t}$ of a 3D parametric surface $\mathbf{C}(u,v)$ is (u,v)). This problem is usually stated as a standard nonlinear least squares problem:

$$\min_{\mathbf{P},\mathbf{t}_1,\ldots,\mathbf{t}_n} \sum_{j=1}^n \|\mathbf{C}(\mathbf{P};\mathbf{t}_j) - \mathbf{X}_j\|^2 \tag{1}$$

Where $\mathbf{t}_j$ is associated with the data point $\mathbf{X}_j$.

There exists vast literature about this problem in mathematics, statistics and computer science. Despite the differences of existing methods in variant contexts, The basics of most methods are the classical optimization theory and the optimization techniques such as decent methods and Gauss-Newton methods [1] appear in different forms.

First we introduce the traditional way of fitting a parametric curve/surface to a given data set in CAGD [2] [3] [4]. The first step is the parametrization which associates the location parameter $\mathbf{t}_j$ to each data point $\mathbf{X}_j$. After substituting $\mathbf{t}_j$ into (1), the second step is solving a linear least squares problem if the shape parameters occur in linear form; for instance, $\mathbf{P}$ are the control points of the B-spline curve/surface. By executing these two steps iteratively, improved location parameters and shape parameters are obtained.

F. Chen and B. Jüttler (Eds.): GMP 2008, LNCS 4975, pp. 384–397, 2008.
© Springer-Verlag Berlin Heidelberg 2008

This approach has been widely used because of its simplicity. However its convergence rate is slow and is proven to be linear [5]. On the other hand, without separating $\mathbf{P}$ and $\mathbf{t}_1,\ldots,\mathbf{t}_n$, the general optimization techniques of course can be applied. One can optimize $\mathbf{P},\mathbf{t}_1,\ldots,\mathbf{t}_n$ simultaneously [6] [7]. Moreover if $\mathbf{P}$ are in the linear form, the separable nonlinear least squares method (variable projection) can be employed and is better than the simultaneous method [5] [8] [9]. But the size of corresponding nonlinear least squares problem becomes larger when n increases. Therefore these methods are not suitable for fitting a large number of data points. In the metrology and pattern recognition communities people prefer the least squares orthogonal distance technique which is an iterative method and considers the relationship between shape parameters and location parameters. We refer the reader to the papers [10] [11] for detailed references. In [12] a curvature-based squared distance minimization(SDM) is proposed for orthogonal distance fitting for B-spline curve fitting. In this paper we consider general parametric curve/surface fitting problems, which are not limited in 2D, 3D curves and surfaces.

Contributions: Inspired by the approaches in [10] [12], we aim to analyze the existing orthogonal distance techniques by rephrasing them into a general optimization framework. We propose two modified methods CDM and GTDM based on geometric and optimizational analysis. We reveal that the existing and our proposed methods have clear geometric meanings. This better understanding will benefit the general parametric models fitting.

The paper is organized as follows: the basic concepts and necessary optimization techniques are introduced in Section 2; In Section 3 the detailed analysis of orthogonal distance fitting is presented including the derivation of the geometric meanings and the modified methods; in Section 4, we illustrate the effectiveness of different methods by numerical examples; finally we close the paper by the conclusion in Section 5.

2   Preliminary

2.1   Notations

Let $\mathbf{C}(\mathbf{P};\mathbf{t}) \subset \mathbb{R}^s$ represent a family of parametric curves or surfaces. A set of points $\{\mathbf{X}_j\}_{j=1}^n \subset \mathbb{R}^s$ are to be approximated by $\mathbf{C}(\mathbf{P};\mathbf{t})$. Here $\mathbf{t} = (\tau_1,\ldots,\tau_m) \in \mathbb{R}^m$ is the location parameter and $\mathbf{P} = (p_1,\ldots,p_r)$ is the shape parameter. For instance, if $m = 1$, $\mathbf{C}(\mathbf{P};\mathbf{t})$ represents a parametric curve. We assume that $\mathbf{C}(\mathbf{P};\mathbf{t})$ has C^2 continuity. In this paper vectors and matrices are denoted by bold face and vectors are in the column format. The first-order partial derivatives of $\mathbf{C}(\mathbf{P};\mathbf{t})$ are denoted as follows:

$$\frac{\partial \mathbf{C}(\mathbf{P};\mathbf{t})}{\partial \mathbf{P}} = \left[\frac{\partial \mathbf{C}(\mathbf{P};\mathbf{t})}{\partial p_1},\ldots,\frac{\partial \mathbf{C}(\mathbf{P};\mathbf{t})}{\partial p_r}\right], \quad \frac{\partial \mathbf{C}(\mathbf{P};\mathbf{t})}{\partial \mathbf{t}} = \left[\frac{\partial \mathbf{C}(\mathbf{P};\mathbf{t})}{\partial \tau_1},\ldots,\frac{\partial \mathbf{C}(\mathbf{P};\mathbf{t})}{\partial \tau_m}\right]$$

$$\nabla_{\mathbf{P}}\mathbf{t} = \begin{bmatrix} \dfrac{\partial \tau_1}{\partial p_1} & \cdots & \dfrac{\partial \tau_m}{\partial p_1} \\ \vdots & \ddots & \vdots \\ \dfrac{\partial \tau_1}{\partial p_r} & \cdots & \dfrac{\partial \tau_m}{\partial p_r} \end{bmatrix}.$$

In many curves and surfaces fitting applications, the initial positions of data points and the model are not aligned well. The data points or the model is allowed to be transformed in the fitting process. By introducing proper transformation, the fitting process can be accelerated and overcome some local minimum cases. The most common transformation is rigid transformation[10], [11]. Combined with rigid transformation, we have shown in [13] that the convergence speed of the fitting algorithm can be faster and high accuracy also can be achieved. Although the transformation can be applied to the data points or the model, for unifying our analysis we assume the transformation is applied on the parametric model, i.e. the shape parameter $\mathbf{P}$ can contain the transformation parameters if needed.

2.2   Nonlinear Least Squares

We consider a standard nonlinear least squares problem which minimizes the objective function $F(\mathbf{X})$:

$$\min_{\mathbf{X}} \frac{1}{2} \sum_{i=1}^{n} f_i^2(\mathbf{X}) \triangleq F(\mathbf{X}) \tag{2}$$

The residual vector is defined as $\mathbf{r}(\mathbf{X}) = (f_1(\mathbf{X}), f_2(\mathbf{X}), \dots, f_n(\mathbf{X}))^T$. The first derivative of $F(\mathbf{X})$ can be expressed in terms of the Jacobian of r: $\mathbf{J}(\mathbf{X}) = \begin{pmatrix} \nabla f_1(\mathbf{X}) \\ \vdots \\ \nabla f_n(\mathbf{X}) \end{pmatrix}$, where $\nabla f_i(\mathbf{X})$ is the gradient of f_i with respect to $\mathbf{X}$. The gradient and Hessian of $F(\mathbf{X})$ have the following forms

$$\nabla F(\mathbf{X}) = \mathbf{J}(\mathbf{X})^T \mathbf{r}(\mathbf{X}); \qquad \mathbf{H} = \nabla^2 F(x) = \mathbf{J}(\mathbf{X})^T \mathbf{J}(\mathbf{X}) + \sum_{i=1}^{n} f_i(\mathbf{X}) \nabla^2 f_i(\mathbf{X})$$

The Gauss-Newton method approximates the Hessian by $\mathbf{J}(\mathbf{X})^T \mathbf{J}(\mathbf{X})$. In practice the line search strategy or the Levenberg-Marquardt method

$$\left(\mathbf{J}(\mathbf{X})^T \mathbf{J}(\mathbf{X}) + \lambda \mathbf{I}\right) \delta \mathbf{X} = -\mathbf{J}(\mathbf{X})^T \mathbf{r}(\mathbf{X})$$

is incorporated with the Gauss-Newton method. The Quasi-Newton type method approximates the Hessian or the inverse of the Hessian by a positive-definite matrix which is updated at each iteration with some specified schemes such as **BFGS** [1]. But in this paper we mainly focus on Gauss-Newton type methods.

2.3   Principal Directions and Curvatures of Parametric Curves and Surfaces

For a smooth parametric curve/surface $\mathbf{C}(\mathbf{t})$, its first-order derivatives $\partial_{\tau_1} \mathbf{C}(\mathbf{t}_p)$, $\dots$, $\partial_{\tau_m} \mathbf{C}(\mathbf{t}_p)$ at point $\mathbf{C}(\mathbf{t}_p)$ span a tangential space $\top_p \mathbf{C}$. Its orthogonal complement defines the normal space $\perp_p \mathbf{C}$. For a given unit normal vector $\mathbf{n}_p \in \perp_p \mathbf{C}$, we can define the principal vectors and curvatures with respect to $\mathbf{n}_p$. The details can be found in Section 2.2 of [14]. Let $\mathbf{T}_1, \dots, \mathbf{T}_m$ be the principle vectors which span $\top_p \mathbf{C}$ and $\kappa_1, \dots, \kappa_m$ be the corresponding principle curvatures with respect to $\mathbf{n}_p$. The orthonormal basis of $\perp_p \mathbf{C}$ are $\mathbf{N}_{m+1}, \dots, \mathbf{N}_s$. One identity about the orthonormal basis will be useful in the paper:

$$\mathbf{I}_s = \mathbf{T}_1 \mathbf{T}_1^T + \cdots + \mathbf{T}_m \mathbf{T}_m^T + \mathbf{N}_{m+1} \mathbf{N}_{m+1}^T + \cdots + \mathbf{N}_s \mathbf{N}_s^T. \tag{3}$$

Where $\mathbf{I}_s$ is a $s \times s$ identity matrix.

Remark: For a 3D parametric curve, the curvature K and curvature direction $\mathbf{N}^0$ are well defined from differential geometry. Since in our discussion $\mathbf{N}$ is not necessarily coincident with $\mathbf{N}^0$, we have $\kappa = K \cdot < \mathbf{N}, \mathbf{N}^0 >$. $< \star, \star >$ is the inner product of two vectors.

3   Orthogonal Distance Fitting

The optimization process of orthogonal distance fitting contains two steps which are executed repeatedly:

1. Reparametrization: compute the foot-point of $\mathbf{X}_j$ on $\mathbf{C}(\mathbf{P}; \mathbf{t})$, i.e, minimize the distance from $\mathbf{X}_j$ to $\mathbf{C}$:

$$\min_{\mathbf{t}_j} \| \mathbf{C}(\mathbf{P}; \mathbf{t}_j) - \mathbf{X}_j \|, j = 1, \dots, n \tag{4}$$

2. minimize one of the following objective functions by applying one step of optimization techniques such as Gauss-Newton methods:

$$\min_{\mathbf{P}} \left\| \left(\| \mathbf{C}(\mathbf{P}; \mathbf{t}_1(\mathbf{P})) - \mathbf{X}_1 \|, \dots, \| \mathbf{C}(\mathbf{P}; \mathbf{t}_n(\mathbf{P})) - \mathbf{X}_n \| \right)^T \right\| \tag{5}$$

or

$$\min_{\mathbf{P}} \left\| \left(\mathbf{C}(\mathbf{P}; \mathbf{t}_1(\mathbf{P}))^T - \mathbf{X}_1^T, \dots, \mathbf{C}(\mathbf{P}; \mathbf{t}_n(\mathbf{P}))^T - \mathbf{X}_n^T \right)^T \right\| \tag{6}$$

Since (5) minimizes the l_2 norm of the residual vector $\mathbf{r}_d$:

$$\mathbf{r}_d = \left(\| \mathbf{C}(\mathbf{P}; \mathbf{t}_1(\mathbf{P})) - \mathbf{X}_1 \|, \dots, \| \mathbf{C}(\mathbf{P}; \mathbf{t}_n(\mathbf{P})) - \mathbf{X}_n \| \right)^T, \tag{7}$$

the corresponding method is called *Distance-based* method; also since (6) minimizes the l_2 norm of the residual vector $\mathbf{r}_c$:

$$\mathbf{r}_c = \left(\mathbf{C}(\mathbf{P}; \mathbf{t}_1(\mathbf{P}))^T - \mathbf{X}_1^T, \dots, \mathbf{C}(\mathbf{P}; \mathbf{t}_n(\mathbf{P}))^T - \mathbf{X}_n^T \right)^T, \tag{8}$$

the corresponding method is called *Coordinate-based* method. By applying nonlinear least squares optimization technique these two methods produce different results. Atieg and Watson present their analysis on *Distance-based* and *Coordinate-based* Gauss-Newton approaches in [10]. We will show the geometry behind these two methods and their variations.

Orthogonality: Because $\mathbf{t}_j$ is the minimizer of (4), the orthogonality condition (9) below always holds in each step, except when the foot-point is at the boundary of $\mathbf{C}$.

$$\left\langle \mathbf{C}(\mathbf{P}; \mathbf{t}_j) - \mathbf{X}_j, \frac{\partial \mathbf{C}(\mathbf{P}; \mathbf{t}_j)}{\partial \tau_k} \right\rangle = 0, \qquad j = 1, \dots, n; k = 1, \dots, m \tag{9}$$

The orthogonality condition (9) plays an important role in parametrization correction and optimization. Many effective foot-point computation methods are available in literature [3] [15] [16]. If the explicit foot-point formula is not available, one can apply Newton-like optimization methods on (9) to obtain the foot-point and corresponding location parameter. But the initial guess $\mathbf{t}^0$ is a key issue in foot-point computation. For complex parametric curves/surfaces, one good strategy is to build a k-D tree from the sample points $\{\mathbf{C}(\mathbf{P};\mathbf{t}_k), k=1,\dots,L\}$ then find the nearest point for $\mathbf{X}_j$ which serves as the initial foot-point.

3.1 Distance-Based Gauss-Newton Method

Distance-based methods are widely used in metrology. Here the l_2 norm of residual vector $\mathbf{r}_d$ is to be minimized. Depending on whether considering the association between the shape parameter $\mathbf{P}$ and the local parameter $\mathbf{t}$, Gauss-Newton distance-based methods can be categorized to two types: the separated method and the standard method.

(1) Separated distance-based Gauss-Newton method

The residual vector $\mathbf{r}_d$ in the separated distance-based Gauss-Newton method is defined as

$$\mathbf{r}_d = (\|\mathbf{C}(\mathbf{P};\mathbf{t}_1) - \mathbf{X}_1\|, \dots, \|\mathbf{C}(\mathbf{P};\mathbf{t}_n) - \mathbf{X}_n\|)^T,$$

where each $\mathbf{t}_j$ is fixed. The first-order total derivative of $\|\mathbf{C}(\mathbf{P};\mathbf{t}_j) - \mathbf{X}_j\|$ with respect to $\mathbf{P}$ is

$$\nabla_{\mathbf{P}}\|\mathbf{C}(\mathbf{P};\mathbf{t}_j) - \mathbf{X}_j\| = \frac{\mathbf{C}(\mathbf{P};\mathbf{t}_j)^T - \mathbf{X}_j^T}{\|\mathbf{C}(\mathbf{P};\mathbf{t}_j) - \mathbf{X}_j\|} \frac{\partial \mathbf{C}(\mathbf{P};\mathbf{t}_j)}{\partial \mathbf{P}}, \tag{10}$$

where it must be assumed that $\mathbf{C}(\mathbf{P};\mathbf{t}_j) \neq \mathbf{X}_j$ such that the derivative exists. Numerical computation can be unstable when $\mathbf{C}(\mathbf{P};\mathbf{t}_j)$ approaches $\mathbf{X}_j$. Notice that if $\mathbf{C}$ is a 2D parametric curve or a 3D parametric surface, the vector $\dfrac{\mathbf{C}(\mathbf{P};\mathbf{t}_j) - \mathbf{X}_j}{\|\mathbf{C}(\mathbf{P};\mathbf{t}_j) - \mathbf{X}_j\|} := \mathbf{N}_j$ actually is the unit normal at $\mathbf{C}(\mathbf{t}_j)$ whose sign may be positive or negative. Thus the instability can be eliminated if we replace it with the unit normal. The Jacobian of $\mathbf{r}_d$ at $\mathbf{C}(\mathbf{P};\mathbf{t}_j)$ can be written as

$$\mathbf{J}_1 = \begin{pmatrix} \mathbf{N}_1^T \dfrac{\partial \mathbf{C}(\mathbf{P};\mathbf{t}_1)}{\partial \mathbf{P}} \\ \vdots \\ \mathbf{N}_n^T \dfrac{\partial \mathbf{C}(\mathbf{P};\mathbf{t}_n)}{\partial \mathbf{P}} \end{pmatrix}.$$

From the normal equation $\mathbf{J}_1^T \mathbf{J}_1 \cdot \delta\mathbf{P} = -\mathbf{J}_1^T \mathbf{r}_d$, we can derive that

$$\sum_{j=1}^{n} \frac{\partial \mathbf{C}(\mathbf{P};\mathbf{t}_j)^T}{\partial \mathbf{P}} \mathbf{N}_j \mathbf{N}_j^T \frac{\partial \mathbf{C}(\mathbf{P};\mathbf{t}_j)}{\partial \mathbf{P}} \cdot \delta\mathbf{P} = -\sum_{j=1}^{n} \frac{\partial \mathbf{C}(\mathbf{P};\mathbf{t}_j)^T}{\partial \mathbf{P}} (\mathbf{C}(\mathbf{P};\mathbf{t}_j) - \mathbf{X}_j), \tag{11}$$

where $\delta\mathbf{P}$ is the increment of the shape parameter $\mathbf{P}$.

Now we show the geometric meaning behind (11). the right hand side of (11) can be rewritten as:

$$\frac{\partial \mathbf{C}(\mathbf{P};\mathbf{t}_j)^T}{\partial \mathbf{P}}(\mathbf{C}(\mathbf{P};\mathbf{t}_j) - \mathbf{X}_j) = \frac{\partial \mathbf{C}(\mathbf{P};\mathbf{t}_j)^T}{\partial \mathbf{P}}\mathbf{N}_j \cdot \mathbf{N}_j^T (\mathbf{C}(\mathbf{P};\mathbf{t}_j) - \mathbf{X}_j) \qquad (12)$$

Now Eqn.(11) actually minimizes the squared distance from data points to their tangent planes at the foot-points:

$$\min_{\mathbf{P}} \sum_{j=1}^{n} \left[\mathbf{N}_j^T \cdot (\mathbf{C}(\mathbf{P};\mathbf{t}_j) - \mathbf{X}_j)\right]^2 \qquad (13)$$

It is easy to verify the normal equation of Eqn.(13) is Eqn.(11) just by applying the Gauss-Newton method on Eqn.(13). We call this kind of geometric minimization TDM (Tangent Distance Minimization) [12].

As we have pointed out, there is no numerical problem for 2D parametric curves and 3D parametric surfaces if we replace $\mathbf{N}_j$ with curves/surfaces' normals. For high dimension parametric curves/surfaces ($m < s - 1$), TDM is not suitable when the data points are almost contained in a low dimension space $\mathbb{R}^l, l < s$. For instance, fitting a 3D parametric curve to a set of points in a plane causes the ill-conditioning of Jacobian matrix [10]. We use a simple example to illustrate this problem. Assume that a 3D curve has the following parametric form (at^2, bt^3, c), where a, b, c are shape parameters and the data points lie in the x-y plane. The third component of $\mathbf{N}_j$ will be always zero. It means that c does not appear in $\mathbf{N}_j^T \cdot (\mathbf{C}(\mathbf{P};\mathbf{t}_j) - \mathbf{X}_j)$. Therefore the normal equations will be singular.

(2) Standard distance-based Gauss-Newton method

With the consideration of the association between $\mathbf{t}$ and $\mathbf{P}$, in the standard distance-based Gauss-Newton method the residual vector $\mathbf{r}_d$ is defined as in (7). The first-order total derivative of each element of $\mathbf{r}_d$ with respect to $\mathbf{P}$ is

$$\nabla_{\mathbf{P}}\|\mathbf{C}(\mathbf{P};\mathbf{t}_j(\mathbf{P})) - \mathbf{X}_j\| = \frac{\mathbf{C}(\mathbf{P};\mathbf{t}_j)^T - \mathbf{X}_j^T}{\|\mathbf{C}(\mathbf{P};\mathbf{t}_j) - \mathbf{X}_j\|}\nabla_{\mathbf{P}}\mathbf{C}(\mathbf{P};\mathbf{t}_j(\mathbf{P}))$$

$$= \frac{\mathbf{C}(\mathbf{P};\mathbf{t}_j)^T - \mathbf{X}_j^T}{\|\mathbf{C}(\mathbf{P};\mathbf{t}_j) - \mathbf{X}_j\|}\left[\frac{\partial \mathbf{C}(\mathbf{P};\mathbf{t}_j)}{\partial \mathbf{P}} + \frac{\partial \mathbf{C}(\mathbf{P};\mathbf{t}_j)}{\partial \mathbf{t}}\nabla_{\mathbf{P}}\mathbf{t}_j\right]$$

$$= \mathbf{N}_j^T \frac{\partial \mathbf{C}(\mathbf{P};\mathbf{t}_j)}{\partial \mathbf{P}}, \qquad (14)$$

where the term $\left(\mathbf{C}(\mathbf{P};\mathbf{t}_j(\mathbf{P}))^T - \mathbf{X}_j^T\right) \cdot \frac{\partial \mathbf{C}(\mathbf{P};\mathbf{t}_j)}{\partial \mathbf{t}}\nabla_{\mathbf{P}}\mathbf{t}_j$ is eliminated due to the orthogonality condition. The result (14) is the same as (10), which means that both separated and standard distance-based approaches produce the same geometric minimization scheme – TDM.

3.2   Coordinate-Based Gauss-Newton Method

Now we consider the coordinate-based Gauss-Newton method based on the objective function (6), which is widely used in pattern recognition community.

(1) Separated coordinate-based Gauss-Newton method

In the separated coordinate-based Gauss-Newton method the residual vector $\mathbf{r}_c$ is defined as

$$\mathbf{r}_c = \left(\mathbf{C}(\mathbf{P};\mathbf{t}_1)^T - \mathbf{X}_1^T, \ldots, \mathbf{C}(\mathbf{P};\mathbf{t}_n)^T - \mathbf{X}_n^T \right)^T.$$

The first-order total derivative of $\mathbf{C}(\mathbf{P};\mathbf{t}_j)^T - \mathbf{X}_j^T$ with respect to $\mathbf{P}$ is $\dfrac{\partial \mathbf{C}(\mathbf{P};\mathbf{t}_j)^T}{\partial \mathbf{P}}$. So the Jacobian $\mathbf{J}_2$ of $\mathbf{r}_c$ is

$$\begin{pmatrix} \dfrac{\partial \mathbf{C}(\mathbf{P};\mathbf{t}_1)}{\partial \mathbf{P}} \\ \vdots \\ \dfrac{\partial \mathbf{C}(\mathbf{P};\mathbf{t}_n)}{\partial \mathbf{P}} \end{pmatrix}.$$

Still from the normal equation $\mathbf{J}_2^T \mathbf{J}_2 \cdot \delta \mathbf{P} = -\mathbf{J}_2^T \mathbf{r}_c$, we obtain

$$\sum_{j=1}^{n} \frac{\partial \mathbf{C}(\mathbf{P};\mathbf{t}_j)^T}{\partial \mathbf{P}} \frac{\partial \mathbf{C}(\mathbf{P};\mathbf{t}_j)}{\partial \mathbf{P}} \cdot \delta \mathbf{P} = -\sum_{j=1}^{n} \frac{\partial \mathbf{C}(\mathbf{P};\mathbf{t}_j)}{\partial \mathbf{P}} \left(\mathbf{C}(\mathbf{P};\mathbf{t}_j) - \mathbf{X}_j \right) \qquad (15)$$

The normal equation actually represents a geometric minimization

$$\min_{\mathbf{P}} \sum_{j=1}^{n} [\mathbf{C}(\mathbf{P};\mathbf{t}_j) - \mathbf{X}_j]^2 \qquad (16)$$

which penalizes the squared distance from data points to foot points, we call this method PDM (**P**oint **D**istance **M**inimization). It is widely used in CAGD community because of its simplicity. Especially when $\mathbf{P}$ is in the linear form in $\mathbf{C}$, one just needs to solve a linear equation and the $\|\mathbf{r}_c\|$ always decreases. However PDM only exhibits linear convergence [12].

(2) Standard coordinate-based Gauss-Newton method

In the standard distance-based Gauss-Newton method the residual vector $\mathbf{r}_c$ is (8), where $\mathbf{t}_j$ is associated with $\mathbf{P}$ through (9). The first-order total derivative of each element with respect to $\mathbf{P}$ is

$$\nabla_{\mathbf{P}} \left(\mathbf{C}(\mathbf{P};\mathbf{t}_j(\mathbf{P})) - \mathbf{X}_j \right) = \frac{\partial \mathbf{C}(\mathbf{P};\mathbf{t}_j)}{\partial \mathbf{P}} + \sum_{k=1}^{m} \frac{\partial \mathbf{C}(\mathbf{P};\mathbf{t}_j)}{\partial \tau_{j,k}} \nabla_{\mathbf{P}} \tau_{j,k}(\mathbf{P}) \qquad (17)$$

In general the explicit expression of $\tau_{j,k}(\mathbf{P})$ with respect to $\mathbf{P}$ is not always available. So we use the implicit procedure presented in [10]. Since the orthogonality condition (9) holds and it is an identity in $\mathbf{P}$, its total derivative with respect to $\mathbf{P}$ is still $\mathbf{0}$. Therefore we have

$$
\begin{aligned}
\mathbf{0} = \nabla_{\mathbf{P}} & \left\langle \mathbf{C}(\mathbf{P};\mathbf{t}_j) - \mathbf{X}_j, \frac{\partial \mathbf{C}(\mathbf{P};\mathbf{t}_j)}{\partial \tau_{j,k}} \right\rangle \\
= & \left\langle \frac{\partial \mathbf{C}(\mathbf{P};\mathbf{t}_j)}{\partial \mathbf{P}} + \sum_{l=1}^{m} \frac{\partial \mathbf{C}(\mathbf{P};\mathbf{t}_j)}{\partial \tau_{j,l}} \nabla_{\mathbf{P}} \tau_{j,l}(\mathbf{P}), \frac{\partial \mathbf{C}(\mathbf{P};\mathbf{t}_j)}{\partial \tau_{j,k}} \right\rangle + \\
& \left\langle \mathbf{C}(\mathbf{P};\mathbf{t}_j) - \mathbf{X}_j, \frac{\partial^2 \mathbf{C}(\mathbf{P};\mathbf{t}_j)}{\partial \tau_{j,k} \partial \mathbf{P}} + \sum_{l=1}^{m} \frac{\partial^2 \mathbf{C}(\mathbf{P};\mathbf{t}_j)}{\partial \tau_{j,k} \partial \tau_{j,l}} \nabla_{\mathbf{P}} \tau_{j,l}(\mathbf{P}) \right\rangle
\end{aligned}
$$

Without loss of generality, suppose $\mathbf{C}(\mathbf{P};\mathbf{t}_j)$ is a local regular parametrization such that $\tau_{j,1}$-, ..., $\tau_{j,m}$- direction vectors are unit principle direction vectors $\mathbf{T}_{j,1}, \ldots, \mathbf{T}_{j,m}$ with respect to $\mathbf{N}_j$ (see Section 2.3). The above equation can be simplified as

$$
\begin{aligned}
\mathbf{0} = & \left\langle \frac{\partial \mathbf{C}(\mathbf{P};\mathbf{t}_j)}{\partial \mathbf{P}} + \sum_{l=1}^{m} \mathbf{T}_{j,l} \nabla_{\mathbf{P}} \tau_{j,l}(\mathbf{P}), \mathbf{T}_{j,k} \right\rangle + \\
& \left\langle \mathbf{C}(\mathbf{P};\mathbf{t}_j) - \mathbf{X}_j, \frac{\partial^2 \mathbf{C}(\mathbf{P};\mathbf{t}_j)}{\partial \tau_{j,k} \partial \mathbf{P}} + \frac{\partial^2 \mathbf{C}(\mathbf{P};\mathbf{t}_j)}{\partial \tau_{j,k}^2} \nabla_{\mathbf{P}} \tau_{j,k}(\mathbf{P}) \right\rangle \\
= & \mathbf{T}_{j,k}^T \frac{\partial \mathbf{C}(\mathbf{P};\mathbf{t}_j)}{\partial \mathbf{P}} + \sum_{l=1}^{m} \mathbf{T}_{j,k}^T \mathbf{T}_{j,l} \nabla_{\mathbf{P}} \tau_{j,l}(\mathbf{P}) + (\mathbf{C}(\mathbf{P};\mathbf{t}_j) - \mathbf{X}_j)^T \frac{\partial^2 \mathbf{C}(\mathbf{P};\mathbf{t}_j)}{\partial \tau_{j,k} \partial \mathbf{P}} + d_j \mathbf{N}_j^T \kappa_{j,k} \mathbf{N}_j \nabla_{\mathbf{P}} \tau_{j,k}(\mathbf{P}) \\
= & \mathbf{T}_{j,k}^T \frac{\partial \mathbf{C}(\mathbf{P};\mathbf{t}_j)}{\partial \mathbf{P}} + (1 + d_j \kappa_{j,k}) \nabla_{\mathbf{P}} \tau_{j,k}(\mathbf{P}) + (\mathbf{C}(\mathbf{P};\mathbf{t}_j) - \mathbf{X}_j)^T \frac{\partial^2 \mathbf{C}(\mathbf{P};\mathbf{t}_j)}{\partial \tau_{j,k} \partial \mathbf{P}},
\end{aligned}
$$

where $d_j = \|\mathbf{C}(\mathbf{P};\mathbf{t}_j) - \mathbf{X}_j\|$, $\kappa_{j,k}$ is the principle curvature along $\mathbf{T}_{j,k}$ with respect to $\mathbf{N}_j$. Then we obtain

$$
\nabla_{\mathbf{P}} \tau_{j,k} = -\frac{\mathbf{T}_{j,k}^T \dfrac{\partial \mathbf{C}(\mathbf{P};\mathbf{t}_j)}{\partial \mathbf{P}} + (\mathbf{C}(\mathbf{P};\mathbf{t}_j) - \mathbf{X}_j)^T \dfrac{\partial^2 \mathbf{C}(\mathbf{P};\mathbf{t}_j)}{\partial \tau_{j,k} \partial \mathbf{P}}}{1 + d_j \kappa_{j,k}} \tag{18}
$$

We can rewrite (17) as

$$
\nabla_{\mathbf{P}}(\mathbf{C}(\mathbf{P};\mathbf{t}_j(\mathbf{P}))) - \mathbf{X}_j) = \frac{\partial \mathbf{C}(\mathbf{P};\mathbf{t}_j)}{\partial \mathbf{P}} - \sum_{k=1}^{m} \frac{\mathbf{T}_{j,k} \mathbf{T}_{j,k}^T \dfrac{\partial \mathbf{C}(\mathbf{P};\mathbf{t}_j)}{\partial \mathbf{P}} + d_j \mathbf{T}_{j,k} \mathbf{N}_j^T \dfrac{\partial \mathbf{T}_{j,k}}{\partial \mathbf{P}}}{1 + d_j \kappa_{j,k}} \tag{19}
$$

In the degenerate case when $1 + d_j \kappa_{j,k} \approx 0$, one can modify the denominator to $1 + d_j |\kappa_{j,k}|$ to improve the condition number of the normal equation. We note that this degenerate case is not addressed in the literature of orthogonal distance fitting, such as [10], [11]. But it can happen in practice. For example, let a parametric circle be $(r \cos t, r \sin t)$ and one data point $\mathbf{X}_j$ be near to the origin. We have $1 + d_j \kappa_{j,k} \approx 1 + r \cdot \frac{-1}{r} = 0$.

(3) Modified standard coordinate-based Gauss-Newton methods

The computational cost of the second-order derivatives $\dfrac{\partial^2 C(P;t_j)}{\partial t^2}$ and $\dfrac{\partial^2 C(P;t_j)}{\partial P \partial t}$ may be high in some applications. So we shall derive two kinds of modified standard Gauss-Newton methods with less computation cost and clear geometric meanings.

First we recall the notations in Section 2.3. At the point $C(t_j)$, $T_{j,1},\ldots,T_{j,m}$ span a tangent vector space $\top_{j,p}C$ and $N_{j,m+1},\ldots,N_{j,s}$ denote the orthonormal basis of $\top_{j,p}C$'s orthogonal complement space $\perp_{j,p}C$. The following identity always holds

$$I = T_{j,1}T_{j,1}^T + \cdots + T_{j,m}T_{j,m}^T + N_{j,m+1}N_{j,m+1}^T + \cdots + N_{j,s}N_{j,s}^T. \tag{20}$$

We will use this identity in our following derivation. By dropping the second-order derivatives from Eqn. 19, we will derive two methods.

1. Drop $\dfrac{\partial^2 C(P;t_j)}{\partial P \partial t}$. This leads to

$$\nabla_P \left(C(P;t_j(P)) - X_j \right) \approx \frac{\partial C(P;t_j)}{\partial P} - \sum_{k=1}^{m} \frac{T_{j,k}T_{j,k}^T \dfrac{\partial C(P;t_j)}{\partial P}}{1 + d_j \kappa_{j,k}}$$

$$= I \cdot \frac{\partial C(P;t_j)}{\partial P} - \sum_{k=1}^{m} \frac{T_{j,k}T_{j,k}^T \dfrac{\partial C(P;t_j)}{\partial P}}{1 + d_j \kappa_{j,k}}$$

$$= \left(\sum_{k=1}^{m} \frac{d_j \kappa_{j,k} T_{j,k}T_{j,k}^T}{1 + d_j \kappa_{j,k}} + \sum_{k=m+1}^{s} N_{j,k}N_{j,k}^T \right) \frac{\partial C(P;t_j)}{\partial P}$$

Substituting the above result into the normal equation $J^T J \cdot \delta P = -J^T r_c$, we obtain

$$\frac{\partial C(P;t_j)^T}{\partial P} \left(\sum_{k=1}^{m} \frac{(d_j \kappa_{j,k})^2 T_{j,k}T_{j,k}^T}{(1 + d_j \kappa_{j,k})^2} + \sum_{k=m+1}^{s} N_{j,k}N_{j,k}^T \right) \frac{\partial C(P;t_j)}{\partial P} \delta P =$$

$$\frac{\partial C(P;t_j)^T}{\partial P} \left(\sum_{k=1}^{m} \frac{d_j \kappa_{j,k} T_{j,k}T_{j,k}^T}{1 + d_j \kappa_{j,k}} + \sum_{k=m+1}^{s} N_{j,k}N_{j,k}^T \right) (C(P;t_j) - X_j).$$

The normal equation represents the following geometric minimization

$$\min_P \sum_{j=1}^{n} \left\{ \sum_{k=1}^{m} \frac{(d_j \kappa_{j,k})^2}{(1 + d_j \kappa_{j,k})^2} \left[T_{j,k}^T \cdot (C(P;t_j) - X_j) \right]^2 + \right.$$

$$\left. \sum_{k=m+1}^{s} \left[N_{j,k}^T \cdot (C(P;t_j) - X_j) \right]^2 \right\} \tag{21}$$

We will call it CDM (Curvature Distance Minimization).

2. Drop $\dfrac{\partial^2 C(P;t_j)}{\partial P \partial t}$ and $\dfrac{\partial^2 C(P;t_j)}{\partial t^2}$. It is easy to verify in this case that the normal equation corresponds to the following geometric minimization

$$\min_P \sum_{j=1}^{n} \left\{ \sum_{k=m+1}^{s} \left[N_{j,k}^T \cdot (C(P;t_j) - X_j) \right]^2 \right\} \tag{22}$$

Since (22) only penalizes the squared distance from data point $\mathbf{X}_j$ to the tangent space $\mathbf{T}_{j,p}(\mathbf{C})$ we call it GTDM (**G**eneralized **T**angent **D**istance **M**inimization). This scheme does not suffer from the ill conditioning problem of high dimension parametric curves/surfaces ($m < s - 1$) which is mentioned before. Still using the same example at the end of the last subsection, let one normal be $\mathbf{N}_j$ and another normal be $\mathbf{N}_z = (0,0,1)^T$. The variable c will appear in $\mathbf{N}_z^T\left(\mathbf{C}(\mathbf{P};\mathbf{t}_j) - \mathbf{X}_j\right)$, so that rank deficiency of the normal equation is avoided.

3.3   SDM - Modified Hessian Approximation

So far our discussion is based on Gauss-Newton methods. Now we look at the Hessian directly. Wang et al. [12] proposed a curvature based squared distance minimization method called SDM where the Hessian is modified to be definite-positive. We do not go into the details and just describe the basic idea here. For each $\mathbf{C}(\mathbf{P};\mathbf{t}_j) - \mathbf{X}_j$, its second-order derivatives $\dfrac{\partial^2 \mathbf{C}(\mathbf{P};\mathbf{t}_j)}{\partial \mathbf{P}^2}$ and $\dfrac{\partial^2 \mathbf{C}(\mathbf{P};\mathbf{t}_j)}{\partial \mathbf{P}\partial \mathbf{t}}$ are dropped. So the modified Hessian is

$$\widetilde{\mathbf{H}} = \sum_{j=1}^{n} \frac{\partial \mathbf{C}(\mathbf{P};\mathbf{t}_j)^T}{\partial \mathbf{P}} \left[\sum_{k=1}^{m} \frac{d_j \kappa_{j,k}}{1 + d_j \kappa_{j,k}} \mathbf{T}_{j,k} \mathbf{T}_{j,k}^T + \sum_{k=m+1}^{s} \mathbf{N}_{j,k} \mathbf{N}_{j,k}^T \right] \frac{\partial \mathbf{C}(\mathbf{P};\mathbf{t}_j)}{\partial \mathbf{P}}. \tag{23}$$

The corresponding geometric minimization is

$$\min_{\mathbf{P}} \sum_{j=1}^{n} \left\{ \sum_{k=1}^{m} \frac{d_j \kappa_{j,k}}{1 + d_j \kappa_{j,k}} \left[\mathbf{T}_{j,k}^T \cdot (\mathbf{C}(\mathbf{P};\mathbf{t}_j) - \mathbf{X}_j) \right]^2 + \sum_{k=m+1}^{s} \left[\mathbf{N}_{j,k}^T \cdot (\mathbf{C}(\mathbf{P};\mathbf{t}_j) - \mathbf{X}_j) \right]^2 \right\}. \tag{24}$$

Remark: If, besides $\dfrac{\partial^2 \mathbf{C}(\mathbf{P};\mathbf{t}_j)}{\partial \mathbf{P}^2}$ and $\dfrac{\partial^2 \mathbf{C}(\mathbf{P};\mathbf{t}_j)}{\partial \mathbf{P}\partial \mathbf{t}}$, we also drop $\dfrac{\partial^2 \mathbf{C}(\mathbf{P};\mathbf{t}_j)}{\partial \mathbf{t}^2}$ from the Hessian, SDM will become GTDM. Thus GTDM also is an approximation of the Hessian.

3.4   Comparisons

We summarize the geometric minimization schemes introduced in previous sections in Table 1 and compare them in several aspects.

Computational cost: The standard coordinate-based Gauss-Newton method (for short, we call it GN) is the most expensive method because of computations of the second-order derivatives $\dfrac{\partial^2 \mathbf{C}(\mathbf{P};\mathbf{t}_j)}{\partial \mathbf{t}^2}$ and $\dfrac{\partial^2 \mathbf{C}(\mathbf{P};\mathbf{t}_j)}{\partial \mathbf{P}\partial \mathbf{t}}$. Since CDM and SDM have similar expressions, their computational cost are the same. GTDM only involves computation of $\dfrac{\partial \mathbf{C}(\mathbf{P};\mathbf{t}_j)}{\partial \mathbf{t}}$ for constructing the normal space if $m < s - 1$. TDM and PDM do not need to compute any derivative of $\mathbf{C}(\mathbf{P};\mathbf{t})$ with respect to $\mathbf{t}$. thus they are more efficient than the others in constructing the approximated Hessian.

Applicability: With proper step-size control or combining Levenberg-Marquardt methods, most methods are suitable for general parametric curve/surface fitting. Only TDM

Table 1. Geometric minimization schemes

Method	Geometric terms
PDM	$\left[\mathbf{C}\left(\mathbf{P};\mathbf{t}_j\right) - \mathbf{X}_j\right]^2$
TDM	$\left[\mathbf{N}_j^T \cdot \left(\mathbf{C}\left(\mathbf{P};\mathbf{t}_j\right) - \mathbf{X}_j\right)\right]^2$
GTDM	$\sum\limits_{k=m+1}^{s} \left[\mathbf{N}_{j,k}^T \cdot \left(\mathbf{C}\left(\mathbf{P};\mathbf{t}_j\right) - \mathbf{X}_j\right)\right]^2$
CDM	$\sum\limits_{k=1}^{m} \dfrac{\left(d_j \kappa_{j,k}\right)^2}{\left(1 + d_j \kappa_{j,k}\right)^2} \left[\mathbf{T}_{j,k}^T \cdot \left(\mathbf{C}\left(\mathbf{P};\mathbf{t}_j\right) - \mathbf{X}_j\right)\right]^2 + \sum\limits_{k=m+1}^{s} \left[\mathbf{N}_{j,k}^T \cdot \left(\mathbf{C}\left(\mathbf{P};\mathbf{t}_j\right) - \mathbf{X}_j\right)\right]^2$
SDM	$\sum\limits_{k=1}^{m} \dfrac{d_j \kappa_{j,k}}{1 + d_j \kappa_{j,k}} \left[\mathbf{T}_{j,k}^T \cdot \left(\mathbf{C}\left(\mathbf{P};\mathbf{t}_j\right) - \mathbf{X}_j\right)\right]^2 + \sum\limits_{k=m+1}^{s} \left[\mathbf{N}_{j,k}^T \cdot \left(\mathbf{C}\left(\mathbf{P};\mathbf{t}_j\right) - \mathbf{X}_j\right)\right]^2$

may have problems in fitting high dimension parametric curves/surfaces, i.e, when $m < s - 1$.

Convergence: Because GN and TDM are standard Gauss-Newton methods, they show quadratic convergence for zero residual problems, super-linear convergence for small residual problems and linear convergence in other cases. For our modified methods CDM and GTDM, they also have the same convergence as TDM. One can see that when $\|\mathbf{C}\left(\mathbf{P};\mathbf{t}_j\right) - \mathbf{X}_j\|$ approaches zero, the second-order derivatives in Eqn. (19) can be ignored so that the Hessian is still well approximated. Unfortunately PDM is an alternating method which is a typical optimization technique for solving a separable nonlinear least squares problem and is known to have only linear convergence [5].

Remark: For high dimension curves/surfaces fitting, i.e. $s > 3$, the principle curvature computation can be expensive. In this case GTDM is a good candidate under the consideration of performance and effectiveness. Also when $m = s - 1$, GTDM is reduced to TDM actually.

4   Numerical Experiments

Now we compare the methods introduced in Section 3: PDM, TDM, CDM, GTDM, GN, SDM. For demonstrating the effectiveness of GTDM, we choose a planar ellipse in 3D space as our parametric models and a point cloud with different scale noises(For general comparison in 2D/3D curve and surface fitting, we refer the reader to the references [10,11,13,12]). In our implementation the Levenberg-Marquardt method is integrated.

Example: We consider fitting an ellipse to 200 data points sampled from an ellipse: $(\cos\frac{2\pi i}{200}, 2\sin\frac{2\pi i}{200}, 0), i = 0, 1, \ldots, 199$ in 3D. The parametric ellipse has the following form which involves rotation and translation

$$\begin{pmatrix} x \\ y \\ z \end{pmatrix} = \mathbf{R}_x \cdot \mathbf{R}_y \cdot \mathbf{R}_z \cdot \begin{pmatrix} a\cos t \\ b\sin t \\ 0 \end{pmatrix} + \begin{pmatrix} c_x \\ c_y \\ c_z \end{pmatrix}$$

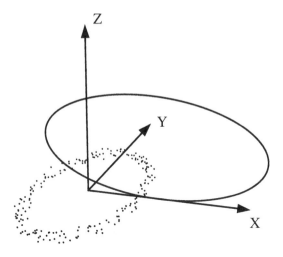

Fig. 1. The initial ellipse and data points of Case 4

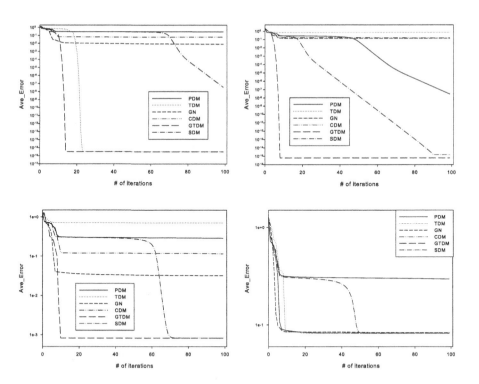

Fig. 2. Comparisons of the six methods on a set of 200 data points. Upper left: Case 1; upper right: Case 2; lower left: Case 3; lower right: Case 4

Where $\mathbf{R}_x = \begin{pmatrix} 1 & 0 & 0 \\ 0 & \cos\alpha & -\sin\alpha \\ 0 & \sin\alpha & \cos\alpha \end{pmatrix}$, $\mathbf{R}_y = \begin{pmatrix} \cos\beta & 0 & -\sin\beta \\ 0 & 1 & 0 \\ \sin\beta & 0 & \cos\beta \end{pmatrix}$, $\mathbf{R}_z = \begin{pmatrix} \cos\gamma & -\sin\gamma & 0 \\ \sin\gamma & \cos\gamma & 0 \\ 0 & 0 & 1 \end{pmatrix}$.

The shape parameters are $\mathbf{P} = [a, b, c_x, c_y, c_z, \alpha, \beta, \gamma]$. We choose four cases to illustrate the convergence of each method, with the following initial values for $\mathbf{P}$

- Case 1: $\mathbf{P} = [3.1, 1.0, 1.0, 2.0, 0.2, 4.0, 1.0, 6.0]$;
- Case 2: $\mathbf{P} = [0.1, 4.0, 2.0, 0.0, 1.0, 1.0, -1.0, 2.0]$;
- Case 3: same $\mathbf{P}$ as in Case 1 but perturb the data points with random noise distributed uniformly in $[-0.001, 0.001]$;
- Case 4: same $\mathbf{P}$ as in Case 1 but perturb the data points with random noise distributed uniformly in $[-0.1, 0.1]$. (See Fig. 1)

Fig.2 shows that the the average error versus the number of iterations of the six methods. The average error is defined as $\sqrt{\frac{\sum_{j=1}^{n} \|\mathbf{C}(\mathbf{P};t_j) - \mathbf{X}_j\|^2}{n}}$. From the figure we find the surprising fact that GTDM is much better than the other methods. It converges very fast and only needs several iterations. The behaviors of TDM in the four cases are different. In Case 2 and 3 TDM is easy to be trapped in the local minimum. In Case 4, since the data points are not nearly planar, TDM shows good performance. In all the cases GN is a little better than CDM but is still slower than GTDM and SDM.

From our experience in 2D/3D curves and surfaces fitting [12,13], actually there is no strong evidence and theoretical guarantee that shows which method (TDM, CDM, GTDM, GN, SDM) is best for most fitting problems since the integrated step-control strategy like line search or the Levenberg-Marquardt method affects the behavior and unexpected local minimum may stop the optimization. Also for large residual problems all the methods exhibit linear convergence which is similar to PDM. For instance, see Case 4 of the example. But in general GTDM is as good as the others at least in most cases. By considering the computational cost and overall performance, we strongly recommend GTDM for general parametric curve and surface fitting including parametric sub-manifold fitting (i.e, when $m < s - 1$) due to its clear geometric meaning and its simplicity since it does not need to compute the principle curvatures and directions.

5   Conclusions

A systematic geometrical and optimizational analysis on least squares orthogonal distance fitting of parametric curves and surfaces is presented in this paper. We give the geometric characterization of existing techniques and propose two modified versions based on geometric meanings. We show how principle curvature and directions are embedded in optimization methods. The presented geometric understanding of optimization techniques will benefit efficient and effective curve/surface fitting. Also for further research, it is interesting to study the geometry behind methods for implicit curve/surface fitting.

Acknowledgements

The work was partially supported by the National Key Basic Research Project of China under 2004CB318000.

References

1. Nocedal, J., Wright, S.J.: Numerical Optimization, 2nd edn. Springer, Heidelberg (2006)
2. Hoschek, J.: Intrinsic parameterization for approximation. Computer Aided Geometric Design 5(1), 27–31 (1988)
3. J. Hoschek and D. Lasser. Fundamentals of Computer Aided Geometric Design. Peters, A. K. (1993)
4. Weiss, V., Andor, L., Renner, G., Várady, T.: Advanced surface fitting techniques. Computer Aided Geometric Design 19(1), 19–42 (2002)
5. Bjorck, A.: Numerical Methods for Least Squares Problems. In: Mathematics Society for Industrial and Applied Mathematics, Philadelphia (1996)
6. Helfrich, H.P., Zwick, D.: A trust region algorithm for parametric curve and surface fitting. Journal of Computational and Applied Mathematics 73(1-2), 119–134 (1996)
7. Speer, T., Kuppe, M., Hoschek, J.: Global reparameterization for curve approximation. Computer Aided Geometric Design 15(9), 869–877 (1998)
8. Golub, G., Pereyra, V.: Separable nonlinear least squares: the variable projection method and its applications. Inverse Problems 19, 1–26 (2003)
9. Borges, C.F., Pastva, T.: Total least squares fitting of Bézier and B-spline curves to ordered data. Computer Aided Geometric Design 19(4), 275–289 (2002)
10. Atieg, A., Watson, G.A.: A class of methods for fitting a curve or surface to data by minimizing the sum of squares of orthogonal distances. Journal of Computational and Applied Mathematics 158, 227–296 (2003)
11. Ahn, S.J.: Least Squares Orthogonal Distance Fitting of Curves and Surfaces in Space. LNCS, vol. 3151. Springer, Heidelberg (2004)
12. Wang, W., Pottmann, H., Liu, Y.: Fitting B-spline curves to point clouds by curvature-based squared distance minimization. ACM Transactions on Graphics 25, 214–238 (2006)
13. Liu, Y., Pottmann, H., Wang, W.: Constrained 3D shape reconstruction using a combination of surface fitting and registration. Computer Aided Design 38(6), 572–583 (2006)
14. Wallner, J.: Gliding spline motions and applications. Computer Aided Geometric Design 21(1), 3–21 (2004)
15. Piegl, L.A., Tiller, W.: The NURBS Book, 2nd edn. Springer, Heidelberg (1996)
16. Saux, E., Daniel, M.: An improved hoschek intrinsic parametrization. Computer Aided Geometric Design 20(8-9), 513–521 (2003)

Shifting Planes to Follow a Surface of Revolution

Eng-Wee Chionh

School of Computing, National University of Singapore
Computing 1, #03-68, Law Link, Singapore 117590
chionhew@comp.nus.edu.sg
http://www.comp.nus.edu.sg/~chionhew

Abstract. A degree n rational plane curve rotating about an axis in the plane creates a degree $2n$ rational surface. Two formulas are given to generate $2n$ moving planes that follow the surface. These $2n$ moving planes lead to a $2n \times 2n$ implicitization determinant that manifests the geometric revolution algebraically in two aspects. Firstly the moving planes are constructed by successively shifting terms of polynomials from one column to another of a spawning 3×3 determinant. Secondly the right half of the $2n \times 2n$ implicitization determinant is almost an n-row rotation of the left half. As an aside, it is observed that rational parametrizations of a surface of revolution due to a symmetric rational generatrix must be improper.

Keywords: implicitization determinants, moving planes, surfaces of revolution, inherently improper parametrizations.

1 Introduction

Surfaces of revolution are ubiquitous in nature and human creation. Familiar examples include circular cones and cylinders, spheres, and torus. Five of the six non-trivial quadric families — ellipsoids, hyperboloids (one-sheet and two-sheet), elliptic paraboloids, and quadric cones — are surfaces of revolution when they are symmetric; the only exception is the family of hyperbolic paraboloids.

A surface of revolution emerges when a plane curve (the generatrix) revolves about a line (the axis of rotation) residing in the same plane. If the algebraic degree of the curve is n, in general the surface degree is $2n$ because a line in the original plane intersects the original curve and its mirror copy about the axis n times each for a total of $2n$ times. If the plane is $X = 0$, the generatrix is $f(Y, Z) = 0$, and the axis of rotation is the Z-axis, the surface is $f(\pm\sqrt{X^2 + Y^2}, Z) = 0$. Thus to obtain the usual implicit surface equation it is necessary to collect odd powers of Y in $f(Y, Z) = 0$ to one side before squaring both sides of the equation to remove the radical symbol. This approach, besides being inelegant, also suffers the absence of an implicit equation in some concise form such as an determinant. This not so satisfactory state of affairs deserves improvements because the implicit equation is such an important representation in geometric processing and computation. It is well-known that the explicit

F. Chen and B. Jüttler (Eds.): GMP 2008, LNCS 4975, pp. 398–409, 2008.

(parametric) representation generates the shape directly and the implicit representation partitions (a point is on, inside, or outside the shape) simply. In some tasks such as finding the intersection of shapes, better algorithms are possible provided one shape is given explicitly and the other implicitly. These facts have motivated much effort in the research of implicitization [6,7].

Fortunately for surfaces of revolution, when the curve $f(Y, Z) = 0$ is rational, it is actually very convenient to express the implicit equation

$$f(\pm\sqrt{X^2 + Y^2}, Z) = 0 \qquad (1)$$

in determinant form using the method of moving surfaces [9,10]. Furthermore, these moving surfaces can be generated from two closed form formulas without the need of solving a system of linear equations, a chore often required when using moving surfaces to implicitize [9,10]. Even more exciting is that the formulas and the resulting $2n \times 2n$ implicitization determinant seem to manifest the geometric action of rotating a curve about a line in two aspects. The formulas generate the moving planes by successively shifting terms of polynomials from one column to another in a spawning 3×3 determinant. The resulting $2n \times 2n$ implicitization determinant is sort of block-circulant: the right-half of the determinant is almost an n-row rotation of the left-half of the determinant.

The aim of this paper is to present the two closed form formulas and to elaborate some delightful properties of the formulas and the $2n \times 2n$ implicitization determinant produced by the formulas. The rest of the paper is structured into the following sections. Section 2 reviews parametrization of surfaces of revolution, the technique of moving surfaces, and Bezout matrices. Their properties are needed for the derivation of the results in the paper. Section 3 presents two formulas. Each formula spawns n moving planes that are used to implicitize a surface of revolution. Section 4 asserts that the $2n \times 2n$ implicitization matrix consists of three Bezout matrices and is almost block circulant. Section 5 observes that rational parametrizations of surfaces of revolution created by symmetric rational curves are inherently improper. Section 6 concludes the paper with several open problems.

2   Preliminaries

This section reviews (a) parametrization of surfaces of revolution, (b) the implicitization technique of moving surfaces, and (c) Bezout matrices.

2.1   Parametrizing Surfaces of Revolution

In the Euclidean 3-space (X, Y, Z), a rational curve in the plane $X = 0$ can be described as

$$Y(t) = \frac{u(t)}{l(t)} = \frac{\sum_{i=0}^{n} u_i t^i}{\sum_{i=0}^{n} l_i t^i}, \quad Z(t) = \frac{v(t)}{l(t)} = \frac{\sum_{i=0}^{n} v_i t^i}{\sum_{i=0}^{n} l_i t^i}. \qquad (2)$$

Taking the rational representations for trigonometric functions

$$\sin(s) = \frac{2s}{1+s^2}, \quad \cos(s) = \frac{1-s^2}{1+s^2}, \tag{3}$$

the surface of revolution obtained by rotating curve (2) about the Z axis can be parametrized as

$$X(s,t) = \frac{x(s,t)}{w(s,t)}, \quad Y(s,t) = \frac{y(s,t)}{w(s,t)}, \quad Z(s,t) = \frac{z(s,t)}{w(s,t)}, \tag{4}$$

where

$$w(s,t) = (1+s^2)l(t), \tag{5}$$
$$x(s,t) = (1-s^2)u(t), \tag{6}$$
$$y(s,t) = 2su(t), \tag{7}$$
$$z(s,t) = (1+s^2)v(t). \tag{8}$$

For example, with spherical coordinates (r, θ, ϕ), we obtain readily the parametrization of a unit sphere centered at the origin as

$$(X(s,t), Y(s,t), Z(s,t)) = \frac{(2(1-s^2)t, 4st, (1+s^2)(1-t^2))}{(1+s^2)(1+t^2)}. \tag{9}$$

2.2   The Technique of Moving Surfaces

A degree $p \times q$ moving plane is a polynomial in the parameter s of degree p and in the parameter t of degree q with coefficients linear in X, Y, Z:

$$P_k(X,Y,Z;s,t) = \sum_{i=0}^{p} \sum_{j=0}^{q} P_{k,ij}(X,Y,Z)s^i t^j \tag{10}$$

where

$$P_{k,ij}(X,Y,Z) = A_{k,ij}X + B_{k,ij}Y + C_{k,ij}Z + D_{k,ij}, \tag{11}$$

$A_{k,ij}$, $B_{k,ij}$, $C_{k,ij}$, and $D_{k,ij}$ are scalars, and k is some index. The moving plane follows surface (4) if

$$P_k\left(\frac{x(s,t)}{w(s,t)}, \frac{y(s,t)}{w(s,t)}, \frac{z(s,t)}{w(s,t)}; s,t\right) \equiv 0. \tag{12}$$

It is known that [9,10] if there are $(p+1)(q+1) = 2n$ linearly independent moving planes that follow (4) then the $2n \times 2n$ determinant

$$\begin{vmatrix} P_{1,00} & \cdots & P_{2n,00} \\ \vdots & \ddots & \vdots \\ P_{1,pq} & \cdots & P_{2n,pq} \end{vmatrix} = 0 \tag{13}$$

is the implicit equation of surface (4).

2.3   Bezout Matrices

The Bezout matrix [1] of two degree n polynomials

$$p(t) = \sum_{i=0}^{n} p_i t^i, \quad q(t) = \sum_{i=0}^{n} q_i t^i, \tag{14}$$

is the $n \times n$ coefficient matrix $BZ(p, q)$ of the polynomial given in quotient form

$$\frac{\begin{vmatrix} p(t) & q(t) \\ p(\tau) & q(\tau) \end{vmatrix}}{t - \tau} = \begin{bmatrix} 1 \cdots t^{n-1} \end{bmatrix} BZ(p, q) \begin{bmatrix} 1 \cdots \tau^{n-1} \end{bmatrix}^T. \tag{15}$$

The Bezout matrix entry is given by

$$BZ_{ij} = - \sum_{k=\max(0,i+j-n-1)}^{\min(i-1,j-1)} p_k q_{i+j-1-k} - p_{i+j-1-k} q_k, \tag{16}$$

for $i, j = 1, \cdots, n$.

3   The Two Spawning Formulas

Two formulas, each spawning n moving planes that follow surface (4), for a total of $2n$ moving planes, are presented in this section.

3.1   The Spawning 3×3 Determinant

First we introduce three variables λ, μ, ν and use the parametric polynomials $l(t)$, $u(t)$, $v(t)$ of (2) to define the spawning 3×3 determinant

$$\pi_k(\lambda, \mu, \nu) = \begin{vmatrix} \sum_{i=0}^{k-1} l_i t^i & \sum_{i=k}^{n} l_i t^{i-k} & \lambda \\ \sum_{i=0}^{k-1} u_i t^i & \sum_{i=k}^{n} u_i t^{i-k} & \mu \\ \sum_{i=0}^{k-1} v_i t^i & \sum_{i=k}^{n} v_i t^{i-k} & \nu \end{vmatrix}, \tag{17}$$

$k = 1, \cdots, n$.

The determinant has the following property.

Lemma 1. $\pi_k(l(t), u(t), v(t)) \equiv 0, \quad k = 1, \cdots, n$.

Proof

After the substitutions, we have

$$C_1 + t^k C_2 = C_3, \tag{18}$$

where C_i is column i, $i = 1, 2, 3$. Thus the determinant vanishes.   □

3.2   The 2n Moving Planes That Follow the Surface of Revolution

The following theorem presents the two formulas that spawn $2n$ moving planes following surface (4).

Theorem 1. *The $2n$ degree $1 \times (n-1)$ moving planes*

$$\phi_k(X, Y, Z; s, t) = \pi_k(1, X, Z) + \pi_k(0, Y, 0)s = 0, \qquad (19)$$
$$\psi_k(X, Y, Z; s, t) = \pi_k(0, Y, 0) + \pi_k(1, -X, Z)s = 0, \qquad (20)$$

$k = 1, \cdots, n$, *follow surface (4)*.

Proof
Recall that surface (4) is

$$X(s,t) = \frac{(1-s^2)u(t)}{(1+s^2)l(t)}, \quad Y(s,t) = \frac{2s\, u(t)}{(1+s^2)l(t)}, \quad Z(s,t) = \frac{(1+s^2)v(t)}{(1+s^2)l(t)}. \qquad (21)$$

Substituting (21) into $\phi_k(X, Y, Z; s, t)$ and applying Lemma 1, we have

$$\phi_k(X(s,t), Y(s,t), Z(s,t); s, t) = \pi_k(1, X(s,t) + Y(s,t)s, Z(s,t)) \qquad (22)$$
$$= \frac{1}{l(t)}\pi_k(l(t), u(t), v(t)) \qquad (23)$$
$$= 0. \qquad (24)$$

That is, the n degree $1 \times (n-1)$ moving planes (19) follow surface (4).
    Similarly, we have

$$\psi_k(X(s,t), Y(s,t), Z(s,t); s, t) = \pi_k(s, Y(s,t) - X(s,t)s, Z(s,t)s) \qquad (25)$$
$$= \frac{s}{l(t)}\pi_k(l(t), u(t), v(t)) \qquad (26)$$
$$= 0. \qquad (27)$$

That is, the n degree $1 \times (n-1)$ moving planes (20) also follow surface (4).   □

3.3   Linear Independence of the 2n Moving Planes

The following theorem ensures that these $2n$ moving planes indeed implicitize surface (4) because they are in general linearly independent.

Theorem 2. *The $2n$ degree $1 \times (n-1)$ moving planes*

$$\phi_1(X, Y, Z; s, t) = 0, \cdots, \psi_n(X, Y, Z; s, t) = 0, \qquad (28)$$

that follow surface (4) are in general linearly independent.

Proof

The $2n$ moving planes can be concisely expressed in matrix form as

$$[\phi_1, \cdots, \phi_n, \psi_1, \cdots, \psi_n] = [1, \cdots, t^{n-1}, s, \cdots, st^{n-1}] M_n. \qquad (29)$$

Consider the particular parametric polynomials

$$l(t) = 1, \quad u(t) = 1, \quad v(t) = t^n. \qquad (30)$$

The corresponding $2n$ degree $1 \times (n-1)$ moving planes (19), (20) are

$$\begin{vmatrix} 1 & 0 & 1 \\ 1 & 0 & X + Ys \\ 0 & t^{n-k} & Z \end{vmatrix} = (1 - X)t^{n-k} - Yst^{n-k} = 0, \qquad (31)$$

and

$$\begin{vmatrix} 1 & 0 & s \\ 1 & 0 & Y - Xs \\ 0 & t^{n-k} & Zs \end{vmatrix} = -Yt^{n-k} + (1 + X)st^{n-k} = 0, \qquad (32)$$

for $k = 1, \cdots, n$.

For these specialized parametric polynomials, the coefficient matrix is

$$M_n = \begin{bmatrix} & & 1-X & & & -Y \\ & \ddots & & & \ddots & \\ 1-X & & & -Y & & \\ & & -Y & & & (1+X) \\ & \ddots & & & \ddots & \\ -Y & & & (1+X) & & \end{bmatrix}. \qquad (33)$$

A straightforward calculation shows that its determinant is

$$|M_n| = (1 - X^2 - Y^2)^n. \qquad (34)$$

Since the matrix is non-singular thus the $2n$ moving planes are linearly independent for the specialized parametric polynomials (30).

Consequently the $2n$ moving planes are linearly independent for general parametric polynomials (2) since they are linearly independent for a particular specialization of these parametric polynomials. □

When generatrix (2) degenerates into a point; that is,

$$l(t) : u(t) : v(t) = \lambda : \mu : \nu \qquad (35)$$

for some constants λ, μ, ν, it is easily checked that M_n is a zero matrix. Thus for arbitrary specialization of the parametric polynomials $l(t)$, $u(t)$, $v(t)$, it is possible that M_n becomes singular and the method fails to produce the implicit equation. Such failure is well known in the matrix-based implicitization technique using resultants. Resultants fail when there are base points because the dimension of the implicitization matrix stays while the degree of the surface drops. This dimension-degree mismatch causes the implicitization matrix to become singular. However, as explained in the Introduction, the degree of a surface of revolution is $2n$ if the degree of the generatrix is n. In other words the degree of the surface never drops. Based on this geometric observation we expect that the method always works if the generatrix is indeed a curve and not a point.

3.4   The $2n \times 2n$ Implicitization Determinant for Surfaces of Revolution

Combining Theorems 1 and 2, we have

Theorem 3. *The implicit equation of surface (4), a surface of revolution, is*

$$|M_n| = 0. \tag{36}$$

To illustrate the theorem, we use the notation

$$|x_i y_j| = x_i y_j - x_j y_i. \tag{37}$$

to express the template 4×4 matrix

$$M_2 = \begin{bmatrix} |v_0 l_1|X + |l_0 u_1|Z + |u_0 v_1| & |v_0 l_2|X + |l_0 u_2|Z + |u_0 v_2| \\ |v_0 l_2|X + |l_0 u_2|Z + |u_0 v_2| & |v_1 l_2|X + |l_1 u_2|Z + |u_1 v_2| \\ |v_0 l_1|Y & |v_0 l_2|Y \\ |v_0 l_2|Y & |v_1 l_2|Y \end{bmatrix}$$

$$\begin{bmatrix} |v_0 l_1|Y & |v_0 l_2|Y \\ |v_0 l_2|Y & |v_1 l_2|Y \\ |l_0 v_1|X + |l_0 u_1|Z + |u_0 v_1| & |l_0 v_2|X + |l_0 u_2|Z + |u_0 v_2| \\ |l_0 v_2|X + |l_0 u_2|Z + |u_0 v_2| & |l_1 v_2|X + |l_1 u_2|Z + |u_1 v_2| \end{bmatrix}. \tag{38}$$

Specialize Equation (2) to

$$l(t) = 1 + t^2, \quad u(t) = 2t, \quad v(t) = 1 - t^2, \tag{39}$$

gives a circle in the plane $X = 0$. Using the template matrix M_2, the implicit equation of the resulting sphere as a surface of revolution is

$$
\begin{vmatrix}
2(Z-1) & 2X & 0 & 2Y \\
2X & -2(1+Z) & 2Y & 0 \\
0 & 2Y & 2(Z-1) & -2X \\
2Y & 0 & -2X & -2(1+Z)
\end{vmatrix} = 16(X^2 + Y^2 + Z^2 - 1)^2. \quad (40)
$$

The square of the expected implicit equation is produced because of improper parametrization. This is clarified further in Section 5.

Specialize Equation (2) to

$$
l(t) = t, \quad u(t) = t^2, \quad v(t) = 1, \quad (41)
$$

gives the hyperbola $YZ = 1$ in the plane $X = 0$. Using the template matrix M_2, the implicit equation of the resulting hyperboloid as a surface of revolution is

$$
\begin{vmatrix}
X & -1 & Y & 0 \\
-1 & Z & 0 & 0 \\
Y & 0 & -X & -1 \\
0 & 0 & -1 & Z
\end{vmatrix} = 1 - Z^2(X^2 + Y^2). \quad (42)
$$

4   Properties of the $2n \times 2n$ Implicitization Determinant

This section demonstrates that the $2n \times 2n$ implicitization matrix M_n composes of several Bezout matrices. The following theorem states this fact precisely.

Theorem 4.

$$
M_n = \begin{bmatrix}
AX + CZ + D & AY \\
AY & -AX + CZ + D
\end{bmatrix}, \quad (43)
$$

where $A = BZ(l(t), v(t))$, $C = BZ(u(t), l(t))$, $D = BZ(v(t), u(t))$, and $l(t)$, $u(t)$, $v(t)$ are the parametric polynomials (2).

Proof

By observing the structures of the moving planes (19) and (20), we see that

$$
M_n = \begin{bmatrix}
A_1 X + C_1 Z + D_1 & B_1 Y \\
B_2 Y & A_2 X + C_2 Z + D_2
\end{bmatrix} \quad (44)
$$

where all the matrices $A_1, \cdots, D_2$ are $n \times n$ matrices.

The (i, j) entry of matrix B_2, $i, j = 1, \cdots, n$, is the coefficient of st^{i-1} of the moving plane

$$\phi_j(X, Y, Z; s, t) = \pi_j(1, X, Z) + \pi_j(0, Y, 0)s. \tag{45}$$

Thus it is the coefficient of t^{i-1} in the polynomial

$$\pi_j(0, Y, 0) = -Y \begin{vmatrix} \sum_{k=0}^{j-1} l_k t^k & \sum_{k=j}^{n} l_k t^{k-j} \\ \sum_{k=0}^{j-1} v_k t^k & \sum_{k=j}^{n} v_k t^{k-j} \end{vmatrix}, \tag{46}$$

which is

$$- \sum_{k_1 < k_2, k_1 + k_2 - j = i - 1} \begin{vmatrix} l_{k_1} & l_{k_2} \\ v_{k_1} & v_{k_2} \end{vmatrix}. \tag{47}$$

From (47) immediately we have

$$k_1 + k_2 = i + j - 1, \tag{48}$$

and

$$0 \leq k_1 \leq j - 1. \tag{49}$$

Applying $j \leq k_2 \leq n$ to (48), we have

$$i + j - n - 1 \leq k_1 \leq i - 1. \tag{50}$$

Combining the above constraints, we conclude

$$\max(0, i + j - n - 1) \leq k_1 \leq \min(i - 1, j - 1). \tag{51}$$

This proves that $B_2 = BZ(l, v)$.

Similarly, we can prove that

$$A_1 = -A_2 = A, \tag{52}$$
$$B_1 = A, \tag{53}$$
$$C_1 = C_2 = C, \tag{54}$$
$$D_1 = D_2 = D. \tag{55}$$

Together these equalities prove the theorem.   □

The theorem substantiates the claim that the $2n \times 2n$ implicitization matrix is almost block circulant. Indeed the rightmost n columns are the leftmost n columns rotate upward (or downward) n rows, except that the sign of the $n \times n$ matrix A has to be flipped.

5   Inherently Improper Parametrization of Symmetric Surfaces of Revolution

As an aside, we remark that any rational parametrization of a surface of revolution due to a symmetric rational generatrix must be improper. Without loss of generality a symmetric surface of revolution is created by a generatrix curve in the $X = 0$ plane that is symmetric with respect to the Z axis. The curve then has an implicit equation of the form

$$g(y^2, z) = 0. \tag{56}$$

Consequently If the curve degree is n, the surface degree is still n with the implicit equation

$$g(x^2 + y^2, z) = 0. \tag{57}$$

However, as discussed earlier, a line in the plane $X = 0$ intersects the surface $2n$ times: n times with the curve and another n times with the 180-degree rotated curve (the mirror copy, which is the original curve due to the symmetry). Thus the parametrized surface has degree $2n$, rather than n. This makes the parametrization improper. The above discussion is formalized in the following theorem

Theorem 5. *Any rational parametrization of a symmetric surface of revolution due to a symmetric generatrix is improper.*

For example, the rational parametrization of the following symmetric non-trivial quadrics must be improper when the parametrization is obtained from formulation (4).

Ellipsoids	$\dfrac{x^2 + y^2}{a^2} + \dfrac{z^2}{c} = 1$	
1-sheet hyperboloids	$\dfrac{x^2 + y^2}{a^2} - \dfrac{z^2}{c^2} = 1$	
2-sheet hyperboloids	$-\dfrac{x^2 + y^2}{a^2} + \dfrac{z^2}{c^2} = 1$	(58)
Elliptic paraboloid	$\dfrac{x^2 + y^2}{a^2} - z = 0$	
Quadric cones	$\dfrac{x^2 + y^2}{a^2} - \dfrac{z^2}{c^2} = 0$	

6   Observations and Open Problems

We conclude the paper with the following observations and attending open problems.

The geometric action of rotating a rational plane curve about an axis in the same plane has algebraic manifestations. This occurs when the method of moving surfaces is applied to implicitize the surface of revolution. The moving planes are spawned by successively shifting terms of a 3×3 spawning determinant from one column to another; the resulting implicitization determinant has almost a block circulant structure.

This way of constructing moving planes is a non-trivial generalization of using Bezout's resultant to implicitize rational plane curves realized with the technique of vector elimination [5]. It might be rewarding to explore if the success here can be repeated in other surface creation schemes.

It is well-known that the Dixon determinant [4,8] implicitizes a general bi-degree parametric surface. Thus the question naturally arises: in between surfaces of revolution and bi-degree parametric surfaces, if there are other families of surfaces that the Bezout-Dixon style of implicitization applies.

The method of moving surfaces is a very powerful implicitization technique. But the method often requires the solution of a linear system. In the context of surfaces of revolution, we have illustrated that actually the implicitization determinant can be constructed with apriori known entries. It is thus worthwhile to find out if there are other surface construction schemes for which an explicit implicitization determinant exists.

Inherently improper surface parametrizations exist algebraically with a suitably sparse monomial support [2]. We have just seen that surfaces of revolution by a symmetric rational generatrix are necessarily parametrized improperly. Again we ask the question if there are other surface schemes that lead to inherently improper parametrizations.

The moving planes given are shown to implicitize a surface of revolution correctly in general and is expected to work in particular as long as the generatrix is indeed a curve. The latter was speculated based on some intuitive geometric reasoning using the surface degree and the implicitization matrix dimension. It would be very useful and interesting either to prove this observation algebraically with rigor or to disprove it with a counter example.

Finally, the matrix dimension is $2n$ since moving planes are used. The matrix dimension will be halved to n if n moving quadrics [3] can be found. This seems to be an immediate sequel that should be pursued.

References

1. Chionh, E.W., Goldman, R.N.: Part 1: Elimination and Bivariate Resultants. IEEE Computer Graphics and Applications 15(1), 69–77 (1995)
2. Chionh, E.W., Gao, X.S., Shen, L.Y.: Inherently improper surface parametric supports. Computer Aided Geometric Design 23, 629–639 (2006)
3. Cox, D., Goldman, R.N., Zhang, M.: On the validity of implicitization by moving quadrics for rational surfaces with no base points. Journal of Symbolic Computation 29, 419–440 (2000)
4. Dixon, A.L.: The eliminant of three quantics in two independent variables. Proc. London Math. Soc. 6, 49–69, 473–492 (1908)
5. Goldman, R.N., Sederberg, T.W., Anderson, D.C.: Vector elimination: A technique for the implicitization, inversion and intersection of planar parametric rational polynomial curves. Computer Aided Geometric Design 1(4), 327–356 (1984)

6. González-Vega, L., Necula, I., Pérez-Díaz, S., Sendra, J., Sendra, J.R.: Algebraic Methods in Computer Aided Geometric Design: Theoretical and Practical Applications. In: Geometric Computation. Lecture Notes Series on Computing, vol. 11, pp. 1–33. World Scientific, Singapore (2004)
7. Kotsireas, I.S.: Panorama of Methods for Exact Implicitization of Algebraic Curves and Surfaces. In: Geometric Computation. Lecture Notes Series on Computing, vol. 11, pp. 126–155. World Scientific, Singapore (2004)
8. Sederberg, T.W., Anderson, D.C., Goldman, R.N.: Implicit representation of parametric curves and surfaces. Computer Vision, Graphics and Image Processing 28, 72–84 (1984)
9. Sederberg, T.W., Chen, F.L.: Implicitization Using Moving Curves and Surfaces. In: Proceedings of SIGGRAPH 1995, pp. 301–308. ACM SIGGRAPH/Addison Wesley, Los Angeles (1995)
10. Zheng, J.M., Sederberg, T.W., Chionh, E.W., Cox, D.A.: Implicitizing rational surfaces with base points using the method of moving surfaces. Contemporary Mathematics 334, 151–168 (2003)

Slit Map: Conformal Parameterization for Multiply Connected Surfaces

Xiaotian Yin[1], Junfei Dai[2], Shing-Tung Yau[3], and Xianfeng Gu[1]

[1] Center for Visual Computing
State University of New York at Stony Brook
{xyin,gu}@cs.sunysb.edu
[2] Center of Mathematics Science
Zhejiang University
jfdai@cms.zju.edu.cn
[3] Mathematics Department
Harvard University
yau@math.harvard.edu

Abstract. Surface parameterization is a fundamental tool in geometric modeling and processing. Different to most existing methods, this work introduces a *linear* method, called *slit map*, to conformally parameterize multiply connected disk-like surfaces with *canonical* domains, namely *circular slit domain* (annulus with concentric circular slits) or *parallel slit domain* (rectangle with parallel slits) equivalently. The construction is based on holomorphic one-forms, which is intrinsic, automatic and efficient. Slit map owns many merits over existing methods. The regularity of the image boundaries simplifies many geometric process, such as surface matching and etc; the full utilization of the parallel slit texture domain improves the packing efficiency for texture mapping; the positions of the slits are completely determined by the surface geometry and can be treated as the finger prints for surface classification. In the paper, both the underlying theory and the algorithm pipeline are explained. Preliminary experimental results are shown for several application purposes.

Keywords: conformal parameterization, slit map, holomorphic one-form, canonical parameter domain.

1 Introduction

Surface conformal parameterization is a fundamental tool in geometric modeling and processing. It is an essential technique for many applications, such as texture mapping, surface matching, registration and tracking, re-meshing, mesh-spline conversion and so on.

Most existing parameterization methods focus on *simply connected* surfaces, namely genus zero surfaces with a single boundary. According to Riemann's mapping theorem, any simply connected surface can be conformally mapped to the unit disk. In other words, they are all *conformally equivalent*, and the unit disk is the canonical representative in common.

F. Chen and B. Jüttler (Eds.): GMP 2008, LNCS 4975, pp. 410–422, 2008.

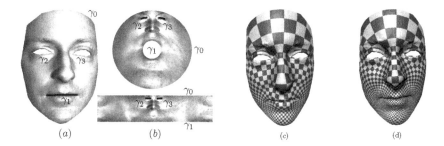

Fig. 1. Slit map. (a) is the original multiply connected surface with four boundary curves. (b) shows the circular slit domain (top) and the parallel slit domain (bottom). (c) and (d) show the conformal texture mapping using the circular and parallel slit map respectively.

But in practice, the surfaces under study are sometimes *multiply connected*; that is, they are genus zero surfaces with multiple boundaries (see figure 1a). According to Riemann surface theory [1], these surfaces can also be mapped to canonical domains conformally. One of such canonical domains is the *circular slit domain*, which is an annulus (figure 1b top) with two prescribed boundaries mapped to concentric circles while the others mapped to concentric arc slits. The other canonical domain is the *parallel slit domain*, which is a rectangle strip (figure 1b bottom) with two prescribed boundaries mapped to the upper and lower straight boundaries while the others mapped to horizontal slits.

Compared to the simply connected case, the canonical mapping for multiply connected surfaces are more complicated. In fact, such surfaces are in general not conformally equivalent. The slit domain for different surfaces will differ by the position and length of the slits. In this sense, the conformal parameterization algorithm for multiply connected surfaces must compute both the mapping and the target domain; whereas the algorithm for simply connected surfaces only needs to compute the mapping, since the domain could be unique.

1.1 Contributions

In this paper we propose a novel computational method of conformal parameterization for multiply connected surfaces. It maps a given surface to either circular slit domain or parallel slit domain. The algorithm is based on finding certain holomorphic one-forms whose integration along boundaries satisfies special constraints. To the best of our knowledge, this is the first linear algorithm to conformally map arbitrary multiply connected disk-like surfaces to canonical domains.

Slit map has many merits for geometric modeling and processing tasks. First, it maps the whole surface to a rectangle, with all the boundaries mapped to parallel slits. This regularity not only helps improve the accuracy for surface

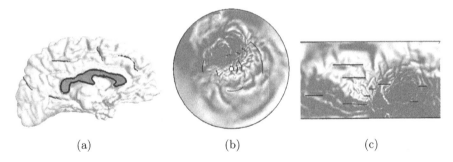

(a)   (b)   (c)

Fig. 2. Slit map for a brain surface model with 12 open landmarks

matching with boundaries, but also makes quad-remeshing, mesh-spline conversion convenient. Second, the whole rectangle in the parallel slit domain is fully occupied without any gap or overlapping, which improves the packing efficiency of texture mapping. Third, the positions of the slits in the regular domain are completely determined by the surface geometry, which can be treated as the finger print to classify surfaces by conformal equivalence. Fourth, the whole pipeline is based on manipulation of one-forms, which is a linear process that is very efficient and stable. Plus, the whole pipeline is very easy to implement.

1.2   Related Work

In recent years, many excellent algorithms have been developed in surface parameterization. In this work, we only briefly review the most related works. For more thorough references on parameterization methods, we refer readers to the following excellent surveys: [9] by Floater et al and [23] by Sheffer et al.

There are many existing methods targeting at simply connected surfaces, either with fixed boundary condition ([8], [6] and etc) or free boundary condition [[16], [3], [22], [15] and etc]. These methods either cannot handle the multiply connected surfaces directly, or they cannot map such surfaces to canonical domains, while our method addresses both issues.

A common way to generalize the parameterization methods for simply connected surfaces is by "cutting-and-packing". This approach has been adapted by Cohen-Steiner et al. [2], Garland et al. [10], Maillot et al. [17] and Sander et al. [21]. All these methods require to cut the surface into multiple patches, parameterize them separately and then pack them together in the texture domain. Because the partitioning process is artificial, global geometric properties of the surfaces will be lost. Furthermore, the conformality can not be achieved along the patch boundaries.

Global methods without segmentation are proposed by Gu et al. [12][13], Ray et al. [20] and Dong et al. [5]. For example, [12] used holomorphic one-forms as the underlying tool. But for multi-hole annuli, this method needs double covering to make the surface close, which will increase the computational cost.

Our slit method, on the other hand, does not require double covering and is more efficient. After all, none of these methods can guarantee a canonical parameter domain as the slit map does.

Another method for multiply connected surfaces is discrete Ricci flow, proposed by Jin et al. [14]. They can prescribe the target curvature of boundaries, so that the target domain has regular shape, such as a unit disk with multiple inner circle boundaries. But it requires non-linear optimization. In our method only linear operations get involved and is therefore much faster and more stable.

Our method is based on one-form, which has been used for many practical purposes, such as vector fields decomposition and smoothing [24] and etc. Discrete one-forms on meshes were studied in [11]. Tong et al. [25] used harmonic one-forms for surface parameterization. They enlarged the space of harmonic one-forms by allowing additional singular points on the surface. Kälberer et al. applied one-forms for surface parameterization combining with branch covering in [19], where the parameter lines were governed by a given frame field. In [7] Fisher et al. used one-forms to design tangent vector fields on surfaces with complicated topologies.

In the rest of the paper, we will first present the theories underlying our method in section 2. Then we give the algorithm pipeline and the implementation details of each step in section 3. The experimental results are reported in 4, followed by a conclusion and a brief discussion on the future direction in 5.

2   Theoretic Background

In this section, we briefly introduce the theoretic background necessary for understanding the algorithm proposed here. Intuitively, a tangent vector field on a surface is *harmonic*, if both its circulation and divergence are zeros. Two harmonic fields form a *holomorphic one-form*, if they are orthogonal everywhere. An intrinsic way to compute the slip map is to search for certain holomorphic one-forms with special behavior on the boundary of the surface.

2.1   Harmonic Function

Suppose S is a surface embedded in $\mathbb{R}^3$ with induced Euclidean metric $\mathbf{g}$. S is covered by an atlas $\{(U_\alpha, \phi_\alpha)\}$. Suppose (x_α, y_α) is the local parameter on the chart (U_α, ϕ_α). We say (x_α, y_α) is *isothermal*, if the metric has the representation

$$\mathbf{g} = e^{2\lambda(x_\alpha, y_\alpha)}(dx_\alpha^2 + dy_\alpha^2).$$

The *Laplace-Beltrami operator* is defined as

$$\Delta_{\mathbf{g}} = \frac{1}{e^{2\lambda(x_\alpha, y_\alpha)}}\left(\frac{\partial^2}{\partial x_\alpha^2} + \frac{\partial^2}{\partial y_\alpha^2}\right).$$

Definition 1 (Harmonic Function). *A function* $f : S \to \mathbb{R}$ *is* harmonic, *if* $\Delta_{\mathbf{g}} f \equiv 0$.

2.2   Holomorphic One-Form

Suppose η is a differential one-form with the representation $f_\alpha dx_\alpha + g_\alpha dy_\alpha$ in the local parameters (x_α, y_α), and $f_\beta dx_\beta + g_\beta dy_\beta$ in the local parameters (x_β, y_β). Then

$$\begin{pmatrix} \frac{\partial x_\alpha}{\partial x_\beta} & \frac{\partial y_\alpha}{\partial x_\beta} \\ \frac{\partial x_\alpha}{\partial y_\beta} & \frac{\partial y_\alpha}{\partial y_\beta} \end{pmatrix} \begin{pmatrix} f_\alpha \\ g_\alpha \end{pmatrix} = \begin{pmatrix} f_\beta \\ g_\beta \end{pmatrix}.$$

η is a *closed one-form*, if on each chart (x_α, y_α)

$$\frac{\partial f}{\partial y_\alpha} - \frac{\partial g}{\partial x_\alpha} = 0.$$

η is an *exact one-form*, if it equals the gradient of some function. An exact one-form is also a closed one-form. If a closed one-form η satisfies

$$\frac{\partial f}{\partial x_\alpha} + \frac{\partial g}{\partial y_\alpha} = 0,$$

then η is a *harmonic one-form*. The gradient of a harmonic function is an exact harmonic one-form.

The so-called *Hodge star operator* turns a one-form η to its *conjugate* $^*\eta$,

$$^*\eta = -g_\alpha dx_\alpha + f_\alpha dy_\alpha.$$

Definition 2 (Holomorphic One-form). *A* holomorphic one-form *is a complex differential form*

$$\eta + \sqrt{-1}^*\eta,$$

where η is a harmonic one-form.

The wedge product of two one-forms $\eta_k = f_k dx + g_k dy, k = 1, 2$ is a two-form

$$\eta_1 \wedge \eta_2 = (f_1 g_2 - f_2 g_1) dx \wedge dy.$$

2.3   Slit Map

Suppose S is an open surface with n boundaries $\gamma_0, \cdots, \gamma_{n-1}$. We can uniquely find a holomorphic one-form $\omega = \eta + \sqrt{-1}^*\eta$ such that η is exact, $^*\eta$ is closed, and

$$Im\left(\int_{\gamma_k} \omega\right) = \begin{cases} 2\pi & k = 0 \\ -2\pi & k = 1 \\ 0 & otherwise \end{cases}$$

where Im takes the imaginary part of the complex valued integration. With such a ω we can define the circular slit map as following.

Definition 3 (Circular Slit Map). *Fix a point p_0 on the surface, for any point $p \in S$, let γ be an arbitrary path connecting p_0 and p, then the circular slit map is defined as*

$$\phi(p) = e^{\int_\gamma \omega}.$$

The parallel slit map can be defined in a similar way.

Definition 4 (Parallel Slit Map). *Let $\bar{S}$ be the universal covering space of the surface S, $\pi : \bar{S} \rightarrow S$ be the projection and $\bar{\omega} = \pi^* \omega$ be the pull back of ω. Fix a point $\bar{p}_0$ on $\bar{S}$, for any point $p \in \bar{S}$, let $\bar{\gamma}$ be an arbitrary path connecting $\bar{p}_0$ and $\bar{p}$, then the parallel slit map is defined as*

$$\bar{\phi}(\bar{p}) = \int_{\bar{\gamma}} \bar{\omega}.$$

The following theorem characterizes the circular slit map. The proof for an equivalent statement can be found in [1].

Theorem 1 (Canonical Domains for Multiply Connected Surface). *The circular slit map ϕ effects a one-to-one conformal mapping of M onto the annulus $e^{\lambda_1} < |z| < e^{\lambda_0}$ minus $n - 2$ concentric arcs situated on the circles $|z| = e^{\lambda_i}, i = 2, \cdots, n - 1$.*

Slit map computes the intrinsic structure of the given surface, which is reflected by the shape of the parameter domain in terms of the size and position of slits. For a given choice of the inner and outer circle, the circular slit map is uniquely determined up to a rotation around the center. Similarly, for a given choice of the upper and lower boundary, the parallel slip map is uniquely determined up to a translation along the horizontal direction.

3   Algorithm Pipeline

In this work, our goal is to compute a conformal mapping of a multiply connected mesh to the slit domain. Suppose the input mesh has $n + 1$ boundaries,

$$\partial M = \gamma_0 - \gamma_1 - \cdots - \gamma_n.$$

Without loss of generality, we map γ_0 to the outer circle of the circular slit domain, γ_1 to the inner circle, and all the others to the concentric slits.
    The following is the algorithm pipeline:

1. Compute the basis for all exact harmonic one-forms; (*section 3.1*)
2. Compute the basis for all closed harmonic one-forms; (*section 3.2*)
3. Compute the basis for all holomorphic one-forms; (*section 3.3*)
4. Construct the slit map. (*section 3.4*).

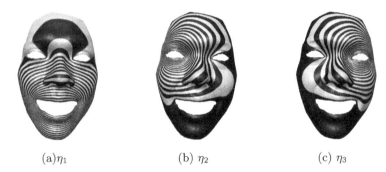

(a)η_1 (b) η_2 (c) η_3

Fig. 3. Basis for the exact harmonic one-forms

3.1   Compute Exact Harmonic One-Form Basis

The first step of the algorithm is to compute the basis for exact harmonic one-forms. Let γ_k be an inner boundary, we compute a harmonic function $f_k : S \to \mathbb{R}$ by solving the following Dirichlet problem on mesh M:

$$\begin{cases} \Delta f_k \equiv 0 \\ f_k|_{\gamma_j} = \delta_{kj} \end{cases}$$

where δ_{kj} is the Kronecker function, Δ is the discrete Laplacian-Beltrami operator using the well-known co-tangent formula proposed in [18].

The exact harmonic one-form η_k can be computed as the gradient of the harmonic function f_k,

$$\eta_k = df_k,$$

and $\{\eta_1, \eta_2, \cdots, \eta_n\}$ form the basis for the exact harmonic one-forms (shown in figure 3).

3.2   Compute Closed Harmonic One-Form Basis

After getting the exact harmonic one-forms, we will compute the closed one-form basis. Let γ_k ($k > 0$) be an inner boundary. Compute a path from γ_k to γ_0, denoted as ζ_k, which cuts the mesh open to M_k, while ζ_k itself is split into two boundary segments ζ_k^+ and ζ_k^- in M_k. As done in [25], define a function $g_k : M_k \to \mathbb{R}$ by solving a Dirichlet problem,

$$\begin{cases} \Delta g_k \equiv 0 \\ g_k|_{\zeta_k^+} = 1 \\ g_k|_{\zeta_k^-} = 0. \end{cases}$$

Compute the gradient of g_k and let $\tau_k = dg_k$, then map τ_k back to M, where τ_k becomes a closed one-form. Then we need to find a function $h_k : M \to \mathbb{R}$, by solving the following linear system:

$$\Delta(\tau_k + dh_k) \equiv 0.$$

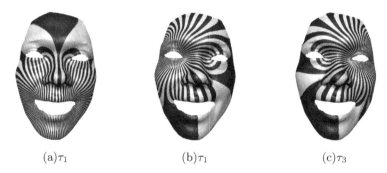

Fig. 4. Basis for the closed but not exact harmonic one-forms

Updating τ_k to $\tau_k + dh_k$, now we have $\{\tau_1, \tau_2, \ldots, \tau_n\}$ as a set of basis for all the closed but not exact harmonic one-forms. Figure 4 shows the closed non-exact harmonic one-form basis for the face model.

With both the exact harmonic one-form basis and the closed non-exact harmonic one-form basis computed, we can construct the harmonic one-form basis by taking the union of them:

$$\{\eta_1, \eta_2, \cdots, \eta_n, \tau_1, \tau_2, \cdots, \tau_n\}.$$

3.3 Basis Holomorphic One-Form Basis

In step 1 we computed the basis for exact harmonic one-forms $\{\eta_1, \cdots, \eta_n\}$. Now we compute their conjugate one-forms $\{^*\eta_1, \cdots, ^*\eta_n\}$, so that we can combine all of them together into a set of holomorphic one-form basis.

First of all, for η_k we compute an initial approximation η_k' by a brute-force method using Hodge star. That is, rotating η_k by $90°$ about the surface normal to obtain η_k'. In practice such an initial approximation is usually not accurate enough. In order to improve the accuracy, we employ a technique utilizing the harmonic one-form basis we just computed. From the fact the η_k is harmonic, we can conclude that its conjugate $^*\eta_k$ should also be harmonic. Therefore, $^*\eta_k$ can be represented as a linear combination of the base harmonic one-forms:

$$^*\eta_k = \sum_{i=1}^{n} a_i \eta_i + \sum_{i=1}^{n} b_i \tau_i.$$

Using the wedge product $\wedge$, we can construct the following linear system,

$$\int_M {}^*\eta_k \wedge \eta_i = \int_M \eta_k' \wedge \eta_i, \int_M {}^*\eta_k \wedge \tau_j = \int_M \eta_k' \wedge \tau_j.$$

We solve this linear system to obtain the coefficients a_i and b_i $(i = 1, 2, \cdots, n)$ for the conjugate one-form $^*\eta_k$. Pairing each base exact harmonic one-form with its conjugate, we get a set of basis for the holomorphic one-form group on M:

$$\{\eta_1 + \sqrt{-1}^*\eta_1, \cdots, \eta_n + \sqrt{-1}^*\eta_n\}$$

Figure 5 demonstrates the holomorphic one-form basis on the mesh.

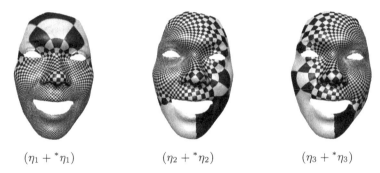

$$(\eta_1 + {}^*\eta_1) \qquad\qquad (\eta_2 + {}^*\eta_2) \qquad\qquad (\eta_3 + {}^*\eta_3)$$

Fig. 5. Holomorphic one-form basis

3.4   Construct Slit Map

After computing the holomorphic one-form basis, we need to find a special holomorphic one-form ω

$$\omega = \sum_{i=1}^{n} \lambda_i(\eta_i + \sqrt{-1}\,{}^*\eta_i)$$

such that the imaginary part of its integration satisfies

$$Im\left(\int_{\gamma_k} \omega\right) = \begin{cases} -2\pi & k = 1 \\ 0 & k > 1 \end{cases}$$

In order to get the coefficients λ_i, solve the following linear system for λ_i, $i = 1, \cdots, n$:

$$\begin{pmatrix} \alpha_{11} & \alpha_{12} & \cdots & \alpha_{1n} \\ \alpha_{21} & \alpha_{22} & \cdots & \alpha_{2n} \\ \vdots & \vdots & \ddots & \vdots \\ \alpha_{n1} & \alpha_{n2} & \cdots & \alpha_{nn} \end{pmatrix} \begin{pmatrix} \lambda_1 \\ \lambda_2 \\ \vdots \\ \lambda_n \end{pmatrix} = \begin{pmatrix} -2\pi \\ 0 \\ \vdots \\ 0 \end{pmatrix}$$

where

$$\alpha_{kj} = \int_{\gamma_j} {}^*\eta_k,$$

It can be proven that this linear system has a unique solution, which reflects the fact that γ_1 is mapped to the inner circle of the circular slit domain. Further, the system implies the following equation

$$\lambda_1\alpha_{01} + \lambda_2\alpha_{02} + \cdots + \lambda_n\alpha_{0n} = 2\pi,$$

which means that γ_0 is mapped to the outer circle in the circular slit domain.

After computing the desired holomorphic one-form ω, we are ready to generate the circular slit map. What we need to compute is a complex-valued function $\phi : M \to \mathbb{C}$ by integrating ω and taking the exponential map. Choosing a base vertex v_0 arbitrarily, and for each vertex $v \in M$ choosing the shortest path γ from v_0 to v, we can compute the map as the following:

$$\phi(v) = e^{\int_\gamma \omega}.$$

Based on the circular slit map ϕ we just computed, we can compute a parallel slit map $\tau : M \to \mathbb{C}$:

$$\tau(v) = \ln \phi(v).$$

Figure 1 shows the results for slip map. In the circular slit domain (figure 1b top), boundary curve γ_0 and γ_1 are mapped to the outer and inner circles, γ_2 and γ_3 mapped to the circular slits. In the parallel slit domain (figure 1b bottom), γ_0 and γ_1 are mapped to the upper and lower lines (single-trip), γ_2 and γ_3 mapped to horizontal slits (round-trip).

4   Applications and Experiments

We implemented our algorithm in C++ on Windows platform. The system has been tested on several real human face surfaces (see figure 7) generated by 3D scanner. Each face mesh has around $15k$ vertices and $30k$ triangles with 4 boundaries. We also tested our system on a brain cortical surface (figure 2), which is reconstructed from MRI images. The cortical surface has $30k$ vertices and $60k$ faces with 12 landmarks sliced on the surface. The harmonic forms are computed using a brute-force implementation of the conjugate gradient method, without using any well-tuned numerical library. The whole processing takes less than 3 minutes. In all experiments the algorithm converges stably, and the final parameterization results are conformal.

As a global conformal parameterization method, slit map can be applied to many applications in geometric modeling and processing. In the current work we applied slit map method for several most direct applications and got some preliminary results.

Texture Mapping. One of most direct applications in computer graphics is texture mapping. As shown in figure 1, the parallel slit domain can serve as the texture domain, which is very regular in shape. More over, the texture domain is fully utilized during the texture mapping, since there is no gap or overlap inside the domain.

Although this work focuses on multiply connected surfaces, in fact our algorithm can be easily generalized to handle high genus closed surfaces, such as the eight model in figure 6. The only extra requirement is to slice the surface open along certain cycles (see figure 6 c). Such cycles can be automatically computed using methods like that proposed by T. Dey et al in [4]. After turning the surface into a multiply connected one, we can carry out the slit algorithm thereafter directly.

Surface Fingerprint. The slit map method computes the conformal invariants of the surface. The shape parameters of the circular slit domain indicate the conformal equivalence class of the surface and can be treated as the fingerprints

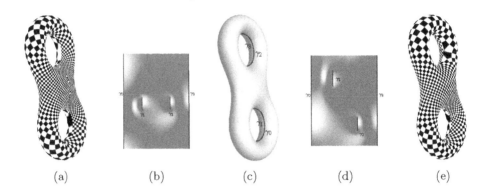

(a)   (b)   (c)   (d)   (e)

Fig. 6. Slit map of closed mesh. The closed eight model (c) is sliced open into an annulus with 4 boundaries. (b) and (d) show the parallel slit domains with different prescribed outer boundaries. (a) and (e) show the texture mapping corresponding to domain (b) and (d) respectively.

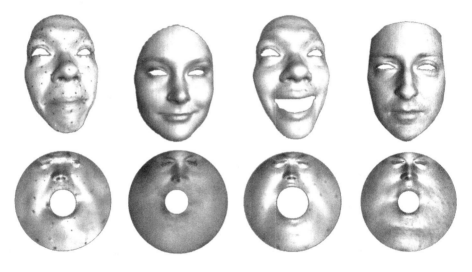

Fig. 7. Slit domain as finger prints. The first row shows four face models with different expression; the second row shows their finger prints using the circular slit domain

of the surface. We test our algorithm for several human faces from different persons with different expressions. The result is illustrated in figure 7. From this figure, it is very clear that the fingerprints of the three calm faces are very similar, whereas the fingerprint of the laughing face is quite different from others. This gives us a way to measure the expression quantitatively.

Brain Mapping. Slit map provides a valuable tool for conformal brain mapping with landmarks. As shown in figure 2, brain surfaces are highly convoluted. It is a great challenge to match two cortical surfaces directly in $\mathbb{R}^3$. Conformal brain mapping flattens the brain surface onto the canonical domains. Special landmarks are labeled on the surface, which are required to be registered across different brain surfaces. By using slit map, all the land marks are mapped to concentric or parallel slits, and the whole brain is mapped to annulus or rectangle. This makes the down stream registration and analysis much easier. In figure 2, the mapping result is shown on the circular slit domain and the parallel slit domain respectively.

As a general method for conformal parameterization, slit map can also benefits many other applications, such as quad-remeshing, geometry image generation, mesh-spline conversion and etc. It is an interesting research direction to explore the potential of our method for these applications.

5   Conclusion

In this work, we presented a novel global conformal parameterization method, called slit map, for multiply connected surfaces. The method is based on computing a holomorphic one-form that has special behaviors on the surface boundaries. The algorithm maps any multiply connected surface to a flat annulus with concentric circular slits (circular slit domain) or to a rectangle with parallel slits (parallel slit domain). The target domain is canonical and reflects the intrinsic conformal structure of the surface; therefore, the shape parameters of the target domains can be used for conformal surface classification. The method can benefit many important applications in geometric modeling and processing, such as texture mapping, surface classification, quad re-meshing, mesh-spline conversion and so on. It can also be applied for conformal brain mapping in medical imaging field. The regularity of the target domain facilitates surface matching with land mark constraints. The method is automatic, efficient, stable and general, which has been shown in our experiments.

In the future, we will explore more applications of slit map. Also, we want to investigate alternative methods of conformal mapping for multiply connected surfaces and compare their performances to that of slit map.

Acknowledgments. This work was supported in part by NSF (CCF-0448399, DMS-0528363, DMS-0626223 and IIS-0713145) and NSF of China (Project Number 60628202). Special thanks to UCLA neurology department for providing the cortex surface model, and to Geometric Informatics Inc. for supplying the 3D face data sets.

References

1. Ahlfors, L.V.: Complex Analysis. Mcgraw-Hill, New York (1953)
2. Cohen-Steiner, D., Alliez, P., Desbrun, M.: Variational Shape Approximation. ACM Transactions on Graphics, 905–914 (2004)
3. Desbrun, M., Meyer, M., Alliez, P.: Intrinsic Parameterizations of Surface Meshes. Computer Graphics Forum 21, 209–218 (2002)
4. Dey, T.K., Li, K., Sun, J.: On Computing Handle and Tunnel Loops. In: Proc. International Conference on Cyberworlds, pp. 357–366 (2007)

5. Dong, S., Bremer, P.T., Garland, M., Pascucci, V., Hart, J.C.: Spectral Surface Quadrangulation. ACM Transactions on Graphics 25(3), 1057–1066 (2006)
6. Eck, M., Derose, T., Duchamp, T., Hoppe, H., Lounsbery, M., Stuetzle, W.: Multiresolution Analysis of Arbitrary Meshes. Computer Graphics 29, 173–182 (1995)
7. Fisher, M., Schröder, P., Desbrun, M., Hoppe., H.: Design of Tangent Vector Fields. ACM Transactions on Graphics 26(3), 56–66 (2007)
8. Floater, M.: Mean Value Coordinates. Computer Aided Geomeric Design 20(1), 19–27 (2003)
9. Floater, M., Hormann, K.: Surface Parameterization: A Tutorial and Survey. In: Advances In Multiresolution For Geometric Modelling, pp. 157–186 (2005)
10. Garland, M., Willmott, A., Heckbert, P.: Hierarchical Face Clustering on Polygonal Surfaces. In: Proc. ACM Symposium on Interactive 3D Graphics, pp. 49–58 (2001)
11. Gortler, S.J., Gotsman, C., Thurston, D.: Discrete One-Forms on Meshes and Applications to 3D Mesh Parameterization. Computer Aided Geometric Design 33(2), 83–112 (2006)
12. Gu, X., Yau, S.-T.: Computing Conformal Structures of Surfaces. Communications in Information and Systems 2(2), 121–146 (2002)
13. Gu, X., Yau, S.-T.: Global Conformal Surface Parameterization. In: Proc. EUROGRAPHICS/ACM Symposium on Geometry Processing, pp. 127–137 (2003)
14. Jin, M., Gu, X.D., Kim, J.: Discrete Surface Ricci Flow: Theory and Applications. In: Martin, R., Sabin, M.A., Winkler, J.R. (eds.) Mathematics of Surfaces 2007. LNCS, vol. 4647, pp. 209–232. Springer, Heidelberg (2007)
15. Karni, Z., Gotsman, C., Gortler, S.: Free-Boundary Linear Parameterization of 3D Meshes in The Presence of Constraints. In: Proc. Shape Modeling and Applications, vol. 00, pp. 268–277 (2005)
16. Lévy, B., Petitjean, S., Ray, N., Maillot, J.: Least Squares Conformal Maps for Automatic Texture Atlas Generation. ACM Transactions on Graphics 21(3), 362–371 (2002)
17. Maillot, J., Yahia, H., Verroust, A.: Interactive Texture Mapping. In: Proc. the 20th Annual Conference on Computer Graphics and Interactive Techniques, pp. 27–34 (1993)
18. Pinkall, U., Polthier, K.: Computing Discrete Minimal Surfaces and Their Conjugates. Experim. Math. 2, 15–36 (1993)
19. Kälberer, F., Nieser, M., Polthier, K.: Quadcover - Surface Parameterization Using Branched Coverings. Computer Graphics Forum 26(3), 375–384 (2007)
20. Ray, N., Li, W., Lévy, B., Sheffer, A., Alliez, P.: Periodic Global Parameterization. ACM Transaction on Graphics 25(4), 1460–1485 (2006)
21. Sander, P., Wood, Z., Gortler, S., Snyder, J., Hoppe, H.: Multi-Chart Geometry Images. In: Proceedings of ACM Symposium on Geometry Processing, pp. 146–155 (2003)
22. Sheffer, A., De Sturler, E.: Parameterization of Faceted Surfaces for Meshing Using Angle Based Flattening. Engineering with Computers 17(3), 326–337 (2001)
23. Sheffer, A., Praun, E., Rose, K.: Mesh Parameterization Methods and Their Applications. Now Publishers (2006)
24. Tong, Y., Lombeyda, S., Hirani, A.N., Desbrun, M.: Discrete Multiscale Vector Field Decomposition. ACM Transactions on Graphics 22(3), 445–452 (2003)
25. Tong, Y., Alliez, P., Cohen-Steiner, D., Desbrun, M.: Designing Quadrangulations with Discrete Harmonic Forms. In: Proc. EUROGRAPHICS/ACM Symposium on Geometry Processing, pp. 201–210 (2006)
26. Xu, G.: Discrete Laplace-Beltrami Operators and Their Convergence. Computer Aided Geometric Design 21(8), 767–784 (2004)

Solving Systems of 3D Geometric Constraints with Non-rigid Clusters

Hilderick A. van der Meiden and Willem F. Bronsvoort

Delft University of Technology
Mekelweg 4, 2628CD Delft
The Netherlands
H.A.vanderMeiden@tudelft.nl,
W.F.Bronsvoort@tudelft.nl
http://graphics.tudelft.nl

Abstract. We present a new constructive solving algorithm for systems of 3D geometric constraints. The solver is based on the cluster rewriting approach, which can efficiently solve large constraint systems, and incrementally handle changes to a system, but can so far solve only a limited class of problems. The new solving algorithm extends the class of problems that can be solved, while retaining the advantages of the cluster rewriting approach. It rewrites a system of constraints to clusters with various internal degrees of freedom. Whereas previous constructive solvers only determined rigid clusters, we also determine two types of non-rigid clusters. This allows us to solve many additional problems that cannot be decomposed into rigid clusters, without resorting to expensive algebraic solving methods.

Keywords: Geometric constraint solving, rewriting, rigidity, clusters.

1   Introduction

In Computer-Aided Design (CAD), geometric constraints are used to specify dimensions such as angles and distances between geometry. Until recently, geometric constraints could only be used in 2D sketches. Some CAD systems now support 3D constraints, e.g. to specify how parts should be assembled. 3D constraints also play a major role in new academic modelling approaches, e.g. [1]. However, solving geometric constraints in 3D is significantly more difficult than it is in 2D.

The most successful solvers are so-called constructive solvers, which determine a decomposition of the problem into generically rigid (well-constrained) subproblems, known as clusters. These clusters are solved independently, and the solutions of the clusters are used to construct a solution for the complete system. The advantages of using constructive solvers in CAD and requirements for such solvers are discussed in [2].

There are two popular approaches to constructive solving: cluster rewriting (the bottom-up approach) and degrees-of-freedom (DOF) analysis (the top-down approach).

F. Chen and B. Jüttler (Eds.): GMP 2008, LNCS 4975, pp. 423–436, 2008.

Some solvers based on the rewriting approach are presented in [3,4,5,6]. In this approach, the solver tries to apply rewrite rules that merge combinations of constraints into clusters, and that merge clusters into a single rigid cluster if the system is well-constrained. Each rewrite rule corresponds to a generically rigid subproblem, for which an efficient and exact solving algorithm is known, resulting in a fast solver. For 3D problems, this approach is not complete, because for any reasonably expressive set of constraints, no set of cluster rewriting rules is known that will reduce all well-constrained systems to a single rigid cluster. In practice this means that some well-constrained systems may not be solved, and may be classified incorrectly as underconstrained.

Solvers with a DOF-based decomposition approach are presented in [7,8,9]. Here, the generic well-constrainedness of subproblems is determined by DOF analysis. Specific solving algorithms for the subproblems that may be found are not known in advance, as was the case in the cluster rewriting approach. Therefore, a symbolic algebraic solver is used to solve subproblems [7,8], which is expensive if the subproblems are large, or a numerical solver is used [9], which is subject to convergence problems. The DOF-based approach is also not complete, because no complete description of generic well-constrainedness in terms of DOF is known for 3D problems. Instead, heuristic rules are used to determine which subproblems are well-constrained. However, recent work [10] has identified a complete characterisation for generic well-constrainedness for rigid clusters with point-incidences. Note that this allows for decomposing systems of distance constraints, but not angle constraints.

The two approaches discussed above have one thing in common: a problem is decomposed into clusters that correspond to subproblems that are generically rigid. In our new solving algorithm, we identify not only rigid clusters (i.e. clusters corresponding to generically rigid subproblems), but also so-called scalable clusters and radial clusters, which represent specific subproblems with some internal DOF. In particular, scalable clusters represent point sets where all angles are known, and radial clusters represent point sets with known angles centred around one point. This allows us to decompose problems that cannot be decomposed further into rigid clusters.

Consider, for example, the 2D constraint problem in Fig. 1. Here, we have a number of points, lying on lines constrained by several angle constraints, and one distance constraint between a pair of points. The whole constraint system is rigid, but there is no subset of the constraints that forms a rigid system. Thus, any solver that tries to make a rigid decomposition must solve this whole system using some generic solving method, e.g a symbolic algebraic solver, whereas our new solver breaks the problem down into smaller subproblems, which are easily solved.

Our new solving algorithm is based on the cluster rewriting approach, which can efficiently solve large systems of constraints. An additional advantage of this approach is that it is relatively easy to write an incremental algorithm that updates the solution(s) when changes are made to the system. This results in an efficient solver that can be used interactively, e.g. in CAD systems.

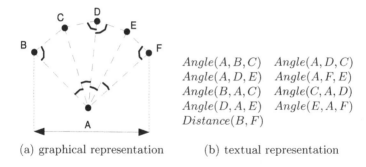

$$Angle(A, B, C) \quad Angle(A, D, C)$$
$$Angle(A, D, E) \quad Angle(A, F, E)$$
$$Angle(B, A, C) \quad Angle(C, A, D)$$
$$Angle(D, A, E) \quad Angle(E, A, F)$$
$$Distance(B, F)$$

(a) graphical representation   (b) textual representation

Fig. 1. System of constraints on points $[A, \dots, F]$. An angle constraint is shown as an arc between dashed lines. A distance constraint is shown by a two-sided arrow.

The rest of the paper is organised as follows. The three clusters types used in our algorithm, rigid clusters, scalable clusters and radial clusters, are described in Section 2. The cluster rewriting approach is explained in Section 3. The details of the solving algorithm are presented in Section 4. Conclusions and suggestions for future work are given in Section 5.

2   Clusters

A cluster is basically a constraint on a set of points that constrains a set of distances and angles on those points. The type of the cluster, *Rigid, Scalable* or *Radial*, determines which distances and angles are constrained. A set of *configu-rations* is associated with the cluster. Each configuration determines a possible set of values for the constrained distances and angles.

The distances $\delta(p, q)$ constrained by clusters are defined as:

$$\delta(p, q) = \sqrt{q - p \cdot q - p}$$

The angles $\angle(p, q, r)$ constrained by clusters are defined as:

$$\angle(p, q, r) = cos^{-1}\left(\frac{q - p}{\delta(p, q)} \cdot \frac{q - r}{\delta(q, r)}\right)$$

where $p, q, r \in \mathbb{R}^2$ or $\mathbb{R}^3$.

In a 2D variant of the solver, signed angles can be used, i.e. a 2D angle $\angle(p, q, r)$ is the angle of rotation to transform a unit-vector $p - q$ to a unit-vector $r - q$. In 3D, $\angle(p, q, r)$ is always unsigned.

A configuration is a set of assignments of coordinates to point variables. For a set of point variables $A = [p_1, \dots, p_n]$, each point p_i is assigned a vector v_i, and for this configuration we write: $c(A) = \{p_1 = v_1, p_2 = v_2, \dots, p_n = v_n\}$.

The actual values of the distances and angles constrained by a cluster are determined by the configurations associated with the cluster. For example, if a

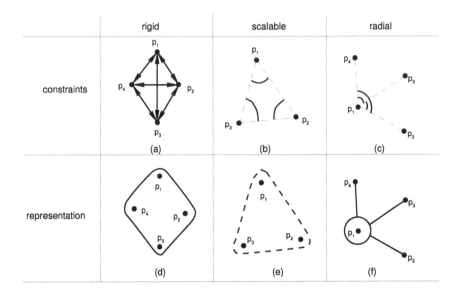

Fig. 2. Constraints (a,b,c) and graphical representation (d,e,f) of cluster types. Left: $Rigid([p_1, p_2, p_3, p_4])$. Middle: $Scalable([p_1, p_2, p_3])$. Right: $Radial([p_1, [p_2, p_3, p_4])$.

cluster specifies that the distance $\delta(p_1, p_2)$ is constrained, and associated with this cluster is a configuration $\{p_1 = (0,0,0), p_2 = (1,1,1)\}$, then there is a constraint $\delta(p_1, p_2) = \sqrt{3}$.

When there are no configurations associated with a cluster, the cluster is considered unsatisfiable, i.e. there are no solutions for the constraints. If there are several configurations associated with a cluster, then these are alternative instantiations of the constraints, i.e. the distance and angle values from one of the configurations must be satisfied.

We distinguish three cluster types: rigid clusters, scalable clusters and radial clusters.

A *rigid cluster* is a constraint on a set of points such that the relative positions of all those points are constrained, i.e. all distances and angles in the point set are constrained (see Fig. 2a). This type of cluster has no internal DOF, but is invariant to translation and rotation, i.e. the constraint is satisfied independently of such transformations. The notation for a rigid cluster on a set of points $[p_1, \ldots, p_n]$ is: $Rigid([p_1, \ldots, p_n])$.

A *scalable cluster* is a constraint on a set of points $[p_1, \ldots, p_n]$ such that all angles $\angle(p_i, p_j, p_k)$ are constrained (see Fig. 2b) The constraint has one internal DOF, namely it may be scaled uniformly, and is further invariant to translation and rotation. The notation for a scalable cluster on this set of points is: $Scalable([p_1, \ldots, p_n])$.

A *radial cluster* is a constraint on a set of points $[p_c, p_1, \ldots, p_n]$ such that all angles $\angle(p_i, p_c, p_j)$ are constrained (see Fig. 2c). Point p_c is called the centre point and points $p_1, \ldots, p_n$ are called radial points. This constraint is invariant to translation and rotation, and has n internal DOF (each point $p_1, \ldots, p_n$ can move along a line though the centre point). The notation for a radial cluster on these points is: $Radial(p_c, [p_1, \ldots, p_n])$.

We use a simple graphical notation for clusters in the figures in this paper, as shown in Fig. 2d, Fig. 2e and Fig. 2f. A point variable is represented by a dot with the name of the corresponding variable next to it. A rigid cluster is represented by a solid curve enclosing the set of points constrained by the cluster. A scalable cluster is represented by a dashed curve enclosing the set of points constrained by the cluster. Finally, a radial cluster is represented by a circle around the centre point and lines connecting the circle to the radial points.

Typically, geometric problems are specified by distances and angles between geometric primitives. Distance and angle constraints on points are easily mapped to clusters and configurations, as follows.

A distance constraint between two points is equivalent to a rigid cluster of two points with one associated configuration. For example, a distance constraint $\delta(p_1, p_2) = 1$ can be represented by a cluster $Rigid(p_1, p_2)$ and a configuration $\{p_1 = (0,0), p_2 = (1,0)\}$. Obviously, the choice for this a particular configuration is somewhat arbitrary: infinitely many different configurations can be used to set the distance value.

An angle constraint on three points is equivalent to a radial cluster with one centre point and two axial points, and one associated configuration. For example, the angle constraint $\angle(p_1, p_2, p_3) = \frac{1}{2}\pi$ can be represented by a cluster $Radial(p_2, [p_1, p_3])$ and a configuration $\{p_1 = (1,0), p_2 = (0,0), p_3 = (0,1)\}$.

A system of distance and angle constraints on points can thus be mapped to a system of clusters. Fig. 3 shows the system of clusters corresponding to the problem in Fig. 1. Note that angle constraints (e.g. $\angle ADC$ and $\angle ADE$) that are mapped to overlapping radial clusters ($Radial(D, [A, C])$ respectively $Radial(D, [A, E])$) have been merged (i.e. $Radial(D, [C, A, E])$) in this figure.

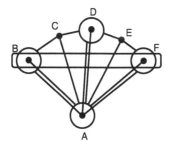

$Radial(A, [B, C, D, E, F])$
$Radial(B, [A, C])$
$Radial(D, [C, A, E])$
$Radial(F, [A, E])$
$Rigid([B, F])$

(a) graphical representation   (b) textual representation

Fig. 3. The system of clusters corresponding to the problem in Fig. 1

3   Cluster Rewriting

To solve a system of clusters, we basically try to rewrite the system to a a single non-rigid cluster, by exhaustively trying to apply a set of rewrite rules.

A *rewrite rule* specifies a *pattern*, describing its *input clusters*, and its *output cluster* in a generic way, and it specifies a *procedure* to determine the configurations of the output cluster from the configurations of the input clusters. If a set of clusters is found in the system that matches the input clusters in a rewrite rule pattern, then the rewrite rule is applied and the corresponding output cluster is added to the system. The rewrite rule also determines the configurations of the output cluster by executing its procedural part, using the configurations of the input clusters as input.

A rewrite rule pattern specifies a number of input clusters of a given type and a number of pattern variables. These pattern variables are bound by the matching algorithm to point variables of clusters, such that the number of variables and the type of the cluster match. If a variable name occurs several times in the pattern, it must be bound to a single point variable, which must therefore occur in several clusters in the constraint system. This is how patterns can specify shared variables between clusters.

A pattern may also specify a cluster with an unknown number of variables, represented by an ellipsis (. . .). Such a pattern can match any cluster with a superset of the given variables, whereas without an ellipsis, a pattern matches only clusters with exactly the given variables.

To determine the configurations of the output cluster of a rewrite rule, the procedural part of the rule is applied for every combination of input cluster configurations. Suppose, for example, that two clusters are used as the input of a rewrite rule, and each cluster has two associated configurations, then four different configurations for the output cluster are computed. Note that some output configurations may be equivalent, thus up to four configurations will be associated with the output cluster.

A set of rewrite rules has been developed for 2D and 3D problems. Of the 15 rules in total, 3 rules are specific for 2D problems, 6 rules are specific for 3D problems, and 6 more can be used in both dimensions. Due to space limitations, the complete set is not given here, but it can be found in [11].

To see how rewrite rules are used to solve a problem, consider the problem in Fig. 3 again. To this problem we can apply the following rule:

Rule 1. *Derive a scalable cluster from two radial clusters:*

Pattern: $A = Radial(p_1, [p_3, p_2, \ldots]) \cup B = Radial(p_2, [p_1, p_3, \ldots])$
$\quad\quad \rightarrow T = Scalable([p_1, p_2, p_3])$
Procedure: $c_T(p_1) = (0, 0, 0)$
$\quad\quad\quad c_T(p_2) = (1, 0, 0)$
$\quad\quad\quad c_T(p_3) = intersection$
$\quad\quad\quad\quad\quad ray\ from\ c_T(p_1)\ direction\ \sphericalangle(c_A(p_3), c_A(p_1), c_A(p_2))$
$\quad\quad\quad\quad\quad ray\ from\ c_T(p_2)\ direction\ \sphericalangle(c_B(p_1), c_B(p_2), c_B(p_3))$

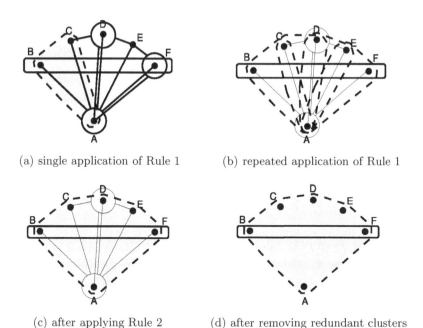

(a) single application of Rule 1   (b) repeated application of Rule 1

(c) after applying Rule 2   (d) after removing redundant clusters

Fig. 4. Intermediate results for solving the system in Fig. 3

This rule can be applied (in 2D or 3D) when two radial clusters (A and B) share three points, including the centre points. When a match is found, a new scalable cluster (T) is added to the system, and a configuration (c_T) is computed by intersecting two rays (directed half-lines).

Applied to the problem at hand, we find the following matches:

$Radial([A, [B, C, D, E, F]) \cup Radial([B, [A, C]) \rightarrow Scalable([A, B, C])$
$Radial([A, [B, C, D, E, F]) \cup Radial([D, [C, A, E]) \rightarrow Scalable([A, C, D])$
$Radial([A, [B, C, D, E, F]) \cup Radial([D, [C, A, E]) \rightarrow Scalable([A, D, E])$
$Radial([A, [B, C, D, E, F]) \cup Radial([F, [A, E]) \rightarrow Scalable([A, E, F])$

The resulting system after applying the rewrite rule corresponding to the first match, is shown in Fig. 4(a). Repeated application of the rule for all the matches listed above, results in the system shown in Fig. 4(b).

When a rewrite rule is applied, the input clusters may become *redundant* and should be removed from the system. A cluster is redundant if all distance and angle constrained by the cluster are also constrained by newer clusters. Thus, an input cluster is removed from the system if all the distances and angles in the cluster are also in the output cluster. If, after applying a rewrite rule, all input clusters are removed, then we say the rewrite rule *merges* the input clusters.

In the system in Fig. 4(a), the cluster $Radial([B, [A, C])$ is redundant and removed, because the angle $\angle ABC$ in this cluster is determined by the new cluster $Scale([A, B, C])$. The cluster $Radial([A, [B, C, D, E, F])$, however, is not removed, even after repeated application of the rewrite rule (result shown in Fig. 4(b)), because it constrains angles that are not in any of the scalable clusters (e.g. $\angle BAF$). The scalable clusters in Fig. 4(b) can be merged using the following rule:

Rule 2. *Merge two scalable clusters with two shared points:*

$$Scalable(X = [p_1, p_2, \ldots]) \cup Scalable(Y = [p_1, p_2, \ldots]) \rightarrow Scalable(X \cup Y)$$

This rule can be applied repeatedly, as follows:

$$Scalable([A, B, C]) \cup Scalable([A, C, D]) \rightarrow Scalable([A, B, C, D])$$
$$Scalable([A, D, E]) \cup Scalable([A, E, F]) \rightarrow Scalable([A, D, E, F])$$
$$Scalable([A, B, C, D]) \cup Scalable([A, D, E, F]) \rightarrow Scalable([A, \ldots, F])$$

This results in the system shown in Fig. 4(c). Now we can remove the clusters $Radial(A, [B, C, D, E, F])$ and $Radial(D, [C, A, E])$, because all angles in those clusters are present in cluster $Scale([A, B, C, D, E, F])$, resulting in Fig. 4(d). Finally the cluster $Scale([A, B, C, D, E, F])$ can be merged with $Rigid([B, F])$ to form the solution cluster $Rigid([A, B, C, D, E, F])$, using the following rule:

Rule 3. *Derive a rigid from a scalable and a rigid with two shared points:*

$$Scalable(X = [p_1, p_2, \ldots]) \cup Rigid([p_1, p_2, \ldots]) \rightarrow Rigid(X)$$

The result of the rewriting process is the *generic solution* of the problem, represented by a directed acyclic graph (DAG) of clusters and rewrite rules. The generic solution of the problem of Fig. 3 is shown in Fig. 5. In this figure, arrows indicate dependencies between clusters created by rewrite rules. The clusters in the generic solution can be classified as problem clusters, i.e. the clusters specified in the original problem, intermediate clusters, and solution clusters, i.e. the clusters that are not used as input for any rewrite rule. In Fig. 5, clusters with no incoming arrows are problem clusters, and clusters with no outgoing arrows are solution clusters.

The configurations associated with the solution cluster are the *particular solutions* of the problem. If there are no configurations associated with the solution cluster, then the problem is *inconsistent*. If there are one or more configurations, then the problem is *consistent*.

From the generic solution we can also determine whether the problem is structurally underconstrained, overconstrained or well-constrained:

- The problem is *structurally underconstrained* if the generic solution has more than one solution cluster or a single non-rigid solution cluster.
- The problem is *structurally overconstrained* if it is *locally overconstrained* and not *generically consistent*. This is elaborated below.
- The problem is *structurally well-constrained* if it is not structurally underconstrained and not structurally overconstrained. Note that these conditions are not mutually exclusive.

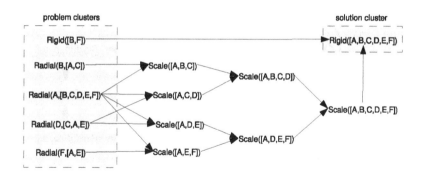

Fig. 5. Generic solution for the problem in Fig. 3

If any distance or angle is constrained by two (or more) clusters, then the system is said to be locally overconstrained. If the overconstrained distances and angles always have the same value in the all clusters, i.e. independently of the actual value, then the system is said to be generically consistent.

To determine whether a locally overconstrained system is generically consistent, we use the following procedure. For each distance or angle that is constrained by two or more clusters, we determine the first clusters in the generic solution that constrain that distance or angle, i.e. the sources of the distance or angle. These sources can be found by following dependencies in the generic solution in the reverse direction, checking for each cluster encountered whether the distance or angle is still constrained. If there is exactly once source for each distance or angle, then the locally overconstrained clusters are generically consistent, because each rewrite rule ensures that all distance/angle constraints in its input clusters are also satisfied in its output clusters. Otherwise, if there is more than one source for a distance or angle, then there is no guarantee that it will have the same value in different clusters, and therefore the system is not generically consistent.

In particular, if two or more problem clusters, i.e. clusters determined by the user, constrain the same distance or angle, then the clusters are not generically consistent, and thus the problem is structurally overconstrained.

However, during the cluster rewriting process, sets of clusters may be created by the solver such that the system is locally overconstrained, but generically consistent. Consider, for example, the system in Fig. 4(c). The system is locally overconstrained, because all the angles in the clusters $Radial(A, [B, C, D, E, F])$ and $Radial(D, [C, A, E])$ are also constrained by the cluster $Scale([A, B, C, D, E, F])$. However, from the generic solution in Fig. 5, we can infer that for each overconstrained angle, there is only one source, in these cases, single problem clusters. Thus, the system is generically consistent.

In other cases, locally overconstrained clusters, even though determined via rewrite rules, are not generically consistent, e.g. in the problem shown in Fig. 6(a). The generic solution for this problem is shown in Fig. 6(b). Here, the clusters $Rigid([p_1, p_3, p_4])$ and $Rigid([p_1, p_2, p_3])$ are locally overconstrained

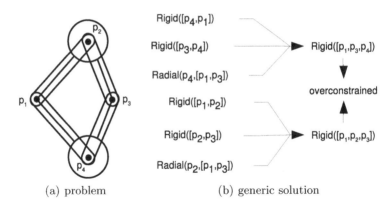

(a) problem   (b) generic solution

Fig. 6. A structurally overconstrained 2D problem and its generic solution

because the distance $\delta(p_1, p_3)$ is constrained by both clusters. This distance is determined twice, by different rewrite rule applications, using different input clusters that do not constrain this distance. Thus, there are two sources for the distance: $Rigid([p_1, p_3, p_4])$ and $Rigid([p_1, p_2, p_3])$. Therefore the problem is not generically consistent, and thus structurally overconstrained.

4   Solving Algorithm

The solving algorithm incrementally updates the generic solution of a constraint problem whenever changes are made to the problem, i.e. clusters are added or removed. The **generic solution** is basically a bi-partite graph data structure, in which nodes are clusters or rewrite rules. Directed edges connect clusters and rewrite rules, i.e. if a cluster is an input cluster of a rewrite rule, then there is an edge from the cluster to the rewrite rule. If a cluster is the output cluster of a rewrite rule, then there is an edge from the rewrite rule to the cluster.

The solving algorithm also keeps track of the set of **active clusters**, i.e. the clusters that represent the problem after all rewriting steps so far.

When the user adds a cluster to the problem (see Algorithm 1), the cluster is added to the generic solution and to the active set of clusters. Because it is not the output of a rewrite rule, it can be identified in the generic solution as a problem cluster. The algorithm then searches for possible rewrite applications on that new cluster, i.e. rewrite rules applications where the cluster is used as input.

When a cluster is removed (see Algorithm 2), all dependent rewrite rules and clusters are also removed from the generic solution. The algorithm must then determine a new set of active clusters. For this purpose, it keeps track of which clusters are removed from the active set after each rewrite rule application. The active set can be restored by re-adding those clusters to the active set when the rewrite rule is removed.

Algorithm 1. Adding a cluster.

```
add cluster c:
   add c to generic solution
   add c to active clusters
   search for rewrite rules on c
```

Algorithm 2. Removing a cluster.

```
remove a cluster c:
   remove c from generic solution
   remove c from active clusters (if it is active)
   remove all dependent clusters and rewrite rules
   re-determine active clusters
```

Searching for possible rewrite rule applications (see Algorithm 3) can be done efficiently because we search only for rewrites on newly added clusters. Since each rewrite rule involves a small number of overlapping clusters (i.e. clusters sharing one or more point variables), we construct a subset of the set of active clusters consisting only of the newly added cluster and the clusters that overlap with it, and search in that subset for possible rewrite rule applications. The pattern matching algorithm thus searches only through a small number of clusters and variables.

The pattern matching algorithm used in our implementation is basically a subgraph matching algorithm that finds all subgraph isomorphisms [12]. The input pattern specified by a rewrite rule is converted into a graph, as is the subset of the active set in which we look for rewrite rule applications. From an isomorphism returned by the graph matching algorithm, we determine which point variable is assigned to which pattern variable, and from that the actual rewrite rule can be instantiated and added to the generic solution.

When a possible rewrite rule application is found, the algorithm first checks whether the rewrite rule application is *progressive*, and only if it is, adds it to the generic solution. A rewrite rule application is progressive if it either increases the number of distances and angles constrained by the active set, or reduces the number of active clusters. The latter condition ensures that locally overconstrained clusters can be merged.

Generally, when a rewrite rule is added to the generic solution, its output cluster becomes part of the set of active clusters, and one or more input clusters may be removed from the active set. A cluster is removed from the active set if it is redundant, i.e. if all distances and angles constrained by it are already constrained by the other clusters in the active set.

To determine whether a cluster is redundant, the algorithm needs to determine whether the set of distances and angles constrained in the cluster is a subset of the set of distances and angles of the other clusters in the active set. Determining these sets explicitly is too expensive. Instead, we determine the number of

---

Algorithm 3. Searching for rewrite rule applications.

```
search for rewrite rules on cluster c:
  subset = c + clusters in active set overlapping with c
  construct reference graph from subset
  for each rewrite rule:
    get pattern graph
    find subgraph isomorphisms
    for each isomorphism:
          rewrite = instantiate rewrite rule from isomorphism
          if rewrite is progressive:
              add output cluster to generic solution and active set
              add rewrite to generic solution
              for each input cluster i:
                  if i is redundant: remove i from active graph
              search for rewrite rules on output cluster (recurse)
```

---

distances and angles constrained by the cluster and the number of distances and angles constrained by each intersection of the cluster with any other overlapping cluster in the active set.

We define the intersection of two clusters as a cluster that constrains only those distances and angles that are constrained in both clusters. The intersection can be determined efficiently using the rules listed in Table 1. The number of distances and angles in a cluster can be determined from Table 2. Note that the number of distances and angles constrained by a cluster is larger than the number of constraints typically needed for a well-constrained system. However, because the values of the distances and angles constrained by a cluster are determined by a configuration, these are always consistent.

If the number of distances and angles constrained by a cluster is larger than the total number of distances and angles constrained by the intersections of the cluster with each other overlapping cluster in the active set, then the cluster is not redundant. Otherwise, the number of distances and angles in the cluster is equal to the total number of distances and angles in the intersections (it cannot be smaller), and the rewrite rule is redundant.

Table 1. Pairwise cluster intersections. If none of the cases listed here matches, then the intersection is empty, i.e. the intersection contains no distances or angles.

intersection	condition
$Rigid(A) \cap Rigid(B) = Rigid(A \cap B)$	$|A \cap B| > 1$
$Rigid(A) \cap Scalable(B) = Scalable(A \cap B)$	$|A \cap B| > 2$
$Rigid(A) \cap Radial(p_c, B) = Radial(p_c, A \cap B)$	$p_c \in A, |A \cap B| > 2$
$Scalable(A) \cap Scalable(B) = Scalable(A \cap B)$	$|A \cap B| > 2$
$Scalable(A) \cap Radial(p_c, B) = Radial(p_c, A \cap B)$	$p_c \in A, |A \cap B| > 2$
$Radial(p_c, A) \cap Radial(p_c, B) = Radial(p_c, A \cap B)$	$|A \cap B| > 2$

Table 2. Number of distance and angles constrained by clusters

cluster	#distances	#angles
$Rigid([p_1, \ldots, p_n])$	$\binom{2}{n}$	$3\binom{3}{n}$
$Scalable([p_1, \ldots, p_n])$	0	$3\binom{3}{n}$
$Radial(p_c, [p_1, \ldots, p_n])$	0	$\binom{2}{n}$

The generic solution of a problem can be used to determine its particular solutions, by evaluating the computation part of each rewrite rule in the generic solution, for each combination of its input clusters' configurations. This may also be done in an incremental way. When the set of configurations associated with a problem cluster is changed, the dependent rewrite rules can be determined from the generic solution, i.e. the rules which use this cluster as input cluster. Only these rewrite rules need to be re-evaluated.

Note that the computation of particular solutions is repeated for every combination of input cluster configurations, and each rewrite rule may in turn have several solutions for each such combination. The number of particular solutions may therefore be very high, and thus expensive to compute, whereas for most applications not all solutions are required or desirable. Thus, a solution selection mechanism is needed to reduce the number of solution and computation time. Our implementation supports two methods of solution selection: by selection constraints, based on [13], and a prototype-based solution selection mechanism, described in [6].

5 Conclusions and Future Work

In this paper we have introduced two new types of clusters, scalable and radial clusters, which can be used, together with traditional rigid clusters, to solve geometric constraint problems. Using a simple cluster rewriting approach, with these new clusters, we can solve a larger class of problems than is possible with only rigid clusters.

The advantages of this approach over more general solving approaches, such as DOF-based decomposition, are that we can solve large problems efficiently, and that an incremental solving algorithm is easy to implement.

Currently, we have not yet done any extensive comparison of our solver to other solving algorithms. It would be interesting to see how it compares in terms of the class of problems that can be solved and in terms of algorithmic complexity.

So far, we have only considered distance and angle constraints on points. Other geometric constraints, defined on other primitives, e.g. lines, planes, cylinders and spheres, can be mapped to systems of clusters. Such mappings have been developed for constraints on lines and circles in 2D, but for 3D primitives this is still work in progress.

The set of rewrite rules for 3D problems we have developed is not complete; there are known well-constrained geometric constructions that cannot be solved with the current rule set. It is not even known whether there exists a complete set of rewrite rules that can be used to solve every 3D system of rigid, scalable and radial clusters. This is related to the more fundamental problem whether a generic, combinatorial description can be given of rigidity in 3D [10].

Altogether, we believe that the use of non-rigid clusters in geometric constraint solving is a promising new approach to solve more complex systems more efficiently.

Acknowledgement

H.A. van der Meiden's work is supported by the Netherlands Organisation for Scientific Research (NWO).

References

1. Bidarra, R., Bronsvoort, W.F.: Semantic feature modelling. Computer-Aided Design 32(3), 201–225 (2000)
2. Hoffmann, C.M., Lomonosov, A., Sitharam, M.: Decomposition plans for geometric constraint systems, Part I: performance measures for CAD. Journal of Symbolic Computation 31(4), 376–408 (2001)
3. Kramer, G.A.: Solving Geometric Constraint Systems: a Case Study in Kinematics. The MIT Press, Cambridge (1992)
4. Hoffmann, C.M., Vermeer, P.J.: Geometric constraint solving in $\mathbb{R}^2$ and $\mathbb{R}^3$. In: Du, D., Hwang, F. (eds.) Computing in Euclidean Geometry, 2nd edn., pp. 266–298. World Scientific Publishing, Singapore (1995)
5. Durand, C., Hoffmann, C.M.: A systematic framework for solving geometric constraints analytically. Journal of Symbolic Computation 30(5), 493–519 (2000)
6. van der Meiden, H.A., Bronsvoort, W.F.: An efficient method to determine the intended solution for a system of geometric constraints. International Journal of Computational Geometry and Applications 15(3), 279–298 (2005)
7. Hoffmann, C.M., Lomonosov, A., Sitharam, M.: Decomposition plans for geometric constraint systems, Part II: new algorithms. Journal of Symbolic Computation 31(4), 409–427 (2001)
8. Sitharam, M., Oung, J.J., Zhou, Y., Abree, A.: Geometric constraints within feature hierarchies. Computer-Aided Design 38(1), 22–38 (2006)
9. Gao, X., Lin, Q., Zhang, G.: A C-tree decomposition algorithm for 2D and 3D geometric constraint solving. Computer Aided Design 38(1), 1–13 (2006)
10. Sitharam, M.: Wellformed systems of point incidences for resolving collections of rigid bodies. International Journal of Computational Geometry and Applications 16(5), 591–615 (2006)
11. van der Meiden, H.A.: Rewrite rules for solving systems of clusters. Technical report, Delft Univerity of Technology (2008), http://graphics.tudelft.nl/~rick/sfo/rwrules.pdf
12. Ullmann, J.: An algorithm for subgraph isomorphism. Journal of the ACM 23(1), 31–42 (1976)
13. Bettig, B., Shah, J.: Solution selectors: a user-oriented answer to the multiple solution problem in constraint solving. Journal of Mechanical Design 125(3), 443–451 (2003)

Space-Time Curve Analogies for Motion Editing

Yuan Wu, Hongxin Zhang, Chao Song, and Hujun Bao

State Key Laboratory of CAD&CG, Zhejiang University, P.R. China
{wuyuan,zhx,songchao,bao}@cad.zju.edu.cn

Abstract. This paper presents a method for analogizing high-dimensional space-time curves, and shows how it can be used to transform a motion sequence into new content and styles according to two accompanied reference sequences. By providing examples, our system can estimate the parameters of a statistical model which incorporates spatial and temporal factors for motion curves. A user can then easily modify existing motion sequences, within these parameters, to have new semantics as examples. This method can also be used to enrich motion capture databases or edit recorded animations in batch mode.

Keywords: motion editing, space-time curve, motion capture.

1 Introduction

With the development of motion capture systems and the growing needs from entertainment industry, complicated motion databases are leveraged in many graphics applications. These databases always contain various characters, contents, styles and their tremendous combinations, and constructing such a database is laborious and quite time-consuming. Moreover, the animators suffer from creating new characters, which means enormous animation clips should be generated in order to match those new characters. It is difficult to rapidly attach new contents and styles to a motion sequence while preserving its original characteristics. To remedy this shortcoming, we propose a method that can fast calculate the difference between two motions and apply this difference to other motions easily.

In this paper, we present a method to learn motion transformation from examples, which alleviates these issues by providing animators a tool for fast motion analogies. With this approach, the animators may choose two reference sequences (say A and A'), our system will capture the difference between them, which includes content, styles, rhythms and pathes. Then, the animator can apply this difference on the third sequence B to obtain a desired motion sequence B'. Mathematically, the above procedure can be formulated as following paradigm:

$$A : A' :: B : B' \qquad (1)$$

For example, one may set a normal walking sequence (A) and a jogging sequence (A') as reference sequences, the difference between the two examples will be extracted immediately, and apply this difference on a normal crouching sequence (B) to acquire a jog-like crouching sequence (B'), see Figure 1. Essentially, we treat motion sequences

F. Chen and B. Jüttler (Eds.): GMP 2008, LNCS 4975, pp. 437–449, 2008.

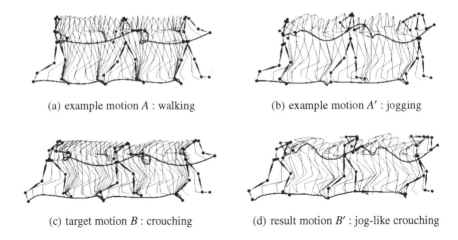

(a) example motion A : walking   (b) example motion A' : jogging

(c) target motion B : crouching   (d) result motion B' : jog-like crouching

Fig. 1. Style analogies: crouching (B) is transformed to jog-like crouching (B'), according to the manner that walking (A) transforms to jogging (A')

as space-time curves and convert our problem of motion editing to the problem of high-dimensional curve processing. So we use the terminology from Hertzmann et al. [10] for reference, and call the approach *motion analogies*.

Our work has three main contributions. First, we show how learning of motion transformation can be formulated as a high dimensional space-time curve analogies problem. Second, we show the way to establish a statistical model, estimate its parameters, and then transfer the target motion sequence with these parameters. Finally, we model and analogize the spatial and temporal transformation in an uniform framework.

The remainder of this paper is organized as follows. Section 2 provides an overview of related work. Section 3 describes our motion analogies algorithm in detail. We demonstrate several results of our analogies experiments in section 4. Lastly, section 5 concludes the paper and gives some promising directions for future work.

2   Related Work

There are several different ways to synthesize motion contents and styles. Retargeting approaches [6,21,27] transfer entire motion from one character to another which has different skeleton size. Graph based approaches [19,16,2,8] search short clips that satisfy constraints specified by user from motion databases and assemble them to synthesize long motion sequences. Blending approaches [24,14,23] interpolate and extrapolate from examples with linear combinations to synthesize motion sequences that have multiple styles. Cut-Paste methods [13,9] cut parts of skeleton from different characters and combine these separate parts into a whole skeleton. Parametric approaches [15,22] can extend motion example space and allow users to adjust motion in parametric space. However, applying these approaches to our problem is difficult, as the sequences generated by these method are unlikely outside the example space and new contents as well as styles cannot be acquired.

Of closet relation to our method are those works that seek to modify the contents and styles by warping motion curves. Witkin et al. [30] adjust motion curves by doing displacement mapping at key-frames. Amaya et al. [1] extract "emotion" component by comparing neutral and "emotional" motions and then apply this emotion component to another neutral motion to generate new sequence with the same "emotion" component. Lee and Shin [18] provide a framework which can hierarchically edit motion sequences by using multilevel B-spline approximation. Our method is similar to [1] on the concept of extracting difference of motions. But our system can transfer not only the "emotion" component but also the semantic and rhythm to the target motion sequences. Moreover, we define difference in a more flexible way than displacement mapping does.

Several researchers focus on extracting style factors from one or more examples. Bruderlin and Williams [4] treat motion data as signal and modifies motion styles by adjusting frequency band. Unuma et al. [28] represent motion data as coefficients of fourier transformation and controls styles on frequent domain. Brand and Hertzmann [3] analyze motion sequences with Hidden Markov Model to extract styles, and these styles can be interpolated, extrapolated and applied to new sequences. Unfortunately, these approaches only change motion sequences mainly on the aspect of styles rather than contents. Our method, on the contrary, has no such limitation, and we can transfer both contents and styles.

Most recently, Urtasun et al. [29] use Principle Component Analysis (PCA) to analyze motion data sets, and use PCA coefficients to generate motion sequences with various styles. However, synthesizing new motions depends on a specified training set such as walking, running, jumping, and these sets must be stored somewhere in advance. Our method differs in that we do not need classifying and training the motion data sets and we can transfer the differences by using only two examples. Shapiro et al. [26] apply Independent Component Analysis (ICA) on one example to extract style components and allow users intuitively choosing the components that they want to transfer. Our method differs in that the users specify the styles for transfer by selecting examples at the beginning, instead of choosing independent components after an analytic procedure. Moreover, we can transfer the differences on cadences which cannot be manipulated by [26].

A few authors have described ways to learn the mapping between input and output motions by examining their relationship. Dontcheva et al. [5] learn to map a low-dimensional acting to a high-dimensional motion by using Canonical Correlation Analysis, and drive character animations by acting widgets movements. Hsu et al. [12] treat the mapping mechanism as a linear time invariable system and transfer styles by estimating dynamic metrics. Their system requires that the input motions must be inside the example space. Our method does not need such requirement, because our intent is exactly to generate motion sequences for new characters which does not exist in the origin motion database.

The paradigms of analogies also exist in other research fields. Hertzmann et al. [10] presents a method for image analogies to filter an image by following the manner of example origin-filtered images pair. Later, Hertzmann et al. [11] analogize stylistic curves from example line drawings for non-photorealistic rendering applications by using texture synthesis technique. Our work is highly inspired by their creative ideas. Our goal is quite different from theirs, although our methods adopt the same paradigms that analogize curves using examples. In our approach, we aim at generating new styles of motion instead of simply modeling styled curves. On the other hand, we emphasis the term *space-time curves* as motion data are characterized not only by geometrical information but also dynamic motion rhythm as well as other time variance factors.

3   Algorithm

Given two example motions A and A', our aim is to learn the transformation from A to A', and apply it to the third motion B to generate a new motion B'. Firstly, the example motions are aligned to the reference motion through a procedure called *time warping*. Our method then use linear model to describe the spatial difference between the warped examples in *spatial estimate* stage. Once the spatial parameters are computed, the model attaches target motion B with new contents and styles to achieve an intermediate motion by *spatial transform* while retains its original characteristics. The procedures of handling temporal factors are quite similar and the details are expatiated in 3.6 . Finally, the intermediate motion is adjusted in temporal domain, through the *inverse time warping* operation, to generate our desirable result. The flow of our method is illustrated in Figure 2.

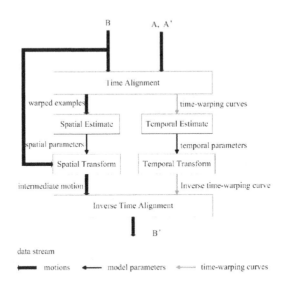

Fig. 2. Diagram of motion analogies

The result motion B' maybe contain artifacts and these can be corrected by an efficient postprocess. In the remainder of section 3, we assume that the skeletons of the three motions, A, A', and B are in the same topology.

3.1 Representation

For the purpose of efficient modeling and general usage, we represent motions in the standard skeletal format. Formally, a motion sequence $M = \{M(t_i)\}$ is a set of frames, which samples a continuous motion curve in time domain. $M(t_i) = (\mathbf{p}_i, \mathbf{q}_i^0, \mathbf{q}_i^1, ..., \mathbf{q}_i^{n-1})$ denotes the configuration of joints at the ith frame, where $\mathbf{p}^i$ is the global position of root, and $\mathbf{q}_i^j$ is the orientation of jth joint in its parent's coordinate system. Especially, $\mathbf{q}_i^0$ is the orientation of root in world coordinate system. Each component of M is called a Degree Of Freedom (DOF).

All joint angles are parameterized by exponential map, since this mapping can handle linear manipulation well on non-linear joint angle and can also avoid the discontinuities problem caused by Euler angle. We refer the readers to Grassia [7] for more discussions on Euler angle, quaternion and exponential map. The root position encodes a global translation, so it is not invariant to different ground-plane coordinate systems. To remove this variability, we represent root position relative to its position in the previous frame.

3.2 Time Alignment

Before modeling the difference between motions, a frame correspondence procedure must be performed to make the example motions, M_A and $M_{A'}$, have the same timing with M_B. Proper time alignment is important to motion analogies; for example, without time alignment the toes in two motions will contact the ground at different time. To align two motions, we construct a time-alignment curve, which maps frames of one motion to correspondence frames in another motion. Formally, the time-alignment curve is defined as follows:

$$S_{A \to B} \triangleq \{(t_i, t_j) : \text{the } i^{th} \text{ frame of } A \text{ corresponds to the } j^{th} \text{ frame of } B\} \quad (2)$$

Moreover, the time-alignment curve must satisfy the constraints of continuity, causality and slope limitation; more details can be retrieved in Kovar et al [14].

To acquire the optimal alignment, first, the user need to select appropriate start frame and the end frame of each sequence in order to make each motion has good alignment on the boundary. Second, the user choose one DOF which can delegate the characteristic of the sequence. The correspondence procedure is then performed on the chosen DOF to compute the time alignment curve. According to our experience, the most closet joints to the root are usually being selected.

With time-alignment curves, we then warp the example motions to the target motion. For the following explanations, we assume that M_A is corresponded to M_B, as shown in Figure 3, and the warped motion is denoted by $\widetilde{M}_A$. There are three types of correspondences between M_A and M_B. The first case is that one $M_A(t_i)$ corresponds to one $M_B(t_j)$, we then set $\widetilde{M}_A(t_j) = M_A(t_i)$. The second case is that multiple $M_A(t_i)$, $(M_A(t_{i+k}), M_A(t_{i+k+1}), ..., M_A(t_{i+k+m}))$, are related one $M_B(t_j)$, the warped frame is set by the average of all correspondent frames, $\widetilde{M}_A(t_j) = \frac{1}{m+1} \sum_{i+k+m}^{i+k} M_A(t_i)$. Lastly, one $M_A(t_i)$ maybe corresponds to multiple $M_B(t_j)$ $(M_B(t_{j+k}), M_B(t_{j+k+1}), ..., M_B(t_{j+k+n}))$. In this case, we employe a cubic B-spline interpolation on $M_A(t_i)$ and its neighbor frames

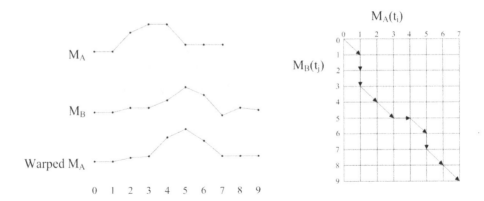

Fig. 3. Time alignment (align M_A to M_B)

to determine the values of $\widetilde{M}_A(t_{j+k}), \widetilde{M}_A(t_{j+k+1}), \ldots, \widetilde{M}_A(t_{j+k+n})$. In addition, the warping procedure, not alignment, is performed on each DOF independently.

The dynamic time warping algorithm we used for alignment is presented by Sederberg et al [25], but there are some other algorithms that also work well, such as Dontcheva et al [5], Kovar et al [14], and Hsu et al [12].

3.3   Analogies Model

In previous section, we warp M_A and $M_{A'}$ in order to align them to M_B, and the three motions are of the same length now. In this section, we use these motions to estimate the parameters of transformation model. To distinguish from the original motions, we denote the warped motions as $\widetilde{M}_A$ and $\widetilde{M}'_A$ and the alignment curves by $S_{A \rightarrow B}$ (aligns A to B) and $S_{A' \rightarrow B}$ (aligns A' to B) respectively.

A motion sequence is a space-time curve in high-dimensional motion space. Since directly handle the analogies problem in such space is difficult, we project motion curve on each DOF and process each DOF independently. On each DOF, we model the transformation from $\widetilde{M}_A$ to $\widetilde{M}'_A$ as follows:

$$\widetilde{M}_{A'}(t_i) = a(t_i)\widetilde{M}_A(t_i) + b(t_i), \tag{3}$$

where t_i is the frame index. The parameters **a** and **b** are all vectors that respectively measure the scale and offset of linear transformation, and we use the two parameters to model the difference between two motions.

The idea of such a model is inspired by the work of Witkin et al. [30]. The main difference between our model and theirs are that we define the two parameters on the whole time domain while they define only on a few key-frames. Their method need to specify the key-frames by users or by automated motion segmentation which is also complicated, and our method is free of this problem.

3.4 Estimation

A naive solution of Equation 3 is derived by directly solving it. However, **a** and **b** solved in this way lack of smoothness and are highly sensitive to data noise. In order to get fine distributions of **a** and **b**, we define and minimize the following objective function:

$$E(\mathbf{a}, \mathbf{b}) = \|\mathbf{a}\widetilde{\mathbf{M}}_{A'} + \mathbf{b} - \widetilde{\mathbf{M}}_A\|^2 + \omega_a\|\mathbf{Sa}\|^2 + \omega_b\|\mathbf{Sb}\|^2, \tag{4}$$

where

$$\mathbf{S} = \frac{1}{2}\begin{pmatrix} -2 & 2 & & & \\ -1 & 0 & 1 & & \\ & \ddots & \ddots & \ddots & \\ & & -1 & 0 & 1 \\ & & & -2 & 2 \end{pmatrix}. \tag{5}$$

Here, **S** is a smooth matrix that approximates the first derivatives, and the smoothness terms $\|\mathbf{Sa}\|^2$ and $\|\mathbf{Sb}\|^2$ are balanced by two weights ω_a and ω_b. Through this equation, we try to make a tradeoff between the accuracy of the linear approximation and the smoothness of the two parameters. In this paper, we set $\omega_a = \omega_b = 2$. To solve this equation, we set $\partial E/\partial \mathbf{a} = \partial E/\partial \mathbf{b} = 0$ and reformulate Equation 4 as follows:

$$\begin{bmatrix} \widetilde{\mathbf{M}}_{A'}^T\widetilde{\mathbf{M}}_{A'} + \omega_a\mathbf{S}^T\mathbf{S} & \widetilde{\mathbf{M}}_{A'}^T \\ \widetilde{\mathbf{M}}_{A'} & \mathbf{I} + \omega_b\mathbf{S}^T\mathbf{S} \end{bmatrix}\begin{bmatrix} \mathbf{a} \\ \mathbf{b} \end{bmatrix} = \begin{bmatrix} \widetilde{\mathbf{M}}_{A'}^T\widetilde{\mathbf{M}}_A \\ \widetilde{\mathbf{M}}_A \end{bmatrix}. \tag{6}$$

By using Gaussian elimination, the above linear system can be easily solved. The form of Equation 4 is similar to the one used in Hsu et al. [12]. They define a similar function with an additional time warping matrix to align motions, and the parameters **a** and **b** are useless at last, in their algorithm. We adopt the definition of the objective function and use these parameters to model the transformation between motions.

3.5 Transformation

In the previous step, we obtain the parameters of transformation model and now we can apply these parameters on M_B to generate a new motion. Because the sequences involves in the transformation procedure are all align to M_B(M_B certainly aligns to itself, $\widetilde{M}_B = M_B$), the result motion also aligns to M_B and we denote it by $\widetilde{M}_{B'}$. The transformation equation can be written as follows:

$$\widetilde{M}_{B'}(t_i) = a(t_i)M_B(t_i) + b(t_i) \tag{7}$$

It should not be surprising that the structure of transformation equation is identical to Equation 3, because the manner of transforming target motion is analogized as the examples. Additionally, we transform each DOF independently, in this step, and finally combine them into a whole motion sequence.

3.6   Inverse Time Alignment

After transforming the target motion M_B, we acquire sequence $\widetilde{M}_{B'}$. However, this sequence is not our desired result motion, as it is align to the target motion M_B, or more intuitively, we can say that $\widetilde{M}_{B'}$ is as long as M_B. Formally, we must inverse the time alignment procedure to recover the original rhythm of $\widetilde{M}_{B'}$.

The key to recover the correct timing is solving the inverse time alignment curves $S_{B \to B'}$ which aligns M_B to $M_{B'}$. Unfortunately, there is a conceptual difficulty that we can't calculate $S_{B \to B'}$ before we own $M_{B'}$. To solve this problem, we treat the time alignment curve as an extra DOF of the motions because it is reasonable that the transformation of timing can also be analogized from the manner that M_A transform into $M_{A'}$. We replace the motion curve terms with time alignment curve terms in Equation 3 and 7 to get equations as below:

$$S_{A' \to B}(t_i) = a(t_i) S_{A \to B}(t_i) + b(t_i), \tag{8}$$

and

$$S_{B' \to B}(t_i) = a(t_i) S_{B \to B}(t_i) + b(t_i). \tag{9}$$

Obviously, we have $S_{B \to B}(t_i) \equiv t_i$. By using the method presented in §3.4 and §3.5, we can solve the time-align curve $S_{B' \to B}$. However, time align curves are not uniform-sampled, which means the capacities of $S_{A \to B}(t_i)$ and $S_{A' \to B}(t_i)$ may be different. So, we align $S_{A \to B}(t_i)$ to $S_{A' \to B}$ with the method presented in §3.2 before we model the temporal transformation.

With $S_{B' \to B}$, we can obtain the inverse time alignment curve $S_{B \to B'}$ according to the following definition:

$$S_{B \to B'} \triangleq \{(t_i, t_j) : \text{where}(t_j, t_i) \in S_{B' \to B}\}. \tag{10}$$

In addition, we refine the curve $S_{B' \to B}$ with uniform quadratic B-spline fitting to satisfy the constraints of continuity, causality and slope limitation, and similar strategy is used by Kovar et al [14]. By warping the intermediate motion $\widetilde{M}_{B'}$ with the inverse time-alignment curve $S_{B \to B'}$, we derive the desired motion $M_{B'}$.

3.7   Postprocessing

Given good example and target motions, our method can generate various stylistic motions, but sometimes a few abnormal frames occurs, which do not satisfy kinematic constraints. The most common phenomenon is footskate, which has been mentioned many times in motion editing and synthesis publications. Fortunately, Kovar et al [17] provide an efficient method to clean up it. Our method on footskate cleaning is similar to theirs, but differs at two points.

First, footskate cleaning problem is also inverse kinematic (IK) problem, and they use analytic method which can solve kinematic constraints of typical skeleton very fast. For general usage purpose, we choose the numerical routine to solve kinematic constraints, which allows our system to handle the case of different skeletal topologies.

Second, in order to avoid sharping adjustments to joint rotations, they allow small changes in the bone lengths. Although the slight changes can hardly be discovered by

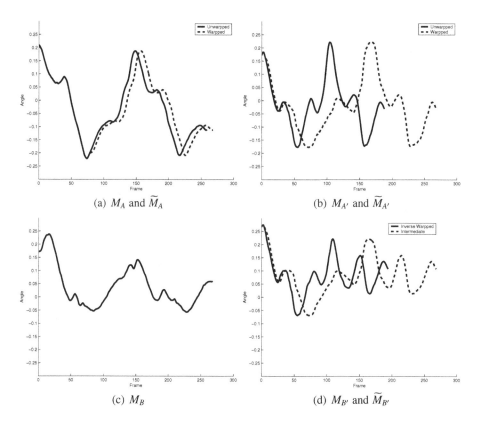

Fig. 4. Example of analogies motion curve of one DOF

the users, the result motions must be saved in the format that support variable bone length. To remedy this shortcoming, we slightly violate the constraints of joints' rotations to match the footprints without introducing visual artifacts, For example, we treat some 1-DOF joints as 3-DOFs joints, and allow these joints rotate slightly on extra DOFs. This technique works well with a lot of data formats, such as BVH, and it can also be seamlessly integrated into traditional algorithms.

4   Results

In this section, we demonstrate some results of our algorithm and all of them can be found in the supplement. We use two motion capture databases from different sources on the internet. One database is shared by http://people.csail.mit.edu/ehsu/, and it was used by Hsu et al [12] in their style translation experiments. The other database is shared by http://www.animazoo.com. The skeleton of Hsu's data has 18 joints and the latter's contains 23 joints.

There are two main factors that determine the speed of our algorithm. One is the the amount of DOFs and the length of motion sequences. As we treat all the DOFs independently, one more DOF only means the estimate and transform procedures are executed one more time and the time cost of each single procedure does not increase. The length of sequences also affect the time cost, as long sequences need more time to be aligned and transformed and the estimate procedure spend more time to optimize parameters through long sequences. According to our experience, the computational expenditure of our method has a near-linear relation with this factor.

The slope limitation of align path is another factor (cf. the right image of Figure 3). With a small slope, the time warping algorithm is faster, but sometimes failed to converge to a global optimal. And a large value of slope means good aligned motions and more time for calculation. In our experiments, we set the slope limitation = 3. In addition, we use the standard time align algorithm without any optimization in this paper because our object is mainly to validate the analogies framework. The discussion on how to accelerate the time warping procedure can be found in [25].

4.1 Style Analogies

For this sample, we show the analogies of motion styles. We choose a walking motion (A) and a jogging motion (A') as examples. Our system captures the difference between them and apply it to a crouching motion (B) to generate a jog-like crouching motion (B'). The analogical procedure transforms the motion style from walking to jogging, according to A and A', while keeping the character's back bending like B, Figure 1.

4.2 Content Analogies

In this sample, we demonstrate the analogies of motion contents. We select a crouching motion (A) and a lateral-crouching motion (A') as examples. We apply the change on a normal forward walking motion (B) to generate a motion that walk in a lateral manner. Notice that the character in B' does not bend its back like it does in A and A', because our analogies model extracts only the difference and certainly eliminates the common features appear in both examples, while most motion blending algorithms are not able to distinguish desired and undesired features.

4.3 Rhythm Analogies

Here, we show a sample of rhythm analogies. We compare a slow-walking motion (A) with a fast-walking motion (A') and apply the rhythmic change on a slow-swaying motion (B). The new motion (B') sways faster while retains other features of (B) except a slight change in arms' swing, according to the examples, see Figure 5.

4.4 Multiple Features Analogies

Lastly we perform a complicated experiment that analogize multiple features. A normal walking motion and a drunk walking motion are be set as A and A' respectively. The two motions are differ in many ways including style, speed and path. We want to transform all of them of a third motion B, in which a man is making a phone call while strolling. The objective was successfully achieved; we get a man in teetering, who is so drunk that he cannot keep the cell phone beside his ear.

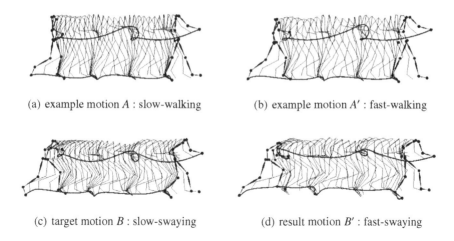

(a) example motion A : slow-walking   (b) example motion A' : fast-walking

(c) target motion B : slow-swaying   (d) result motion B' : fast-swaying

Fig. 5. Rhythm analogies: slow-swaying (B) is transformed to fast-swaying (B'), according to speed change between slow-walking (A) and fast-walking (A'). Notice that the content of walking motion, in A and A', is not be transformed to B, because the feature of walking is removed by our analogies model by comparing the two examples.

5   Conclusion

In this paper, we present a novel method for example based motion editing. We treat motion sequences as high-dimensional space-time curves and solve the motion analogies problem as a space-time curve analogies problem. Given two examples , our system can extract the difference between them and apply it to the other motions to generate sequences with new contents and styles while retain their original characteristics.

With our method, animators can easily modify a batch of existing sequences to attach one or more new characteristics. In addition, the animators can also generate animations with highly sophisticated styles by layered transferring original data. Actually, it is not our objective to design a brand new system to replace current motion editing and synthesis approaches. Our goal is to provide an efficient tool which can be integrated into larger animation systems.

As nothing is perfect, our method still has following limitations. Like most motion editing and synthesis methods, our system cannot guarantee that the result motions are valid under physical principle. We believe that physically-based techniques can enhance the reality of motions in postprocessing stage. For example, Liu et al. [20] present a physical model to handle constraints that cannot be solved by kinematic routines. Furthermore, applying our method in cartoon animations maybe encounter some problems. Because the actions of cartoon characters are very exaggerated sometimes and these actions cannot be recorded by motion capture system. Whether our system does work in such cases has not been proved yet. We hope to validate our method on non-human motion data in the future.

Acknowledgements. Motion data are courtesy of `http://people.csail.mit.edu/ehsu/` and `http://www.animazoo.com`. This project is supported in partial by 973 Program of China (No.2002CB312102), 863 Programs of China (No.2006AA01Z335) and NSFC (No.60505001).

References

1. Amaya, K., Bruderlin, A., Calvert, T.: Emotion from motion. In: Graphics Interface 1996, pp. 222–229 (1996)
2. Arikan, O., Forsyth, D.A., O'Brien, J.F.: Motion synthesis from annotations. ACM Transactions on Graphics 22(3), 402–408 (2003)
3. Brand, M., Hertzmann, A.: Style machines. In: ACM SIGGRAPH 2000, pp. 183–192 (2000)
4. Bruderlin, A., Williams, L.: Motion signal processing. In: SIGGRAPH 1995, pp. 97–104 (1995)
5. Dontcheva, M., Yngve, G., Popović, Z.: Layerd acting for character animation. ACM Transactions on Graphics 22(3), 409–416 (2003)
6. Gleicher, M.: Retargetting motion to new characters. In: ACM SIGGRAPH 1998, pp. 33–42 (1998)
7. Grassia, F.S.: Practical parameterization of rotation using the exponential map. Journal of Grahpics Tools 3(3), 29–48 (1998)
8. Heck, R., Gleicher, M.: Parametric motion graphs. In: Symposium on Interactive 3D Graphics and Games (2007)
9. Heck, R., Kovar, L., Gleicher, M.: Splicing upper-body actions with locomotion. Computer Graphics Forum 24(3), 459–466 (2006)
10. Hertzmann, A., Jacobs, C.E., Oliver, N., Curless, B., Salesin, D.H.: Image analogies. In: ACM SIGGRAPH 2001, pp. 327–340 (2001)
11. Hertzmann, A., Oliver, N., Curless, B., Seitz, S.M.: Curve analogies. In: Proceedings of the Thirteenth Eurographics Workshop on Rendering, pp. 233–245 (2002)
12. Hsu, E., Pulli, K., Popović, J.: Style translation for human motion. ACM Transaction on Graphics 24(3), 1082–1089 (2005)
13. Ikemoto, L., Forsyth, D.A.: Enrich a moton collection by transplanting limbs. In: The 2004 Symposium on Computer animation, pp. 99–108 (2004)
14. Kovar, L., Gleicher, M.: Flexible automatic motion blending with registration curves. In: ACM SIGGRAPH/Eurographics Symposium on Computer Animation, pp. 214–224 (2003)
15. Kovar, L., Gleicher, M.: Automated extraction and parameterization of motions in large data sets. ACM Transactions on Graphics 23(3), 559–568 (2004)
16. Kovar, L., Gleicher, M., Pighin, F.: Motion graphs. ACM Transactions on Graphics 21(3), 473–482 (2002)
17. Kovar, L., Schreiner, J., Gleicher, M.: Footskate cleanup for motion capture editing. In: ACM SIGGRAPH Symposium on Computer Animation, pp. 97–104 (2002)
18. Lee, J., Shin, S.Y.: A hierarchical approach to interactive motion editing for human-like figures. In: SIGGRAPH 1999, pp. 39–48 (1999)
19. Li, Y., Wang, T., Shum, H.-Y.: Motion texture: A two-level staticstical model for character motion synthesis. ACM Transactions on Graphics 21(3), 465–472 (2002)
20. Liu, C.K., Hertzmann, A., Popović, Z.: Learning physics-based motion styl with nonlinear inverse optimization. ACM Transcations on Graphics 24(3), 1071–1081 (2005)
21. Monzani, J.S., Baerlocher, P., Boulic, R., Thalmann, D.: Using an intermediate skeleton and inverse kinematics for motion retargeting. In: Eurograhics 2000 (2000)

22. Mukai, T., Kuriyama, S.: Geostatistical motion interpolation. ACM Transactions on Grahics 24(3), 1062–1070 (2005)
23. Park, S.I., Shin, H.J., Kim, T.H., Shin, S.Y.: On-line motion blending for real-time locomotion generation. Computer Animation and Virtual Worlds 15(3-4), 125–138 (2004)
24. Rose, C., Cohen, M.F., Bodenheimer, B.: Verbs and adverbs: Multidimensional motion interpolation. IEEE Computer Graphics and Applications 18(5), 32–41 (1998)
25. Sederberg, T., Greenwood, E.: A physically-based approach to 2-d shape blending. In: SIGGRAPH 1992, vol. 26, pp. 26–34 (1992)
26. Shapiro, A., Cao, Y., Faloutsos, P.: Style components. In: Graphics Interface 2006, pp. 33–39 (2006)
27. Shin, H.J., Lee, J., Gleicher, M., Shin, S.Y.: Computer pupperty: An importance-based approach. ACM Transactions on Graphics 20(2), 67–94 (2001)
28. Unuma, M., Anjyo, K., Tekeuchi, R.: Fourier principles for emotion-based human figure animation. In: SIGGRAPH 1995, pp. 91–96 (1995)
29. Urtasun, R., Glardon, P., Boulic, R., Thalmann, D., Fua, P.: Style-based motion synthesis. Computer Graphics Forum 23(4), 1–14 (2004)
30. Witkin, A., Popović, Z.: Motion warping. In: SIGGRAPH 1995, pp. 105–108 (1995)

Variational Skinning of an Ordered Set of Discrete 2D Balls

Greg Slabaugh[1], Gozde Unal[2], Tong Fang[1], Jarek Rossignac[3],
and Brian Whited[3]

[1] Siemens Corporate Research, Princeton NJ 08540 USA
[2] Sabanci University, Istanbul Turkey
[3] Georgia Institute of Technology, Atlanta, GA 30332 USA

Abstract. This paper considers the problem of computing an interpolating skin of a ordered set of discrete 2D balls. By construction, the skin is constrained to be C^1 continuous, and for each ball, it touches the ball at a point and is tangent to the ball at the point of contact. Using an energy formulation, we derive differential equations that are designed to minimize the skin's arc length, curvature, or convex combination of both. Given an initial skin, we update the skin's parametric representation using the differential equations until convergence occurs. We demonstrate the method's usefulness in generating interpolating skins of balls of different sizes and in various configurations.

1 Introduction

In this paper, we consider the geometric problem of *ball skinning*, which we define to be the computation of a continuous interpolation of a discrete set of balls; an example appears in Figure 1. This problem arises in numerous applications, including character skinning, molecular surface model generation, and modeling of tubular structures. The balls can have different radii, be configured in different positions, and may or may not overlap. In our formulation of the problem, we require that there exist a series of contact points arranged along the skin, so that each ball has a point of contact with the skin and that the skin and ball are tangent to each other at the point of contact. The skin then rests on and interpolates the underlying balls.

For a given configuration of balls, there exist an infinite number of possible solutions to this problem as expressed above. To formulate the problem so that it is well-posed, we seek the skin that has minimal arc length, curvature, or combination of both. We achieve this by deriving, and solving, differential equations that minimize an energy, composed of arc length and curvature terms, based on this constrained variational problem. By minimizing this energy, the method provides an optimal constrained interpolation of the balls. In this paper we consider both *one-sided* and *two-sided* skins. A one-sided skin is a contour that rests on one side (left or right) of a collection of balls such as that portrayed in Figure 2 (a), while a two-sided skin defines an interpolating region that is composed of both left and right skins, and has an inside and outside, as demonstrated in Figure 1.

F. Chen and B. Jüttler (Eds.): GMP 2008, LNCS 4975, pp. 450–461, 2008.

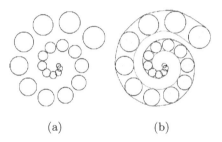

(a)   (b)

Fig. 1. An example ball skinning. Given an ordered sequence of balls (a), we produce a skin that optimally interpolates the balls (b). This skin consists of two splines (green and blue) and is computed using differential equations.

1.1 Related Work

The problem of skinning appears in various contexts. In computer animation, often an articulated object or character is constructed using a layered representation consisting of a skeletal structure and a corresponding geometric skin [1]. The skeleton has fewer degrees of freedom and is simpler to adjust by an animator. Given a new skeletal pose, the skinning algorithm is responsible for deforming the geometric skin to respond to the motion of the underlying skeleton. The skinning problem is a special case of the problem of computing the envelopes of families of quadrics, which have been investigated by Peternell [2] via the use of cyclographic maps.

The problem of ball skinning appears frequently in the context of computational chemistry and molecular biology, when generating surface meshes for molecular models [3] [4] [5]. Several algorithms exist to skin a molecular model to produce a C^1 continuous surface that is tangent smooth and has high mesh quality. These methods are typically either based on Delaunay triangulation [3] or by finding the isosurface of an implicit function [5]. The work of [5] derives a special subset of skins that is piece-wise quadratic. When dealing with a continuous family of balls, the skin is the envelope of the infinite union of the circles of intersection of two consecutive pearls of infinitely close center. While the surfaces generated by these methods are tangent to the balls and have smoothness at the point of tangency, none of these existing methods provide an optimally smooth skin, unlike the method we present here.

In our application, we are interested in modeling the geometry of a blood vessel that has been identified using a 2D variant of Pearling [6], a ball packing algorithm that places numerous balls of different radii so that they fit snugly inside an imaged blood vessel. Given these balls, we would like to find a smooth, C^1 skin that smoothly interpolates the balls. This surface can then be used for visualization of the blood vessel as well as measurements such as volume or surface area.

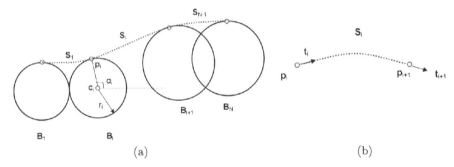

Fig. 2. Depiction of the problem. In (a), we would like to find the skin (dotted curve) that interpolates the ordered set of balls. In (b), we show a depiction of a segment of the skin.

1.2   Our Contribution

We model the skin as a C^1 spline, which, by construction, must touch each ball at a point of contact and be tangent to the ball at the point of contact. We then provide two novel derivations, one for deforming this constrained spline to minimize its arc length; and a second derivation for minimizing its curvature. The result of these derivations are differential equations, which we then solve to update a given spline to its optimal position. We then show experimental examples of how these differential equations are used perform *optimally smooth* skinning of balls.

2   Skin Representation

In this paper we consider the problem of *smoothly* interpolating between a discrete ordered set of balls. Our objective is to find a skin **S**, that satisfies several geometric criteria:

1. The skin should be modeled by a point of contact with each ball.
2. The skin should be tangent to a ball at the point of contact.
3. The skin should optimize an energy functional composed of arc length and curvature terms.

A depiction of the problem is presented in Figure 2 (a), where the desired skin is rendered as a dotted line. The skin **S** is composed of a set of segments, $\mathbf{S}_i$, for $i = 1 \ldots N - 1$, where N is the total number of balls.

2.1   Segment Representation

Various spline representations (such as Catmull Rom, 4-point, etc.) are possible for modeling the segments; in this paper we choose to model each segment $\mathbf{S}_i$ using a spline that starts at point $\mathbf{p}_i$ in direction $\mathbf{t}_i$, and ends at point $\mathbf{p}_{i+1}$ in direction $\mathbf{t}_{i+1}$, as depicted in Figure 2 (b). We model the segment using a cubic polynomial curve

$$\mathbf{S}_i = \mathbf{A}_i t^3 + \mathbf{B}_i t^2 + \mathbf{C}_i t + \mathbf{D}_i, \tag{1}$$

since the four constraints require four degrees of freedom. For the ith segment, $\mathbf{A}_i$, $\mathbf{B}_i$, $\mathbf{C}_i$, and $\mathbf{D}_i$ are coefficients, and $t \in [0, 1]$ is a time variable that parameterizes the curve.

Each segment $\mathbf{S}_i$ of the skin is defined by the Hermite interpolation of the boundary conditions defined by $\mathbf{p}_i$, $\mathbf{t}_i$, $\mathbf{p}_{i+1}$, and $\mathbf{t}_{i+1}$, specifically, $\mathbf{S}_i|_{t=0} = \mathbf{p}_i$, $\frac{d\mathbf{S}_i}{dt}|_{t=0} = \mathbf{t}_i$, $\mathbf{S}_i|_{t=1} = \mathbf{p}_{i+1}$, and $\frac{d\mathbf{S}_i}{dt}|_{t=1} = \mathbf{t}_{i+1}$. With these constraints, and the derivative of the segment,

$$\mathbf{S}'_i = \frac{d\mathbf{S}_i}{dt} = 3\mathbf{A}_i t^2 + 2\mathbf{B}_i t + \mathbf{C}_i, \tag{2}$$

we obtain a system of four equations for the four coefficients: $\mathbf{D}_i = \mathbf{p}_i$, $\mathbf{C}_i = \mathbf{t}_i$, $\mathbf{A}_i + \mathbf{B}_i + \mathbf{C}_i + \mathbf{D}_i = \mathbf{p}_{i+1}$, and $3\mathbf{A}_i + 2\mathbf{B}_i + \mathbf{C}_i = \mathbf{t}_{i+1}$, which is easily solved, yielding

$$\begin{aligned}
\mathbf{A}_i &= -2\mathbf{p}_{i+1} + 2\mathbf{p}_i + \mathbf{t}_i + \mathbf{t}_{i+1} \\
\mathbf{B}_i &= 3\mathbf{p}_{i+1} - 3\mathbf{p}_i - 2\mathbf{t}_i - \mathbf{t}_{i+1} \\
\mathbf{C}_i &= \mathbf{t}_i \\
\mathbf{D}_i &= \mathbf{p}_i
\end{aligned} \tag{3}$$

2.2 Endpoints

We now have a way to model each segment of the skin. But we have not yet described how to determine the endpoints $\mathbf{p}_i$, $\mathbf{p}_{i+1}$ and their respective tangents, $\mathbf{t}_i$, $\mathbf{t}_{i+1}$ of each segment. As shown in Figure 2 (a), we can represent the point of contact $\mathbf{p}_i$ on the ith ball as

$$\mathbf{p}_i = \mathbf{c}_i + \begin{bmatrix} r_i \cos \alpha_i \\ r_i \sin \alpha_i \end{bmatrix}, \tag{4}$$

where r_i is the radius of ball, $\mathbf{c}_i$ is its center, and α_i is an angle. In addition, we can represent the tangent

$$\mathbf{t}_i = \begin{bmatrix} -a_i \sin \alpha_i \\ a_i \cos \alpha_i \end{bmatrix}, \tag{5}$$

where a_i is a stiffness coefficient that controls the influence of the tangential constraint. Each a_i is fixed to be half the distance between the next and previous ball centers (for the first and last balls, it is the distance between the ball center and its neighbor ball center). Note that both the point $\mathbf{p}_i$ and the tangent $\mathbf{t}_i$ are only a function of the angle α_i, since the radius, center, and stiffness coefficient of the ball are fixed.

We now have a way to represent the skin $\mathbf{S}$ as a set of segments $\mathbf{S}_i$, where each segment $\mathbf{S}_i$ interpolates between the points of contact $\mathbf{p}_i, \mathbf{p}_{i+1}$ with balls $\mathbf{B}_i, \mathbf{B}_{i+1}$, subject to tangent conditions $\mathbf{t}_i, \mathbf{t}_{i+1}$ respectively.

By construction of the problem, the angle α_i affects only the segment $\mathbf{S}_i$ as well as the segment $\mathbf{S}_{i-1}$, as can be easily seen in Figure 2 (a). Finally, we note that the skin is fully parameterized by the balls and the spline angles α_i. Since the balls are fixed, our objective will be to compute the angles α_i that form the optimal skin.

3   Energy Minimization

There are an infinite number of skins that are modeled by a contact point on each ball and have a direction tangent to the ball at the point of contact. To further constrain the problem, we require that the skin have minimal arc length and/or be smooth. We achieve this by finding the angles α_i that optimize an energy functional. First, we derive equations used to compute the skin with minimal arc length, then we consider curvature.

3.1   Arc Length Minimization

In this section, we consider the minimization of arc length. This will result in the shortest skin that satisfies the geometric constraints imposed by our ball representation. That is, we would like to find the angles α_i that minimize

$$E_a = \int |\mathbf{S}'| dt, \tag{6}$$

where $\mathbf{S}'$ is the derivative of $\mathbf{S}$ with respect to t. Since the skin is represented as a set of segments, this is equivalent to

$$E_a = \sum_{i=1}^{N-1} \int |\mathbf{S}'_i| dt, \tag{7}$$

Next, we take the derivative of the energy with respect to the angle α_i. As stated above, the ith angle only affects the segments $\mathbf{S}_{i-1}$ and $\mathbf{S}_i$. Therefore,

$$\frac{\partial E_a}{\partial \alpha_i} = \frac{\partial}{\partial \alpha_i} \left(\int |\mathbf{S}'_i| dt \right) + \frac{\partial}{\partial \alpha_i} \left(\int |\mathbf{S}'_{i-1}| dt \right) \tag{8}$$

First term. Let us consider the first term of Equation 8. Propagating the derivative with respect to α_i through the integral, it is easy to show that

$$\frac{\partial}{\partial \alpha_i} \left(\int |\mathbf{S}'_i| dt \right) = \int \frac{\partial}{\partial \alpha_i} < \mathbf{S}'_i, \mathbf{S}'_i >^{\frac{1}{2}} dt$$

$$= \int < \mathbf{S}'_i, \mathbf{S}'_i >^{-\frac{1}{2}} < \mathbf{S}'_i, \frac{\partial \mathbf{S}'_i}{\partial \alpha_i} > dt,$$

where $<>$ denotes an inner product. Next, we derive an expression for the $\frac{\partial \mathbf{S}'_i}{\partial \alpha_i}$ terms using Equation 2, yielding

$$\frac{\partial \mathbf{S}'_i}{\partial \alpha_i} = 3t^2 \frac{\partial \mathbf{A}_i}{\partial \alpha_i} + 2t \frac{\partial \mathbf{B}_i}{\partial \alpha_i} + \frac{\partial \mathbf{C}_i}{\partial \alpha_i} \tag{9}$$

The derivatives $\frac{\partial \mathbf{A}_i}{\partial \alpha_i}$, $\frac{\partial \mathbf{B}_i}{\partial \alpha_i}$ and $\frac{\partial \mathbf{C}_i}{\partial \alpha_i}$ can be derived using Equation 3, as

$$
\frac{\partial \mathbf{A}_i}{\partial \alpha_i} = 2\frac{\partial \mathbf{p}_i}{\partial \alpha_i} + \frac{\partial \mathbf{t}_i}{\partial \alpha_i}
$$
$$
\frac{\partial \mathbf{B}_i}{\partial \alpha_i} = -3\frac{\partial \mathbf{p}_i}{\partial \alpha_i} - 2\frac{\partial \mathbf{t}_i}{\partial \alpha_i}
$$
$$
\frac{\partial \mathbf{C}_i}{\partial \alpha_i} = \frac{\partial \mathbf{t}_i}{\partial \alpha_i} \tag{10}
$$

Finally, the derivatives $\frac{\partial \mathbf{p}_i}{\partial \alpha_i}$ and $\frac{\partial \mathbf{t}_i}{\partial \alpha_i}$ can be derived from Equations 4 and 5 as

$$
\frac{\partial \mathbf{p}_i}{\partial \alpha_i} = \begin{bmatrix} -r_i \sin \alpha_i \\ r_i \cos \alpha_i \end{bmatrix}
$$
$$
\frac{\partial \mathbf{t}_i}{\partial \alpha_i} = \begin{bmatrix} -a_i \cos \alpha_i \\ -a_i \sin \alpha_i \end{bmatrix} \tag{11}
$$

We now have all the derivatives needed to compute the first term in Equation 8.

Second term. Now let us consider the first term of Equation 8, which has a very similar derivation. Propagating the derivative with respect to α_i through the integral

$$
\frac{\partial}{\partial \alpha_i}\left(\int |\mathbf{S}'_{i-1}| dt\right) = \int <\mathbf{S}'_{i-1}, \mathbf{S}'_{i-1}>^{-\frac{1}{2}} <\mathbf{S}'_{i-1}, \frac{\partial \mathbf{S}'_{i-1}}{\partial \alpha_i}> dt, \tag{12}
$$

As before, we derive an expression for the $\frac{\partial \mathbf{S}'_{i-1}}{\partial \alpha_i}$ terms using Equation 2, yielding

$$
\frac{\partial \mathbf{S}'_{i-1}}{\partial \alpha_i} = 3t^2\frac{\partial \mathbf{A}_{i-1}}{\partial \alpha_i} + 2t\frac{\partial \mathbf{B}_{i-1}}{\partial \alpha_i} + \frac{\partial \mathbf{C}_{i-1}}{\partial \alpha_i} \tag{13}
$$

Next, the derivatives $\frac{\partial \mathbf{A}_{i-1}}{\partial \alpha_i}$, $\frac{\partial \mathbf{B}_{i-1}}{\partial \alpha_i}$ and $\frac{\partial \mathbf{C}_{i-1}}{\partial \alpha_i}$ can be derived using Equation 3, as

$$
\frac{\partial \mathbf{A}_{i-1}}{\partial \alpha_i} = -2\frac{\partial \mathbf{p}_i}{\partial \alpha_i} + \frac{\partial \mathbf{t}_i}{\partial \alpha_i}
$$
$$
\frac{\partial \mathbf{B}_{i-1}}{\partial \alpha_i} = 3\frac{\partial \mathbf{p}_i}{\partial \alpha_i} - \frac{\partial \mathbf{t}_i}{\partial \alpha_i}
$$
$$
\frac{\partial \mathbf{C}_{i-1}}{\partial \alpha_i} = 0 \tag{14}
$$

We now have all the derivatives needed to compute the second term of Equation 8.

3.2   Curvature Minimization

We now consider the problem of minimizing curvature. Since curvature can be positive or negative, we choose to minimize the squared curvature, i.e, we would like to find the angles α_i that minimize

$$E_c = \int \kappa^2(t)dt, \tag{15}$$

where $\kappa(t)$ is the curvature of $\mathbf{S}$ at point t. Since the skin is represented as a set of segments, this is equivalent to

$$E_c = \sum_{i=1}^{N-1} \int \kappa_i^2(t)dt, \tag{16}$$

where $\kappa_i(t)$ is the curvature at point t along segment $\mathbf{S}_i$. Next, we take the derivative of the energy with respect to the angle α_i. As stated above, the ith angle only affects the segments $\mathbf{S}_{i-1}$ and $\mathbf{S}_i$. Therefore,

$$\frac{\partial E_c}{\partial \alpha_i} = \frac{\partial}{\partial \alpha_i}\left(\int \kappa_i^2(t)dt\right) + \frac{\partial}{\partial \alpha_i}\left(\int \kappa_{i-1}^2(t)dt\right) \tag{17}$$

Recall that the curvature is given by

$$\kappa_i = \frac{|\mathbf{S}_i' \times \mathbf{S}_i''|}{|\mathbf{S}_i'|^3} = \frac{< \mathbf{S}_i', J\mathbf{S}_i'' >}{< \mathbf{S}_i', \mathbf{S}_i' >^{\frac{3}{2}}}, \tag{18}$$

where $J = \begin{bmatrix} 0 & 1 \\ -1 & 0 \end{bmatrix}$ is a 90 degree rotation matrix. Using these equations, Equation 17 becomes

$$\frac{\partial E_c}{\partial \alpha_i} = \frac{\partial}{\partial \alpha_i}\left(\int \left[\frac{< \mathbf{S}_i', J\mathbf{S}_i'' >}{< \mathbf{S}_i', \mathbf{S}_i' >^{\frac{3}{2}}}\right]^2 dt\right) + \frac{\partial}{\partial \alpha_i}\left(\int \left[\frac{< \mathbf{S}_{i-i}', J\mathbf{S}_{i-i}'' >}{< \mathbf{S}_{i-i}', \mathbf{S}_{i-i}' >^{\frac{3}{2}}}\right]^2 dt\right) \tag{19}$$

First term. Let us consider the first term of Equation 19. Propagating the derivative with respect to α_i through the integral, we see that

$$\frac{\partial}{\partial \alpha_i}\left(\int \left[\frac{< \mathbf{S}_i', J\mathbf{S}_i'' >}{< \mathbf{S}_i', \mathbf{S}_i' >^{\frac{3}{2}}}\right]^2 dt\right) =$$

$$\int 2\left[\frac{< \mathbf{S}_i', J\mathbf{S}_i'' >}{< \mathbf{S}_i', \mathbf{S}_i' >^{\frac{3}{2}}}\right]\left(\frac{\frac{\partial}{\partial \alpha_i}< \mathbf{S}_i', J\mathbf{S}_i'' >}{< \mathbf{S}_i', \mathbf{S}_i' >^{\frac{3}{2}}} - \frac{3}{2}\frac{< \mathbf{S}_i', J\mathbf{S}_i'' >\frac{\partial}{\partial \alpha_i}< \mathbf{S}_i', \mathbf{S}_i' >}{< \mathbf{S}_i', \mathbf{S}_i' >^{\frac{5}{2}}}\right) dt$$

For this, we need the derivatives $\frac{\partial}{\partial \alpha_i}< \mathbf{S}_i', J\mathbf{S}_i'' >$ and $\frac{\partial}{\partial \alpha_i}< \mathbf{S}_i', \mathbf{S}_i' >$. It is easy to show that these derivatives are

$$\frac{\partial}{\partial \alpha_i}< \mathbf{S}_i', J\mathbf{S}_i'' > = < \frac{\partial \mathbf{S}_i'}{\partial \alpha_i}, J\mathbf{S}_i'' > + < \mathbf{S}_i', J\frac{\partial \mathbf{S}_i''}{\partial \alpha_i} >$$

$$\frac{\partial}{\partial \alpha_i}< \mathbf{S}_i', \mathbf{S}_i' > = 2 < \mathbf{S}_i', \frac{\partial \mathbf{S}_i'}{\partial \alpha_i} > \tag{20}$$

Equation 9 gives an expression for $\frac{\partial \mathbf{S}'_i}{\partial \alpha_i}$, and from this we see $\frac{\partial \mathbf{S}''_i}{\partial \alpha_i} = 6t\frac{\partial \mathbf{A}_i}{\partial \alpha_i} + 2\frac{\partial \mathbf{B}_i}{\partial \alpha_i}$. The derivatives $\frac{\partial \mathbf{A}_i}{\partial \alpha_i}$, $\frac{\partial \mathbf{B}_i}{\partial \alpha_i}$ and $\frac{\partial \mathbf{C}_i}{\partial \alpha_i}$ are given in Equation 10. We now have all the derivatives needed to compute the first term in Equation 19.

Second term. The first term of Equation 19 is very similar the second term derived above. Propagating the derivative with respect to α_i through the integral, we see that

$$
\frac{\partial}{\partial \alpha_i}\left(\int \left[\frac{< \mathbf{S}'_{i-1}, J\mathbf{S}''_{i-1} >}{< \mathbf{S}'_{i-1}, \mathbf{S}'_{i-1} >^{\frac{3}{2}}} \right]^2 dt \right) = \int 2\left[\frac{< \mathbf{S}'_{i-1}, J\mathbf{S}''_{i-1} >}{< \mathbf{S}'_{i-1}, \mathbf{S}'_{i-1} >^{\frac{3}{2}}} \right] \cdot
$$
$$
\left(\frac{\frac{\partial}{\partial \alpha_i} < \mathbf{S}'_{i-1}, J\mathbf{S}''_{i-1} >}{< \mathbf{S}'_{i-1}, \mathbf{S}'_i >^{\frac{3}{2}}} - \frac{3}{2}\frac{< \mathbf{S}'_{i-1}, J\mathbf{S}''_{i-1} > \frac{\partial}{\partial \alpha_i} < \mathbf{S}'_{i-1}, \mathbf{S}'_{i-1} >}{< \mathbf{S}'_{i-1}, \mathbf{S}'_{i-1} >^{\frac{5}{2}}} \right) dt \tag{21}
$$

For this, we need the derivatives $\frac{\partial}{\partial \alpha_i} < \mathbf{S}'_{i-1}, J\mathbf{S}''_{i-1} >$ and $\frac{\partial}{\partial \alpha_i} < \mathbf{S}'_{i-1}, \mathbf{S}'_{i-1} >$. It is well known that the derivative of the cross product is

$$
\frac{\partial}{\partial \alpha_i} < \mathbf{S}'_{i-1}, J\mathbf{S}''_{i-1} > = < \frac{\partial \mathbf{S}'_{i-1}}{\partial \alpha_i}, J\mathbf{S}''_{i-1} > + < \mathbf{S}'_{i-1}, J\frac{\partial \mathbf{S}''_{i-1}}{\partial \alpha_i} >
$$
$$
\frac{\partial}{\partial \alpha_i} < \mathbf{S}'_{i-1}, \mathbf{S}'_{i-1} > = 2 < \mathbf{S}'_{i-1}, \frac{\partial \mathbf{S}'_{i-1}}{\partial \alpha_i} > \tag{22}
$$

Equation 13 gives an expression for $\frac{\partial \mathbf{S}'_{i-1}}{\partial \alpha_i}$, and from this we see $\frac{\partial \mathbf{S}''_{i-1}}{\partial \alpha_i} = 6t\frac{\partial \mathbf{A}_{i-1}}{\partial \alpha_i} + 2\frac{\partial \mathbf{B}_{i-1}}{\partial \alpha_i}$. The derivatives $\frac{\partial \mathbf{A}_{i-1}}{\partial \alpha_i}$, $\frac{\partial \mathbf{B}_{i-1}}{\partial \alpha_i}$ and $\frac{\partial \mathbf{C}_{i-1}}{\partial \alpha_i}$ are given in Equation 14. Thus, all the derivatives needed to compute the second term of Equation 19 have been derived.

In these equations, we evaluate the integrals in Equations 8 and 19 for each angle α_i. However, for the first ball, $i = 1$, there is no segment $\mathbf{S}_{i-1}$, so we ignore the integral this term. Likewise, for the last ball, $i = N$, there is no segment $\mathbf{S}_i$, so we ignore the integral this term.

3.3   Discussion

To summarize, we have derived the gradient of energy functionals E_a and E_c with respect to angles, α_i. The derivation inherently consisted of several steps via the chain rule, as the energy is a combination of the arc length and squared curvature, which in turn are functions of the skin, which in turn is a function of the segment coefficients $\mathbf{A}_i, \mathbf{B}_i, \mathbf{C}_i, \mathbf{D}_i$ and $\mathbf{A}_{i-1}, \mathbf{B}_{i-1}, \mathbf{C}_{i-1}, \mathbf{D}_{i-1}$, which in turn are functions of the angles α_i.

3.4   Implementation

We combine the energies E_a and E_c together, as

$$
E = (1 - k)E_a + kE_c \tag{23}
$$

where k is a constant used to weight the arc length minimization relative to the curvature minimization. Convex combinations of the two can be selected using $k \in [0, 1]$. Therefore, the combined energy minimization is given by

$$\frac{\partial E}{\partial \alpha_i} = (1 - k)\frac{\partial E_a}{\partial \alpha_i} + k\frac{\partial E_c}{\partial \alpha_i} \tag{24}$$

where $\frac{\partial E_a}{\partial \alpha_i}$ is given in Equation 8 and $\frac{\partial E_c}{\partial \alpha_i}$ is provided in Equation 17. In all of the experiments in this paper, we fix $k = 0.9$, to encourage smoother solutions.

These equations are a set of differential equations that can be used in a gradient descent procedure to optimize the skin by manipulating the angles $\alpha = [\alpha_1, \ldots \alpha_N]^T$. Let α_i^n be the ith angle at iteration n. We can then update the angles by moving them in the negative gradient direction, i.e.,

$$\alpha^{n+1} = \alpha^n - \Delta t \nabla E_{\alpha^n}, \tag{25}$$

where Δt is a time step and $\nabla E_{\alpha^n} = [\frac{\partial E}{\partial \alpha_1^n}, \frac{\partial E}{\partial \alpha_2^n}, \ldots, \frac{\partial E}{\partial \alpha_N^n}]^T$.

The computational complexity of the algorithm depends on the number of balls N and the number of points L on a segment where the points and derivatives are evaluated. For each iteration of the gradient descent procedure, the computational complexity is $O(NL)$. The number of iterations required depends on the time step Δt as well as how close the initial skin is to the final solution.

4   Results

A simple example is provided in Figure 3. Here, four balls of radius 50, 75, 50, and 25 pixels, respectively were set along the x-axis. The initial angles for this experiment were 0.57, 1.07, 1.57 and 2.07 radians, respectively; the initial skin is shown in part (a) of the figure. The angles were iteratively updated using Equation 25. An intermediate solution after 50 iterations in shown in (b), at this stage, the skin is considerably smoother while still satisfying the constraints of the problem. We show the result after 100 iterations in (c), at which point the energy has reached a minimum and the angles have converged. The solution (all 100 iterations) is computed in 47 milliseconds using C++ code compiled on a machine with a 3.0 GHz single-core processor.

Figure 4 shows a slightly more complicated example for which some balls overlap and others do not. The initial skin is shown in (a), an intermediate result after 70 iterations in (b), and the final result upon convergence after 140 iterations in (c). The solution (all 140 iterations) is computed in 143 milliseconds.

In Figure 5 we show an example of generating an interpolating region for a collection of balls. In this case, we have two skins, one defining the interior boundary of the region (rendered in green), and another defining the exterior boundary (rendered in blue). For each ball, there are two points of contact: one from the interior skin and one for the exterior skin; however, we constrain these points of contact to be separated by 180 degrees. Therefore, for each ball there is only one angle α_i to be determined as in the examples above. We solve for the

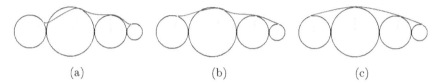

Fig. 3. Example ball skinning. The initialization is shown in (a), and the result after 50 iterations is shown in (b), and the converged result after 100 iterations is shown in (c). The skin is rendered in a blue color.

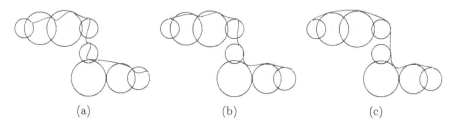

Fig. 4. Another skinning of a set of balls. Initialization (a), intermediate result (b) after 70 iterations, and final result upon convergence (c) after 140 iterations.

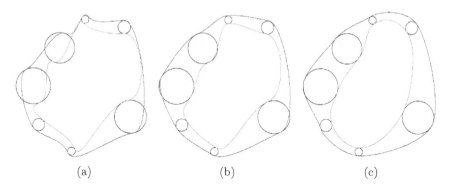

Fig. 5. Generating a smooth interpolating region between a set of balls. Initialization (a), intermediate result (b) after 50 iterations, and final result upon convergence (c) after 100 iterations.

angle for all the balls, with each skin contributing a term in Equation 25. In part (a) of Figure 5 we show an initialization, in (b) and intermediate result after 50 iterations, and in (c) the final converged result after 100 iterations. Convergence for this example occurs in 190 milliseconds.

More examples are provided in Figures 6 and 1. In Figure 6, the balls are arranged on a sine wave and have a variable radius. Convergence of the skinning algorithm, starting from a set of angles far from the optimal result, takes 775

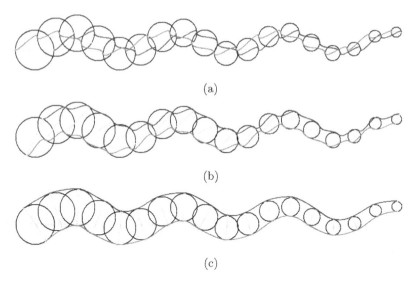

Fig. 6. Generating a smooth interpolating region between a set of balls. Initialization (a), intermediate result (b) after 80 iterations, and final result upon convergence (c) after 160 iterations.

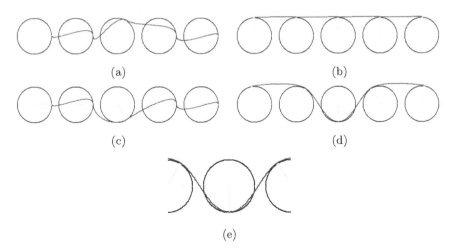

Fig. 7. Solution depends on initial condition. Different solutions (b) and (d) are possible depending on the initial condition (a) and (c), respectively. In (e), we show that the skin may pass through a ball.

milliseconds. In Figure 1, the variable radius balls are arranged in a spiral. The skin is generated in 2.5 seconds.

We note that our gradient descent approach only guarantees a locally optimal solution; the particular solution depends on the convexity of the energy functional as well as the initial condition, as demonstrated in Figure 7. In (a) we show an initial skin, and in (b) the result of our approach. The initial condition in (c) is identical to (a) except the middle ball has a different angle drawing the skin down. In (d), we show the result of our approach starting from (c), resulting in a different solution. In this example and many others in the paper, the initial skins are chosen to be far from the final solution to demonstrate the effect and robustness of the differential equations. However, in practice, it is typically easy to determine a good initialization based on the choosing an angle for each ball that is along the ray orthogonal to the centerline connecting adjacent ball centroids. Finally, we note that the skin our method generates may pass through a ball (shown in Figure 7 (e)) since it is only constrained to be tangent to the ball at one point of incidence. In our application of modeling blood vessels, this is an acceptable solution since ball itself is a geometric proxy of the local vessel geometry.

5 Conclusion

In this paper, we presented a method for optimally skinning an ordered set of 2D balls. Our formulation of the problem requires that the skin be modeled by a point of contact with each ball and at the point of contact, be tangent to the ball. We have presented novel derivations resulting in differential equations that minimize a convex combination of the arc length and curvature of a third order polynomial spline subject to these constraints. Starting with an initial skin, we evolve the skin's parameters until convergence. Experimental results demonstrate the viability of the method.

For future work, we are interested in extending the approach to interpolate balls in R^3. In this case, the point of contact for a ball will be modeled with two angles, and the derivation will result in a coupled set of partial differential equations for these two angles. The skin will then be a surface that interpolates the balls, constrained by the points of contact.

References

1. Singh, K., Kokkevis, E.: Skinning Characters using Surface Oriented Free-Form Deformations. In: Graphics Interface, pp. 35–42 (2000)
2. Peternell, M.: Rational Parametrizations for Envelopes of Quadric Families. PhD thesis, University of Technology, Vienna, Austria (1997)
3. Cheng, H., Shi, X.: Quality Mesh Generation for Molecular Skin Surfaces Using Restricted Union of Balls. In: IEEE Visualization (2005)
4. Edelsbrunner, H.: Deformable smooth surface design. Discrete and Computational Geometry 21(1), 87–115 (1999)
5. Kruithov, N., Vegter, G.: Envelope Surfaces. In: Annual Symposium on Computational Geometry, pp. 411–420 (2006)
6. Whited, B., Rossignac, J., Slabaugh, G., Fang, T., Unal, G.: Pearling: 3D Interactive Extraction of Tubular Structures from Volumetric Images. In: Interaction in Medical Image Analysis and Visualization, held in conjunction with MICCAI (2007)

Part II
Short Papers

3D Mesh Segmentation Using Mean-Shifted Curvature

Xi Zhang[1], Guiqing Li[1,*], Yunhui Xiong[1], and Fenghua He[2]

[1] School of CSE, South China University of Technology, Guangzhou, China
[2] Nanchang Land Army College
Brain.zhang@hotmail.com, {ligq,yhxiong}@scut.edu.cn,
wskmqwxf@163.com

Abstract. An approach to segmentation of a 3D mesh is proposed. It employs mean-shift curvature to cluster vertices of the mesh. A region-growing scheme is then established for collecting them into connected subgraphs. The mesh faces consisting of vertices in the same subgraph constitute a patch while faces whose vertices are in different subgraphs are split and then lined out to near patches to complete the segmentation. To produce pleasing results, several ingredients are introduced into the segmentation pipeline. Firstly, we enhance the original model before mean-shifting and then transfer the curvature of the enhanced mesh to the original one in order to make the features distinguishable. To rectify the segmentation boundaries, the min-cut algorithm is used to repartition regions around boundaries. We also detect sharp features.

Keywords: Mesh segmentation, mean-shift, curvature, graph cut.

1 Introduction

Recently, mesh segmentation has been applied to many research areas such as 3D shape retrieval, compression, texture mapping, deformation, and simplification. 3D Mesh segmentation methods can generally be categorized into two different kinds [1,2]: patch-type and part-type. The first one partitions meshes into disk-like patches while the other divides 3D models into meaningful parts.

In essence, segmentation approaches cluster primitives with similar attributes into large regions. There are three popular schemes to do this [1]. At the first place is the region-growing scheme, a greedy approach focusing on local optimization. Its key problems are start seed selection and growing stop criteria. One may use either vertex [3,4,5] or face [6] as the basic growing element, and then generate start seeds according to some regulations [6-9]. The second is hierarchical-clustering which assumes that each face is initially a cluster and then merges clusters with some minimum cost [10,11]. The last is the k-means based iterative-clustering scheme which starts from k representatives of k clusters and assigns each element to one of the clusters. An iterative process updates the representatives until they stop change [11].

Depending on applications, the features of a mesh used as segmentation criteria may be either scalar or vector functions, such as curvature [1,8,12], normal [7] and geodesic distance [13,14], or even symmetry [15].

---

* Corresponding author.

F. Chen and B. Jüttler (Eds.): GMP 2008, LNCS 4975, pp. 465–474, 2008.

This paper proposes a part-type and region-growing mesh segmentation algorithm. Its contributions are two folds. Firstly, a new segmentation pipeline is proposed in which several optimization procedures are introduced to improve the correctness. Secondly, a non-parametric method based on the mean-shift technique is established by constructing a graph feature space to simplify the cluster procedure. In addition, the approach exhibits efficiency in segmenting transitive parts of CAD models.

2   Related Works

The mean-shift clustering technique has successfully been used in image segmentation, image blurring and video object tracking. Shamir et al. [16] firstly investigated its application to feature space analysis of 3D meshes and demonstrated its applications on feature-extraction, data exploration and model partitioning. In order to combine spatial information and other feature of the mesh into a uniform feature space, a moving geodesic parameterization has to be introduced in their work. Yamauchi et al. employed mean-shifted face normals to partition 3D meshes [7]. Their approach can only be patch-type due to face normal used and therefore is not suitable for semantic segmentation. James and Twigg used a non-parametric mean-shift algorithm to reveal the near-rigid structure of given meshes [17].

Adjacent meaningful parts of a model are generally bordered by deep concavities [18]. Curvature is a good attribute for measuring these features. Yamauchi et al. [19] employed Gaussian curvature to design a patch-based segmentation algorithm.

Our approach is part-based and different types of curvature are employed. To combine spatial information and curvature into a feature space, we employ a generalized density function definition which uses the product of kernels to filter two feature components respectively. Especially, we substitute the mesh for a graph naturally induced from it in order to make the approach non-parametric. In addition, our method is rotation-and-scale invariant due to curvature used.

3   Overview of the Proposed Approach

The algorithm mainly consists of three phases: curvature estimation, curvature mean-shifting and region growing. We enhance the initial mesh in order to obtain a more discernable curvature map as preprocessing and optimize boundaries after region growing as post-processing. Fig.1 illustrates the outline of the algorithm.

Preprocessing. Given mesh is enhanced by intensify its feature parts so as to make sure the natural boundary of the geometry can be recognized during segmentation. Concrete steps for enhancement are described in Section 4.

Curvature computation. Vertex curvature of the enhanced mesh is then estimated using a fitting method and then transferred to the given mesh. Both the mean-curvature and the combined curvature are tested in our experiments (See Section 5).

Curvature mean-shifting. The curvature of all vertices is then mean-shifted iteratively until the stop condition are met to lead to a curvature distribution compatible with the structure of the shape. More details are described in Section 6.

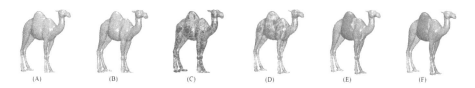

Fig. 1. The pipeline of the proposed algorithm: (A) Original model; (B) Enhanced model; (C) Curvature map for the enhanced model; (D) Mean-shifted curvature map with enhancing curvature; (E) Region-growing. (F) The post-processing result.

Vertices clustering. Vertices are clustered by a region-growing algorithm in terms of the mean-shifted curvature in this phase. Section 7 describes the details.

Postprocessing. The segmented boundaries of patches are rectified in the step to diminish zigzag cut lines using the minimum cut algorithm. A solution is designed for dealing with special cases for CAD models with transitive parts and sharp features.

4   Features Enhancement

The underlying idea feature enhancement is to enhance the interface effect of different regions such as large convexities and concavities so as to make it more efficient to pick feature parts out. As the mean-shift procedure, the body of the proposed approach, may be viewed as a curvature filter, we enhance the geometry instead of the curvature directly in this preprocessing step. Given an original mesh $M =<V,F>$, we firstly employ a low-passing filter described in [20], which is an iterative, anti-shrinkage fairing algorithm, to get a smoothed version of M, denoted by $M_S =<V_s,F>$. A set D of detail coefficients are then defined as follows

$$D = \{\mathbf{d}_i = v_i - v_{s,i} : v_i \in V, v_{s,i} \in V_s\}$$

A vertex $v_i^E \in V_E$ of the enhanced mesh $M_E =<V_E,F>$ is evaluated as

$$v_i^E = v_i + \lambda \mathbf{d}_i \qquad (1)$$

where λ is a relaxation scalar.

Let $\kappa(v)$ be the curvature of vertex v (See Section 5 for curvature estimation). We define the enhancing curvature of $v_i \in M$ as $\kappa(v_i) = \kappa(v_i^E)$. In Fig.1, (C) and (D) give curvature comparison between the original and enhanced models. Fig. 2 shows two segmentation results using original (Fig. 2 Left)

Fig. 2. Segmentations without (Left) and with enhancement (Right)

and enhancing (Fig. 2 Right) curvature, respectively. It can be observed that the latter gives a more precise segmentation.

5 Curvature Estimation

We estimate the principle curvature of vertices of mesh M using the approach by Chen and Schmitt [21,22]. Let $N_F(v)$ be the set of all triangles adjacent to vertex $v \in V$, and $\mathbf{n}(f)$ and $\mathbf{c}(f)$ respectively unit normal and centroid of face f. The vertex normal $\mathbf{n}(v)$ is defined as the weighted average of normals of its adjacent faces. For $v_j \in N_V(v)$, the set of adjacent vertices of v, a unit tangent $\mathbf{t}_j$ associated with v_j is defined as the normalization of the projection of $v_j - v$ onto the tangent plane of v. The normal curvature of v along $\mathbf{t}_j$ is approximated by

$$\kappa_n(\mathbf{t}_j) = -<v_j - v, \mathbf{n}(v_j) - \mathbf{n}(v)> / |v_j - v|^2, \text{ for } \mathbf{t}_j \ (j = 1,2,...,|N_V(v)|).$$

Without loss of generality, assume $\kappa_n(\mathbf{t}_1)$ is the maximum among all these values. Denote θ_j the angle between $\mathbf{t}_j$ and $\mathbf{t}_1$, we then express $\kappa_n(\mathbf{t}_j)$ as

$$\kappa_n(\mathbf{t}_j) = a \cos^2 \theta_j + b \cos \theta_j \sin \theta_j + c \sin^2 \theta_j,$$

where $a = \kappa_n(\mathbf{t}_1)$, and b and c can be estimated by least square fitting. This leads to the Gaussian curvature, mean curvature and two principle curvatures of v as follows

$$\kappa_G = ac - \frac{1}{4}b^2, \ \kappa_m = \frac{1}{2}(a+c), \text{ and } \kappa_{1,2} = \kappa_m \pm \sqrt{\kappa_m^2 - \kappa_G}. \quad (2)$$

Finally, we also use the following curvature combination [23] to amplify the difference between convex and concave regions:

$$\kappa_c = \frac{2}{\pi} \arctan \frac{\kappa_1 + \kappa_2}{\kappa_1 - \kappa_2} \quad (3)$$

6 Curvature Mean-Shift

Given a sample of a feature space, a mean-shift algorithm firstly estimates the density of the feature space, then evaluates the gradient of the density function, and finally moves the sample points along their gradient direction. A standard mean-shift algorithm is generally guaranteed to be convergent [24]. Our situation is more complex since there are two components in the feature space, the 2D manifold and the 1D curvature.

Our feature space χ comprises two components: $\chi = \{(v, \kappa(v)) : v \in V, \kappa(v) \in K\}$. A generalized mean-shift density function [7] is defined as:

$$\hat{f}(v, \kappa(v)) = \frac{1}{|V|\hat{h}^2 h} \sum_{v_j \in V} \hat{K}\left(\frac{v - v_j}{\hat{h}}\right) K\left(\frac{\kappa(v) - \kappa(v_j)}{h}\right), \quad (4)$$

where $\hat{h}$ and h are constants called bandwidth, and $\hat{K}(\bullet)$ and $K(\bullet)$ are two kernels. To simplify the density function, we set $\hat{K}(\bullet)$ to be a flat kernel: $\hat{K}(x) = 1$ for $\|x\| \le 1$ and 0 otherwise. Here threshold 1 is chosen since we want to update the curvature of a vertex with that of its 1-ring vertices. Let $\hat{h} = 2$ and $\|v - v_j\|$ be the number of edges of the shortest path from v to v_j. The density function of Eq. (4) is then reduced to

$$\hat{f}(\kappa(v)) = \frac{1}{4|V|h} \sum_{v_j \in N_2(v)} K\left(\frac{\kappa(v) - \kappa(v_j)}{h}\right) \tag{5}$$

Furthermore, define $K(\bullet) = pk(\|\bullet\|^2)$, where p is a normalization constant assuring that $K(\bullet)$ integrates to 1 and $k(x)$ is a profile. The gradient of the curvature density of Eq. (5) then has the following form [25]:

$$\nabla \hat{f}(\kappa(v)) = \frac{pm(\kappa(v))}{2|V|h^3}\left[\sum_{v_j \in N_2(v)} g\left(\left\|\frac{\kappa(v) - \kappa(v_j)}{h}\right\|^2\right)\right] \tag{6}$$

where $g(\bullet) = -k'(\bullet)$, $N_2(v)$ denotes the set of 2-neighboring vertices of v, and

$$m(\kappa(v)) = -\kappa(v) + \sum_{v_j \in N_2(v)} \kappa(v_j)g\left(\left\|\frac{\kappa(v) - \kappa(v_j)}{h}\right\|^2\right) \Big/ \sum_{v_j \in N_2(v)} g\left(\left\|\frac{\kappa(v) - \kappa(v_j)}{h}\right\|^2\right) \tag{7}$$

is the curvature mean-shift at vertex v. Note that parameter p and the total vertex number $|V|$ of Eq. (6) do not appear in Eq. (7). Parameter h plays a key role for obtaining good results. $h = 10^{-5}$ is used for all examples in the paper.

In the $(t+1)$-th iteration, we renew the curvature of v with

$$\kappa^{t+1}(v) = \kappa^t(v) + m(\kappa^t(v)) \qquad t = 0,1,2,\cdots, \tag{8}$$

where $\kappa^0(v) = \kappa(v^E)$ is the enhancing curvature at v.

Experimental comparison (Gaussian kernel, uniform kernel, biweight kernel etc.), shows that Epanechnikov kernel [7,24] may yield best results. The mean-shift procedure then repeats Eq. 8 until $error(v) = |\kappa^{t+1}(v) - \kappa^t(v)| < \varepsilon$, where ε is a user specified threshold.

7 Region-Growing, Boundary Rectification and Thin Region Removing

Region-growing. The region-growing scheme is responsible for clustering vertices with similar feature. Initially, the vertex with maximal curvature is taken as a start seed. A patch is initialized with the seed and grows from its boundary. The growing strategy is greedy in the sense that one adjacent vertex is added only if the difference between its curvature and the average curvature of the current patch is less than a specified threshold. If no new vertex satisfies the condition, a new patch is launched by selecting a vertex with maximal curvature from unprocessed vertices until no vertex remains. In our experiment, the threshold is set to 10^{-1} for combined curvature and 10^{-4} for mean curvature. After vertex region growing is completed, faces are classified into three kinds according to the status of their vertices. Triangles in the same patch are internal faces, triangles whose vertices belong to two patches are boundary triangles, and triangles shared by three patches are not regarded as a boundary one. A strip of boundary triangles forms a boundary.

Boundary rectification. Segmented results still exhibit serrated boundaries at this stage. A graph min-cut procedure is devised to straighten dividing lines [13,14,26], but instead of building dual graphs like previous work, we directly apply it to original sub-meshes. For given boundary b, the associated graph is made up of two additional points (source S and sink T) and vertices in a search region of b which is defined as the sub-mesh consisting of b and τ layers of triangles on its two sides, where τ is specified by the user ($\tau = 3$ in our experiments). A flow network is then constructed by connecting S and T to the points in the outmost layers as shown in Fig. 3.

A new capacity for reflecting vertex distance is defined as follows. Let v be a vertex of valence k in a flow network and $N_E(v)$ be the set of its adjacent edges. Denote θ_e the angle between normals of two faces sharing $e \in N_E(v)$. We define [13]

$$w(v) = \frac{1}{|N_E(v)|} \sum_{e \in N_E(v)} \lambda_e (1 - \cos \theta_e),$$

where $\lambda_e = 1$ for concave edge e and is set to a small positive number for convex edges. The capacity of edge (v_i, v_j) is then defined as

$$Cap(v_i, v_j) = \frac{1}{1 + |w(v_i) - w(v_j)|}.$$

The capacity of edges with source/sink is set to be infinite. Fig. 4 depicts that the boundary smoothing made the result more pleasing.

Fig. 3. Source S, sink T and the yellow region form a flow network

Fig. 4. Segmentation results of Ilk model with (Top) and without (Bottom) boundary smoothing

Thin region removing. There exist some slender regions with sharp features in their center (see Fig. 5) since curvature of sharp features is disseminated to near regions. It is more natural to cut this kind of models along sharp features. Sharp edges are

Fig. 5. Segmented results before (left) and after (right) the post-processing step

detected by examining the dihedral angle of two adjacent faces. The regions containing sharp edges are then decomposed to several parts along the creases which are finally merged into adjacent patches.

8 Experimental Results

This section gives some experimental results. Comparison with some previous methods is also performed. Different regions are depicted by pseudo-color. Figs. 6 are some examples based on mean curvature and combined curvature, respectively. It demonstrates that the proposed algorithm may yield qualified segmentation for a variety of models. Fig. 7 demonstrates that our approach exhibits strong advantage in extracting transitive parts for CAD models. Fig. 8 presents a comparison of our proposed method with the face based approach using mean-shifted normals [7]. It can be observed from Fig. 8 that our approach indeed produces more meaningful results for manifold geometries and more reasonable partition for CAD models. Using

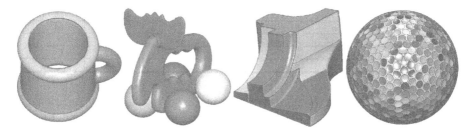

Fig. 6. Segmentation results of Cup, Elk, Fandisk models using mean-curvature and the Golf ball model based on combined curvature

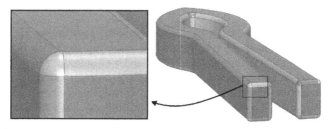

Fig. 7. The transitive parts are fairly partitioned for a U-shape CAD part

Fig. 8. Segmentation comparison between the approach described in [7] (Top row) without boundary rectification and our algorithm (Bottom row) for models of Elk, Pear, Chessman, and Fandisk models

Table 1. Performance of the proposed algorithm (M: Mean curvature, C: combined curvature)

Model	Num. of vertices	Mean shift (sec.)	Region growth (sec.)	Patch No.	Curvature Type
Cup	5668	1.53	1.31	5	M
Elk	5194	1.03	0.99	6	M
U-shape	3132	0.76	0.67	27	M
Fandisk	6475	1.48	1.32	20	M
Pear	10754	3.79	7.27	6	M
Chessman	33664	37.61	30.78	18	M
Golf ball	122882	440.08	580.86	643	C
Octopus	16942	5.76	15.39	10	C
Camel	34506	35.46	32.85	11	C

curvature scalar instead of normal vector make our method more efficiency in memory and running time. Finally, Table 1 summarizes the running time for segmentation of the present models. All examples were obtained on a PC with an Intel(R) Core(TM) 1.86GHz CPU.

9   Conclusion and Future Work

The paper proposes a new framework for meaningfully segmenting 3D mesh models. The essence of the framework is a mean-shift clustering algorithm over a curvature space which is a non-parameterization method and well suited for large uncertain data analysis. It is able to partition models with salient part preserved and is competent for recognizing transitive parts of model. A model enhancement technique is introduced as preprocessing to make the segmentation more robustly. In addition, a strategy of min-cut algorithm based on a new graph capacity is introduced for fairing the segmentation results.

Acknowledgement. The work is supported by NSF of China (60673005, 60573019), NSF of Guangdong Prov. (05006540), and Open Fund of the State Key Lab of CAD & CG, Zhejiang Univ.. All models are courtesy of http://dsw.aimatshape.net.

References

1. Shamir, A.: A formulation of boundary mesh segmentation. In: Proc. of 2nd Symposium on 3D Data Processing, Visualization and Transmission, pp. 82–89 (2004)
2. Attene, M., Katz, S., Mortara, M., Patane, G., Spagnuolo, M., Tal, A.: Mesh segmentation - A comparative study. In: Proc. of Shape Modeling International (SMI 2006), pp. 14–25 (2006)
3. Mangan, A., Whitaker, R.: Partitioning 3D surface meshes using watershed segmentation. IEEE Trans. on Visualization and Computer Graphics 5(4), 308–321 (1999)
4. Vieira, M., Shimada, K.: Surface mesh segmentation and smooth surface extraction through region growing. Computer Aided Geometric Design 22(8), 771–792 (2005)
5. Ji, Z., Liu, L., Chen, Z., And Wang, G.: Easy mesh cutting. Computer Graphics Forum (Eurographics 2006) 25(3), 283–291 (2006)
6. Pan, X., Ye, X., Zhang, S.: 3D Mesh segmentation using a two-stage merging strategy. In: CIT 2004, pp. 730–733 (2004)
7. Yamauchi, H., Lee, S., Lee, Y., Ohtake, Y., Belyaev, A., And Seidel, H.-P.: Feature sensitive mesh segmentation with mean shift. In: Proc. of Shape Modeling and Applications, pp. 238–245 (2005)
8. Guillaume, L., Florent, D., Atilla, B.: Curvature tensor based triangle mesh segmentation with boundary rectification. In: Proceedings of the Computer Graphics International (CGI 2004), pp. 10–17 (2004)
9. Sheffer, A.: Model simplification for meshing using face clustering. CAD 33, 925–934 (2001)
10. Várady, T., Facello, M.A., Terék, Z.: Automatic Extraction of Surface Structures in Digital Shape Reconstruction. In: Kim, M.-S., Shimada, K. (eds.) GMP 2006. LNCS, vol. 4077, pp. 1–16. Springer, Heidelberg (2006)
11. Lai, Y.-K., Zhou, Q.-Y., Hu., S.-M., Martin, R.R.: Feature sensitive mesh segmentation. In: Proc. of ACM Symposium on Solid and Physical Modeling (SPM 2006), pp. 17–25. ACM Press, New York (2006)
12. Guillaume, L., Florent, D., Atilla, B.: A new CAD mesh segmentation method, based on curvature tensor analysis. CAD 37(10), 975–987 (2005)
13. Katz, S., Tal, A.: Hierarchical mesh decomposition using fuzzy clustering and cuts. ACM Transactions on Graphics 22(3), 954–961 (2003)
14. Katz, S., Leifman, G., Tal, A.: Mesh segmentation using feature point and core extraction. The Visual Computer (PG 2005) 21(8-10), 649–658 (2005)
15. Mitra, N.J., Guibas, L.J., Pauly, M.: Partial and approximate symmetry detection for 3D geometry. ACM Trans. on Graphics 25(3), 560–568 (2006)
16. Shamir, A., Shapira, L., Cohen-Or, D.: Mesh analysis using geodesic mean-shift. The Visual Computer 22(2), 99–108 (2006)
17. James, D.L., Twigg, C.D.: Skinning mesh animations. ACM Trans. on Graphics (ACM SIGGRAPH 2005) 24(3), 399–407 (2005)
18. Biederman, I.: Recognition-by-Components: a theory of human image understanding. Psychological Review 94(2), 115–147 (1987)

19. Yamauchi, H., Gumhold, S., Zayer, R., Seidel, H.-P.: Mesh segmentation driven by Gaussian curvature. The Visual Computer (PG 2005) 21(8-10), 659–668 (2005)
20. Li, G., Bao, H., Ma, W.: A unified approach for fairing arbitrary polygonal meshes. Graphical Models 66(3), 160–179 (2004)
21. Dong, C., Wang, G.: Curvatures estimation on triangular mesh. Journal of Zhejiang University (SCIENCE), 128–136 (2005)
22. Chen, X., Schmitt, F.: Intrinsic surface properties from surface triangulation. In: Proc. 2nd European Conf. On Computer Vision, pp. 739–743 (1992)
23. Koenderink, J.J., Van Doorn, A.J.: Surface shape and curvature scales. Image and Vision Computing 10(8), 557–565 (1992)
24. Cheng, Y.: Mean shift, mode seeking, and clustering. IEEE Trans. on Pattern Analysis and Machine Intelligence 17(8), 790–799 (1995)
25. Fukunaga, K., Hostetler, L.D.: The estimation of the gradient of a density function, with Applications in pattern recognition. IEEE Trans. Information Theory 21, 32–40 (1975)
26. Yan, D., Liu, Y., Wang, W.: Quadric Surface Extraction by Variational Shape Approximation. In: Kim, M.-S., Shimada, K. (eds.) GMP 2006. LNCS, vol. 4077, pp. 73–86. Springer, Heidelberg (2006)

Convex Surface Interpolation

Malik Zawwar Hussain and Maria Hussain

Department of Mathematics
University of the Punjab
Lahore-Pakistan
malikzawwar@math.pu.edu.pk

Abstract. This work is a contribution towards the graphical display of 3D data over a regular grid when it is convex. A piecewise rational bi-cubic function [8] has been utilized for this objective. Simple sufficient data dependent conditions are derived on free parameters in the description of rational bi-cubic function to preserve the shape of data. The presented method applies equally well to data or data with derivatives. The developed scheme is not only local and computationally economical but also visually pleasing.

Keywords: Surface interpolation; Convexity; Rational bi-cubic function; Shape parameters; Free parameters.

1 Introduction

In recent years, a good amount of work has been published [1-12] that focuses on shape preserving curves and surfaces. Convexity, monotonicity and positivity are the shape properties of data. However, this paper has addressed the first one i.e. convexity. Convexity is an important shape property and plays a vital role in nonlinear programming which arises in engineering, scientific applications such as design, optimal control, parameter estimation and approximation of function.

Some of the work on convexity was discussed in [1, 3, 5, 6, 9, 10]. The scheme of Asaturyan [1] divided each grid rectangle into nine sub rectangles to generate the convex surfaces. This scheme was not local i.e. by changing data in x direction of one edge of a sub rectangle there was a change through out the grid for all sub rectangle's edges located in the original x direction. Constantini and Fontanella [3] developed a semi global scheme to preserve the shape of convex data. The drawback of scheme was that in some rectangular patches degree of interpolant became too large and polynomial patches tend to be linear in x and/or y, and sometimes the corresponding surfaces were not visually pleasing. Dodd et al [5] produced quadratic splines along the boundary of each grid rectangle. These splines were used to define functional and partial derivatives on the boundaries of the rectangles. The scheme preserved the convexity along the grid lines but failed to preserve the convexity inside. Floater [6] derived sufficient conditions for the convexity of tensor-product Bézier surface. The convexity condition

F. Chen and B. Jüttler (Eds.): GMP 2008, LNCS 4975, pp. 475–482, 2008.
© Springer-Verlag Berlin Heidelberg 2008

was generalized to C^1 tensor product B-spline surfaces. These sufficient conditions were in the form of inequalities which involved control points. Kouibia and Pasadas [9] presented a shape preserving method for the scattered data. They defined a k-convex interpolating spline function in a Sobolev space. The shape preservation of the data was assured by making the derivative function of order k positive. Renka [10] described a Fortran 77 software package for constructing a convex surface that interpolates arbitrarily distributed convex data. In this paper, we have presented an efficient and economical method to preserve the shape of 3D regular convex data. Rational bi-cubic function [8] with shape parameter is worthy for convex surface interpolation due to its increase flexibility and ability to incorporate the data having singularities, whereas, ordinary polynomials cannot.

2   Rational Bi-cubic Function

The rational bi-cubic function [8] used in this paper is written as:

Let $(x_i, y_j, F_{i,j})$, $i = 0, 1, 2, \ldots, m$; $j = 0, 1, 2, \ldots, n$ be given set of data points defined over the rectangular domain $D = [a, b] \times [c, d]$, where $\pi : a = x_0 < x_1 < x_2 < \cdots < x_m = b$ be partition of $[a, b]$ and $\tilde{\pi} : c = y_0 < y_1 < y_2 < \cdots < y_n = d$ be partition of $[c, d]$. The rational bi-cubic function $S(x, y)$ is defined over each rectangular patch $[x_i, x_{i+1}] \times [y_j, y_{j+1}]$ where $i = 0, 1, 2, \ldots, m - 1$; $j = 0, 1, 2, \ldots, n - 1$ as:

$$S(x, y) = \frac{R(\theta) Z R^T(\phi)}{q_{i,j}(\theta) \hat{q}_{i,j}(\phi)}, \tag{1}$$

where
$R(\theta) = \left[(1 - \theta)^3 \ (1 - \theta)^2 \theta \ (1 - \theta)\theta^2 \ \theta^3 \right]$, $\quad Z = [Z_{i,j}]_{1 \leq i,j \leq 4}$,
with

$$
\begin{aligned}
Z_{11} &= F_{i,j}, \ Z_{12} = \hat{\alpha}_{i,j} F_{i,j} + \hat{h}_j F_{i,j}^y, \ Z_{13} = \hat{\beta}_{i,j} F_{i,j+1} - \hat{h}_j F_{i,j+1}^y, \ Z_{14} = F_{i,j+1}, \\
Z_{21} &= \alpha_{i,j} F_{i,j} + h_i F_{i,j}^x, \ Z_{22} = \hat{\alpha}_{i,j}(\alpha_{i,j} F_{i,j} + h_i F_{i,j}^x) + \hat{h}_j(\alpha_{i,j} F_{i,j}^y + h_i F_{i,j}^{xy}), \\
Z_{23} &= \hat{\beta}_{i,j}(\alpha_{i,j} F_{i,j+1} + h_i F_{i,j+1}^x) - \hat{h}_j(\alpha_{i,j} F_{i,j+1}^y + h_i F_{i,j+1}^{xy}), \\
Z_{24} &= \alpha_{i,j} F_{i,j+1} + h_i F_{i,j+1}^x, \ Z_{31} = \beta_{i,j} F_{i+1,j} - h_i F_{i+1,j}^x, \\
Z_{32} &= \hat{\alpha}_{i,j}(\beta_{i,j} F_{i+1,j} - h_i F_{i+1,j}^x) + \hat{h}_j(\beta_{i,j} F_{i+1,j}^y - h_i F_{i+1,j}^{xy}), \\
Z_{33} &= \hat{\beta}_{i,j}(\beta_{i,j} F_{i+1,j+1} - h_i F_{i+1,j+1}^x) - \hat{h}_j(\beta_{i,j} F_{i+1,j+1}^y - h_i F_{i+1,j+1}^{xy}), \\
Z_{34} &= \beta_{i,j} F_{i+1,j+1} - h_i F_{i+1,j+1}^x, \ Z_{41} = F_{i+1,j}, \ Z_{42} = \hat{\alpha}_{i,j} F_{i+1,j} + \hat{h}_j F_{i+1,j}^y, \\
Z_{43} &= \hat{\beta}_{i,j} F_{i+1,j+1} - \hat{h}_j F_{i+1,j+1}^y, \ Z_{44} = F_{i+1,j+1}, \\
q_{i,j}(\theta) &= (1 - \theta)^3 + \alpha_{i,j}(1 - \theta)^2\theta + \beta_{i,j}(1 - \theta)\theta^2 + \theta^3, \\
\hat{q}_{i,j}(\phi) &= (1 - \phi)^3 + \hat{\alpha}_{i,j}(1 - \phi)^2\phi + \hat{\beta}_{i,j}(1 - \phi)\phi^2 + \phi^3, \\
\theta &= \tfrac{x - x_i}{h_i}, \ \phi = \tfrac{y - y_j}{\hat{h}_j}, \ h_i = x_{i+1} - x_i, \ \hat{h}_j = y_{j+1} - y_j.
\end{aligned}
$$

$F_{i,j}^x$ and $F_{i,j}^y$ are first order partial derivatives, $F_{i,j}^{xy}$ is the first order mixed derivative, also known as twist vector at the data site. If the values of derivatives are not provided, then $F_{i,j}^x$, $F_{i,j}^y$ and $F_{i,j}^{xy}$ can be approximated by derivative approximation schemes e.g. arithmetic mean and geometric mean choice of derivatives. In this paper, derivatives at the data site are approximated by arithmetic mean choice of derivatives written in [8]. $\alpha_{i,j}$, $\beta_{i,j}$, $\hat{\alpha}_{i,j}$ and $\hat{\beta}_{i,j}$ are variable weights(free parameters) which provide local control on the shape of surface. The shape of the surface can be modified by assigning different values to these parameters.

3   Convex Surface Interpolation

Michael S. Floater [6] defined the convexity as:

Definition 1. *A function $S(x,y)$ of two variables is convex iff its Hessian matrix*

$$H = \begin{pmatrix} S_{xx} & S_{xy} \\ S_{xy} & S_{yy} \end{pmatrix}$$

is positive semi-definite for all x and y. This is equivalent to the property

$$S_{xx}\lambda_1^2 + 2S_{xy}\lambda_1\lambda_2 + S_{yy}\lambda_2^2 \geq 0, \tag{2}$$

for any λ_1, $\lambda_2 \in \Re$ for all x and y. It is also known from linear algebra that the inequality

$$a\lambda_1^2 + 2b\lambda_1\lambda_2 + c\lambda_2^2 \geq 0, \tag{3}$$

holds for all λ_1, $\lambda_2 \in \Re$ iff

$$a \geq 0, \, c \geq 0, \, ac \geq b^2. \tag{4}$$

Thus the convexity of S is equivalent to

$$S_{xx} \geq 0, \, S_{yy} \geq 0, \, S_{xx}S_{yy} \geq S_{xy}^2. \tag{5}$$

$\square$

Let $(x_i, y_j, F_{i,j})$, $i = 0, 1, 2, \ldots, m$; $j = 0, 1, 2, \ldots, n$ be convex data defined over the rectangular grid $D = [x_0, x_m] \times [y_0, y_n]$ such that

$$\Delta_{i,j} \leq \Delta_{i+1,j}; \; F_{i,j}^x \leq F_{i+1,j}^x; \; \Delta_{i,j} = \frac{F_{i+1,j} - F_{i,j}}{h_i},$$

$$i = 0, 1, 2, \ldots, m-1; \; j = 0, 1, 2, \ldots, n.$$

$$\hat{\Delta}_{i,j} \leq \hat{\Delta}_{i,j+1}; \; F_{i,j}^y \leq F_{i,j+1}^y; \; \hat{\Delta}_{i,j} = \frac{F_{i,j+1} - F_{i,j}}{\hat{h}_j},$$

$$i = 0, 1, 2, \ldots, m; \; j = 0, 1, 2, \ldots, n-1.$$

$$F_{i,j}^x \leq \Delta_{i,j} \leq F_{i+1,j}^x, i = 0, 1, 2, \ldots, m-1; \; j = 0, 1, 2, \ldots, n.$$

$$F_{i,j}^y \leq \hat{\Delta}_{i,j} \leq F_{i,j+1}^y, \; i = 0, 1, 2, \ldots, m; \; j = 0, 1, 2, \ldots, n-1.$$

According to the above definition, the rational bi-cubic function (1) preserves convexity if

$$S_{xx}(x,y) > 0, \; S_{yy}(x,y) > 0 \; and \; S_{xx}(x,y)S_{yy}(x,y) - S_{xy}^2(x,y) > 0, \; \forall (x,y) \in D.$$

$$S_{xx}(x,y) = \frac{\sum_{i=0}^{7}(1-\theta)^{7-i}\theta^i A_i}{h_i(q_i(\theta))^3 \hat{q}_j(\phi)}, \; S_{yy}(x,y) = \frac{\sum_{i=0}^{3}(1-\theta)^{3-i}\theta^i B_i}{\hat{h}_j(\hat{q}_j(\phi))^3 q_i(\theta)},$$

$$S_{xy}(x,y) = \frac{\sum_{i=0}^{5}(1-\theta)^{5-i}\theta^i C_i}{\hat{h}_j(q_i(\theta))^2(\hat{q}_j(\phi))^2}.$$

A_i, B_i and C_i are the terms involving $\alpha_{i,j}$, $\beta_{i,j}$, $\hat{\alpha}_{i,j}$, $\hat{\beta}_{i,j}$ and derivative values.
$S_{xx}(x,y) > 0$ if

$$\hat{q}_j(\phi) > 0, \; (q_i(\theta))^3 > 0, \; \sum_{i=0}^{7}(1-\theta)^{7-i}\theta^i A_i > 0.$$

$\hat{q}_j(\phi) > 0$ if
$$\hat{\alpha}_{i,j} > 0, \; \hat{\beta}_{i,j} > 0.$$

$(q_i(\theta))^3 > 0$ if
$$\alpha_{i,j} > 0, \; \beta_{i,j} > 0.$$

$\sum_{i=0}^{7}(1-\theta)^{7-i}\theta^i A_i > 0$ if
$$A_i > 0, \; i = 0,1,2,\ldots,7.$$

$A_i > 0, \; i = 0,1,2,\ldots,7$ if

$$\alpha_{i,j} > Max\left\{D_k, 1 \le k \le 4, k \in Z^+\right\}, \; \beta_{i,j} > Max\left\{D_k, 5 \le k \le 8, k \in Z^+\right\},$$

where

$$D_1 = \frac{F_{i+1,j}^x - F_{i,j}^x}{\Delta_{i,j} - F_{i,j}^x}, \; D_2 = \frac{F_{i+1,j+1}^x - F_{i,j+1}^x}{\Delta_{i,j+1} - F_{i,j+1}^x}, \; D_3 = \frac{\Delta_{i,j} - F_{i,j}^x}{-\Delta_{i,j} + F_{i+1,j}^x},$$

$$D_4 = \frac{\Delta_{i,j+1} - F_{i,j+1}^x}{-\Delta_{i,j+1} + F_{i+1,j+1}^x}, \; D_5 = \frac{F_{i+1,j}^x - F_{i,j}^x}{F_{i+1,j}^x - \Delta_{i,j}}, \; D_6 = \frac{F_{i+1,j+1}^x - F_{i,j+1}^x}{F_{i+1,j+1}^x - \Delta_{i,j+1}},$$

$$D_7 = \frac{F_{i+1,j}^x - \Delta_{i,j}}{\Delta_{i,j} - F_{i,j}^x}, \; D_8 = \frac{F_{i+1,j+1}^x - \Delta_{i,j+1}}{\Delta_{i,j+1} - F_{i,j+1}^x}.$$

$S_{yy}(x,y) > 0$ if

$$(\hat{q}_j(\phi))^3 > 0, \; q_i(\theta) > 0, \; \sum_{i=0}^{3}(1-\theta)^{3-i}\theta^i B_i > 0.$$

$(\hat{q}_j(\phi))^3 > 0$ if

$$\hat{\alpha}_{i,j} > 0, \ \hat{\beta}_{i,j} > 0.$$

$q_i(\theta) > 0$ if

$$\alpha_{i,j} > 0, \ \beta_{i,j} > 0.$$

$\sum_{i=0}^{3}(1-\theta)^{3-i}\theta^i B_i > 0$ if

$$B_i > 0, \ i = 0, 1, 2, 3.$$

$B_i > 0, \ i = 0, 1, 2, 3$ if

$$\hat{\alpha}_{i,j} > Max\left\{D_k, 9 \le k \le 12, k \in Z^+\right\}, \ \hat{\beta}_{i,j} > Max\left\{D_k, 13 \le k \le 16, k \in Z^+\right\},$$

where

$$D_9 = \frac{F^y_{i,j+1} - F^y_{i,j}}{\hat{\Delta}_{i,j} - F^y_{i,j}}, \ D_{10} = \frac{F^y_{i+1,j+1} - F^y_{i+1,j}}{\hat{\Delta}_{i+1,j} - F^y_{i+1,j}}, \ D_{11} = \frac{\hat{\Delta}_{i,j} - F^y_{i,j}}{F^y_{i,j+1} - \hat{\Delta}_{i,j}},$$

$$D_{12} = \frac{\hat{\Delta}_{i+1,j} - F^y_{i+1,j}}{-\hat{\Delta}_{i+1,j} + F^y_{i+1,j+1}}, \ D_{13} = \frac{F^y_{i,j+1} - F^y_{i,j}}{F^y_{i,j+1} - \hat{\Delta}_{i,j}}, \ D_{14} = \frac{F^y_{i+1,j+1} - F^y_{i+1,j}}{F^y_{i+1,j+1} - \hat{\Delta}_{i+1,j}},$$

$$D_{15} = \frac{F^y_{i,j+1} - \hat{\Delta}_{i,j}}{\hat{\Delta}_{i,j} - F^y_{i,j}}, \ D_{16} = \frac{-\hat{\Delta}_{i+1,j} + F^y_{i+1,j+1}}{\hat{\Delta}_{i+1,j} - F^y_{i+1,j}}.$$

$S_{xx}(x,y)S_{yy}(x,y) - S^2_{xy}(x,y) > 0$ if

$$\hat{\beta}_{i,j} > Max\left\{D_k, 17 \le k \le 20, k \in Z^+\right\},$$

where

$$D_{17} = \frac{2(F^y_{i,j+1} - F^y_{i,j}) + \hat{h}_j F^{xy}_{i,j}}{2(\hat{\Delta}_{i,j} - F^y_{i,j})}, \ D_{18} = \frac{2(F^y_{i,j+1} - F^y_{i,j}) + \hat{h}_j F^{xy}_{i,j+1}}{2(F^y_{i,j+1} - \hat{\Delta}_{i,j})},$$

$$D_{19} = \frac{2(F^y_{i+1,j+1} - F^y_{i+1,j}) + \hat{h}_j F^{xy}_{i+1,j}}{2(\hat{\Delta}_{i+1,j} - F^y_{i+1,j})}, \ D_{20} = \frac{2(F^y_{i+1,j+1} - F^y_{i+1,j}) + \hat{h}_j F^{xy}_{i+1,j+1}}{2(F^y_{i+1,j+1} - \hat{\Delta}_{i+1,j})}.$$

The above can be summarized as:

Theorem 1. *The piecewise rational bi-cubic interpolant $S(x,y)$, defined over the rectangular mesh $D = [x_0, \ x_m] \times [y_0, \ y_n]$, in (1), is convex if the following sufficient conditions are satisfied*

$$\alpha_{i,j} > Max\left\{0, D_k, 1 \le k \le 4, k \in Z^+\right\}, \ \beta_{i,j} > Max\left\{0, D_k, 5 \le k \le 8, k \in Z^+\right\},$$

$$\hat{\alpha}_{i,j} > Max\left\{0, D_k, 9 \le k \le 12, k \in Z^+\right\}, \ \hat{\beta}_{i,j} > Max\left\{0, D_k, 13 \le k \le 20, k \in Z^+\right\}.$$

The complete expressions of D_k, $1 \le k \le 20$, $k \in Z^+$ are written on pages 478–479. The above constraints can be rearranged as:

$$\alpha_{i,j} = l_{i,j} + Max\left\{0, D_k, 1 \le k \le 4, k \in Z^+\right\}, \quad l_{i,j} > 0.$$
$$\beta_{i,j} = m_{i,j} + Max\left\{0, D_k, 5 \le k \le 8, k \in Z^+\right\}, \quad m_{i,j} > 0.$$

Table 1.

y/x	-3	-2	-1	0	1	2	3
-3	18.1	13.1	10.1	9.1	10.1	13.1	18.1
-2	13.1	8.1	5.1	4.1	5.1	8.1	13.1
-1	10.1	5.1	2.1	1.1	2.1	5.1	10.1
0	9.1	4.1	1.1	0.1	1.1	4.1	9.1
1	10.1	5.1	2.1	1.1	2.1	5.1	10.1
2	13.1	8.1	5.1	4.1	5.1	8.1	13.1
3	18.1	13.1	10.1	9.1	10.1	13.1	18.1

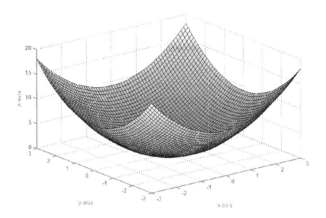

Fig. 1. Convex rational bi-cubic function with $l_{i,j} = m_{i,j} = n_{i,j} = o_{i,j} = 0.01$

$$\hat{\alpha}_{i,j} = n_{i,j} + Max\left\{0, D_k, 9 \leq k \leq 12, k \in Z^+\right\}, \quad n_{i,j} > 0.$$
$$\hat{\beta}_{i,j} = o_{i,j} + Max\left\{0, D_k, 13 \leq k \leq 20, k \in Z^+\right\}, \quad o_{i,j} > 0.$$

4   Demonstration

First example: Take a set of convex data in Table 1, generated by $F = x^2 + y^2 + 0.1$. The shape of the data in Table 1 is shown in Figure 1 that is produced by using the convexity preserving scheme developed in Section 3 with $l_{i,j} = m_{i,j} = n_{i,j} = o_{i,j} = 0.01$. It is clear from Figure 1, that the shape of the convex data in Table 1 is preserved, smooth and visually pleasing.

Second example: The convex data in Table 2 is generated from the function $F = e^{\sqrt{x^2 + y^2}}$.

The shape of the data in Table 2 is shown in Figure 2. Figure 2 is generated by using the convexity preserving scheme developed in Section 3, with $l_{i,j} = m_{i,j} = n_{i,j} = o_{i,j} = 0.1$. It is clear from Figure 2 that the convexity preserving scheme developed in Section 3 attains the desired results.

Table 2.

y/x	1	2	3	4	5	6
1	4.1	9.4	23.6	61.8	163.9	438.2
2	9.4	16.9	36.8	87.5	218.1	558.1
3	23.6	36.8	69.6	148.4	340.7	819.1
4	61.8	87.5	148.4	286.2	603.7	1354.4
5	163.9	218.1	340.7	603.7	1177.4	2465.7
6	438.2	558.1	819.1	1354.4	2465.7	4843.0

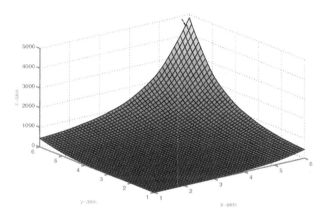

Fig. 2. Convex rational bi-cubic function with $l_{i,j} = m_{i,j} = n_{i,j} = o_{i,j} = 0.1$

5   Conclusion

In this study, we have considered the generation of convex surfaces through convex data using rational bi-cubic function (1). Data dependent sufficient conditions are derived on four parameters in the description of rational bi-cubic function to preserve the shape of convex data. The convexity preserving scheme of this paper is simple and easy to implement than the schemes developed in [1, 3, 5, 6]. The methods developed in [1, 3] for the shape preserving convex data were not local, whereas, the scheme developed in this paper is local. Thus the presented method is more flexible. Further, unlike [5,6], no derivative constraints are imposed so that the presented method applies equally to data or data with derivatives. However in this paper derivatives are approximated by arithmetic mean choice of derivatives.

References

1. Asaturyan, S.: Shape Preserving Surface Interpolation, Ph.D. Thesis, Department of Mathematics and Computer Science, University of Dundee, Scotland, UK (1990)
2. Carnicer, J.M., Garcia-Esnaola, M., Peña, J.M.: Convexity of Rational Curves and Total Positivity. Journal of Computational and Applied Mathematics 71, 365–382 (1996)

3. Constantini, P., Fontanella, F.: Shape Preserving Bivariate Interpolation. SIAM Journal of Numerical Analysis 27(2), 488–506 (1990)
4. Delgado, J., Peña, J.M.: Are Rational Bézier Surfaces Monotonicity Preserving. Computer Aided Geometric Design 24, 303–306 (2007)
5. Dodd,, McAllister, S.L., Roulier, J.A.: Shape Preserving Spline Interpolation for Specifying Bivariate Functions of Grids. IEEE Computer Graphics and Applications 3(6), 70–79 (1983)
6. Floater, M.S.: A Weak Condition for the Convexity of Tensor-Product Bézier and B-spline Surfaces. Advances in Computational Mathematics 2(1), 67–80 (1994)
7. Goodman, T.N.T.: Refinable Spline Functions and Hermite Interpolation. In: Lyche, T., Schumaker, L. (eds.) Mathematical Methods for Curves and Surfaces, Oslo, pp. 147–161. Vanderbilt University Press, Nashville (2001)
8. Hussain, M.Z., Hussain, M.: Visualization of Surface Data Using Rational Bi-cubic Spline. PUJM 38, 85–100 (2006)
9. Kouibia, A., Pasadas, M.: Approximation by Shape Preserving Interpolation Splines. Applied Numerical Mathematics 37, 271–288 (2001)
10. Renka, R.J.: Interpolation of Scattered Data with a convexity-Preserving Surface. ACM Transactions on Mathematical Software 30(2), 200–211 (2004)
11. Roulier, J.A.: A Convexity Preserving Grid Refinement Algorithm for Interpolation of Bivariate Functions. IEEE Computer Graphics and Applications 7, 57–62 (1987)
12. Sarfraz, M., Hussain, M.Z.: Data Visualization Using Rational Spline Interpolation. Journal of Computational and Applied Mathematics 189, 513–525 (2006)

Deformation and Smooth Joining of Mesh Models for Cardiac Surgical Simulation

Hao Li[1], Wee Kheng Leow[1], Ing-Sh Chiu[2], and Shu-Chien Huang[2]

[1] Dept. of Computer Science, National University of Singapore,
Computing 1, Singapore 117590
{lihao,leowwk}@comp.nus.edu.sg
[2] Dept. of Surgery, National Taiwan University Hospital,
No. 7 Chung San South Road, Taipei, Taiwan, R.O.C.
ingsh@ntu.edu.tw, dtsurg99@yahoo.com.tw

Abstract. This paper focuses on an important aspect of cardiac surgical simulation, which is the deformation of mesh models to form smooth joins between them. A novel algorithm based on the Laplacian deformation method is developed. It extends the Laplacian method to handle deformation of 2-manifold mesh models with 1-D boundaries, and joining of 1-D boundaries to form smooth joins. Test results show that the algorithm can produce a variety of smooth joins common in cardiac surgeries, and it is efficient for practical applications.

Keywords: predictive surgical simulation, Laplacian mesh deformation, smooth join, discrete differential geometry.

1 Introduction

3D modeling and model deformation are indispensable components in computer simulation of surgery. In *reactive* surgical simulation [1,5,10], 3D models of body tissues deform in real-time according to applied forces provided by user inputs. Reactive simulation is suitable for surgical training and preoperative planning that involves only basic surgical operations.

In *predictive* surgical simulation [2,4,15], 3D models of body tissues may deform according to applied forces or geometrical constraints on the body tissues. The simulation system predicts surgical results given minimum user inputs. In this way, a surgeon can perform preoperative planning of complex surgical procedure without going through the entire procedure in detail.

To achieve the goal of predictive simulation of complex cardiac surgery, it is necessary to devise novel 3D model manipulation algorithms that satisfy the requirements of cardiac surgery and the physical properties of cardiac tissues. One of these important algorithms is the deformation of 3D models of cardiac tissues to form smooth joins, which is the focus of this paper.

This paper develops a novel algorithm that extends the Laplacian deformation method [8] to handle deformation of 2-manifold mesh models with 1-D boundaries, and joining of 1-D boundaries to form smooth joins. Test results show

F. Chen and B. Jüttler (Eds.): GMP 2008, LNCS 4975, pp. 483–490, 2008.

that the algorithm can produce a variety of smooth joins common in cardiac surgeries, and is efficient for practical applications.

2   Related Work

Four main approaches exist for 3D model deformation: free-form deformation (FFD), finite element method (FEM), mass spring model (MSM), and differential geometry (DG) approach. FFD [7, 12] does not work directly on the geometric shape of an object. So it is difficult for FFD to manage geometrical constraints defined on the mesh in surgical simulation. FEM [2, 4, 15] offers accurate and realistic modeling of soft tissue deformation. However, it is computationally too expensive for real-time or near-real-time simulation. MSM [1, 5, 10] is efficient and easy to implement, so it is useful for real-time surgical simulation. However, its simulation results are not necessarily accurate because its behavior is highly dependent on the mesh topology and resolution.

DG approach directly computes the positions of the mesh points after deformation, given the initial configuration of the object and predefined constraints on the geometric properties of the mesh surface [8, 9, 13, 16]. DG offers more accurate and stable simulation results than MSM and is computationally less expensive than FEM. The standard DG method applies Laplacian operators to estimate the mean-curvature normals of the surface. As the Laplacian operators are ill-defined on mesh boundaries, existing Laplacian methods assume that the models are 2-manifold surfaces without boundaries. In contrast, our applications require the joining of meshes with boundaries.

One DG method for achieving smooth joins is to apply Laplacian filtering to smoothen the joined meshes [6]. This method is, however, not favored in our application because surface smoothing or fairing may cause the models to violate physical properties of cardiac tissues (see Section 3 for physical constraints).

Another DG method is to model the mesh surfaces using Poisson method [11, 16], which describes the differential properties of mesh surfaces using continuous partial differential equations. The Poisson method requires boundary conditions on all boundaries, which may not be known a priori in our applications.

3   Simulation of Smooth Joins

Suturing of cardiac tissues to form smooth joins are common operations in cardiac surgery. Figure 1 illustrates two typical scenarios: the suturing of coronary button to the aorta (Fig. 1(a)), and end-to-end joining of two arteries (Fig. 1(b)).

In this paper, the two anatomical parts to be joined are modeled as 3D meshes with zero surface thickness. One of the two parts is modeled as a fixed, rigid object called the *host H*. The other one is modeled as a flexible object called the *guest G* that is deformed to fit the host to form a smooth join.

The boundary curves of G and H to be joined, denoted as U and V, are identified by corresponding starting points $\mathbf{b}_s$ and $\mathbf{q}_s$, and end points $\mathbf{b}_e$ and $\mathbf{q}_e$ (Fig. 1). For closed curves, the starting points are also the end points. The curves

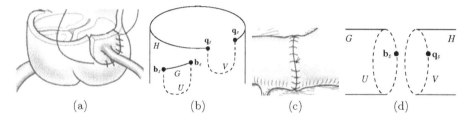

Fig. 1. Two kinds of smooth joins in cardiac surgery. (a, b) Partial patching. (c, d) End-to-end join. (a, c) Schematic drawings from [3]. Joins are indicated by the stitch marks. (b, d) Computational models. Dashed curves denote boundaries to be joined.

U and V may have different shape and length. The host is fixed and the guest is allowed to stretch or shrink. The minimum amount of deformation required to produce a smooth join is computed and displayed. In this way, a surgeon can assess whether the amount of deformation is acceptable empirically.

Now, we can define the joining problem as follows: Given two triangulated 2-manifold mesh models with 1-D boundaries, of which one is a rigid host H and the other a flexible guest G, and their corresponding boundary curves V and U to be joined, determine the shape of the deformed guest that forms smooth, continuous, and seamless join with the host. That is, the shape and length of the deformed U, denoted as U', should be identical to those of V.

The deformation of the guest model G is subjected to three soft constraints:

A. The shape of the deformed guest should be similar to that of the original guest, and its surface should be smooth.
B. The stretching or shrinking of the deformed guest should be minimized.
C. The join should be smooth and continuous.

4   Smooth Joining Algorithm

The algorithm for smoothly joining two mesh models consists of two steps:

1. Map boundary points $\mathbf{b}$ on U to the new positions $\mathbf{b}'$ on U'.
2. Deform G into G'.

Step 1 maps points $\mathbf{b}$ on U to the new positions $\mathbf{b}'$ on U' (Section 4.1). This mapping is imposed as positional hard constraints in the algorithm. Step 2 deforms the guest model G subject to the positional hard constraints and the three soft constraints. The algorithm is extended from the Laplacian method [8] (Section 4.2) to handle minimum stretching and shrinking (Section 4.3) and smooth joining of meshes (Section 4.4).

4.1   Mapping Corresponding Points

This step determines the corresponding point of each mesh point $\mathbf{b}$ on U. First, the lengths $l(U)$ of U and $l(V)$ of V are measured, and a scaling factor k is computed as the ratio $l(U)/l(V)$. Then, each point $\mathbf{b}$ on U is mapped to a position $\mathbf{p}$ on V such that

$$d_U(\mathbf{b}_s, \mathbf{b}) = k\, d_V(\mathbf{q}_s, \mathbf{p}). \tag{1}$$

So, the new position $\mathbf{b}'$ of $\mathbf{b}$ is $\mathbf{p}$.

In real application, there may not be an existing mesh point at position $\mathbf{p}$ on V. To ensure a close fit between U and V, each mesh point $\mathbf{q}$ on V is also mapped to a corresponding position $\mathbf{a}$ on U according to Eq. 1. A new point and the associated edges are added to G at position $\mathbf{a}$.

4.2   Laplacian Mesh Deformation

In the Laplacian method of mesh deformation, the hard constraints on the positions of the mesh points on U', i.e., $\mathbf{b}' = \mathbf{q}$, are represented by the equation:

$$\mathbf{H}\mathbf{x} = \mathbf{d} \tag{2}$$

where $\mathbf{x}$ is a vector of all the mesh points in G' and $\mathbf{d}$ is a vector of the desired coordinates of $\mathbf{b}'$, and $\mathbf{H}$ is a matrix that relates $\mathbf{x}$ and $\mathbf{d}$.

Shape constraint (Constraint A) is imposed by minimizing the difference between the surface normals before and after deformation. The surface normal $\mathbf{n}(\mathbf{p})$ at mesh point $\mathbf{p}$ can be approximated by the Laplacian operator $\mathbf{l}(\mathbf{p})$:

$$\mathbf{l}(\mathbf{p}) = \sum_{\mathbf{v}_i \in N(\mathbf{p})} w_i\,(\mathbf{p} - \mathbf{v}_i) \tag{3}$$

where $N(\mathbf{p})$ is the set of neighboring vertices of $\mathbf{p}$, and $w_i = 1/|N(\mathbf{p})|$.

Shape constraint is imposed by minimizing the squared difference between the mean-curvature normals before and after deformation, i.e.,

$$\|\mathbf{l}(\mathbf{p}') - \mathbf{l}(\mathbf{p})\| \tag{4}$$

4.3   Minimum Stretching and Shrinking

Minimum stretching and shrinking (Constraint B) is achieved by preserving the distances between neighboring connected points. For each pair of connected points $\mathbf{p}_i$ and $\mathbf{p}_j$ in G, the stretching (and shrinking) energy to be minimized is:

$$k_{ij}\,(\|\mathbf{p}'_i - \mathbf{p}'_j\| - \|\mathbf{p}_i - \mathbf{p}_j\|)^2 \tag{5}$$

where k_{ij} is the stiffness coefficient. It is equivalent to minimizing the following:

$$k_{ij}\,\|\mathbf{d}_{ij} - \mathbf{s}_{ij}\|, \quad \text{where } \mathbf{d}_{ij} = \mathbf{p}'_i - \mathbf{p}'_j,\ \mathbf{s}_{ij} = \mathbf{d}_{ij}\|\mathbf{p}_i - \mathbf{p}_j\|/\|\mathbf{p}'_i - \mathbf{p}'_j\|. \tag{6}$$

4.4   Smooth Join

Our method of imposing Constraint C is to match the surface tangents of the corresponding boundary points on U' and V. As there is more than one surface tangent at a point, the one that is normal to the boundary curve is chosen.

Let $\mathbf{b}$ be a boundary point on U and $\mathbf{b}_1, \ldots, \mathbf{b}_n$ denote the neighboring points of $\mathbf{b}$ such that $\mathbf{b}_1$ is also a boundary point. Among these connected neighbors, there is a neighbor $\mathbf{b}_t$ such that $\mathbf{b} - \mathbf{b}_t$ is (approximately) normal to the surface normal and U. We call this point the *tangential point*.

The tangential point $\mathbf{b}_t$ of $\mathbf{b}$ is computed as follows. First, the normal $\mathbf{n}(\mathbf{b})$ at $\mathbf{b}$ is estimated by the weighted mean of the normals of the neighboring faces of $\mathbf{b}$. Next, the required surface tangent $\mathbf{t}(\mathbf{b})$ is computed as the cross-product $\mathbf{n}(\mathbf{b}) \times (\mathbf{b} - \mathbf{b}_1)$. The difference vector $(\mathbf{b} - \mathbf{b}_1)$ approximates the tangent of U at $\mathbf{b}$, which is normal to the required surface tangent. Then, the tangential point $\mathbf{b}_t$ of $\mathbf{b}$ is the point whose difference vector $\mathbf{b} - \mathbf{b}_t$ is most parallel to $\mathbf{t}(\mathbf{b})$. This tangential point is used as an approximation of the corresponding tangential point of $\mathbf{b}'$, whose position is to be computed by the deformation algorithm.

The desired surface tangent $\mathbf{w}(\mathbf{b}')$ at a boundary point $\mathbf{b}'$ is given by the tangent $\mathbf{t}(\mathbf{q})$ at the corresponding boundary point $\mathbf{q}$ on V. This tangent $\mathbf{t}(\mathbf{q})$ is computed in the same way as described above.

Now, the smooth join constraint can be specified as

$$\mathbf{b}' - \mathbf{b}_t = \mathbf{w}(\mathbf{b}'), \quad \text{where } \mathbf{w}(\mathbf{b}') = -\mathbf{t}(\mathbf{q})\|\mathbf{b} - \mathbf{b}_t\|/\|\mathbf{t}(\mathbf{q})\|. \tag{7}$$

4.5   Constrained Minimization

Assembling Eq. 4, 6 and 7 gives the total energy to be minimized subjected to the hard constraints described in Eq. 2:

$$\min_{\mathbf{x}} \|\mathbf{A}\mathbf{x} - \mathbf{c}(\mathbf{x})\| \quad \text{such that } \mathbf{H}\mathbf{x} = \mathbf{d} \tag{8}$$

where $\mathbf{c}(\mathbf{x})$ collects the $\mathbf{l}(\mathbf{p})$, $\mathbf{s}_{ij}$ and $\mathbf{w}(\mathbf{b}')$ terms, and $\mathbf{A}$ relates $\mathbf{x}$ and $\mathbf{c}(\mathbf{x})$.

This is a non-linear optimization problem, which can be solved iteratively using Gauss-Newton method, as demonstrated in [14], in conjunction with the equality-constrained least square method described in [8].

5   Experiments and Discussion

This section evaluates the effectiveness and efficiency of the proposed algorithm in producing smooth joins of mesh models. Two test scenarios were constructed manually based on the application examples illustrated in Fig. 1. Two variants of the algorithm was tested for comparison: with and without smooth join constraints. Three test samples were constructed for each scenario, with the length of the guest's boundary smaller than, equal to, and greater than that of the host's boundary. For partial patching, the guest models were flat surfaces, which agree with real applications. The amount of deformation (stretching or shrinking) of the guest model was computed by measuring the amount of change in the areas of the triangles before and after deformation.

Figures 2–3 show that the smooth join constraint is necessary for the algorithm to produce a smooth join between the guest and the host. The constraint ensures that the algorithm achieves a balance between smooth joining and preservation of the orientation and shape of the guest model.

When the guest's boundary was shorter than the host's boundary, the guest was stretched significantly to match the host (column (a)). When they had the

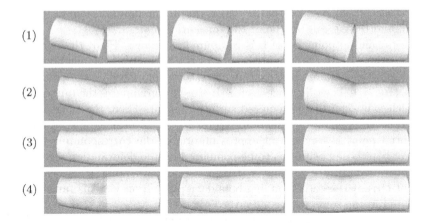

Fig. 2. Results of end-to-end joint. (1) Initial configurations, (2) without smooth join constraint, (3, 4) With smooth join constraint. The guest's boundary is (a) shorter than, (b) equal to, and (c) longer than the host's boundary. Red color indicates the amount of deformation of the guest model (white: no change, dark red: large change).

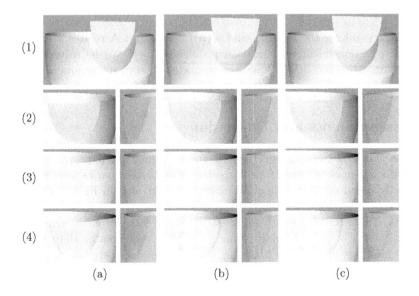

Fig. 3. Results of partial patching. (1) Initial configurations. (2) Without smooth join constraint. (3, 4) With smooth join constraint. The guest's boundary is (a) shorter than, (b) equal to, and (c) longer than the host's boundary. Red color indicates the amount of deformation of the guest model (white: no change, dark red: large change).

same length, the guest was minimally deformed (column (b)). When the guest's boundary was longer than the host's boundary (column (c)), the guest did not shrink significantly due to the minimum shrinking constraint. Instead, it bulged to match the shape of the host. This is an expected result in real applications.

The computation time of the algorithm was measured in an Intel Pentium 4 PC with 3GHz CPU and 1GB RAM to assess its efficiency. The algorithm took 70 seconds to solve the end-to-end join scenario (Fig. 2(4)), which was an optimization problem with 2,494 points, 522 hard constraints, and 29,688 soft constraints. For partial patching (Fig. 3(4)), with 1509 points, 1,218 hard constraints, and 17,967 soft constraints, the algorithm took 13 seconds to solve.

6 Conclusion

This paper presented a novel algorithm that extends the Laplacian deformation method to handle deformation of 2-manifold mesh models with 1-D boundaries, and joining of 1-D boundaries to form smooth joins. In addition to the positional and shape constraints in the original method, two other constraints are incorporated. The smooth join constraint ensures that the guest model is deformed properly to fit the host model and to produce a smooth join between their joining boundaries. The minimum stretching and shrinking constraint ensures that the deformation of the guest model is minimum. Test results show that the algorithm can produce a variety of smooth joins common in cardiac surgeries. Moreover, the algorithm can execute efficiently for practical applications.

References

1. Castañeda, M.A.P., Cosío, F.A.: Deformable model of the prostate for TURP surgery simulation. Computers & Graphics 28(5), 767–777 (2004)
2. Crouch, J.R., Merriam, J.C., Crouch, E.R.: Finite element model of cornea deformation. In: Duncan, J.S., Gerig, G. (eds.) MICCAI 2005. LNCS, vol. 3750, pp. 591–598. Springer, Heidelberg (2005)
3. Gardner, T.J., Spray, T.L.: Operative Cardiac Surgery, 5th edn. Arnold (2004)
4. Gladilin, E., Ivanov, A., Roginsky, V.: Generic approach for biomechanical simulation of typical boundary value problems in cranio-maxillofacial surgery planning. In: Barillot, C., Haynor, D.R., Hellier, P. (eds.) MICCAI 2004. LNCS, vol. 3217, pp. 380–388. Springer, Heidelberg (2004)
5. Kuhnapfel, U., Cakmak, H., Maass, H.: Endoscopic surgery training using virtual reality and deformable tissue simulation. Computers & Graphics 24, 671–682 (2000)
6. Liepa, P.: Filling holes in meshes. In: Proceedings of Eurographics/ACM SIGGRAPH symposium on Geometry processing, pp. 200–205 (2003)
7. MacCracken, R., Joy, K.I.: Free-form deformations with lattices of arbitrary topology. In: Proc. of ACM SIGGRAPH, pp. 181–188 (1996)
8. Masuda, H., Yoshioka, Y., Furukawa, Y.: Interactive mesh deformation using equality-constrained least squares. Computers & Graphics 30(6), 936–946 (2006)
9. Meyer, M., Desbrun, M., Schröder, P., Barr, A.H.: Discrete differential-geometry operators for triangulated 2-manifolds. In: Hege, H.-C., Polthier, K. (eds.) Visualization and Mathematics III, pp. 35–57. Springer, Heidelberg (2002)

10. Mosegaard, J.: LR-spring mass model for cardiac surgical simulation. In: Proc. of 12th Conf. on Medicine Meets Virtual Reality, pp. 256–258 (2004)
11. Park, S., Guo, X., Shin, H., Qin, H.: Surface completion for shape and appearance. The Visual Computer 22, 168–180 (2006)
12. Sederberg, T.W., Parry, S.R.: Free-form deformation of solid geometric models. In: Proc. of ACM SIGGRAPH, pp. 151–160 (1986)
13. Sorkine, O., Lipman, Y., Cohen-Or, D., Alexa, M., Rössl, C., Seidel, H.-P.: Laplacian surface editing. In: Proc. of Eurographics/ACM SIGGRAPH Symposium on Geometry Processing, pp. 175–184 (2004)
14. Weng, Y., Xu, W., Wu, Y., Zhou, K., Guo, B.: 2D shape deformation using nonlinear least squares optimization. The Visual Computer 22(9), 653–660 (2006)
15. Williams, C., Kakadaris, I.A., Ravi-Chandar, K., Miller, M.J., Patrick, J.C.W.: Simulation Studies for Predicting Surgical Outcomes in Breast Reconstructive Surgery. In: Ellis, R.E., Peters, T.M. (eds.) MICCAI 2003. LNCS, vol. 2878, pp. 9–16. Springer, Heidelberg (2003)
16. Yu, Y., Zhou, K., Xu, D., Shi, X., Bao, H., Guo, B., Shum, H.-Y.: Mesh editing with Poisson-based gradient field manipulation. ACM Trans. on Graphics 23(3), 644–651 (2004)

Digital Design for Functionally Graded Material Components Rapid Prototyping Manufacturing

Su Wang[1], Yuming Zhu[1], Chin-Sheng Chen[2], and Xinxiong Zhu[1]

[1] School of Transportation Science & Engineering, Beijing University of Aeronautics & Astronautics, 100083 Beijing, China
[2] Department of Industrial and Systems Engineering, Florida International University, 33174 Florida, USA
wangsu2000@buaa.edu.cn, zhuyuming2008@163.com, chenc@fiu.edu, xxzhu33@yahoo.com

Abstract. The paper presents an adaptive slicing algorithm of functionally graded material (FGM) components for RP. The algorithm is studied by considering both geometry information and material distribution information. The layer thickness is computed according to the area variation rate between two adjacent slices, the gradient of material distribution and material resolution. After slicing, the material information in each slice is stored discretely by material composition function and material resolution. Finally, an example is presented.

Keywords: Rapid prototyping manufacturing, Adaptive slice, Functionally graded material components.

1 Introduction

The request of modern manufacturing industry to the product performance is higher and higher, the components which are composed of homogeneous material do not usually satisfy the request. In FGM, the variation of the material composition and the structure is continuous and implicit interface inside, so the performance of FGM also varies continuously. FGM meets the request of manufacturing industry to material. But it is hard for traditional methods to process FGM components; RP technology provides a valid way to process the kind of components.

RP technology involves deposition of material to manufacture an object; it can manufacture the components whose internal structure is very complicated. The fabrication of a FGM component is achieved by selectively depositing various materials on point by point technology to vary the continuous material composition throughout a component [1].

2 Representation of CAD Models for FGM Components

For fabricating FGM components by RP, a CAD model of the components is required which not only has the geometry information, but also the material information at

F. Chen and B. Jüttler (Eds.): GMP 2008, LNCS 4975, pp. 491–497, 2008.

each point of the components. Some modeling and representing schemes have been reported in the literature [2, 3, 4, 5]. In this paper, the 'source-based' FGM components modeling [5] is used, so f is a function of distance from the geometric point to the grading source. The grading source is a geometric feature of the CAD model (such as a point, line, and plane). At any point in FGM components, the volume fraction for the ith primary material is computed:

$$v_i = \begin{cases} M_{si} & f(D)<0 \\ f(D)\times(M_{ei}-M_{si})+M_{si} & 0\le f(D)\le 1 \quad i=1,2\cdots n \\ M_{ei} & f(D)>1 \end{cases} \tag{1}$$

Where M_s and M_e are defined as the material composition array which the grading source starts and ends respectively. M_{si} is ith element of M_s, M_{ei} is ith element of M_e, D represents the perpendicular distance from the geometric point to the grading source.

3   Adaptive Slicing Algorithms of FGM Components

Because of the various materials information in FGM components, adaptive slicing of FGM components generates not only the geometric staircase effect, but also the material staircase effect. In this paper, for the computation of the layer thickness, not only geometry information but also material information of FGM components is considered. The layer thickness is computed according to the area variation rate between two adjacent slices, the gradient of material distribution and material resolution.

3.1   The Computation of Geometry-Based Layer Thickness

The computation of geometry-based thickness is almost the same as homogeneous components. The geometry-based layer thickness can be computed based on STL model [6] or based on parametric algebraic surfaces model [7].

In this paper, the method based on the area variation rate from Zhao, Z. W. and Laperriere, L. U. C. [8] is applied to compute geometry-based layer thickness. It is an indirect method, which is achieved by computing the area variation rate between two adjacent slices. If the relative area variation rate between two adjacent slices is larger, the variation of the geometric features are considered to be obvious, the layer thickness should be minished; if the relative area variation rate between two adjacent slices is smaller, the geometric features are considered to be similar. An area deviation ratio δ is defined based on the request of precision. The next slice area is A_{i+1} at height $Z_i + t$, the current slice area is A_i at height Z_i (Fig.1).

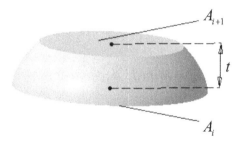

Fig. 1. The principle of the geometry-based layer thickness

$$\left|\frac{A_i - A_{i+1}}{A_i}\right| \leq \delta \tag{2}$$

In general, if the layer thickness of the next slice does not satisfy the condition (2), the layer thickness should be changed to slice newly until the condition is satisfied. The smaller the area deviation ratio δ is, the smaller each layer thickness is, and so the geometric stair effect generated is also smaller. The maximal layer thickness which satisfied the condition (2) is chosen as the geometry-based layer thickness d_g of the next slice.

3.2 The Computation of Material-Based Layer Thickness

The material-based layer thickness can be computed by the gradient of material distribution and material resolution. In this paper, the build direction is set to be the z direction. When the materials are sliced, the material staircase effect is generated for each material composition. For any certain value Δz, the larger the volume fraction of each material composition changes, the more obvious the material staircase effect is, and so the larger error is generated. So for each primary material, the maximum absolute value of the variation rate in volume fraction along the direction for the entire slice is first computed:

$$v_i' = \left|\frac{\partial}{\partial z}V(x)\right| = Max\left(\left|\frac{\partial}{\partial z}v_i(x)\right|\right) \quad i = 1,2 \cdots n \quad x \in \text{current slice} \tag{3}$$

Where v_i denotes the volume fraction of a primary material in a component, and $v_i \in V \subset M^n$, v_i is computed by equation (1).

This maximum variation for each ith material then compared with the material resolution Δv to obtain the ith material-based layer thickness d_{mi} :

$$d_{mi} = \frac{\Delta v}{v_i'} \tag{4}$$

The minimum of these d_{mi} values is chose as the material-based layer thickness d_m for the layer.

$$d_m = Min(d_1, d_2, \cdots, d_n) \tag{5}$$

3.3 Selection of Layer Thickness

The geometry-based and material-based layer thickness values are compared and the minimum of those two is chosen as the layer thickness for the slice, namely $d = Min(d_g, d_m)$. The next slice is obtained by slicing the FGM components model at the height $z + d$ where z is the height of the current slice.

3.4 The Geometry and Material Information in the Slice

Each slice is saved by STEP standard format, and the geometry contour of the slice is generated by reading the format in turn. If the contour which is in the upper section of the layer is read as the contour of the entire slice, or the contour which is in the lower section of the layer is read as the contour of the entire slice, the negative material tolerance or the positive material tolerance will be generated in a component. The negative material tolerance and the positive material tolerance have been explained in the literature [7].

In the slicing process, the continuous material distribution in FGM components is dispersed and now it is very difficult for RP technology to fabricate the FGM component whose material distribution is completely continuous. So the material distribution in each slice is described in a discrete way.

4 Example

A hemisphere shown in Fig. 2 composed of two materials A and B. Its radius is $50\,mm$. The grading source is the centre of the hemisphere. In the grading direction, the material composition array which the grading source starts and ends for the material A and B is $M_s = (0,1)$ and $M_e = (1,0)$ respectively. The material composition function is $f(D) = \left(\dfrac{D}{50}\right)^{\frac{1}{4}}$, The maximum allowable layer thickness is $2\,mm$, the minimum allowable layer thickness is $0.1\,mm$. The area deviation ratio δ is 0.03, the material resolution Δv is 0.02.

After adaptive slicing, the number of the layers is 126; the contour which is in the lower section of the layer is read as the contour of the entire slice, so the positive material tolerance will be generated in the component. The material distribution in first slice is shown in Fig. 3.

Fig. 2. The FGM hemisphere mode

Fig. 3. The adaptively sliced model of the hemisphere

In Fig. 4, the curve 1 shows the variation of the layer thickness which is affected by the geometric feature and the curve 2 shows the variation of the layer thickness which is affected by the material distribution.

The variation of the layer thickness which is affected by the geometric features and the material distribution together is shown in Fig. 5.

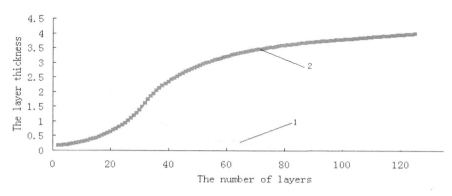

Fig. 4. The variation of geometry-based layer thickness and material-based layer thickness respectively

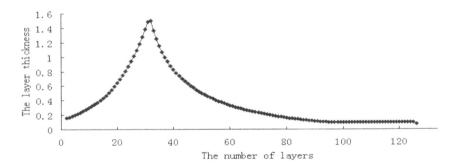

Fig. 5. The variation of the final layer thickness

5 Conclusions

The paper presents an approach for adaptive slicing of FGM components, in which layer thickness is computed by considering not only the geometric features but also the material distribution information. The geometry-based layer thickness is computed based on the area variation rate between two adjacent slices and the method is simple, practical, and suitable for various kinds of geometry models. The material-based layer thickness is computed by the gradient of material distribution and material resolution, and the material information in each slice is stored discretely by material composition function and material resolution. Therefore, much memory space can be saved to store the material information of FGM components.

To manufacture FGM components using RP technology, apart from adaptive slicing, orientation, support generation and tool path generation should also be studied by considering the geometry and material information inside FGM components.

Acknowledgments. We gratefully acknowledge the financial support from National Natural Science Foundation of China (NSFC) (grant 60773154).

References

1. Sihn, K.H., Dutta, D.: Process-Planning for Layer Manufacturing of Heterogeneous Objects Using Direct Mental Deposition. Journal of Computing and Information Science in Engineering 2(4), 330–344 (2002)
2. Jackson, T.R., Liu, H., Partikalakis, N.M., et al.: Partikalakis NM, et al.Modeling and designing functionally graded material components for fabrication with local composition control. Materials and Design 20(2–3), 63–75 (1999)
3. Bhashyam, S., Shin, K.H., Dutta, D.: An integrated CAD system for design of heterogeneous objects. Rapid Prototyping Journal 6(2), 119–135 (2000)
4. Pegna, J., Sali, A.: CAD modeling of multi-modal structures for free-form fabrication. In: Solid Freeform Fabrication Symposium, Austin, Tx (1998)
5. Siu, Y.K., Tan, S.T.: 'Source-based' heterogeneous solid modeling. Computer-Aided Design 34(1), 41–45 (2002)

6. Yang, R., Guo, D.M., Xu, D.M., et al.: An Adaptive Slicing Algorithm for Digital Manufacturing of Ideal Functional Material Components. China Mechanical Engineering, (in Chinese) 14(9), 770–772 (2003)
7. Zhou, M.Y.: Adaptive slicing of functionally graded material objects for rapid prototyping. International Journal of Advanced Manufacturing Technology 24(5-6), 345–352 (2004)
8. Zhao, Z.W., Laperriere, L.U.C.: Adaptive direct slicing of the solid model for rapid prototyping. International Journal of Production Research 38(1), 69–83 (2000)

Layer-Based Mannequin Reconstruction and Parameterization from 3D Range Data

Weishu Wei, Xiaonan Luo, and Zheng Li

Computer Application Institute, Sun Yat-Sen University, Guangzhou,
510275, P.R. China
Weishu Wei
Tel.: +86-84112291-307; +86-84112291-303
wei-ws@163.com

Abstract. Personalize mannequin design are the basic element in virtual garment simulation and visualization. This paper addresses the problem of mannequin construction from scanned point cloud and regenerates their shapes by inputting variant dimension. Layer-based simplification and parameterization algorithm is presented to preserve the mannequin's features as far as possible. The regenerated models are watertight and used for clothes design.

Keywords: mannequin, scan point, construction, parameterization.

1 Introduction

Reverse engineering (RE) refers to creating a CAD model from an existing physical object, which can be utilized as a design tool for producing geometrical and physical models in computer. In RE, the physical object shape can be rapidly captured by optical non-contact measuring techniques, e.g. 3D laser scanner. Usually, the point cloud data set produced is arbitrary scattered. Base on the RE concept, this paper presents a layer-based human model reconstruction and parameterization approach for automatic clothes design. Since in garment design, various body shapes require design of different size and style clothes, various manual-measure key feature dimensions are input to generate individual mannequins. We take one human scan point cloud as an example template to generate different individual figures for various precision using the parameterization method presented. The point clouds are obtained from Telmat 3D human body scanner; it contains approximately 4 hundred thousands points in each point cloud. The point cloud obtained from laser scanner contains noise and multi-silhouette, hence such problems must be preprocessed to ensure the completeness of the input data. At the first step, a manual-measure human body database is established. After that, it is possible to generate 3D parameterized human models from input dimensions without scanning. The parameterization of human model is the essential step for developing automation system of clothes design products.

F. Chen and B. Jüttler (Eds.): GMP 2008, LNCS 4975, pp. 498–504, 2008.

2 Related Works

The human body modeling from real scan data is a part of Reverse Engineering. It has important application in virtual reality, virtual clothes design, and surgical simulations. According to the methodology, the human model reconstruction is divided into surface reconstruction, tissue reconstruction, and multi-images based reconstruction.

Since human surface model is widely used in various domains, many works are related to the geometrical model reconstruction. In the feature-based modeling field [1-3], object semantics, which defined on geometry models, are systematically represented for a specific application. Wang [1] brought forward a feature-based wire frame to construct human models. It has the advantage to unify all the models in the same feature framework, but it is hard to construct various precision models under the restriction of the wire framework and the distribution of point density was not even. The Gregory patch was used to construct the cell mesh, which can make the model smooth. The mannequin processed by this scheme is an ideal model with no noises, and the points' distribution was sparse and well proportioned. Feature extraction was the critical step during modeling. Parts of the feature points were marked automatically on the model [1, 3] to hold functional symbolic information. Wang [2] utilize fuzzy logic concept to locate "sharpness" points. Setfan et al. [4] used point penalty functions and Minimum Spanning Graph to extract the crease lines and borderlines directly on the point cloud. Douros and Dekker [5] used B-spline curve to build the horizontal contour and local B-spline patches to build the quadrangular cell mesh. But B-spline curve and surface is an approximate method in CAGD, it is hard to create precise model cling to the original point cloud. The model surface satisfies G1 continuity in conjunction boundaries. But this technique does not conveniently and automatically modify the reconstructed models into different shapes following the user intends.

Example-based shape modeling technique can overcome this disadvantage. Seo and Thalmann [6] consider the body geometric model as two distinct entities: rigid and elastic components. Individual physique is determined through deforming rigid part by linear approximation. The detailed body shape is depicted in elastic parts using spring mesh over quad-patch. This runtime modeling technique can generate a wide range of different body models for real-time animation.

Related to parameterization of models and unorganized point cloud, Barhak and Fischer [7] presented a PDE-based method about the parameterization for reconstruction of 3D freeform objects from range point cloud. Praun et al [8] utilized a same connectivity wire-frame to generated various body shapes fast from a template model or synthesize an average model when given a set of specified models. Allen et al [9] reconstruct and synthesize high-resolution individual models from range scans. All the approaches mentioned above are geometry-oriented and feature modeling technique.

In multi-view based modeling field, human model is constructed from multi-images or from a series of video sequences. Agarwal and Triggs [10] use a sparse Bayesian regression method for recovering 3D human body motion directly from silhouettes extracted from monocular video sequences. It is not realistic to recover an exquisite human model by this method unless inexhaustible video clips are provided.

3   Layer-Based Mannequin Reconstruction and Parameterization

This section illustrates the layer-based mannequin reconstruction and parameterization from noisy and coarse scanned point clouds. The layer-based process has the advantage of keeping the many features well, for the reasons that the girth dimensions are related to a certain layer points directly. And also, it avoids the cross section with the meshed model. As a result, it increases the feature veracity on the model surface. There is an assumption that the mannequin datum being processed is in upright posture. The whole reconstruction scheme consists of series steps: (1) eliminate the noises and adjust the model to the proper coordinate in a uniform format; (2) find the feature points on human model surface and segment the model into isolated parts; (3) simplify the model by point cloud simplification algorithm and construct the model by triangular mesh; (4) generate individual figures according to the measure database. Various precision constructed models are obtained according to different simplification densities. In order to construct watertight models, calvaria, soles and fingertips points are added at corresponding places. Their positions are computed by averaging all the points in neighbor-layer. Simultaneously, the multi-contour and holes problem from original point cloud are rectified in simplification phase. The resultant model constructed by the above steps is used as the template model for parameterization. As a result, various figures are generated according to different height and girth dimension from database.

3.1   Point Cloud Pre-process

The range data from 3D scanner contains numerous noisy points and the model position is not quite accurate. Point vacancies are formed when scanning some shaded areas. Before reconstruction, noisy points must be removed and the model must be adjusted to standard position. The model is normalized in Descartes coordinate system by multiplying transform matrix and rotation matrix. After the adjustment, the body face direction is z-axis, stature direction is y-axis and the side direction is x-axis, since the models were required to stand in upright posture when scanning. The noisy points obtained in scan phase are caused by dust and the laser scanner. They are classified into background noise and edge noise. Background noises are eliminated by the clustering points into different groups according to the Euclid distances. The largest group is the result we need, other groups regards as background noise that must be removed.

3.2   Feature Layer and Segmentation

The purpose of finding the feature layers (contains feature points) is to make the point cloud apart and to measure the mannequin's key-dimensions. They are critical part in the following parameterization process. Furthermore, the feature layers are correlated to height and girth in garment design. Feature points on human model surface includes the neck point, the armpit points, the breast points, the under-breast points, the navel point, the crotch point, etc. Feature points on human body surface are classified into two types: (1) branch points, and (2) other feature points. All the feature points (layers) shall be gained before they lose in point cloud simplification.

Indeed, only branch points and neck point are helpful in human body segmentation. The algorithm adopted here is to project the point cloud on a plane and compute the average points distance on certain layer to find out the branch points. If the gap is larger than a pre-establish value, it regards as the feature layer that contains branch points. Dekker's method [3] is adopted here to find the feature points. Nearly 80 percent of the feature points are founded accurately in the actual experimental point cloud. The validity decreases when processing on some special figure point clouds, such as large waist-line human model. The remnant feature points that cannot be found precisely are marked on the point cloud manually.

3.3 Layer-Based Point Cloud Simplification and Multi-silhouette Elimination

Because of the unorganized and dense point cloud, it is essential to simplify the points and arrange them in order. Layer-based point cloud simplification method is presented to make points well-organized. The whole point cloud simplification algorithm includes:

- Vertical simplification by extracting layers
- Layer iterative simplification

In vertical direction, various layer-extraction schemes can be adopted here. We must ensure that all the key-features must be preserved along the whole processing.

In horizontal direction, the general idea of layer simplification algorithm is as follows. Firstly, points are separated into layers, and points in different layer are considered as isolated groups to be dealt with. Take one layer points as an example. The arbitrary non sharp-angle point is taken as a starting point as well as initial direction is indicated. Secondly, circle regions are set up to compute the track points along the contour. The central points of the circle regions are the track points (simplified points). We analyze the principle direction in every circle region, and the cross-point of circle edge and radial (from circle centre) is regarded as the next track point. The circle radius R is a simplification precision parameter controlled by user. Compute next track points until all the rest points in this layer are processed. Finally, a horizontal contour in one layer is formed, as is shown in Fig 2. The points in other layers are processed in the same way. As a result, various precision models can be generated according to the simplification radius parameter.

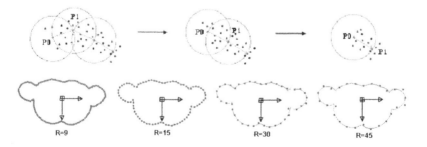

Fig. 1. The upper row indicates the process of finding track points; the lower row exhibits the results

To construct a watertight model, triangular mesh is applied in the surface construction. Since the body is segmented into several parts, each part is meshed independently using parallel tracks mesh algorithm [11]. And all the isolate parts are united through constructing a compound contour [11] in different parts. The results are shown in Fig 3. There is a limitation that the simplification precision cannot exceed the mesh algorithm that used in the experiment. And the veracity of the iterative simplification algorithm is not high, when the simplification precision is too low.

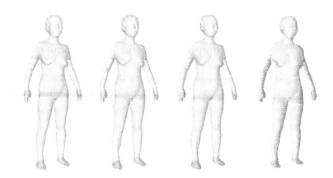

Fig. 2. Various precision mannequins are constructed base on the simplification method

3.4 Layer-Based Model Parameterization

For the requirement of clothes simulation and visualization, the critical part is to utilize the mannequins' key-feature parameters to generate various figures. The mannequin reconstructed above is considered as a template for parameterization. In general, a template model is computed by averaging many models with adding weights to every model [1, 9]. For the precision of key-feature measurement, approximation-type curves are abandoned. Here, compound interpolations are utilized to implement model parameterization. Points on the models are classified into feature layer points and non-feature layer points. The former are transformed directly by a scale factor $S = D_{Input} / D_{Template}$, where D is the Dimension of height or girth. The latter are transformed through an interpolation matrix defined as follows,

$$\begin{bmatrix} x' \\ y' \\ z' \\ 1 \end{bmatrix} = \begin{bmatrix} s & 0 & 0 & 0 \\ 0 & 0 & 0 & h \\ 0 & 0 & s & 0 \\ 0 & 0 & 0 & 1 \end{bmatrix} \bullet \begin{bmatrix} x \\ y \\ z \\ 1 \end{bmatrix} \tag{1}$$

Where h is the linear interpolation height between two feature layers, $h=S(H_{High} - H_{Low})+H_{Low}$; s is Cardinal spline interpolation [12], the four control points are in four neighboring feature layers. The interpolation map is shown in Fig 4. Fig 5 shows the resultant models generated by the Table 1 input size dimensions. In parameterization application, there are 800 manual-measured data of mannequins in the human model

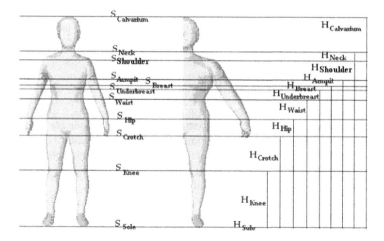

Fig. 3. Scale factor and Height of the feature layers and non-feature layers

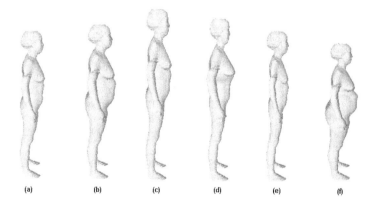

Fig. 4. Models generated according to the parameters list in Table1

Table 1. Input parameters in generating model that are listed in Fig18. (H:Height, G:Girth)

Model	Height (cm)	Neck		Chest		Under-breast		Waist		Hip	
		H	G	H	G	H	G	H	G	H	G
(a)	165	142	31	124	104	117	75	106	65	85	97
(b)	173	149	35	130	120	123	90	111	90	89	121
(c)	190	164	30	143	108	135	78	122	71	98	99
(d)	181	156	25	136	123	128	90	116	60	93	90
(e)	168	145	28	126	85	119	65	108	59	87	78
(f)	154	133	29	116	109	109	77	99	92	79	130

database, which contains 36 items of details measured by manual anthropometric method. Only parts of the data, such as girth and height, are utilized to generate parametric models.

4 Conclusion and Future Work

This paper discusses the whole process of mannequin reconstruction and parameterizetion from scan point cloud. The major contributions of this paper are the layer-based simplification and parameterization methods. It keeps the girth feature dimension as far as possible, which is critical in garment design. The present simplification method eliminates multi-silhouette and make point in-order. Further more, the parameterization method make the model satisfy C1continuity on the scaling parts.

Hereby considering the model precision, curve and surface will be introduced in model construction in the future work. To compute an average human model from many models is the next step in consideration. And, to establish digital mannequins and measurement database are significant works in human model researches.

References

1. Charlie, C.L.: Wang, Parameterization and parametric design of mannequins. Computer-Aided Design 37, 83–98 (2005)
2. Wang, C.C.L., Cheng, T.K.K., Yuen, M.M.F.: From laser-scanned data to feature human model: a system based on fuzzy logic concept. Computer-Aided Design 35(3), 241–253 (2003)
3. Dekker, L., Douros, I., Buxton, B.F., Treleaven, P.: Building symbolic information for 3D human body modeling from range data. In: Proceeding of the second international conference on 3-D digital imaging and modeling, pp. 388–397. IEEE Computer Society, Los Alamitos (1999)
4. Gumhold, S., Wang, X., Macleod, R.: Feature extraction from point clouds. In: Proceedings of 10th International Meshing Roundtable, pp. 293–305 (2001)
5. Douros, I., Dekker, L., Buxton, B.: An improved algorithm for reconstruction of the surface of the human body from 3D scanner data using local B-spline patches. In: People Workshop Proceedings, IEEE-ICCV 1999, pp. 29–36 (1999)
6. Seo, H., Magnenat-Thalmann, N.: An example-based approach to human body manipulation. Graph Models 66(1), 1–23 (2004)
7. Barhak, J., Fischer, A.: Parameterization for reconstruction of 3D freeform objects from laser-scanned data based on a PDE method. Vis Comput 2001 17(6), 353–369 (2001)
8. Praun, E., Sweldens, W., Schroder, P.: Consistent mesh parameterizations. In: Proceedings of the 28th annual conference on Computer graphics and interactive techniques, pp. 179–184 (2001)
9. Allen, B., Curless, B., Popovie, Z.: The space of human body shapes: reconstruction and parameterization from range scans. ACM SIGGRAPH 2003 22(3), 587–594 (2003)
10. Agarwal, A., Triggs, B.: Learning to track 3D human motion from silhouettes. In: Proceedings of the twenty-first international conference on machine learning ICML 2004, pp. 9–16 (2004)
11. Bajaj, C.L.: Arbitrary topology shape reconstruction from planar cross sections. Graphical Models and Image Processing 58(6), 523–543 (1996)
12. Ahlberg, J.H.: The theory of splines and their applications, vol. 38. Academic Press, New York (1972)

Manifoldization of β-Shapes by Topology Operators

Donguk Kim[1], Changhee Lee[2], Youngsong Cho[1], and Deok-Soo Kim[2]

[1] Voronoi Diagram Research Center, Hanyang University,
17 Haengdang-dong, Seongdong-gu, Seoul 133-791, Korea
{donguk,ycho}@voronoi.hanyang.ac.kr
[2] Department of Industrial Engineering, Hanyang University,
17 Haengdang-dong, Seongdong-gu, Seoul 133-791, Korea
chlee@voronoi.hanyang.ac.kr, dskim@hanyang.ac.kr

Abstract. It is well known that the geometric structure of a protein is an important factor to determine its functions. In particular, the atoms located at the boundary of a protein are more important since various physicochemical reactions happen in the boundary of the protein. The β-shape is a powerful tool for the analysis of atoms located at the boundary since it provides the complete information of the proximity among these atoms. However, β-shapes are difficult to handle and require heavy weight data structures since they form non-manifold structure. In this paper, we propose topology operators for converting a β-shape into a manifold. Once it is converted, compact data structures for representing a manifold are available. In addition, general topology operators used for manifold structures can also be available for various applications.

Keywords: β-shapes, β-complex, non-manifold, manifoldization.

1 Introduction

In many real world problems in science and engineering, neighboring information among particles often plays an important role in solution processes. One of emerging application areas is the structural analysis of molecules such as proteins since the geometry of a molecule is important in determining the functions of the molecule [2,5,9,10,11,14]. Among many geometric properties of a molecule, finding atoms which is placed on a boundary and representing in a well-defined data structure is of importance. In this problem area, various geometric structures such as Voronoi diagrams of either atom centers or the atoms themselves, Delaunay triangulations, α-shapes, power diagrams, regular triangulations, etc. have been frequently used.

Recently, a geometric structure called the β-shape has been proposed [8]. The theory of the β-shape is based on the Voronoi diagram of spheres. It turns out that various geometric problems for a molecule such as the computation of the correct molecular surface [12] or the computation of pockets on the boundary of a protein can be very efficiently computed once the β-shape of a molecule is

F. Chen and B. Jüttler (Eds.): GMP 2008, LNCS 4975, pp. 505–511, 2008.

available [6,7]. In those applications, it is known that the boundary information of a β-shape plays important roles. Moreover, the traversal along the boundary can be an important part in many applications.

It is known that the boundary of a β-shape consists of vertices, edges, and triangular faces and can be represented as a mesh structure. However, the boundary of a β-shape does not form a manifold polyhedral surface in general, and there can be a lot of dangling faces and edges. Due to this fact, the traversal among boundary entities becomes complicated. Although we can use general data structures for non-manifold structures [1,15,16], they are generally heavier than data structures for manifolds and there are many special cases to handle depending on applications.

In this paper, we propose a method to convert the β-shape into a manifold object so that traversal along boundary simplices of the β-shape can be easily and efficiently done. Once the boundary of β-shape is converted into a manifold, many concise but rich data structure such as a winged-edge data structure or a half-edge data structure can be applied to represent the surface. Guéziec et al. [4] reported an approach to convert a non-manifold polyhedral surface which has holes and degenerate vertices by cutting and stitching non-manifold topological elements.

This paper is organized as follows: In Section 2, β-shape which is the input non-manifold structure is briefly explained. In Section 3, we categorize non-manifold configurations appearing in β-shapes. In Sections 4 and 5, topology operators to handle non-manifold configurations and conversion method into a manifold by using the operators are discussed. Then, we conclude this paper.

2 β-Shapes

The β-*hull* can be defined in a way similar to the well-known α-hull [3] since the β-hull generalizes the α-hull as the generators are generalized from points to spheres. The point set given in α-hull is now replaced by a set of three dimensional spherical atoms. As in the α-hull, think of a Styrofoam filled $\mathbb{R}^3$ with some spherical rocks of varying radii scattered within the Styrofoam. Carving out the Styrofoam with an omnipresent and empty spherical eraser whose radius is β called a β-probe will result in a shape which we call a β-hull.

Suppose that we have a β-hull of an atom set A. We straighten the surface of the β-hull by substituting straight edges for the circular ones and triangles for the spherical caps. The straightened object bounded by planar facets is the β-*shape* of A. In this paper, we consider a β-shape is connected. Otherwise, each connected component of the β-shape is separately handled as discussed in this paper.

Shown in Fig. 1(a) is a set of thirteen two-dimensional atoms and the β-hull corresponding to a β-probe with a radius of β. The corresponding β-shape is shown in Fig. 1(b) where the shaded region is the interior of the β-shape. The dangling edge corresponds to a pair of atoms that both are exposed to or touched by the β-probe. As shown in Fig. 1(b), β-shapes are not manifolds since they

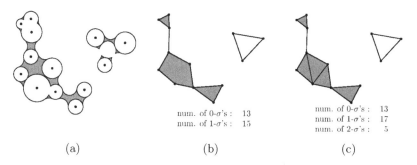

num. of 0-σ's :   13
num. of 1-σ's :   15

num. of 0-σ's :   13
num. of 1-σ's :   17
num. of 2-σ's :   5

(a)   (b)   (c)

Fig. 1. β-hull, β-shape, and β-complex for the atoms in $\mathbb{R}^2$: (a) a set of thirteen two-dimensional atoms and the β-hull, (b) the corresponding β-shape, and (c) β-complex

have edges without incident face and vertices whose degree is more than two. Similar to the case in 2D, β-shapes in 3D are not necessarily manifolds.

Shown in Fig. 1(c) is the β-complexe of the β-shape in Fig. 1(b). As shown in the figure, each β-shape is tessellated into a set of simplices. A simplex σ in a β-complex takes one of the three states: singular, regular, or interior. If σ belongs to the boundary of β-shape and does not bound any higher-dimensional simplex in the corresponding β-complex, then σ has a *singular* state. If σ belongs to the boundary of β-shape and bounds a higher-dimensional simplex in the corresponding β-complex, σ has a *regular* state. If σ does not belong to the boundary of β-shape, then the state of σ is *interior*. Details on the β-complex can be found in [13].

From the above definition, all the singular edges and faces of a β-complex corresponding to the β-shape form a non-manifold since there is no bounding higher dimensional simplex. On the other hand, in the case of regular simplices, some of them form a non-manifold. In this paper, we call simplices at which a non-manifold structure occurs a *non-manifold simplices*. Therefore, all the singular edges and faces are called non-manifold edges and non-manifold faces, respectively. In addition, the regular edges and vertices which result in a non-manifold also called non-manifold edges and non-manifold vertices, respectively.

3   Non-manifold Configurations in β-Shapes

In order to convert a β-shape into a manifold, all possible configurations where a non-manifold case occurs have to be identified. In this section, we will show the non-manifold configurations in a β-shape assuming that it is represented by corresponding β-complex. Our strategy of enumerating configurations is to find all possible situations where non-manifold simplices exist among the simplices constituting the β-complex by considering simplex type.

A non-manifold simplex corresponding to the boundary of a β-shape may have a state either singular or regular. As mentioned earlier, all the singular simplices are non-manifold simplices. On the other hand, not all the regular simplices are non-manifold simplices. According to the relative position of neighboring simplices, some regular simplices become non-manifold simplices. Therefore, we need to consider the situations where two other simplices intersect at the regular simplex as a non-manifold configuration. Suppose that a singular simplex σ_i is connected with another simplex σ_j regardless the state of σ_j, then the simplex at contact locations which is either a vertex or an edge becomes a non-manifold simplex and those simplices have a regular state.

The combination of simplex pairs where each pair causes a non-manifold case can be enumerated as follows: EE, EF, ET, FF, FT, and TT, where E, F, and T denote edge, face, and tetrahedron, respectively. Note that we do not need to consider vertices since the contact location between the vertex and another simplex becomes the vertex itself. It is important that the state of edges and faces in the case is singular. This is due to the fact that only singular edges and faces cause non-manifold configurations.

For each combination, we should consider contact location. In the case of FF, it is possible that two singular faces contact at a vertex or an edge. For the purpose of considering contact location, we use XY_z to denote that the contact between simplex types of X and Y occurs at a simplex type z. For example, FF_v means the configuration that two singular faces contact at a vertex and FF_e means that both faces contact at an edge. We call each of FF_v, FF_e, etc. a non-manifold *configuration*. In the cases of EE, EF, and ET, it is easy to verify that edge contact is not available.

4   Topology Operators for Manifoldization

As discussed in the previous section, there are nine exclusive non-manifold configurations: EE_v, EF_v, ET_v, FF_v, FF_e, FT_v, FT_e, TT_v, and TT_e. In this section, we discuss how to convert each configuration into a manifold. In each configuration, we added one artificial vertex and the numbers of topological entities are increased after each operation. Cases of non-manifold vertex, i.e. EE_v, EF_v, ET_v, FF_v, FT_v, and TT_v, are converted into those of non-manifold edge. In addition, non-manifold edge configurations are converted into a manifold.

$O(EE_v)$. EE_v is the configuration that two singular edges are connected at a vertex, v_n. Our strategy is to convert the non-manifold vertex v_n into a non-manifold *edge* by adding an artificial vertex v_a at the same location as v_n. Note that the length of the new-born edge is zero. The cardinality of the topological elements at the boundary of β-shape changes due to this operation. The numbers of vertices, edges, and faces are increased by one, three, and two, respectively. Note that the resulting configuration after handling EE_v is not a manifold but still a non-manifold case of FF_e. This means that we have to handle FF_e after handling EE_v to convert into a manifold.

$O(EF_v)$. EF_v is the configuration that a singular edge and a singular face have a contact at a vertex, v_n. Similar to the EE_v case, an artificial vertex v_a is inserted at the location of v_n. Then, by adding an artificial edge connecting v_n and v_a and three more artificial edges, we convert the singular edge into a singular face and the singular face into a tetrahedron. After the procedure of this operation, EF_v becomes a FT_e configuration. The numbers of vertices, edges, and faces are increased by one, four, and four, respectively.

$O(ET_v)$. ET_v is the configuration that a singular edge and a tetrahedron have a contact at a vertex v_n. Similar to the previous configuration, an artificial vertex v_a is added to create an edge connecting v_n and v_a. Then, one of the faces which bound the tetrahedron and share v_n makes a tetrahedron with v_a. A singular face incident to the singular edge is created by adding artificial edges. Now, ET_v is transformed into a FT_e configuration. The numbers of vertices, edges, and faces on the boundary of β-shape are increased by one, four, and three, respectively.

$O(FF_v)$. FF_v is the configuration that two singular faces have a contact at a vertex v_n. In this case, an artificial vertex v_a is added at the vertex v_n to create an artificial zero-length edge e_a. Then, two singular faces become two incident tetrahedron at the edge e_a. After the procedure, FF_v becomes TT_e

Table 1. Summary of the topology operators

Operator	Figure		Output Config.	Cardinality		
	Input	Output		Input	Output	Delta
$O(EE_v)$			FF_e	V : 3 E : 2 F : 0	V : 4 E : 5 F : 2	ΔV : +1 ΔE : +3 ΔF : +2
$O(EF_v)$			FT_e	V : 4 E : 4 F : 1	V : 5 E : 8 F : 5	ΔV : +1 ΔE : +4 ΔF : +4
$O(ET_v)$			FT_e	V : 5 E : 7 F : 4	V : 6 E :11 F : 7	ΔV : +1 ΔE : +4 ΔF : +3
$O(FF_v)$			TT_e	V : 5 E : 6 F : 2	V : 6 E :11 F : 8	ΔV : +1 ΔE : +5 ΔF : +6
$O(FF_e)$			manifold	V : 5 E : 5 F : 2	V : 5 E : 9 F : 6	ΔV : +1 ΔE : +4 ΔF : +4
$O(FT_v)$			TT_e	V : 6 E : 9 F : 5	V : 7 E :14 F :10	ΔV : +1 ΔE : +5 ΔF : +5
$O(FT_e)$			manifold	V : 5 E : 8 F : 5	V : 6 E :12 F : 8	ΔV : +1 ΔE : +4 ΔF : +3
$O(TT_v)$			TT_e	V : 7 E :12 F : 8	V : 8 E :17 F : 12	ΔV : +1 ΔE : +5 ΔF : +4
$O(TT_e)$			manifold	V : 6 E :11 F : 8	V : 7 E :15 F :10	ΔV : +1 ΔE : +4 ΔF : +2

configuration. From the operation, the numbers of vertices, edges and faces are respectively increased by one, five, and six.

Note that the other operators $O(FF_e)$, $O(FT_v)$, $O(FT_e)$, $O(TT_v)$, and $O(TT_e)$ are similarly explained and we will omit details due to page limit. All the operators are summarized in Table 1 with representative figures.

5   Manifoldization Via Topology Operators

First of all, we identify all the configurations induced by singular edges and faces connected by another simplex, i.e. EE_v, EF_v, ET_v, FF_v, FT_v, FF_e, and FT_e. This process can be done by visiting all the edges and faces in the β-complex. In addition, TT_e case can be identified by visiting all the regular edges and checking the number of incident faces with regular state, and this process can be done in linear time. In the case of TT_v, all regular vertices have to be visited to detect, and this process takes linear time.

After detecting all the non-manifold configurations for initial β-complex, we handle the configurations one by one by topology operators described in the above section. During each operation, some non-manifold configuration will be changed into another configuration at the simplices having contacts with involving simplices in the operation. Therefore we have to update such configuration changes during operations. This problem can solved if the configuration type is stored and maintained in each non-manifold simplex as an additional information. Then the processing takes a linear time with respect to the number of simplices. It is possible that more than one configuration are associated with one single vertex or edge. Even in this mixed case, the above procedure will properly handle the case if we update the configuration change after each operator.

6   Conclusions

In this paper, we proposed a method to convert the β-shape into a manifold by identifying non-manifold configurations and handling by proposed topology operators. Manifoldized β-shape surface enable us to traverse boundary simplices efficiently and simplify algorithms in application problems. Once the boundary of β-shape is converted into a manifold, many concise but rich data structures widely used in computer graphics and geometric modeling community can be used to represent the surface. Therefore, we expect that the proposed method will be used importantly in many application problems dealing with the surface of a protein such as a pocket recognition on a boundary of a protein and molecular surface construction.

Acknowledgements

This research was supported by the Korea Science and Engineering Foundation (KOSEF) through the National Research Lab. Program funded by the Ministry of Science and Technology, Korea (No. R0A-2007-000-20048-0).

References

1. Choi, Y.: Vertex-Based Boundary Representation of Non-Manifold Geometric Models. PhD thesis, Carnegie-Mellon University, USA (1989)
2. Edelsbrunner, H., Facello, M., Liang, J.: On the definition and the construction of pockets in macromolecules. Discrete Applied Mathematics 88, 83–102 (1998)
3. Edelsbrunner, H., Mücke, E.P.: Three-dimensional alpha shapes. ACM Transactions on Graphics 13(1), 43–72 (1994)
4. Guéziec, A., Taubin, G., Lazarus, F., Horn, B.: Cutting and stitching: Converting sets of polygons to manifold surfaces. IEEE Transactions on Visualization and Computer Graphics 7(2), 136–151 (2001)
5. Heifets, A., Eisenstein, M.: Effect of local shape modifications of molecular surfaces on rigid-body protein-protein docking. Protein Engineering 16(3), 179–185 (2003)
6. Kim, D., Cho, C.-H., Cho, Y., Ryu, J., Bhak, J., Kim, D.-S.: Pocket extraction on proteins via the Voronoi diagram of spheres. Journal of Molecular Graphics & Modelling (in press, 2007) doi:10.1016/j.jmgm.2007.10.002
7. Kim, D.-S., Cho, C.-H., Kim, D., Cho, Y.: Recognition of docking sites on a protein using β-shape based on Voronoi diagram of atoms. Computer-Aided Design 38(5), 431–443 (2006)
8. Kim, D.-S., Seo, J., Kim, D., Ryu, J., Cho, C.-H.: Three-dimensional beta shapes. Computer-Aided Design 38(11), 1179–1191 (2006)
9. Lee, B., Richards, F.M.: The interpretation of protein structures: Estimation of static accessibility. Journal of Molecular Biology 55, 379–400 (1971)
10. Liang, J., Edelsbrunner, H., Fu, P., Sudhakar, P.V., Subramaniam, S.: Analytical shape computation of macromolecules: I. molecular area and volume through alpha shape. PROTEINS: Structure, Function, and Genetics 33, 1–17 (1998)
11. Peters, K.P., Fauck, J., Frömmel, C.: The automatic search for ligand binding sites in protein of known three dimensional structure using only geometric criteria. Journal of Molecular Biology 256, 201–213 (1996)
12. Ryu, J., Park, R., Kim, D.-S.: Molecular surfaces on proteins via beta shapes. Computer-Aided Design 39(12), 1042–1057 (2007)
13. Seo, J., Cho, Y., Kim, D., Kim, D.-S.: An efficient algorithm for three-dimensional β-complex and β-shape via a quasi-triangulation. In: Proceedings of the ACM Symposium on Solid and Physical Modeling 2007, pp. 323–328 (2007)
14. Shoichet, B.K., Kunts, I.D.: Protein docking and complementarity. Journal of Molecular Biology 221, 327–346 (1991)
15. Weiler, K.: The radial edge structure: A topological representation for non-manifold geometric boundary modeling. In: Wozny, M., McLaughlin, H., Encarnacao, J. (eds.) Geometric Modeling for CAD Applications, North Holland, Elsevier Science Publishers B.V, pp. 3–36 (1988)
16. Yamaguchi, Y., Kimura, F.: Nonmanifold topology based on coupling entities. IEEE Computer Graphics and Applications 15(1), 42–50 (1995)

A Mesh Simplification Method Using Noble Optimal Positioning

Han Kyun Choi, Hyun Soo Kim, and Kwan H. Lee

Dept. of Mechatronics at GIST, 1 Oryong-dong,
Puk-gu, Gwangju 500-712, Korea
{korwairs,hskim,khlee}@gist.ac.kr
http://ideg.gist.ac.kr

Abstract. This paper proposes a mesh simplification method by finding the optimal positions of vertices. It generates a Bézier patch around the collapsed edge and finds the optimal position based on visual importance and curvature. It successfully maintains the geometry and topology of the model even when the size of the model is reduced to less than 5 % of the original model. Our method uses QEM for the error measure. It can be applied to usual mesh simplification but also to mesh parameterization and remeshing.

Keywords: Mesh simplification, Multi-resolution modeling, Level of detail, Point-normal triangle, Optimal position.

1 Introduction

As the application areas of computer graphics have become more widespread, highly detailed models with a large file size are often used in many fields such as virtual reality and 3D games. And the need of real-time interaction has been continuously increased as well. Consequently, many researchers have focused their work on improving the rendering efficiency that can reduce the computation time. One of the most popular approaches is the mesh simplification methods. The mesh simplification methods have been used for parameterization and remeshing as well as reducing the number of polygonal mesh. For these applications, the approximated mesh needs to preserve both topology and geometry of the original mesh model while satisfying the specified error level. Most mesh simplification methods have concentrated on finding a cost function for error estimation; few researchers have made an effort to find an optimal position of a collapsed vertex. However, if the optimal positioning is not applied, well defined cost functions do not necessarily guarantee the preservation of topology and geometry due to the accumulated error when the mesh is simplified drastically.

Overall Procedure

In this research, the candidates for the optimal position are determined by the vertices that are produced as a result of subdividing the neighboring triangles around the collapsed edge using a PN triangle [1]. The PN triangle is defined

F. Chen and B. Jüttler (Eds.): GMP 2008, LNCS 4975, pp. 512–518, 2008.
© Springer-Verlag Berlin Heidelberg 2008

as a Bèzier patch and the collapse priority is calculated by QEM [2]. Then we select the optimal position among various candidates on a curved triangle using three parameters: normal variance, curvedness and compactness.

2   Previous Work

Many researchers have come up with simplification schemes that show different ideas on finding cost function and optimal position. We briefly reviewed the previous methods as below.

One of the popular simplification algorithms was the vertex decimation originally proposed by Schroeder et al. [3] before the edge collapse method was generally used. It estimated the approximation error iteratively using the distance between the vertices and the average plane. Some methods developed from vertex decimation [4] [5] used more accurate error metric such as the localized Hausdorff error.

The edge contraction method was developed next, and it became the popular method in mesh approximation. It continuously collapses vertex pairs (v_i, v_j) and generates a new vertex $(\bar{v})$ at each iteration until the desired error is satisfied. Hoppe proposed the progressive mesh [6] based on minimizing the energy function. It measures the error as the distances between the sampled points on an original surface and the closest points on the approximated surface. The algorithm was later improved by Guéziec [7] who defined the tolerance volume around the approximated mesh model and thereby guaranteed the maximum geometric error within that volume.

A moore enhanced method which used quadric error metric (QEM) was then developed by Garland and Heckbert. It defines the error as the squared distances to sets of planes. This method is relatively fast and accurate although it requires 4×4 symmetric metrics for each vertex. After this edge collapse method, some researchers have sought the vertex position methods that minimize the quadric error by assuming the gradient of quadric error metric equals zero. These methods are further developed to handle additional attribute errors such as texture coordinates and colors [8]. Some researchers propose a more general method that expands the error metric to any dimension [9]. Lindstrom and Turk [10] developed the memoryless algorithm that uses linear constraints based on the conservation of volume. It only considers the current approximation error alone. In other words, there is no information about the original shape of model. In fact, "volume optimization" is almost the same as the deviation of quadric error [11].

3   Determination of an Optimal Position

Finding an optimal position in each iteration from many candidate positions requires a significant amount of computation time. Our proposed method uses four metrics: normal variance, Gaussian and mean curvatures and compactness. These metrics are relatively easy to compute while they are sensitive to the change of geometry and topology. They are briefly described as follows.

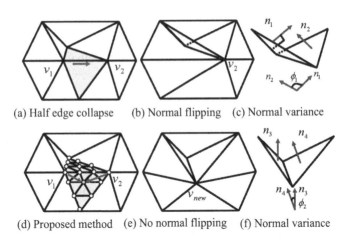

(a) Half edge collapse   (b) Normal flipping   (c) Normal variance

(d) Proposed method   (e) No normal flipping   (f) Normal variance

Fig. 1. Normal variance

Normal variance. Normal variance can be calculated by measuring the difference of the normal values of corresponding triangles before and after the edge collapse. It has been widely used for its efficiency in geometry preservation and removal of non-manifolds. However, previous methods usually skip the edge collapse and proceeds to the next step when the value of normal variance exceeds the threshold. But the proposed method can contract edges without skipping the current step since it uses many candidates. This helps to avoid normal flipping as shown in Figure 1.

Curvedness using Gaussian and mean curvature. Gaussian curvature K is originally expressed as multiplication of principal curvatures κ_1 and κ_2. In discrete geometry, it is related with angles and areas of neighboring triangles of a vertex. Mean curvature H is defined as the average of the principal curvatures. In a polygonal model, it is related to dihedral angle and edge lengths. We can use both Gaussian and mean curvature when the optimal position is determined. However, if we use two components simultaneously, the function of optimal positioning will be more complex. So, we use curvedness that contains the property similar to both Gaussian and mean curvature. Originally, the curvedness of a vertex can be defined as $R = \sqrt{\kappa_1{}^2 + \kappa_2{}^2}/\sqrt{2}$ and it can be expressed using Gaussian and mean curvature $R = \sqrt{2H^2 - K}$.

Compactness. The compactness is another measure that we use to determine the optimal position. We use this to improve the mesh quality by removing sliver triangles. The sliver triangle, ones which contains a very small angle, gives undesirable effects for many applications. Guéziec [12] defines the compactness as bellow.

$$\gamma = \frac{4\sqrt{3}A}{l_1{}^2 + l_2{}^2 + l_3{}^2} \tag{1}$$

where l is the length of edges and A is the area of the triangle. The compactness γ varies from 0 to 1 according to the shape of the triangle. It goes close to 1 when the shape of the triangle is regular, and it approaches to 0 for the case of sliver triangles. In the proposed method, it is used to penalize candidate positions which generate a poor mesh using the threshold value. In practice, the predefined threshold value is assigned heuristically.

An optimal positioning function. As we mentioned before, an optimal positioning function uses three components: normal variance, curvedness and compactness. Normal variance is used for consistency check and making the smooth model. Curvedness is applied to preserve geometry and geometric feature. Finally, the compactness is also used to prevent collapsing topology and to improve mesh quality. The optimal positioning function using normal variance ϕ, curvedness R and compactness γ can be explained as follows

$$
P_1(e(v_i, v_j)) = \sum_i \alpha\phi_i + \beta R_i + w\gamma_i, \, P_2(e(v_j, v_i)) = \sum_j \alpha\phi_j + \beta R_j + w\gamma_j
$$
$$
P_{total} = P_1 + P_2
$$
(2)

where α, β and w are the weight of variance, curvedness and compactness respectively. The three weight factors are user defined values, so that the user can provide different values for different applications.

4 Experimental Results

We demonstrate the performance of our method by comparing it with the QEM method in terms of visual comparison and geometric error. The visual comparison allows the user to recognize the difference between the original model and the approximated models. The geometric error is compared by the normal distance error which is acquired by the measurement of squared distance along the normal direction between the original model and approximated models. It quantitatively shows the difference in geometry as well as in volume shrinkage.

Visual Comparison. The use of three-dimensional surface models for visualization has been continuously increased. The models now need to be rendered without any visual artifacts by using a relatively small number of polygons, especially, in the game industry. Since visual comparison has become important as described above, we visually compare the result of our method and that of the QEM with the original model as shown by Figure 2 and 3.

Figure 2 shows the IGEA model which contains relatively complex geometric features. As observed in Figure 2, our proposed method gives a better result with respect to the overall shape. Figure 3 shows a visual comparison a foot model that has relatively simple features. As we can see, the proposed method also gives better results.

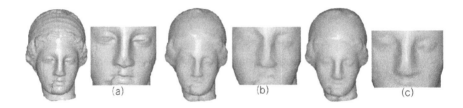

Fig. 2. Visual comparison of approximated IGEA models: (a) An original model (Number of vertex: 50,000), (b) QEM (Number of vertex: 1500), (c) The proposed method (Number of vertex: 1500)

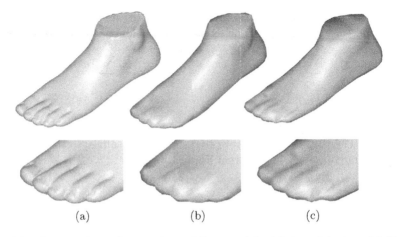

Fig. 3. Visual comparison of approximated Foot models: (a) An original model (Number of vertex: 5,200), (b) QEM (Number of vertex: 250), (c) The proposed method (Number of vertex: 250)

Table 1. Normal distance error and time cost

Model	Number of vertices	Average Normal Distances(mm) (Time cost(sec))	
		QEM(sec)	Proposed Method(sec)
Dragon	2,000	0.11441(0.86)	0.11379(8.56)
(# of vertices: 25,000)	1,500	0.16366(0.88)	0.14254(8.75)
IGEA	2,000	0.07724(2.592)	0.07923(18.195)
(# of vertices: 50,000)	1,500	0.09167(2.753)	0.08977(18.542)
Foot	600	0.11959(0.16)	0.10832(1.798)
(# of vertices: 5,200)	300	0.21465(0.17)	0.20516(1.844)

The Geometric Error. The geometric error can be estimated by calculating the deviations of the normal distance error between the original and approximated models. We evaluate our algorithm by comparing it with QEM shown in

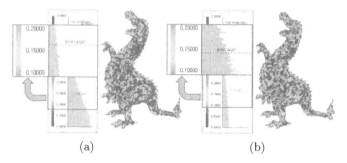

Fig. 4. The visualization of normal distance error, (a) QEM(Number of vertex: 1,500), (b) The proposed method(Number of vertex: 1,500)

Table 1 using the normal distance error. Three models are used for the test: a dragon model, the IGEA model and a foot model. These models have 25,000, 50,000 and 5,200 vertices with 150mm diagonal length of the bounding box, respectively. The table shows that the proposed algorithm gives a better result in terms of the average normal distance error. Especially, it gives a distinguishable result compare to QEM for the highly approximated models. The normal distance error for the dragon model is visualized using a color map in Figure 4.

5   Conclusion

We have developed a QEM-based mesh simplification method that utilizes optimal positioning of collapsed vertices. The method gives a good appearance preservation result for highly approximated models. The proposed method uses a PN triangle and curvedness to achieve a more accurate estimation of quadric error that considers color attributes and vertex replacement to reduce the accumulated error. The proposed method is evaluated using visual comparison and the normal distance error. The visual quality of the mesh obtained using the proposed method shows better result than that of the QEM. The proposed method also yields better results with regard to the normal distance error. However, this method requires additional computation time due to calculating the curvature and a PN triangle.

Acknowledgments. This research is supported by the NRL (National Research Laboratory) funded by the Ministry of Science and Technology of Korea and the CTI development project supported by the MCT and KOCCA in S. Korea.

References

1. Boyd, C., Peters, J., Mitchell, J.L., Vlachos, A.: Curved PN triangles. In: Proc.2001 Symposium on Interactive 3D graphics, pp. 159–166. ACM Press, New York (2001)
2. Garland, M., Heckbert, P.S.: Surface simplification using quadric error metrics. In: Proc. SIGGRAPH 1997, pp. 209–216 (1997)

3. Schroeder, W.J., Zarge, A., Lorensen, W.E.: Decimation of triangle mesh. In: Proc. SIGGRAPH 1992, pp. 65–70 (1992)
4. Ciampalini, A., Cigonini, P., Montani, C., Scopigno, R.: Multiresolution decimation based on global error. The Visual Computer, 228–246 (1997)
5. Kobbelt, L., Campagna, S., Seidel, H.P.: A general framework for mesh decimation. In: Proc. Graphics Interface 1998, pp. 43–50 (1998.4)
6. Hoppe, H.: Progressive meshes. In: Proc. SIGGRAPH 1996, pp. 99–108 (1996)
7. Guéziec, A.: Surface of irregular surface meshes in 3D medical images. In: IEEE Visualization 1998 Conference Proceedings, pp. 271–278 (1998)
8. Garland, M., Heckbert, P.S.: Simplifying surfaces with color and texture using quadric error metrics. In: IEEE Visualization 1998 Conference Proceedings, pp. 263–269 (1998)
9. Garland, M., Zhou, Y.: Quadric-based simplification in any dimension. ACM Trans. Graphics 24(2), 209–239 (2005)
10. Lindstrom, P., Turk, G.: Fast and memory efficient polygonal simplification. In: IEEE Visualization 1998 Conference Proceedings, pp. 279–286 (1998)
11. Heckbert, P.S., Garland, M.: Survey of polygonal surface simplification algorithm. In: Multiresolution Surface Modeling Course Note. ACM SIGGRAPH (1997)
12. Guèziec, A.: Surface simplification with variable tolerance. In: Second Int. Symp. on Medical Robotics and Computer Assisted Surgery (MRCAS 1995), pp. 132–139 (1995)

Narrow-Band Based Radial Basis Functions Implicit Surface Reconstruction

Xiaojun Wu[1], Michael Yu Wang[2], and Jia Chen[1]

[1] Shenzhen Graduate School, HIT, Shenzhen, China
[2] The Chinese University of Hong Kong, Shatin, NT, Hong Kong, China
{wuxj,chenjia}@hitsz.edu.cn,
yuwang@mae.cuhk.edu.cn

Abstract. We propose a narrow-band based RBFs implicit surface reconstruction method which can substantially reduce the computational complexity compared with other RBFs implicit surface reconstruction techniques. Our scheme only deals with a narrow-band subdomains, rather than the traditional whole computational domain. A criteria for polygonization is presented for correctly extracting iso-surfaces from RBFs implicits. Experiments show that our method can offer a very effective RBFs based surface reconstruction algorithm.

Keywords: Radial basis functions, implicit surface reconstruction, narrow-band based method.

1 Introduction

Surface reconstruction is a classical topic in computer graphics and geometry processing community. There are lots of algorithms to create surfaces from scattered point set. As a complex shape can be described in a compact form and boolean operations can be easily performed on implicit models, implicit shape representations attract more and more attentions[8] recently. There are conventional implicit models, such as blobby model [7], level-set based method[14], moving least square based method[5][6] and multi-level of partition of unity[11].

Radial basis functions (RBFs) is one of the most accurate and stable interpolation methods. Savchenko et al.[9], Carr et al.[1] and Turk and O'Brien [10] use globally supported radial basis functions to reconstruct smooth surfaces, while Morse, et al.[2], Ohtake, et al[3]. utilize compactly supported RBFs. Whereas, these methods utilize the whole domain as computational domain, it is time and storage consuming. Ohtake et al.[11] and Tobor et al[12] use hierarchy structure to organize the scattered point set and employ the partition of unity (POU) technique to decompose the whole computational domain into smaller pieces with mild overlap, which mitigate the computational load. Traditionally, the evaluation of RBFs is very time and storage consuming especially to reconstruct implicit surface from a large amount of samples or even failure since the computational cost of $O(N^3)$ to solve the linear system and $O(N^2)$ in memory. In [1], Carr et al. use a iterative strategy on the basis of *Fast Multipole Method*,

F. Chen and B. Jüttler (Eds.): GMP 2008, LNCS 4975, pp. 519–525, 2008.

FMM, of Greengard and Rohklin to reduce the evaluation cost drops from $O(N)$ to $O(1)$ after a $O(N \log N)$ setup, and the computational cost to solve the linear system drops from $O(N^3)$ to $O(N \log N)$ and the storage consuming to $O(N)$ from $O(N^2)$. Though this method is attractive, it is not understandable and complicated to implement.

There are two reasons to hamper further improve the speed of RBFs based method. Firstly, the solving and evaluation of RBFs are much time consuming. Secondly, every point in the computational domain must be evaluated to ensure the correctness of isosurface extraction, see the left figure in Figure 1. In order to accelerate the RBFs based approach, we propose the narrow-band based RBFs reconstruction method. We take advantage of the fact that the zero level-set of implicit function intersects only with a small portion of the marching cube grid, shown in the right figure of Figure 1, which can save lots of computation resources. Narrow-band based strategy was used firstly in level set method to accelerate the surface evaluation [13].

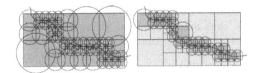

Fig. 1. Traditional computation scheme and narrow band based method

Fig. 2. Adaptive octree subdivision and subdomain computation

The rest of this paper is organized as follows. Section 2 describes the method of radials basis functions and partition of unity. In section 3, we present details of our narrow-band based surface reconstruction method. In section 4, we show reconstructed results of our approach and discuss the limitations. Section 5 is the conclusion.

2   Radial Basis Functions and Partition of Unity

RBFs is utilized in multidimensional interpolation or approximation. In general, a RBFs can be expressed as follows:

$$\phi(\boldsymbol{x}) = \sum_{i=1}^{N} \alpha_i g(\|\boldsymbol{x} - \boldsymbol{x}_i\|) + P(\boldsymbol{x}) \tag{1}$$

where $\boldsymbol{x}_i$ are called RBFs centers, and α_i are coefficients corresponding to each center, $g(r)$, $r = \|\boldsymbol{x} - \boldsymbol{x}_i\|$, is basis functions with real value unbounded $[0, \infty)$. $P(\boldsymbol{x})$ is a polynomial of low degree. More about RBFs please refers to [1][2][10] and literatures therein.

The main idea of the partition of unity method is to divide the global domain of interest into smaller overlapping subdomains where the problem can be solved locally on a small scale. The local solutions are combined together by using blending functions to obtain the global solution. The smoothness of the global solution in the overlap regions of two subdomains can be guaranteed by a polynomial stitching function.

The global domain Ω is first divided into M overlapping subdomains $\{\Omega_i\}_{i=1}^M$ with $\Omega \subseteq \cup_i \Omega_i$. For a partition of unity on the set of subdomains $\{\Omega_i\}_{i=1}^M$, we then need to define a collection of non-negative blending functions $\{w_i\}_{i=1}^M$ with limited support and with $\sum w_i = 1$ in the entire domain Ω. For each subdomain Ω_i, the data set of the points within the subdomain is used to compute a local reconstruction function ϕ_i that interpolates the data points. The global reconstruction function Φ is then defined as a combination of the local functions: $\Phi(\boldsymbol{x}) = \sum_{i=1}^M \phi(\boldsymbol{x}) w_i(\boldsymbol{x})$. $w_i(\boldsymbol{x})$ is called blending function, defined as $w_i(\boldsymbol{x}) = \frac{W_i(\boldsymbol{x})}{\sum_{j=1}^n W_j(\boldsymbol{x})}$. The weight function is defined as $W_i(\boldsymbol{x}) = b(\frac{3|\boldsymbol{x} - c_i|}{2R_i})$ [11]. In our implementation, the subdomain is defined as a sphere centered at c_i and with support radius R_i, shown in Figure 2.

3   Narrow Band-Based RBFs Method

3.1   Adaptive Space Subdivision and Subdomain Creation

Just like Ohtake's MPU[11], we also use an adaptive space subdivision method based on octree to decompose the point set $\mathcal{P}$. Two threshold values T_{min} and T_{max} with $T_{min} < T_{max}$ are defined to control the number of points in computational subdomains each of which associates with a non-empty leaf node of octree, illustrated in Figure 2. The support radius R_i is defined as $R_i = \beta d_i$, where d_i are the diagonal length of octree cell and β is set 0.6 in our implementation. In order to confine the number of points in each subdomain in the range of thresholds, we use the following iterative procedures to tune the support radius.

$$\begin{cases} \hat{R} = \hat{R} - \lambda_1 R, & N > T_{max} \\ \hat{R} = \hat{R} + \lambda_2 R, & N < T_{min} \end{cases} \tag{2}$$

where $\hat{R} = R$ initially, N is the number of point in current subdomain, λ_1 and λ_2 are two different coefficients to accelerate the convergence, $\lambda_1 = 0.005$ and $\lambda_2 = 0.0045$ in our implementation. When $T_{min} \leqslant N \leqslant T_{max}$ but $\hat{R} < (0.5 + \epsilon)d_i$ (ϵ is a tiny value for robustness) during the iteration, the ith cell should be subdivided further. Figure 3 demonstrates a real point cloud and the corresponding narrow band sumdomains.

When we set up the octree and the corresponding subdomain, the RBFs coefficients can be calculated by using equation (1) locally. As the point set is organized by octree structure, the point is reduced to a small scale that can be solved by simple methods. In our implementation, we use LU-decomposition based linear system solving approach.

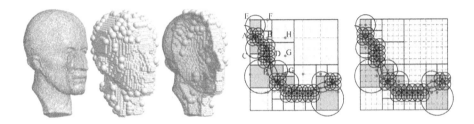

Fig. 3. The point cloud and the corresponding narrow band subdomains

Fig. 4. Determination of the MC grid size

3.2   Polygonization of Implicit Surface

In our scheme, we employ a modified marching cube algorithm to extract the
iso-surface. If every point in the octree space could be evaluated by the implicit
function, the iso-surface could be extracted safely from the volumetric grid.
However, in our case, the local surface patches are only computed in a narrow
band. Therefore, some new strategies are needed to avoid the artifacts.

We employ a 2D example to clarify our tricks as illustrated in Figure 4, where
only a portion of octree cells, the surface and the subdomains are displayed. The
shaded cells indicates the narrow band. The circles represents the computational
subdomains, and the dashed lines are the marching cube (MC) grid. In the left
figure of Figure 4, the size of marching grid is d, $d > L$, L the smallest edge size of
octree cell. The red dots locating the subdomains can be evaluated, however, the
green dots outside the subdomains can not find implicit function values. That
is these grids can not produce correct triangle facets during marching cube, see
edges of AB, BD, and DG in the grids ABCD, ABEF, and BDGH in left figure.
Thus, the final iso-surface would appear some artifacts. When we reduce the size
of the marching cube grid as revealed in the right figure, all the vertex of MC
grid intersecting with surfaces can be evaluated. That is we can get the correct
reconstructed iso-surfaces. When the size of the marching cube reduced further,
this claim is also hold. We can conclude that the marching cube size d and the
smallest octree cell size L should meed the requirement of $d \leqslant L$ if the correct
iso-surface can be extracted. Otherwise, some artifacts can occur, see the left
figure in Figure 5.

4   Experimental Results and Discussion

All the experimental results in this section were performed on a notepad computer with Intel Pentium 1.5GHz with 512MB of RAM. Some surfaces are shown
in Figure 5 and 6. There are some holes on the middle left surface in Figure 5 as
the marching cube grid size is larger than the smallest size of octree leaf node.

Table 1 presents the processing time of different point cloud. N denotes the
number of points from which the surfaces are reconstructed, $Nleaf$ the amount
of leaf nodes of octree, $Nenode$ the number of non-empty leaf nodes of octree

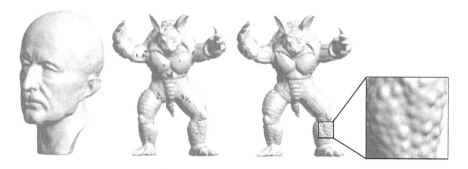

Fig. 5. Surface reconstruction of the Armadillo point set, left is the artifact model, the middle is the fine result and a zoomed out displayed in the right figure

Fig. 6. Reconstructed surfaces of Isis, Lucy and blade consisting of187.6K, 262.9K points and 883K points respectively. Total reconstruction time is 108.8sec, 196 sec and 494.9 sec using RBFs interpolation.

Table 1. Reconstruction time with narrow-band based method

Model	N	$Nleaf$	$Nenode$	Ttree	Trec	Tpoly	Ttotal
Maxplanck	48.5K	18K	8K	0.4	2.7	11.8	14.9
Armadillo	173K	66K	28K	1.2	9.8	55.3	66.3
Isis	187.6K	18K	36.6K	2.8	13.4	92.6	108.8
Lucy	262.9K	91K	43K	3.3	16.2	176.6	196.1
Blade	883K	314K	151K	11.5	46.6	436.8	494.9

that is also the quantity of sumdomains, Ttree the time of set up the octree and creation of the subdomains, Trec the reconstruction time that is time of solve the RBFs linear equations, Tpoly the polygonization time in seconds, Ttotal the total time of processing. To compare the efficiency of our method, we list the processing time of Ohtake's adaptive compactly supported radial basis functions

Table 2. Processing time of adaptive compactly supported RBFs [4]

Model	N	PU	RBF	T_{SA}	M
Squirrel	46K	6	4	5.0×10^{-5}	1.6K
Bunny	362K	60	42	2.0×10^{-6}	23K
Dragon	2.11M	778	380	2.0×10^{-6}	36K
Armadillo	2.37M	861	590	2.0×10^{-6}	41K

in Table 2, where N means the number of the original point set, PU the time cost of adaptive PU method in seconds, RBF the time consumption of RBF scheme in seconds, T_{SA} a scale factor, M the quantity of points to approximate the surfaces[4]. A point must be noticed is that we utilize all the original points $Npoints$ to interpolate the implicit surface, however the surfaces are approximated by selected fewer centers M in [4]. With the comparable number of points in model Maxplanck, 48.5K in ours and 41K in PU+RBF, the computation time is 14.97sec and 1451sec respectively. The adaptive PU+RBF were performed on a 1.6GHz Mobile Pentium 4PC with 1GB RAM.

Though our approach is quite fast, there is a limitation. Specifically, when the distribution of the point set is heavily nonuniform, the hole on the point set can not be filled automatically in the areas of lacking samples since the computational sumdomains are constructed only in a way of narrow band. When the hole size is so large that the boundary subdomains can not cover the hole, the point outside of the sumdomain can not be evaluated, which leads to the incapability of hole filling. One solution to this problem is point resampling which can redistribute the point set more evenly. These techniques can be borrowed from point based surfaces [5].

5   Conclusions

In this paper, we present a new surface reconstruction method for large scattered point set based on radial basis functions, partition of unity and octree hierarchy space subdivision. In this approach, we construct some narrow-band computational subdomains corresponding to the non-empty leaf nodes of octree, which can save the time storage tremendously. A criterion for isosurface extraction is proposed as well, which guarantees the vertex of MC grid are located in subdomains. In contrast to other RBFs based methods, where the whole octree space, empty and non-empty, are covered by subdomains, which blocks the speedup of RBFs based surface reconstruction methods. Though there is one limitation in our narrow-band based approach, it can be solved by using point resampling strategy to even the distribution of points. One important point is worth noting that we use interpolation instead of approximation scheme to reconstruct RBFs surfaces. Because of the good approximation characteristic of radial basis functions, the point set used in the experiments can be considered as down-sampled from a large scale point data, that is we can get fine surfaces from tremendous scale of point set.

Acknowledgments. This work was funded in part by the China Postdoctoral Science Foundation. The models are courtesy of the Stanford 3D Scanning Repository (Lucy, Armadillo , and bunny), Cyeberware (Blade), Max-Planck-Institut für Informatic (Head of Max Planck).

References

1. Carr, J.C., Beatson, R.K., Cherrie, J.B., Mitchell, T.J., Fright, W.R., McCallum, B.C., Evans, T.R.: Reconstruction and representation of 3d objects with radial basis functions. In: Computer Graphics Proceedings. Annual Conference Series, pp. 67–76 (2001)
2. Morse, B.S., Yoo, T.S., Rheingans, P., Chen, D.T., Subramanian, K.R.: Interpolating implicit surfaces from scattered surfaces data using compactly supported radial basis functions. In: Proceedings of Shape Modeling International (SMI 2001), pp. 89–98 (2001)
3. Ohtake, Y., Belyaev, A., Seidel, H.P.: A multilevel approach to 3D scattered data interpolation with compactly supported basis function. In: Proceedings of Shape Modeling International (SMI 2003), pp. 153–161 (2003)
4. Ohtake, Y., Belyaev, A., Seidel, H.P.: 3D scattered data approximation with adaptive compactly supported radial basis functions. In: Proceedings of the Shape Modeling International 2004 (SMI 2004), pp. 31–39 (2004)
5. Alexa, M., Behr, J., Cohen-or, D., Fleishman, S., Levin, D., Silva, C.T.: Computing and rendering point set surfaces. IEEE Transactions on Visualization and Computer & Graphics 9(1), 3–15 (2003)
6. Adamson, A., Alexa, M.: Approximating and interesting surfaces from points. In: Symposium on Geometry Processing, pp. 245–254 (2003)
7. Blinn, J.F.: A generalization of algebraic surface drawing. ACM Transactions on Graphics 1(3), 235–256 (1982)
8. Bloomenthal, J., Bajaj, C., Blinn, J., Cani-Gascuel, M.P., Rockwood, A., Wyvill, B., Wyvill, G.: Introduction to implicit surfaces. Morgan Kaufman, San Francisco (1997)
9. Savchenko, V.V., Wendland, H.: Adaptive greedy techniques for approximate solution of large RBF systems. Numerical Algorithms 24, 1115–1138 (2000)
10. Turk, G., O'Brien, J.: Modelling with implicit surfaces that interpolate. ACM Transactions on Graphics (SIGGRAPH 2002) 21(4), 855–873 (2002)
11. Ohtake, Y., Belyaev, A., Peter Seidel, H.P.: Multi-level partition of unity implicits. ACM Transactions on Graphics(SIGGRAPH 2003) 22(3), 463–470 (2003)
12. Tobor, I., Reuter, P., Schlick, C.: Multiresolution reconstruction of implicit surfaces with attributes from large unorganized point sets. In: Proceedings of Shape Modeling International (SMI 2004), pp. 19–30 (2004)
13. Sethian, J.A.: Level Set Methods and Fast Marching Methods: Evolving Interfaces in Computational Geometry, Fluid Mechanics, Computer Vision, and Materials Science, 2nd edn. Cambridge University Press, Cambridge (1999)
14. Zhao, H.K., Osher, S., Merriman, B., Kang, M.: Implicit and non-parametric shape reconstruction from unorganized points using variational level set method. Computer Vision and Image Understanding 80, 295–319 (2000)

Progressive Interpolation Using Loop Subdivision Surfaces

Fuhua (Frank) Cheng[1], Fengtao Fan[1], Shuhua Lai[2],
Conglin Huang[1], Jiaxi Wang[1], and Junhai Yong[3]

[1] Depart. of Computer Science, University of Kentucky, Lexington, KY 40506, USA
cheng@cs.uky.edu, {ffan2,Conglin.Huang,Jiaxi.Wang}@uky.edu
[2] Depart. of Mathematics and Computer Science, Virginia State University,
Petersburg, VA 23806, USA
slai@vsu.eud
[3] School of Software, Tsinghua University, Beijing 100084, P.R. China
yongjh@tsinghua.edu.cn

Abstract. A new method for constructing interpolating Loop subdivision surfaces is presented. The new method is an extension of the progressive interpolation technique for B-splines. Given a triangular mesh M, the idea is to iteratively upgrade the vertices of M to generate a new control mesh $\bar{M}$ such that limit surface of $\bar{M}$ interpolates M. It can be shown that the iterative process is convergent for Loop subdivision surfaces. Hence, the method is well-defined. The new method has the advantages of both a local method and a global method, i.e., it can handle meshes of any size and any topology while generating smooth interpolating subdivision surfaces that faithfully resemble the shape of the given meshes.

Keywords: Geometric Modeling, Loop Subdivision Surface, Interpolation.

1 Introduction

Subdivision surfaces are becoming popular in many areas such as animation, geometric modeling and games because of their capability in representing any shape with only one surface. A subdivision surface is generated by repeatedly refining a control mesh to get a limit surface. Hence, a subdivision surface is determined by the way the control mesh is refined, i.e., the *subdivision scheme*. A subdivision scheme is called an *interpolating scheme* if the limit surface interpolates the given mesh. Otherwise, it is called an *approximating scheme*. Popular subdivision schemes such as Catmull-Clark [2], Doo-Sabin [1], and Loop schemes [8] are approximating schemes while the Butterfly [6], the improved Butterfly [9] and the Kobbelt schemes [18] are interpolating schemes.

An interpolating subdivision scheme generates new vertices by performing local affine combinations on nearby vertices. This approach is simple and easy to implement. It can handle meshes with a large number of vertices. However, since no vertex is ever moved once it is computed, any distortion in the early stage of

F. Chen and B. Jüttler (Eds.): GMP 2008, LNCS 4975, pp. 526–533, 2008.

the subdivision will persist. This makes interpolating subdivision schemes very sensitive to irregularity in the given mesh. In addition, it is difficult for this approach to interpolate normals or derivatives.

On the other hand, even though subdivision surfaces generated by approximating subdivision schemes do not interpolate their control meshes, it is possible to use this approach to generate a subdivision surface to interpolate the vertices of a given mesh. One method, called *global optimization*, does the work by building a global linear system with some fairness constraints to avoid undesired undulations [3,4]. The solution to the global linear system is a control mesh whose limit surface interpolates the vertices of the given mesh. Because of its *global property*, this method generates smooth interpolating subdivision surfaces that resemble the shape of the given meshes well. But, for the same reason, it is difficult for this method to handle meshes with a large number of vertices.

To avoid the computational cost of solving a large system of linear equations, several other methods have been proposed. A two-phase subdivision method that works for meshes of any size was presented by Zheng and Cai for Catmull-Clark scheme [5]. A method proposed by Lai and Cheng [15] for Catmull-Clark subdivision scheme avoids the need of solving a system of linear equations by utilizing the concept of similarity in the construction process. Litke, Levin and Schöder avoid the need of solving a system of linear equation by quasi-interpolating the given mesh [7]. However, a method that has the advantages of both a local method and a global method is not available yet.

In this paper a new method for constructing a smooth Loop subdivision surface that interpolates the vertices of a given triangular mesh is presented. The new method is an extension of the progressive interpolation technique for B-splines [10,12,13]. The idea is to iteratively upgrade the locations of the given mesh vertices until a control mesh whose limit surface interpolates the given mesh is obtained. It can be proved that the iterative interpolation process is convergent for Loop subdivision surfaces. Hence, the method is well-defined for Loop subdivision surfaces. The limit of the iterative interpolation process has the form of a global method. But the control points of the limit surface can be computed using a local approach. Therefore, the new technique enjoys the advantages of both a local method and a global method, i.e., it can handle meshes of any size and any topology while generating smooth interpolating subdivision surfaces that faithfully resemble the shape of the given meshes.

The remaining part of the paper is arranged as follows. The concept of progressive interpolation for Loop subdivision surfaces is presented in Section 2. The convergence of this iterative interpolation process is proven in Section 3. Implementation issues and test results are presented in Section 4. Concluding remarks are given in Section 5.

2 Progressive Interpolation Using Loop Surfaces

Given a 3D triangular mesh $M = M^0$. To interpolate the vertices of M^0 with a Loop subdivision surface, one needs to find a control mesh $\bar{M}$ whose Loop surface passes through all the vertices of M^0. Instead of a direct approach, we use an iterative process to do the job.

First, we consider the Loop surface $\mathbf{S}^0$ of M^0. For each vertex $\mathbf{V}^0$ of M^0, we compute the distance between this vertex and its limit point $\mathbf{V}^0_\infty$ on $\mathbf{S}^0$, $\mathbf{D}^0 = \mathbf{V}^0 - \mathbf{V}^0_\infty$, and add this distance to $\mathbf{V}^0$ to get a new vertex $\mathbf{V}^1$, $\mathbf{V}^1 = \mathbf{V}^0 + \mathbf{D}^0$. The set of all the new vertices is called M^1. We then consider the Loop surface $\mathbf{S}^1$ of M^1 and repeat the same process.

In general, if $\mathbf{V}^k$ is the new location of $\mathbf{V}^0$ after k iterations of the above process and M^k is the set of all the new $\mathbf{V}^k$'s, then we consider the Loop surface $\mathbf{S}^k$ of M^k. We first compute the distance between $\mathbf{V}^0$ and the limit point $\mathbf{V}^k_\infty$ of $\mathbf{V}^k$ on $\mathbf{S}^k$

$$\mathbf{D}^k = \mathbf{V}^0 - \mathbf{V}^k_\infty. \tag{1}$$

We then add this distance to $\mathbf{V}^k$ to get $\mathbf{V}^{k+1}$ as follows:

$$\mathbf{V}^{k+1} = \mathbf{V}^k + \mathbf{D}^k. \tag{2}$$

The set of new vertices is called M^{k+1}.

This process generates a sequence of control meshes M^k and a sequence of corresponding Loop surfaces $\mathbf{S}^k$. $\mathbf{S}^k$ converges to an interpolating surface of M^0 if the distance between $\mathbf{S}^k$ and M^0 converges to zero. Therefore the key task here is to prove that $\mathbf{D}^k$ converges to zero when k tends to infinity. This will be done in the next section.

Note that for each iteration in the above process, the main cost is the computation of the limit point $\mathbf{V}^k_\infty$ of $\mathbf{V}^k$ on $\mathbf{S}^k$. For a Loop surface, the limit point of a control vertex $\mathbf{V}$ with valence n can be calculated as follows:

$$\mathbf{V}_\infty = \beta_n \mathbf{V} + (1 - \beta_n)\mathbf{Q} \tag{3}$$

where

$$\beta_n = \frac{3}{11 - 8 \times \left(\frac{3}{8} + \left(\frac{3}{8} + \frac{1}{4}\cos\frac{2\pi}{n}\right)^2\right)}, \qquad \mathbf{Q} = \frac{1}{n}\sum_{i=1}^{n}\mathbf{Q}_i.$$

$\mathbf{Q}_i$ are adjacent vertices of $\mathbf{V}$. This computation involves nearby vertices only. Hence the progressive interpolation process is a local method and, consequently, can handle meshes of any size.

Another point that should be pointed out is, even though this is an iterative process, one does not have to repeat each step strictly. By finding out when the distance between M^0 and $\mathbf{S}^k$ would be smaller than the given tolerance, one can go directly from M^0 to M^k, skipping the testing steps in between.

3   Convergence of the Iterative Interpolation Process

We need the following lemma for the proof.

Lemma 1. *Eigenvalues of the product of positive definite matrices are positive.*

The proof of Lemma 1 follows immediately from the fact that if P and Q are square matrices of the same dimension, then PQ and QP have the same eigenvalues (see, e.g., [16], p.14).

To prove the convergence of the iterative interpolation process for Loop subdivision surfaces, note that at the $(k+1)$st step, the difference $\mathbf{D}^{k+1}$ can be written as:

$$\mathbf{D}^{k+1} = \mathbf{V}^0 - \mathbf{V}^{k+1}_{\infty} = \mathbf{V}^0 - \left(\beta_n \mathbf{V}^{k+1} + (1 - \beta_n)\mathbf{Q}^{k+1}\right)$$

where $\mathbf{Q}^{k+1}$ is the average of the n adjacent vertices of $\mathbf{V}^{k+1}$

$$\mathbf{Q}^{k+1} = \frac{1}{n}\sum_{i=1}^{n} \mathbf{Q}^{k+1}_i$$

By applying (2) to $\mathbf{V}^{k+1}$ and each $\mathbf{Q}^{k+1}_i$,

$$\mathbf{Q}^{k+1}_i = \mathbf{Q}^k_i + \mathbf{D}^k_{\mathbf{Q}^k_i}$$

we get

$$\mathbf{D}^{k+1} = \mathbf{V}^0 - \left(\beta_n \mathbf{V}^k + (1-\beta_n)\mathbf{Q}^k\right) - \left(\beta_n \mathbf{D}^k + \frac{1-\beta_n}{n}\sum_{i=1}^{n}\mathbf{D}^k_{\mathbf{Q}_i}\right)$$

$$= \mathbf{D}^k - \left(\beta_n \mathbf{D}^k + \frac{1-\beta_n}{n}\sum_{i=1}^{n}\mathbf{D}^k_{\mathbf{Q}_i}\right)$$

where $\mathbf{Q}^k$ is the average of the n adjacent vertices of $\mathbf{V}^k$. In matrix form, we have

$$\left[\mathbf{D}^{k+1}_1, \mathbf{D}^{k+1}_2, \ldots, \mathbf{D}^{k+1}_m\right]^T = (I - B)\begin{bmatrix} \mathbf{D}^k_1 \\ \mathbf{D}^k_2 \\ \vdots \\ \mathbf{D}^k_m \end{bmatrix} = (I - B)^{k+1}\begin{bmatrix} \mathbf{D}^0_1 \\ \mathbf{D}^0_2 \\ \vdots \\ \mathbf{D}^0_m \end{bmatrix}$$

where m is the number of vertices in the given matrix, I is an identity matrix and B is a matrix of the following form:

$$\begin{pmatrix} \beta_{n_1} & \cdots & \frac{1-\beta_{n_1}}{n_1} & \cdots \\ \vdots & \ddots & & \\ \frac{1-\beta_{n_i}}{n_i} & \cdots & \beta_{n_i} & \cdots \\ \vdots & & & \beta_{n_m} \end{pmatrix}$$

The matrix B has the following properties:

(1) $b_{ij} \geq 0$, and $\sum_{j=1}^{n} b_{ij} = 1$ (hence, $\|B\|_\infty = 1$);
(2) There are $n_i + 1$ positive elements in the i-th row, and the positive elements in each row are equal except the element on the diagonal line.
(3) If $b_{ij} = 0$, then $b_{ji} = 0$;

Properties (1) and (2) follow immediately from the formula of $\mathbf{D}^{k+1}$ in eq. (3). Property (3) is true because if a vertex $\mathbf{V}_i$ is an adjacent vertex to $\mathbf{V}_j$ then $\mathbf{V}_j$ is obviously an adjacent vertex to $\mathbf{V}_i$. Due to these properties, we can write the matrix B as

$$B = DS$$

where D is a diagonal matrix and S is a symmetric matrix:

$$D = \begin{pmatrix} \frac{1-\beta_{n_1}}{n_1} & 0 & \cdots & 0 \\ 0 & \frac{1-\beta_{n_2}}{n_2} & \cdots & 0 \\ \vdots & & \ddots & \\ 0 & & & \frac{1-\beta_{n_m}}{n_m} \end{pmatrix}, \quad S = \begin{pmatrix} \frac{n_1\beta_{n_1}}{1-\beta_{n_1}} & \cdots & 1 & \cdots \\ \vdots & \ddots & & \\ 1 & \cdots & \frac{n_i\beta_{n_i}}{1-\beta_{n_i}} & \cdots \\ \vdots & & & \frac{n_m\beta_{n_m}}{1-\beta_{n_m}} \end{pmatrix}.$$

D is obviously positive definite. We will show that the matrix S is also positive definite, a key point in the convergence proof.

Theorem 1. *The matrix S is positive definite.*

Proof: To prove S is positive definite, we have to show the quadric form

$$f(x_1, x_2, \ldots, x_m) = \mathbf{X}^T S \mathbf{X}$$

is positive for any non-zero $\mathbf{X} = (x_1, x_2, \ldots, x_m)^T$.

Note that if vertices $\mathbf{V}_i$ and $\mathbf{V}_j$ are the endpoints of an edge e_{ij} in the mesh, then $s_{ij} = s_{ji} = 1$ in the matrix S. Hence, it is easy to see that

$$f(x_1, x_2, \ldots, x_n) = \sum_{e_{ij}} 2x_i x_j + \sum_{i=1}^m \frac{n_i\beta_{n_i}}{1-\beta_{n_i}} x_i^2$$

where e_{ij} in the first term ranges through all edges of the given mesh. On the other hand, if we use f_{ijr} to represent a face with vertices $\mathbf{V}_i$, $\mathbf{V}_j$ and $\mathbf{V}_r$ in the mesh, then since an edge in a closed triangular mesh is shared by exactly two faces, the following relationship holds:

$$\sum_{f_{ijr}} (x_i + x_j + x_r)^2 = \sum_{e_{ij}} 4x_i x_j + \sum_{i=1}^m n_i x_i^2$$

where f_{ijr} on the left hand side ranges through all faces of the given mesh. The last term in the above equation follows from the fact that a vertex with valence n is shared by n faces of the mesh. Hence, $f(x_1, x_2, \ldots, x_n)$ can be espressed as

$$f(x_1, x_2, \ldots, x_n) = \sum_{f_{ijr}} \frac{1}{2} (x_i + x_j + x_r)^2 + \sum_{i=1}^m \left(\frac{n_i\beta_{n_i}}{1-\beta_{n_i}} - \frac{n_i}{2} \right) x_i^2$$

From eq. (3), it is easy to see that $\frac{n\beta_n}{1-\beta_n} \geq \frac{3}{5}n$ for $n \geq 3$. Hence, $f(x_1, x_2, \ldots, x_m)$ is positive for any none zero $\mathbf{X}$ and, consequently, S is positive definite.

Based on the above lemma and theorem, it is easy to conclude that the iterative interpolation process for Loop subdivision is convergent.

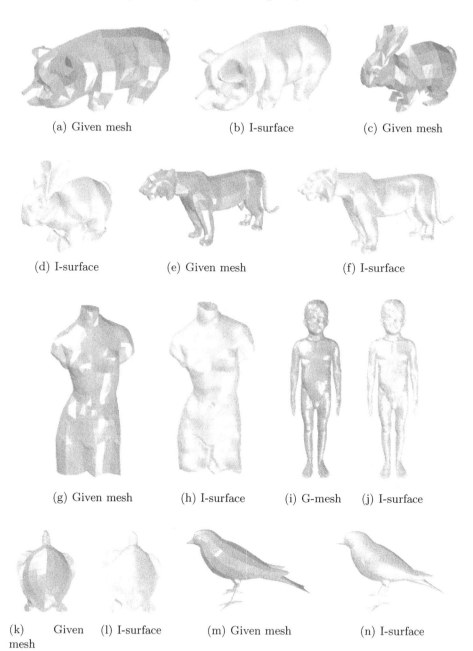

(a) Given mesh   (b) I-surface   (c) Given mesh

(d) I-surface   (e) Given mesh   (f) I-surface

(g) Given mesh   (h) I-surface   (i) G-mesh   (j) I-surface

(k)   Given   (l) I-surface   (m) Given mesh   (n) I-surface
mesh

Fig. 1. Examples of progressive interpolation using Loop subdivision surfaces (G-mesh ≡ given mesh; I-surface ≡ interpolating Loop surface)

Theorem 2. *The iterative interpolation process for Loop subdivision surface is convergent.*

Proof: The iterative process is convergent if and only if absolute value of the eigenvalues of the matrix $P = I - B$ are all less than 1, or all eigenvalues λ_i, $1 \le i \le m$, of B are $0 < \lambda_i \le 1$.

Since $\|B\|_\infty = 1$, we have $\lambda_i \le 1$. On the other hand, since B is the product of two positive definite matrices D and S, following Lemma 1, all its eigenvalues must be positive. Hence, the iterative process is convergent.

4   Results

The progressive interpolation process is implemented for Loop subdivision surfaces on a Windows platform using OpenGL as the supporting graphics system. Quite a few cases have been tested. Some of them (a hog, a rabbit, a tiger, a statue, a boy, a turtle and a bird) are presented in Figure 1. All the data sets are normalized, so that the bounding box of each data set is a unit cube. For each case, the given mesh and the constructed interpolating Loop surface are shown. The sizes of the data meshes, numbers of iterations performed, maximum and average errors of these cases are collected in Table 1.

Table 1. Loop surface based progressive interpolation: test results

Model	# of vertices	# of iterations	Max Error	Ave Error
Hog	606	10	0.000870799	0.000175255
Rabbit	453	13	0.000932857	0.000111197
Tiger	956	9	0.000720453	0.00014148
Statue	711	11	0.000890806	0.000109163
Boy	17342	6	0.000913795	0.000095615
Turtle	445	10	0.000955765	0.0001726
Bird	1129	9	0.000766811	0.000088345

From these results, it is easy to see that the progressive interpolation process is very efficient and can handle large meshes with ease. This is so because of the expotential convergence rate of the iterative process. Another point that can be made here is, although no fairness control factor is added in the progressive iterative interpolation, the results show that it can produce visually pleasing surface easily.

5   Concluding Remarks

A progressive interpolation technique for Loop subdivision surfaces is presented and its convergence is proved. The limit of the iterative interpolation process has the form of a global method. Therefore, the new method enjoys the strength of a global method. On the other hand, since control points of the interpolating surface can be computed using a local approach, the new method also enjoys the strength of a local method. Consequently, we have a subdivision surface based interpolation technique that has the advantages of both a local method and a global method. Our next job is to investigate progressive interpolation for Catmull-Clark and Doo-Sabin subdivision surfaces.

Acknowledgments. This work is supported by NSF (DMI-0422126). The last author is supported by NSFC(60625202, 60533070). Triangular meshes used in this paper are downloaded from the Princeton Shape Benchmark [17].

References

1. Doo, D., Sabin, M.: Behaviour of recursive division surfaces near extraordinary points. Computer-Aided Design 10, 356–360 (1978)
2. Catmull, E., Clark, J.: Recursively generated B-spline surfaces on arbitrary topological meshes. Computer-Aided Design 10, 350–355 (1978)
3. Halstead, M., Kass, M., DeRose, T.: Efficient, fair interpolation using Catmull-Clark surfaces. In: Proc.SIGGRAPH, pp. 47–61 (1993)
4. Nasri, A.H.: Surface interpolation on irregular networks with normal conditions. Computer Aided Geometric Design 8, 89–96 (1991)
5. Zheng, J., Cai, Y.Y.: Interpolation over arbitrary topology meshes using a two-phase subdivision scheme. IEEE Trans. Visualization and Computer Graphics 12, 301–310 (2006)
6. Dyn, N., Levin, D., Gregory, J.A.: A Butterfly Subdivision Scheme for Surface Interpolation with Tension Control. ACM Trans. Graphics 9, 160–169 (1990)
7. Litke, N., Levin, A., Schröder, P.: Fitting Subdivision Surfaces. Proc. Visualization 319-324 (2001)
8. Loop, C.: Smooth Subdivision Surfaces Based on Triangles. Master'thesis. Dept. of Math., Univ. of Utah (1987)
9. Zorin, D., Schroder, P., Sweldens, W.: Interpolating Subdivision for Meshes with Arbitrary Topology. Computer Graphics, Ann. Conf. Series 30, 189–192 (1996)
10. de Boor, C.: How does Agee's method work? In: Proc. 1979 Army Numerical Analysis and Computers Conference ARO Report, Army Research Office, vol. 79, pp. 299–302 (1979)
11. Lin, H., Bao, H., Wang, G.: Totally positive bases and progressive iteration approximation. Computer & Mathematics with Applications 50, 575–585 (2005)
12. Lin, H., Wang, G., Dong, C.: Constructing iterative non-uniform B-spline curve and surface to fit data points. Science in China (Series F) 47, 315–331 (2004)
13. Qi, D., Tian, Z., Zhang, Y., Zheng, J.B.: The method of numeric polish in curve fitting. Acta Mathematica Sinica, (in Chinese) 18, 173–184 (1975)
14. Delgado, J., Peña, J.M.: Progressive iterative approximation and bases with the fastest convergence rates. Computer Aided Geometric Design 24, 10–18 (2007)
15. Lai, S., Cheng, F.: Similarity based Interpolation using Catmull-Clark Subdivision Surfaces. The Visual Computer 22, 865–873 (2006)
16. Magnus, I.R., Neudecker, H.: Matrix Differential Calculus with Applications in Statistics and Econometrics. John Wiley & Sons, New York (1988)
17. Shilane, P., Min, P., Kazhdan, M., Funkhouser, T.: The Princeton Shape Benchmark. Shape Modeling Int'l (2004)
18. Kobbelt, L.: Interpolatory Subdivision on Open Quadrilateral Nets with Arbitrary Topology. Comput. Graph. Forum 5, 409–420 (1996)

Protein Surface Modeling Using Active Contour Model

Junping Xiang and Maolin Hu

Key Laboratory of Intelligent Computing & Signal Processing, Ministry of Education, and
School of Mathematics and Computational Science, Anhui University, China
jpingxiang@gmail.com, hml@ahu.edu.cn

Abstract. A PDE (partial differential equation) based method, 3D active contour model, is presented to model the surface of protein structure. Instead of generating a single molecular surface, we create a series of surfaces associated with the atomic energy inside the protein, which describe different resolutions of molecular surface. Our results indicate that the surfaces we generated are suitable for shape analysis and visualization. So, when the solvent-accessible surface is not enough to represent the features of protein structure, the evolution surface sequence may be an alternative choice. Besides, if the initial surface is smooth enough, the generated surfaces will preserve this property partly, because the evolution of the surfaces is controlled by the PDE.

Keywords: Chan-Vese method, solvent-accessible surface, evolution, partial differential equation, finite difference, atomic energy.

1 Introduction

The representation and display of protein surfaces are useful in many areas of molecular modeling, and surface shape is particularly important in the analysis of protein-ligand and protein-protein interactions. There are several types of surfaces that are used in molecular modeling. The van der Waals surface is the surface area of the volume formed by placing van der Walls spheres at the center of each atom in a molecule. Many solvent accessibility programs use the surface defined by Richards that is computed by rolling a spherical water molecule probe over the surface [1]. The trajectory of the center of the solvent sphere defines the solvent-accessible surface. Whereas, the solvent-excluded surface is defined as the trajectory of the boundary of the solvent sphere in contact with the van der Waals surface.

Numerous methods have been developed to compute molecular surfaces. One of the earliest algorithms was proposed by Connolly [2]. A grid-based algorithm was described by Nicholls et al. [3] and used in the program GRASP [4]. Sanner et al. [5] developed a method that relies on the reduced surface or alpha shape for computing the molecular surfaces. An implementation of the method is used by the UCSF [6] Chimera molecular graphics program. Can et al. [7] proposed a level-set method to identify solvent-accessible surface and solvent-excluded surface.

In this paper, we developed a new surface generation method based on the active contour without edges model proposed by Chan [8] for image processing. Active contours are used to detect objects in a given image u_0 using techniques of curve

F. Chen and B. Jüttler (Eds.): GMP 2008, LNCS 4975, pp. 534–540, 2008.
© Springer-Verlag Berlin Heidelberg 2008

evolution. The Chan-Vese active contour model without edges use the Mumford-Shah segmentation techniques to find the boundary. This model has several advantages: it detects edges both with and without gradient; it automatically detects interior contours; the initial curve does not necessarily have to start around the objects to be detected and instead can be placed anywhere in the image; it gives in addition a partition of the image into two regions, the first formed by the set of the detected objects, while the second one gives the background; finally, there is no need for an a priori noise removal. In this paper, instead of generating a single solvent-accessible surface, we create a series of surfaces which describe different resolution details of molecular surface using different atomic energy functions. Our results indicate that the final surfaces we generated provide an alternative way to understand protein surface, and they are suitable for shape analysis (such as curvatures calculation) and visualization (such as Delaunay Triangulation).

2 Methods

In this section, we formulate the 3d active contour model in order to make our experiments repeatable, adopting the similar notations used in the Chan-Vese method.

The inputs to our method are the atomic coordinates of the 3d protein structures as a PDB file. The size of an atom is enlarged with the van der Waals radius of this atom, and the van der Walls radii are: C atom-1.7, H atom-1.2, N atom-1.55A, O atom-1.52 and S atom-1.85 (in Å). The union of the surfaces of atom spheres is known as the van der Walls surface. If inflate each atom by the van der Walls radius plus the probe radius (solvent radius), the union of the enlarged sphere surfaces is known as the solvent-accessible surface (SAS). So protein structure is treated as a set of spheres. The molecular structure is then placed on a 3d orthogonal grid with $M{\times}N{\times}P$ cells, which is determined by the size of this molecular. Now we can consider these $M{\times}N{\times}P$ cells as an "image", which is composed of the molecular structure cells and background cells (outside protein structure), and the boundary C_0 between these two regions is the protein surface to be detected.

Let $u_0(m,n,p)$ be the "image", which is formed by two regions of approximatively piecewise constant intensities, of distinct values u_0^i and u_0^o, defined in the set $\Omega = \{(m,n,p) \mid m = 1, 2, ..., M; n = 1, 2, ..., N; p = 1, 2, ..., P\}$.

Then we have $u_0(m,n,p) \approx u_0^i$ inside the protein (or inside C_0) and $u_0(m,n,p) \approx u_0^o$ outside the protein (or outside C_0). The "image" u_0 is defined as follows: If the set $s(m,n,p)$ is not empty, $u_0(m,n,p) = \sum_{i \in s(m,n,p)} E(i)$; else, $u_0(m,n,p) = 0$, where $E(i)$ is one kind of energy of atom i, and $s(m,n,p)$ is an atom set whose enlarged regions cover the grid point (m,n,p). There are three kinds of energies are used in this paper, which are Molecular Electrostatic Potential, Electric Field Energy and Polarization Energy. See Supplemental Material S3 for the definitions of these energy functions. Different forms of energy will generate different surfaces, especially, if E is a constant function, the surface generated will be the van der Walls surface or solvent-accessible surface (decided by the radii adopted).

2.1 3D Model of the Chan-Vese Method

Consider the "energy": $F_1(C)+F_2(C)=\int_{inside(C)}|u_0-u_1|^2dxdydz+\int_{outside(C)}|u_0-u_2|^2dxdydz$, where C is any other variable surface, and the constants c_1, c_2 depending on C, are the averages of u_0 inside C and respectively outside C. It is obvious that C_0 is the minimizer of the "energy": $\inf_C\{F_1(C) + F_2(C)\} = 0 \approx F_1(C_0) + F_2(C_0)$.

When adding some regularizing terms, like the area and/or the volume inside C, the energy is: $F(C,c_1,c_2) = \mu\cdot(area(C)) + \nu\cdot(volume(C)) + \lambda_1\int_{inside(C)}|u_0-c_1|^2dxdydz + \lambda_2\int_{outside(C)}|u_0-c_2|^2dxdydz$, where c_1 and c_2 are constants unknown, and μ, $\nu\geq0$, λ_1, $\lambda_2>0$ are fixed parameters. Therefore, we consider the minimization problem: $\inf_{C,c1,c2}F(C,c_1,c_2)$, which can be implemented using the level-set method. In the level-set method, C is represented by the zero level set of a Lipschitz function $\phi:R^3\to R$, such that: $C=\{x \in R^3:\phi(x)=0\}$, 'inside'$C=\{x \in R^3:\phi(x)>0\}$ and 'outside'$C=\{x \in R^3:\phi(x)<0\}$.

Using the Heaviside function H and the one-dimensional Dirac measure δ concentrated at 0, the terms in the energy are expressed in the following way: $F(\phi,c_1,c_2) = \mu\cdot(\int_\Omega\delta(\phi)|\nabla\phi|dxdydz) + \nu\cdot(\int_\Omega H(\phi)dxdydz) + \lambda_1\int_\Omega|u_0-c_1|^2H(\phi)dxdydz + \lambda_2\int_\Omega|u_0-c_2|^2(1-H(\phi))dxdydz$. Here, If $w\geq0$, $H(w) = 1$, else $H(w) = 0$; $\delta(w) = dH(w)/dz$.

To find the minimum, we need to consider the problem that the functional F is not Gateaux differentiable with respect to the third variable. The reason is simply that the Heavisible function is not differentiable. It is then classical to regularize the problem by changing H to H_τ and changing δ to δ_τ, such that $\delta_\tau = H_\tau'$: $H_\tau = 1/2(1+(2/\pi)\arctan(w/\tau))$, $\delta_\tau(w) = (1/\pi)\cdot(\tau/(\tau^2+w^2))$.

Denote the associated regularized F be the functional F_τ. So, to minimize F_τ with respect to ϕ, c_1 and c_2, we need to solve the equations

$$\partial\phi/\partial t = \delta_\tau(\phi)[div(\nabla\phi/|\nabla\phi|) - \nu - \lambda_1(u_0-c_1)^2 + \lambda_2(u_0-c_2)^2] \text{ in } \Omega,$$
$$c_1(\phi)=(\int_\Omega u_0H_\tau(\phi)dxdydz)/(\int_\Omega H_\tau(\phi)dxdydz),$$
$$c_2(\phi)=(\int_\Omega u_0(1-H_\tau(\phi))dxdydz)/(\int_\Omega(1-H_\tau(\phi))dxdydz),$$
$$\phi(t,x,y,z) = \phi_0(x,y,z), (\delta_\tau(\phi)/|\nabla\phi|)\cdot(\partial\phi/\partial n) = 0 \text{ in } \partial\Omega.$$

$$(1)$$

Notice that we do not need H_τ for calculating c_1 and c_2. These equations can then be implemented using standard finite differences described below. The proof will be found in the Supplemental Material S1.

2.2 The Numerical Approximation

Let h be the step space, and $(x_i, y_j, z_k)=(ih, jh, kh)$ be the gird points, for $1\leq i\leq M$, $1\leq j\leq N$, $1\leq k\leq P$. Let $\phi^n_{i,j,k}=\phi(n\Delta t, x_i, y_j, z_k)$ be an approximation of $\phi(t, x, y, z)$ with $\phi^0=\phi_0$. The centered differences are: $\Delta^x\phi^n_{i,j,k} = \phi^n_{i-1,j,k}-\phi^n_{i+1,j,k}$, $\Delta^y\phi^n_{i,j,k} = \phi^n_{i,j-1,k}-\phi^n_{i,j+1,k}$, $\Delta^z\phi^n_{i,j,k} = \phi^n_{i,j,k-1}-\phi^n_{i,j,k+1}$

Knowing ϕ^n, we first compute $c_1(\phi^n)$ and $c_2(\phi^n)$, then ϕ^{n+1} is computed by the following equation:

$$\phi^{n+1}_{i,j,k} = \phi^n_{i,j,k} +$$
$$\Delta t\delta_k(\phi^n_{i,j,k})[\mu\nabla\cdot(\nabla\phi^n_{i,j,k}/|\nabla\phi^n_{i,j,k}|)-\nu-\lambda_1(u_{0,i,j,k}-c_1(\phi^n))^2+\lambda_2(u_{0,i,j,k}-c_2(\phi^n))^2]$$

$$(2)$$

where: $\nabla \cdot (\nabla \phi^{n}_{i,j,k}/|\nabla \phi^{n}_{i,j,k}|) = \Delta^{x}(\Delta^{x}\phi^{n}_{i,j,k}/((\Delta^{x}\phi^{n}_{i,j,k})^{2}+(\Delta^{y}\phi^{n}_{i,j,k})^{2}+(\Delta^{z}\phi^{n}_{i,j,k})^{2})^{1/2})/(2h)+$
$\Delta^{y}(\Delta^{y}\phi^{n}_{i,j,k}/((\Delta^{x}\phi^{n}_{i,j,k})^{2}+(\Delta^{y}\phi^{n}_{i,j,k})^{2}+(\Delta^{z}\phi^{n}_{i,j,k})^{2})^{1/2})/(2h)+\Delta^{z}(\Delta^{z}\phi^{n}_{i,j,k}/((\Delta^{x}\phi^{n}_{i,j,k})^{2}+(\Delta^{y}\phi^{n}_{i,j,k})^{2}+$
$\Delta^{z}\phi^{n}_{i,j,k})^{2})^{1/2})/(2h)$.

Similar to the situation in the original Chan–Vese method, we need at each step to reinitialize ϕ to be the signed distance function to its zero-level surface, because we work with the regularized function δ_r. This procedure is standard [9], and prevents the level set function to become too flat, or it can be seen as a rescaling. See the Supplemental Material S2 for the details. The outline of the algorithm is:

(1) Initialization: $\phi^{0}=\phi_{0}$, $n=0$; (2) Compute $c_1(\phi^{n})$ and $c_2(\phi^{n})$ using formula (1); (3) Solve the PDE from formula (2), to obtain ϕ^{n+1}; (4) Re-initialization: replace ϕ^{n+1} by the signed distance function to $\{\phi^{n+1}\}$; (5) Check whether the solution is stationary. If not, $n=n+1$ and repeat.

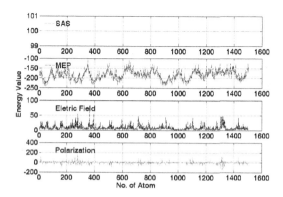

Fig. 1. The four energy distributions of 8DFR. From the top to the bottom, the energy functions are: the constant function with value 100, Molecular Electrostatic Potential function, Electric Field Energy function and Polarization Energy function. The last three energy functions used here are defined in Supplemental Material, and the constant function is corresponding to "SAS".

3 Results

We conducted three experiments to evaluate the performance and utility of our method. Recall that the protein structure is represented by a set of spheres. Here we show that the active contour model mentioned above can generate the solvent-accessible surface (however, it should be noted that the van der Waals surface can be generated in the similar way using our algorithm if we let the radius of probe sphere zero). The radius of probe sphere used in this paper is 1.4 Å (corresponding to the water molecule radius), and $\mu=300$, $\nu=1$, $\lambda_1=2$, $\lambda_2=2$, $t=0.1$, $h=1$, $u_0^i=1$, $u_0^o=0$. The initial surface C_0 is a sphere in the center of the protein with the radius related to the size of the protein. We choose three proteins for the experiments, which have diverse structures and in particular include all α (1ECD, Globins, 1048 atoms), all β (1PLC, Plastocyanin, 1547 atoms) and α/β (8DFR, Dihydrofolate Reductases, 1504 atoms) proteins.

In order to show that different energy functions will generate different surfaces, we adopt four kinds of energy functions for each protein: Constant Energy

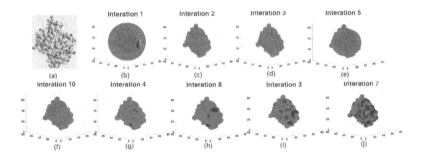

Fig. 2. (a) The spacefill rendering of the structure of 8DFR generated by RasWin. (b) The initial surface of iteration and the dot structure of 8DFR. The dot protein structure is marked by blue points and the dot surface is marked by red points. The size of the grid space is 79 by 79 by 83 and the radius of initial surface is 34. (c)~(d) The evolution of the surface of 8DFR_ based on the constant function: the generated surfaces in 2nd iteration and final (3rd) iteration respectively. (e)~(f) The evolution of the surface of 8DFR_ based on the Molecular Electrostatic Potential function: the generated surfaces in 5th iteration and final (10th) iteration respectively. (g)~(h) The evolution of the surface of 8DFR_ based on the Electric Field Energy function: the generated surfaces in 4th iteration and final (8th) iteration respectively. (i)~(j) The evolution of the surface of 8DFR_ based on the Polarization Energy function: the generated surfaces in 3rd iteration and final (7th) iteration respectively.

(corresponding to the SAS), Molecular Electrostatic Potential, Electric Field Energy and Polarization Energy. The CPU time is about 20~50s for one surface evolution in a 125×125×125 domain on a personal computer using MatLab. See Figures 1~4 and their legends for the results of 8DFR, and see the Supplemental Material S4-S5 for the results of the other two proteins.

Fig 1 shows the four energy distributions of 8DFR, and Fig 2 shows the generated surfaces based on these four kinds of energies and Fig 3 show the histograms of

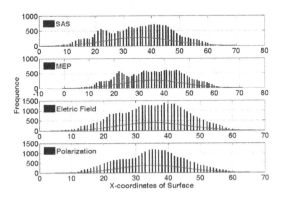

Fig. 3. The histograms of *x*-coordinates of the four generated surfaces. Here we plot the histograms of the values of the coordinate using a number of bins equal to the square root of the number of grid points on the surfaces and then superimpose the fitted normal distributions. The corresponding histograms of *y* and *z* corrdinates are similar and not showed here. From the figure, we can obviously see that the surfaces generated using different energy functions are different.

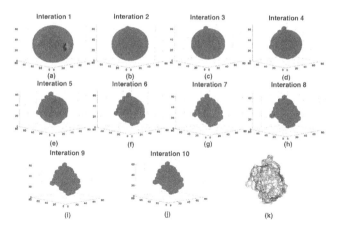

Fig. 4. The evolution of the surface of 8DFR_ based on the Molecular Electrostatic Potential function, which is the detail evolution sequence of Fig 2 (e)~(f). The surface is stationary in 10 iterations. (a)~(j): the generated surfaces from 1st iteration to last(10th) iteration respectively. The dot protein structure is marked by blue points and the dot surface is marked by red points. (k) The Delaunay Triangulation of the sub-fig (j) and the marks by curvature: mark the high curvature regions by red markers. The surface shapes showed in (j) and (k) maybe have some differences which are caused by the different view angles.

x-coordinates of the four generated surfaces. From these figures, we could conclude that different forms of energy produce different surfaces. Obviously, the classical SAS can be generated using active contour method, which is expected. Besides, maybe more important, the other three kinds of surfaces, which are associated with the energy inside the protein, provide an alternative choice for analyzing protein surface.

Fig 4 shows that we can generate the protein surface in a few of iterations, and different surfaces generated in different iterations represent the various resolution details of the surface. In fact, in particular iteration, it is only the shape of the protein structure itself determines what the sharp of the generated surface is. So, when the solvent-accessible surface is not enough to represent the features of protein structure, the evolution surfaces sequence may be an alterative choice. Besides, if the initial surface is smooth enough, the generated surfaces will preserve this property partly, because the evolution of the surfaces is controlled by the PDE.

From the figures, we can also conclude that the surfaces we generated are suitable for shape analysis and visualization. The characterization of molecular surface shape is useful in molecular modeling because shape complementarily is an important aspect of protein-protein interactions. In our implementation, surface features can be analyzed at different spatial resolution (the concept of spatial resolution is not part of classical surface analysis). The deep study of the relationships between the evolution surfaces and protein-protein (ligand) interactions is our ongoing work.

4   Conclusions

In this paper, we proposed a new protein surface modeling method, and did some experiments to show the efficiency of our PDE based model. However, the proposed active contour model is not mature, further study is in process. The evolution time should be decreased and the accuracy should be improved for more significant implementations. Code Optimization, C language programming and larger "image" will be helpful. Some potential improvements and implements are: (1) By using different energy function, the "cavities" (both empty and water-containing) within protein structures can be detected, which may have several important functions contributing to protein stability. (2) When protein molecular is endowed with some fields (such as potential energy field), we can "segment" the field to some similar domains.

References

1. Richards, F.M.: Areas, volumes, packing and protein structure. Annu. Rev. Biophys. Bioeng. 6, 151–176 (1977)
2. Connolly, M.L.: Analytical molecular surface calculation. J. Appl. Crystallogr. 548–558 (1983)
3. Nicholls, A., Sharp, K., Honig, B.: Protein folding and association: Insights from the interfacial and thermodynamic properties of hydrocarbons. Proteins 11(4), 281–296 (1991)
4. Nicholls, A.: GRASP: Graphical Representation and Analysis of Surface Properties. Columbia University, New York (1992)
5. Sanner, M.F., Olson, A.J., Spehner, J.C.: Reduced surface: An efficient way to compute molecular surfaces. Biopolymers. 38(3), 305–320 (1996)
6. Pettersen, E.F., Goddards, T.D., Huang, C.C., Couch, G.S., Greenblatt, D.M., Meng, E.C., Ferrin, T.E.: UCSF chimera–a visualization system for exploratory research and analysis. J. Comput. Chem. 25(13), 1605–1612 (2004)
7. Tolga, C., Chao-I, C., Yuan-Fang, W.: Efficient molecular surface generation using level-set methods. Journal of Molecular Graphics and Modeling 25, 442–454 (2006)
8. Chan Tony, F., Vese Luminita, A.: Active Contours Without Edges. IEEE Transactions on image progressing 10(2) (2001)
9. Sussman, M., Smereka, P., Osher, S.: A Level Set Approach for Computing Solutions to Incompressible Two Phase Flow. J. Comput. Phys. 119, 146–159 (1994)

Quasi-interpolation for Data Fitting by the Radial Basis Functions

Xuli Han and Muzhou Hou

School of Mathematical Sciences and Computing Technology,
Central South University, 410083 Changsha, China
xlhan@mail.csu.edu.cn

Abstract. Quasi-interpolation by the radial basis functions is discussed in this paper. We construct the approximate interpolant with Gaussion function. The suitable value of the shape parameter is suggested. The given approximate interpolants can approximately interpolate, with arbitrary precision, any set of distinct data in one or several dimensions. They can approximate the corresponding exact interpolants with the same radial basis functions. The given method is simple without solving a linear system. Numerical examples show that the given method is effective.

Keywords: interpolation, approximation, radial basis function.

1 Introduction

Consider a set of ordered pairs $(x_i, f_i), i = 0, 1, \cdots, n$, where $x_i \in \mathbb{R}^d$, $f_i \in \mathbb{R}, i = 0, 1, \cdots, n$. A radial basis function is a function

$$\phi_i(x) = \phi(a\|x - x_i\|),$$

which depends on the shape parameter a and the distance between $x \in \mathbb{R}^d$ and a fixed point $x_i \in \mathbb{R}^d$. Each function ϕ_i is radially symmetric about the center x_i. The interpolating RBF approximation is

$$F(x) = \sum_{i=0}^{n} c_i \phi_i(x), \tag{1}$$

where the expansion coefficients, c_i, are chosen so that

$$F(x_i) = f_i, \ i = 0, 1, \cdots, n.$$

That is, they are obtained by solving the linear system. For the RBFs that we have considered in this paper $\phi(t) = e^{-t^2}$, the interpolation matrix can be shown to be invertible for distinct points [1,2].

Despite the fact that the interpolation matrix can be shown to be invertible, the linear system may often be very ill-conditioned and it may be impossible to solve accurately using standard floating point arithmetic. The condition number

F. Chen and B. Jüttler (Eds.): GMP 2008, LNCS 4975, pp. 541–547, 2008.

of the interpolation matrix is influenced by the number of centers, the minimum
separation distance of the centers, as well as values of parameters.

The shape parameter affects both the accuracy of the approximation and
the conditioning of the interpolation matrix. Many researchers (e.g., [3,4]) have
attempted to develop algorithms for selecting optimal values of the shape parameter. By optimal choice of the shape parameter is still an open question.
Several different strategies see [5], have been somewhat successful in reducing
the ill-conditioning problem when using RBF methods in PDE problem. The
strategies include: variable shape parameters, domain decomposition, precondi-
tioning the interpolation matrix, and optimizing the center location. Often, more
than one of these strategies are used together.

When the number of samples for fitting functions is large, the interpolation
methods present the typical drawbacks of global methods, since each interpolated
value is influenced by all the data. Moreover, the numerical condition of the
interpolation matrix heavily depends on the data density and the smoothness
of the radial basis functions. This leads to unstable solutions or unacceptable
computational costs, see [6]. Compactly supported positive definite radial basis
functions have been introduced to overcome these problems and to provide local
methods, see [7,8] Among local methods, the method given in [9,10] seems to
match both requirements of efficiency and reproduction quality.

Quasi-interpolation, one of approximation method, possesses some advantages, such as less computation time. In [11], [12] and [13], some methods are
discussed for the approximation of a function with a neural network whose ac-
tivation function is sigmoidal. In [14], the univariate quasi-interpolants are discussed. In this paper we give a method for approximate interpolation. We show
a constructive method for obtaining a family of approximate interpolation.

2   Quasi-interpolants

A quasi-interpolant is constructed by letting c_i in (1) are some given finite linear
combinations of f_j. We consider a simple interpolant of type

$$s(x) := \sum_{i=0}^{n} f_i \varphi_i(x), \tag{2}$$

where

$$\varphi_i(x) = \frac{\phi(a_i \|x - x_i\|)}{\sum_{j=0}^{n} \phi(a_j \|x - x_j\|)}, \quad a_i = \frac{\sqrt{\lambda}}{\min_{k \neq i} \|x_k - x_i\|}.$$

Here, we denote by λ a shape parameter.

Remark. If we consider

$$g(x) = f_{i-1}\phi_{i-1}(x) + f_i\phi_i(x) + f_{i+1}\phi_{i+1}(x),$$

with $x_{i+1} - x_i = x_i - x_{i-1} = h > 0$, then $g'(x_i) = (f_{i+1} - f_{i-1})/(2h)$ when
$e^\lambda = 4\lambda$. This leads to $\lambda \approx 2.153292$ and $\lambda \approx 0.357403$.

Therefore, for the shape-preserving property, we may choose $\lambda \approx 2$ for suitable
shape properties of approximation.

3   Approximability

Corresponding to the quasi-interpolant (2), we consider interpolant as follows

$$s_0(x) := \sum_{i=0}^{n} c_i \varphi_i(x), \tag{3}$$

where the expansion coefficients, c_i, are chosen so that

$$s_0(x_i) = f_i, \; i = 0, 1, \cdots, n. \tag{4}$$

The system (4) can be written as an $(n+1) \times (n+1)$ linear system in vectorial form

$$Mc = f,$$

where $c = (c_0, c_1, \cdots, c_n)^T$, $f = (f_0, f_1, \cdots, f_n)^T$, and the elements of the interpolation matrix are

$$m_{ij} = \varphi_j(x_i), \; i, j = 0, 1, \cdots, n.$$

If the matrix M is invertible, then the coefficients are obtained by solving the linear system.

Theorem 1. If $\lambda > \ln(n)$, then the matrix M is invertible.

Proof. The matrix can be written as $M = D^{-1}B$,

$$D = diag \left(\sum_{j=0}^{n} b_{0,j}, \sum_{j=0}^{n} b_{1,j}, \cdots, \sum_{j=0}^{n} b_{n,j} \right).$$

The entries of the matrix B are

$$b_{ij} = \phi(a_j \|x_i - x_j\|) = e^{-a_j^2 \|x_i - x_j\|^2}.$$

From this we obtain $b_{ii} = 1$, $m_{ij} \le e^{-\lambda} (i \ne j)$. Thus

$$\sum_{j \ne i} |b_{ij}| = \sum_{j \ne i} b_{ij} \le n e^{-\lambda}.$$

If $\lambda > \ln(n)$, then

$$\sum_{j \ne i} |b_{ij}| < |b_{ii}|, \; i = 0, 1, \cdots, n.$$

This implies that matrix B is a strictly diagonally dominant matrix, and then matrix M is invertible. □

The estimation in Theorem 1 is conservative. Specially, we have the following result.

Theorem 2. Let $x_i \in \mathbb{R}, x_{i+1} - x_i = h$ for all nodes. If $\lambda \ge \ln(3)$, then the matrix M is invertible.

Proof. For the entries of the matrix B, we have

$$\sum_{j \ne i} |b_{ij}| = \sum_{j \ne i} e^{-\lambda |i-j|^2} < \frac{e^{-\lambda}(2 - e^{-\lambda i} - e^{-\lambda(n-i)})}{1 - e^{-\lambda}}.$$

Therefore, If $\lambda \geq \ln(3)$, then

$$\sum_{j \neq i} |b_{ij}| < 1 = b_{ii}.$$

This implies the theorem.   □

Since $\ln(3) \approx 1.098612$, we may choose $\lambda \approx 2$ for the two factors of approximability and shape-preserving property.

We note that $s_0(x)$ and $s(x)$ differ only in the expansion coefficients and $s_0(x_i) = f_i$ for $i = 0, 1, \cdots, n$. We will prove that $s(x)$ is an approximate interpolant that is arbitrarily near of the corresponding $s_0(x)$ when the parameter λ increases.

Theorem 3. If $\lambda > \ln(n)$, then

$$|s(x) - s_0(x)| < \frac{2ne^{-\lambda}}{1 - ne^{-\lambda}} \|f\|_\infty. \tag{5}$$

Proof. For $\lambda > \ln(n)$, the matrix M is invertible. We can get c by solving the linear system (4). Thus, we have

$$c - f = (I - M)c = (I - M)(c - f) + (I - M)f,$$

where I is an unit matrix. Therefore

$$\|c - f\|_\infty \leq \frac{\|I - M\|_\infty}{1 - \|I - M\|_\infty} \|f\|_\infty.$$

From (4) , we have

$$\|I - M\|_\infty = \max_{0 \leq i \leq n} 2\left(1 - \frac{1}{\sum_{j=0}^n b_{ij}}\right)$$

$$= \max_{0 \leq i \leq n} \frac{2\sum_{j \neq i} b_{ij}}{1 + \sum_{j \neq i} b_{ij}} \leq \frac{2ne^{-\lambda}}{1 + ne^{-\lambda}}.$$

For $\lambda > \ln(n)$, we have

$$\frac{2ne^{-\lambda}}{1 + ne^{-\lambda}} < 1.$$

Then, considering that the function $g(t) = t/(1 - t)$ is strictly increasing on $(-\infty, 1)$, it follows that

$$\|c - f\|_\infty \leq \frac{2ne^{-\lambda}}{1 - ne^{-\lambda}} \|f\|_\infty.$$

On the other hand

$$|s(x) - s_0(x)| = |\sum_{i=0}^n (c_i - f_i)\varphi_i(x)|$$

$$\leq \|c - f\|_\infty \sum_{i=0}^n \varphi_i(x) = \|c - f\|_\infty.$$

From this we obtain (5) immediately.   □

4   Numerical Examples and Conclusions

Example 1. We consider the approximate interpolation corresponding to the function $f(x) = e^{-(x-3)^2} \sin(x)$. Figure 1 shows the curves (dashed lines) of the interpolant (2) with the dimension $d = 1$ and $\lambda = 2$.

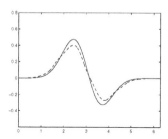 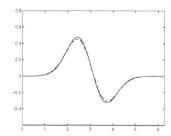

Fig. 1. Approximate interpolation curves with $n = 10$ (left) and $n = 20$ (right)

Example 2. We consider the approximate interpolant (2) corresponding to the function

$$f(x, y) = \sin(\sqrt{x^2 + y^2})/\sqrt{x^2 + y^2} \tag{6}$$

on the 64 points $S = \{(i, j) : -8 \leq i \leq 8, -8 \leq j \leq 8\}$. We have set $error = \max_{x \in S} |f(x) - s(x)|$. The results are given in Table 1. Figure 2 shows the different surfaces of the approximate interpolant for the function (6) with $\lambda = 2$ (left) and $\lambda = 5$ (right).

Table 1. The approximate errors for the function (6)

λ	1	2	6	12	15
error	0.4322	0.3030	0.0660	0.0037	8.2187×10^{-4}

Example 3. We consider the approximate interpolant (2) corresponding to the function

$$f(x, y) = ye^{-x^2 - y^2} \tag{7}$$

on the 64 points $S = \{(i, j) : -3 \leq i \leq 3, -3 \leq j \leq 3\}$. The results are given in Table 2. Figure 3 shows the different surfaces of the approximate interpolant for the function (7) with $\lambda = 2$ (left) and $\lambda = 5$ (right).

Stated numerical examples above show that the given method is effective. With the Gaussian function, the given approximate interpolant can approximately interpolate, with arbitrary precision, any set of distinct data in one or several dimensions without solving a linear system. It can approximate the corresponding exact interpolant with the same radial basis functions. The suitable value of the shape parameter is suggested. We give a rigorous bound of the interpolation error.

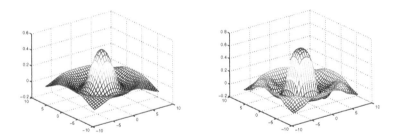

Fig. 2. Approximate interpolation surfaces for the function (6)

Table 2. The approximate errors for the function (7)

λ	1	2	5	10	15
error	0.2358	0.1765	0.0619	0.0061	5.0486×10^{-4}

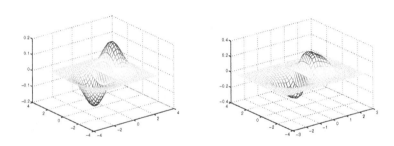

Fig. 3. Approximate interpolation surfaces for the function (7)

For the two factors of approximability and shape-preserving property, the suitable value of the shape parameter is suggested. Compare with [13], the given approximant is a simple explicit expression and effective for multivariate functions.

References

1. Micchelli, C.: Interpolation of scattered data: Distance matrices and conditionally positive definite function. Constr. Approx. 2, 11–22 (1986)
2. Wendland, H.: Piecewise polynomial, positive definite and compactly supported radial functions of minimal degree. Adv. Comput. Math. 4, 389–395 (1995)
3. Carlson, R.E., Foley, T.A.: The parameter in multiquadric interpolation. Comput. Math. Appl. 21, 29–42 (1991)
4. Rippa, S.: An algorithm for selecting a good parameter c in radial basis function interpolation. Adv. Comput. Math. 11, 193–210 (1999)
5. Kaansa, K., Hon, Y.C.: Circumventing the ill-conditioning problem with multiquadric radial basis function: Applications to elliptic partial differential equations. Comput. Math. Appl. 39, 123–137 (2000)

6. Schaback, R.: Creating surfaces from scattered data using radial basis function. In: Daehlen, M., Lyche, T., Schumacker, L. (eds.) Mathematical Methods for curve and Surfaces, pp. 477–496. Vanderbilt University Press, Nashville (1995)
7. Wu, Z.: Multivariate compactly supported positive definite radial functions. Adv. Comput. Math. 4, 283–292 (1995)
8. Wendland, H.: Piecewise polynomial, positive definite and compactly supported radial basis functions of minimal degree. Adv. Comput. Math. 4, 359–396 (1995)
9. Lazzaro, D., Montefusco, L.B.: Radial basis functions for the multivariate interpolation of large scattered data sets. J. Comput. Appl. Math. 140, 521–536 (2002)
10. Davydov, O., Morandi, R., Sestini, A.: Local hybrid approximation for scattered data fitting with bivariate splines. Computer Aided Geometric Design 23, 703–721 (2006)
11. Debao, C.: Degree of approximation by superpositions of a sigmoidal function. Approx. Theory & its Appl. 9, 17–28 (1993)
12. Mhaskar, H.N., Michelli, C.A.: Approximation by superposition of sigmoidal and radial basis functions. Adv. Appl. Math. 13, 350–373 (1992)
13. Lianas, B., Sainz, F.J.: Constructive approximate interpolation by neural networks. J. Comput. Appl. Math. 188, 283–308 (2006)
14. Zhang, W., Wu, Z.: Shape-preserving MQ-B-Splines quasi-interpolation. In: Proceedings Geometric Modeling and Processing, pp. 85–92. IEEE Computer Society Press, Los Alamitos (2004)

A Shape Feature Based Simplification Method for Deforming Meshes

Shixue Zhang[1] and Enhua Wu[1,2]

[1] Dept. of Computer and Information Science, University of Macau, Macao, China
[2] State Key Lab. of Computer Science, Institute of Software,
Chinese Academy of Sciences, China
{ya57406,EHWu}@umac.mo

Abstract. Although deforming surfaces are frequently used in numerous domains, only few works have been proposed until now for simplifying such data. In this paper, we propose a new method for generating progressive deforming meshes based on shape feature analysis and deformation area preservation. By computing the curvature and torsion of each vertex in the original model, we add the shape feature factor to its quadric error metric when calculating each QEM edge collapse cost. In order to preserve the areas with large deformation, we add deformation degree weight to the aggregated quadric errors when computing the unified edge contraction sequence. Finally, the edge contraction order is slightly adjusted to further reduce the geometric distortion for each frame. Our approach is fast, easy to implement, and as a result good quality dynamic approximations with well-preserved fine details can be generated at any given frame.

Keywords: Deforming mesh, LOD, Mesh simplification, Shape feature.

1 Introduction

In computer graphics and virtual reality, polygonal mesh is most commonly used to construct three-dimensional models. Especially, more and more time-varying surfaces, which are also called deforming meshes are frequently used from scientific applications (simulation) to animation (movie, games). In many cases, some of the details of them might be unnecessary especially when viewing from a distance. Mesh simplification algorithms are investigated more often with a single static mesh, but very little work has been proposed to address the problem of accurate approximation for time-varying surfaces. In this paper, we propose a new method for generating progressive deforming meshes.

Our method is based on shape feature analysis and deformation area preservation. And it is a better tradeoff between the temporal coherence and geometric distortion, i.e. we try to maximize the temporal coherence while minimizing the visual distortion during the simplification process. We propose the use of shape feature which is calculated using curvature and torsion of the surface, to improve the quadric error metric of edge collapse operation. A deformation weight is added to the aggregated edge contraction cost, so areas with large deformation can be preserved. Finally, we

F. Chen and B. Jüttler (Eds.): GMP 2008, LNCS 4975, pp. 548–555, 2008.

adjust the unified edge collapse sequence for each frame to further reduce geometric distortion. We demonstrate that this provides an efficient means of multiresolution representation of deforming meshes over all frame of an animation.

The rest of this paper is organized as follows: Section 2 will review the previous works in the related fields. Section 3 will mainly introduce the procedure of our algorithm and discuss its advantage. Section 4 will show the experimental results. Finally, conclusion and some future work will be given in Section 5.

2 Related Work

Simplification and LOD. The mesh simplification methods can be roughly divided into five categories: vertex decimation [21], vertex clustering [12], region merging [3], subdivision meshes [11], and iterative edge contraction [4, 5, 6, 7, 8, 9]. A complete review of the methods has been given in [2, 13, 16].

Shape error metric. Yan [9] proposed a new method based on hierarchical shape analysis. Sun [18] described a mesh simplification algorithm using a discrete curvature norm. Bin-Shyan [1] proposed a new method that uses torsion detection to improve the quadric error metric.

Approximation of time-varying surfaces. Shamir et al. [17, 19] designed a global multiresolution structure named Time-dependant Directed Acyclic Graph (TDAG) which merges each individual simplified model of each frame into a unified graph. Unfortunately this scheme is very complex, and can not be easily handled. Mohr and Gleicher [14] proposed a deformation sensitive decimation (DSD) method, which directly adapt the Qslim algorithm [4] by summing the quadrics errors over each frame of the animation. Kircher and Garland [10] proposed a multiresolution representation with a dynamic connectivity for deforming surfaces. This method seems to be particularly efficient because of its connectivity transformation. Recently, Huang et al. [5] proposed a method based on the deformation oriented decimation and dynamic connectivity update. They used vertex tree to further reduce geometric distortion by allowing the connectivity to change.

3 Our Algorithm

Our algorithm consists of three parts: (1) Use the shape feature factor to improve the quadric error metric. (2) Add the deformation information when computing the unified edge collapsing sequence. (3) Adjust the edge collapse sequence for each frame to further reduce the approximation distortion.

3.1 Shape Feature Definition

According to the definition of [15], we can express the shape feature of a vertex as following form:

$$S(v) = -\nabla_V U$$

In this equation, U is unit normal field which is composed of normal of the vertex v and its neighbor vertices, $\nabla_V U$ is the gradient along with curve V. Further, for calculation of shape feature, we need the Frenet-Serret Formulas [20]:

On the curve $r(s)$ belonging to category C^3 based on the parameter of arc s, the derivatives of three unit vectors t, n, and b on each regular point, we have the following equations:

$$\begin{pmatrix} t' \\ n' \\ b' \end{pmatrix} = \begin{pmatrix} 0 & k & 0 \\ -k & 0 & \tau \\ 0 & -\tau & 0 \end{pmatrix} \cdot \begin{pmatrix} t \\ n \\ b \end{pmatrix}$$

where t is the tangent vector of the curve, n is principle normal vector, b is binormal vector, k is the curvature of a point on the curve and τ is the torsion.

On the curve $r(s)$ belonging to category C^3 based on the parameter of arc s, each regular point on the curve owns one torsion τ and one curvature k:

$$k = \left\| \frac{dt}{ds} \right\| = \|t'\|, \ \tau = \left\| \frac{db}{ds} \right\| = \|b'\|$$

We can rewrite $S(v)$ as following form:

$$S(v) = \|-k \cdot t + \tau \cdot b\| = \sqrt{k^2 + \tau^2}$$

In this equation, the torsion [1] of each regular point on the curve records the binormal vector variation of the points while the curvature of the point records the normal vector variation.

3.2 Estimating Shape Feature

Some papers define the normal vector of vertex v_i as the normalized weight sum of normals of the incident faces, with weights proportional to the surface areas of the faces. But if the shapes of the incident triangles are largely different as **Fig. 1** shows, only the area can't represent this difference. So we have the equation:

$$N_{v_i} = \frac{\sum\limits_{f_k \in F^i} \gamma_k A_{f_k} N_{f_k}}{\left\| \sum\limits_{f_k \in F^i} \gamma_k A_{f_k} N_{f_k} \right\|}$$

where f_k is face that is incident to v_i, N_{fk} is the normal of the face of face f_k, A_{fk} is the area of the face f_k, γ_k is one of the internal angle of f_k which is adjacent to v_i.

The tangent and binormal of vertex v_i on the curve (v_i, v_j) can be calculated by:

$$T_{ij} = \frac{(I - N_{v_i} N_{v_i}^t)(v_i - v_j)}{\left\| (I - N_{v_i} N_{v_i}^t)(v_i - v_j) \right\|}, \ B_{ij} = T_{ij} \otimes N_{v_i}$$

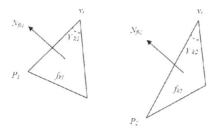

Fig. 1. Triangles adjacent to vertex v_i with same areas and normal vectors, but different shapes

According to the above equation, the curvature and torsion of v_i can be obtained as:

$$k_{v_i} = \frac{\sum_{j=1}^{k} \|v_j - v_i\| \cdot k_{ij}}{\sum_{j=1}^{k} \|v_j - v_i\|} \, , \, \tau_{v_i} = \frac{\sum_{j=1}^{k} \|v_j - v_i\| \cdot \tau_{ij}}{\sum_{j=1}^{k} \|v_j - v_i\|}$$

where k_{ij} and τ_{ij} are respectively the curvature and torsion on the curve (v_i, v_j), thus the shape feature is:

$$S(v) = \sqrt{k_{v_i}^2 + \tau_{v_i}^2}$$

We add this shape feature factor to the vertex's quadric error metric computation: $Q(v_i) = S(v_i) \cdot \sum k_p$, where P denote the triangle plane that contains the point v_i, and k_p represents 4*4 metric of this plane[4].

3.3 Deformation Measurement

The deformation sensitive decimation (DSD) algorithm addresses this problem by summing QEM costs across all frames [14]. We add an additional deformation weight to the DSD cost which is measured by the change of frame-to-frame edge collapse cost. For areas with large deformation, the change of the collapsing cost must be prominent, while collapsing cost may change slightly in areas with small deformation. We append this additional deformation weight to the DSD cost:

$$cost\_final_{ij} = DSD_{ij} + k_1 * \sum_{t=1}^{f} \left| ec_{ij}^t - \overline{ec}_{ij} \right| = \sum_{t=1}^{f} ec_{ij}^t + k_1 * \sum_{t=1}^{f} \left| ec_{ij}^t - \overline{ec}_{ij} \right|$$

where ec_{ij}^t is the collapse cost of edge (v_i, v_j) in frame t, $\overline{ec}_{ij}$ is the average collapse cost of edge (v_i, v_j) over all of the frames, and k_1 is a user-specified coefficient to adjust the influence the deformation degree. In our experiment, we set k_1 to around 1.

3.3 Distortion Adjustment

Based on the final collapse cost described above, we can finally get the overall edge-collapse sequence. This unified collapse sequence can maintain the connectivity of the whole mesh sequence unchanged. However, the approximations can't be very satisfying on every frame. So our algorithm finally adjusts this sequence slightly for each frame to get better approximation and preserve more features.

For a certain frame, the independent QEM method can generate the optimal edge-collapse sequence. So we can first do a high percentage (μ) collapse operations based on the unified sequence, then do the rest collapse operation based on the shape feature embedded QEM cost. Since most of the operation is based on static connectivity, the temporal coherence can still be maintained. We have tested different μ during the animation, and we found that set μ to around 95% can generate elegant result.

3.4 Algorithm Outline

Our algorithm can be summarized into the following steps:

1. Calculate the shape feature factor for all the vertices.
2. Use the shape feature to calculate the quadric error metric of each vertex.
3. Measure the deformation degree to obtain the final edge collapsing cost.
4. For each frame, slightly adjust the edge contraction order.
5. Iteratively perform the edge collapse operation until the desired resolution is reached.

Fig. 2. Horse-gallops animation simplification with 48 frames. Original sequence: 8431v (upper), simplified models: 800v (bottom).

4 Experimental Results

We test the result of our algorithm on a computer with Pentium4 3.2G CPU and 2G memory. The simplification of the horse gallop animation is shown in the **Fig. 2**. Compared to original models on the upper row, the bottom shows our simplified results when 90% of its components are reduced. We could see that most of the features in the original models are preserved. **Fig. 3** shows another example on a facial-expression animation. Even after removing 95% of the vertices, the simplified meshes can still be rendered faithfully to the rendering of the original ones.

We compared our method with the [10]'s method in **Fig. 4**. The left is [10]'s result of the last frame in the horse-to-human morphing sequence, while the right is

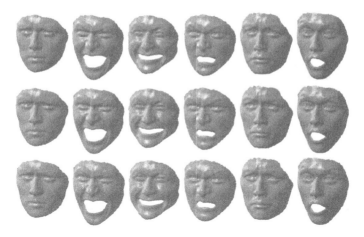

Fig. 3. A facial expression animation with 192 frames. Top: The original sequence with 23725 vertices and 46853 faces. Middle: 3201-vertices and 5825-faces approximation(80% reduced) using our method. Bottom: 1200-vertices and 1874-faces approximation(95% reduced).

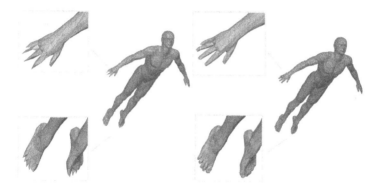

Fig. 4. Comparison of [10]'s method (left) and our method (right) applied on the horse_to_human morphing sequence(we only show the last frame), the approximated versions contains 3200 vertices and 6396 triangles.

our result. By enlarging the detail of some human features, we could see that result preserves the hands and feet better.

5   Conclusion and Future Work

In this paper, we propose a simplification method for deforming surfaces based on shape feature analysis and deformation area preservation. Given a sequence of meshes representing time-varying 3D data, our method produces a sequence of simplified meshes that are good approximation of the original deforming surface for a given frame. Our method is easy to implement and can produce better approximations than other previous method.

There are certainly further improvements that could be made to our algorithm. For example, we believe that there must be a way to extend our algorithm to be view-dependent.

Acknowledgments. We should give thanks to the anonymous reviewers for their valuable comments. We also wish to thank Scott Kicher for providing the horse-to-man sequence. The work has been supported by the Studentship & Research Grant of University of Macau.

References

[1] Bin-Shyan, J., Juin-Ling, T., Wen, H.: An Efficient and Low-error Mesh Simplification Method Based on Torsion Detection. The Visual Computer 22(1), 56–67 (2006)

[2] Garland, M.: Multiresolution Modeling: Survey & future opportunities. In: Proceedings of Eurographic 1999, Milano, pp. 49-65 (1999)

[3] Garland, M., Willmott, A., Heckbert, P.S.: Hierarchical Face Clustering on Polygonal Surfaces. In: Proc. ACM Symp. Interactive 3D Graphics, pp. 49–58 (2001)

[4] Garland, M., Heckbert, P.S.: Surface Simplification using Quadric Error Metrics. In: ACM SIGGRAPH 1997 Conference Proceedings, pp. 209–216 (1997)

[5] Huang, F.C., Chen, B.Y., Chuang, Y.Y.: Progressive Deforming Meshes Based on Deformation Oriented Decimation and Dynamic Connectivity Updating. In: Proceedings of ACM SIGGRAPH/Eurographics Symposium on Computer Animation, pp. 53–62 (2006)

[6] Hoppe, H.: Progressive meshes. In: ACM SIGGRAPH 1996 Conference Proceedings, pp. 99–108 (1996)

[7] Hoppe, H.: View-dependent Refinement of Progressive Meshes. In: ACM SIGGRAPH 1997 Conference Proceedings, pp. 189–198 (1997)

[8] Hoppe, H.: New Quadric Metric for Simplifying Meshes with Appearance Attributes. In: Proc. IEEE Visualization 1999, pp. 59–66 (1999)

[9] Jingqi, Y., Pengfei, S.: Mesh Simplification with Hierarchical Shape Analysis and Iterative Edge Contraction. IEEE Tran. on Visual. and Computer Graphics 10(2), 142–151 (2004)

[10] Kircher, S., Garland, M.: Progressive Multiresolution Meshes for Deforming Surfaces. In: Proceedings of ACM SIGGRAPH/Eurographics Symposium on Computer Animation, pp. 191–200 (2005)

[11] Lee, A., Moreton, H., Hoppe, H.: Displaced Subdivision Surfaces. In: SIGGRAPH 2000 Conference Proceedings, pp. 85–94 (2000)

[12] Low, K.L., Tan, T.S.: Model Simplification Using Vertex-Clustering. In: Proc. ACM Symp. Interactive 3D Graphics, pp. 75–82 (1997)

[13] Luebke, D., Reddy, M., Cohen, J.: Level of Detail for 3-D Graphics. Morgan Kaufmann, San Francisco (2002)

[14] Mohr, A., Gleicher, M.: Deformation Sensitive Decimation. Tech. rep., University of Wisconsin (2003)

[15] Neill, B.O.: Elementary Differential Geometry. Academic Press, London (1997)

[16] Oliver, M., van, K., Hélio, P.: A Comparative Evaluation of Metrics for Fast Mesh Simplification. Computer Graphics Forum 2006 25, 197–210 (2006)

[17] Shamir, A., Bajaj, C., Pascucci, V.: Multi-resolution Dynamic Meshes with Arbitrary Deformations. In: IEEE Visualization 2000 Conference Proceedings, pp. 423–430 (2000)

[18] Sun-Jeong, K., Chang-Hun, K., Levin, D.: Surface Simplification Using a Discrete Curvature Norm. Computers&Graphics 26(5), 657–663 (2002)
[19] Shamir, A., Pascucci, V.: Temporal and Spatial Level of Details for Dynamic Meshes. In: Proceedings of ACM Symp. on Virtual Reality Software and Technology, pp. 77–84 (2001)
[20] Spivak, M.: A Comprehensive Introduction to Differential Geometry, 3rd edn., vol. 3. Publish or Perish (1999)
[21] Schroeder, W.J., Zarge, J.A., Lorensen, W.E.: Decimation of Triangle Meshes. In: ACM Computer Graphics (SIGGRAPH 1992 Conference Proceedings), vol. 26(2), pp. 65–70 (1992)

Shape Representation and Invariant Description of Protein Tertiary Structure in Applications to Shape Retrieval and Classification

Dong Xu[1,2], Hua Li[1,2], and Tongjun Gu[3]

[1] Key Lab. of Computer System and Architecture, Chinese Academy of Sciences
[2] National Research Center for Intelligent Computing Systems, Institute of Computing Technology, Chinese Academy of Sciences
[3] Center for Systems Biology, Institute of Biophysics, Chinese Academy of Sciences
xudong@ict.ac.cn, lihua@ict.ac.cn, gutongjun@ncic.ac.cn

Abstract. Each protein tertiary structure is represented by three sorts of geometric models, which are polyline curves, triangulated surfaces and volumetric solids. Moment invariants are employed to describe the shapes of the three kinds of protein models and form a multidimensional feature vector for each model. We further analyze the influence of the three representations in applications to protein shape retrieval and classification.

Keywords: triangulated surface, shape retrieval, moment invariants.

1 Introduction

There are more than 48,000 protein structures in the Protein Data Bank (PDB) [1] by Jan. 2008, which are created with the technologies of X-ray crystallography and nuclear magnetic resonance spectroscopy. However, the totally unique structures are no more than 5,000. Hence, it is worthwhile to find shape similarities between proteins and classify them correctly.

Protein structures have been grouped together in several well-known databases. SCOP (**S**tructural **C**lassification **of P**roteins) [2] classifies proteins to reflect both structural and evolutionary relatedness. The principal levels are family, superfamily and fold. CATH [3] is a hierarchical classification of protein domain structures, which clusters proteins at four major levels, Class(**C**), Architecture(**A**), Topology(**T**) and Homologous superfamily (**H**). The assignments of structures to fold groups and homologous superfamilies are made by sequence and structure comparisons. FSSP (**F**amilies of **S**tructurally **S**imilar **P**roteins) [4] classification scheme is based on an exhaustive pairwise alignment of protein structures.

To an unknown protein structure, we may want to retrieve the most similar ones quickly. Protein structure retrieval systems have already appeared recently [5]. They use coefficients of Zernike moment and Fourier transform as invariant features.

In this paper, we introduce the three kinds of macromolecular surfaces and represent each protein tertiary structure by 3D polyline curve, triangulated surface and volumetric solid in part 2. Moment invariants are then used as global

F. Chen and B. Jüttler (Eds.): GMP 2008, LNCS 4975, pp. 556–562, 2008.

shape descriptors in applications to protein shape retrieval and structure classification in part 3. Conclusions and future work are given in part 4.

2   Shape Representation and Invariant Description of Protein Tertiary Structure

Protein tertiary structure contains coordinate information of all the atoms. Each atom has its own atomic mass and van der Waals radius. Protein structure can be represented by polyline curve, triangulated surface and volumetric solid. The alpha-carbon (Cα) backbone of each protein is a point set which contains the 3D coordinates of all the Cα atoms. Since the backbone also has a sequential order, it could be connected to form a 3D polyline curve. Some methods have made use of the coordinate information and atomic weights in the periodic table for structure classification and shape retrieval. Protein surfaces are very important because they reflect the interactions within protein complexes or between a protein and its surroundings. The part which is wrapped in the protein surface forms a solid. Next, we will discuss how to generate them in detail.

2.1   Macromolecular Surfaces Generation

The definitions of the three macromolecular surfaces are given in [6]. The van der Waals surface is the topological boundary of a set of overlapping spheres, each having a van der Waals radius. The solvent-accessible surface is defined as the area traced out by the center of a probe sphere as it is rolled over the van der Waals surface. The molecular surface is also called the solvent-excluded surface, which is the boundary of the union of all possible probes which do not overlap with the molecule. We generate the three surfaces by using Euclidean distance transform and vertex-connected marching cube. The three surfaces are shown in Fig. 1 (a), (b) and (c). The voxels covered by the molecular surface is in Fig. 1(d).

All of the three kinds of surfaces have been used for shape representation and feature extraction. Van der Waals surface in [7] and solvent-accessible surface in [5] are extracted and each protein is treated as a 3d model for shape retrieval.

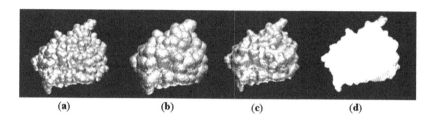

(a)   (b)   (c)   (d)

Fig. 1. (a) van der Waals surface of 1aaj, (b) solvent-accessible surface of 1aaj, (c) molecular surface of 1aaj, (d) volumetric solid of 1aaj covered by the molecular surface

2.2   A Secondary Structure Database

Each protein structure in the PDB may contain several domains. The number of total residues also alters greatly, from dozens to tens of thousands. Hence, each protein is divided according to its domains before comparison. The original database we use is from the protein challenge track of SHREC 2007 [8] which contains 647 test structures and 30 query structures. Each of the test structure is a single domain of a protein structure in the PDB. SCOP classification information of them is also given. For example, the hierarchical SCOP id for 1aaj is b.6.1.1, where b, 6, 1, 1 stand for the names or numbers of class, fold, superfamily, family separately.

We build a secondary database which contains the three kinds of representations of protein tertiary structures based on this primary database. For each protein structure in the database, we extract its $C\alpha$ backbone and save the 3D coordinates of the $C\alpha$ atoms and line segments to wrl format. Both the triangulated molecular surface and the volumetric solid which is enveloped in the molecular surface are saved in off format. The molecular surface contains the 3D coordinates of the vertices and the list of all the triangular patches. The volumetric solid in off file contains the 3D coordinates of all the voxels. The contents of the three kinds of files can be opened as text files and the 3D models they stand for can be viewed intuitively by many 3D model viewers.

Van der Waals surface and solvent-accessible surface which are much easier to construct are not considered here. Molecular surface plays a key role in protein functional analysis, so it is more useful to further research.

This secondary database is available upon request.

2.3   Moment Invariants

Feature extraction is a necessary step for shape comparison and recognition. Since protein structures may have different orientations in 3D space, it is better to extract features which are independent of rotation and translation. Principal Component Analysis (PCA) could be used for pose normalization. However, proteins with different sizes and conformations cannot be always transformed to the same orientation by PCA.

Furthermore, we need some shape descriptors which could describe all the three kinds of geometric models. Moment invariants which don't suffer from these problems will be used as shape descriptors in this paper. The generation of moment invariants for 3D solids is introduced in [9]. The definition of moment invariants also could be extended from solid to surface and curve. In the next part, we will use 54 volume moment invariants (VMIs), surface moment invariants (SMIs), curve moment invariants (CMIs) to represent volumetric solid, triangulated molecular surface and protein backbone separately. All the moment invariants are less than fifth order which are validated to be very robust to noise and distortion.

3   Retrieval and Classification of Protein Structures

Some methods use exhaustive structure alignment algorithms to compare each pair of proteins. It is accurate but very time-consuming for we need to compare the query protein with all the proteins in the database during the query process. If we use moment invariants as global shape descriptors, the feature vectors of all the test proteins in the database could be calculated offline. Then, we only need to compare the feature vectors between proteins to avoid pairwise alignment.

3.1   Protein Shape Retrieval

The feature vector of each shape representation of protein structure is extracted by using its corresponding moment invariants. We use Euclidean distance to measure the similarity of every two 54-dimensional feature vectors between the 30 proteins in the query set and the 647 proteins in the test set. For each query protein, we sort the test proteins with descent orders according to the calculated Euclidean distances.

We retrieve each query protein using all the three kinds of shape representations. In Fig. 2, we show the retrieval results of the polyline curve of 1mi3 chain A, the molecular surface of 1jzm chain A and the volumetric solid of 1g26.

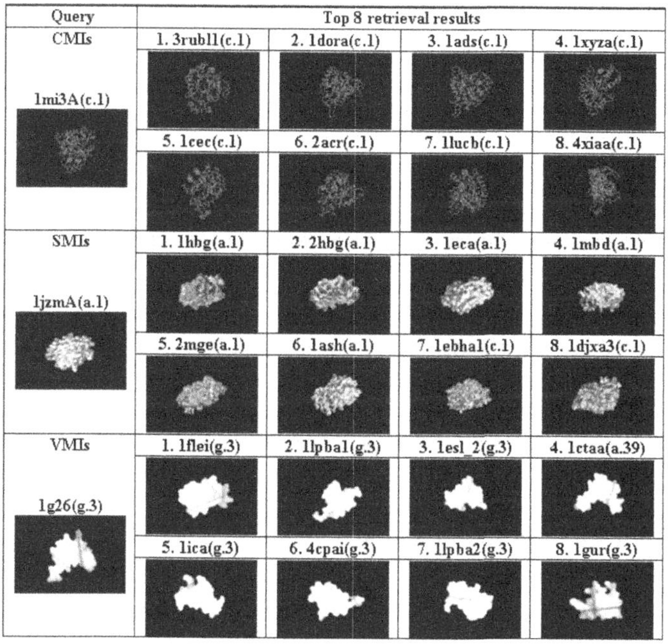

Fig. 2. Protein shape retrieval of the three kinds of shape representations

The most similar 8 retrieval results are given, each of which is represented by a typical snapshot of its corresponding 3D representation. The first two codes of SCOP id are also given. If the two codes of the retrieved protein are the same as the query protein, then they belong to the same fold. If only the first code is the same, they are regarded as proteins in the same class. Otherwise, they are of different classes.

The precision-recall plot is used to evaluate the average retrieval result. Precision is the ratio of the relevant retrieved models to the total retrieved models while recall is the division of the relevant retrieved models by the total relevant models. In Fig. 3, we compare the three shape representations for protein shape retrieval. We can see that 3D polyline curve is the best representation for protein tertiary structure for it can achieve the best retrieval result. Since the shapes of triangulated molecular surface and the volumetric solid it covers are the same, the retrieval results are very similar to them.

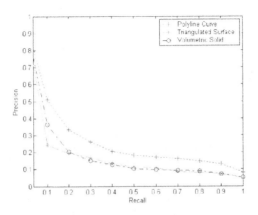

Fig. 3. Precision-recall plot of the three kinds of protein shape representations

Shape retrieval by using moment invariants as shape descriptors is purely based on the shape similarity. Since the classification of SCOP integrates many factors in it besides the overall shape similarity, there are maybe some proteins which are not in the same class, but have high similarities. For example, when we use volumetric representation for retrieval, 1ctaa ranks fourth, which means it is more similar to 1g26 than 1ica, 4cpai and 1lpba2 etc. 1ctaa even ranks first for the query 1g26 when we represent it by a polyline curve, which also proves the high similarity between 1ctaa and 1g26. However, they belong to different folds in the SCOP database.

3.2   Protein Structure Classification

Protein structure classification is more rigorous than protein shape retrieval for we need to estimate the class each protein belongs to as accurately as possible.

Many classification methods which are commonly used in research areas of artificial intelligence and data mining are adopted for protein structure classification, such as nearest neighbor classification [10].

We still use the database which contains SCOP classification information to test the classification accuracy by using moment invariants as features. As we have summarized above, pure shape descriptors are not enough to classify protein structures correctly. Similar to [7], we also add some statistical information (SI) to describe protein backbone besides curve moment invariants. It contains: (1) Primary structure information, including percentages of 20 kinds of residue; percentage of the hydrophobic residues; number of the alpha-carbon atoms; number of the total atoms. (2) Secondary structure information, including percentages of α-helix, β-sheet, β-turn and random coil; distribution of the two dihedral angles Φ and Ψ. (3) Tertiary structure information, including the radius of the bounding sphere for protein backbone. This information is also independent of the orientations of protein structures.

We use nearest neighbor classification to estimate the class of the 30 query proteins. That is to say, we consider each query protein to have the same class as the most similar protein among the 647 test proteins. If the estimated class has the same fold as its real SCOP fold, then the score is 2. If the estimated class has the same class as its real SCOP class, then the score is 1. Otherwise the score is 0. In addition to the three kinds of invariant shape descriptors, we also integrate curve moment invariants with statistical information to describe protein backbone and get an improved method called CMIs+SI.

Table 1. Comparison of protein classification accuracy

Group/Method	# (Wrong classification)	# (Correct SCOP class only)	# (Correct SCOP fold)	Total score
Purdue	5	5	20	5*1+20*2=45
ITI(Trace)	15	8	7	8*1+7*2=22
ITI(Graph)	14	3	13	3*1+13*2=29
LMB	2	4	24	4*1+24*2=52
CMIs	11	6	13	6*1+13*2=32
SMIs	17	8	5	8*1+5*2=18
VMIs	12	7	11	7*1+11*2=29
CMIs+SI	1	7	22	7*1+22*2=51

In Table 1, we compare our methods with that of the four groups attending the SHREC 2007. From the table, we can see that 3D curve representation is most suitable for classification, which is coincident with the conclusion from protein shape retrieval. If we only use curve moment invariants, the classification accuracy is better than two of the others' methods. We achieve nearly the best result by using the combined method. Since the classification of folds in SCOP is mainly based on the secondary structures, the statistical information could improve the classification accuracy greatly. Only one query protein (1gqz chain A, d.19) has the wrong estimated class (c.2) by us. However, groups ITI(Graph) and LMB also attribute it to SCOP fold c.2. Maybe the reason is that the test database is short of adequate test proteins whose fold is c.2 (only 2 against 647).

4   Conclusions and Future Work

In this paper, we build a secondary structure database which represents protein structure as polyline curve, triangulated surface and volumetric solid. Moment invariants are employed as shape descriptors for feature extraction. They are then used in applications to protein shape retrieval and classification.

In the future, we can train the test dataset by Neural Network or Support Vector Machine which could achieve higher classification accuracy. Molecular surface can be used for protein docking and other functional analysis. The retrieval performance could be improved by relevance feedback. In order to search similar structures quickly, we can also index the feature vectors by k-d tree etc.

Acknowledgments. This work is supported by NKBRP (2004CB318006) and NSFC (60573154 and 60533090). Thank the Freiburg track on protein challenge for SHREC 2007, which is supported by AIM@SHAPE, for providing data.

References

1. Berman, H.M., Westbrook, J., Feng, Z., Gilliland, G., Bhat, T.N., Weissig, H., Shindyalov, I.N., Bourne, P.E.: The Protein Data Bank. Nucleic Acids Research 28, 235–242 (2000)
2. Murzin, A.G., Brenner, S.E., Hubbard, T., Chothia, C.: SCOP: a Structural Classification of Proteins for the Investigation of Sequences and Structures. Journal of Molecular Biology 247, 536–540 (1995)
3. Orengo, C.A., Michie, A.D., Jones, S., Jones, D.T., Swindells, M.B., Thornton, J.M.: CATH- a Hierarchic Classification of Protein Domain Structures. Structure 5, 1093–1108 (1997)
4. Holm, L., Ouzounis, C., Sander, C., Tuparev, G., Vriend, G.: A Database of Protein Structure Families with Common Folding Motifs. Protein Science 1, 1691–1698 (1992)
5. Yeh, J.-S., Chen, D.-Y., Chen, B.-Y., Ouhyoung, M.: A Web-based Three-Dimensional Protein Retrieval System by Matching Visual Similarity. Bioinformatics 21, 3056–3057 (2005)
6. Sanner, M.F., Olson, A.J., Spehner, J.-C.: Reduced Surface: An Efficient Way to Compute Molecular Surfaces. Biopolymers 38, 205–320 (1996)
7. Chen, S.-C., Chen, T.: Protein Retrieval by Matching 3D Surfaces. In: Proceedings of the Workshop on Genomic Signal Processing and Statistics, vol. CP2-09, pp. 1–4 (2002)
8. Veltkamp, R.C., ter Haar, F.B.: SHREC, 3D Retrieval Contest. Technical Report UU-CS-2007-015 (2007)
9. Xu, D., Li, H.: Geometric Moment Invariants. Pattern Recognition 41, 240–249 (2008)
10. Ankerst, M., Kastenmuller, G., Kriegel, H., Seidl, T.: Nearest Neighbor Classification in 3D Protein Databases. In: ISMB, pp. 34–43 (1999)

The Structure of V-System over Triangulated Domains

Ruixia Song[1], Xiaochun Wang[2], Meifang Ou[1], and Jian Li[3]

[1] College of Sciences, North China University of Technology, Beijing, China
[2] College of Sciences, Beijing Forestry University, Beijing, China
[3] Faculty of Information Tech., Macau University of Science and Tech., Macau, China
songrx880@sohu.com

Abstract. The V-system on $L_2[0, 1]$ constructed in 2005 is a complete orthogonal system. It has multiresolution property. This paper further studies the V-system of two variables. The orthogonal V-system of degree k defined over triangulated domains is presented. With the orthogonal V-system over triangulated domains, all the application of the V-system on $L_2[0, 1]$ can be generalized onto the surface. Especially, the triangulated surface represented by piecewise polynomial of two variables of degree k with multi-levels discontinuities can be precisely reconstructed by finite terms of the V-series.

Keywords: Triangular domain, Triangulation, Orthogonal functions, Multiresolution, Haar system, V-system, U-system.

1 Introduction

The V-system of degree k is constructed by authors in 2005 [1], which is a class of complete orthogonal function system on $L_2[0, 1]$, and closely related to the U-system proposed by Qi Dongxu and Feng Yuyu [2]. The U-system of degree 0 is just Walsh system and the V-system of degree 0 is just Haar system. Compared with the other orthogonal systems, V-system has some distinctive characteristics: (1) The V-system is composed of both smooth functions and discontinuous piecewise polynomials at multi-levels. (2) The V-system has multiresolution property and local support. (3) Using partial sum of the V-series, the geometric information expressed by piecewise polynomials can be precisely reconstructed without Gibbs phenomenon. Please be noted that almost all the well-known orthogonal function system, such as Fourier orthogonal system, continuous wavelets, will inevitably cause Gibbs phenomenon when they are applied to reconstruction of geometric models with discontinuities. Because of this distinctive characteristic of the V-system, it is feasible to apply the frequency spectrum analysis method on graphic groups for its overall characteristic extraction. In addition, we have already tried to apply the V-system to the areas of reconstruction of graphics group, pattern recognition, points cloud fitting as well as digital watermark, and obtained some satisfying results [3-6].

The orthogonal multi-variable function systems have broad practical applications. Furthermore, orthogonal system over triangular domain is more significant than rectangular (tensor) domain in applications. However, constructing an orthogonal multivariable function system over triangulated domain is usually a remarkable and challenging work. So the study on interpolation and approximation over triangulated domain is more complicated and important. The Bernstein polynomials and Bezier surface over triangulated domain have been well studied [7-8], and many beautiful results about Walsh

F. Chen and B. Jüttler (Eds.): GMP 2008, LNCS 4975, pp. 563–569, 2008.

functions, Haar functions and wavelets over triangulated domain have also been obtained [9-13]. In this paper we generalize the V-system on $L_2[0, 1]$ to triangulated domain.

2   Triangulated Domain and Triangulation

Given a triangular domain G, we divide G as follows: First, connect the midpoint of each side, the domain G is divided into four subtriangles in a manner shown in Figure 1, denoted by $G_{1,1}, G_{2,1}, G_{2,2}, G_{2,3}$, we call the first step division triangulation at level 1. The intersection line of two different subtriangles (we call it partition line segment) is denoted by $I_j, j=1,2,3$. Second, each of these four subtriangles is subdivided into four smaller subtriangles denoted by $G_{i,j}, i = 1, 2, 3, 4, j = 1, 2, \cdots, 2i-1$, shown in Figure 2. Accordingly, this process is called triangulation at level 2. Go on with this triangulation process, in the triangulation at level m there are 4^m subtriangles represented by $G_{i,j}, i = 1, 2, \cdots, 2^m, j = 1, 2, \cdots, 2i - 1, m = 1, 2, \cdots$.

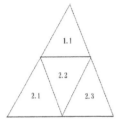

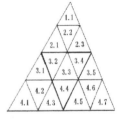

Fig. 1. The labeling order scheme for subtriangles at level 1

Fig. 2. The labeling order scheme for subtriangles at level 2

It is noted that coordinate transformation from one triangular domain to another doesn't change the orthogonality since the transformation Jacobi determinant is nonzero. Therefore, we choose a right triangle domain G with vertices $(0,0)$, $(1,0)$, $(0,2)$ as base triangle, it is a simple triangle with area 1. The inner product of $f(x, y)$ and $g(x, y)$ on G is defined as

$$< f, g >= \iint\limits_{G} fg dx dy = \int_0^1 dx \int_0^{2(1-x)} fg dy \tag{1}$$

Without losing the generality, in the following we concentrate our discussion on triangle domain G.

3   The V-System of Degree k over Triangulated Domain

The V-system of degree k over triangulated domain G, like the V-system defined on $L_2[0, 1]$, is also constructed by groups and classes. Let's take account of degree $k = 1$ first. Since the polynomial of two variables of degree 1 has three coefficients, it is necessary to construct three orthogonal polynomials of two variables of degree 1. They can be easily obtained by applying Schmidt orthogonalization method to any three linear independent polynomials. The followings are three simple independent polynomials chosen.

$$f_1(x, y) = 1, f_2(x, y) = x, f_3(x, y) = y,$$

Using the inner product defined in Equation (1), three orthonormal functions defined on G can be computed

$$V^1_{1,1}(x, y) = 1, \; V^2_{1,1}(x, y) = 3\sqrt{2}x - \sqrt{2}, \; V^3_{1,1}(x, y) = \sqrt{6}x + \sqrt{6}y - \sqrt{6} \tag{2}$$

These functions constitute the first group.

Besides three functions in Equations (2), we need to construct other 9 piecewise polynomials of two variables of degree 1, since the dimension of linear space, which consists of piecewise polynomials of two variables of degree 1 defined on G under triangulation at level 1, is 12. We call these 9 functions the generators of degree 1. The nine generators should satisfy the following conditions: (i) They are piecewise polynomials of two variables of degree 1 with norm 1 defined on domain G under triangulation at level 1 and orthogonal to each other. (ii) They are also orthogonal to the functions in Equations 2. (iii) On each of the three partition line segments of domain G under triangulation at level 1, there is at least one generator is discontinuous, and each generator should be discontinuous on at least one of the partition line segments.

According to the above conditions, nine generators of degree 1 can be determined. We dispense with the complicated calculation, just write down the functions below

$$V^1_{1,2} = \begin{cases} -3y + 5, & (x, y) \in G_{1,1} \\ -3y + 1, & (x, y) \in G_{2,1} \\ -3y + 1, & (x, y) \in G_{2,2} \\ -3y + 1, & (x, y) \in G_{2,3} \end{cases}, \qquad V^2_{1,2} = \begin{cases} 6x + 3y - 5, & (x, y) \in G_{1,1} \\ 6x + 3y - 1, & (x, y) \in G_{2,1} \\ 6x + 3y - 5, & (x, y) \in G_{2,2} \\ 6x + 3y - 5, & (x, y) \in G_{2,3} \end{cases}$$

$$V^3_{1,2} = \begin{cases} -6x + 1, & (x, y) \in G_{1,1} \\ -6x + 1, & (x, y) \in G_{2,1} \\ -6x + 1, & (x, y) \in G_{2,2} \\ -6x + 5, & (x, y) \in G_{2,3} \end{cases}, \qquad V^4_{1,2} = \begin{cases} \sqrt{2}(6x - 1), & (x, y) \in G_{1,1} \\ -\sqrt{2}(6x - 1), & (x, y) \in G_{2,1} \\ \sqrt{2}(6x - 2), & (x, y) \in G_{2,2} \\ -\sqrt{2}(6x - 4), & (x, y) \in G_{2,3} \end{cases}$$

$$V^5_{1,2} = \begin{cases} -\sqrt{2}(6x - 1), & (x, y) \in G_{1,1} \\ \sqrt{2}(6x - 1), & (x, y) \in G_{2,1} \\ \sqrt{2}(6x - 2), & (x, y) \in G_{2,2} \\ -\sqrt{2}(6x - 4), & (x, y) \in G_{2,3} \end{cases}, \qquad V^6_{1,2} = \begin{cases} \sqrt{2}(-3y + 4), & (x, y) \in G_{1,1} \\ \sqrt{2}(-3y + 1), & (x, y) \in G_{2,1} \\ \sqrt{2}(3y - 2), & (x, y) \in G_{2,2} \\ \sqrt{2}(3y - 1), & (x, y) \in G_{2,3} \end{cases}$$

$$V^7_{1,2} = \begin{cases} \sqrt{3}(4x + 4y - 6), & (x, y) \in G_{1,1} \\ \sqrt{3}(-4x - 4y + 2), & (x, y) \in G_{2,1} \\ 0, & (x, y) \in G_{2,2} \\ 0, & (x, y) \in G_{2,3} \end{cases}, \qquad V^8_{1,2} = \begin{cases} 0, & (x, y) \in G_{1,1} \\ 0, & (x, y) \in G_{2,1} \\ -\sqrt{3}(4x + 4y - 4), & (x, y) \in G_{2,2} \\ \sqrt{3}(4x + 4y - 4), & (x, y) \in G_{2,3} \end{cases}$$

$$V^9_{1,2} = \begin{cases} \sqrt{6}(4x + y - 2), & (x, y) \in G_{1,1} \\ \sqrt{6}(4x + y - 1), & (x, y) \in G_{2,1} \\ \sqrt{6}(-4x - y + 2), & (x, y) \in G_{2,2} \\ \sqrt{6}(-4x - y + 3), & (x, y) \in G_{2,3} \end{cases}$$

To define the value of above nine generators at I_j, we assume I_j,j=1,2,3is $G_{2,2}$ inside. These 9 generators constitute the second group of the V-system of degree 1. Let

$$V_{1,3}^{i,1} = \begin{cases} 2V_{1,2}^i(2x, 2(y-1)), & (x,y) \in G_{1,1} \\ 0, & others \end{cases} , V_{1,3}^{i,2} = \begin{cases} 2V_{1,2}^i(2x, 2y), & (x,y) \in G_{2,1} \\ 0, & others \end{cases}$$

$$V_{1,3}^{i,3} = \begin{cases} 2V_{1,2}^i(-2(x-\frac{1}{2}), -2(y-1)), & (x,y) \in G_{2,2} \\ 0, & others \end{cases}$$

$$V_{1,2}^{i,4} = \begin{cases} 2V_{1,2}^i(2(x-\frac{1}{2}), 2y), & (x,y) \in G_{2,3} \\ 0, & others \end{cases}$$

$$i = 1, 2, \cdots 9.$$

There are totally 9×4 functions. These 9×4 functions are divided into 9 classes, and each class consists of four functions. All these 36 functions constitute the third group of the V-system of degree 1. Figure 3 is the sketch map of a class of the third group (the white represents the function value 0).

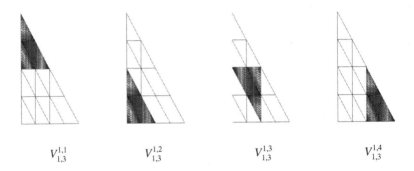

$$V_{1,3}^{1,1} \qquad\qquad V_{1,3}^{1,2} \qquad\qquad V_{1,3}^{1,3} \qquad\qquad V_{1,3}^{1,4}$$

Fig. 3. The sketch map of a class of the third group of V-system of degree 1

Go on with above process over and over again. In step m-1, let $G_{i,j}$ is the subtriangles under the triangulation at level $m-2 (i = 1, 2, \cdots, 2^m, j = 1, 2, \cdots, 2i-1, m = 3, 4, \cdots)$, we can obtain $9 \times 4^{m-2}$functions according to the following formulas, which form the m-th group:

$$V_{1,m}^{i,j}(x,y) = \begin{cases} 2^{m-2}V_{1,2}^i(2^{m-2}(x - \frac{\beta}{2^{m-2}}), 2^{m-2}(y - \frac{2^{m-1}-2\alpha}{2^{m-2}})), & (x,y) \in G_{\alpha,2\beta+1} \\ 0, & others \end{cases}$$

$$j = (\alpha - 1)^2 + 2\beta + 1$$

$$V_{1,m}^{i,j}(x,y) = \begin{cases} 2^{m-2}V_{1,2}^i(-2^{m-2}(x - \frac{\beta}{2^{m-2}}), -2^{m-2}(y - \frac{2^{m-1}-2\alpha+2}{2^{m-2}})), & (x,y) \in G_{\alpha,2\beta} \\ 0, & others \end{cases}$$

$$j = (\alpha - 1)^2 + 2\beta$$

Here $m = 3, 4, \cdots, i = 1, 2, \cdots, 9, \alpha = 1, 2, \cdots, 2^{m-2}, \beta = 0, 1, 2, \cdots, \alpha - 1.$

The function set: $\{V_{1,1}^1, V_{1,1}^2, V_{1,1}^3, V_{1,2}^1, V_{1,2}^2, \cdots, V_{1,2}^9, V_{1,m}^{i,j}, m = 3, 4, \cdots, i = 1, 2, \cdots, 9,$ $j = 1, 2, \cdots, 4^{m-2}\}$ is called the V-system of degree 1 over triangulated domain.

From the construction process of the V-system of degree k=1, we can easily find the method to construct the V-system of degree k over triangulated domain. They are also ordered by classes and groups. The number of functions is relative to k. There are $\frac{1}{2}(k+1)(k+2)$ functions in the first group obtained by applying Schmidt orthogonalization method to any $\frac{1}{2}(k+1)(k+2)$ linear independent polynomials, $\frac{3}{2}(k+1)(k+2)$ functions in the second group, which are called generators. The functions in the m-th group are obtained by squeezing and duplicating the functions in the second group. The m-th group is divided into $\frac{3}{2}(k+1)(k+2)$ classes, and each class contains $4^{m-2}(m = 3, 4, \cdots)$ functions. We have omitted further details because of the length limitation of the paper.

From the constructing process of the V-system of degree k over triangulated domain, we can obtain the following properties.

Theorem 1 (Orthonormality) The V-system of degree k over triangulated domain is a orthonormal function system, and composed of both smooth functions and discontinuous piecewise polynomials at multi-levels.

Theorem 2 (Reproducibility) If $f(x, y)$ is a piecewise polynomial of two variables of degree k, and its discontinuous line segment is just the partition line segment of G, then $f(x, y)$ can be exactly represented by the finite functions of the V-system of degree k over triangulated domain G.

4   Application Illustration

Figure 4 is a graphics group which is composed of two totally separate polyhedrons. It can be triangulated into 16 triangular pieces. We can calculate its frequency spectrum of the V-system of degree 1 over triangulated domain. Figure 5 depicts the frequency spectra of three coordinates. In addition, Figure 6 is a model of a plane, and notice that the aerofoil is separate from its airframe. It is regarded as a graphics group, and triangulated into 16 triangular pieces. Calculate its frequency spectrum of the V-system of degree 1 over triangulated domain. The frequency spectra of three coordinates are depicted in Figure 7. The significance of these two groups of frequency spectra is that we can further analyze the characteristics of the graphics group of polyhedrons and plane using them. What's more important is that these two geometric models are exactly

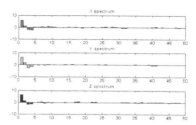

Fig. 4. A graphics group of two polyhedrons   **Fig. 5.** The frequency spectra of figure 4

Fig. 6. A model of a plane

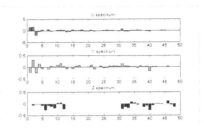

Fig. 7. The frequency spectra of the plane

reconstructed without Gibbs phenomenon using these frequency spectra. The result is guaranteed by theorem 2.

5   Conclusion

Constructing an orthogonal function system over triangulated domain is a challenge work in mathematics. The same as the V-system defined on $L_2[0, 1]$, the V-system over triangulated domain constructed in this paper is also an orthogonal function system. It contains both smooth and discontinues functions, and has compactly local support and multiresolution property. These properties are extraordinary useful in CAGD. Using V-system the signals represented by piecewise polynomials on triangulated domain can be precisely reproduced, in other words, Gibbs phenomena can be eliminated, while it is inevitable when continuous orthogonal function or continuous wavelet is used. It means that the powerful frequency spectrum analysis method, which is used in image processing and signal analysis, can be applied to geometry models for their overall characteristic extraction. Our future work is exploring the application of the V-system over triangulated domain and extending V-system to n-dimensional simplex.

Acknowledgments

This research is partially supported by NSF-CHINA Project (No.10631080, No.10771002, No.10671002), National Key Basic Research Project of China Grant (No. 2004CB 318000), Open Project of State Key Lab of CAD&CD of Zhejiang University(No.A0503), and Science and Technology Development Fund of Macao SAR (No. 045/2006/A).

References

1. Song, R., Ma, H., Wang, T., Qi, D.: Complete Orthogonal V-system and Its Applications. Communications on Pure and Applied Analysis 6(3), 853–871 (2007)
2. Feng, Yu-yu. Qi, Dong-xu: A Sequence of Piecewise Orthogonal Polynomials. SIAM J.Math, Anal. 15(4),834-844 (1984)

3. Song, R., Wang, X., Ma, H., Qi, D.: V-descriptor and Shape Similarity Measurement between B-spline Curves. In: Proc. of the First International Symposium on Pervasive Computing and Application, Urumchi, China, pp. 486–490 (2006)
4. Liang, Y.Y., Song, R.X., Qi, D.X.: Complete orthogonal function system V and points cloud fitting. Journal of System Simulation, (in Chinese) 18(8), 2109–2113 (2006)
5. Liang, Y.Y., Song, R.X., Wang, X.C., Qi, D.X.: Application of a New Class of Orthogonal Function System in Geometrical Information, J. of Computer-Aided Design&Computer Graphics, (in Chinese) 19(7), 871–875 (2007)
6. Song, R., Liang, Y., Wang, X., Qi, D.: Elimination of Gibbs phenomenon in Computational Information based on the V-system. In: Proc. of The Second International Conference on Pervasive Computing and Applications., Birmingham, UK, pp. 337–341 (2007)
7. Chang, G.: The Mathematics of Surfaces. Hunan Education Press (1995)
8. Gang, X., Guozhao, W.: Harmonic B-B Surfaces over the Triangular Domain. Chinese Journal of Computers, (in Chinese) 29(12), 2180–2185 (2006)
9. Feng, Y., Qi, D.: On the Haar and Walsh System on a Triangle. J. of Computational Math. 1(3) (1983)
10. Nielson, G.M., Il-Hong, J., Junwon, S.: Haar wavelets over triangular domains with applications to multiresolution models for flow over a sphere. In: Proceedings of Visualization 1997, pp. 143–149 (1997)
11. Schroeder, P., Sweldens, W.: Spherical Wavelets: Efficiently Representing Functions on the Sphere. In: SIGGRAPH: Proceedings of the 22nd annual Conference on Computer Graphics and Interactive Techniques, pp. 161–172. ACM Press, New York (1995)
12. Rosca, D.: Haar Wavelets on Spherical Triangulations. In: Dodgson, N.A., Floater, M.S., Sabin, M.A. (eds.) Advances in Multiresolution for Geometric Modelling. Mathematics and Visualization, pp. 405–417. Springer, Heidelberg (2005)
13. Bonneau, G.-P.: Optimal Triangular Haar Bases for Spherical Data. In: VIS: Proceedings of the Conference on Visualization, pp. 279–284. IEEE Computer Society Press, Los Alamitos (1999)

Tool Path Planning for 5-Axis Flank Milling Based on Dynamic Programming Techniques

Ping-Han Wu, Yu-Wei Li, and Chih-Hsing Chu*

Department of Industrial Engineering and Engineering Management
National Tsing Hua University, Hsinchu 300, Taiwan
chchu@ie.nthu.edu.tw

Abstract. This paper proposes a novel computation method for tool path planning in 5-axis flank milling of ruled surfaces. This method converts the path planning (a geometry problem) into a curve matching task (a mathematical programming problem). Discrete dynamic programming techniques are applied to obtain the optimal matching with machining error as the objective function. Each matching line corresponds to a cutter location and the tool orientation at the location. An approximating method based on z-buffer is developed for a quick estimation of the error. A set of parameters is allowed to vary in the optimization, thus generating the optimal tool paths in different conditions. They reveal useful insights into design of the tool motion pattern with respect to the surface geometry. The simulation result of machining different surfaces validates the proposed method. This work provides an effective systematic approach to precise error control in 5-axis flank milling.

Keywords: 5-axis machining, flank milling, ruled surface, dynamic programming, optimization.

1 Introduction

5-axis CNC machining has found wide applications in manufacture of aerospace, automobile, air-conditioning, and mold parts since the 90s. The two additional degrees of freedom provide several advantages [1, 2] over 3-axis machining such as higher productivity and better quality. There are two different milling methods in five-axis machining. In point milling (or end milling), the cutting edges near the end of a tool performs the action of material removal [3]. On the contrary, the cylindrical part of a tool does the main cutting in flank milling [4]. To generate the tool path that achieves the specified surface quality is a major concern in point milling. Elimination of tool interference is another challenge in use of five-axis machining. The situation in 5-axis flank milling is more complicated. To completely avoid tool interference is difficult, if not impossible, in most cases except for machining of simple shapes such as cylindrical, conical, and developable surfaces [5]. The following literature review discusses some previous work related to 5-axis flank milling.

---

* Corresponding author.

F. Chen and B. Jüttler (Eds.): GMP 2008, LNCS 4975, pp. 570–577, 2008.

Liu [6] offset the points corresponding to the parameter values 0.25 and 0.75 of a surface ruling with a distance of the tool radius in the tool path generation. The direction determined by the offset points becomes the tool orientation. Bohez et al. [7] determined the tool orientation in a similar manner. The offset direction is chosen as the average of the surface normal at the sample points on the boundary curves. This approach can reduce the amount of tool interference compared to Liu's method. Lee and Suh [8] proposed an interference-free tool path planning method for twisted ruled surfaces. This method can minimize undercut volume without global tool interference by adjusting the rotation angle together with the offset distance. Lartigue et al. [9] proposed the envelope surface to model the tool swept volume in space. They also proposed an evaluation method of the machining error that can be applied to compare the quality of different tool paths. Tsay et al. [10, 11] studied the influence of the yaw and tilt angles on the amount of tool interference in 5-axis flank milling from the machining error estimation of various examples. The result works as look-up tables integrated in a tool path generation algorithm for B-spline ruled surface. Bedi et al. [12] looked into the relationship between the tool orientation and the amount of undercut/overcut in flank milling of ruled surfaces. A cylindrical cutter initially makes a contact with a boundary curve of the surface with the tool orientation as the surface ruling. A nonlinear optimization scheme [13] is developed for minimizing the tool interference at discrete positions by locally adjusting the tool position and orientation around the contact ruling. Wang and Tang [14] developed a technique of optimal triangulation for interpolating two polygonal chains which approximate the boundary curves of a ruled surface.

Previous studies have employed optimization methods in adjustment of the tool orientation at discrete cutter locations. However, they are all based on a greedy approach, which can only give local optimal. An implicit assumption is made by these approaches: the global optimum equals to the sum of local optimums, which is generally not true. Further, the previous methods neglect the possibility of skipping some contact points along the boundary curves in the tool path generation. To overcome these problems, we propose a new scheme for the tool path generation in 5-axis flank milling of ruled surfaces. A geometric problem (tool path generation) is transformed into a mathematical programming problem (curve matching). The machining error serves as the objective function to be minimized in a global optimization process. In addition, the method extends the boundary curves in different extents for enlarging the solution space of tool contact lines. A z-buffer method is developed for a quick estimation of the machining error in the optimization. A set of parameters is allowed to vary and thus generate the optimal tool paths in different conditions. Important findings are obtained regarding the optimal patterns of the tool motion. This work provides a simple, effective and systematic approach to precise error control in 5-axis flank milling of ruled surfaces.

2 Optimization Tool Planning

2.1 Tool Path Planning in 5-Axis Flank Milling

A systematic approach to tool path generation in 5-axis flank milling consists of contact line generation and cutter axis determination (see Fig. 1). The former step determines a series of tool contact lines which bridge the boundary curves of the machined surface. The next step is to choose a cutter axis at each cutter contact line. One straightforward solution is to offset the end points of each contact line along the surface normal vectors at the points, with a distance equal to the cutter radius. CAM/NC systems can then simulate the machining process, estimate the machining error, and eliminate the potential errors prior to real cutting. The tool path planning is a critical step determining the machining quality, time, and cost.

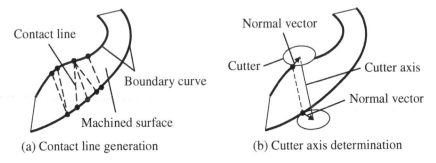

(a) Contact line generation   (b) Cutter axis determination

Fig. 1. Tool path generation in 5-axis flank milling

2.2 Optimization Scheme Based on Dynamic Programming

In this work, an optimization scheme based on discrete dynamic programming is adopted to solve the problem. The scheme contains five steps. First, the boundary curves of a machined surface are discretized into discrete points, designated as DP_{m-n}. m identifies the boundary curve to which the discrete point belongs, and n represents the point index on the boundary curve. Thus, there are two sets of discrete points for a machined surface. Second, users select one boundary as the focal curve used for creating stages in dynamic programming. Each discrete point on the focal curve represents a stage. The third step is to create states in each stage. Each state in a stage consists of several connection lines which connect a discrete point on the focal curve with discrete points on the other boundary. The number of the states in a stage and the connection lines are determined by two user-defined parameters (SR and LS). After creating the states in each stage, the next step is to make linkages between states in adjacent stages. The determination of these linkages is based on the two criteria. The first one is to avoid intersection between the connection lines within adjacent states. The second one prevents the number of skipped discrete points on the non-focal boundary from violating the LS value. Finally, dynamic programming is adopted to obtain an optimal route while minimizing machining error. The optimal route comprises a series of connection lines which connect the two boundary curves. Each line represents a cutter location on the machined surface. Consequently, the optimization scheme provides a systematic method for the tool path planning and for achieving a globally optimal solution with respect to the minimization of machining error.

2.3 Parameters of the Optimization Scheme

Four parameters, the expansion ratio (ER), the sampling size (SS), the scanning range (SR), and the leaping step (LS), define the optimization scheme and provide users a flexible way to configure the optimization scheme so that their domain knowledge can be compatibly integrated into the scheme in order to improve tool path generation.

Previous studies assume that the tool contact lines are generated from the boundary curves of a machined surface. This study proposes to enlarge the solution space by expanding the boundary curves in different extents and use the expanded curves in the generation process. Accordingly, the term ER is coined to represents the expansion ratio of the control points on boundary curves (CP_i and CP_j). Based on the ER values, the control points are moved to the adjusted positions (CP_i^{new} and CP_j^{new}) by using $CP_i^{new} = CP_i + (CP_j - CP_i) \times ER_i$ and $CP_j^{new} = CP_j + (CP_i - CP_j) \times ER_j$.

SS defines the number of the discrete points obtained from the boundary curves. Discretization can be performed with either an equal curve parameter or an arc length. Moreover, the number of discrete points on the two boundary curves can be different from each other in the optimization scheme.

SR restricts the number of states in a stage and connection line patterns by locating a range on the non-focal boundary. The discrete points within the range are eligible for forming the connection lines. For example, to create states in the stage corresponding to DP_{1-m}, the points to be considered on the non-focal boundary is ranged from $DP_{0-(m-SR)}$ to $DP_{0-(m+SR)}$. The discrete points within the range correspond to a set with $(2 \times SR+1)$ elements. After the set is determined, the next step is to generate combinations of the set. Each of the valid combinations represents an "original" state in the stage which belongs to DP_{1-m}. On the other hand, each stage has "dummy" states if the LS value for the focal boundary is non-zero. The purpose of introducing "dummy" states in a stage is to allow for the skipping of discrete points on the focal boundary.

LS defines the number of skipped discrete points between the adjacent tool contact lines on each of the boundary curves. For example, given LS_0 and LS_1, and the end points of the two adjacent lines ((DP_{0-i}, DP_{1-j}) and (DP_{0-m}, DP_{1-n})), $(m - i) - 1 \leq LS_0$ and $(n - j) - 1 \leq LS_1$ should be satisfied. Therefore, some of the combinations are deleted based on these two constraint specified by the LS values.

3 Machining Error Estimation

The optimization scheme employs dynamic programming for the generation of tool paths with the objective function as minimizing the machining error. To estimate the machining error is a complicated issue. In addition to the precision in the estimation, we need to take into account the computation load in the optimization. We propose a method based on the z-buffer technique. This method simulates the shape change of the stock material during the machining. Fig. 2 illustrates its major steps. The machining error, depending on the application, can be calculated as the sum of the lengths of the trimmed straight lines or as the maximum length of the trimmed straight lines. The latter is employed in this paper. For verification purpose, the result generated by our method is compared with that of commercial simulation software NC VERICUT™. Fig. 3 demonstrates that our method tends to estimate the machining error in a way consistent to that obtained from the software, which is considered de facto in practice.

(a)   (b)

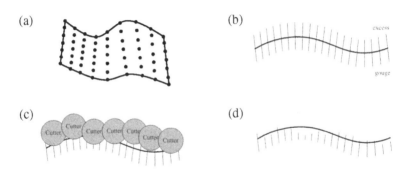

(c)   (d)

Fig. 2. Z-buffer estimation method

4 Implementation Results

This section demonstrates several tests to show how SS, SR, and LS affect the obtained optimal solution. Each test is designed for only one parameter and the remaining parameters remain identical, as presented in Table 1. The boundary curves of the test surface are both cubic Bézier curves whose control points are (7, 7, -22), (21, 28, -22), (35, 35, -22), and (7, 49, -22) and (7, 7, -22), (21, 21, -22), (35, 42, -22), and (7, 49, -22). In practice, one way of reducing machining error is to increase the discrete points on boundary curves so that the cutter can follow the denser surface rulings. Increasing the SS value can attain the same result. However, the greater the SS value, the more computation time the optimization scheme takes. To consider both efficiency and quality, SS should be properly selected. The results shown in Table 2

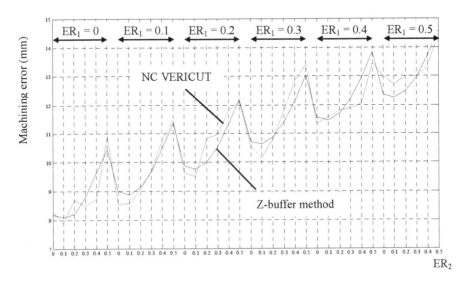

Fig. 3. Comparison between the proposed estimation method and NC VERICUT™

illustrates that the machining error is reduced with the increasing SS value. SR is one of the factors that determine the number of possible solutions to be considered in the optimization. When the SR value increases, the number of the states on each stage also increases and the possible number of different tool paths is thus enhanced. Table 3 shows a series of test results with different SR values. For a machined surface, the SR has a critical value which guarantees that the obtained solution is globally optimal. Any smaller value than the critical one will yield a suboptimal solution, while setting a greater value will reduce the efficiency of the operation of the optimization scheme. Similarly, the LS has a critical value which ensures obtaining a globally optimal solution. The critical value varies according to the surface properties. Setting the LS value improperly can result in missing an optimal solution or obtaining the optimal in an unaccepted computation time. The result shown in Table 4 indicates that allowing the cutter to skip points on the boundary curves leads to better tool paths in terms of smaller machining errors. This is an important finding that cannot be discovered by local optimization methods.

Table 1. Test parameters

Cutter Radius	ER (Focal/Non-focal)	SS (Focal/Non-focal)	SR	LS (Focal/Non-focal)
2.0 mm	0.0/0.0	40/40	1	2/2

Table 2. Experiment with SS adjustments

SS (Focal/Non-focal)	20/20	40/40	60/60
Machining error (mm)	2.5798	1.5473	0.852
Generated tool paths			

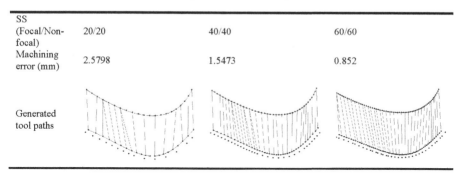

Table 3. Experiment with SR adjustments

SR	1	2	3
Machining error (mm)	1.5474	1.5474	1.5474
Generated tool paths			

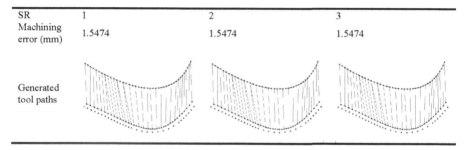

Table 4. Experiment with LS adjustments

LS (Focal/Non-focal)	1/1	2/2	3/3
Machining error (mm)	1.6024	1.5474	1.36
Generated tool paths			

5 Conclusion

This paper presents a computation scheme that generates the tool path in 5-axis flank milling of ruled surfaces in a novel manner. The path planning problem is transformed into a curve matching task, which is then solved by dynamic programming techniques. The machining error is used as the objective function to be minimized. Such a global optimization method overcomes several limitations in previous research. First, it provides a more systematic method for the machining error control. Second, a larger solution space is produced by extending the boundary curves of the machined surface for generation of the tool contact lines. This leads to a better optimal solution and the resultant tool path. In addition, several important findings are obtained from the results generated by varying several parameters in the scheme. They reveal insights into the patterns of the optimal tool motion in reducing the machining error. Triangular contacts lines occur when the surface area exhibits larges curvatures or twist of the boundary curves. Moreover, to skip the tool contact lines helps reduce the error in some cases. This work provides an effective systematic approach to precise error control in 5-axis flank milling. The current method is limited in single pass tool path. It can be extended into flank milling of multiple passes. The machined surface needs be decomposed into strips. The decomposition process becomes another optimization problem in addition to the tool path generation within each strip, which has been solved by this work. We are concurrently working on this.

References

1. Vickers, G.B., Guan, K.W.: Ball-Mills Versus End-Mills for Curved Surface Machining. ASME Journal of Engineering for Industry 111, 22–26 (1989)
2. Wu, C.Y.: Arbitrary Surface Flank Milling of Fan, Compressor, and Impeller Blades. Transactions of ASME Journal of Engineering for Gas Turbines and Power 117(3), 534–539 (1995)
3. Choi, B.K., Park, J.W., Jun, C.S.: Cutter-Location Data Optimization in 5-Axis Surface Machining. Computer-Aided Design 25(6), 377–386 (1993)

4. Li, S.X., Jerard, R.B.: 5-Axis Machining of Sculptured Surfaces with a Flat-End Cutter. Computer-Aided Design 26(3), 165–178 (1994)
5. Chu, C.H., Chen, J.T.: Automatic Tool Path Generation for 5-axis Flank Milling based on Developable Surface Approximation. International Journal of Advanced Manufacturing Technology 29(7–8), 707–713 (2006)
6. Liu, X.W.: Five-Axis NC Cylindrical Milling of Sculptured Surfaces. Computer-Aided Design 27(12), 87–94 (1995)
7. Bohez, E.L.J., Ranjith Senadhera, S.D., Pole, K., Duflou, J.R., Tar, T.: A Geometric Modeling and Five-Axis Machining Algorithm for Centrifugal Impellers. Journal of Manufacturing Systems 16(6), 422–436 (1997)
8. Lee, J.J., Suh, S.H.: Interference-Free Tool Path Planning for Flank Milling of Twisted Ruled Surface. The International Journal of Advanced Manufacturing Technology 14, 795–805 (1998)
9. Lartigue, C., Duc, E., Affouard, A.: Tool Path Deformation in 5-Axis Flank Milling Using Envelope Surface. Computer-Aided Design 35, 375–382 (2003)
10. Tsay, D.M., Her, M.J.: Accurate 5-Axis Machining of Twisted Ruled Surfaces. ASME Journal of Manufacturing Science and Engineering 123, 731–738 (2001)
11. Tsay, D.M., Chen, H.C., Her, M.J.: A Study on Five-axis Flank Machining of Centrifugal Compressor Impellers. ASME Journal of Engineering for Gas Turbines and Power 124, 177–181 (2002)
12. Bedi, S., Mann, S., Menzel, C.: Flank Milling with Flat End Milling Cutters. Computer-Aided Design 35, 293–300 (2003)
13. Menzel, C., Bedi, S., Mann, S.: Triple Tangent Flank Milling of Ruled Surfaces. Computer-Aided Design 36, 289–296 (2004)
14. Wang, C.C., Tang, K.: Optimal Boundary Triangulations of an Interpolating Ruled Surface. Journal of Computing and Information Science in Engineering 5, 291–301 (2005)

Trimming Bézier Surfaces on Bézier Surfaces Via Blossoming*

Lian-Qiang Yang and Xiao-Ming Zeng

Department of Mathematics, Xiamen University, Xiamen 361005, China
ylqylq@yahoo.cn(Lian-Qiang Yang), xmzeng@xmu.edu.cn(Xiao-Ming Zeng)

Abstract. The problem of trimming Bézier surfaces on Bézier surfaces contains many cases, such as the subdivision, conversion and conjoining. Different methods have been given for some special cases. In this paper, by means of blossoming and parameter transformation, a united approach is given for this problem. The approach can be extended to trim Bézier patches on any polynomial or rational surfaces naturally.

Keywords: Triangular Bézier patch, Rectangular Bézier patch, reparametrization, Blossoming, Trimming.

1 Introduction

Triangular Bézier patch and rectangular Bézier patch are most widely used surfaces in CAGD. For conveniently, we call them TBP(Triangular Bézier Patch) and RBP(Rectangular Bézier Patch) in the following of this paper. Both TBP and RBP play key roles in modeling systems, then the relationship between them arouses many researchers' attention. A class of interesting questions is:

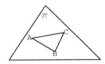   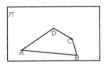

| Fig. 1.1. | Fig. 1.2. | Fig. 1.3. | Fig. 1.4. |

Question 1 (Fig 1.1): Given one TBP defined on the triangular domain π, the triangular subdomain $\triangle ABC \subseteq \pi$. Then, whether the subpatch defined on $\triangle ABC$ can be represented as a TBP, if it is true, how to compute its Bézier control points?

Question 2 (Fig 1.2): Given one RBP defined on the rectangular domain π. Then the same question as question 1 can be proposed;

Question 3 (Fig 1.3): Given one TBP defined on the triangular domain π, the convex quadrilateral subdomain $\square ABCD \subseteq \pi$. Then, whether the subpatch defined on $\square ABCD$ can be represented as a RBP, if it is true, how to compute its

---

* This work is supported by NSFC under Grant 10571145.

F. Chen and B. Jüttler (Eds.): GMP 2008, LNCS 4975, pp. 578–584, 2008.

Bézier control points ? (Note, the sign "□" only means convex quadrilateral in this paper, but not a square.)

Question 4 (Fig 1.4): Given one RBP defined on the rectangular domain π. Then the same question as question 3 can be proposed.

All the four questions can be expressed as one problem: How to trim Bézier surfaces on Bézier surfaces? Furthermore, each question contains some special cases, these cases are related to important algorithms such as subdivision, trimming, connection or conversion of TBP and RBP. Many works have been accomplished for these cases, we illustrate some of them with figures(Fig1.1a-1.4a).

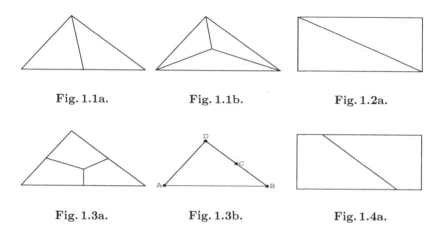

Fig. 1.1a.   Fig. 1.1b.   Fig. 1.2a.

Fig. 1.3a.   Fig. 1.3b.   Fig. 1.4a.

1. (Fig 1.1a and Fig 1.1b, the cases of question 1.)   Which mean the subdivision of TBP. This question's detailed answer can be found in [5,3](Goldman, 1983; Farin, 1993);

2. (Fig 1.2a, the case of question 2.)   Which means how to subdivide one RBP into two TBP. This is the work of [6](Goldman and Filip, 1987);

3. (Fig 1.3a and Fig 1.3b, the case of question 3.)   Fig1.3a points to the subdivision of one TBP into three RBP, which is discussed in [10](Hu, 1996); Fig 1.3b implies how to convert one TBP into a RBP, which is studied in [1,9,12](Brueckner,1980; Hu, 1993; Hu, 2001);

4. (Fig 1.4a, the case of question 4.)   This case is about the generalized subdivision of RBP, the work is investigated by [11](Hu, 1996).

Moreover, [4](Feng and Peng, 1999) researched the question 2 and 3 by shifting operator; [13](Lasser, 2002) gave an explicit formula and a geometric algorithm for RBP on TBP, where the parametric domain of this RBP is a planar RBP in the triangular parameteric domain of the TBP. The method is extened to the case of TBP on RBP in [14](Lasser, 2007); Another correlative profound work is [2](Derose, Goldman, Hagen, Mann, 1993), which provided a frame for solving these questions from the viewpoint of functional composition.

As mentioned above, these four questions are dispersively researched via different ways. In this paper, a united solution is given for them. We prove that the bilinear reparametrization of a convex quadrilateral is well defined, and the explicit formula for blossoming bivariate polynomial is given. With the two tools, all the four questions can be solved through the same mode.

2   Reparametrization of Convex Quadrilateral

Given one convex quadrilateral (or degenerate convex quadrilateral as Fig 1.3b) over Euclidean plane $\mathbb{R}^2$, with vertices $A(a_1, a_2)$, $B(b_1, b_2)$, $C(c_1, c_2)$, $D(d_1, d_2)$, denote the close domain enclosed by $\square ABCD$ as $\mathcal{D}(u, v)$.

Theorem 1. *Let mapping* $\phi : \overline{\mathcal{D}}(s, t) \rightarrow \mathcal{D}(u, v), \overline{\mathcal{D}} = [0, 1] \otimes [0, 1] \subset \mathbb{R}^2,$ *by*

$$\begin{cases} u = (1 - s)(1 - t)a_1 + (1 - s)tb_1 + stc_1 + (1 - t)sd_1 \\ v = (1 - s)(1 - t)a_2 + (1 - s)tb_2 + stc_2 + (1 - t)sd_2 \end{cases} \tag{1}$$

If $\square ABCD$ *is a convex quadrilateral, then the mapping* ϕ *is a regular parameter transformation;*
If $\square ABCD$ *is a degenerate convex quadrilateral as Fig 1.3b, then the mapping* ϕ *is a regular parameter transformation except at the unique singular point* $(s, t) = (1, 1)$.

It can be proved that ϕ is a differentiable homeomorphism except at the unique singular point in degenerate cases. After this transformation, the convex quadrilateral is then defined on a unit square. Moreover, the latter part of Theorem 1 means that even a regular TBP can only be represented as a degenerate RBP with a unique singular point.

3   Blossomings of Bivariate Polynomials

Blossoming has many important applications in geometric modeling[8](Goldman, 2003) . In this paper, we focus on the blossoming of bivariate polynomial, one explicit formula for blossoming bivariate polynomial is given, which is the unique tool for computing the Bézier control points of a surface patch. All notations and terms about blossoming are adopted from [7](Goldman, 2002).

Theorem 2. *Given bivariate polynomial*

$$P_{ij}(u, v) = u^i v^j, \qquad 0 \leq i \leq m, \quad 0 \leq j \leq n, \quad i + j \leq k.$$

Let

$$M = \{1, 2, 3, \cdots, m\}, \quad N = \{1, 2, 3, \cdots, n\}, \quad K = \{1, 2, 3, \cdots, k\},$$

2.1 Consider it as bivariate polynomial of total degree k, then the blossoming of $P_{ij}(u,v)$ is

$$b_{ij}^{\triangle}((u_1,v_1),\cdots,(u_k,v_k))$$

$$= \sum_{\substack{\{\alpha_1,\alpha_2,\cdots,\alpha_i\}\subseteq K \\ \{\beta_1,\beta_2,\cdots,\beta_j\}\subseteq K\backslash\{\alpha_1,\alpha_2,\cdots,\alpha_i\}}} \frac{u_{\alpha_1}u_{\alpha_2}\cdots u_{\alpha_i}v_{\beta_1}v_{\beta_2}\cdots v_{\beta_j}}{\binom{k}{i,j}}; \quad (2)$$

2.2 Consider it as bivariate polynomial of bidegree (m,n), then the blossoming of $P_{ij}(u,v)$ is

$$b_{ij}^{\square}(u_1,\cdots,u_m;v_1,\cdots,v_n)$$

$$= \sum_{\{\alpha_1,\alpha_2,\cdots,\alpha_i\}\subseteq M} \frac{u_{\alpha_1}u_{\alpha_2}\cdots u_{\alpha_i}}{\binom{m}{i}} \sum_{\{\beta_1,\beta_2,\cdots,\beta_j\}\subseteq N} \frac{v_{\beta_1}v_{\beta_2}\cdots v_{\beta_j}}{\binom{n}{j}}. \quad (3)$$

where,

$$\binom{k}{i,j} = \frac{k!}{i!j!(k-i-j)!}, \quad \binom{m}{i} = \frac{m!}{i!(m-i)!}, \quad \binom{n}{j} = \frac{n!}{j!(n-j)!}.$$

The similar expressions have been appeared in (Goldman, 2002) without proof, we reformulate them and the proof can be completed via the three axioms of blossoming.

Because blossoming is a linear operator, for any bivariate polynomial

$$P(u,v) = \sum_{(i,j)} c_{ij}u^iv^j, \quad 0 \le i+j \le k, \quad 0 \le i \le m, \quad 0 \le j \le n, \quad (5)$$

its blossoming is

$$b^{\triangle}((u_1,v_1),\cdots,(u_k,v_k)) = \sum_{(i,j)} c_{ij}b_{i,j}^{\triangle}((u_1,v_1),\cdots,(u_k,v_k)); \quad (6)$$

$$b^{\square}(u_1,\cdots,u_m;v_1,\cdots,v_n) = \sum_{(i,j)} c_{ij}b_{i,j}^{\square}(u_1,\cdots,u_m;v_1,\cdots,v_n). \quad (7)$$

One wonderful character of blossoming is the dual functional property [7,8] (Goldman, 2002, 2003), which provides the Bernstein coefficients(Bézier control points) of the polynomial surface patch. For any polynomial surface patch $Q(u,v) = \sum_{(i,j)} c_{ij}u^iv^j$ of total degree k defined on triangular domain $\triangle ABC$, then the Bézier control points Q_{fgh} of the surface patch are

$$Q_{fgh} = b^{\triangle}(\underbrace{A,\cdots,A}_{f},\underbrace{B,\cdots,B}_{g},\underbrace{C,\cdots,C}_{h}), \quad f+g+h=k; \quad (8)$$

For any polynomial surface patch $Q(u, v) = \sum_{(i,j)} c_{ij} u^i v^j$ of bidegree (m, n) defined on rectangular domain $[a, b] \otimes [c, d]$, then the Bézier control points Q_{fg} of this surface patch are

$$Q_{fg} = b^{\square}(\underbrace{a, \cdots, a}_{m-f}, \underbrace{b, \cdots, b}_{f}; \underbrace{c, \cdots, c}_{n-g}, \underbrace{d, \cdots, d}_{g}). \tag{9}$$

4   Applications

In this section, concrete algorithms are provided for the questions from 1 to 4. The steps for questions 1 and 2 are same, and the steps for questions 3 and 4 are only one step more than the former.

Question 1(2): Fig 1.1 (1.2), Given one TBP(RBP) $P(u, v)$ defined on the triangular (rectangular) domain π, the triangular subdomain $\triangle ABC \subseteq \pi$. Evaluate the Bézier control points Q_{fgh} of the triangular subpatch $Q(u, v)$ defined on $\triangle ABC$.

Steps 1:   Rewrite $P(u, v)$ as the form of (5);
Steps 2:   Generate the blossoming $b^{\triangle}((u_1, v_1), \cdots, (u_k, v_k))$ of $P(u, v)$ via (6);
Steps 3:   Evaluate the Bézier control points Q_{fgh} of $Q(u, v)$ via (8).

Question 3(4): Fig 1.3 (1.4), Given one TBP(RBP) $P(u, v)$ defined on the triangular (rectangular) domain π, the convex quadrilateral subdomain $\square ABCD \subseteq \pi$. Evaluate the Bézier control points Q_{fg} of the rectangular subpatch $Q(s, t)$ defined on $\square ABCD$.

Steps 1:   Substitute (1) into $P(u, v)$, get $Q(s, t) = P(\phi(s, t))$;
Steps 2:   Rewrite $Q(s, t)$ as the form of (5);
Steps 3:   Generate the blossoming $b^{\square}(u_1, \cdots, u_m; v_1, \cdots, v_n)$ of $Q(s, t)$ via (7);
Steps 4:   Evaluate the Bézier control points Q_{fg} of $Q(s, t)$ via (9) :
$$Q_{fg} = b^{\square}(\underbrace{0, \cdots, 0}_{m-f}, \underbrace{1, \cdots, 1}_{f}; \underbrace{0, \cdots, 0}_{n-g}, \underbrace{1, \cdots, 1}_{g}).$$

5   Examples

Example 1: Fig 2.1: Given quadratic TBP defined on the parametric domain $\triangle ABC$, where $A(0, 1), B(0, 0), C(1, 0)$, with control points $\{P_{200}, P_{110}, P_{101}, P_{020}, P_{011}, P_{002}\} = \{(0,0,-3),(0,1,0),(1,0,0),(0,4,-2),(1,1,0),(3,3,-4)\}$; Fig 2.2: Regard $\triangle ABC$ as a degenerate convex quadrilateral $\square ABCD$, where $D(1/2, 1/2)$, with the algorithms for *Question 3*, the TBP is converted into a degenerate biquadratic RBP with control points $\{P_{00}, P_{01}, P_{02}, P_{10}, P_{11}, P_{12}, P_{20}, P_{21}, P_{22}\} = \{(3, 3, -4),(1, 0,0),(0,0,-3),(1,1,0),(1/2,3/4,0),(0,1/2,-3/2),(0,4,-2),(0,5/2,-1),(0,3/2,-5/4)\}$; Fig 2.3: Trimmed three RBPs on the TBP, control points are $\{(3, 3, -4),(7/3, 7/3, -8/3),(16/9, 20/9, -2),(7/3, 2, -8/3),(31/18, 3/2, -14/9),(11/9, 14/9, -10/9),(16/9, 4/3, -19/9),(11/9, 1, -11/9),(7/9, 11/9, -1)\}$, $\{(7/9, 23/9, -4/3),(1/3, 3, -4/3),(0, 4, -2),(7/9, 16/9, -8/9),(1/3, 19/9, -7/9),(0, 3, -4/3),(7/9, 11/9, -1),(1/3, 13/9, -7/9),(0, 20, -11/9)\}$ and $\{(7/9, 11/9, -1),(1/3, 8/9, -8/9),(0, 8/9, -14/9),(7/9, 2/3, -10/9),(1/3, 1/3, -7/6), (0, 1/3, -2),(7/9, 1/3, -16/9),(1/3, 0, -2),(0, 0, -3)\}$; Fig 2.4: Trimmed letter

"H"(made up of three TBPs) on the TBP, the TBPs' control points are {(56/25, 107/50, -261/100),(103/50, 101/50,-231/100),(189/100,39/20, -207/100),(38/25, 59/50, -29/20),(34/25, 28/25, -123/100),(121/100, 111/100,-107/100),(24/25, 7/10, -141/100), (41/50, 7/10, -127/100), (69/100, 3/4, -119/100)}, { (561/400, 111/80, -591/400), (499/400, 109/80, -519/400),(441/400, 111/80, -471/400),(499/400, 489/400, -533/ 400), (439/400, 97/80, -469/400),(383/400, 501/400, -429/400),(441/400, 87/80, -503/400), (383/400, 437/400, -447/400),(329/400, 459/400, -83/80)} and {(39/25, 93/50, -33/20), (7/5, 46/25, -147/100),(5/4, 187/100, -27/20),(23/25, 57/50, -81/100), (39/50, 59/50, -71/100),(13/20, 127/100, -67/100),(11/25, 9/10, -109/100),(8/25, 1, -107/100),(21/100, 23/20, -111/100)} .

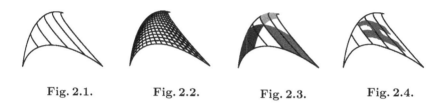

Fig. 2.1.   **Fig. 2.2.**   **Fig. 2.3.**   **Fig. 2.4.**

Example 2: Fig 3.1: We subdivide the square $[0,1]^2$ into a *Tangram*(contains subdomains from 1 to 7); Fig 3.2: The RBP of degree (2,1) defined on the square with control points {(0, 0, 0),(0, 2, 0),(1, 0, 1),(1, 2, 1),(2, 0, 0),(2, 2, 0)} ; Fig 3.3: Correspondingly, the RBP is subdivided into five quadratic TBPs and two biquadratic RBPs, their control points are 1: {(0, 2, 0),(0, 1, 0),(1/2, 3/2, 1/2),(0, 0, 0),(1/2, 1/2, 1/2),(1, 1, 1/2)}; 2: {(0, 2, 0),(1/2, 3/2, 1/2),(1, 2, 1),(1, 1, 1/2),(3/2, 3/2, 1/2),(2, 2, 0)}; 3: {(1/2, 1/2, 3/8),(1/4, 1/4, 1/4),(3/4, 1/4, 1/2),(0, 0, 0),(1/2, 0, 1/2),(1, 0, 1/2)}; 4: {(3/2, 3/2, 3/8),(5/4, 5/4, 1/2),(3/2, 1, 3/8),(1, 1,1/2),(5/4, 3/4, 1/2),(3/2, 1/2, 3/8)}; 5: {(2, 1, 0),(3/2, 1/2, 1/2),(2, 1/2, 0),(1, 0, 1/2),(3/2, 0, 1/2),(2, 0, 0)}; 6: {(1/2, 1/2, 3/8),(3/4, 1/4, 1/2),(1, 0, 1/2),(3/4, 3/4, 1/2),(1, 1/2, 3/8),(5/4, 1/4, 1/8),(1, 1, 1/2),(5/4, 3/4, 1/8),(3/2, 1/2,-3/8)}; 7: {(3/2, 3/2, 3/8),(7/4, 7/4, 1/4),(2, 2, 0),(3/2, 1, 3/8),(7/4, 5/4, 1/4),(2, 3/2, 0),(3/2, 1/2, 3/8),(7/4, 3/4, 1/4),(2, 1, 0)}; Fig 3.4: Trimmed letter "T" (made up of two biquadratic RBPs)on the RBP, the RBPs' control points are {(2/5, 1, 8/25),(3/10, 11/10, 13/50),(1/5, 6/5, 9/50), (7/10, 13/10, 1/2),(3/5, 7/5, 47/100),(1/2, 3/2, 21/50), (1, 8/5, 1/2),(9/10, 17/10, 1/2),(4/5, 9/5, 12/25)} and {(7/5, 1/5, 21/50),(1, 7/10, 29/50),(3/5, 6/5, 21/50),(3/2, 2/5, 19/50),(11/10, 17/20, 29/50),(7/10, 13/10, 23/50), (8/5, 3/5, 8/25),(6/5, 1, 14/25),(4/5, 7/5, 12/25)}.

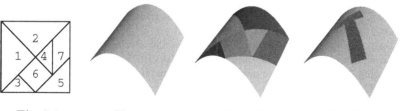

Fig. 3.1.   **Fig. 3.2.**   **Fig. 3.3.**   **Fig. 3.4.**

6   Conclusions

Using blossoming and parameter transformation, the united approach is given for the problem of free trimming Bézier surfaces on Bézier surfaces. And this work can be extended to the following cases naturally:

First, trimming the surface patches out and along the side of the parametric domain, then this is the work correlative with the conjoining of two Bézier surfaces; Second, with homogeneous coordinates, the approach is also worked for rational cases; Fianlly, this method can be used for trimming Bézier patches on polynomial or rational surfaces defined over any shaped parametric domains, whether they are represented as Bézier forms or not. For example, we can trim TBPs and RBPs on the multisided patches such as S-patches and *Toric* patches.

References

1. Brueckner, I.: Construction of Bézier points of quadrilaterals from those of triangles. Computer-Aided Design 12, 21–24 (1980)
2. DeRose, T.D., Goldman, R.N., Hagen, H., Mann, S.: Functional composition algorithms via blossoming. ACM Transactions on Graphics 12, 113–135 (1993)
3. Farin, G.: Curves and Surfaces in Computer Aided Geometric Design, 3rd edn. Academic Press, San Diego (1993)
4. Feng, J.Q., Peng, Q.S.: Functional compositions via shifting operators for Bézier patches and their applications. Journa of Software 10, 1316–1321 (1999)
5. Goldman, R.N.: Subdivision algorithms for Bézier triangles. Computer-Aided Design 15, 159–166 (1983)
6. Goldman, R.N., Filip, D.: Conversion from Bézier rectangles to Bézier triangles. Computer-Aided Design 19, 25–27 (1987)
7. Goldman, R.N.: Pyramid Algorithms: A dynamic Programming Approach to Curves and Surfaces for Geometric Modeling. Morgan Kaufmann, San Francisco (2002)
8. Goldman, R.N.: Polar forms in geometric modeling and algebraic geometry. In: Contemporary mathematics:Topics on algebraic geometry and geometric modeling, vol. 334, pp. 3–24. AMS (2003)
9. Hu, S.M.: Conversion between two classes of Bézier surfaces and geometric continuity jointing. Applied Mathematics: A Journal of Chinese Universities 8, 290–299 (1993)
10. Hu, S.M.: Conversion of a triangular Bézier patch into three rectangular Bézier patches. Computer Aided Geometric Design 13, 219–226 (1996)
11. Hu, S.M., Wang, G.Z., Jin, T.G.: Generalized subdivision of Bézier surfaces. Graphical Models and Image Processing 58, 218–222 (1996)
12. Hu, S.M.: Conversion between triangular and rectangular Bézier patches. Computer Aided Geometric Design 18, 667–671 (2001)
13. Lasser, D.: Tensor product Bézier surfaces on triangle Bézier surfaces. Computer Aided Geometric Design 19, 625–643 (2002)
14. Lasser, D.: Triangular subpatches of rectangular Bézier surfaces, Computers and Mathematics with Applications (2007), doi:10.1016/j.camwa.2007.04.049
15. Ramshaw, L.: Blossoms are polar forms. Systems Research Center Reports 34, Palo Alto, Califorinia (1989)
16. Ramshaw, L.: Blossoming: A connect-the-dots approach to splines. Systems Research Center Reports 19, Palo Alto, Califorinia (1987)

A Volumetric Framework for the Modeling and Rendering of Dynamic and Heterogeneous Scenes

Duoduo Liao[1] and Shiaofen Fang[2]

[1] Department of Computing Science, George Washington University
Washington, DC. 20052, U.S.A.
dliao@gwu.edu
[2] Department of Computer and Information Science, Indiana University-Purdue University
Indianapolis, IN 47202, U.S.A.
sfang@cs.iupui.edu

Abstract. This paper presents a novel unified solution -- a general volumetric framework for 3D modeling, interaction, and visualization applications with complex volumetric scenes composed of virtually any types of conventional and unconventional object representations, potentially for a wide range of volumetric applications. As an example, this framework is applied to a general problem of volume modeling, and the experimental results demonstrate the effectiveness, flexibility and generality of our volume fusion algorithms for 3D interactive volumetric applications.

Keywords: Volume Fusion, Volume Rendering, Volumetric Scene Tree, Voxelization, Volume Modeling.

1 Introduction

A great challenge in computer graphics today comes from the emergence of a new wave of applications that require the synthesis of extremely complex scenes from multiple data sources. This trend has been largely inspired by the rapid advances in 3D data acquisition. Biomedical scanning devices, such as CT, MRI, and confocal microscopes, produce large volumes of biomedical image data that may be used by a wide range of applications such as surgical simulation, medical diagnosis, and biomedical education and training. It is not hard to envision that future computer graphics will be shaped by applications where synthetic models and characters, scanned real objects, volumes scans of real human, and even microscopic scans of biological structures can interact with each other within the same scene. The increasing number of such applications involving unconventional representations provides a strong motivation for a systematic study of complex scenes and their interaction techniques.

Our goal is to develop a unified solution for the interaction of 3D heterogeneous scenes composed of virtually any types of conventional and unconventional object representations. Although many different interaction and rendering techniques have been developed in recent years, these techniques are mostly designed for one specific type of datasets under certain conditions. Generalization to heterogeneous and complex scenes is difficult. In the work described in this paper, we developed an

F. Chen and B. Jüttler (Eds.): GMP 2008, LNCS 4975, pp. 585–591, 2008.

efficient and unified framework for volume fusion, including novel volumetric scene expression design methods, and fast voxelization algorithms for scene expression evaluation. This framework is applied to the general problem of volume modeling to demonstrate its flexibility and related implementation details, in particular, the design of a volumetric scene expression with a specific set of blending or filtering functions to generate the required volumetric information.

2   Related Work

A popular technique in scene composition is scene graph. It has been widely used in many 3D graphics systems such as Open Inventor, VRML and Java 3D. Recently, this idea has been extended to volume scene graph for the composition of complex volumetric scenes from multiple volume datasets and space-filling functions [9]. However, the evaluation of a volume scene graph is done in a brute-force manner, i.e. for each point in volume space, recursively computing the value of the scene graph starting from the root. This is a very expensive procedure, and can only be used as a preprocessing step, which is not practical for interactive applications. A similar technique often used for 3D model composition is the Constructive Solid Geometry (CSG) representation. A natural extension of the CSG method is to allow volume datasets to be included as primitives in the CSG construction process, as described independently in several publications [3][6]. A Volume Scene Tree (VST) was proposed in [5], in which each leaf node represents an input dataset or synthetic geometric model, and each interior node represents an operator such as Boolean operators or blending.

The slice sweeping algorithm and VST tree pruning optimization techniques presented in [5] is used for volumetric CSG modeling. They will be extended for scene evaluation in volume fusion. The basic idea of slice sweeping is to generate a slice for each object in the scene first, and then apply the operation on the slice in postfix order of the VST. Since heterogeneous representations are used, the computation process in this algorithm differs from various object representations.

Traditional voxelization [8][7] approaches are not suitable for interactive volume operations due to the lack interactive speed. Several fast voxelization algorithms for various object types (i.e. curve, surface, solids, etc.) have been proposed recently based on hardware acceleration in [1][2][5]. These algorithms have markedly improved the voxelization speed at interactive time. All of them will be integrated into a unified framework for volume fusion.

3   Volume Fusion

3.1   Volumetric Scene Representation

Volume fusion is concerned with the representation, manipulation and rendering of a volumetric scene. A volumetric scene expression is represented as a sequence of objects and operators that are carried out in a certain order. It describes the composition process of a complex volumetric scene, and can be best represented by a Volume Scene Tree (VST). The VST is extended by more complex operators (i.e. fusion functions) as interior nodes and primitives as leaf nodes. In a VST, each leaf node represents an input dataset or synthetic geometric model, and each interior node represents an operator such as blending or filtering, or even an unconventional operator defined for a certain special application. This kind of data structure can represent a volumetric scene as well as the entire computation process of volume fusion.

Volume fusion is the process of designing and evaluating a volumetric scene expression. The designing process, called scene design, involves the selection of appropriate blending, filtering, or other operations and their operational orders in the scene expressions, which represent certain volumetric computation tasks such as scene modeling and collision detection. The execution of the operations requires a volumetric evaluation of the scene expression as called scene evaluation. The evaluation process is normally done by computing the value of each point in the volume space through either a top-down or bottom-up traversal of the scene expression tree. The evaluated result of a volumetric scene expression is a field defined over a volume space. Conceptually, however, each object represents a volumetric field, defined by its attribute values. The evaluation process will extract volumetric information from different objects by voxelization and perform corresponding fusion operations. In practice, the volume fusion approach described in this paper includes a volumetric scene representation method and an efficient scene evaluation algorithm based on slice-sweeping (i.e. voxelization). It provides a unified framework for general volumetric scene manipulation.

3.2 Volumetric Scene Evaluation

Volume fusion functions for specific applications are represented by the interior nodes in a VST. Theoretically, VST interior nodes can represent any types of fusion function operators, such as filtering, blending, and transformation. A filtering function is an operator that changes the attribute values of the data, similar to image filters. Typical filters include thresholding, boundary detection, color coding, sharpening, blurring, and other image processing operations. A blending function is an operator with multiple operands, which combines multiple attribute values from different objects at each point in the volume space, similar to the image composition process.

Objects in a scene expression can be transformed by transformation operations, filtered by filtering functions, and blended by blending functions. Most commonly used transformation operations are simple affine transformations such as translation, rotation and scaling. However, non-linear transformations like volume deformation may also be applied to any node of the VST. The result of this process is a deformation function [4] that can be applied to any volume dataset, and generate and unstructured point set. In addition, transform functions such as Fourier Transform and Discrete Cosine Transform (DCT) may be used for some special applications through the general volume fusion framework.

The design and manipulation of the blending, filtering, and transformation operators constitute the main process of scene manipulation and design. One or more parameters can be defined in each operator, and may be modified dynamically by the users during interaction. A complex volumetric scene can be constructed and composed using various types of geometric and volumetric objects such as curve, surface, solids and CT dataset. One of the main motivations of using the volumetric method is to provide a unified data format for objects and datasets of heterogeneous representations. Thus how to convert a complex volumetric scene into a unified volume model is critical. This process is called volume scene evaluation, and is carried out by a slice sweeping algorithm [5].

The slice sweeping algorithm proceeds by moving a slice plane with a constant step size in volume space. For each new slice, all primitives are rendered into the slice buffer (i.e., frame buffer) using standard OpenGL rendering pipeline. This algorithm needs a stack data structure to store the intermediate result slices. For each slice, stored as a 2D texture image, the algorithm reads one node at a time from the postfix expression of the VST, and pushes slices of leaf nodes (objects) into the stack, and pops slices from the stack for operator nodes. At the end, the last slice in the stack will be the final evaluated result for this slice.

The volume scene evaluation algorithm can be described in the following pseudo-code:

```
VolumeSceneEvaluation (VST Tree)
{
    for ( each slice S) {
        Read the first node P from Tree;
        while (P is not NULL) {
            if( P is a leaf node) {
                Voxelize P into the slice S;
                Push (S);
            }
            else if ( P is a function operator) {
                for (each operator)
                    T[i] = Pop(); // i = 0,1, …, k
                Slice = Func(T[0], …, T[k], P);
                Push (S);
            }
            P = next node in Tree;
        }
        S = Pop();
        Output final slice S into volume memory;
        Move to the next slice;
    }
}
```

4 Volume Modeling

A volumetric CSG model provides a constructive way of designing 3D models using intermixed geometric and volumetric objects. Several publications ([3][1]) have provided the definitions of volumetric CSG tree and volumetric Boolean operations. These volumetric Boolean operations can also be used as fusion functions for volume modeling based on VST. The standard definition of volumetric Boolean operations is straightforward and has been discussed in [1][5]. They can also be directly represented by frame buffer blending functions in a 3D graphics API such as OpenGL. More specifically, a frame buffer blending function in OpenGL, when enabled, is applied between the incoming pixel (source) color and the pixel color at the corresponding location in the frame buffer (destination). Let C_s denote the incoming pixel color (source color) and C_d denote the corresponding pixel color in the frame buffer (destination color). And C is the output color to be written back to the frame buffer. Thus, volumetric Boolean operations in the volumetric CSG tree can be implemented with the replacement of blending functions in the scene expression by the following combinations:

Union: $C = \max(C_s, C_d)$

Or: $C = k_s C_s + k_d C_d$

Intersection: $C = \min(C_s, C_d)$

Difference: $C = \min(C_d - C_s, 0)$

It is quite easy to design a VST for volume modeling or turn a volumetric CSG tree into a VST under the volume fusion framework. By VST definition [5], a VST can be any kind of a tree, which allows a parent node to have any number of children but at least one. So the VST of the volumetric CSG tree may also have more than two operands for an OR (Union) blending operator, but the filtering operator has only one operand by conventions.

5   Implementation Results

The methods and techniques described in this paper have been implemented on a standard PC with Nvidia Quadro FX3000 accelerator with 256MB texture memory and SGI Onyx2 with a single processor and 64MB texture memory. The program is written in C++ and OpenGL.

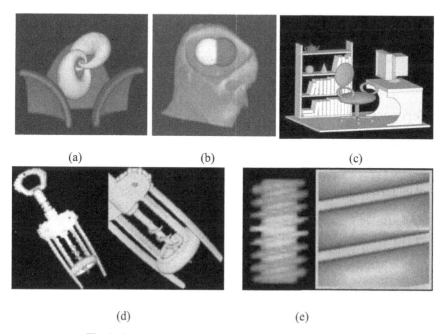

(a)   (b)   (c)

(d)   (e)

Fig. 1. Some Experimental Results for Volume Modeling

Figure 1 shows several volume modeling examples. Fig.1(d) and Fig.1(e) show the zoom-in effects that represent the multi-scale capabilities of the system. Each reconstructed volume has a resolution of 128^3, and is volume rendered by 3D texture mapping. The timing of the operations depends on the scene complexity, framebuffer size and the number of slices. The model in Fig.1(c) takes 1.5 seconds, all the rest of the models take between 0.1 to 1 second.

6 Conclusions

This paper presents a general volume fusion framework for 3D interactions with volumetric scenes using scene expression and scene evaluation, potentially for a wide range of volumetric applications. The major contributions of this paper are as follows:

1) Developed a new volumetric 3D interaction technique using volume fusion for a volumetric scene evaluation. A slice-sweep approach is employed to provide a general computational framework that extracts volumetric information of the scene expression on a 2D slice.
2) The design processes and algorithms for volume modeling using the volume fusion framework are described in detail.
3) The design of the entire volume fusion system is largely platform independent. Our experimental results show that volume fusion process can run at interactive speeds on standard PCs with 3D graphics accelerators.

In the future, we will continue improving the volume fusion optimization techniques (e.g. tree pruning and slice cache) for better scene evaluation speed based on more efficient GPU programming. We are exploring more applications by using this technology in several domains: surgical simulation, medical imaging, and geospatial information visualization. We hope that the availability of this integrated framework and fast algorithms will spur more researches in new and innovative interaction techniques for heterogeneous data.

References

1. Fang, S., Chen, H.: Hardware Accelerated Voxelization. In: Volume Graphics, Ch. 20, pp. 301–315. Springer, Heidelberg (2000)
2. Fang, S., Liao, L.: Fast CSG Voxelization by Frame Buffer Pixel Mapping. In: Proceedings of IEEE/ACM Symposium on Volume Visualization, pp. 43–48 (2000)
3. Fang, S., Srinivasan, R.: Volumetric CSG – A Model-Based Volume Visualization Approach. In: Proceedings of Sixth International Conference in Central Europe on Computer Graphics and Visualization, pp. 88–95 (1998)
4. Fang, S., Srinivasan, S., Huang, S., Raghavan, R.: Deformable Volume Rendering by 3D Texture Mapping and Octree Encoding. In: Proceedings of IEEE Visualization 1996, pp. 73–80 (1996)
5. Liao, D., Fang, S.: Fast CSG Modeling For Volumetric Scene Using Standard Graphics system. In: Proceedings of 7th ACM Symposium on Solid Modeling and Application (2002)

6. Hable, J., Rossignac, J.: GPU-based rendering of Boolean combinations of free-form triangulated shapes. ACM Transactions on Graphics 24, 1024–1031 (2005)
7. Huang, J., Yagel, R., Filippov, V., Kurzion, Y.: An Accurate Method for Voxelizing Polygon Meshes. In: Proceedings of IEEE/ACM Symposium on Volume Visualization, pp. 119–126 (1998)
8. Kaufman, A.: Efficient Algorithms for 3D Scan-conversion of Parametric Curves, Surfaces, and Volumes. In: Proceedings of ACM SIGGRAPH 1987, pp. 171–179 (1987)
9. Nadeau, D.: Volume Scene Graphs. In: Proceedings of IEEE/ACM Symposium on Volume Visualization, pp. 49–56 (2000)

Geometric Calibration of Projector Imagery on Curved Screen Based-on Subdivision Mesh

Jun Zhang[1], BangPing Wang[1], and XiaoFeng Li[1,2]

[1] Institute of Image & Graphic Department of Computer Science
Sichuan University, Chengdu, 610064, China
[2] Key Laboratory of Fundamental Synthetic Vision Graphics and Image for National Defense
Sichuan University, Chengdu, 610064, China
{mecca_zj,wbangping}@163.com, lixiaofengflc@sina.com

Abstract. A novel method based on subdivision mesh for geometric calibration problem is proposed, which can correct the projector imagery on smooth curved screen with respect to the screen geometry and the observer's position. Using sparse initial mesh extracted from the image of camera and a special designed subdivision algorithm, we can subdivide the initial mesh into arbitrary precision. Hence, the one-one correspondence of pixels between the frame buffer of projector and image of camera can be established. Using this correspondence, we can transform the projector imagery into any shape through the image of camera. So the geometric calibration can be done easily by mapping all pixels of frame-buffer image to an undistorted mesh on curved screen specified by a laser theodolite. Theoretical analysis and experimental results show that the method can automatically, accurately and rapidly display an undistorted image on curved screen, without any projector and camera registrations or prior knowledge of screen.

Keywords: geometric calibration, subdivision mesh, curved surface projection, computer vision.

1 Introduction

Large area multi-projector displays are becoming increasingly popular for various applications, changing the way we interact with our computing environment [1]. Accurate estimation of geometric relationship between overlapping projectors is the key for achieving seamless displays. They also influence the rendering and intensity blending algorithms [2].

Generally, the algorithms for geometric calibration for undistorted display need one or more cameras to capture the information of geometric distortion. The commercial systems such as PixelFlex2 designed by Aditi[1, 3] can correct the imagery of projector to show an undistorted and seamless image on plane by multi-projectors. Rasker [2, 4] exploited a quadric image transfer function to achieve sub-pixel registration while interactively displaying two or three-dimensional datasets for a head-tracked user. Tardif

F. Chen and B. Jüttler (Eds.): GMP 2008, LNCS 4975, pp. 592–600, 2008.
© Springer-Verlag Berlin Heidelberg 2008

[5] presented an approach allowing one or more projectors to display an undistorted image on curved surface of unknown geometry. They computed the correspondence of each pixel between the projector image and the camera image by using structured light patterns and piecewise linear interpolation. Hartville's method [6] can get mapping of screen-to-camera on developable surfaces by triangle meshes and texture mapping. However, it's quite difficult to model the image distortion by linear or piecewise linear method, because it is actually the composition of several nonlinear mappings that occur in the process of projecting an image on the curved screen, such as lens of projectors and cameras, curvature of surface and so on. Meanwhile, the hypothesis of quadric surface isn't always true because of the limitation of process precision facing the large scaled screen. All of these make the geometric calibration algorithms need dense sampling and time-consuming.

We present a novel geometric calibration method to correct the image of projector's frame-buffer without any prior information of projector or geometry of surface and camera's registration. Using camera to capture the projected chessboard image on screen, we can get the sparse initial mesh and use it to generate arbitrary precision mesh to establish the one-one correspondence between each pixel of a projector image and the camera image. To do this, we design a special subdivision algorithm, which can self-adapting subdivide the mesh holding the geometric shape and density distribution characters of original mesh. Then we do geometric calibration on images of camera and use the established correspondence to do same geometric calibration on frame-buffer image of projectors. The presented method reduces the amount of sampling dramatically and has no use for the prior information of projector or geometry of surface, which is a main contribution of our method.

2   Construction of the Subdivision Mesh

Motivated by techniques for the discrete generation of B-spline curves initiated by Lane and Riesenfeld (1980) as well as Cohen et al. (1980), various universal subdivision frameworks or schemes have been developed [7]. But most researches were focused on the performance of subdivision surface and the higher-order continuity of limit surface.

We present a 2D subdivision mesh which is composition of different families of curves along the two orthogonal directions based on the classic four-point interpolatory subdivision curve [8]. The subdivision mesh is designed specially for the geometric calibration. Face points are used to balance the displacement of subdivision curves along different directions.

2.1   Mode of Subdivision Curves Along the Two Orthogonal Directions

Any 2D grid on rectangular domain can be viewed as composition of different families of curves along the two orthogonal directions:

$$U_i^k = \{P_{i0}^k \quad P_{i1}^k \quad \cdots \quad P_{in}^k\}(i=0,1\cdots m) \quad V_j^k = \{P_{0j}^k \quad P_{1j}^k \quad \cdots \quad P_{mj}^k\}(j=0,1\cdots n)$$

Where U_i^k denotes i th u curve, V_j^k denotes j th v curve and M^k denotes the whole grid, k denotes the times of subdivision.

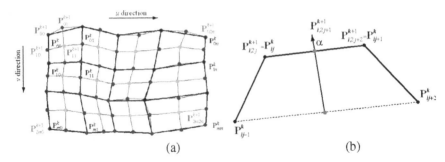

Fig. 1. (a) A subdivision grid on rectangular domain: the blue points are new edge points and the green points are face points. (b) Illustration of four-point interpolatory subdivision curve.

We subdivide the initial mesh by method of four-point interpolatory subdivision curve along u/v directions. The blue points in Fig. 1 can be computed recursively.

u Curves: $U_{2i}^{k+1} = \{P_{2i,0}^{k+1}\quad P_{2i,1}^{k+1}\quad \cdots\quad P_{2i,2n}^{k+1}\}(i=0,1\cdots m)$ v Curves:

$V_{2j}^{k+1} = \{P_{0,2j}^{k+1}\quad P_{1,2j}^{k+1}\quad \cdots\quad P_{2m,2j}^{k+1}\}(j=0,1\cdots n)$

$$\begin{cases} P_{2i,2j}^{k+1} = P_{i,j}^{k} \\ P_{2i,2j+1}^{k+1} = (1+\alpha)\left(\dfrac{P_{i,j}^{k}+P_{i,j+1}^{k}}{2}\right)-\alpha\left(\dfrac{P_{i,j-1}^{k}+P_{i,j+2}^{k}}{2}\right) \end{cases} \begin{cases} P_{2i,2j}^{k+1} = P_{i,j}^{k} \\ P_{2i+1,2j}^{k+1} = (1+\alpha)\left(\dfrac{P_{i,j}^{k}+P_{i+1,j}^{k}}{2}\right)-\alpha\left(\dfrac{P_{i-1,j}^{k}+P_{i+2,j}^{k}}{2}\right) \end{cases} \quad (2.1)$$

Where α is a parameter of tension.

2.2 Computation of Face Points

Using (2.1), we can refine the curves of initial grid along u/v directions. To get a refined mesh, we need add a new point to each face in the original grid.

To balance the displacement of subdivision curves along different directions and hold the density distribution characters of original mesh, we hope that the added point can construct four parallelograms with its 8-neighborhood. As Fig. 2 (a) shows, we expect face point $P_{2i+1,2j+1}^{k+1}$ to construct parallelograms with

$\{P_{2i,2j}^{k+1}, P_{2i,2j+1}^{k+1}, P_{2i+1,2j}^{k+1}\}$, $\{P_{2i,2j+2}^{k+1}, P_{2i,2j+1}^{k+1}, P_{2i+1,2j+2}^{k+1}\}$, $\{P_{2i+2,2j}^{k+1}, P_{2i+2,2j+1}^{k+1}, P_{2i+1,2j}^{k+1}\}$,

$\{P_{2i+1,2j+2}^{k+1}, P_{2i+2,2j+2}^{k+1}, P_{2i+2,2j+1}^{k+1}\}$,

like $\{Q_1,Q_2,Q_3,Q_4\}$.

So, we take the mean of $\{Q_1,Q_2,Q_3,Q_4\}$ as the new face point:

$$P_{2i+1,2j+1}^{k+1} = (Q_1+Q_2+Q_3+Q_4)\Big/ 4 = \frac{1}{2}\left(P_{2i+1,2j}^{k+1}+P_{2i,2j+1}^{k+1}+P_{2i+1,2j+2}^{k+1}+P_{2i+2,2j+1}^{k+1}\right)-\frac{1}{4}\left(P_{2i,2j}^{k+1}+P_{2i,2j+2}^{k+1}+P_{2i+2,2j}^{k+1}+P_{2i+2,2j+2}^{k+1}\right) \quad (2.2)$$

2.3 Disposal of Boundary

The mesh discussed in this paper is open mesh, so the boundary is important to disposal peculiarly. To maintain consistency of subdivision mode and also to hold the geometric shape and density distribution characters of original mesh, we adopt the algorithm that mirrors inner points of mesh to the virtual border of mesh.

Without loss of generality, let the left three end points of some u curve be $P_{i1}^k, P_{i2}^k, P_{i3}^k$, we construct a virtual point P_{i0}^k, which is the mirror of P_{i3}^k about the perpendicular bisector of $P_{i1}^k P_{i2}^k$:

$$P_{i0}^k = P_{i3}^k + (P_{i1}^k - P_{i2}^k) + \frac{2\left|(P_{i2}^k - P_{i3}^k, P_{i1}^k - P_{i2}^k)\right|}{\left(\left\|P_{i1}^k - P_{i2}^k\right\|_2\right)^2}(P_{i1}^k - P_{i2}^k) \tag{2.3}$$

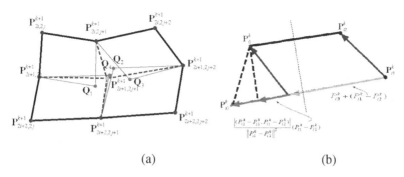

$$(a) \qquad\qquad\qquad\qquad (b)$$

Fig. 2. The geometric significance of (2.2) and (2.4). (a) Illustration of new face point (the blue points are 8-neighborhood of face point, the red points $\{Q_1, Q_2, Q_3, Q_4\}$ are expected points of face point, green point $P_{2i+1,2j+1}^{k+1}$ is face point) (b) Geometric significance of equation (2.5) (the virtual point P_{i0}^k is the mirror point of P_{i3}^k about the perpendicular bisector of $P_{i1}^k P_{i2}^k$).

Where $(*,*)$ denotes inner product, $\left\|*\right\|_2$ denotes 2-norm of vector. The geometric significance of (2.6) is shown in Fig. 2 (b).

2.4 Analysis of Convergence

To ensure that the subdivision mesh approaches to each pixel of image and keeps the characters of original mesh, we explore the convergence of our subdivision algorithm. In fact, the image is a discrete lattice of pixels, so the subdivision mesh can approach each pixel of image if it is continuous. We use the following theorem to show the convergence of our subdivision mesh.

Theorem 1. The subdivision mesh constructed by (2.1)(2.2) is convergent if $0 < \alpha < 1$. Each u curves and v curves of the subdivision mesh constructed by (2.1)(2.2) is C^1 continuous if $0 < \alpha < \frac{\sqrt{5}-1}{4}$.

3   Correction of Projector Imagery on Curved Surface Based-on Subdivision Mesh

The crucial problem of multi-projectors geometric calibration method is to find the corresponding relationship between each pixel of a projector image and the camera image. The current mainstream geometric calibration methods used cameras to capture the special structure image projected on the surface. The corresponding relationship was established through intensive sampling from camera images.

We also use a camera to capture a special projected "checkerboard" image on curved surface. By a feature extraction method which has been research extensively [12-13], we can get the information of projected image deformation and the initial mesh. Using the subdivision algorithm depicted in section 2, we subdivide the mesh into enough precision mesh to establish the one-one mapping from projector-to-camera. In addition, we can adjust α to adapt the plane, quadric surface and other continuous surface. This is the biggest advantage of our algorithm.

3.1   Acquisition of Deformation Features

Suppose (cx, cy) to be a pixel of camera image, $(px, py, pz)^T$ to be its projected point on surface. Because of the deformation of photographic process, there is a nonlinear mapping (Fig. 3):

$$f_1 : R^3 \to R^2 \quad (cx \quad cy)^T = f_1(px, py, pz) \tag{3.1}$$

Without linear assumption to f_1, we project a special "checkerboard" image on surface and extract the x-corner points $\left\{(cx_{ij}, cy_{ij})\right\}_{i=1, j=1}^{m,n}$ to be the initial mesh M^0, which capture the deformation features of surface and camera.

Using formula (2.1)(2.2), we can subdivide M^0 into enough precision (the resolution of camera) M^k.

3.2   Establishing Corresponding Relationship between Camera and Projector

Let (bx, by) be a pixel of frame-buffer image of projector, $(px, py, pz)^T$ to be its projected point on surface. Because of the deformation of projection on surface, there is a nonlinear mapping (Fig. 3):

$$f_2 : R^2 \to R^3 \quad (px \quad py \quad pz)^T = f_2(bx, by) \tag{3.2}$$

If we don't know the inner parameter and position of projector, we can't know f_2. Fortunately we needn't know f_2. Combining (3.1) and(3.2), we can get:

$$(bx, by)^T = f_2\left[(px \quad py \quad pz)^T\right] = f_2\left\{f_1^{-1}\left[(cx \quad cy)^T\right]\right\} = g(cx, cy) \tag{3.3}$$

Where $g = f_2 f_1^{-1}$.

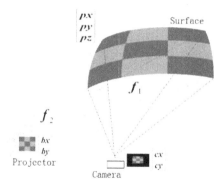

Fig. 3. Illustration of our method to acquisition of deformation features (the checkerboard image is projected on curved surface by a non-linear function f_2 related the projector, and then a camera captures the projected image by another non-linear function f_1 related the camera. The sparse initial mesh can be extracted from the camera image by x-corner detection method, which contains the sparse sampling information of f_1 and f_2).

Observed that image pixel lattice is a regular rectangular domain structure, the order of $(cx, cy)^T$ and $(bx, by)^T$ is the same. So we can establish one-one mapping from $(cx, cy)^T$ to $(bx, by)^T$ without any prior information of g .

Let $I_{buffer} = \{I_{buffer}(bx, by)\}_{bx=0, by=0}^{m,n}$ be the frame-buffer image of projector and $I_{camera} = \{I_{camera}(cx, cy)\}_{cx=0, cy=0}^{m', n'}$ be the image of camera. The final subdivision mesh is $M^k = \{P_{ij}^k\}_{i=0, j=0}^{m,n}$.

Then we have:

$$I_{buffer}(bx, by) \leftrightarrow I_{camera}(cx, cy)$$
$$(bx, by)^T = (i, j) \leftrightarrow P_{ij}^k = (cx, cy)^T \tag{3.4}$$

So far, we have established the corresponding relationship between each pixel of a projector image and the camera image without knowing any information of projectors, camera and curved screen.

3.3 Algorithm of Geometric Calibration

Once the corresponding relationship has been established, the camera can be used to be a tool to measure the deformation of image projection. Finally, we can compute the undistorted frame-buffer image inversely according the established corresponding relationship.

We use laser theodolite to specify an initial mesh which is the expected region to project undistorted image. The laser theodolite is controlled by computer through a Pan & Tilt shown in Fig. 4 (right). We let it project a series of points with equal interval of azimuth and elevation from the objector's eye (we will hereafter call this as *equal visual angle*). Our subdivision algorithm can also subdivide it into arbitrary precision.

Using formula (3.4) we can transform pixels in frame-buffer image of projector to the exactly points that are specified by laser theodolite.

Algorithm Steps:

1. Project the "checkerboard" image on surface and capture this image by camera.
2. Extract M^0 the initial mesh from camera image.
3. Use formula (2.1)(2.2) to subdivide the initial mesh into pixel-level precision M^k.
4. Use laser theodolite to specify an initial mesh $\widehat{M}^0$ divided by *equal visual angle* and Use formula (2.1)(2.2) to subdivide it into $\widehat{M}^k$ according to the precision of M^k.
5. Use formula(3.4) to establish the corresponding relationship between each pixel of a projector image and the camera image.
6. Compute the undistorted frame-buffer image inversely according the established corresponding relationship.

4   Results

We implement our algorithm on the large-scale approximate cylindrical surface respectively to verify our geometric calibration method. All experiments were run on notebook computer with PM1.6GHz CPU and 512 MB of memory. The programming language is Matlab. Other computers only used to control projectors to show image. The setup of our algorithm spent in 10 minutes on single projector. The projection screen is the exhibition hall of the airport control tower simulator in Sichuan University which is a 10-meter-diameter approximate cylinder screen (Fig 4. left). The projectors are NEC LT240K+ and the camera is OLYMPUS SP350 (Fig 4. center).

For two or more projectors, we establish the corresponding relationship between each pixel of a projector image and the camera image one by one. Fig.5 shows the results of geometric correction on approximate cylinder screen with $\alpha = 1/16$. One can see that the corrected image has no deformation within it and two geometric corrected images achieve precise stitching in the boundary.

Fig. 4. Experimental environment: curved screen (left), projectors (center) and Pan & Tilt (right)

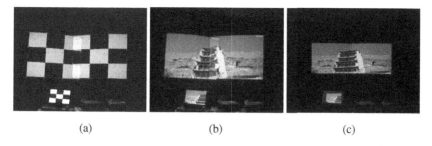

(a)   (b)   (c)

Fig. 5. The results of geometric correction on approximate cylinder screen with two projectors. (a) Projected "checkerboard" image with laser points (red points) and extracted initial mesh (red lines) on it. (b) The projected image before geometric correction. (c) The projected image after geometric correction.

Fig. 6. A 360-degree virtual airport control tower simulation system. The green lines are geometric calibration mesh and pink lines are also geometric calibration mesh to blend images of two adjacent projectors. All these calibration mesh are computed by our method presented in this paper.

The small images in Fig. 5 (b) and (c) are the geometric corrected frame-buffer images of left projector controlled by notebook computer. Since the main purpose of this paper is to present a multi-projectors geometric correction method without any camera calibration and information of projector and screen, we don't do any edge fusion and color calibration. These methods can be found in Wang Xiu Hui [14] and Brown [3]. Our method can make edge fusion and color calibration much easier than before. A practical application of our method is given in Fig. 6, which is a 360-degree virtual airport control tower simulation system. Our method presented in this paper is the foundation to display a seamless virtual airport scene with 10 projectors.

5   Conclusion

In this paper, we presented a new geometric calibration method for the problem of multi-projectors display wall. The method can be used to correct frame-buffer images of projectors to show an undistorted image on any continuous curved screen without any information of camera, projector and screen. The precision of our algorithm is a single pixel. The method is based on a simple, stable and accurate subdivision mesh. It subdivides the initial feature sampling mesh into precision of single pixel. Compared to the traditional intensive sampling method, our method reduces the amount of initial data dramatically. Meanwhile, due to the subdivision mesh of this paper can maintain the geometric shape and density distribution characters of original mesh, it avoids the piecewise linear interpolation error brought by traditional methods of texture mapping.

The value of α can influence the subdivision mesh notably. Using a group of hierarchical patterns to decide the value of α automatically according the curved screen are the emphases of our future research

References

1. Raij, A., Gill, G., Majumder, A., Towles, H., Fuchs, A.H.: PixelFlex2: A Comprehensive, Automatic, Casually-Aligned Multi-Projector Display. In: Proc. IEEE Int'l Workshop Projector-Camera Systems (2003)
2. Rasker, R., van Baar, J., Willwacher, T.: Quadric Transfer for Immersive Curved Display, Mitsubishi Electric Research Laboratories, Inc., Massachusetts, TR-2004-34 (2004)
3. Brown, M., Majumder, A., Yang, R.: Camera-Based Calibration Techniques for Seam-less Multiprojector Displays. IEEE Transactions on Visualization and Computer Graphics 11, 193–206 (2005)
4. Raskar, R., van Baar, J., Beardsley, P.: iLamps: Geometrically Aware and Self-Configuring Projectors. In: ACM SIGGRAPH 2003 Proceedings, pp. 809–818 (2003)
5. Tardif, J., Roy, S., Trudeau, M.: Multi-projectors for arbitrary surfaces without explicit calibration nor reconstruction. In: Proc. 2003 Fourth International Conference on 3-D Digital Imaging and Modeling, pp. 217–224 (2003)
6. Harville, M., Culbertson, B., Sobel, I., Gelb, D., Fitzhugh, A., Tanguay, D., et al.: Practical Methods for Geometric and Photometric Correction of Tiled Projector Displays on Curved Surfaces. In: Proceedings of the 2006 Conference on Computer Vision and Pattern Recognition Workshop (2006)
7. Li, G., Ma, W.: Composite √2 subdivision surfaces. Computer Aided Geometric Design 24, 339–360 (2007)
8. Dyn, N., Levin, D.: A Butterfly Subdivision Scheme for Surface Interpolation with Tension Control. ACM Transactions on Graphics 9, 160–169 (1990)
9. Zorin, D.: A Method for analysis of C1-Continuity of Subdivision surface. Numerical Analysis 37, 1677–1708 (2000)
10. Zorin, D.: Smoothness of Stationary Subdivision on irregular meshes. Constructive Approximation 16, 359–397 (2000)
11. Reif, U.: A unified approach to subdivision algorithms near extraordinary vertices. Computer Aided Geometric Design 12, 153–174 (1995)
12. Chen, D., Zhang, G.: A New Sub-pixel Dector for X-Corners in Camera Calibration Targets. In: Proc. WSCG 2005 Short Papers Proceedings, pp. 97–101 (2005)
13. J. Bouguet, Camera Calibration Toolbox for Matlab,
   http://www.vision.caltech.edu/bouguetj/
14. Xiuhui, W., Wei, H., Hujun, B.: Global Color Correction for Multi2Projector Tiled Display Wall. Journal of Computer Aided Design & Computer Graphics 19, 96–101 (2006)

Part III

A Comment

A Comment on 'Constructing Regularity Feature Trees for Solid Models'

F.C. Langbein, M. Li, and R.R. Martin

School of Computer Science, Cardiff University, Cardiff, UK
{F.C.Langbein,M.Li,R.R.Martin}@cs.cf.ac.uk

In [2] we presented an algorithm for decomposing a boundary representation model hierarchically into regularity features by recovering broken symmetries. The algorithm adds new recoverable edges and faces, which can be constructed from existing geometry. This generates positive and negative volumes giving simple, more symmetric sub-parts of the model. The resulting regularity feature tree may be utilised for regularity detection to describe a model's design intent in terms of regularities such as symmetries and congruencies.

In that paper we cited Leyton [1]. It has been brought to our attention that we could have more fully acknowledged this important work in which Leyton extensively discusses design intent, based on the idea that constructing a geometric model can be described as a sequence of symmetry breaking operations. A generative model of an object's shape can be built based on these operations. He argues that the asymmetries present in a shape represent its design intent. Hence, design intent is represented in a generative model by a sequence of symmetry breaking opertions, which can be recovered from the generative representation.

Our approach starts with only a boundary representation model. As there are infinitely many construction sequences for such a model [3], we detect design intent only as regularities of the final model, and cannot infer its construction history. The regularity feature tree does not represent a construction history, but is only constructed to aid regularity detection. Leyton's work is mainly concerned with the representation of shape and the fact that design intent as a symmetry-breaking construction history can be recovered from a generative representation without the need for detection algorithms.

References

1. Leyton, M.: A generative theory of shape. In: Leyton: M. (ed.) A Generative Theory of Shape. LNCS, vol. 2145, Springer, Heidelberg (2001)
2. Li, M., Langbein, F.C., Martin, R.R.: Constructing Regularity Feature Trees for Solid Models. In: Kim, M.-S., Shimada, K. (eds.) GMP 2006. LNCS, vol. 4077, pp. 267–286. Springer, Heidelberg (2006)
3. Shapiro, V., Vossler, D.: Separation for boundary to CSG conversion. ACM Trans. Graphics 12(1), 35–55 (1993)

F. Chen and B. Jüttler (Eds.): GMP 2008, LNCS 4975, p. 603, 2008.

Author Index

Lecture Notes in Computer Science

Sublibrary 1: Theoretical Computer Science and General Issues

For information about Vols. 1– 4664
please contact your bookseller or Springer

.